AN INTRODUCTION TO
HUMAN
EVOLUTIONARY
ANATOMY

AN INTRODUCTION TO

HUMAN

EVOLUTIONARY

ANATOMY

Dep... Group,

Department... ge London, UK

This book is printed on acid-free paper

Copyright © 2002, Elsevier Ltd. All rights reserved.
Sixth printing of the paperback 2002
Reprinted 2004

Elsevier Academic Press
84 Theobald's Road, London WC1X 8RR, UK
http://www.elsevier.com

Elsevier Academic Press
525 B Street, Suite 1900, San Diego, California 92101-4495, USA
http://www.elsevier.com

British Library Cataloguing in Publication Data
A catalogue record for this book is available from the British Library

ISBN 0-12-045590-0
      0-012-045591-9 (Pbk)

Typeset by Photo graphics, Honiton, UK
Printed and bound in China

CONTENTS

CONTENTS _____

FOREWORD

My initial response to the authors' invitation to introduce this book was to undertake a modest comparative study in order to establish how a 'good' foreword should be constructed. Regrettably, the results provided no clear guidance, but there was a definite pattern to the variation. Some forewords explicitly offered the equivalent of 'covering fire' to shield a text that was innovative, controversial, or sometimes both. Others took the form of an intellectual 'warranty', implicitly underwriting, or vouching for, the veracity and relevance of what followed. A third category comprised the 'testimonial' variety in which a respected figure attests to the worthiness of the text, much as a purveyor of quack medicines reels off the names of social notables who are prepared to testify to the success of a hair restorer, or tonic. Fortunately, this book has no need for any of these three varieties of foreword, nor, for that matter, does it require any other form of support or validation.

The authors were perspicacious enough to have identified the need for a textbook which, for student and researcher alike, would present the anatomical background to the study of human evolution. Fossil hominids and hominoids are unique, and there has developed an increasing realisation that 'Whiggish' methods of analogy are outmoded as the basis for research strategies aimed at solving the puzzle of human evolution. We have all faced, at one time or another, the problem of describing or comparing a piece of hominid anatomy, and ending up using an unhappy amalgam of 'Martin', 'Raven', and others, in order to find adequate words and measurements to describe what, in detail, is a novel structure. While it must always be the case that human and other higher primate anatomy will provide the starting point for attempts to understand hominid form and function, researchers are increasingly seeking to interpret form in terms of structural and functional 'rules' which have a much wider application. Hominids will need to be described and explained on their own terms, and this book introduces and explains the language that should be used for those tasks.

On the face of it, 1990, the beginning of a new decade, seems to be a singularly inopportune time to be offering a classical text of form and structure. Surely the language of splicing and cloning is more relevant to the run up to the twenty first century, than the details of hypoconulids, frontal trigones and third trochanters? We must all acknowledge and accommodate to the increasing role that molecular biology has played, and will play, in the task of exploring the relationships between taxa. However, recent developments in phylogenetic analytical techniques, and increasingly sophisticated programmes of functional analysis, make it more important than ever that the developmental and comparative contexts of hominid anatomy are well understood. This book is designed to facilitate such research programmes.

The complementary interests of Leslie Aiello and Christopher Dean have enabled them to face, and rise to, the challenge of providing a balanced and systematic account of a wide range of anatomy relevant to both functional analysis and phylogenetic reconstruction. The authors have chosen an unashamedly analytical presentation, being careful to provide the 'chapter and verse' behind each proposition. Particular care has been taken to commission 'intelligent' artwork. All too often diagrams are an afterthought to the text, so that they often make no more than a marginal contribution. In this book, the care taken to plan the textfigures is evident, and the artist deserves particular praise for her efforts.

We must be grateful that Leslie Aiello and Christopher Dean were equipped with the vision to conceive of, and the industry to prepare, such a useful text as *An Introduction to Human Evolutionary Anatomy*. Their reward will be a book that is a worthy companion to Napier and Napier's *A Handbook of Living Primates*. My prediction is that it will have a long and useful life, and that future generations will wonder how their predecessors ever managed without it.

Bernard Wood
The University of Liverpool

ACKNOWLEDGEMENTS

We must first acknowledge and thank our students who initially recognized the need for this book. They convinced us that we should attempt to clarify the sometimes rather esoteric comparative anatomy that makes up at least part of the literature in human evolution. As a result, we have done our best to bring together, in a functional and a comparative context, the comparative anatomy of living humans and apes with that of the early fossil hominids. We have tried to do this in plain English and in a format that will provide a well illustrated and easy-to-use companion for students, teachers and researchers in human evolution, who have until now had no single source to draw on at this level. We would also like to think that there is much within this book that will be interesting and useful for others in archaeology, dentistry, medicine, speech science and zoology, who pursue issues that touch on aspects of hominoid anatomy and human evolution.

In attempting this task we must first thank our own teachers, Bernard Campbell, Michael Day, Geoffrey Harrison, and Bernard Wood, who introduced us to human evolution and successfully passed on to us some of their own knowledge of and enthusiasm for the subject. We are especially grateful to Bernard Wood who read and commented on the entire manuscript. He is one of the few people in the field today whose knowledge extends into every chapter of this book and we thank him for undertaking the task of reading the manuscript as well as for his friendship, support and constant encouragement in this project. We also owe special thanks to Clark Howell, Rob Foley, David Pilbeam and Bob Martin who each offered support for this book at the very beginning.

Over the past three years a great number of our friends and colleagues have read parts of the manuscript and have made constructive suggestions. Without their help our task would have been much more difficult and we owe the following a great debt of gratitude: Fred Anapol, Peter Andrews, Baruch Arensburg, Tim Arnett, Jan Austin, David Beynon, Tim Bromage, Andrew Chamberlain, Dave Daegling, Michael Day, Bill Hylander, Sheila Jones, Bill Jungers, Gail Kennedy, Bill Kimbel, Anne MacLarnon, Theya Molleson, Monte Montemurro, Tom Naiman, John Pegington, Yoel Rak, Pat Smith, Chris Stringer, Gen Suwa, and Russ Tuttle.

While writing this book many of our students in the Department of Anatomy and Developmental Biology and in the Department of Anthropology at University College London have helped by clarifying (and sometimes by discovering) important information that is not immediately obvious in the literature, and also by carrying out original research in areas that had been overlooked by the field. Here we would like to acknowledge

Heather Baldwin, Stephen Baycroft, Sean Blaney, Rogan Corbridge, Helen Liverside, Mark Thompson, and Jason Spenser. We would also like to thank the staff of our own departments for their encouraging support and also the staff and students of the Department of Anthropology, Yale University, who offered their hospitality to one of us (LA) during the 1987–1988 academic year when the majority of the chapters on the postcranial anatomy were written. Especial thanks go to Mark Birchette, Jon Marks, Alison Richard, and Marion Swartz and also to Laura Bishop, Nora Giles, Danny Povinelli, Patricia Princehouse, Todd Preuss, Melissa Remis, Jeff Rogers, David Sprague, Eleanor Sterling, and Andrew Young.

We are also very grateful to the following who have allowed us to reproduce photographs of fossils, figures, radiographs and micrographs of bone and teeth in this book: Peter Abrahams, David Beynon, Alan Boyde, Michael Day, Sheila Jones, Bill Jungers, Gail Kennedy, OJ Lewis, Lawrence Martin, Yoel Rak, Mike Rose, Holly Smith, Jack Stern, Chris Stringer, Randy Susman, Russ Tuttle, and Alan Walker. We are indebted to Joanna Cameron for her tireless skill in producing the illustrations for this book. Her talent speaks for itself and has brought the whole subject to life. We are also grateful to Mike Gilbert, Jane Pendjiky, and Chris Sims for their help with the photographic work and to Joy Pollard for her help with the photographs of the postcrania. Special thanks also go to Robert Kruszynski for providing helpful access to the hominid and cast collection of the British Museum (Natural History) as well as to Molly Badham and Twycross Zoo for supply of the nonhuman primate specimens upon which much of the original research on ape soft tissue anatomy was conducted.

We would also like to offer sincere thanks to Andrew Richford and Carol Parr and to the rest of the editorial staff at Academic Press for their unfailing support, enthusiasm and hard work in bringing this project to completion. We also are indebted to Anne Ginger who gave up her Christmas holiday to help correct the proofs and to Richard Bruce for support throughout the project. Not least we are grateful to our families and friends without whose help and support we would not have been able to complete this project.

Leslie Aiello
Christopher Dean

CHAPTER ONE

AN INTRODUCTION TO CLASSIFICATION, PHYLOGENETIC RECONSTRUCTION AND THE FOSSIL RECORD

Anatomy is the science of the structure of animals, and comparative anatomy provides the main basis upon which our knowledge of the course of human evolution is based. In order to interpret the fossils, to determine what hominid fossils were like in life, it is necessary to compare the structure of their fossil bones and teeth to those of humans, apes and other primates. This can help us to determine not only that they were on the human line, but also details about their function, how they moved and what they ate. Only by such analogy with modern humans and non-human primates can we have confidence in our conclusions about the nature of our evolutionary ancestors.

Comparative anatomy also has further uses to the palaeoanthropologist. By understanding the ways in which bones and teeth grow and develop in humans and other primates it is becoming possible to determine the speed and the course of their individual growth and development to adulthood, their **ontogeny** (ontos (Gk) = really; gennan (Gk) = to produce). Comparative anatomy also pro-

vides the basis for the reconstruction of the path taken by a particular animal or group of animals during evolutionary history. Such an exercise in evolutionary inference is known as **phylogenetic reconstruction**, which literally means the reconstruction of the generation or development of a tribe (phylon (Gk) = a tribe; genesis (Gk) = generation). The utility of comparative anatomy does not stop here, however. It also forms the basis of our system of classification, or the naming and categorizing of fossil animals. Because it is impossible to apply the test of breeding compatibility to fossils, fossil species are defined in the first instance on the basis of their anatomy. This exercise is normally guided by the rule that fossils can be included in a single species if they vary no more from each other than do the members of an average modern species.

A good working understanding of comparative anatomy is therefore essential for the palaeoanthropologist, and to this end the following chapters are primarily concerned with an introduction to and interpretation of the anatomy of humans, of apes and of the major fossils on the human evolutionary line. These chapters are oriented primarily towards the first two uses of comparative anatomy, towards the interpretation of structure and function and towards the interpretation of the ontogeny, or growth and development of the hominids. However, before embarking on this excursion into comparative anatomy a few more detailed comments about the uses of comparative anatomy in phylogenetic reconstruction and in classification will be helpful to students who are working with, or anticipating working with, fossil material.

Classification and phylogenetic reconstruction

Biological classification makes no claim about evolution or evolutionary relationships. It simply involves the ordering of organisms into groups on the basis of their anatomical similarities and differences. The system that is currently in use is known as the Linnean Hierarchy and dates in its (near) present form to 1758 and the 10th edition of Linnaeus' *Systema Naturae*. The Linnean Hierarchy is a pyramidal structure where each higher category includes a nest, or set, of one or more lower, or subordinate, categories. At the minimum there are seven levels to this hierarchy but in practice modern applications may include 20 or more separate hierarchical levels (Fig. 1.1).

A **taxon** (taxis (Gk) = arrangement) is a group of organisms at any level of the hierarchy (such as a genus or a species or a family) and **taxonomy** (taxis (Gk) = arrangement; nomous (Gk) = law) is literally the theoretical study of the laws, procedures and principles of the arrangement of organisms. We will shortly show that there is considerable difference of opinion over the manner in which taxa (= plural of taxon) should be grouped in the hierarchical system. However, there is no possibility for difference of opinion over how various taxa are defined and named. **Nomenclature** (nomenclatos (L)

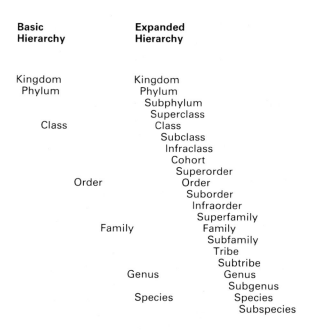

Basic Hierarchy	Expanded Hierarchy
Kingdom	Kingdom
Phylum	Phylum
	Subphylum
	Superclass
Class	Class
	Subclass
	Infraclass
	Cohort
	Superorder
Order	Order
	Suborder
	Infraorder
	Superfamily
Family	Family
	Subfamily
	Tribe
	Subtribe
Genus	Genus
	Subgenus
Species	Species
	Subspecies

FIGURE 1·1 *The Linnean Hierarchy.*

= listing of names), the application of names to the taxa in the hierarchy, is strictly governed by the **International Code of Zoological Nomenclature**. This code is based on the concept of name-bearing types, which are particular specimens or taxa that provide an objective standard of reference against which other specimens or taxa must be compared before they are included in the rank of interest. For example, each **family** or **subfamily** has a type **genus**, each genus a type **species** and each species a type **specimen**. The code stipulates that a family or subfamily name is formed by adding to the stem of the name of the type genus the latinized suffix '–idae' for a family name and '–inae' for a subfamily name. Moreover, it recommends that the suffix '–oidea' be added to the stem for the name of a superfamily and '–ini' for the name of a tribe. For example, in the classification of the human species, the genus *Homo* gives its name to the tribe Hominini to the subfamily Homininae to the family Hominidae and to the superfamily Hominoidea (see Figs 1.6 and 1.7).

The code also specifies that all species (and only species) should be referred to by two names, their genus name and their species name. Such a double name is known as a **binomen** and is the reason why this naming system is known as the system of **Binominal Nomenclature**. The genus name always begins with an upper-case letter and the species name with a lower-case one as, for example, *Homo sapiens*, the binomen for modern humans. Frequently with fossil material it may not be possible to assign a particular specimen to a particular species, but it may be possible to assign it to the next highest taxonomic level, the genus. In this case the binomen for that specimen would be, for example, *Homo* sp. indet. which identifies the particular specimen as a member of the genus *Homo* but of an indeterminate species.

Sometimes the term 'cf.' (= confer) may also be included in a binomen as, for example, *Homo* cf. *erectus*. This indicates that there is some doubt about the formal referral of the particular specimen to the species *erectus*. A species name might also be qualified as, for example, *Homo habilis* s.s. (= *sensu stricto*) or *Homo habilis* s.l. (= *sensu lato*). The use of these terms is not governed by the International Code. However *sensu stricto* is generally used in the sense of the type of the species, referring specifically to those fossil specimens that are formally included in the type series of the taxon (see below). Likewise, *sensu lato* refers to the broader (and perhaps more variable) group of specimens that might have been referred to the species subsequent to its formal establishment. Such a group of referred specimens is known as the **hypodigm**.

As a last general word about the International Code of Zoological Nomenclature it is also common to see various fossil specimens referred to as **holotypes** or **paratypes**. A new species must have a designated specimen or series of specimens as the name-bearing type(s) for that species. These terms refer to particular categories of name-bearing type that are important in the definition of a new species. The holotype is the single specimen designated as the name-bearing type of a species or subspecies when it was established. Each specimen of a type series other than the holotype is known as a paratype. A type series consists of all the specimens eligible to be name-bearing types included in the original description of a new species.

The International Code of Zoological Nomenclature lays out specific rules for both the establishment and the naming of taxa, but other areas of classification engender potential controversy and deserve further discussion. These areas are: (1) the definition of and recognition of species; (2) the grouping of taxa into an hierarchical system; and (3) the assignment of taxa to their proper levels in the Linnean Hierarchy.

The definition and recognition of species

Species are the lowest level in the Linnean Hierarchy. However, this definition of species is not of much use to the practical palaeontologist or palaeoanthropologist who is trying to define species or to assign particular fossil specimens to one or another species. A number of other definitions of the species are more useful for this purpose. Perhaps the most common definition of species is the **biospecies** definition of Mayr (1940: 254). Here biospecies are defined as groups of actually or potentially interbreeding natural populations, which are reproductively isolated from other such species. This definition is directly applicable only to contemporaneous living organisms and has been criticized by Simpson (1961) for not explicitly addressing the evolutionary duration, or time-depth, of a species. Consequently, he proposes a definition for **evolutionary species** in which 'an evolutionary species is a lineage (an ancestral–descendant sequence of populations) evolving separately from others and with its own unitary evolutionary role and tendencies' (Simpson, 1961: 153). This definition emphasizes reproductive continuity through time and is similar to a more modern definition proposed by Wiley (1978: 18) which defines a species as '. . . a single lineage of ancestral descendant populations of organisms which maintains its identity from other such lineages and which has its own evolutionary tendencies and historical fate'.

These species definitions have one very basic problem when applied to extinct organisms. Any inference about breeding continuity in the fossil record must be based on continuity of anatomical characters in the fossils, assuming an equivalence between anatomical similarity and breeding compatibility. Indeed, Simpson has suggested that if the ranges of population variation in anatomical features of two or more samples overlap for all observable characters it is likely that the samples come from one interbreeding population, or biospecies. Conversely, if the ranges of variation do not overlap it is likely that the samples come from two distinct biospecies. However, things are never as clear cut as this in the fossil record, where there are frequently ambiguous ranges of overlap between fossil samples. Moreover, there is no evidence that there is any hard and fast relationship between speciation and anatomical change (Vrba, 1980; Tattersall, 1986; Turner and Chamberlain, 1989). In human palaeontology this issue becomes very important in relation to the later-occurring fossils such as Neanderthals. The question at issue here is whether the anatomical differences between Neanderthals and modern humans justify species distinction between these hominids, e.g. *Homo neanderthalensis* and *Homo sapiens*, and if so whether this species distinction also implies breeding isolation between these species with the obvious implication that, by definition, Neanderthals could not have contributed to the modern human gene pool.

When it comes down to it, palaeontologists are, in the first instance, confined to working with **morphospecies**, which are species defined on the basis of morphological, or anatomical, similarity, regardless of other considerations (Simpson, 1961). It must be clearly recognized that any inferences about breeding compatibility are just hypotheses based on anatomical similarity and that there is always considerable room for doubt about the boundaries of fossil species.

Other definitions of species that are applicable to fossil species are **palaeospecies** and **chronospecies**. These terms both refer to temporally successive species in a single lineage (Simpson, 1961). The problem here is to identify where in a continuous lineage one species ends and the next begins. For example, in human evolution, if one posits a linear relationship between, say, *Homo erectus* and *Homo sapiens*, where does the former end and the latter begin? Simpson (1961: 165) suggests that 'Successive species should be so defined as to make the morphological difference between them at least as great as sequential differences among contemporaneous species of the same group or closely allied groups'. There are some taxonomists, particularly those adhering to the school of phylogenetic systematics (see below), who believe that there is no cause to divide continuous, non-branching, lineages into palaeospecies or chronospecies. In other words, these workers argue that if there is a continuous lineage from, say, *Homo habilis* to *H. sapiens*, that does not give rise to side branches, this lineage should be included in one evolutionary species (i.e. *H. sapiens*) (Eldredge and Cracraft, 1980; Bonde, 1981). In this case *H. sapiens* would be the species name with historical priority.

The grouping of taxa into an hierarchical system

The grouping of taxa into an hierarchical system involves the formation of nested sets, where say a group of similar species is placed in one set and a number of such sets are grouped together into a higher set and so on up the hierarchy. Such hierarchical grouping is based on the one simple principle of anatomical similarity. However, as might be expected there is a serious problem with this simple principle. The problem is that all similarities are not equal. Some anatomical features are similar by virtue of their presence in a common ancestor of the animals concerned. These features are known as **homologous features**, or **homologues** (homos (Gk) = same; logos (Gk) = a treatise). These features have a structural similarity to each other but do not necessarily need to have a functional similarity (or identity) in the animals concerned. The classic example of a homologous feature in mammals is the wing of a bat and the forelimb of, say, humans. Although the bat wing is adapted to flight the same bones make up the wing as make up the human forelimb which has an entirely different function (manipulation). The bones in the two animals differ primarily in size and proportion.

Other anatomical features can be similar in two or more animals without being homologous. These features are lumped under the category of **homoplastic features** (plasis (Gk) = moulding). Most homoplastic features are similar by virtue of function rather than by inheritance from a common ancestor. For example, the wings of bats, birds and insects are homoplasies. They are superficially similar because they serve the same function, but they are structurally very different. This type of homoplasy results from either **parallelism**, **convergence** or **analogy**. The difference between these forms of homoplasy is basically one of degree. Parallelism refers to the development of homoplastic features in animals with a fairly recent common ancestry where the characteristics are based on, or are channelled by, that common ancestry (Simpson, 1961). Convergence refers to the parallel development of features in more remotely related animals, while analogy refers to such development in very remotely related animals where the feature is not related to any community of common descent. The wing of a bird and that of an insect is a good

example of analogy. Homoplastic features can also be similar, particularly in insects, by virtue of **mimicry**. In this case the animals achieve some advantage by mimicking the body form, colouring or patterning of another animal. The last category of homoplasy is **chance**. Here, similar characteristics result from independent causes and without causal relationship involving the similarity in each (Simpson, 1961).

In hierarchical groupings that are meant to reflect evolutionary relationships it is essential to base inferences on homologous features rather than on homoplastic features, since it is the homologous features that are present in animals by virtue of descent from a common ancestor. However, it is not always so easy to distinguish them, particularly with parallelisms from recent common ancestry where the similarity in structure is also close.

Unfortunately, there is also another problem. It is not good practice to base the degree of relationship between species on the absolute number of homologies present. It is not uncommon that homologous features disappear, or be otherwise modified, during the course of evolution of a particular taxon. Absolute number may, therefore, produce entirely false hierarchical relationships (Fig. 1.2). There is, however, a solution to

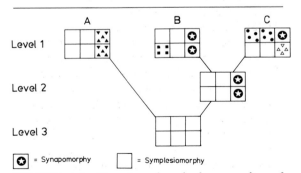

FIGURE 1·2 *Counting the absolute number of homologies shared between any two organisms (boxes) at Level 1 would give a false idea of relatedness. Organisms A and B share more homologies (three) than do organisms B and C (two). Only organisms B and C share a synapomorphic feature indicating relatedness.*

this problem. As all similarities are not equal, all homologies are also not equal. Some homologies may stem from a very remote common ancestor while others stem from a much more recent common ancestor, and the solution lies in distinguishing these homologies. For example, if you have three species and the problem is to determine which two are more closely related to each other than they are to the third, your conclusion needs to be based only on those homologies that were present in the common ancestor of the two most closely related taxa and not in the common ancestor of all three taxa. In other words, the problem is to distinguish a closely related set from a more remotely related set. In the jargon of **cladistic analysis** (clade from klados (Gk) = branch) these homologies shared by closely related taxa are called **synapomorphic features**, or shared derived features (syn (Gk) = together; apo (Gk) = from; morphos (Gk) = form). The more remote homologies are called **symplesiomorphic features**, or shared ancestral features (syn (Gk) = together; plesio (Gk) = near; morphos (Gk) = form). Those features that are unique to a particular taxon are called **autapomorphic features** (auto (Gk) = self).

The most common way to distinguish synapomorphic features from symplesiomorphic features is to compare the frequency of the occurrence of particular traits, or character-states, in a larger number of relatively closely related taxa than those of immediate concern to the analysis. Such a comparative group is called an **outgroup** and the assumption upon which this approach is based is simply that the most frequently occurring features, or character-states, in a wide group of taxa will be those that were present in the common ancestor. Eldredge and Cracraft (1980) provide a simple and clear example of this technique (Fig. 1.3). They begin with five hypothetical species A, B, C, D and E, and their problem is to determine whether A is most closely related to B or to C. Species D and E provide the outgroup. Each of the five species has two

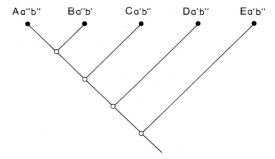

FIGURE 1.3 *An outgroup comparison. See text for discussion. After Eldredge and Cracraft (1980).*

different characters, 'a' and 'b', and each of these characters can have two forms, or character-states (a′ or a″ and b′ or b″). In hominoid evolution, for example, character 'a' might be enamel thickness and a′ would be thick dental enamel and a″ would be thin dental enamel. These character-states are distributed through the five species in the following way:

A	B	C	D	E
a″ b″	a″ b′	c′ b″	a′ b″	a′ b″

In this example, a′ is more frequently found than is a″ and is therefore assumed to be the primitive form of character 'a'. Likewise, b″ is more frequently found than is b′ and is taken to be the primitive form of this feature. Species A and B are the only two species that share a derived, or less frequently occurring character-state (a″), and are therefore assumed to be more closely related to each other than they are to species C by virtue of the common occurrence of this character-state in the two species. The assumption is that this derived, or synapomorphic feature, would have been inherited from the inferred common ancestor of species A and B. Trait b′ only occurs in one species, species B, and would be an autapomorphic, or unique, feature of this species. Traits a′ and b″, because of their common and widespread occurrence, would then be the symplesiomorphic, or shared primitive, features.

This exercise of determining which particular form of a trait is more primitive or more derived is called **determining the polarity of the morphocline**. A morphocline is the sequence presumed to reflect the probable pathway of change among the character-states (Eldredge and Cracraft, 1980). 'Determining the polarity' refers to the determination of the direction of change in these traits. In other words, this would be the determination of which of the character-states were primitive and which were advanced, or derived. This is perhaps the most controversial area in modern evolutionary analysis today, because while there is little argument over the definition or description of different character-states, there is considerable disagreement over the inferred direction of change and the resulting composition of the nested sets.

The outgroup comparison is perhaps the most reliable way of establishing the polarity of a morphocline, but in practice it is not always as easy as described above. It frequently produces conflicting results when different characters are introduced into the analysis or when different researchers carry out the analysis. Another way of determining the polarity of the morphocline is by looking at the fossil record. However, here there is an additional problem. It is not always reliable to assume that the particular character-state found in a fossil is necessarily primitive in relation to the character-state observed in a later-occurring species. Fossil species can be very specialized in their character-states, just as can modern species. Figure 1.4, also taken from Eldredge and Cracraft (1980), illustrates a situation in which a living species is primitive in relation to its fossil antecedents. In this case, if polarity had been determined by age, the polarities would simply have been wrong.

This example also brings up the differences between branching diagrams and phylogenies (evolutionary trees) (Tattersall and Eldredge, 1977). A branching diagram, or cladogram, merely expresses the inferred pattern of relationships between species, based on the analysis of character states. It does not include any information about time and, moreover, fossil species are included in the analysis on equal footing with living species. A cladogram also does not include any information about ancestor–descendant relationships, or more specifically about who is directly ancestral to whom. These features are in the realm of the phylogeny.

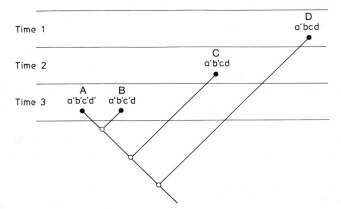

FIGURE 1·4 *An example of a later occurring species (species D) that is primitive in relation to its fossil antecedents. This illustrates the danger of assuming that character-states found in fossil species are necessarily primitive in relation to the character-states found in later occurring species. (Prime marks indicate derived character-states.) After Eldredge and Cracraft (1980).*

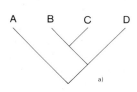

FIGURE 1·5 *Some of the possible phylogenies that can be derived from a single cladogram. (a) Cladogram; (b) all species have autapomorphic features and therefore no species can be directly ancestral to any other species; (c) species D is primitive in relation to its descendent species and can therefore be linearly related to them; and (d) species D is primitive in relation to species C which is primitive in relation to species B.*

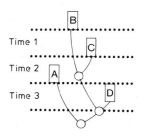

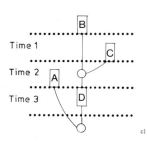

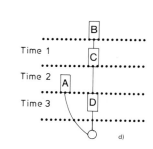

To transform a cladogram into a phylogeny, the first step is either to stretch or to shrink its branches to fit the known time distributions of the species concerned (see Fig. 1.4). An important caution here is that the known age of a fossil taxon is not necessarily the age of the first appearance, or of the initial evolution, of that taxon. The particular species could have been in existence long before the dated sample from the fossil record. However, even with this proviso, the second step in transforming a cladogram into a phylogeny can be much more problematic (Fig. 1.5). In strict cladistic analysis, the determination of specific ancestral-descendant relationships involves the (rather controversial) assumption that a species that is directly (linearly) ancestral to another cannot have any character-states that are derived in relation to the character-states found in its inferred descendant species. This is the same thing as saying that evolutionary reversals (from derived to primitive) are not likely to occur in the fossil record. Almost all known species, either living or fossil, have at least some uniquely specialized features (autapomorphies). As a result, the logical conclusion must be that virtually no known species (which admittedly are only a small percentage of the total species that must have actually existed through time) are directly and linearly ancestral to other species. This results in branching evolutionary trees that resemble tangled bushes to a far greater degree than they do more simple, linear sequences. Many would argue that this is a more accurate reflection of the actual course and process of evolution, taking the point of view that evolution proceeds more frequently by **cladogenesis**, or branching, rather than by **anagenesis**, or slow gradual transformation in one evolving line.

Even with the many problems inherent in the hierarchical grouping of taxa and in phylogenetic reconstruction, the important point to remember is that comparative anatomy is the main basis we have for unravelling the path of human evolution through time.

The assignment of taxa to their proper levels in the Linnean Hierarchy

The assignment of taxa to their proper levels in the Linnean Hierarchy is perhaps the most controversial area in classification. There are basically two rival schools of thought on this matter, **grade-based** classification (also known as **evolutionary systematics**) and **clade-based** classification (also known as

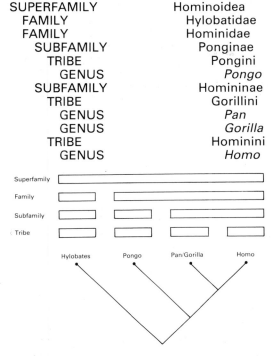

SUPERFAMILY	Hominoidea
FAMILY	Hylobatidae
FAMILY	Hominidae
SUBFAMILY	Ponginae
TRIBE	Pongini
GENUS	*Pongo*
SUBFAMILY	Homininae
TRIBE	Gorillini
GENUS	*Pan*
GENUS	*Gorilla*
TRIBE	Hominini
GENUS	*Homo*

FIGURE 1·6 *A clade-based classification of humans and apes. (a) The classification; (b) the relationship between the classification and the cladogram.*

phylogenetic systematics). Systematics is defined by Simpson (1961: 7) as the scientific study of the kinds and diversity of organisms and of any and all relationships among them. The clade-based systematists, or cladists (e.g. Hennig, 1966; Eldredge and Cracraft, 1980; Bonde, 1977, 1981; Andrews and Cronin, 1982) argue that the Linnean Hierarchy should accurately reflect the pattern of relationships revealed in the branching diagram, or cladogram. Because branches of the cladogram, or sister groups, stem from an inferred common ancestral species, equivalent ranking in this system also reflects relative age of divergence of the branches (Eldredge and Cracraft, 1980). An example of such a cladistic classification of humans and apes, with its corresponding cladogram, is given in Fig. 1.6) (Andrews, pers. comm., 1990). All

taxa in a clade-based classification must be **monophyletic**. This means that they must include the common ancestor of the descendant taxa as well as all of those descendant taxa. The unacceptable alternatives to monophyly are **polyphyletic** taxa and **paraphyletic** taxa. A polyphyletic taxon does not include the single common ancestor of the descendant taxa (although it may include some ancestors of the descendant taxa). It is a taxon that has arisen from two (or more) separate stems, or evolutionary lines. A paraphyletic taxon is one that does not include all of the descendant taxa derived from a single common ancestor.

Grade-based (evolutionary) systematists (e.g. Simpson, 1961; Mayr, 1969; Bock, 1977), on the other hand, argue that a classification should not only be consistent with the evolutionary history of a group of organisms, but also should reflect other things

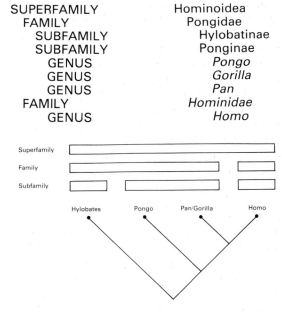

SUPERFAMILY	Hominoidea
FAMILY	Pongidae
SUBFAMILY	Hylobatinae
SUBFAMILY	Ponginae
GENUS	*Pongo*
GENUS	*Gorilla*
GENUS	*Pan*
FAMILY	Hominidae
GENUS	*Homo*

FIGURE 1·7 *A grade-based classification of humans and apes after Simpson (1963). (a) The classification; (b) the relationship between the classification and the cladogram. The family Pongidae and the subfamily Ponginae are paraphyletic taxa.*

about the included organisms than just the branching pattern of a cladogram. Their classifications tend to be based on **grades** (groups of animals similar in general levels of organization) rather than on **clades** (groups of animals of common genetic origin). Grade-based systematists do not reject the criterion of monophyly completely but argue that minimum monophyly, or monophyly at the genus level, is sufficient. Polyphyletic or paraphyletic taxa at higher levels of the hierarchy are allowed (Simpson, 1961). An example of a grade-based classification of humans and the apes is given in Fig. 1.7 (Simpson, 1963).

Hominids and human ancestors

The term **hominid** comes from the family name Hominidae and under the Simpsonian classification includes modern humans and our evolutionary ancestors back to the separation of the human line from that leading to the living African apes. Despite the different compositions of the family Hominidae under rival classifications (e.g. Bonde 1977; Andrews and Cronin 1982), the term hominid, in common usage, still refers to living humans and our exclusive ancestors. It contrasts with the word **pongid** (derived from the family name Pongidae) that refers to the living apes and their exclusive evolutionary ancestors. The term **hominoid** (from the superfamily Hominoidea) refers to both living humans and apes and the evolutionary ancestors of both these groups.

Traditionally, three general features have been recognized as being characteristic of the hominids (Le Gros Clark, 1964; Pilbeam, 1972). The first is bipedal locomotion (with its associated anatomical features), the second is a relatively large brain size in relation to body size, and the third is a reduced dentition and particularly a reduced anterior dentition. Although these features distinguish clearly modern humans from modern apes, they do not serve equally well when applied to the fossil hominids. Dental reduction

(particularly of the canine) is at best ambiguous when used to distinguish the early fossil hominids. For example, the dentition of our earliest known ancestors, *Australopithecus afarensis*, is much more similar to the dentition of the Miocene apes (dryopithecines and sivapithecines) than it is to the dentition of modern humans. Equally problematic is relative brain size. There are not many crania known for our earliest ancestors from which absolute brain size can be determined (see Chapter 10). Moreover, there are also a number of uncertainties involved in predicting body weight for these early hominids (see Chapter 14). However, the current estimates suggest that *A. afarensis* had a brain size relative to body weight that was approximately the same as that of the orang-utan (*Pongo pygmaeus*). Indeed these data show that all the australopithecines and paranthropines had relative brain sizes within the range observed for living monkeys and apes and it is only members of the genus *Homo* that exceed the observed range of variation in these living primates (see Chapter 10 and Fig. 10.17).

Of the three general features that distinguish living humans from living apes, bipedal locomotion is the only one that clearly defines the early australopithecines as hominids. The remarkably preserved footprints from Laetoli, Tanzania (ca. 3.6–3.75 million years ago) clearly demonstrate that hominids of this period habitually moved on two legs. However, the limb bones of at least some of these early hominids suggest that this bipedalism was of a different form than that found in modern *Homo sapiens* and may

---

FIGURE 1·8 *The ages of the hominid fossils. (a) The ages of the main Plio-Pleistocene hominid fossils and the sites at which they were found. The shading indicates the hominid-bearing levels at the sites. The following abbreviations are used in the specimen numbers: AL = Hadar; LH = Laetoli; MLD = Makapansgat, Sts = Sterkfontein; O and L both = Omo; WT = West Turkana; ER = Koobi Fora; TM = Transvaal Museum, Kromdraai; SK = Swartkrans; and OH = Olduvai. Based on Feibel et al. (1989), Grine (1988), Vrba (1982). Fig. 1.8(b) overleaf.*

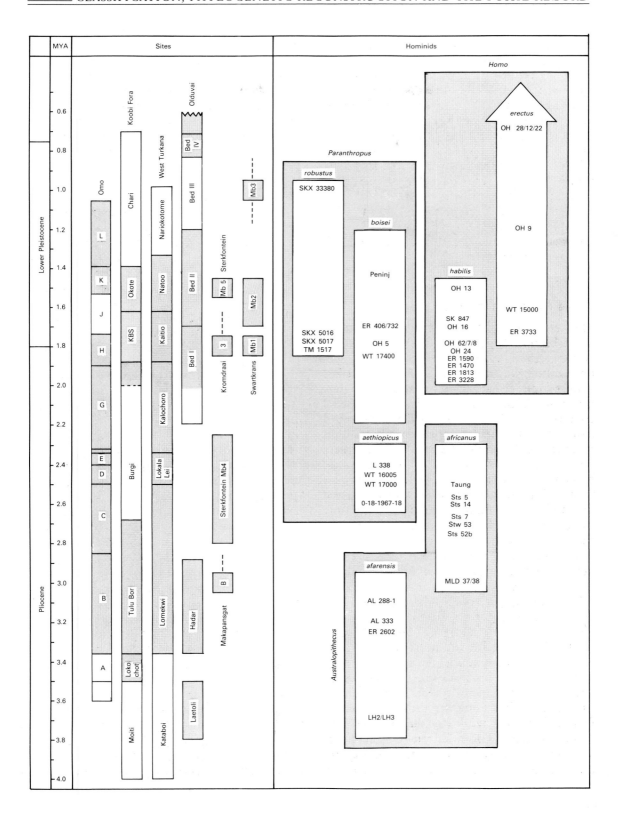

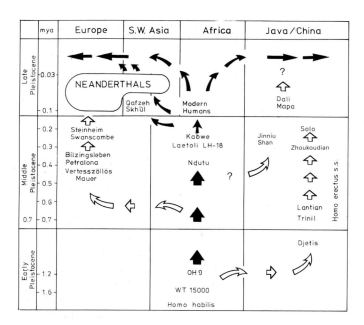

FIGURE 1·8 (b) *The age and geographical location of the main Middle and Late Pleistocene hominid fossils. Dark arrows indicate the evolution of modern humans in Africa and their movement out of Africa in the Late Pleistocene. Open arrows indicate probable earlier migrations and subsequent evolutionary development.*

even not have been their exclusive form of locomotion. Indeed, a fully modern postcranial skeleton (implying fully modern bipedal locomotion) is not found in the fossil record until the appearance of anatomically modern *Homo sapiens* approximately 100 000 years ago (see Chapter 21).

Molecular comparisons of living primates suggest that the hominids split off from the line leading to the African apes sometime between about 6 and 8 million years ago (Sibley and Alhquist, 1984; Andrews, 1986a, b). Known hominid fossils extend back in time almost that far in East Africa to about 5.6 million years ago (Hill and Ward, 1988). However, fossils older than the Laetoli footprints (and other Laetoli fossils) (ca. 3.6 million years ago) are few and very fragmentary and for the most part are considered to be hominids on the basis of a few anatomical similarities to later-occurring hominids. The majority of the hominids discussed in this book are those that extend in time from Laetoli to the present day (Fig. 1.8a, b). Three different genera are recognized in this group: *Australopithecus*, *Paranthropus* and *Homo*. The earliest of

these, *Australopithecus*, includes two separate species. *Australopithecus afarensis*, dates from 3.75 to about 2.8 million years ago and presently includes material from Laetoli (Tanzania), Hadar (Ethiopia), Koobi Fora, (Kenya) and Omo (Ethiopia) (Boaz, 1988). The second is *Australopithecus africanus* which is known from the southern African sites of Taung, Makapansgat and Sterkfontein and dates between approximately 3 and 2.5 million years ago.

The second genus, *Paranthropus*, also includes (for purposes of our discussions) two species, *Paranthropus robustus*, from South Africa, and *Paranthropus boisei* from East Africa. These two species are frequently included in the genus *Australopithecus* (*A. robustus* and *A. boisei*); however, we are of the opinion that there is sufficient difference between these hominids and those included in *Australopithecus*, particularly in the anatomy of the skull and dentition, to justify distinction at the generic (genus) level.

The third genus, *Homo*, is perhaps the most difficult to discuss because it is the most variable. We have chosen to discuss three species, *Homo habilis*, *Homo erectus* and

Homo sapiens (including archaic _sapiens_, Neanderthal and anatomically modern _sapiens_). _Homo habilis_, dating between about 2.2 and 1.6 million years ago, may eventually prove to include fossils belonging to more than one species. Our anatomical discussions directly address this interesting problem of variation, particularly as it applies to the cranium and mandible. Once hominids spread out of Africa (most probably in the later Lower Pleistocene) there is an exceedingly large amount of morphological variation, again primarily in the anatomy of the skull. Because of this variation we have chosen to restrict the taxon _H. erectus_ primarily to the hominids from Java and China that span the Middle Pleistocene time period between about 750 000 and 125 000 years ago (following: Andrews, 1984; Stringer, 1984a; Wood, 1984) (Fig. 1.8b). Nevertheless, we

also have included Early Pleistocene specimens such as KNM-ER 3733, KNM-ER 3883, OH 9, KNM-WT 15000) in this taxon. However, Middle Pleistocene hominids from Africa and Europe are discussed (according to the currently accepted convention) as 'archaic _H. sapiens_'. The Neanderthals of Europe and south-western Asia could also be included under this heading; however, because of their extreme and rather remarkable morphology we have chosen to discuss them under the separate heading of Neanderthal, refraining from referring to them by either a specific taxonomic designation (_Homo neanderthalensis_) or a sub-specific designation (_Homo sapiens neanderthalensis_).

The geographical distribution of the relevant hominid fossil sites are given in Fig. 1.9. There is no present consensus over the phylogenetic relationships of hominid taxa,

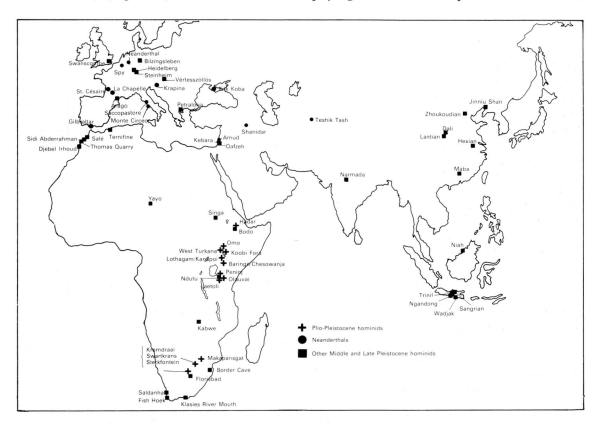

FIGURE 1·9 _The geographical location of the main hominid fossil sites._

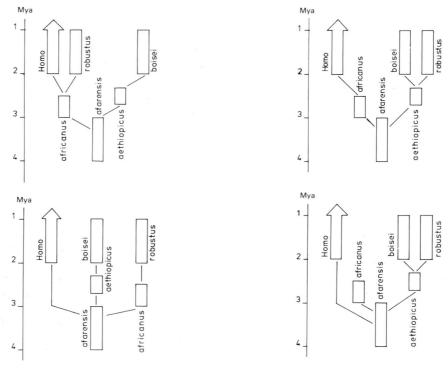

FIGURE 1·10 *Some possible phylogenies for the Plio-Pleistocene hominids.*

particularly in the Plio-Pleistocene period (before the beginning of the Middle Pleistocene). However, Fig. 1.10 illustrates some possible phylogenies for this period, while Fig. 1.8b gives our preferred phylogeny for the Middle and Later Pleistocene. This figure places the origin of modern humans in Africa in the later part of the Middle Pleistocene followed by a movement of anatomically modern humans out of Africa in the early Late Pleistocene. Whether the various Middle Pleistocene hominid populations in Africa, East Asia and Europe were actually separate biospecies in the sense that they could not interbreed and whether anatomically modern humans were able to interbreed with indigenous peoples they encountered as they spread during the Late Pleistocene are questions that we feel cannot be answered conclusively at present.

CHAPTER TWO

ANATOMICAL NOMENCLATURE

Because anatomy began as a descriptive science in the days when Latin was the universal scientific language, early anatomists described structures they saw in that language, comparing them with common or familiar objects or borrowing terms from Greek and Arabic scholars before them. While the official list of anatomical terminology is still written in Latin, all countries liberally translate it to a less-formal version for ease of teaching and learning. Nevertheless, some descriptions still seem odd and obscure. Sometimes we feel it helps to know a little more about the origin of a word that on the face of it may seem a ridiculous way of describing a bone or a muscle. To this end we have quoted freely from Field and Harrison (1968) in the text of this book in the hope that it may provide a little insight (and perhaps some amusement).

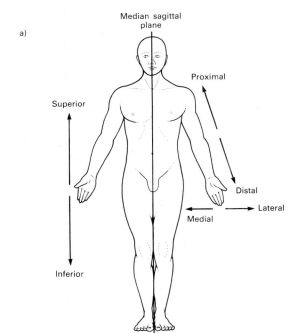

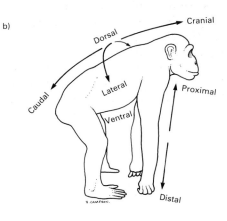

FIGURE 2·1 *(a) The anatomical position. (b) The habitual position in a quadropedal primate with equivalent terms indicated.*

15

Anatomical language depends upon a set of rules and reference planes that define how the body and its parts shall be described. The planes, positions and movements that are constantly used in this and most other anatomical texts are explained and described here. Some of them, especially some of the definitions of movements that occur in the limbs, will be emphasized again later in the appropriate chapters of the book.

The anatomical position

Anatomical descriptions are always made with the body positioned in the **anatomical position** (Fig. 2.1). Sometimes this has to be imagined, because the bone or specimen under consideration is actually lying on a table or is still half buried in the ground. The anatomical position is that assumed by a standing person with the upper limbs at the side and with the face, palms of the hands and feet pointing forwards. All the anatomical descriptions in this book will be made in the anatomical position as defined for modern humans, even if some of the hominoids under consideration might not have naturally assumed that posture as well as others.

The planes of the body

A series of imaginary planes are used in anatomical descriptions in order to provide a reference framework for terms relating to direction (Fig. 2.2). An imaginary plane that divides the body into right and left halves from top to bottom is called the **median sagittal plane**. Planes that run parallel with this median sagittal plane are described as **parasagittal** planes and, theoretically, there are an infinite number of these. Often, we simply refer to a structure lying in, or running in, the sagittal plane. The example you should remember now is the sagittal suture of the skull from which the sagittal plane gets its name.

The plane that runs from the top of the head to the soles of the feet and divides the front of the body from the back of the body is called the **coronal plane** and, again, there are theoretically an infinite number of coronal planes or sections although usually we simply say that a structure lies or runs in the coronal plane. The coronal suture of the skull runs in the coronal plane.

Planes that divide the body into upper and lower portions are described as **transverse planes** wherever they occur along the length of the body. Transverse planes can equally easily be made or imagined through the head, the pelvis or the foot.

Anatomical terms that describe direction

When a person stands in the anatomical position, the nearer a structure is towards the top of the head the more **superior** it is said to be. Likewise, the nearer a structure is to the soles of the feet the more **inferior** it is. The terms **cranial** (or **above**) and **caudal** (or **below**) are synonymous with superior and inferior. Often, in less-formal descriptions, 'above' and 'below' are used instead of superior and inferior.

The more a structure lies towards the front or **ventral** aspect of the body in the anatomical position, the more **anterior** it is said to be, and the more a structure lies towards the back or **dorsal** aspect of the body the more **posterior** it is said to be.

Likewise, the closer a structure is to the

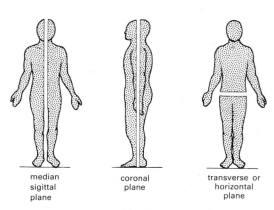

median sigittal plane   coronal plane   transverse or horizontal plane

FIGURE 2·2 *The planes of the body.*

median sagittal plane the more **medial** it is and the further away it is from the median sagittal plane the more **lateral** it is said to be. Some structures of the body are closer to the external environment than others and are then said to be more **superficial** than those that lie nearer the centre of the body, or a part of the body, which is then said to be **deeper**.

The further along a limb a structure is from the attachment of that limb to the trunk, the more **distal** it is said to be. Similarly, the nearer a part of the limb is to its attachment the more **proximal** it is said to be.

Joints

Unless bones have fused together, some sort of soft-tissue-filled gap exists between them. Not all these gaps or joints permit movement between bones, but many joints are designed especially to permit motion between two bones.

Joints between bones that are filled with cartilage are called **primary cartilaginous joints** or **synchondroses**. They do not permit movement but are often important growth centres. Joints that in addition contain fibro-cartilage are called **secondary cartilaginous** joints and a little movement or separation between bones can sometimes occur here. Secondary cartilaginous joints are found in the mid-line of the body; for example, the pubic symphysis, the joints between the bodies of the vertebrae or the joint at the sternal angle between the body and the manubrium of the breast bone. The joints between the vault bones of the skull are filled with fibrous tissue only and these are known as **sutures**. Elsewhere in the body these are simply known as **fibrous joints**. In adults no movement occurs at sutures.

Joints designed for movement are almost exclusively **synovial joints**. These consist of a capsule that joins the ends of two adjacent bones. Within this capsule a synovial membrane secretes a fluid known as **synovial fluid** into the joint compartment. The ends of the bones facing the gap are covered and cushioned with **articular cartilages** and these slip over one another when movement occurs at the joint. Synovial joints are also designed to resist dislocation and for this reason have strong ligaments around them. The ligaments and bone morphology at synovial joints also restrict active movements determined by muscle attachments to one or two planes and only rarely do active movements at synovial joints occur in more than two planes.

Movements at joints

General rules exist for naming movements at joints which are described with reference to one of the planes of the body. Movements of the thumb and parts of the lower limb can be confusing and their movements need to be discussed separately.

The movement of any part of the body from its location in the anatomical position to a more anterior position is usually called **flexion**. However, perhaps a better definition of flexion is the approximation of two developmentally ventral surfaces. These movements occur about a transverse axis and in a sagittal plane. It follows that bending the head or the back forwards is flexion and bending the fingers anteriorly so that they grip tightly is flexion. But flexion of the thumb sweeps it medially across the palm of the hand.

During embryological development the lower limb rotates so the equivalent of flexion at the elbow where the forearm moves anteriorly is reversed at the knee. In the lower limb, flexion at the knee is actually a posterior movement of the leg! Nonetheless, again we take note of this embryological rotation and remember that flexion at the knee approximates two developmentally ventral surfaces. You will recall that the knee of a bird bends the other way to your own and this will help you to remember that yours has rotated during development! Another complication arises because the foot is at a right angle to the leg and flexion of the foot occurs when we stand on our toes and the foot rotates backwards. This third exception to the rule is described as **plantarflexion**.

Extension is the exact opposite movement to flexion and with the now obvious exceptions of the thumb, leg and foot, occurs when movement carries a part of the body posterior with respect to the anatomical position. It follows that returning to the anatomical position from an already flexed position is also called extension. In order to remind ourselves that extension of the foot is defined oddly, we sometimes use the term **dorsiflexion** to denote an upwards movement of the ball of the foot.

Abduction is a movement that carries any part of the body away from the median sagittal plane. In fact we use left or right **lateral flexion** to describe this movement of the vertebral column and it will also become clear that this sort of movement at the knee and elbow cannot occur at all unless the joint is injured. Describing abduction of the fingers and toes presents a problem, so we describe a spreading of the fingers relative to a neutral middle finger as abduction and a spreading of the toes about a neutral second toe as abduction. Both left and right movements of these neutral digits is then called abduction, but is distinguished in the hand by referring to the movement as **radial** or **ulnar abduction** and as lateral or medial abduction in the foot. Again the thumb is a problem; the movement of the thumb that is equivalent to abduction of the fingers occurs when there is anterior movement away from the palm of the hand in the anatomical position.

Adduction is the opposite movement to abduction and brings parts of the body nearer to the median sagittal plane. This is most easily thought of as the movement that returns abducted limbs and digits, as we have just described them, to the anatomical position. You will realize that with the feet together in the anatomical position it is only possible to adduct the lower limb further across the midline by avoiding the stationary limb.

Rotation at a joint is any movement that occurs around any longitudinal axis. Rotation of the head and neck to the left or right is easy to imagine, but rotation of the limbs can also occur in two ways on either side of the body and needs further clarification. If the anterior surface of a limb rotates towards the body, the movement is described as **medial rotation**. If the anterior surface of a limb rotates away from the body or laterally, the movement of the limb is described as **lateral rotation**. Special terms exist for the forearm: medial rotation is called **pronation** and lateral rotation is called **supination**. Rotatory movements at the foot are difficult to carry out, but it is possible to raise or elevate the medial border of the foot which is called **inversion** or, to raise or elevate the lateral border of the foot which is called **eversion**. Inversion and eversion can only really occur when the foot is off the ground. These same movements made when the feet are in contact with the ground are called **pronation** instead of eversion and **supination** instead of inversion although several other definitions exist. Remember that embryological rotation of the hindlimb confuses any comparisons of supination and pronation with the forearm.

CHAPTER THREE

THE MICROANATOMY OF MUSCLE AND BONE

Muscle

This introduction to the microanatomy of muscle has two aims. The first is to provide some background to the mechanism of muscle contraction and to **electromyography** (EMG), now an important and widely used technique in many comparative studies of primate locomotion and mastication. The second is to explain something of how the size and internal structure of muscles relate to the forces they generate and the amount of movement they can bring about.

Muscles exert a force by contracting and so bring about the movements of the skeleton, e.g. the contractions of the heart, the contractions of the walls of blood vessels, intestine, bladder and uterus, and in so doing produce heat. The broadest classification of muscle tissue is based upon the presence or absence of regular **cross-striations** that can be seen with a light microscope. Skeletal muscle and heart muscle both appear **striated**. Muscles of the internal organs and of blood vessel walls are unstriated and are usually described as **smooth muscle**. Heart muscle, while striated, is often considered to be a separate type of muscle in that its fibres differ from skeletal muscle by being branched and connected somewhat like a meshwork, so that a single contraction can spread more swiftly through the entire organ. In smooth muscle the thick and thin filaments that make up all muscle tissue are not arranged in such a regular pattern as they are in cardiac and skeletal muscle so that no cross-striations can be seen with a light microscope. The rate of contraction of smooth muscle is also very much slower than that of striated muscle.

Muscle microstructure

It is probable that the mechanism of muscle contraction is similar in all types of muscle. In order to understand this, something of the structure of a muscle must be described. This is done here with reference to Fig. 3.1. A skeletal muscle is composed of bundles of muscle **fibres** that are easily visible to the naked eye (these bundles can be teased apart in cooked meat). The individual muscle fibres

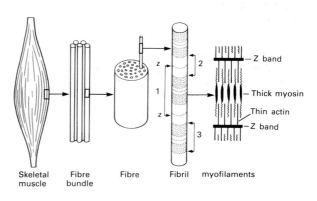

Skeletal muscle   Fibre bundle   Fibre   Fibril   myofilaments

FIGURE 3·1 *The microstructure of muscle. 1 = sarcomere (the unit between two Z lines); 2 = I band; 3 = A band.*

within each bundle may be many centimetres long and may run through the entire length of the muscle. Each fibre is between 10 and 100 μm in diameter. These fibres are, in turn, made up of thinner **fibrils** or **myofibrils** which comprise closely packed contractile protein structures known as **myofilaments** which give the fibre and fibril a banded appearance. In transmitted light microscopy (TLM), the strongly birefringent (**anisotropic**) striations are darker than the weakly birefringent (**isotropic**) striations. Accordingly, they have been named **A bands** and **I bands**. In the middle of each I band there is a thin dark strip known as the **Z line**. The unit between two Z lines is about 2 μm and is called the **sarcomere**. The sarcomere represents the smallest functional unit of the myofibril.

The fine structure of the sarcomere can be resolved further by using transmission electron microscopy (TEM). Z lines link adjacent thin myofilaments of **actin**. In the central section of the sarcomere, thick myofilaments of **myosin** are positioned between the thin filaments. Actin and myosin are elongated protein molecules that are connected together by a system of molecular cross linkages. When the muscle contracts and shortens, these cross linkages are rearranged so that the thick filaments slide between the thin filaments, thus reducing the distance between the Z lines. As this happens the ends of the myosin filaments approach the Z lines and the I bands consequently become narrower (see also Fig. 3.4).

The energy supply for muscle contraction

Adenosine triphosphate (ATP), the immediate source of energy for muscle contraction, is only present in muscle in very small amounts, enough perhaps for only 10 rapid contractions. Another organic phosphate molecule, **creatine phosphate**, is present in greater amounts and its phosphate group is transferred to **adenosine diphosphate** (ADP) as supplies of ATP become exhausted. Ulti-

mately, the oxidation of carbohydrates, stored in muscle in the form of **glycogen**, replenishes the creatine phosphate supplies within the muscle. In fact between 0.5% and 2% of the wet weight of a muscle is stored glycogen.

The innervation of muscle fibres

Muscle contraction is usually initiated when a nerve impulse arrives at a neuromuscular junction, or **motor end-plate**, in the muscle. The impulse spreads rapidly as an electric depolarization which extends over the surface of the muscle fibres. The muscle-cell membrane or **sarcolemma** that covers each fibre connects with the deeper parts of the muscle via a system of **transverse tubules** that runs across the muscle cells near the Z lines (Fig. 3.2). This system of tubules is called the T system. Each muscle fibre is also enveloped by a sleeve of flattened vesicles called the **sarcoplasmic reticulum** that contains calcium ions. Each wave of depolarization spreads through the T system and causes an increase in the permeability of the

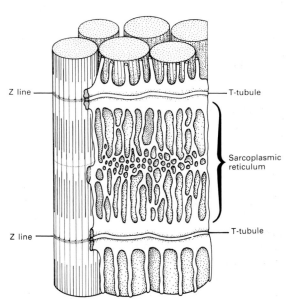

FIGURE 3·2 *The system of T-tubules and the sarcoplasmic reticulum which surrounds the fibres of striated muscle.*

membrane of the sarcoplasmic reticulum to calcium ions. These calcium ions are released and then interact with the thin myofilaments altering their configuration and permitting an interaction of the actin molecules with the myosin molecules, which is the basis of muscle contraction.

The number of nerve fibres that innervate a muscle is smaller than the number of muscle fibres. Within the muscle, nerve fibres branch and innervate several muscle fibres. The nerve and the group of muscle fibres that are simultaneously innervated by it are called a **motor unit**. Motor units vary in size. In densely innervated muscles such as those which move the eyeball, each motor unit may contain an average of about seven muscle fibres, whereas a motor unit in the leg may contain more than 1000 muscle fibres (Schmidt, 1978). The excitatory impulses of the motor units can be recorded as an electromyogram (EMG). The EMG is a recording of the extracellular potential of the muscle. The electrodes are either placed on the skin over the muscle or are inserted into the muscle between individual muscle fibres. In a fully relaxed muscle no change in potential is recorded, but with increasing force of contraction extracellularly recorded action potentials, or impulses, show up in the EMG.

The control of muscle contraction

Whereas cardiac muscle and smooth muscle have both an excitatory and an inhibitory nerve supply to control and modify the degree of contraction, skeletal or striated muscle has only an excitatory nerve supply. Nevertheless, skeletal muscle is under precise control. Two kinds of major sensory nerve fibres come from small sensory organs located in the tendons (**tendon organs**) and in **intrafusal** muscle fibres within the muscle (**muscle spindles**). Tendon organs seem to sense the deformation produced by tension in the tendon and send information about the degree of force of muscle contraction back to the central nervous system. Muscle

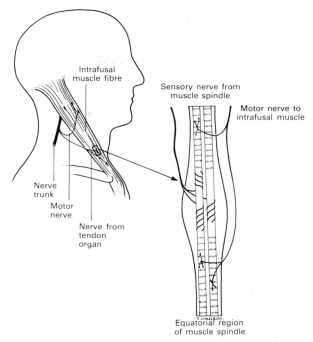

FIGURE 3·3 _Intrafusal muscle fibres and muscle spindles together with tendon organs pass important information about muscle length and force of contraction back to the central nervous system_

spindles are bundles of specialized muscle fibres that have nerve endings wrapped around their centre, where they lack the usual contractile apparatus (Fig. 3.3). Any unexpected load to a muscle causes the central region of the muscle spindle to stretch and to send nerve impulses to the spinal cord. These in turn initiate motor nerve impulses that return to the rest of the muscle and cause sufficient contraction to compensate for the initial stretching. Muscle spindles are themselves contractile and when they contract the central region is again stretched, thereby initiating motor activity to the surrounding muscle fibres until the muscle spindle is restored to its unstretched condition. In this situation they act like a mechanical servosystem such as the power steering of a large vehicle. Feedback information from tendon organs and muscle spindles allows a graded speed of contraction to variable loads on the skeleton and is vital in maintaining normal posture and integrating muscular movements during locomotion. As we shall see, some muscles have more muscle spindles than others. Tendon organs and muscle spindles provide information about the position of the body in space known as **proprioceptive** feedback information.

The force of muscle contraction

We have seen that during muscle contraction, the thin actin filaments are drawn between the thicker myosin filaments. However, the extent to which this basic process manifests itself as a force at the attachments of the muscle depends upon other circumstances. The duration of muscle fibre contraction is not uniform within whole muscles. In fact all muscles are made up of a variety of types of fibre. Some muscle fibres take only 80 ms to attain their maximum tension, whereas others take more than twice this time. Different types of muscle fibres are active in different situations. **Fast** muscle fibres are active during rapid movements and **slow** muscle fibres are active during prolonged contractions where maintenance of a low force is important. Fast-contracting muscle fibres fatigue quickly but generate high forces, whereas slow contracting fibres are more fatigue resistant but generate lower forces.

Muscles can be stretched to about 1.5 times their resting length without being damaged. At this length the filaments of actin and myosin no longer overlap and, therefore, no cross linkages can form between them and no contractile force can be developed. Muscles develop maximum force at about their resting length, but this force is again increasingly reduced if a muscle is shorter than its resting length. This is because the actin filaments are drawn in so far between the myosin filaments that they interfere with each other in the centre and, in addition, the myosin filaments eventually run up against the Z discs (Fig. 3.4).

Whether or not shortening of a muscle

occurs during contraction depends upon whether its points of attachment are able to move. Muscles that are fixed at both ends remain constant in length but exert tension under **isometric** conditions. If, on the other hand, a muscle is free to move at one end, the load remains the same throughout the contraction. This is then called **isotonic** contraction. When muscles contract naturally, the contraction is rarely completely isotonic or isometric but it is usual to define and use these two conditions in experimental situations because different properties of muscles can be revealed under these circumstances. To illustrate these points it is worth pointing out that when our jaw muscles are completely at rest our teeth are no more than

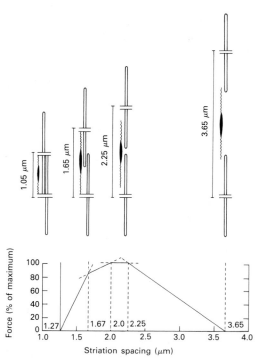

FIGURE 3·4 *The force developed by a contracting skeletal muscle (Y axis) in relation to its initial length (X axis). Force is expressed as a percentage of the maximal isometric force the muscle can develop. The amount of overlap between the thick and thin filaments is shown above the corresponding place on the force curve. After Gordon* et al. *(1966).*

1–2 mm apart. When we then clench our teeth together very little isotonic contraction occurs and the majority of the bite force generated in these circumstances is under isometric conditions but with the muscles close to their ideal resting length. A third form of muscle contraction is known as **eccentric** contraction when muscle tension is generated as a muscle lengthens, for example, contraction of the quadriceps femoris as you squat.

The maximum force a muscle of any given kind can exert is directly related to its cross-sectional area, but not to its length. It is the number of muscle filaments that can be packed within the cross-sectional area of a muscle that determines the maximum force of contraction. Measured in this way the maximum force of muscles from many different animals is about the same: 4–6 kg per cm² of cross-sectional area (Schmidt-Nielsen, 1979).

Some muscles are **pennate** (penna (L) = a feather) so that the anatomical arrangement of their fibres is not parallel with the muscle pull but at an angle to it. Muscles can be described as **bipennate** or **multipennate** and we shall come across examples of these later. Increasing the number of muscle fibres in this way increases the maximum force that the muscle can generate, but because the length of the fibres is reduced, this is at the expense of the distance over which the attachment can be moved (Fig. 3.5).

The external work performed by a muscle is the product of the force it generates and the distance that its free attachment moves. Therefore, the external work performed is also related to the length of the muscle fibres since most muscles can contract to, say, about one-third of their resting length. Two muscles of the same cross-sectional area will have the same force of contraction, but if one is twice as long as the other it will be able to contract further and, therefore, perform more external work. It follows that the external work performed by a pennate muscle cannot be greater than that of a parallel fibred muscle despite its increased force, because the distance moved at its insertion is greatly reduced.

Bone

Bones are organs made up of a tissue called bone as well as of cartilage and marrow. We are all familiar with dead or even fossilized bones but living bone bleeds and hurts when it is cut or broken and is in a continual state of change. Bone is essentially a highly vascular, constantly changing, mineralized connective tissue that is remarkable for its hardness and resilience. While all living bone consists of cells embedded in an organic matrix that is permeated by inorganic bone salts, its fine structure varies widely with age, site and natural history of the tissue. Because of this, cut sections of bone have been used to estimate the age at death of modern humans (Thompson, 1979) and fossil hominids (Thompson and Trinkaus, 1987) and the surfaces of bone have even been used to study the mechanisms of facial growth in early hominids (Bromage, 1985, 1986, 1989). To understand how these sorts of studies are possible, something of the development and the microstructure of bone must be understood.

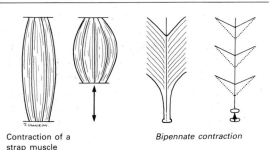

Contraction of a strap muscle

Bipennate contraction

FIGURE 3·5 *Muscle contraction in a strap muscle, where the fibres run its whole length, is approximately equal to one half of its resting length. Muscle contraction in a bipennate muscle results in much less movement at its insertion because the fibres are much shorter. However, in this situation there are many more muscle fibres and while the distance moved is reduced, the force exerted by the bipennate muscle is greater.*

Bone first forms either by **intramembranous ossification** or by **endochondral ossification**. In other words, it either develops surrounded by a connective tissue 'membrane' among a cluster of cells that produce a **matrix** which then subsequently mineralizes or, more usually, it replaces a cartilaginous precursor of the bone whose shape already resembles the bone in some crude way. (A matrix (L), literally means a female animal kept for breeding but in this sense it denotes something sheltered or enclosed, i.e. the bone cells, by their matrix.) A useful way of describing the microanatomy of bone is to follow these two processes of ossification through and then to describe the various remodelling processes that eventually result in the typical histological picture we recognize in cut sections of adult bone as well as on the surfaces of bone. This way of describing the structure of bone emphasizes the dynamic nature of bone tissue in life, as well as the initial processes that contribute to its formation.

Intramembranous ossification

Many bones of the skull as well as the clavicle develop 'in membrane' and so have no cartilage precursor. Mineralization first begins in bones at a **primary centre of ossification** and secondary centres that may appear later eventually coalesce with them to form a single bone. The appearance of ossification centres throughout the skeleton occurs in a regular sequence in a given species. Clusters of cells within a connective-tissue membrane begin to secrete a matrix called **osteoid** which consists mainly of collagen fibres together with a complex mixture of 'ground substance' proteins, including osteocalcin, osteonectin, sialoproteins and proteoglycans. This substance quickly mineralizes and the cells that have just formed it become the first **osteoblasts** (bone-forming cells) on the surface of this mineralized spicule. The mineral component of the spicule is largely hydroxyapatite (approximately $Ca_{10}(PO_4)_6 \cdot (OH)_2$). The collagen fibres in this first-formed bone form a random feltwork which gives it another name: **woven bone** (Fig. 3.6). Continued activity by these osteoblasts results in further **appositional growth** of bone in the form of layers of **lamellae**. Some of the bone-producing cells quickly become enclosed by their own deposits and are thereby established within the bone and are known as **osteocytes**. These cells communi-

---

FIGURE 3·6 *Scanning electron micrograph (stereo pair) illustrating the random feltwork of collagen fibres laid down in rapidly formed woven bone. Foetal human mandible, wall of tooth crypt (× 1180). After Boyde and Hobdell (1969). Courtesy of Alan Boyde.*

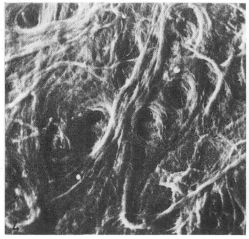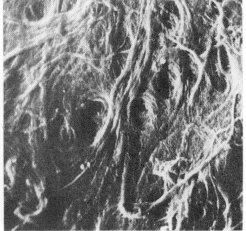

cate with their neighbouring cells via long cell processes that run in minute canals, known as **canaliculi**, through the bone. The chamber within which the osteocyte is enclosed is known as an **osteocyte lacuna** (lacuna (L) = a pond). As bone formation continues, in for example the membrane bones of the skull vault, the enlarging primary centre of ossification differentiates into two types of bone. The first is a middle region where **trabeculae** (diminutive of trabes (L) = beam of a ship or battering ram) have formed from earlier-formed bone that has been partially eaten away or resorbed so that the spaces between them containing blood vessels and connective tissue have enlarged. The cells that resorb bone are called **osteo-clasts** (klasis (Gk) = a breaking). Under the membrane that encloses the forming bone (known as the **periosteum**) the bone forms more densely. The outer and inner regions of the bone become more compact as lamella bone is laid down at the surface and around blood vessels. Large blood vessels that become incorporated at the surface of the bone create **primary vascular canals** that are surrounded by lamella bone. In this way a region of trabecular (sometimes called **cancellous**, cancelli (L) = a lattice work) bone becomes sandwiched between an inner and outer region of more compact (or cortical) bone. Before we describe further remodelling processes that occur, we shall outline the process of endochrondral ossification because, once bone has been laid down, remodelling processes are common to all bone, however ossified.

Endochrondral ossification

Young cartilage has the advantage of being able to grow **interstitially** as well as in an appositional manner which is the reason it is retained in the growing skeleton. In other words cartilage cells enclosed by their matrix can divide and for a while occupy the same lacuna. Soon, however, the cells become separated by fine matrix partitions which subsequently thicken as the cells move apart from one another. In this way a mass of young cartilage can grow more quickly in volume than by appositional growth alone. Nevertheless, cartilage also grows in an appositional manner at the surface. Here, continual surface increments are added with some of the cells that secrete cartilage matrix becoming enclosed in lacunae. Larger volumes of cartilage can be formed by a combination of interstitial and appositional growth than by appositional growth alone as is the case with bone.

The first indication of the appearance of a primary ossification centre in a long bone occurs in the cartilage cells placed deep within the centre of the primitive shaft or **diaphysis** (dia (Gk) = apart; physis (Gk) = growth in the sense that the shaft holds the growing ends apart). These cells begin to enlarge and eventually become so big that their adjacent lacunae become confluent. Eventually the cartilage cells appear to perish and the thin walls of the lacunae calcify. The centre of the bone at this time is then formed of calcified cartilage. Calcified cartilage is very highly mineralized. At the same time, the young osteoblasts of the outer membrane surrounding all the cartilage precursor of the bone, (the **perichondrium**) form a collar of young trabecular bone around this core of calcified cartilage in exactly the same way as we described above for intramembranous ossification. Sprouts of blood vessels now arise from the perichondrium and invade right through the outer bony collar into the central region of calcified cartilage by way of large vascular channels. Osteoclasts that have invaded via the blood vessels resorb some of the calcified cartilage and so enlarge the confluent lacunae which then become known as **medullary spaces**. Osteoblasts then lay down new bone on top of what remains of the calcified cartilage columns and, eventually, through a continual process of resorption by osteoclasts and new bone formation by osteoblasts, the shaft of the long bone is formed. This process gradually spreads

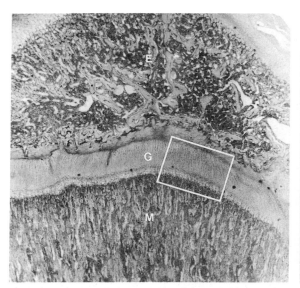

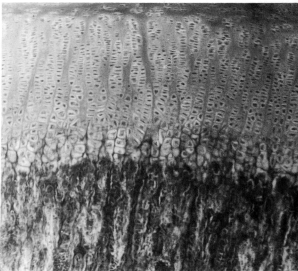

FIGURE 3·7   *Endochondrial ossification adjacent to the growth plate cartilage of a long bone. Columns of cartilage cells (chondrocytes) increase in size within their lacunae and eventually appear to die. Osteoblasts then replace the calcified cartilage matrix with newly formed bone. E is the epiphysis, G is the growth plate and M the metaphysis of a growing long bone. On the right hypertrophying chondrocytes can be seen from a section of the growth plate at higher magnification (× 80).*

towards the ends of the bone and the newer formed parts of the shaft become known as the **metaphysis** (meta (Gk) = after).

The ends of long bones develop secondary centres of ossification that are separated from the metaphysis by a plate of cartilage that continues to grow in the manner described above until the length of the bone is completed (Fig. 3.7). The plate of cartilage is known as the **growth plate**. The cells of the growth plate continue to enlarge and die, their matrix continues to calcify and be invaded by blood vessels and is then resorbed by osteoclasts and replaced by new bone formed by osteoblasts. The secondary centres of ossification develop into the bony **epiphyses** (epi (Gk) = upon; physis (Gk) = growth) that occur at the ends of bones.

Remodelling of bone during growth and development

We noted earlier that bone differs from cartilage in that it cannot grow interstitially.

Bone growth at a growth-plate cartilage or at an ossification centre is on its own, insufficient to form the shapes of complex bones. Constant remodelling occurs within bone tissue and also at the internal and external bone surfaces in the form of bone deposition and bone resorption.

Bone surfaces that face the direction of growth show bone deposition and, conversely, bone surfaces which face away from the direction of growth tend to undergo bone resorption. In this way the whole cortex of a bone can become translocated as one side remains resorptive and the other continues to lay down new bone (Enlow, 1968). These two types of bone surface look different. We have seen that osteoclasts resorb bone and when whole bone surfaces are being actively resorbed they are scalloped with the large lacunae that house each osteoclast; these are commonly called **Howship's lacunae** (after John Howship (1781–1841) the surgeon who first described them). Forming bone surfaces look entirely different. There are no How-

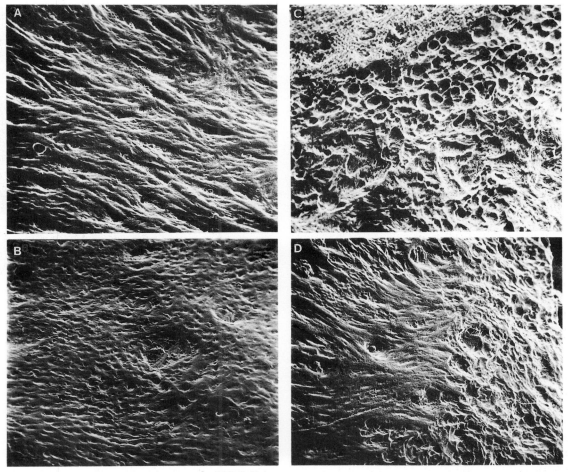

FIGURE 3·8 *Forming and resorbing periosteal bone surfaces look different. A and B are scanning electron micrographs made from epoxy replicas of mandibles SK 64 and Sts 24 respectively. They show forming bone surfaces; fieldwidths 1.5 mm for SK 64 (A) and 0.75 mm for Sts 24 (B). Resorbing periosteal bone surfaces are characterized by scalloped Howship's lacunae and contrast with the mineralizing collagen fibre pattern of the forming or resting bone surface. C is a region of periosteal resorption in a modern human (fieldwidth 207 μm) and D is the periosteal bone surface of the SK 64 mandible showing a reversal line with forming bone on the left of the field and resorbing bone on the right of the field (fieldwith 1.3 mm). Courtesy of Tim Bromage.*

ship's lacunae and mineralizing collagen fibres can be seen all over the surface (Fig. 3.8).

The bone surface on the face of some juvenile fossil hominids is preserved sufficiently well to reveal resorbing and forming surfaces with scanning electron microscopy (SEM) (Bromage, 1985, 1986, 1989). The face of juvenile hominids attributed to *Australopithecus* show forming bone surfaces compatible with the relatively prognathic faces of the adults of this group. The face of individuals attributed to *Paranthropus* show resorbing surfaces which are compatible with the relatively flatter faces of some adult *Paranthropus* crania.

Bone formation at primary and secondary sites of ossification is in the first instance very rapid. It consists of either woven bone or lamella bone that surrounds blood-vessel canals or that has become incorporated into the growing bone at a surface. When lamella bone surrounds a blood vessel that has become incorporated into new bone, the amount of bone relating to and nourished by this blood vessel is called a **primary osteon**. Three-dimensionally, osteons resemble cylinders of bone arranged around a blood vessel. In the life history of a bone, large regions of lamella bone, woven bone and calcified cartilage, where it remains, are resorbed as erosion spaces develop. These spaces become invaded by new regions of bone growth within the older tissue. The resorption is carried out by osteoclasts which leave a characteristic scalloped border where cells have taken bites out of the bone. The scalloped line which is visible in sections of bone where this has occurred is sometimes known as a **reversal line** because new bone begins to form again at this limit of bone resorption. Scalloped reversal lines are characteristic of the perimeter of **secondary osteons** (but not primary osteons).

Secondary osteons then, are the product of an internal remodelling process which secondarily alters the structure of the blood-vessel canals associated with the primary osteons. A secondary osteon, however, still represents the amount of bone relating to the territory of a single (but now new) nutrient blood vessel. The formation of new bone within an erosion space occurs much more slowly than the first-formed woven bone. Such new bone also forms in concentric layers or lamellae that fill in the hole, so that histological sections of secondary osteons appear to be made up of a series of rings arranged around a central blood-vessel canal (Figs 3.9 and 3.10). Similar slower-growing lamellae develop at the circumference of bones from the periosteum, where the surface of the bone is smooth and uninterrupted. These lamellae are known as **circumferential lamellae** when they overlay a proportion of the bone surface. The collagen fibres within a single lamella have the same orientation but

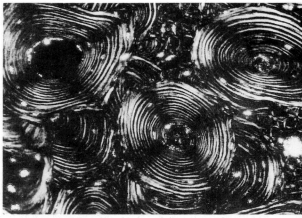

FIGURE 3.9 *Polarized light micrograph of secondary osteons in cortical bone. Concentric lamellae can be seen in each osteon as can the central blood vessel canal. Fieldwidth 540 μm. Courtesy of Chris Riggs.*

FIGURE 3·10 *Scanning electron micrograph (backscattered image that reveals mineral density) of two human parietal bones (a) a full thickness specimen of a one day old infant and (b) a full thickness specimen of a 4.5 year old child (× 48). In (a) the most dense (whitest) band and cores are mineralized woven bone. Elsewhere, lamella bone can be seen infilling primary osteons. Occasionally, within the woven bone, regions where osteoclasts have cut cones can be seen; these appear as black holes and have scalloped borders (unlike the periphery of the infilled primary osteons). The parietal bone in (b) is older and much thicker (× 40). Circumferential lamella bone was being laid down on the ectocranial and endocranial surfaces. A sequence of internal remodelling was underway in the cortex and osteoclasts have cut cones into the circumferential lamellae which in most cases were infilling (or have infilled) with secondary osteons. The darkest secondary osteons are the least dense (therefore the youngest) and the whitest are the most dense (the oldest). Extensive evidence of previous bone remodelling is apparent from the internal structure of the trabecular bone. This is evidence for the cortical origin of the trabeculae. Resorption and deposition of established trabecular bone occurs at the surface of trabeculae and not through further secondary osteon formation. A few regions of resorbing and forming bone can be seen at the surface of some trabeculae in this micrograph. After Boyde et al. (1900). Courtesy of Alan Boyde, Sheila Jones, Elaine Maconnachie and Springer-Verlag.*

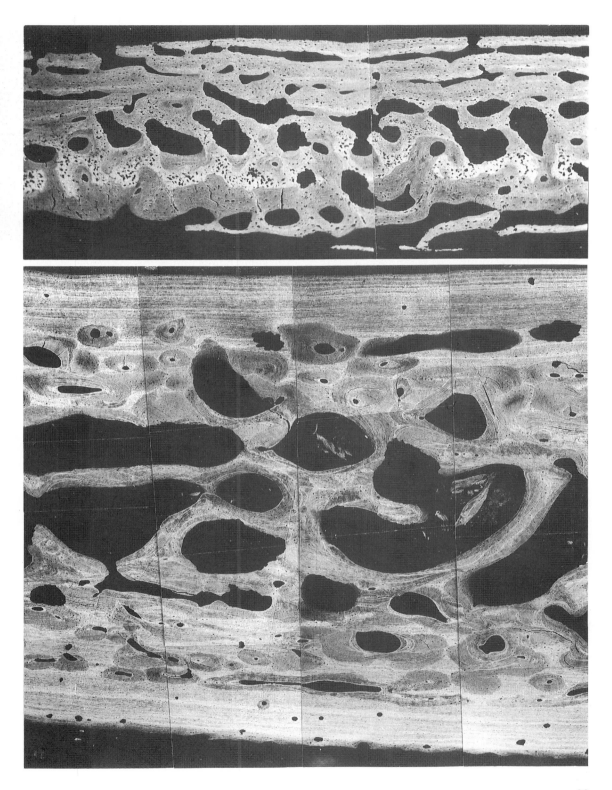

the collagen fibres in immediately adjacent lamellae may, but do not always, have a different orientation; this contrast is responsible for their layered structure, apparent in light microscopy (LM) and especially in polarized light microscopy (PLM).

Adult bone remodelling

Secondary osteons are sometimes called **Haversian systems** (because they were first described by Clopton Havers in 1691). Secondary osteons continue to replace bone that is resorbed throughout the life of an individual and many generations of secondary osteons can be seen in sections of older human bone. Estimates of the number of secondary osteons have been used to provide estimates of the age at death of Neanderthals (Thompson and Trinkaus, 1987) and comparisons with modern humans suggest that none were much older than 35 years at death. Remnants of woven bone, primary osteons and earlier formed secondary osteons that fail to become resorbed constitute the **interstitial bone**. This occupies the crevices between the subsequently formed secondary osteons of cortical bone.

In mature adult bones the trabeculae of the cancellous or spongy regions of a bone represent more fragmentary interconnected systems of bone tissue with large spaces between them that contain blood vessels, bone cells and marrow. Trabecular bone tissue contains no vascular canals because of its close proximity to such large volumes of marrow space. In adult humans the cortical bone turnover rate is about 2–3% year^{-1}, whereas roughly 25% of trabecular bone is renewed annually (Eriksen, 1986). A major function of bone tissue is to act as a reservoir for calcium, phosphorus, magnesium, sodium, citrate, carbonate and hydroxyl ions and it is primarily resorption and deposition of trabecular bone that facilitates this. In cortical bone it is secondary osteon formation that is responsible for this turnover. Exactly the same amount of bone is replaced by a new Haversian system as was resorbed prior to its formation. In cancellous or spongy bone the trabeculae are alternately resorbed away by osteoclasts and built up again by osteoblasts in an appositional manner. However, because osteoblasts slightly underfill the resorption cavities of osteoclasts during trabecular bone turnover, the trabecular plates become progressively thinner with increasing age. On the surface of a bone, however, just beneath the periosteum, osteoblasts tend to overfill resorption cavities. The net effect with age is for a bone to increase in diameter and for the spongy cancellous region to become increased in size and be composed of thinner trabeculae.

Besides being under the influence of well known systemic factors like vitamin D, parathyroid hormone and calcitonin, bone resorption is now known to be effected by other powerful factors such as the sex hormones, prostaglandins, and interleuken 1 (Raisz, 1988). Each of these factors is important in the aetiology of diseases, such as senile osteoporosis, that form an important aspect of palaeopathology.

Lever systems

While not to do with the microanatomy of muscle or bone, lever systems are central to the mechanics of movement. It is fitting to review some of the basic principles relating to the mechanics of levers at the end of this chapter, because they describe how muscle forces are applied to bone.

A **lever** is regarded simply as a rigid bar with no account taken of its shape or structure. While most long bones are easily conceived as rigid bars the bones of the skull, for example, that come in many shapes and sizes are not, and yet they can still act as levers. A **fulcrum** is the point around which a lever rotates. That part of the lever between the fulcrum and the point of force application is known as the **force arm** or **power arm**. That point between the fulcrum and the point of load application is known as the **load arm**.

TABLE 3·1 Sequence of epiphyseal union in the long bones of the limbs. The stage at which the permanent dentition becomes complete is indicated by horizontal dotted lines (after Schultz, 1956).

Sequence	Elbow	Hip	Ankle	Knee	Wrist	Shoulder	Elbow	Hip	Ankle	Knee	Wrist	Shoulder
1	Hum.d.	—	—	—	—	—	Hum.d.	—	—	—	—	—
2	Epicon.	—	—	—	—	—	Epicon.	—	—	—	—	—
3	Ulno p.	s.Tro.	—	—	—	—	Ulna p.	—	—	—	—	—
4	—	g.Tro.	—	—	—	—	Rad.p.	s.Troc.	—	—	—	—
5	Rad.p.	Fem.h.	—	—	—	—	—	Fem.h.	—	—	—	—
6	—	—	Tib.d.	—	—	—	—	g.Troc.	—	—	—	—
7	—	—	Fib.d.	—	—	—	—	—	Tib.d.	—	—	—
8	—	—	—	Fem.d.	—	—	—	—	Fib.d.	—	—	—
9	—	—	—	Tib.p.	—	—	—	—	—	Fib.p.	—	—
10	—	—	—	Fib.p.	—	—	—	—	—	Fem.d.	—	—
11	—	—	—	—	—	Hum.p.	—	—	—	Tib.p.	—	—
12	—	—	—	—	Rad.d.	—	—	—	—	—	—	Hum.p.
13	—	—	—	—	Ulna d.	—	—	—	—	—	Ulna d.	—
14	CAPUCHIN AND					—	GUENON AND			—	Rad.d.	—
15	SPIDER MONKEY					—	MANGABEY					—
1	Hum.d.	—	—	—	—	—	Hum.d.	—	—	—	—	—
2	Epicon	—	—	—	—	—	Ulna p.	—	—	—	—	—
3	Ulna p.	—	—	—	—	—	Epicon.	—	—	—	—	—
4	—	s.Tro	—	—	—	—	—	Fem.h.	—	—	—	—
5	Rad.p.	g.Tro	—	—	—	—	—	s.Tro	—	—	—	—
6	—	Fem.h.	—	—	—	—	Rad.p.	g.Tro	—	—	—	—
7	—	—	Tib.d.	—	—	—	—	—	Fib.d.	—	—	—
8	—	—	Fib.d.	—	—	—	—	—	Tib.d.	—	—	—
9	—	—	—	Fib.p.	—	—	—	—	—	Fib.p.	—	—
10	—	—	—	Tib.p.	—	—	—	—	—	Tib.p.	—	—
11	—	—	—	Fem.d.	—	—	—	—	—	Fem.d.	—	—
12	—	—	—	—	Rad.d.	—	—	—	—	—	—	Hum.p.
13	—	—	—	—	Ulna d.	—	—	—	—	—	Rad.d.	—
14	GIBBON					Hum.p.	ORANG-UTAN				Ulna d.	—
15						—						—
1	Humd.	—	—	—	—	—	Hum.d.	—	—	—	—	—
2	Epicon.	—	—	—	—	—	Epicon.	—	—	—	—	—
3	Ulna p.	—	—	—	—	—	Ulna p.	—	—	—	—	—
4	—	s.Tro.	—	—	—	—	Rad.p.	—	—	—	—	—
5	—	g.Tro.	—	—	—	—	—	Fem.h.	—	—	—	—
6	Rad.p.	—	—	—	—	—	—	s.Tro.	—	—	—	—
7	—	Fem.h.	Tib.d.	—	—	—	—	g.Tro.	—	—	—	—
8	—	—	Fib.d.	—	—	—	—	—	—	—	—	Hum.p.
9	—	—	—	Fem.d.	—	—	—	—	Tib.d.	—	—	—
10	—	—	—	Tib.p.	—	—	—	—	Fib.d.	—	—	—
11	—	—	—	Fib.p.	—	—	—	—	—	Fem.d.	—	—
12	—	—	—	—	Rad.d.	Hum.p.	—	—	—	Fib.p.	—	—
13	—	—	—	—	Ulna d.	—	—	—	—	Tib.p.	—	—
14	CHIMPANZEE					—	GORILLA			—	Rad.d.	—
15						—				—	Ulna d.	—
1	Hum.d.	—	—	—	—	—	Hum.d.	—	—	—	—	—
2	Ulna p.	—	—	—	—	—	Ulna p.	—	—	—	—	—
3	Rad.p.	—	—	—	—	—	Rad.p.	—	—	—	—	—
4	Epicon.	—	—	—	—	—	Epicon.	—	—	—	—	—
5	—	Fem.h.	—	—	—	—	—	Fem.h.	—	—	—	—
6	—	s.Tro	—	—	—	—	—	s.Tro.	—	—	—	—
7	—	g.Tro	—	—	—	—	—	g.Tro.	—	—	—	—
8	—	—	Tib.d.	—	—	—	—	—	Tib.d.	—	—	—
9	—	—	Fib.d.	—	—	—	—	—	Fib.d.	—	—	—
10	—	—	—	Tib.p.	—	—	—	—	—	Fib.p.	—	—
11	—	—	—	Fib.p.	—	—	—	—	—	Fib.p.	—	—
12	—	—	—	Fem.d.	—	—	—	—	—	Fem.d.	—	—
13	—	—	—	—	Rad.d.	—	—	—	—	—	Rad.d.	—
14	ESKIMO AND			—	Ulna d.	—	EUROPEAN			—	Ulna d.	—
15	AMERICAN INDIAN			—		Hum.p						Hum.p.

Hum. = humerus; Epicond. = medial epicondyle of humerus; Rad. = radius; Fem. = femur; s.Tro. = small trochanter; g.Tro. = great trochanter; Tib. = tibia; Fib. = fibula; h. = head; p. = proximal; d. = distal.

There are different classes of levers that are defined depending upon the different arrangements of the fulcrum, force arm and load arm (Fig. 3.11). **First class levers** have the fulcrum positioned between the force arm and the load arm. The usual example of a first class lever that is most easily conceived is a see-saw. First class levers are used in balancing weight and/or changing the direction of pull. There is usually no gain in mechanical advantage with a first class lever. One example of a first class lever is given in Fig. 3.11 but another important example relates to bipedal locomotion. When standing on the right leg the right hip joint acts as a fulcrum, the load being body weight applied medial to this hip joint and the force applied being provided by the contraction of the right gluteus medius and minimus muscles lateral to the joint. This example is developed further and illustrated in Chapter 19.

Second class levers have the fulcrum at one end of the lever and the applied force at the other, with the load situated between them. This is the principle on which weight is lifted in a wheelbarrow. The example illustrated in Fig. 3.11 shows how the foot and calf muscles can act as a second class lever. Second class levers gain mechanical advantage thereby allowing large loads to be moved but only at the expense of a loss of speed.

Third class levers are most common within the musculoskeletal system. Third class levers

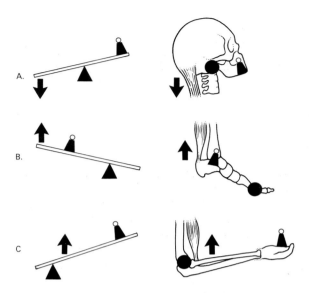

FIGURE 3·11 *Examples of (a) a first class lever; (b) a second class lever; and (c) a third class lever as described in the text.*

have the force applied between the fulcrum and the load. They work at a mechanical disadvantage moving less weight but often at great speed. The biceps muscle working across the elbow is a good example of this class of lever. Bear in mind that lifting 50 kg with the arms in this way requires considerably more effort than lifting 50 kg of body weight by rising onto the toes as in the example of a second class lever illustrated in Fig. 3.11.

CHAPTER FOUR

THE BONES OF THE SKULL

Each of the bones that comprise the modern human skull are described here in detail since they form an essential framework for appreciating the comparative anatomy of other hominoid skulls. Chapter 5 provides an overview of the comparative anatomy of great ape and fossil hominid mandibles and crania and builds on many details about the bones of the modern human skull set out in this chapter. Some details of individual bones are not described in this chapter because they figure more appropriately elsewhere in the book (for example, the internal morphology of the occipital bone is described in Chapter 9 on the intracranial region). Where this is the case other sections of the book are cross-referenced in the text.

The **mandible** and the **cranium** together are known as the **skull** (Fig. 4.1). When the mandible is removed the cranium alone comprises the mid and upper **facial skeleton**, the **calvarium** that surrounds the sides and top of the brain (calvus (L) = bald) and the bones of the **cranial base** (that lie beneath the brain). Sometimes all the bones that surround the brain are referred to together as the **neurocranium** or **brain case**. Another common way of describing the calvarium is to refer to the individual bones of the neurocranium or brain case that form the walls and roof of the cranial cavity as the cranial **vault bones**. Occasionally, especially among physical anthropologists, the parts of the bones that form only the roof of the

FIGURE 4·1 *Skull in lateral view.*

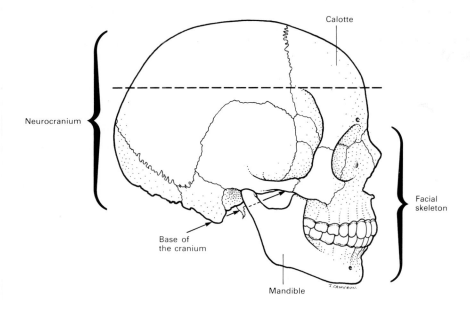

cranial vault are referred to together as the **calotte** (which simply means the skull cap). Besides these broad descriptive regions of the skull several cavities exist which in life house and protect the brain and the eyes or surround the structures of the nose and the mouth. These are the cranial cavity, the two orbital cavities, the nasal cavity and the oral cavity.

The mandible

The mandible is one of the first bones in the body to begin to ossify (two others being the clavicles). At birth the two halves of the mandible are separated anteriorly by a mid-line **symphysis** or **fibrocartilaginous joint** (see Fig. 4.12) but begin to fuse together between six and nine months of age (Marshall, 1986) so that in skeletal material older than this both halves of the mandible are joined together. The **alveolar bone** of the mandible provides bony support for the lower teeth but the mandible also provides skeletal attachment for some of the muscles of the tongue and some of the muscles in the floor of the mouth. Precise occlusal relations with the upper dentition, both static and dynamic, are possible because the **muscles of mastication** also attach to the mandible and are able between them to control both fine and powerful movements of the lower jaw. The complicated morphology of the **temporomandibular** joints are adapted to permit all mandibular movements, not only during chewing but, for example, while speaking, singing or yawning. The detailed structure of this joint and the anatomy of the muscles that move the mandible are described in Chapter 6.

The **body** or **corpus** of the mandible is horse-shoe shaped when viewed from above. The vertical or ascending **rami** (ramus (L) = a branch) project upwards from the posterior ends of the body. The external surface of the body (also known as the **labial** (lip) or **buccal** (cheek) surface), is marked anteriorly in the mid-line by a **mental protuberance** (mentum (L) = chin) which begins below the alveolar

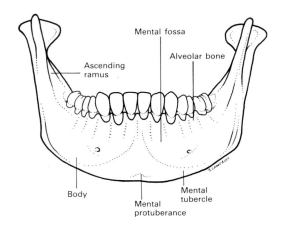

FIGURE 4·2 *Anterior view of the mandible.*

bone of the central incisor teeth and spreads downwards and outwards to the lower border of the mandible to form a raised triangular area, the **mental trigone** (Fig. 4.2). The lowest, most lateral and usually most prominent part of this raised area is known as the **mental tubercle**. Above the mental trigone the alveolar bone is hollowed to form a **mental fossa** either side of the mid-line. Together, these features give the human chin its characteristic appearance.

The anterior border of the ascending ramus of the mandible joins the body and sweeps forwards as an **external oblique line** or ridge on the buccal aspect of the body (Fig. 4.3). The **mental foramen**, the anterior opening of the **mandibular canal** that runs below the teeth, is another prominent feature on the external surface of the mandible. The mental foramen is usually situated below and between the position of the root apices of the premolar teeth. Because of the way the mandible grows the opening of the mental foramen usually points backwards in modern humans.

The internal or lingual surface of the mandibular body is marked by a ridge, the **mylohyoid line** that indicates the line of attachment of the mylohyoid muscle that lies

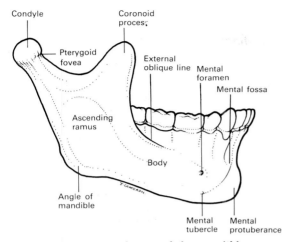

FIGURE 4·3 *Lateral view of the mandible.*

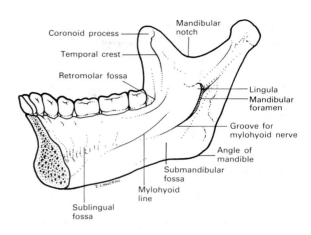

FIGURE 4·4 *Medial aspect of the mandible.*

across the floor of the mouth (Fig. 4.4). This line becomes less distinct in the premolar and canine region. It runs from a position **above** the level of the root apices of the last molar tooth, obliquely downwards, to end *below* the root apices of the premolar teeth. Two of the salivary glands in the floor of the mouth are closely related to the lingual surface of the mandible and create two depressions in the mandible. One of these depressions, the **submandibular fossa**, lies below the posterior end of the mylohyoid line and the other, the **sublingual fossa**, lies above the mylohyoid line (but still below the level of the roots of the premolar teeth). A prominent groove on the lingual aspect of the mandible marks the position in life of the blood vessels and nerves which supply the mylohyoid muscle (see Chapter 6). This **mylohyoid groove** begins at the mandibular foramen (the entrance of the mandibular canal) runs forwards below the mylohyoid line and fades away in the submandibular fossa.

Behind the chin, in the mid-line of the lingual surface of the mandible, there are two raised pairs of bony tubercles known as the **genial tubercles** (geneion (Gk) = chin) (Fig. 4.5). These tubercles mark muscular

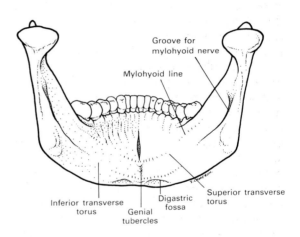

FIGURE 4·5 *Posterior view of the mandible.*

attachments for the genioglossus and geniohyoid muscles and often have adjacent hollows associated with them. On the lower border of the mandible either side of the mid-line are two depressions for attachment of the anterior bellies of the digastric muscles, the **digastric fossae**. Two faint bars of bone run horizontally around the inside of the chin, the superior and inferior **transverse tori**. These tori are not usually very prominent in modern human mandibles. The

35

superior transverse torus runs just above the genial tubercles and the inferior transverse torus just above the digastric fossae. In great apes (see Chapter 5) the inferior transverse torus is extended at the lower border of the mandible as a shelf of bone a short distance back along the floor of the mouth. This torus is sometimes referred to as a 'simian shelf'.

The vertical or ascending ramus of the mandible joins the horizontal body of the mandible in the region of the **retromolar fossa** behind the last molar tooth. The **angle** of the mandible, marks the junction of the horizontal inferior border of the body of the mandible with the near-vertical posterior border of the ascending ramus. Two prominent processes, the **condyle** and the **coronoid process**, are separated along the superior border of the ramus by a notch, the **sigmoid** or **mandibular notch**. The constricted region beneath the condyle is known as the **neck** of the mandible. More details of the mandibular condyle are described with the temporomandibular joint in Chapter 6.

The lateral surface of the ascending ramus of the mandible is usually roughened (especially in male mandibles), for the attachment of the masseter muscle. The medial surface of the angle of the mandible is also roughened by raised ridges that are associated with the attachment of the medial pterygoid muscle. Higher up the medial surface of the ascending ramus is the mandibular foramen. This lies at the level of the occlusal plane of the teeth and in human mandibles is midway between the anterior and posterior borders of the ascending ramus. A spur of bone, the **lingula** (lingula (L) = a little tongue), extends backwards from the anterior aspect of the mandibular foramen. In life a ligament, the **sphenomandibular ligament**, connects it to the spine of the sphenoid bone (see the section on the sphenoid bone in this chapter and Chapter 13). A **temporal crest** or **internal oblique line** on the medial aspect of the ascending ramus of the mandible runs from the coronoid process to the lingual aspect of the alveolar bone around the last molar tooth and is associated with the attachment of the temporalis tendon.

The parietal bone

The paired parietal bones (paries (L) = a wall) form most of the sides of the cranial vault. The parietal bones form in membrane and the parietal **eminence** of adult crania marks the region of the original centre of ossification. Occasionally, this may remain pronounced and is then referred to as a parietal **boss**. The parietal bones articulate with each other along the **sagittal suture** (Fig. 4.6). They also form two Y-shaped articulations, one anteriorly at *bregma* where the sagittal suture meets the **coronal suture** at the frontal bone, and another posteriorly at *lambda* where the sagittal suture meets the lambdoidal sutures at the occipital bone.

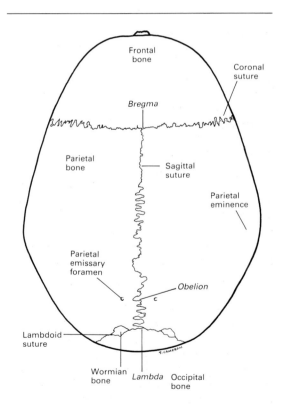

FIGURE 4·6 *Superior view of the cranium.*

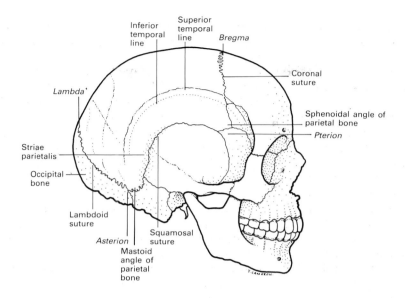

FIGURE 4·7 *Sutures and landmarks of the cranium.*

Along the lambdoidal suture and especially at the site of the **posterolateral fontanelle** (one of the deficiences in the bony vault covered only by tough fibrous tissue at birth) there are commonly small additional sutural or **Wormian bones**.

Although rounded to follow the contours of the vault, the parietal bones are quadrilateral having near-parallel superior and inferior borders and near-parallel anterior and posterior borders. The external surface is marked by parts of the **superior** and **inferior temporal lines** and also often by the **striae parietalis** (Fig. 4.7). These radiate from the temporal bone and extend over the inferior aspect of the parietal, parallel with the overlying fibres of the temporalis muscle. The internal surface of the parietal bone is deeply scored by grooves for the middle meningeal vessels and also near the sagittal suture by deep pits, once occupied by arachnoid granulations (Fig. 4.8). Large **diploic veins** form between the inner and outer tables of adult parietal bones and are easily recognized as **parietal spiders** on lateral skull radiographs (Fig. 4.9; see also Chapter

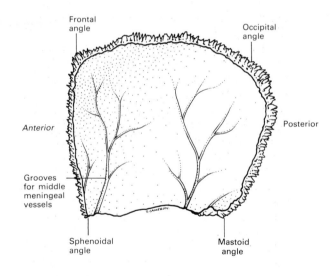

FIGURE 4·8 *Endocranial surface of the parietal bone.*

9). Typically, there are two parietal emissary veins that pass right through this bone and mark *obelion* (the mid-line point on the sagittal suture between the parietal emissary

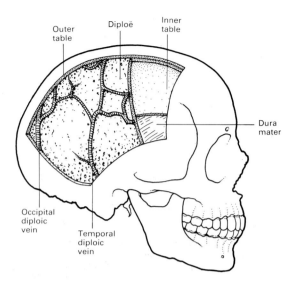

FIGURE 4·9 *The diploë of the vault bones.*

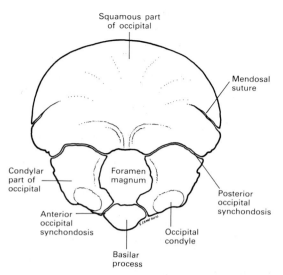

FIGURE 4·10 *Occipital bone of a neonate.*

foramina; Fig. 4.6). The four corners of the parietal bone are known as the occipital, frontal, sphenoidal and mastoid **angles**, respectively. The mastoid angle may be quite pronounced and creates a notch (known as the parietal notch) in the squamous temporal bone above the mastoid region. The confluence of the **lambdoid, occipitomastoid** and **parietomastoid** sutures here is known as *asterion* (because it looks like a star; aster (Gk) = a star).

The occipital bone

In the majority of cases this bone which forms the back of the cranium presents first at birth and it is reputed that because of this the bone was named the **occipital** (occipio (L) = I begin or commence). In neonates there are four distinct parts to the occipital bone that are joined to each other by **synchondroses** or primary cartilaginous joints; the **squamous** part, two **condylar** portions and a **basilar** part (Fig. 4.10). The squamous part behind the foramen magnum is incompletely divided in immature bones

by two sutures (the left and right **mendosal sutures**). These run laterally in the future position of the superior nuchal line and mark the junction between the upper part of the squamous occipital that forms in membrane from the rest of the occipital bone that is preformed in cartilage. Failure of the mendosal suture to fuse results in the formation of an extra bone at the junction of the lambdoid and sagittal sutures. The high prevalence (over 30%; Martin and Saller, 1959; Brues, 1977) of this bone in Peruvian populations has led to it being named the **Inca bone**. The two condylar parts of the occipital bone form the sides of the foramen magnum and the basilar part forms the portion of the occipital bone anterior to the foramen magnum. By six years of age the human occipital has fused to form a single bone and usually no sutures or synchondroses remain between any of the parts.

The internal aspect of the squamous part of the occipital bone is cupped to support the posterior part of the brain. It is marked on its external surface by the **external occipital protuberance** (Fig. 4.11). The

superior nuchal line runs laterally around the bone, on either side of this protuberance and divides the area below it that is roughened by the attachment of the nuchal and suboccipital musculature from a smoother region above it. Two **jugular processes** of the occipital bone extend laterally on either side of the occipital condyles. These form the posterior margin of the **jugular foramen** the boundary of which is completed anteriorly by the temporal bone. Behind the condyles there are **posterior condylar emissary foramina** and above the articular surface of each condyle, another canal, the **hypoglossal** or **condylar canal**, passes through the occipital bone. The basilar part of the occipital bone projects forwards in front of the foramen magnum to form a very important primary cartilaginous joint with the body of the sphenoid bone. This joint, the **sphenooccipital synchondrosis** is an important site of growth in the vertebrate cranium and has attracted special attention in hominoids. Internal markings on the occipital bone are described in Chapter 9.

The frontal bone

The frontal bone (frons (L) = brow), develops from two intramembranous centres of ossification whose positions are marked by the frontal **eminences** of the skull (Fig. 4.12). Before two years of age, the frontal bone is always divided by the **metopic suture** (metopon (Gk) = forehead, the space between the eyes). This usually closes with the **anterior fontanelle** (the largest of the deficiencies in the bony vault that exist at birth), although persistent metopic sutures are common in some groups of adult crania. The **vertical** portion of the frontal bone forms the anterior part of the cranial vault. The **horizontal** part or **orbital plates** of the frontal bone form the roof of the orbits (Fig. 4.13). The external surface of the vertical part of the frontal bone is marked in the mid-line by a prominence above the nose known as *glabella* (glaber (L) = smooth or

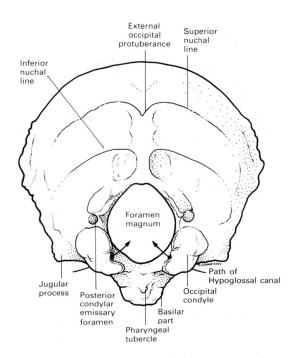

FIGURE 4·11 *Inferior view of the occipital bone.*

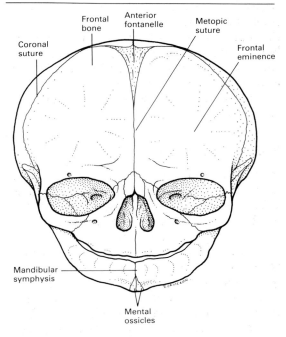

FIGURE 4·12 *Anterior view of a neonatal skull.*

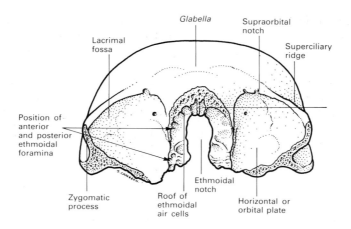

FIGURE 4·13 *Inferior view of the frontal bone.*

without hair) and laterally on each side by two **superciliary ridges** above the orbits. The supraorbital margin of the frontal bone is traversed by a **supraorbital foramen** or, commonly, is simply interrupted by a **supraorbital notch**. The internal and external **angular processes** of the supraorbital arch mark the medial and lateral extremities of the frontal bone overlying each orbit.

Laterally, the superior and inferior temporal lines extend backwards over the frontal portion of the vault. These mark the most anterior attachment of the temporalis muscle. (More details about what forms each line are given in Chapter 6.) Internally, the frontal bone, like the parietal bone, is marked by grooves for meningeal vessels and pits for **arachnoid granulations** (the sites where cerebrospinal fluid (CSF) returns to the venous sinuses of the intracranial region; see Chapter 9). As well as these markings there is a mid-line groove on the inner surface of the frontal bone for the superior sagittal sinus (see Chapter 9). Anteriorly this groove becomes a crest, the **frontal crest** (Fig. 4.14).

The orbital plates of the frontal bone are divided by an **ethmoidal notch** into which the ethmoid bone articulates. When disarticulated, openings into the right and left **frontal air sinuses** can be seen behind *glabella* in the inferior aspect of the frontal bone. The frontal air sinuses begin to develop from the ethmoidal air sinuses after the age of two

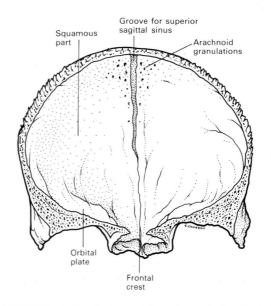

FIGURE 4·14 *Endocranial surface of the frontal bone.*

years and are very variable in size. The frontal bone forms the bony roof overlying the ethmoidal air sinuses. The top of these air cells can easily be seen along the medial walls of the orbital part of a disarticulated frontal bone. Two small foramina, the **anterior** and **posterior ethmoidal foramina** squeeze through the medial wall of the orbit here between the frontal bone and the ethmoid bone.

The sphenoid bone

The sphenoid bone is wedged between the face and the brain and indeed this is how it gets its name (sphen (Gk) = a wedge). The centrally-positioned part of the sphenoid bone is preformed in cartilage from many centres of ossification. The more lateral portions of the sphenoid bone are formed from membrane. Fusion of the **mid-sphenoidal synchondrosis** (which like the spheno-occipital synchondrosis is an important growth centre before birth in primates) occurs shortly after birth in man and represents the last event in the coalescence of these centres. After this time the whole of the sphenoid continues to develop as a single bone.

The central part of the sphenoid bone is known as the **body**. The superior aspect of the body forms the **pituitary fossa** and is bound at each corner by anterior and posterior **clinoid processes** (Figs 4.15–4.17). This arrangement was thought by some to give the fossa the appearance of a high-backed Turkish saddle when viewed from the side so that we still commonly call the pituitary fossa the **sella turcica** of the sphenoid bone. The four clinoid processes of the body of the sphenoid bone are also strangely named since they were thought to resemble the posts of a four-poster bed! It is because of this (kline (Gk) = a bed) that we now refer to them as clinoid processes. The optic nerves pass into the orbits through the **optic canals** which lie just medial to the anterior clinoid process. Further details of this region are given in Chapter 9.

By six years of age, the **sphenoidal air sinus** has begun to hollow out the body of the bone and in adults this part of the sphenoid is little more than a shell of bone with a thin mid-line partition. As already noted, the body of the sphenoid forms an important joint or synchondrosis with the basilar process of the occipital bone that fuses in late puberty when growth ceases in this region (Figs 4.16 and 4.17). A similar primary cartilaginous joint at the front of the body with the ethmoid bone fuses earlier, at around five or six years of age.

Four **wings** of the sphenoid bone extend laterally from the body, two on each side. The two lesser wings form the whole of the posterior part of the floor of the anterior cranial fossa and have a sharp free posterior margin. In modern human crania the lesser wings of the sphenoid bone extend outwards towards a region on the side of the vault where the sphenoid, temporal, frontal and parietal bones all meet forming an H-shaped confluence of sutures called the *pterion* (Fig. 4.7). This is the place in Greek mythology where Hermes had his wings and it is from this that the *pterion* took its name (pteryx (Gk) = a wing). *Pterion* is another site in the cranium where sutual bones commonly form. In this region they are known as **epipteric bones** and there is some evidence that they may be more common on the left side of the modern human cranium and, therefore, possibly linked with asymmetries of the vault bones (Aitchison, 1960). In great apes and some fossil hominids the lesser wings of the sphenoid bone are not so extensive and form a smaller proportion of the floor of the anterior cranial fossa and so do not reach the *pterion*. The lesser wings are separated from the **greater wings** of the sphenoid bone by the **superior orbital fissures** below them (Figs 4.15 and 4.17). It is easiest to see both the lesser and greater wings of the sphenoid bone together in the orbit or in the middle cranial fossa where the superior orbital fissure lies between them.

The greater wings, as their name suggests,

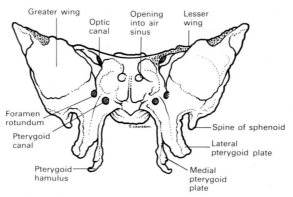

FIGURE 4·15   *Anterior view of the sphenoid bone.*

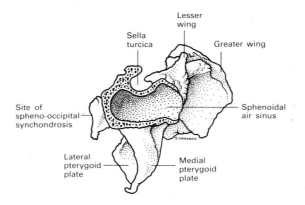

FIGURE 4·16   *Midline section through the sphenoid bone.*

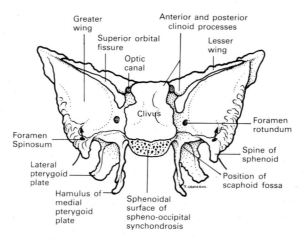

FIGURE 4·17   *Posterior view of the sphenoid bone.*

are more extensive than the lesser wings of the sphenoid bone and form the floor of the middle cranial fossa, part of the roof of the infratemporal fossa on the other side of this and a small portion of the wall of the vault of the cranium. The most posterior part of the greater wing of the sphenoid bone, which forms the floor of the middle cranial fossa, is sharp when disarticulated. For this reason it is called the **spine** of the sphenoid and the **foramen spinosum** that passes through it is named after it. This is the foramen through which the middle meningeal artery passes into the cranial cavity to supply the vault bones with blood (see Chapter 9).

Hanging down from the undersurface of the body of the sphenoid bone are two pairs of bony plates, the medial and lateral **pterygoid plates** (Figs 4.15–4.17). The lateral pterygoid plate provides the origin of the pterygoid muscles which move the mandible. The medial pterygoid plate forms part of the bony framework for the muscles of the pharynx and soft palate. A small hook, the **pterygoid hamulus**, turns outwards at the lower extremity of the medial pterygoid plate. Between the roots of the pterygoid plates at their origin from the body of the sphenoid bone there is a small shallow scalloped depression called the **scaphoid fossa** where a muscle of the soft palate takes origin before it drops down and winds around the hamulus into the palate. The significance of these bony landmarks associated with the medial pterygoid plate are discussed in Chapter 13.

The temporal bone

The temporal bones, so called because greying hair at the temple was one of the first signs of ageing before birth certificates were invented (tempus (L) = time), form both part of the vault and part of the cranial base. Each temporal bone consists of **squamous**, **petrous** and **mastoid portions**. The temporal bone also has other distinct parts such as the **tympanic plate** and **styloid process** which

will be described with the appropriate part of the bone.

The flat squamous (squama (L) = the scale of a fish or serpent) part of the temporal bone forms part of the lateral wall of the vault. On each side of the cranium a **zygomatic process** (zygoma (Gk) = a yoke for oxen), extends forwards from its **roots**, on the squamous temporal bone to articulate with zygomatic bone. There are three roots of the zygomatic process (Figs 4.18 and 4.19): the **anterior root** turns inwards to form the **articular eminence** at the anterior end of the **glenoid fossa**, the **posterior root** runs as a ridge over the external auditory meatus (see below), and the **middle root** again turns inwards but this time forms the **post-glenoid process** behind the glenoid fossa (this is very well developed in great apes). The post-glenoid process marks the posterior boundary of the joint capsule of the temporomandibular joint which forms between the glenoid fossa and the head of the condyle of the mandible.

The external surface of the adult squamous and mastoid parts of the temporal bone are marked by several crests and ridges of bone that represent muscle attachments to this region (Figs 4.18 and 4.19). The lowermost fibres of the temporalis muscle create a **supramastoid crest** that extends backwards from the posterior root of the zygomatic process. Above the external auditory meatus it is known as the **suprameatal crest**. The sternocleidomastoid muscle creates another crest, the **mastoid crest**, which is lower down on the mastoid process and is continuous posteriorly with the superior nuchal line on the occipital bone. Between these two crests there may be a **supramastoid sulcus**. When these two crests coalesce, as they do in great apes and some fossil hominids, the supramastoid sulcus is obliterated and the resulting crest is known as a **compound temporal–nuchal crest**. We noted earlier in the description of the parietal bone that a deep notch, the **parietal notch** of the temporal bone, articulates with the **mastoid angle** of the parietal bone above the mastoid

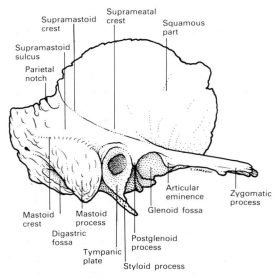

FIGURE 4·18   *Lateral view of the temporal bone.*

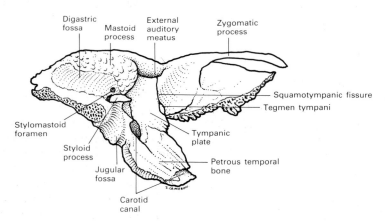

FIGURE 4·19   *Inferior view of the temporal bone.*

process. Rather than interdigitating like many bones of the cranium do when they articulate, the squamous temporal overlaps the parietal bone considerably. The importance of this type of sutural articulation and its relationship to strong masticatory stresses are reviewed further in Chapter 6.

A denser (hence petrosus (L) = stony or petrified) **petrous pyramid** of the temporal bone projects from the squamous portion medially across the cranial base and ends in a **petrous apex** between the basioccipital and sphenoid bone (Fig. 4.19). The petrous and most of the mastoid portions of the temporal bone are preformed in cartilage, yet at birth no projecting mastoid process has developed.

The squamous portion, like the rest of the vault wall, develops in membrane. At birth only a **tympanic ring** of bone supports the tympanic membrane or ear drum. Surprisingly, the tympanic ring, the ear ossicles (the **malleus, incus** and **stapes**) and the semicircular canals of the vestibular apparatus that are housed in the petrous bone are all fully grown even at this early age. The bone of the ear ossicles, unlike all bone elsewhere in the body, is not replaced or 'turned over' after this time so that any damage to them subsequent to birth is permanent.

The **internal carotid artery** that supplies most of the blood to the brain passes into the petrous temporal bone via the **carotid canal** through which it travels to emerge at the apex of the petrous bone above the **foramen lacerum**. The petrous temporal bone also completes the anterior part of the jugular foramen in the posterior cranial fossa with the jugular process of the occipital bone forming the posterior margin of this foramen.

During the first five or six years of life the tympanic ring grows laterally into a **tympanic plate** of bone that forms the anterior part of the **external auditory meatus** and extends along the undersurface of the petrous temporal bone. A central region of the plate remains deficient until about five years of age (occasionally throughout life) and is called the **foramen of Hushcke**. The tympanic plate also grows around the base of the **styloid process** of the temporal bone where it is called the **vaginal process** of the tympanic plate. A small part of the petrous portion of the temporal bone that forms the roof over the middle ear becomes trapped on the external aspect of the base of the cranium during development in modern humans and is known as the **tegmen tympani**. This small lip of bone projects between the tympanic plate and the glenoid fossa into a well-marked fissure the **squamotympanic** or **Glaserian fissure**. The tegmen tympani is not normally visible in ape crania because the sutures, which are all that demarcate it, fuse very early in juvenile specimens. More details of

this region are given in Chapter 6. Another more prominent projection of the temporal bone that is however much more prominent in great ape crania than in modern human crania is the **post-glenoid process**. This represents the middle root of the zygomatic process of the temporal bone and forms the posterior limit of the glenoid fossa and joint capsule of the temporomandibular joint.

The petrous and mastoid parts of the temporal bone become pneumatized during development from the cavity of the middle ear. Other air sinuses or **antra** in the cranial bones pneumatize more directly as outgrowths from the nose and pharynx. Unlike these small air cells in the petrous temporal bone and the other air sinuses of the skull, the **mastoid antrum** is present and well formed at birth in modern humans. The mastoid antrum is an air cavity that extends upwards and backwards from the middle-ear cavity.

The ethmoid bone

The ethmoid bone is situated between the two orbital cavities and articulates with the frontal bone in its ethmoidal notch (Fig. 4.20). Above the nose in the floor of the anterior cranial fossa only the **cribriform plate** of the ethmoid bone and a vertically projecting plate of bone, the **crista galli**, are visible. The sieve-like cribriform plate of the ethmoid is responsible for the name of the bone (ethmos (Gk) = a sieve). The human cribriform plate is fully grown early in the growth period, by two years of age. The ethmoidal air cells that make up much of the space between the orbital cavities and the lateral wall of the nose expand most rapidly after birth and continue to grow more slowly until puberty. The part of the ethmoid bone that contributes to the medial wall of the orbit is particularly frail and egg-shell like. A central **perpendicular plate** of the ethmoid bone contributes to the mid-line septum of the nose and the two uppermost **turbinate**

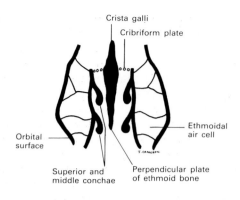

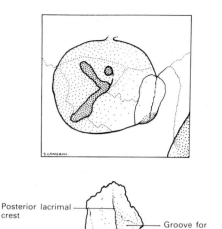

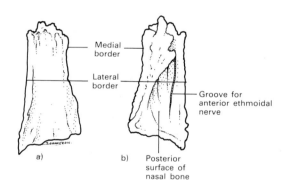

FIGURE 4·21 *The lacrimal bone in the medial wall of the orbit (a), and disarticulated (b).*

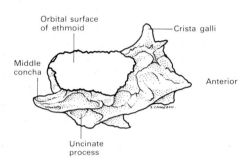

FIGURE 4·20 *Coronal section through the ethmoid bone (a), and lateral aspect of ethmoid bone (b) as it appears in the medial wall of the orbit.*

bones (or **conchae** when they are covered with mucous membrane) are also parts of the ethmoid bone (Fig. 4.20a and 20b). Further details about the sutural patterns of the bones of the medial wall of the orbit and of the ethmoidal air cells are described in Chapter 11.

The lacrimal bone

The lacrimal bone articulates with the frontal bone, the maxilla and the ethmoid bone in the medial wall of the orbit (Fig. 4.21, see also Fig. 11.5). It forms the posterior portion of the first part of the **nasolacrimal duct** that drains tears from the lacrimal sac and orbital cavity into the nose beneath the inferior concha. The anterior portion of the orbital

FIGURE 4·22 *The nasal bones (a), anterior surface, (b), posterior surface.*

opening into this canal is formed by the **frontal process** of the **maxilla**.

The nasal bone

The paired nasal bones (Fig. 4.22) articulate with the frontal bone and also with the orbital processes of the maxillae and with each other.

They are particularly variable both in their orientation to one another and the degree to which their width flares superiorly and inferiorly. Usually the bone is narrowest midway along its length in modern human crania. The undersurface of the bone is always grooved.

The palatine bone

Despite its name, much of the palatine bone (the **vertical or perpendicular plate**) is sandwiched between the medial pterygoid plate of the sphenoid bone and the posterior part of the maxilla and so takes no part in the formation of the palate at all. As a result, this part of the bone is hidden from view in an articulated cranium. It is the **horizontal** part of the palatine bone that forms the posterior aspect of the hard palate (Figs 4.23, 4.24 and 4.26). The **greater palatine**

foramina open onto the hard palate through the horizontal plate of the palatine bone. Together, the horizontal plates form a **posterior nasal spine** in the mid-line (Fig. 4.26). The vertical plate of the palatine bone is notched superiorly by a **sphenopalatine notch**. This divides the most superior aspect of the vertical plate of the palatine bone into an **orbital process** anteriorly and a **sphenoidal process** posteriorly (Fig. 4.26). The notch is turned into a foramen by its articulation with the sphenoid bone above so that it becomes the **sphenopalatine foramen** in an articulated cranium. The orbital process of the palatine bone is just visible on the posteromedial wall of the orbital cavity and so the palatine bone is also one of the bones that make up the medial wall of the orbital cavity. Articulation of the vertical plate of the palatine bone with the maxilla creates a canal between the two that communicates with the region of the

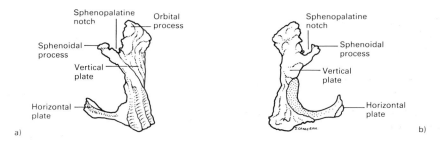

FIGURE 4·23   *Posterior (a), and anterior (b), aspects of the palatine bone.*

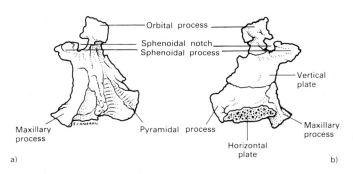

FIGURE 4·24   *Lateral (a), and medial (b), aspects of the palatine bone.*

sphenopalatine foramen and descends to open onto the hard palate as the greater and lesser palatine foramina.

The vomer

This mid-line, perpendicular plate of bone contributes to the nasal septum posteriorly where it divides the posterior nasal aperture into two, (the **choanae**). Shaped like a plough at its base (Fig. 4.25), its name is derived from the Roman name for a plough that 'threw up' earth either side (vomere (L) = to vomit or throw up). Its base in modern humans articulates with the undersurface of the body of the sphenoid bone where two small flanges of bone, the **vaginal processes** of the sphenoid bone, extend over the **alae** or wings of the expanded base of the vomer. In adult great apes, where the cranial base is less flexed in the sagittal plane, the base of the vomer is positioned further anteriorly. For this reason a patent **sphenoidal canal** that communicates with the pituitary fossa is often visible on the undersurface of the body of the sphenoid bone in ape crania (Sprinz and Kaufman, 1987; see Chapter 9), but is covered over by the base of the vomer in human crania. The growing vomer has been implicated as underlying the appearance of a mid-line **torus palatinus**, a raised ridge of bone that occasionally appears on the roof of the mouth. The vomer is also thought by some to be an important site of growth in the downward growing human face.

The zygomatic bones

The zygomatic bones are occasionally also referred to as **malar**, **jugal** or simply cheek-bones. Because they join the temporal and maxillary bones they were, as we noted earlier, thought to resemble the yoke used to join oxen together. Three processes of the zygoma, the **maxillary**, **frontal** and **temporal**, form sutures with the adjacent bones of the same names (Fig. 4.26). The

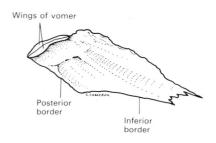

FIGURE 4·25   *Lateral view of the vomer.*

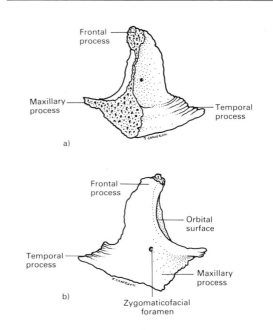

FIGURE 4·26   *Medial (a), and lateral (b), aspects of the zygomatic bone.*

zygoma forms the inferolateral border of the bony orbit and additionally provides a great deal of the bony origin for the masseter muscle.

The maxilla

The maxilla forms the bulk of the floor of the orbital cavity, much of the floor and lateral wall of the nasal cavity (as well as a large part of the roof of the mouth). Most importantly, however, each of the maxillary teeth develop within the maxilla and are then supported in the alveolar bone of the maxilla, while in functional occlusion. The **median palatine suture** between the two palatal processes of the maxillae in the hard palate does not fuse like the mandibular symphysis but remains patent well into adult life. The maxilla has three processes that extend from the pneumatized body to articulate with other bones of the same name: the **frontal process**, the **zygomatic process** and the contralateral **palatal process** of the maxilla (Figs 4.27 and 4.28). In addition, there is an **alveolar process** which, as noted, supports the maxillary dentition but resorbs when the teeth are lost.

Before birth in modern humans the **os incisivum**, the supposed homologue of the premaxilla in other primates, is completely overgrown by bone anteriorly such that there is little evidence of a separate bone in modern

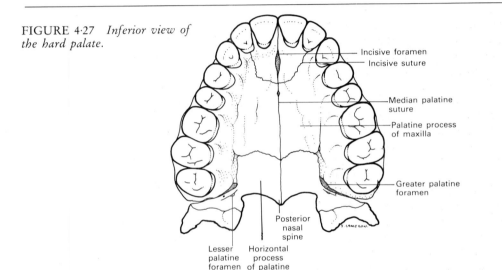

FIGURE 4·27 *Inferior view of the hard palate.*

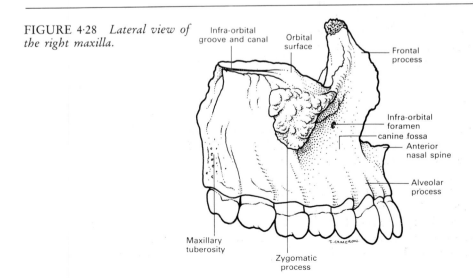

FIGURE 4·28 *Lateral view of the right maxilla.*

humans. The only possible remnants of a premaxillary ossification centre in modern humans are the **incisive sutures** that run from the incisive canal behind the incisor teeth to fade out between the lateral incisors and canines. These are, however, usually only visible in the crania of very young individuals. In great apes the premaxillary sutures remain clearly visible, both on the palate and anterior aspect of the face, well into the growth period.

Only the rudiment of a **maxillary antrum** or **sinus** is present at birth, what cavity there is being occupied by unerupted deciduous teeth in their bony crypts. Gradually, as the maxilla grows in height, the sinus begins to invade the body of the bone. In the mid-line at the anterior end of the floor of the nose the maxilla forms an **anterior nasal spine**. Below this and above the incisor teeth on the anterior aspect of the maxilla, is a shallow depresson called the **incisive fossa**. Lateral to this over the canine root a **canine eminence** separates this incisive fossa from a deeper **canine fossa** that lies some way beneath the lower border of the orbital margin. Above the canine fossa is the **infra-orbital foramen**, the anterior end of the **infra-orbital canal**. Rarely in modern humans, but commonly in great apes, this canal may open as more than one foramen onto the face.

The inferior concha

This turbinate bone or **concha** when covered with mucous membrane (concha (L) = an oyster shell), unlike the other two which are part of the ethmoid, is a separate bone (Fig. 4.29). Surprisingly, unlike all the other bones of the facial skeleton, it is preformed in cartilage. The inferior turbinate bone is situated in the lateral wall of the nose and articulates with the maxilla and palatine bone to occlude the otherwise wide entrance into the maxillary antrum. In so doing, this bone forms a large part of the lateral wall of the nose.

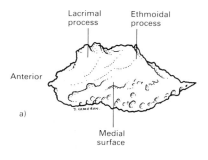

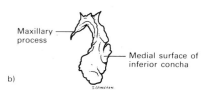

FIGURE 4·29 *Medial aspect of the inferior concha (a), and coronal section through the inferior concha (b).*

Anthropometric landmarks

Various anthropometric landmarks have been defined on the skull and many of these are commonly used by physical anthropologists, orthodontists and craniofacial biologists both to describe points on the skull and to make measurements from and between points on the skull. Some of the most important of these are shown in Figs. 4.30–4.32 and these and more are listed below.

Glabella: the most anterior point in the mid-sagittal plane between the superciliary arches.

Bregma: the point at which the sagittal and coronal sutures meet.

Obelion: the mid-line point on the sagittal suture between the two parietal emissary foramina.

Lambda: the point at which the lambdoid and sagittal sutures meet.

Opisthocranion: the mid-line point of the cranium that projects farthest posteriorly.

FIGURE 4·30 _Anthropometric landmarks on the facial skeleton._

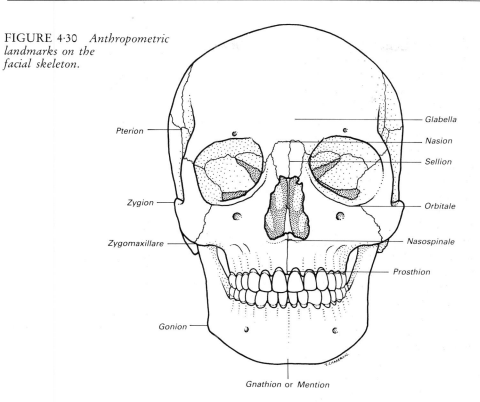

FIGURE 4·31 _Anthropometric landmarks on the external surface of the base of the cranium._

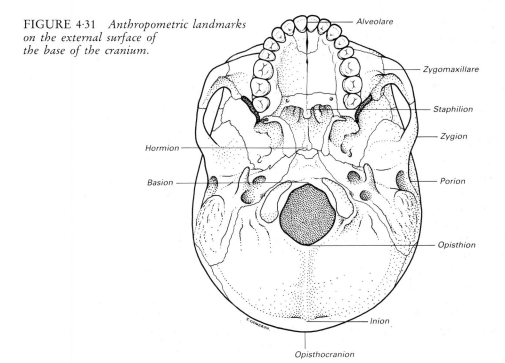

FIGURE 4.32 *Anthropometric landmarks on the lateral aspect of the skull.*

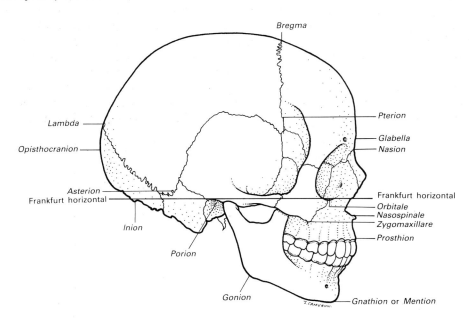

Inion: the point in the mid-line of the cranium that crosses a tangent to the upper convexities of the superior nuchal lines.

Asterion: the sutural point at which the parietal, occipital and temporal bones meet.

Pterion: the midpoint of the sphenoparietal suture, or, alternatively, the confluence of sutures where the frontal, parietal, temporal and sphenoid bones meet.

Porion: the most lateral and most superior point on the roof of the bony external auditory meatus.

Opisthion: the mid-line point on the posterior border of the foramen magnum.

Basion: the mid-line point on the anterior margin of the foramen magnum.

Sphenobasion: the position (estimated in adult crania) of the spheno-occipital synchondrosis in the mid-line of the cranium.

Hormion: the point on the sphenoid bone in the mid-line that crosses a line between the two most posterior points of the alae of the vomer.

Sella: The centre of the sella turcica as determined by examination of lateral skull radiographs or CT scans in the mid-line.

Pituitary point: the mid-line point on the raised tuberculum sella of the body of the sphenoid bone.

Prosphenion: the junction between the sphenoid and ethmoid bones in the mid-line.

Nasion: the point where the internasal and frontonasal sutures meet in the mid-line.

Sellion: the deepest point in the hollow beneath **glabella** in the mid-line.

Dacryon: the apex of the lacrimal fossa as it impinges on the frontal bone.

Nasospinale: the midpoint on a line that connects the lowest points of the border of the piriform aperture on either side at the base of the anterior nasal spine.

Alveolare: the mid-line point on the most inferior point of the tip of the alveolar septum between the right and left upper central incisors.

Prosthion: that point on the maxillary alveolar process that projects most anteriorly in the mid-line.

Orbitale: the lowest point on the infraorbital margin.

Zygomaxillare: the lowermost point of the zygomaticomaxillary suture.

Zygion: the most lateral projection of the zygomatic arch.

Gnathion or **Menton**: the lowest point of the mandible in the mid-line.

Gonion: the apex or point of maximum curvature at the mandibular angle.

Condylion, laterale or mediale: the most lateral or most medial point of the poles of the mandibular condyle.

Frankfurt Horizontal: the plane of a line drawn between **porion** and **orbitale** with the cranium viewed in **norma lateralis**.

CHAPTER FIVE

THE COMPARATIVE ANATOMY OF THE HOMINOID MANDIBLE AND CRANIUM

In this chapter we review the comparative morphology of the mandible and cranium in the living great apes and in fossil hominids. With the exception of teeth, mandibles are the most numerous of all fossil hominid finds so that the morphology of the hominoid mandible can be described in some detail. A whole chapter of this book is devoted to the comparative anatomy of the face in great apes and hominids, so that no detailed overview of facial morphology appears in this chapter. However, other major anatomical features that distinguish great ape and hominid crania are summarized here in four orientations: *norma lateralis*, *norma basilaris*, *norma occipitalis* and *norma verticalis*. These summaries are intended to provide a background framework, so that aspects of comparative hominoid morphology can be expanded and developed separately later in the book, some in the form of regional comparisons others as part of discussions of broader functional issues.

The great ape mandible

While usually larger than the human mandible in its dimensions, the great ape mandible also differs from the human mandible in its overall proportions. More often than not the coronoid process is taller than the condyle and the mandibular notch between them is less deep (Fig. 5.1). This is the reverse of the situation in most modern human mandibles where the condyle is taller than the coronoid process (very young or very old individuals being an exception). The ascending ramus of the great ape mandible is relatively wider anteroposteriorly than in modern humans. The body of the ape mandible is also much longer, since apes are relatively prognathic compared with modern humans. This together with the broad ascending ramus emphasizes the relatively low mandibular body in apes despite their absolutely longer tooth roots. As a consequence of this, the mental foramen opens lower down on the outer aspect of the body of the mandible in apes and is also commonly multiple.

Viewed from above (Fig. 5.2) the adult great ape mandible is U-shaped and not so evenly curved in outline as are modern human mandibles. This means that anteriorly the ape mandible is narrower than modern human mandibles and that, relatively speaking, the condyles are not so widely separated as they are in modern humans. A consequence of this is that the floor of the mouth is narrower

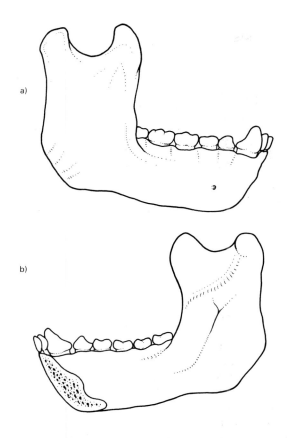

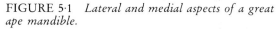

FIGURE 5·1   _Lateral and medial aspects of a great ape mandible._

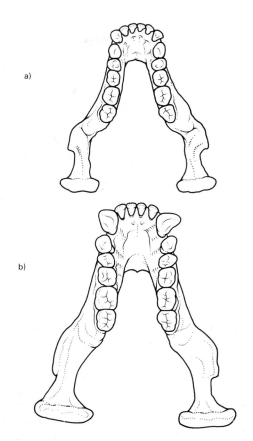

FIGURE 5·2   _Superior views of female (a) and male (b) great ape mandibles._

in apes than it is in modern humans and so does not need so much muscular support. More details about this and the shape of the dental arcade are presented in Chapters 6 and 8.

While the mandibular foramen lies well above the level of the occlusal plane in some large specimens of _Gorilla_ and _Pongo_ it is still positioned midway between the anterior and posterior borders of the ascending ramus. Smaller great ape mandibles may occasionally differ from those of larger apes in that the mandibular foramen remains nearer the level of the occlusal plane of the teeth in its position on the medial aspect of the ascending ramus. No chin or mental fossa exist in great

apes as the basal and alveolar bone, in the mid-line anteriorly, curve smoothly round to become continuous with the inferior transverse torus or 'simian shelf' (Fig. 5.3). Rather than always arising from tubercles, two muscles in the floor of the mouth, the geniohyoid and the genioglossus, often take origin from deep fossae which are also commonly associated with one or several neurovascular foramina on the internal aspect of the mandible anteriorly. No digastric fossae exist on the anterior–inferior border of the mandible in the orang-utan as the muscle has no anterior belly. Instead the posterior belly inserts at the angle of the mandible (see Chapter 6). In all three apes,

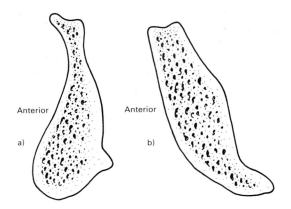

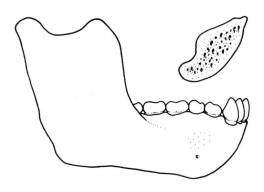

FIGURE 5·3 *Midline section through the mandibular symphysis of a modern human (a) and an orangutan (b) showing the form of the chin and the inferior transverse torus or simian shelf. After Scott, (1967).*

FIGURE 5.4 *Lateral view of a reconstructed Australopithecus afarensis mandible together with a symphyseal cross section of an Australopithecus afarensis mandible based upon AL 288-1.*

however, the lower border of the mandible is flattened below the molar and premolar teeth and in this region there is an extensive attachment for platysma, one of the muscles of facial expression.

The mandibles of fossil hominids

Australopithecus afarensis

Some 25 adult and juvenile mandibular remains from Hadar and Laetoli make up the sample of *Australopithecus afarensis*. The general features of these mandibles have been described and discussed by Johanson et al. (1978), White et al. (1981), Johanson and White (1985) and Kimbel et al. (1984, 1985). The mandible of *A. afarensis* has a broad low ascending ramus that slopes somewhat posteriorly and is surmounted by a broad condyle (Fig. 5.4). The body of the mandible is deeper anteriorly than it is posteriorly at the junction with the ascending ramus and is hollowed on the buccal surface in the region of the mental foramen. The mental foramen itself tends to open low on the lateral aspect of the body of the mandible. The mandibular symphysis is rounded and bulbous with an axis that inclines downwards and backwards.

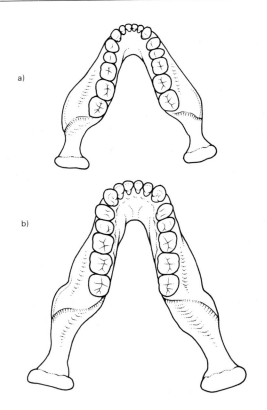

FIGURE 5·5 *Superior view of reconstructed male (a) and female (b) Australopithecus afarensis mandibles. After White et al. (1981).*

The superior transverse torus is moderately well developed and the inferior transverse torus is low and rounded rather than shelf like. Occlusally, the posterior teeth can be seen to form straight rows and the anterior part of the dental arch is narrow (Fig. 5.5). There is a tendency in smaller mandibles attributed to *A. afarensis*, as there is in great apes, for the last molar tooth to lie wide of the dental arcade which accentuates the near-V-shaped arch of some specimens.

Australopithecus africanus

There are far fewer mandibular remains of *Australopithecus africanus* than of *A. afarensis*, the most complete being a sub-adult specimen (Sts 52b) from Sterkfontein. Others include adult and juvenile mandibular remains from Makapansgat (MLD 2, 18, 34, 40), and Taung. Kramer (1986) has summarized the differences that exist between *A. afarensis* and *A. africanus* mandibles. Judged by Sts 52b (Fig. 5.6), the ascending mandibular ramus of *A. africanus* is probably relatively taller than that of *A. afarensis*, but, similarly, the ascending ramus also slopes slightly backwards. (The ascending ramus of the juvenile specimen from Taung is also relatively high, suggesting that this feature was established early in the growth period.) The mandibular notch in *A. africanus* is shallow

and the anterior border of the ramus curves convexly, ending in a strong external oblique ridge on the lateral surface of the body. At the junction of the ascending ramus with the body there is a broad and deep **extramolar sulcus**. Two mental foramina are present below the root of the PM3 of Sts 52b but other mandibles (e.g. MLD 40) have single mental foramina which are always positioned above the mid-corpus height. There is a pronounced superior transverse torus (MLD

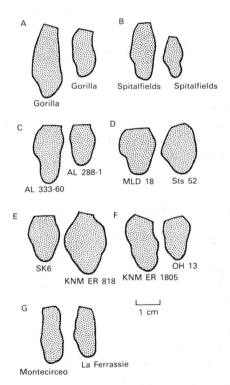

FIGURE 5·7 *Cross-sections of hominoid mandibular corpora made in the region of M1 and excluding any tooth crowns present. (A) Male and female Gorilla; (B) male and female modern Homo sapiens from Spitalfields, London; (C) two mandibles attributed to* A. afarensis *AL 333-60 and AL 288-1; (D) two mandibles attributed to* A. africanus *MLD 18 and Sts 52; (E) two mandibles attributed to Paranthropus, one to P. robustus SK 6 and one to P. boisei KNM ER 818; (F) two mandibles attributed to early Homo KNM ER 1802 and OH 13; (G) two Neanderthals, Montecirceo and La Ferrassie.*

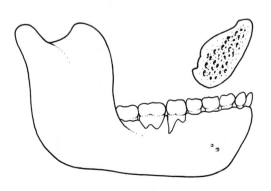

FIGURE 5·6 *Lateral view of a mandible attributed to* Australopithecus africanus *together with a cross section of the symphyseal region. Based on Sts 52b.*

57

18, 40, Sts 7) at the symphysis of the mandible but less of an inferior transverse torus than in *A. afarensis*. The overall size of the mandible, as reflected by the cross-sectional area measured at M1 (Fig. 5.7) is greater than that of *A. afarensis* (Chamberlain and Wood, 1985). Adult mandibles from Makapansgat also share the straight diverging posterior tooth rows typical of *A. afarensis*.

Paranthropus

Mandibular remains of *Paranthropus* are well represented in the fossil record. They are typically massive with extremely tall vertical ascending rami that reflect the increased height of the temporomandibular joint above the occusal plane of the teeth (Fig. 5.8). The ascending ramus is also broad with a shallow mandibular notch and a tall coronoid process. The width between the mandibular condyles in *Paranthropus* is great, reflecting the great width of the cranium. A large bony buttress arises from the **crista pharyngea** (a crest on the lingual aspect level with the last molar tooth; Day, 1986) and continues to rise upwards towards the condyle on the medial aspect of the ramus.

The body of the mandible is also massive and thickened buccolingually more than in any other hominid, being especially wide superiorly which gives it a diamond-shaped outline in cross-section in large specimens (Fig. 5.7). The body of the mandible in *Paranthropus* increases in width posteriorly but increases in height anteriorly towards the mid-line. While there may be a hint of a chin in some specimens, or a lateral prominence (e.g. KNM-ER 729), the symphysis typically recedes on the buccal aspect and is marked lingually by both superior and inferior transverse tori (Fig. 5.8). The inferior transverse torus is well above the mandibular base in *Paranthropus* (for example some 15 mm in KNM-ER 404; Leakey et al. (1971)) and can, like the superior transverse torus, extend as far posteriorly as M1. A deep fossa and or genial spine is a feature of the mandible below the lingual alveolar planum and must

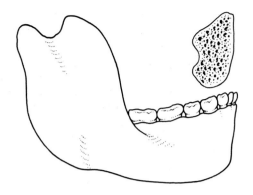

FIGURE 5·8   *Lateral view of a mandible attributed to* Paranthropus robustus *together with a cross-section of the symphyseal region. Based on SK 23 from Swartkrans.*

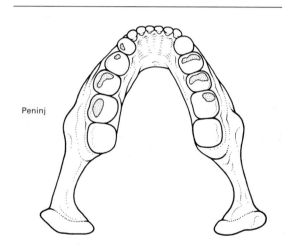

Peninj

FIGURE 5·9   *Superior view of the Peninj mandible attributed to* Paranthropus boisei.

mark the attachment of the genioglossus superiorly and the geniohyoid inferiorly.

The shape of the adult mandibular dental arch in *Paranthropus* is more rounded posteriorly than it is in *Australopithecus* (Fig. 5.9) and there is a tendency for the last one or two molar teeth to curve inwards towards each other rather than to form a straight diverging line as they do in the mandibular arch of *Australopithecus afarensis*. This leaves a very wide extramolar sulcus (as great as 1 cm in some specimens of *Paranthropus boisei*) between the ramus and the buccal aspect of the molar teeth. Indeed, a wide extramolar sulcus is essential to accommodate transverse movements of the last lower molar tooth across the last maxillary tooth, both of which lie posterior to the ascending ramus of the mandible in *Paranthropus*.

Anteriorly, the dental arch is variable (Tobias, 1967), but despite the small size of the anterior teeth the dental arch tends to be more crowded and flattened than that of *Australopithecus*. The anterior teeth of *Paranthropus* are set more vertically in the alveolar bone than in *Australopithecus* where a degree of alveolar prognathism may alleviate any crowding due to lack of space in the dental arch. Despite their small anterior teeth, several specimens of *Paranthropus* show evidence of a crowded anterior dental arch, good examples being the SK 23 mandible and the maxilla of KNM-WT 17400. The shape of the dental arch of the juvenile *Australopithecus* specimen from Taung differs from that of the juvenile *Paranthropus* specimens from Swartkrans in being less rounded and more U-shaped with parallel posterior tooth rows (Fig. 5.10).

Early Homo

Even though there is debate about the specific status of many of the mandibular remains attributed to early *Homo* and considerable variation in morphology amongst this large group, some general morphological features characterize them all. In overall build, the mandibles of *Homo* are less robust than those of other hominids; i.e. they are less wide across the body relative to the height of the body in a given position such as at M1 (Fig. 5.7). The distance between the mandibular condyles and the ascending rami is relatively wider than in other fossil hominids and the dental arcade is reduced in length and wider anteriorly (Figs 5.11 and 5.12). Tobias (1980) has noted that compensatory occlusal curves of Spee (see Chapter 6) first

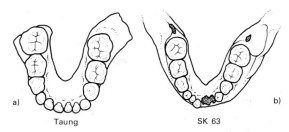

FIGURE 5·10 *The form of the dental arch in two juvenile fossil hominids (a) Taung (*Australopithecus africanus*) and (b) SK 63 from Swartkrans (*Paranthropus robustus*).*

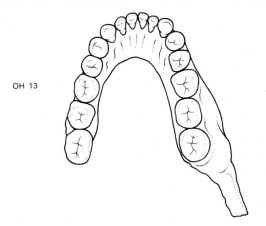

FIGURE 5·11 *Superior view of OH 13 from Bed II at Olduvai, attributed to* Homo habilis. *Note that the width across the floor of the mouth in the premolar region is greater than the width of the corpus in this region.*

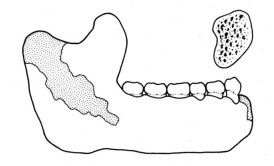

FIGURE 5·12 *Lateral view of mandible KNM ER 992, attributed to early* Homo, *(left side reversed) together with a cross-section of the symphyseal region which has been partially reconstructed.*

feature in the dentitions of *Homo habilis*, although Smith (1986b) has argued that these can occur in most hominids and result from an increased axial inclination of the molar teeth, which may lean increasingly inwards towards the lingual aspect as the back of the mandibular tooth row is approached. The ascending ramus of early *Homo* mandibles is less tall than that in *Australopithecus* and *Paranthropus* and is typically broad anteroposteriorly, even more so than in populations of anatomically modern *Homo sapiens* (Fig. 5.11). In fact, some mandibles of early *Homo* have broader ascending rami than do specimens of *Paranthropus*, the Mauer mandible from Heidelberg (a specimen now often attributed to archaic *H. sapiens*) being a case in point (Wells, 1958; Fig. 5.13). Others, however, while still having ascending rami that are not tall are very variable in their breadth (for example the sample of *Homo erectus* mandibles from Zhoukoudian). The root of the ascending ramus of the mandible in *Homo* is generally not so far forward along the tooth row as it is in earlier hominids. Interestingly, Chamberlain and Wood (1985) have demonstrated that early *Homo* mandibles from East Africa are comparable in cross-sectional area at M1 with early mandibles from Europe, North Africa and Asia, but have a higher index of robusticity ((Br/Ht) × 100). None of the mandibles attributed to early *Homo* have a chin but a few (KNM-ER 730 and Ternifine mandibles 2 and 3 for example) have a mental trigone. The inner contour of the mandible of early *Homo* has been described by Leakey et al. (1964), Tobias (1966) and Robinson (1965, 1966) as being distinct from earlier hominids. According to Robinson, the width between the second premolars across the floor of the mouth is wider in *Homo erectus* ('*Telanthropus*') than the width of the corpus at this point, whereas in *Australopithecus* the width of the mandibular corpus at PM4 is greater than the width across the floor of the mouth at this point. This is one of the characters that Leakey et al. (1964) and

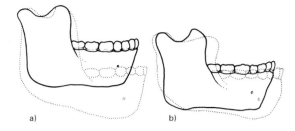

FIGURE 5·13 *Profiles of mandibles from Zhoukoudian* (Homo erectus) *and Swartkrans* (Paranthropus robustus) *(a) with a profile of the mandible from Heidelberg (Mauer mandible, now commonly attributed to archaic* Homo sapiens*) and the Zhoukoudian mandible (b). After Wells (1958).*

Tobias (1966) noted as being diagnostic of *Homo habilis*, although Robinson (1965, 1966) was unhappy about the diagnosis of this (and other) characters in the type specimen of *H. habilis*.

Weidenreich (1936) noted several features of the '*Sinanthropus*' (*H. erectus*) mandibles that set them apart from modern human mandibles as well as from other mandibles attributed to early *Homo* (Fig. 5.14). Weidenreich noted a degree of sexual dimorphism greater than that he had observed in a modern Chinese sample and similar to that in chimpanzees. However, it is not at all clear that this degree of mandibular sexual dimorphism would exceed that in other modern human populations. All specimens in this *H. erectus* sample also possessed two or more mental foramina and a **torus mandibularis** (a bony swelling) along the lingual aspect of the body. Besides a low ascending ramus and a great width between the mandibular condyles, Weidenreich measured differences in the mandibular angle between specimens of *H. erectus* and modern mandibles. The mandibular angle is formed between the posterior aspect of the ascending ramus and the inferior border of the body of the mandible and is generally greater in very young and old edentulous mandibles than it

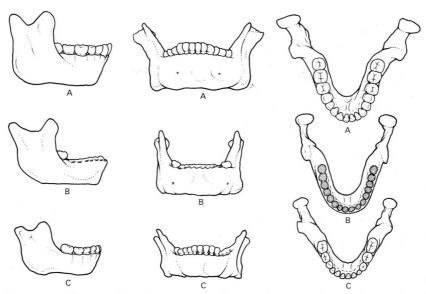

FIGURE 5·14   _Three mandibles attributed to_ Homo erectus, _a supposed male (A); a supposed female (B); and a juvenile (C), from Zhoukoudian, each illustrated in lateral superior and frontal views. After Weidenreich (1936)._

is in adult modern mandibles. In both the juvenile and adult _H. erectus_ mandibles this angle is smaller than in modern humans; (97–108° in the adult '_Sinanthropus_' sample and 107–112° in the juveniles, whereas in modern humans the average value is 125°). All specimens in the '_Sinanthropus_' sample, whether male, female or juvenile (as judged by Weidenreich), possessed flared mandibular angles and an alveolar plane parallel with the lower border of the mandibular corpus. This latter feature is atypical of other fossil hominids where there is a reduction in corpus height distally. It can be argued that all these features are more typical of great ape mandibles than of those of modern humans, but on the other hand smaller overall size, a very rounded anterior dental arch and genial tubercles rather than a fossa are especially modern features.

Mandibles identified as Neanderthal have some additional and important distinguishing characteristics. Besides being more robust than modern human mandibles Neanderthal mandibles tend to be especially broad anteri-

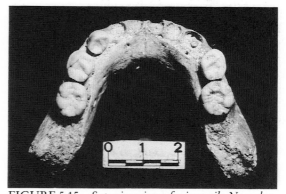

FIGURE 5·15   _Superior view of a juvenile Neanderthal mandible, Archi, Italy, illustrating the wide symphyseal region. Courtesy of Chris Stringer._

orly in the region of the mandibular symphysis (Fig. 5.15). This is true even in the juvenile specimens and it may be that this relates to the large dimensions of the incisor teeth in Neanderthals. Neanderthal mandibles are also characterized by an extended retromolar space between the last molar tooth and the junction of the ascending ramus with the body of the mandible and, on the whole, by lack of a chin (Fig. 5.16). This is in contrast with earlier australopithecines where the last one or two molars are distal to the anterior border of the ascending ramus. Wolpoff (1979) has noted that eruption of the third permanent molar teeth in Neanderthals may have been advanced in comparison with some modern human populations, possibly erupting at 15 years of age, and it may well be that this increased retromolar space permitted their early eruption. That Neanderthal mandibles are large is clear from the fact that in four measurements (bicondylar breadth, bigonial breadth, bimental foramen breadth and maximum body length) the Kebara mandible from Israel is more than three standard deviations beyond the means of a sample of modern Bedouin mandibles (Arensburg et al., 1989).

The great ape and fossil hominid cranium *in* norma lateralis

The sagittal profile of great ape crania varies between two extremes (largely as a consequence of body size). These can be represented on the one hand by the gorilla and on the other by the pygmy chimpanzee (Fig. 5.17). Nevertheless, the great ape profile is distinct from that of hominids and this is mostly due to the combined effects of an absolutely smaller brain and neurocranium and an absolutely larger facial skeleton and masticatory system in great apes.

A mid-line sagittal section through a modern human cranium and an adult gorilla cranium (Fig. 5.17) reveals more about the relationship of the brain and cranial cavity to the face. Much more of the modern human face is rotated backwards and downwards underneath the brain case and more of the modern human brain has overgrown the top of the facial skeleton.

One result of this is that the bones of the modern human cranial base, those which lie between the face and the brain, have a larger bend in them than do those of the great apes.

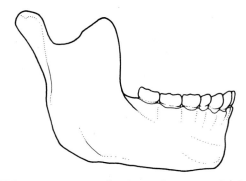

FIGURE 5·16 *Lateral view of a Neanderthal mandible based on Amud I from Israel. Compare the retromolar space of this mandible with the condition in SK 23, Figure 5·8. Note that this specimen has more of a chin than many Neanderthals do.*

The usual way of describing this is to say that the cranial base of modern humans is more flexed than that of great apes. While this is a simple concept it is difficult to find a series of truly homologous landmarks by which to measure the degree of flexure. Two points are fairly reliable: the most anterior point on the cribriform plate (i.e. the **foramen caecum**); and the most anterior point, in the mid-line, on the anterior margin of the foramen magnum *basion*. Defining a third that coincides with the site of the bend in the cranial base has been problematical, but the *pituitary point*, the mid-line point on the tuberculum sella, or even the centre point of the sella turcica seen in lateral view (called *sella*) are as good as any other reference points chosen for hominoids. The angle between these three points is called the **cranial base angle**. This angle serves to emphasize how flat the great ape basioccipital is with respect to the anterior cranial fossa and how flexed is the human basioccipital (Fig. 5.17). Similar angles have been devised on the external surface of the cranium in the mid-sagittal plane which measure **exocranial base flexion**. These angles show how the basioccipital has become more angled to, for example, the mid-line of the palate. One thing that obviously affects the size of this

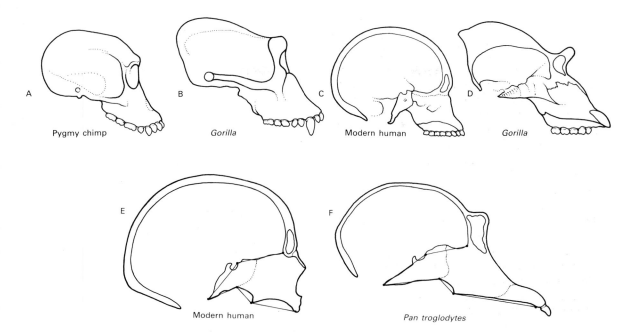

FIGURE 5·17 *(A), (B) Crania of* Pan paniscus *and* Gorilla *demonstrating the range of variation that exists in the proportions of the face and neurocranium among great apes. (C), (D) Midsagittal sections through a modern human and a* Gorilla *cranium illustrating the position and proportions of the cranial cavity and the facial skeleton in each. (E), (F) Also midsagittal sections of* Homo sapiens *and* Pan troglodytes *crania with the cranial base angle (foramen caecum – pituitary point – basion) shown on each together with the exocranial base angle devised by Laitman (1977); (prosthion – staphilion – hormion – basion).*

angle is how far back the foramen magnum is (or indeed how long the basioccipital bone is). Mid-line sections of crania also show the true position and size of the brain. Large crests for muscle attachment and tori above the orbits can make it difficult to judge the position of the brain relative to the face in complete dried skulls and many early comparisons of fossil hominids with great apes and humans are misleading because of this.

One feature that distinguishes hominids from great apes, and which is apparent in lateral profile, is the level of the occipital torus or superior nuchal line in modern human crania. In all early hominids this lies well below the level in great apes and, although there is variation, this feature is always associated with a relatively more forward position of the foramen magnum and occipital condyles. The long axes of the occipital condyles in early hominids are, generally speaking, orientated either parallel with, or close to the plane of the Frankfurt Horizontal and not orientated more posteriorly as they are in great apes.

Early fossil hominids are an extremely variable group and few generalizations can be made about cranial form in any one taxon. Nevertheless, when account is taken of the cranial crests in hominids, such that they do not exaggerate measurements made on crania, it can be shown that the whole of the calvaria in *Paranthropus* and the great apes is hafted onto the facial skeleton at a lower level than it is in *Australopithecus* and early *Homo* (Tobias, 1967). As a result, the vault of the cranium rises above the upper margin of

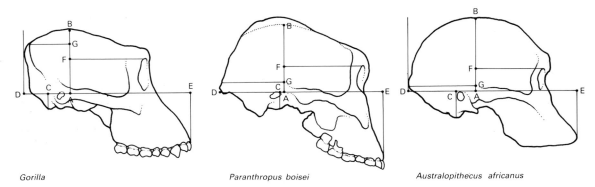

Gorilla

Paranthropus boisei

Australopithecus africanus

FIGURE 5·18 *Lateral craniograms of (a)* Gorilla *(female), (b)* Paranthropus boisei, *OH 5, and (c)* Australopithecus africanus, *Sts 5. The indices illustrated are nuchal height index AG/AB × 100;* *supraorbital height index FB/AB × 100 and condylar position index CD/CE × 100. After Le Gros Clark (1971) and Tobias (1967).*

the orbital cavities to different extents in hominids, quite independent of the size of the brain.

Figure 5.18 illustrates three crania (*Gorilla*, *A. africanus* and *Paranthropus*) where the percentage ratio between the height of the calotte above the upper margin of the orbit and total calvarial height (FB/AB × 100) demonstrates this point. For example, the value for Sts 5 is 61%, OH 5 52%, SK 48 50%, and *H. erectus* from Zhoukoudian between 63% and 67%. Values for modern great apes are between 49% and 54% and values for modern humans between 63% and 77%. Clearly, of these hominids *A. africanus* (Sts 5) resembles modern humans more closely in this respect than do the other early fossil hominids represented here. Other ratios that can be calculated from Fig. 5.18 are the occipital condyle position index (CD/CE × 100) and the nuchal area height index (AG/AB × 100) and these emphasize the point that the foramen magnum and occipital condyles are further forward in hominids than apes and that the nuchal area is swung down markedly in hominids as compared with the gorilla shown in the figure (Tobias, 1967).

A great deal has been said about the general proportions of the face in early fossil hominids and, in general, it is true that the hominid facial skeleton is reduced in size and less prognathic than it is in great apes. Within each early hominid taxon, however, there is great variation apparent in *norma lateralis* (Fig. 5.19). Specimens of *Australopithecus* exist which are very prognathic (Sts 5) but others have much more retruded facial skeletons (Sts 71 and Sts 19). Specimens of *Paranthropus* exist which are very prognathic (KNM-WT 17000) or which have much more flattened (or **orthognathic**) faces (KNM-WT 17400 and SK 48). Even within early *Homo*, one hallmark of which is a reduced facial skeleton, some specimens (KNM-ER 1470) have extremely flat faces with respect to the brain case and others (KNM-ER 1813) are more prognathic (although not as markedly so as many specimens of *Australopithecus*).

The bones of the cranial vault of great apes and hominids also reveal some interesting features when their external surface is examined in *norma lateralis*. Some variation exists among primates in the pattern of sutures at *pterion*, for example, between New World and Old World monkeys. Among hominoids there is also variation in the sutural pattern at *pterion* and it is useful to review this here. In modern humans, the pygmy chimpanzee (*Pan paniscus*) and the orang-utan it is most common for the greater wing of the sphenoid and the parietal bone to prevent the temporal and frontal from articulating at *pterion*

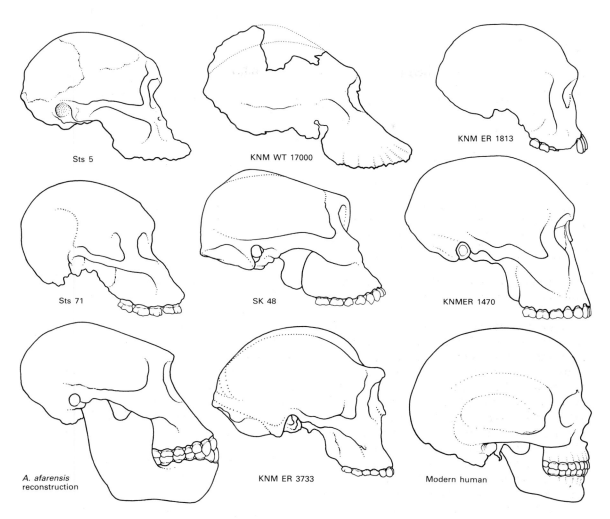

Sts 5

KNM WT 17000

KNM ER 1813

Sts 71

SK 48

KNMER 1470

A. afarensis
reconstruction

KNM ER 3733

Modern human

FIGURE 5·19 _Lateral views of eight early hominid crania and one of modern_ Homo sapiens _illustrating different degrees of facial prognathism and expansion of the neurocranium._

(Fig. 5.20). However, in the African apes it is more common for the temporal and frontal bones to prevent the parietal and sphenoid from suturing (Ashley-Montagu, 1933). Little phylogenetic significance can be attached to this as both sutural patterns can occur in all extant hominoids; nevertheless, the most common modern human pattern occurs consistently in all early fossil hominids.

The squamosal suture in apes lacks the

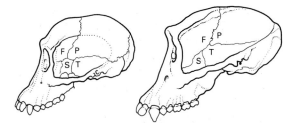

FIGURE 5·20 _Crania of_ Pan paniscus _(a) and_ Pan troglodytes _(b) illustrating how the confluence of sutures at_ pterion _varies among hominoids._

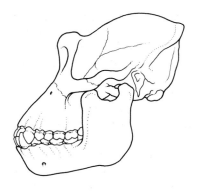
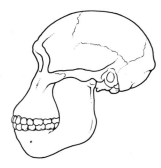
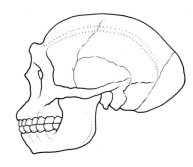

FIGURE 5·21 *Skull of* Gorilla *together with reconstructed skulls of* Australopithecus africanus *and* Homo erectus *illustrated in lateral view to show the proportions of the frontal, parietal and occipital bones that make up the vault as well as the arched contour of the hominid squamosal suture.*

strong curvature and height of modern human and fossil squamosal sutures (something that is apparent even in human foetal skulls) and in this respect all hominids contrast with the flat squamosal suture of great apes (Fig. 5.21). Measurements of the sagittal arc length of the modern human parietal bone are greater than that of the occipital bone and this modern human parietal predominance over the occipital is again shared with *Australopithecus, Paranthropus* and Neanderthals. Tobias (1967) notes, however, that this is not the case in *H. erectus* from Java and Beijing judged by measurements made by Weidenreich (1943). In these crania the arc lengths of the occipital bone predominate over the parietal one. Great ape crania are somewhat intermediate in this respect with near equal parietal and occipital sagittal arc lengths so that it could be argued that the group of Asian *H. erectus* crania appear to be derived in this respect compared with other hominids.

The sagittal contours of *H. erectus* crania are typically **platycephalic** or flattened (Fig. 5.22). Even though the vault rises above the superior margin of the orbit more than it does in *Australopithecus*, these long crania have little vertical height to the vault when compared with later hominids. In fact in the occipital region there is a very marked angle between the squamous occipital and the nuchal region of the bone below the superior nuchal line that shelves strongly horizontally inwards, giving little or no height to the back of the cranium. In *H. erectus* crania *inion* (the mid-line point level with the top of the superior nuchal lines), is coincident with *opisthocranion* (the most posterior point in the mid-line of the cranium). In many later

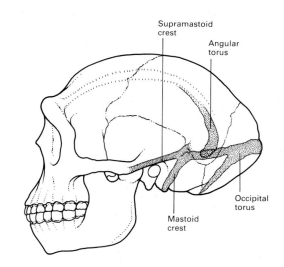

Supramastoid crest

Angular torus

Occipital torus

Mastoid crest

FIGURE 5·22 *Reconstructed skull of* Homo erectus *from Zhoukoudian with several bony markings characteristic of* Homo erectus *crania, (crests and tori), highlighted on the cranium.*

hominid crania the superior nuchal lines fall anteroinferior to *opisthocranion*. In *H. erectus* crania the cranial base angles, measured in the mid-sagittal plane between the *basion*, *sella* and *foramen caecum* are flattened or 'opened out and unflexed', more so than is typical of any modern human populations (Maier and Nkini, 1984). Earlier fossil hominid crania are very variable in this respect and it is virtually impossible to make accurate measurements in many of them. Nevertheless, the consistent condition in *H. erectus* (and Neanderthals) should dispel any notion of a simple trend in increased cranial base flexure from apes through the early fossil hominids to modern humans.

A sagittal keel is common in *H. erectus* crania but not, however, in the form of a crest related to the superior temporal lines and the temporalis muscles, but rather as simply a raised mid-line bony torus on the frontal and parietal bones. Marked supra-orbital tori anteriorly and an extensive nuchal region behind bounded superiorly by another marked bony ridge, the **occipital torus**, are also characteristics of *H. erectus* as are small mastoid processes. Some *H. erectus* crania have an **angular torus** which is the enlarged posterior aspect of the superior temporal line as it approaches the mastoid angle of the parietal bone.

Neanderthal crania are also platycephalic in their sagittal contour (Fig. 5.23) and have flattened unflexed cranial base angles. They also have marked and uninterrupted supra-orbital tori which, however, unlike *H. erectus* crania, can be strongly arched over each orbit rather than straight and 'bar like'. Unlike the condition in *H. erectus* crania, the supra-orbital torus in Neanderthals is pneumatized by enlarged frontal air sinuses. Neanderthal crania have an occipital torus but also a pronounced occipital region (sometimes referred to as a **bun** or **chignon**). Just above the middle of the occipital torus many Neanderthal crania have a slight depression known as the **supra-iniac depression** or **fossa** in the occipital bone. Neanderthals are also

marked by a greatly inflated nasomaxillary region and enlarged orbital and nasal cavities that contribute further to the great length of the cranium as do the elongated greater wings of the sphenoid in the temporal fossa, anteroposteriorly (see Chapter 11).

Despite the flattened contours of the vault, the cranial capacity of Neanderthal crania exceeds that of *H. erectus* and may equal or exceed that of many modern *H. sapiens* crania. It is clear that Neanderthal crania are notably 'full' in the occipital region and that this contributes to the length of the crania.

An awkward clutch of fossils has become known as archaic *H. sapiens* (Stringer, 1974) which includes fossils showing features transitional between *H. erectus* and anatomically modern *H. sapiens* on the one hand, and Neanderthals on the other. The group includes the Swanscombe skull, the Heidelberg jaw, the Petralona skull, the Steinheim skull, the Vertesszöllös remains from Hungary and others from Africa (Kabwe, Ngaloba and Bodo) and China (Da-li). All these remains lack the special derived characteristics typical of Neanderthals. The Petralona skull, for example (Fig. 5.24), has a continuous ridge across the orbits, angled parietal bones

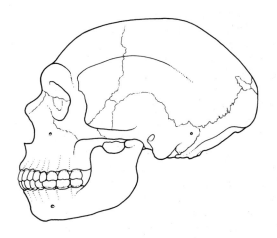

FIGURE 5·23 *Lateral view of the reconstructed Neanderthal skull from Amud, Israel, illustrating the general features and contours of Neanderthal crania.*

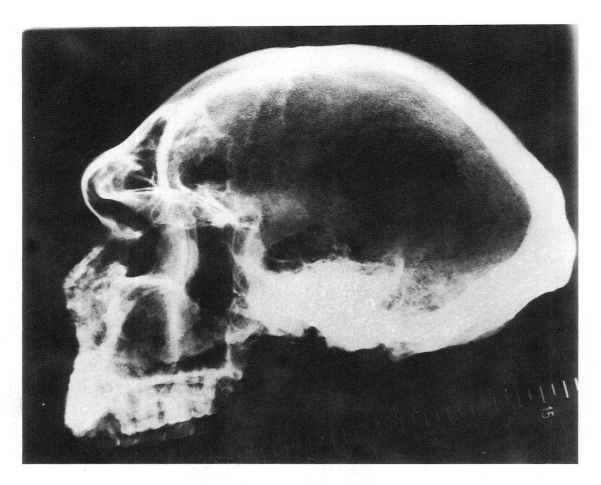

FIGURE 5·24 *Lateral skull radiograph of the Petralona cranium. Note the thick bone in the occipital region and the well-developed frontal air sinuses. Courtesy of C.B. Stringer, J.K. Melentis, A. Christoforidis and P. Papadopoulous.*

with a slight keel in the mid-line, an occipital torus strongly angled in the mid-line posteriorly (and which comprises extremely thick bone) as well as a flattened frontal. All these are characteristics found in *H. erectus* crania but, unlike *H. erectus* crania, Petralona has large frontal sinuses, expanded parietal regions that take the maximum breadth of the cranium above the level of the expanded base, and an occipital torus which is most marked in the mid-line, fading away towards the more lateral aspects of the cranium.

The hominoid cranium in norma basilaris

The morphology of the base of the great ape cranium varies surprisingly little between the gorilla, chimpanzee and orang-utan. While the bizygomatic breadth is wide (especially so in the orang-utan) and the distance between the extremities of the tympanic plates equally far apart, most of the other landmarks on the cranial base that can be identified as homologous among hominoids fall into a long and narrow pattern that contrasts markedly with the broad and short cranial base of modern *Homo sapiens*.

Many of the foramina on the cranial base of great apes, through which nerves, arteries and veins pass, are closer to the mid-line

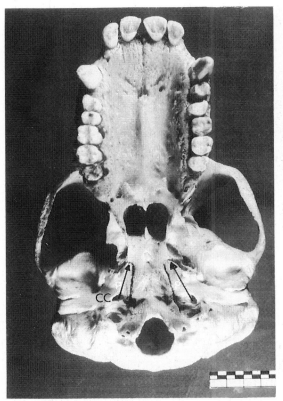

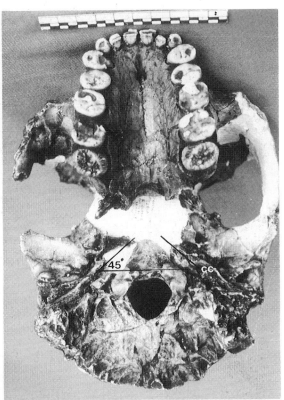

FIGURE 5·25   *The cranial base of* Pan troglodytes *(a) and* Paranthropus boisei, *OH 5 (b). The foramen magnun is positioned more posteriorly in* Pan *than in the early hominids. The petrous temporal bones in monkeys, great apes and* Australopithecus *are orientated more parallel to the sagittal plane rather than at 45 degrees to a line connecting both carotid canals as in* Homo *and* Paranthropus. *Note also the heart-shaped foramen magnum of OH 5.*

than they are in modern man so that the distance between bilateral landmarks is less than in humans (Fig. 5.25 and see Chapter 12). The foramen magnum in great apes lies behind the bitympanic line (even in very young apes) and the basioccipital bone is relatively long. The length of the basioccipital and the position of the foramen magnum are linked, such that in large male gorillas the basioccipital is *very* long and the foramen magnum *very* far behind the bitympanic line. In contrast, the pygmy chimpanzee, *Pan paniscus*, has a much shorter basioccipital and a foramen magnum nearer the bitympanic line. The eventual adult position of the foramen magnum in great apes, therefore,

depends on how much growth there is at the spheno-occipital synchondrosis. In all great apes, however, a relatively large area of the occipital bone behind the foramen magnum is devoted to attachment of the nuchal muscles and this, like the foramen magnum and the occipital condyles, is orientated more vertically in the apes than it is in hominids.

The petrous parts of the temporal bones are orientated with their axes pointing forwards in apes, closer to the sagittal plane than they are in modern humans (about 60° to the bicarotid line). This is in keeping with the long narrow cranial base in apes and the posteriorly positioned foramen magnum.

The cranial base of *Australopithecus* is in

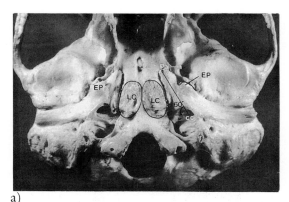

a)

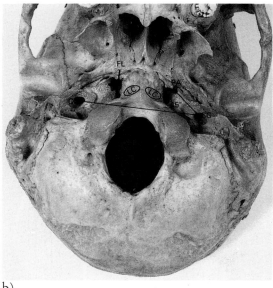

b)

FIGURE 5·26   *The cranial base of (a)* Pan troglodytes *illustrating (i) the petrous bones orientated at 60° to the coronal plane drawn between the carotid canals (CC–CC–PA), (ii) markings for the longus capitis muscles (LC) and (iii) the eustachian processes (EP). (Compare this with Figure 13.4, the same view of MLD 37/38 attributed to* Australopithecus africanus*). (b) Illustrates the cranial base of a modern human with smaller markings for the longus capitis muscles on the basioccipital and a petrous angle (CC–CC–PA) of 45°. Note also the foramen lacerum*

(FL arrowed) anterior to the apex of the petrous temporal bone in the modern human cranium. This is not present in apes.

many ways similar to the cranial base of great apes (see Fig. 5.26 and also Fig. 13.4). However, the foramen magnum is further forward relative to the bitympanic line and there has been some widening between the landmarks across the posterior part of the cranial base (such as the distance between the styloid processes). Much less area is devoted to the nuchal musculature behind the foramen magnum in *Australopithecus* than is the case in great ape crania. The basioccipital, however, is still, as in apes, strongly marked by the large insertions for the longus capitis muscles in MLD 37/38 and Sts 5. The petrous temporal bones are also orientated some 60° to the coronal plane much as they are in apes and monkeys. Two club-like eustachian processes, similar to those typical of African apes and not present on human crania are also prominent in Sts 5, MLD 37/38 (as well as on the cranium of TM 1517 the type specimen of *P. robustus* and *H. erectus* described by Weidenreich (1943).

The outline of the cranium of *Paranthropus*

in *norma basilaris* (Fig. 5.26) is extremely wide across the tympanic plates and the zygomatic bones. This is a simple reflection of the large masticatory system but other features of the cranial base vary independently of this. The foramen magnum in *Paranthropus* is further forward than in both *Australopithecus* and *Homo* being well in front of the bitympanic line. In specimens attributed to *Paranthropus boisei* the foramen magnum is characteristically heart-shaped. Many of the markings for the nuchal muscles are very pronounced in large specimens, but even they are still smaller in area than those of similar-sized apes. Reduced areas for muscle markings on the basioccipital and absent eustachian processes (excepting, as noted above, TM 1517) set the majority of specimens attributed to *Paranthropus* apart from great apes or *Australopithecus*, although it must be said that the very early cranium attributed to *P. boisei* from West Turkana (KNM-WT 17000) has a comparatively long basioccipital with a large area for the insertion of longus capitis. Nevertheless, the petrous temporal bones in all specimens of *Paranthropus*, where they are preserved, are aligned like those of *Homo*, across the cranial base at approximately 45° to the coronal plane, being much more coronal in their orientation than those of *Australopithecus* or great apes.

The cranial base morphology of both early and modern *Homo* is much more similar to each other and distinct from the typically long and narrow cranial base of the apes. Posteriorly, the occipital squama is expanded greatly because of the enlarged brain, but bears only relatively small areas of attachment for the nuchal muscles. The foramen magnum does not lie as far forward as it does in *Paranthropus* but is well forward in comparison with great apes. In some hominids the shape of the foramen magnum is quite characteristic, for example in *P. boisei*. As noted already, it tends to be heart-shaped and in Neanderthals it tends to be oval and elongated, possibly reflecting the long cranium of Neanderthals.

As a consequence of the great reduction in the size of the muscles of mastication, the bizygomatic and bitympanic breadths are also much reduced across the modern human cranium but are wider in early *Homo* where the masticatory system remains better developed. All the nervous and vascular foramina that are close to the mid-line in the great apes are wider apart on the cranial base of *Homo*. Neanderthals and modern *Homo* crania are characterized by a long bony styloid process; however, in apes and some early hominids this process is small and easily falls out to leave only a pit in the cranial base to mark its position. The overall appearance of the human cranial base is short and wide and a major contrast to the long and narrow cranial base of great apes.

The hominoid cranium in norma occipitalis

Many features of the hominoid crania described in *norma lateralis* and *norma basi-*laris are also apparent in *norma occipitalis* (Fig. 5.27). A shift in the position of the foramen magnum and an increase in the size of the brain, together with a reduction in the area of attachment of the nuchal musculature, can easily be seen in the series of crania representing great apes, *Australopithecus*, *Paranthropus* and *Homo*. The full rounded contour of the vault posteriorly in specimens of modern *Homo sapiens* contrasts with the appearance of apes and some specimens of *Australopithecus afarensis* and *Paranthropus*. In these groups the inflated mastoid region and prominent temporal and nuchal crests form the widest part of the cranium. The coronal outline of the vault bones in other specimens of *Australopithecus* and early *Homo* has steeper parietotemporal walls that rise almost vertically before rounding over to the mid-line. While it is easy to attribute these differences in outline to the relative sizes of the brain and the relative sizes of the nuchal and masticatory muscles, it is not so easy to establish whether such differences are due to the allometric effects of differences in body size or to real evolutionary shifts towards an increase in brain size or a reduction in the masticatory system. The interplay between all these factors is a particular problem among specimens attributed to *Australopithecus*, *Paranthropus* and *Homo habilis*, for example. The relative size of the brain and the development of the masticatory system together also determine the pattern of cranial cresting in hominoids and these patterns are most variable in the occipital region.

Large individuals with small brains, such as KNM-WT 17000, the early specimen attributed to *Paranthropus boisei* (Leakey and Walker, 1988) from West Turkana and KNM CH 304 a late *P. boisei* specimen from Chesowanga, have compound temporal/nuchal crests with no **bare area** of bone remaining at their confluence (Walker and Leakey, 1988) (Fig. 5.28). Other crania with both sagittal and nuchal crests have bigger brains and so do not form such extensive

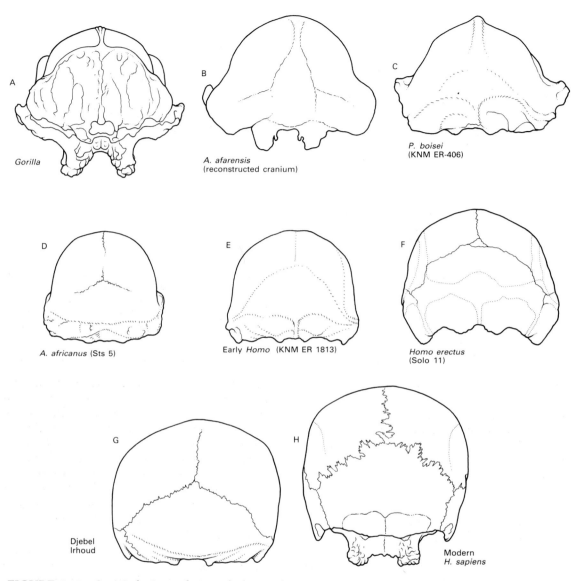

FIGURE 5·27 *Occipital views of six early hominid crania; compared with* Gorilla *(A) and modern* Homo sapiens *(H).*

compound temporal/nuchal crests in this area (KNM-ER 1805). The sagittal crest in great apes or the temporal lines posteriorly in *Australopithecus afarensis* arise more posteriorly on the vault due to the relative enlargement of (and so arguably functional emphasis of) the posterior temporalis muscle. In *Paranthropus* there has been a trend towards greater enlargement of and functional emphasis on the anterior temporalis muscle, so that the superior temporal lines (or sagittal crest in most specimens) come together more anteriorly. This has increased the potential for a bare area of bone between the nuchal

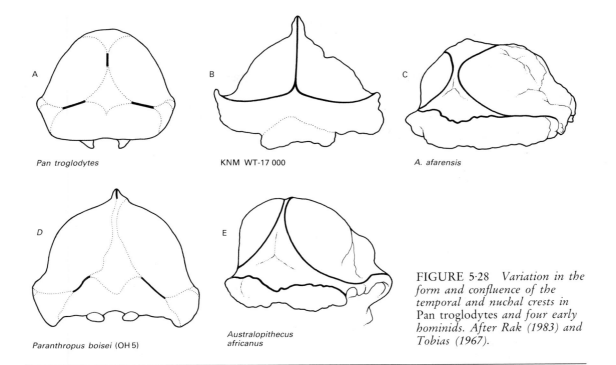

A Pan troglodytes

B KNM WT-17 000

C A. afarensis

D Paranthropus boisei (OH 5)

E Australopithecus africanus

FIGURE 5·28 *Variation in the form and confluence of the temporal and nuchal crests in* Pan troglodytes *and four early hominids. After Rak (1983) and Tobias (1967).*

and temporal crests or lines posteriorly but has acted anteriorly to restrict the post-orbital breadth of the cranium.

The widest part of all early *Australopithecus* and *Paranthropus* crania in the occipital region is across the supramastoid crests or mastoid processes. This is also true of *Homo erectus* crania that are widest low down in the temporal region. Neanderthal crania, however, are much more rounded in the occipital view than are crania of *H. erectus*, but lack the vertical height to the occipital outline that modern human crania show. Their greatest breadth is, like modern human crania, higher on the vault wall than in other early hominids.

The hominoid cranium in norma verticalis

Seen in *norma verticalis*, many of the features of hominid crania discussed in *norma occipitalis* again present clearly (Fig. 5.29). The majority of adult great ape crania have an outline dominated by the temporal/nuchal crests, zygomatic arches and projecting facial skeleton. In extreme contrast is the outline of the modern human cranium, which is entirely dominated by the outline of the vault bones that exceed even the zygomatic bones and facial skeleton in their dimensions.

The crania of *Australopithecus* and *Paranthropus robustus* are long and ovoid in superior view but have relatively less of a post-orbital constriction than do crania attributed to *Paranthropus boisei*. In these latter specimens there is greater emphasis on the anterior temporalis muscle than there is in *Australopithecus*, and *P. robustus*. This has acted both to narrow further the post-orbital constriction of the cranium and to flare the zygomatic arches to produce a more rounded outline (Tobias, 1967; Rak, 1983). The outline of the early specimen attributed to *Paranthropus boisei* (KNM-WT 17000) is more similar to *Australopithecus* crania: it is longer, has a less rounded zygomatic profile than other *P. boisei* specimens and a more

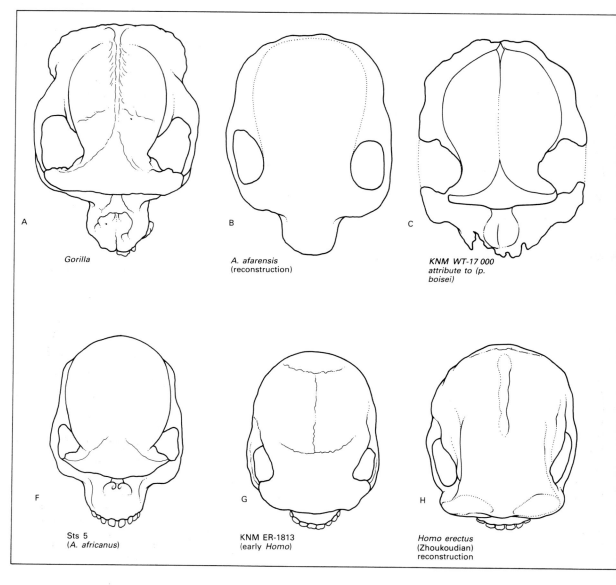

A. Gorilla

B. A. afarensis (reconstruction)

C. KNM WT-17 000 attribute to (p. boisei)

F. Sts 5 (A. africanus)

G. KNM ER-1813 (early Homo)

H. Homo erectus (Zhoukoudian) reconstruction

prognathic facial skeleton that projects beyond the brain case. Some post-orbital constriction is characteristic of later fossil hominids that are also attributed to *Homo*. Specimens attributed to *Homo habilis* have quite a marked post-orbital constriction, but along with specimens attributed to *H. erectus* they are considerably more expanded in the occipital and parietal regions than are earlier fossil hominids. One special feature of

H. erectus crania that can usefully be illustrated in a section cut through the bones of cranium is the great thickness of the vault bones, especially posteriorly (Fig. 5.30). Further details about the form of the supra-orbital and post-orbital region appear in Chapter 11 and additional information about the muscle markings on the nuchal and mastoid regions appears in Chapter 12.

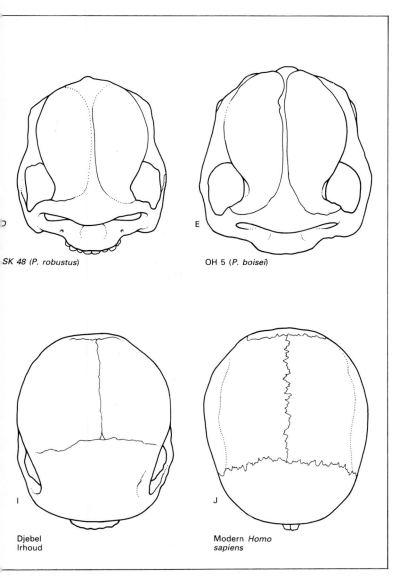

SK 48 (P. robustus)

E

OH 5 (P. boisei)

I

Djebel
Irhoud

J

Modern _Homo
sapiens_

FIGURE 5·29 _Superior views of_ Gorilla _(A) and modern_ Homo sapiens _(J) together with eight early fossil hominids. Note the form of the temporal foramen and zygomatic arch as well as the size of the neurocranium in each._

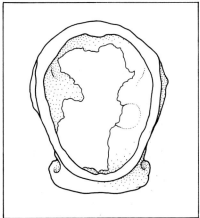

FIGURE 5·30 _Superior view of_ Homo erectus _cranium sectioned in the transverse plane to illustrate the thickness of the vault bones. After Weidenreich (1943)._

CHAPTER SIX

THE MASTICATORY SYSTEM OF HOMINOIDS

In this chapter we shall consider the comparative anatomy of the muscles that bring about movements of the mandible together with the comparative anatomy of the temporomandibular joint and its ligaments. Some simple movements at the mandibular condyles are also discussed as well as some of the forces that act on the mandible during mastication as these relate to morphological features that can be made use of when interpreting fossil hominid remains. In the final part of this chapter we shall review some of the more general features of the skull that relate to masticatory strategies in fossil hominids and consider the morphology of the glenoid fossa of fossil hominids in some detail.

The four major pairs of muscles that bring about movements of the mandible are known collectively as the **muscles of mastication**. They are attached both to the mandible and the cranium and are named the **temporalis**, the **masseter**, the **medial pterygoid** and the **lateral pterygoid** muscles. Muscles that raise the mandible and occlude the teeth are known as **elevator** muscles of the jaw and, with one exception (the lateral pterygoid), all the muscles of mastication act primarily as elevators of the mandible. Other muscles that act to open the mouth by lowering the jaw are known as **depressor** muscles of the mandible. One of the muscles of mastication (the lateral pterygoid) is active when the mouth is opened but there are also other depressor muscles. The **digastric muscles** as well as another group of muscles that run directly between the mandible and the hyoid bone and continue as a 'strap-like' group of muscles to the sternum and pull on the mandible from below (the **infrahyoid muscles**) are all important depressor muscles of the mandible. Clearly then, not all muscles that move the mandible have been included under the common descriptive title of 'muscles of mastication' so that in this chapter we shall also describe several additional muscles in the head and neck.

Each group of muscles not only play an important role in opening and closing movements of the mandible but also in **protrusive**, **retrusive** and **lateral** movements of the mandible. To enable precise occlusal contacts to be made between the maxillary and mandibular teeth during mastication, co-ordination and control of these muscles is such that the muscles on each side of the head and neck may be active at different times. The way in which this usually occurs during chewing will be described later. Co-ordination of complicated jaw movements develops during childhood but at birth the muscles of mastication are poorly developed. In fact infantile suckling and swallowing movements are largely restricted to simple opening and closing movements that are in part brought about by the muscles of facial expression. Sperber (1981) notes that after birth the facial muscles increase only four times in weight, whereas the muscles of mastication increase seven times before adulthood.

Muscles that move the mandible

The temporalis muscle

The temporalis is a fan-shaped muscle that extends upwards and backwards from the coronoid process of the mandible to take origin from the side of the cranial vault (Fig. 6.1). Two elevated bony markings, the **superior** and **inferior temporal lines**, circumscribe an area overlying the frontal, sphenoid, temporal and parietal bones that provide attachment for the muscle on each side of the skull. Two extensions of the muscle continue downwards from the coronoid process along the anteromedial aspect of the ramus of the mandible. These become tendinous and end in the region of the retromolar fossa, just behind the third molar tooth. If the temporalis muscle is cut in coronal section it can be seen that it is bipennate (see Chapter 2). A central tendinous sheet that inserts onto the coronoid process and passes backwards through the middle of the muscle mass to take origin from the inferior temporal line (Batson, 1953) (Fig. 6.2). Beneath this central tendinous sheet muscle fibres arise from the side of the cranial vault and insert obliquely into its undersurface. Some of the deepest muscle fibres of temporalis in this region originate from the infratemporal crest on the greater wing of the sphenoid bone. Often this raises a spur of bone, the **infratemporal spine** that is a prominent feature of many hominoid crania.

Arising from the superior temporal line on the side of the cranial vault is the **temporalis fascia**. This is a tough fibrous investing layer and overlies muscle fibres that are lateral to the central tendinous sheet. These more superficial fibres take origin from the temporalis fascia and then pass obliquely downwards and forwards from it to insert into the central tendinous sheet. In modern humans, muscle fibres do not usually arise from the whole of the outer aspect of the temporalis fascia, so that when it is removed a part of the central tendon of the muscle remains bare

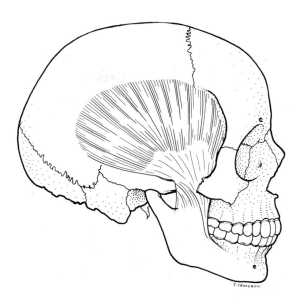

FIGURE 6·1   *The temporalis muscle.*

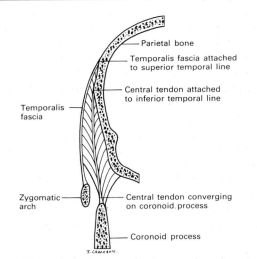

FIGURE 6·2   *The temporalis muscle in coronal section, After Batson (1953).*

and is visible in the region of the coronoid process. The more muscle fibres that arise from the lower part of the temporalis fascia the smaller the bare area of the central tendon that remains visible, such that in great apes, muscle fibres cover it all and none of the central tendon is visible when the temporalis fascia is removed. The temporalis fascia splits inferiorly and attaches to both the inner and the outer aspects of the superior border of the zygomatic arch. It performs an additional function here as a bracing suspensory sling that supports the zygomatic arch against the pull of the masseter muscle (Eisenberg and Brodie, 1965).

Because temporalis is a bipennate muscle (Fig. 6.2) the individual muscle fibres are much shorter than they appear to be when the muscle is viewed laterally (and therefore are also more numerous). We noted in Chapter 2 that muscles are not damaged by being stretched to about 1.5 times their resting length but additional stretching tears the muscle fibres. Most of us can open our mouths to about three-fingers breadth between our upper and lower incisor teeth but more than this is usually difficult and many people cannot even manage this (Fig. 6.3). Temporalis muscle fibres in modern humans are only associated with about some 15 mm of movement at the coronoid process of the mandible (and not the 30 mm

or so one might predict by measuring the apparent length of the muscle fibres in lateral view (Batson, 1953)). As the power of a muscle is related to the number of muscle fibres, however, the temporalis is a more powerful muscle than it would be were it not bipennate.

Fibres of the temporalis muscle on both the inside and the outside of the central tendon converge towards the opening between the zygomatic arch and the lateral surface of the cranial vault. The apex of the coronoid process of the mandible is positioned in the centre of this opening. When seen in lateral view, the most anteriorly positioned fibres (which form the bulk of the temporalis muscle), run vertically downwards towards the coronoid process. The fibres in the middle part of the muscle become increasingly obliquely orientated towards the coronoid process and the most posterior fibres run horizontally forwards but then bend sharply downwards in front of the articular eminence at the root of the zygoma. In this region the most posterior muscle fibres of the temporalis become tendinous but, like the anterior fibres, still pass vertically downwards to insert onto the edge of the lowest point of the mandibular notch. It follows that when the condylar head is in the 'rest position' (see later), the horizontal fibres of the posterior temporalis in fact pull vertically (and not as might be expected backwards) and so pull the condyle upwards and forwards into firm contact with the articular eminence of the temporal bone (DuBrul, 1980).

The masseter muscles

The masseter muscles are the most superficial of the muscles of mastication and overlie the ascending rami of the mandible (Fig. 6.4). The masseter can be incompletely divided into a deep part and a superficial part which, although separable posteriorly, become fused together in a single muscle mass anteriorly (very occasionally three parts are discernible).

The deeper part of the muscle is inseparable from the underlying vertically orientated

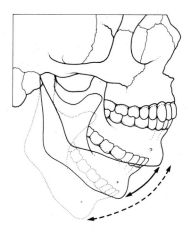

FIGURE 6·3 *The commonly observed range of movement of the mandible compared with the impossible range of mandibular movement that can be predicted if the temporalis muscle fibres were as long as they appear to be in lateral view. After Batson (1953).*

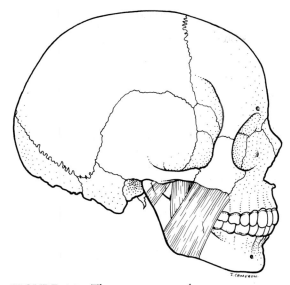

FIGURE 6·4   *The masseter muscle.*

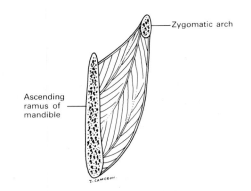

FIGURE 6·5   *In coronal section the masseter muscle can be seen to be multipennate.*

aspect of the zygomatic arch as far back as the articular eminence and insert onto the uppermost aspect of the lateral surface of the mandibular ramus. Apart from a small triangular area in front of the temporomandibular joint, the whole of the deep portion of the muscle is hidden beneath the most superficial part of the muscle.

The superficial part of the muscle is multipennate (presenting a herring-bone pattern (Fig. 6.5) in coronal section (DuBrul, 1980)). The origin of the superficial part is from the inferior edge of the zygomatic arch but only as far back as the zygomaticotemporal suture. Anteriorly, the superficial head of the masseter may occasionally attach to the outer aspect of the zygomatic process of the maxilla. This tendinous origin extends downwards and backwards to insert into the angular region of the mandible a third of the way up the ramus and anteriorly as far forward as the level of the second molar tooth. The oblique orientation of the superficial part of the masseter muscle runs at right angles to the posteriorly ascending occlusal plane of the molar teeth (this is known as the **curve of Spee**).

The medial pterygoid muscles

The medial pterygoid muscles are situated on the inner aspects of the ascending mandibular rami and are similar to the superficial heads of the masseters both in the oblique orientation of their muscle fibres (that are also at right angles to the occlusal plane of the molar teeth) and in their multipennate structure (Figs 6.6 and 6.7). The majority of the medial pterygoid muscle takes its origin from the medial aspect of the lateral pterygoid plate but a small anterior portion arises from the lateral and inferior part of the pyramidal process of the palatine bone and, occasionally, even the most posterolateral part of the maxilla. Unlike the masseter, which inserts onto a large area of the lateral aspect of the mandibular ramus, the medial pterygoid muscle inserts into a smaller triangular area at the medial aspect of the mandibular angle.

anterior fibres of the temporalis muscle which run in the same direction. A part of the deep portion of the masseter muscle that is fused with superficial fibres of anterior temporalis and which runs from the inner border of the zygomatic arch to attach to the base of the coronoid process, is easily distinguishable in many animals and is known as the **zygomaticomandibularis muscle**. The rest of the deep portion of the masseter muscle fibres take origin from the whole of the inner

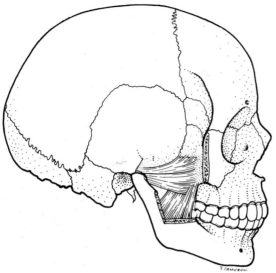

FIGURE 6·6 *The medial and lateral pterygoid muscles in lateral view with part of the zygomatic arch and ascending ramus of the mandible removed.*

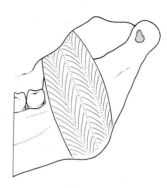

FIGURE 6·7 *When viewed medially, the multipennate structure of the medial pterygoid muscle is apparent.*

The lateral pterygoid muscles

The lateral pterygoid muscles each arise as two separate heads, one from the inferior surface of the greater wing of the sphenoid (the smaller of the two heads), and the other from the lateral aspect of the lateral pterygoid plate (Figs 6.6 and 6.8). Both heads then pass laterally towards each condyle to become inseparable at their insertion into the depression or **fovea** in the neck of the condyle. Some of the deepest and most superior of the muscle fibres of the upper head of the lateral pterygoid insert into the joint capsule and the articular disc.

To reach the condylar neck from the lateral pterygoid plate or from the infratemporal surface of the greater wing of the sphenoid,

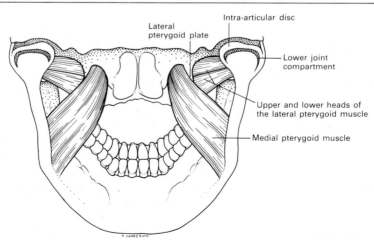

Lateral pterygoid plate

Intra-articular disc

Lower joint compartment

Upper and lower heads of the lateral pterygoid muscle

Medial pterygoid muscle

FIGURE 6·8 *Posterior view of the medial and lateral pterygoid muscles, illustrating that they run in different planes to one another.*

muscle fibres must travel laterally at nearly 45° (Fig. 6.9). It follows that when both lateral pterygoid muscles contract at the same time, they also act to pull the condyles towards each other, while they pull each condyle and disc forwards. The resulting forces that act to bend the mandible are concentrated at the chin. Muscle fibres of the lateral pterygoid, whose origin is higher in the infratemporal fossa than the level of the condylar fovea, act to pull the condylar head upwards and medially against the articular eminence. These fibres of the upper head are active during closing of the jaws and maintain contact between the articular surface of the condyle, the articular disc and the articular eminence, as the mandible returns to the rest position (MacNamara, 1973; Juniper, 1981). Muscle fibres whose origin is from the lowermost part of the lateral pterygoid plate pass upwards and outwards to the neck of the condyle and so act to pull the condyle downwards and forwards onto the articular eminence during opening and protrusion of the mandible. Because of the two quite different functions of the upper and lower heads of the lateral pterygoid muscle, some authorities now refer to the upper head as the **superior pterygoid muscle** and to the lower head only as the lateral pterygoid (Juniper, 1981; Stern, 1988).

Despite its complicated and important actions, there is some evidence that the lateral pterygoid muscle has fewer muscle spindles than do the other muscles of mastication (Gill, 1971). It follows that the proprioceptive feedback (see Chapter 2) from this muscle may be poorer than that from the other muscles that move the mandible, so that the lateral pterygoid may be a less finely co-ordinated muscle of mastication. This fact may go some way to explaining its implication in painful temporomandibular joint syndromes.

The digastric muscles

In many mammals the principal depressor of the mandible is a double-bellied muscle that

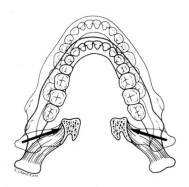

FIGURE 6·9 *The action of the lower head of the lateral pterygoid is to open and protrude the lower jaw. (In fact according to Juniper (1981) and Stern (1988) the upper head, which is active during elevation of the jaw, has such a different action to that of the lower head illustrated here that these authors are in favour of the upper head being named the superior pterygoid muscle).*

passes from the cranial base to the angle of the mandible and which, for this reason, is often called the **occipito-mandibularis**. In man the digastric muscles are homologous with these and along with the geniohyoids are the most important depressors of the mandible. However, they run between the cranial base and the two digastric fossae of the mandible. Each digastric muscle consists of a posterior belly and an anterior belly joined by a strong round tendon (Fig. 6.10). This **intermediate tendon** is bound down to the side of the hyoid bone by a loop of fascia that allows the muscle to contract and slip through the sling without pulling the hyoid bone backwards or forwards. In this way, the chin can be depressed independent of the position of the hyoid bone (Last, 1955).

The posterior belly of the digastric muscle originates from a deep elongated groove in the temporal bone called the **mastoid notch** or **digastric fossa** (see also Chapter 12). This lies immediately medial to the mastoid process of the temporal bone. The two digastric fossae of the cranial base are widely separated from each other but the posterior bellies of the muscle run downwards, forwards and

FIGURE 6·10 *Lateral view of the strap muscles together with the omohyoid, digastric and stylohyoid muscles.*

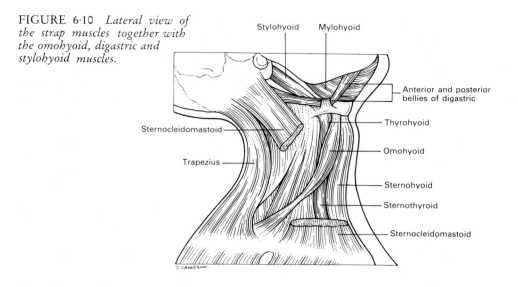

medially to the hyoid bone. The anterior bellies pass upwards and even further medially to insert into the digastric fossae at the lower border of the mandible, either side of the mid-line.

A major function of the digastric muscle is to open the mouth by depressing the mandible. It is also probable that the digastric muscles assist in retruding the jaw as far back as it will go into what is known as the **retruded contact position**. In fact it is not possible to carry out such retrusion of the mandible without feeling activity in the infrahyoid muscles above the manubrium which strongly suggests that the digastric muscle is active during retrusion (see also Last, 1955). Some authorities have argued that it is the posterior fibres of temporalis that retrude the mandible but, as we have noted above, DuBrul (1980) emphasizes that these muscle fibres turn 90° around the root of the zygomatic process of the temporal bone to insert (and presumably pull) vertically onto the mandibular notch and coronoid process. While fibres of posterior temporalis may well be active to some degree in active retrusion (otherwise the jaw would open) the digastric muscle is likely to be the prime retractor.

The mylohyoid muscle

The mylohyoid muscle forms the diaphragm or floor of the mouth and extends between the mylohyoid line on the inner aspect of the mandibular body and the anterior part or body of the hyoid bone (Fig. 6.10 and 6.11).

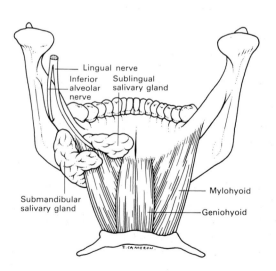

FIGURE 6·11 *Posterior view of the floor of the mouth illustrating the geniohyoid and mylohyoid muscles together with the submandibular and sublingual salivary glands.*

Fibres from the two sides form a mid-line **raphe** (or seam) and only the most posterior muscle fibres that originate from the mylohyoid line in the region of the third molar tooth insert into the hyoid bone directly. These posterior fibres are able to depress the mandible weakly, or, alternatively, raise the hyoid bone if the mandible is fixed. When the more anterior fibres contract, the floor of the mouth is raised, thereby raising the structures above the mylohyoid muscle up into the oral cavity.

The geniohyoid and stylohyoid muscles

At the most anterior end of the mylohyoid line, two small strap-like muscles, the geniohyoids, arise from the inferior genial tubercles and adjacent mylohyoid line and pass backwards and downwards to the body of the hyoid bone (Figs 6.10 and 6.11). Together they exert a downwards and backwards pull on the chin or, alternatively, with the mandible fixed by other muscles, the geniohyoid muscles are able to pull the hyoid bone forwards and upwards.

The stylohyoid muscles act to retract the hyoid bone. Together, with the mylohyoid and geniohyoid muscles they are able to oppose the downwards pull of the infrahyoid muscles (Last, 1955). The two stylohyoid muscles arise and extend from the styloid processes of the temporal bone on the cranial base, just anteromedial to the origin of the posterior bellies of the digastric muscles (Fig. 6.10). They insert onto the greater cornua (or horns) of the hyoid bone by splitting to pass either side of the intermediate tendon of the digastric muscle.

The infrahyoid muscles

Although the infrahyoid or **strap muscles** of the neck do not act directly on the mandible, they act with the suprahyoid muscles to stabilize the hyoid bone and so provide a stable base for the digastric and mylohyoid muscles to depress the mandible (Last, 1955).

The two deepest pairs of infrahyoid muscles are the **sternothyroid** and **thyrohyoid** muscles. The sternothyroid muscles run in turn from the posterior–superior aspect of the manubrium to the thyroid cartilage of the larynx. The thyrohyoid muscles run from the **oblique line** of the thyroid cartilage to the hyoid bone. Overlying these two deeper pairs of strap muscles are the **sternohyoid** muscles which, as their name suggests, run directly to the hyoid bone from the sternum (Fig. 6.10). The origin of the sternohyoid muscles differs from that of the sternothyroid muscles in that they are separated in the mid-line at their origin and may extend laterally onto the cartilage of the first rib. In contrast, the origin of the sternothyroid muscles at the manubrium meet in the mid-line and may even overlap each other.

The omohyoid muscle, like the digastric muscle, has two bellies and passes obliquely across the neck between the upper border of the scapula and the lower border of the hyoid bone (Fig. 6.10). Its central tendinous part is tied down to the clavicle by a tough loop of fascia. Contraction of the omohyoids aid in depressing or fixing the hyoid bone. However, with the hyoid and clavicle fixed it has been suggested that the omohyoid may also help to prevent the apex of the lung bulging out above the thoracic inlet. In addition, because the **omoclavicular fascia** that binds the central tendon to the clavicle also covers the internal jugular vein, movements of the omohyoid muscle are said to widen the lumen of this important vein. However, contraction of the omohyoid muscle can be observed most commonly in the neck during speaking and swallowing when the hyoid is most mobile.

Muscles that move the mandible in the great apes

The masticatory system of great apes is large so that the muscles associated with movements of the mandible are also larger in apes than they are in modern humans. This, together with the fact that the brain in great apes is smaller than the brain in modern humans (which as we have seen in Chapter 5 results in a reduced amount of the skull vault being available for the origin of several massive muscles), has an effect both upon the structure of the muscles that act on the mandible and upon the form of their bony attachments. It has been noted that, whereas the jaws and face of primates scale in a positive allometric fashion with body weight, dry weights of muscles of mastication from primates scale isometrically (Cachel, 1984).

The temporalis muscle

In each of the great apes the temporalis muscle may take origin from a **sagittal crest** or **keel** as well as from the lateral aspect of the cranial vault. A sagittal crest is usually present in adult male gorillas and adult male orang-utans but is rare in male chimpanzees and unknown in female chimpanzees and orang-utans. Where there are cranial crests in great apes (either nuchal or sagittal) both the nuchal muscles and the temporal muscles always rise proud of their crests of origin. In living great apes that have a bony crest there is actually commonly a groove or furrow on the top of the head and not a bony ridge (see also Chapter 12). In some individuals the nuchal and temporal crests meet and fuse to form a compound temporal–nuchal crest posteriorly. In these cases, however, the temporalis muscle does not extend to the outer margin of this compound crest. Laterally, in the region of the external auditory meatus, there may be as much as 1 cm between the temporalis muscle and the edge of the supramastoid crest. Posteriorly, the temporalis encroaches closer to the lip of the temporal–nuchal crest in apes, but there is always another more superficial muscle attached to the crest. The **occipito-frontalis** (a muscle of facial expression) inserts onto the upper aspect of the nuchal crest and so separates the origin of temporalis from that of trapezius (Sakka, 1972, 1984).

Although the fibres of the temporalis originate from almost the whole of the side

of the cranium in the great apes, behind the lower part of the orbit there is a considerable amount of fat where the muscle does not lie directly against bone. The investing temporalis fascia, which gives origin to muscle fibres that insert onto the outer aspect of the central tendon of the muscle, covers the whole of the temporalis muscle in apes. When it is dissected off, no part of the central tendon is exposed as it is in modern humans because the outer portion of the muscle is

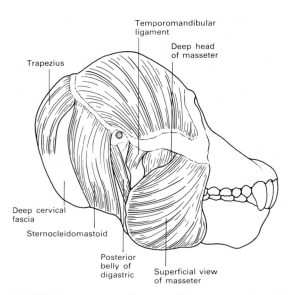

FIGURE 6.12 *Lateral view of the temporalis and masseter muscles in an orang-utan. After Bluntschli (1929).*

FIGURE 6.13 *Serial sections cut through the temporalis muscle of a chimpanzee reveal that it is bipennate and that there is a fleshy part of the muscle over the coronoid process. After Baycroft (1988).*

so well developed in apes (Fig. 6.12). The insertion of temporalis onto the coronoid process of the mandible continues downwards as two strong tendinous extensions along the inner anterior border of the mandibular ramus, as far as the retromolar fossa. Sections cut coronally through the temporalis muscles of great apes reveal that there can be a separate fleshy part of the muscle over the crest of the coronoid process around which the central tendon divides (Fig. 6.13).

The masseter muscle

The masseter muscle is more easily separable into superficial and deep parts in the great apes. In the chimpanzee there can be a strong aponeurotic sheet between the two parts that gives origin to muscle fibres of both heads, but commonly these heads are only separated by loose connective tissue. Muscle fibres forming part of the deep head of the masseter in great apes, also arise from an additional attachment, the outer aspect of the temporalis fascia in the infratemporal fossa. In fact, as in modern humans, in this region the muscle fibres of both the temporalis and the masseter run in the same direction and attach to adjacent regions near the base of the coronoid

process. Therefore, these muscles are very difficult to separate here. The superficial part of masseter has a tough tendinous insertion at the lower border of the mandibular ramus and is close to the insertion of the medial pterygoid here. Sections cut through the masseter muscle in great apes (Fig. 6.14) reveal that the superficial part of the muscle is multipennate with two primary tendinous septa running superoinferiorly in the sagittal plane. One of the septa originates from the zygomatic arch and fades out inferiorly. The other septa arises from the lower border of the mandible and fades out as it extends superiorly through the muscle. Thus the superficial head is divided into three 'compartments' (Baycroft, 1988).

The medial and lateral pterygoid muscles

On the inner aspect of the angle of the mandible in great apes, there are several raised bony ridges that radiate in the direction of the lateral pterygoid plate and represent the bony attachments of several tendinous sheets within the medial pterygoid muscle which is then also a multipennate muscle. The tendinous septa of the medial pterygoid muscle, however, are arranged at right angles to those within the masseter muscle. They run superoinferiorly, but more nearly in the coronal plane. In this way the muscle fibres of the masseter and medial pterygoid muscles are arranged so that they pull in different planes as they elevate the mandible (Baycroft, 1988).

The lateral pterygoid muscle arises from the outer aspect of the lateral pterygoid plate and that part of the undersurface of the sphenoid bone that is continuous with the lateral pterygoid plate in this region. When the lateral pterygoid plate is very extensive in great apes, the most posterior part may in fact not give origin to muscle fibres but have the mandibular division of the trigeminal nerve closely applied to its edge here as it exits the cranial cavity via the foramen ovale.

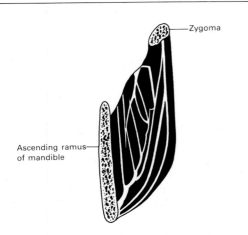

FIGURE 6·14 *Coronal section through the masseter muscle of an orang-utan illustrating the direction of primary and secondary tendinous septa within the superficial part of the muscle. After Baycroft (1988).*

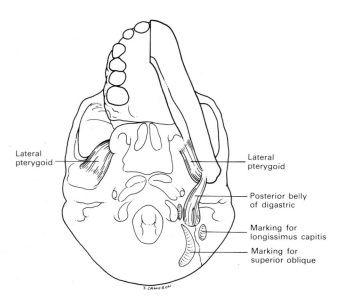

Lateral
pterygoid

Lateral
pterygoid

Posterior belly
of digastric

Marking for
longissimus capitis

Marking for
superior oblique

J. CAMERON

FIGURE 6·15   *Inferior view of the right and left lateral pterygoid muscles together with the posterior belly of the digastric muscle, which in the orang-utan inserts at the angle of the mandible. After Bluntschli (1929).*

A dense sheet of connective tissue arises from the posterior margin of the lateral pterygoid plate and sweeps round covering the underlying tissues, before it fuses with the capsule of the temporomandibular joint posteriorly. The insertion of the upper and lower heads of the lateral pterygoid muscle in great apes at the head of the mandibular condyle differs slightly from that in modern humans. Rather than attaching only to a discrete fovea anteriorly, the great ape lateral pterygoid attaches to the whole of the medial aspect of the neck and also to the undersurface of the medial pole of the condyle (Fig. 6.15). The upper head of the lateral pterygoid and upper part of the lower head of the lateral pterygoid muscle in great apes insert into the disc or meniscus of the temporomandibular joint. The point of attachment, however, is more medial than in modern humans, close to the medial pole of the head of the mandibular condyle (Baycroft, 1988).

In modern man, the bone of the undersurface of the greater wing of the sphenoid is thin and the temporal lobe of the brain in the middle cranial fossa lies immediately above it. However, in apes, the temporal lobe of the brain is much smaller; the bone in this region is pneumatized to expand the external area of the outer table of the sphenoid bone such that sufficient area of bone is available to act as origin for the lateral pterygoid and temporalis muscles. The temporal lobe of the brain may be 2 or even 3 cm above the infratemporal crest in large apes. During growth an extensive sphenoidal air sinus invades this region as far laterally as the infratemporal crest, between the inner and outer tables of the sphenoid bone.

Powerful heavy mandibles also require powerful depressor muscles because gravity alone is insufficient to bring about speedy opening movements of the jaws during mastication. In addition, fine control of mandibular movements results from synergistic contraction of muscles with opposing actions. These are especially important during unilateral excursive movements of the mandible. The digastric muscle is an important depressor of the mandible and in the chimpanzee and gorilla and, like in modern humans, it is double bellied. The digastric muscle originates from the cranial base of great apes from a wide area lateral to the rectus capitis lateralis and medial to the sternocleidomastoid and splenius capitis muscles (Fig. 6.15). Occasionally, the anterior bellies of the digastric may be fused in the chimpanzee (Sonntag, 1923, 1924). The orang-utan is unique amongst hominoids in having no anterior belly of the digastric at all. The posterior belly of the digastric in the orang-utan is especially large at the cranial base, but quickly converges to form a thick round tendon that inserts onto the angle of the mandible between the medial pterygoid and the masseter. Therefore, the digastric muscle in the orang-utan functions like the craniomandibularis in other mammals.

The mylohyoid muscle that forms the floor of the mouth together with the geniohyoid muscles and anterior bellies of the digastric muscles (except in the orang-utan) is probably also active during opening movements of the mandible. The narrow distance between the mandibular bodies in the apes results in the mylohyoid being less well developed than in modern humans where the floor of the mouth is wider (Duckworth, 1904). Some reports suggest that the mid-line raphe (or seam) of the mylohyoid muscle is less prominent in great apes than it is in modern humans.

Platysma, the most superficial muscle of the face and neck, is a muscle of facial expression. It is much better developed in the great apes than in modern humans, sometimes even meeting in the mid-line at the back of the neck and so enveloping all underlying structures. The platysma inserts into a large area of the inferior border of the mandible in the apes. This bony attachment may extend along the whole of the lower border of the mandible, anterior to the attachment for masseter, as far forwards as the region of the symphysis and even include the posterior border of the inferior transverse

torus or simian shelf (Lightoller, 1929; Winkler, 1989). This large mandibular attachment far exceeds the thin line of attachment to the mandible in modern humans and may be important when the platysma retracts the lips of apes to expose the teeth. Moreover, the extensive mandibular attachment may also be indicative of some activity during depression of the mandible in apes.

The modern human temporomandibular joint

The temporomandibular joint is a bilateral joint which like the bilateral facet joints between vertebrae has to move at the same time on each side. Because of this the movements of the mandible are complicated. The bony morphology of the head of the condyle and the glenoid fossa are not good guides to the movements and the functional morphology of the temporomandibular joint because: firstly, both are covered with a layer of dense fibrous connective tissue of varying thickness; and secondly, an interposing **articular disc** (or **meniscus**) further modifies the contours actually travelled by the condylar heads during mandibular movements. Radiographs reveal that, with the teeth in occlusion (or slightly apart with all muscles relaxed, when the mandible is said to be in the **rest position**), the head of the condyle abuts the posterior slope of the articular eminence, but never lies in the depth of the glenoid fossa. In this position the fibrous articular disc is firmly sandwiched between the head of the condyle and the articular eminence of the temporal bone. Muscle fibres of the upper head of the lateral pterygoid and of the horizontal, posterior part of the temporalis that turn vertically downwards help to hold the condyle in this position (DuBrul, 1980; MacNamara, 1973, 1974).

The glenoid fossa
The bony morphology of the modern human **condylar** or **glenoid fossa** is dominated by the steeply sloping anterior **articular**

FIGURE 6.16 *Anatomical landmarks in the region of the modern human glenoid fossa.*

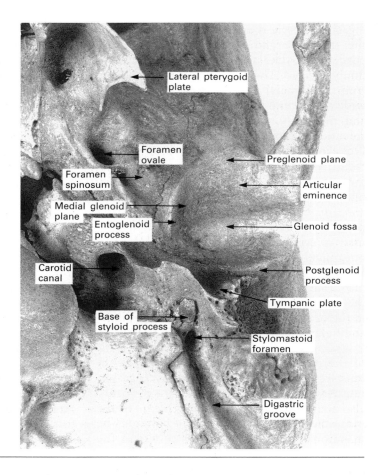

Lateral pterygoid plate

Foramen ovale

Foramen spinosum

Preglenoid plane

Articular eminence

Medial glenoid plane

Entoglenoid process

Glenoid fossa

Carotid canal

Postglenoid process

Tympanic plate

Base of styloid process

Stylomastoid foramen

Digastric groove

eminence (Fig. 6.16). This becomes continuous in front with the more nearly horizontal **preglenoid plane**. Medially, the glenoid fossa narrows and is bounded by a bony wall, the **entoglenoid process**, that is continuous with a **medial glenoid plane** at its inferior edge. This process can occasionally be elevated into a prominent bony spine, the **temporal spine** (not to be confused with the infratemporal spine on the greater wing of the sphenoid bone that gives rise to fibres of the temporalis muscle). In fact, the temporal spine probably got its name because it articulates anteriorly with the spine of the sphenoid bone. Posteriorly, a lip of bone marks the line of attachment of the joint capsule around the rim of the glenoid fossa and becomes more prominent laterally as it forms the thickened

post-glenoid process that projects downwards in front of the **tympanic plate**. Behind the rim of the fossa and the post-glenoid process at the back, the squamotympanic fissure creates a clearer demarcation between the tympanic plate and the glenoid fossa. At this point it is worth mentioning that the small lip of petrous bone, which we described in Chapter 1 as the **tegmen tympani**, protrudes **into** the squamotympanic fissure and so divides it into a **petrotympanic fissure** posteriorly and a **petrosquamous fissure** anteriorly (Fig. 4.19). These names sometimes appear in descriptions of fossil hominid crania and can be confusing, but really they simply describe a fissure that splits in one part of its course to travel around an isolated part of the petrous temporal bone.

The head of the mandibular condyle is orientated at right angles to the line of the posterior tooth row. If the long axes of each condyle are extended medially across the human cranial base, they intersect in the region of the anterior border of the foramen magnum (at *basion*, the mid-line point on the anterior margin of the foramen magnum). This means that when simple hinge movements take place during minimal opening of the mandible, the axis of rotation, that passes straight across the cranial base from one condyle to the other, does not coincide with the axes of the condylar heads. The point of articulation of the condyle with the glenoid fossa then, moves inwards over the head of the condyle as opening continues and in this way pressure is not maintained in any one spot for very long. The medial articular aspect of the head of the condyle is more rounded and follows the contours of the entoglenoid process and medial glenoid plane that faces it.

The development of the glenoid fossa

The morphology of the adult glenoid fossa establishes itself gradually throughout the growth period as the permanent teeth are erupting into occlusion (Kozam, 1985). The glenoid fossa of young individuals is shallow and there is much less articular eminence anteriorly. In adult modern humans, two extremes in glenoid fossa morphology can be recognized which may result partly from the interplay between the occlusion of the teeth and the degree of constraint this imposes upon mandibular movement. A deep incisor overbite in adults is often associated with a deep glenoid fossa together with well-developed fibres of the *inferior* portion of the lower head of the lateral pterygoid muscle that are necessary for pulling the head of the condyle down the steep articular eminence (Kozam, 1985). This strong muscle action may also result in an anterior tilting of the head of the condyle. Lateral excursion in individuals with a deep incisor overbite is constrained and the predominant movements

are elevation and depression of the mandible. On the other hand, adults with reduced overbites or an edge-to-edge occlusion of the anterior teeth are able to make wide excursive movements more easily and during the growth period usually retain shallow glenoid fossae and a flatter than average incline to the articular eminence. The *upper* fibres of the lower part of the lateral pterygoid muscle are well-developed in these individuals and act to protrude the mandible (Kozam, 1985).

Traditionally, it was held that little or no remodelling of the temporomandibular joint occurred after the establishment of occlusion in any individual, even following major changes to the occlusion later in life. Recently, however, data have been published which demonstrate that with the loss of teeth or with severe attrition, the angle of the slope of the articular eminence decreases and the glenoid fossa becomes more shallow (Granados, 1979). It is also possible that a reduction in the height of the mid-face or in the degree of incisor overbite between different modern human populations may also be associated with more shallow glenoid fossae and a reduced mandibular plane angle (Fletcher, 1985).

The capsule and compartments of the temporomandibular joint

The **articular disc** or meniscus of the temporomandibular joint completely divides the **synovial joint capsule** into upper- and lower-joint compartments (for a detailed account of this see Hylander, 1979c). Two thickened regions of the disc that run coronally in an arc from pole to pole of the condyle are known as the **anterior** and **posterior bands** of the disc (Fig. 6.17). Between these is an **intermediate zone** that is not as thickened as the two bands. Behind the posterior band the disc splits into two layers: a loose fibroelastic upper part that attaches to the **squamotympanic fissure** posteriorly; and a lower part that attaches to the back of the mandibular condyle. During protrusive

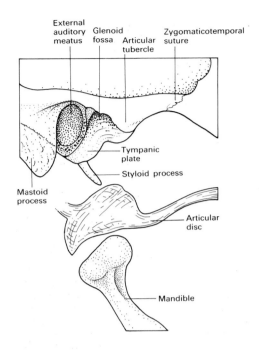

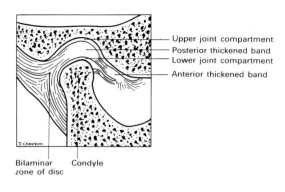

FIGURE 6·17   *The structure of the temporomandibular joint and the articular disc in modern humans.*

movements of the condyle and disc, this loose, more vascular, bilaminar region becomes sucked into the back of the joint capsule. In front, the disc attaches to both the anterior edge of the articular eminence and the head of the condyle. At the sides, the disc blends with the joint capsule and along with it attaches to the poles of the condyle. It is this attachment to the poles of the condyle that causes the disc to travel with the condyle when it slides forwards.

The ligaments that limit movements of the temporomandibular joint

The **temporomandibular ligament** is a tough dense sheet of collagenous tissue, devoid of any elastic fibres. It is inseparably blended with the joint capsule and lies on the lateral aspect of the joint only (Fig. 6.18). (A medial ligament is unnecessary as the ligament of the opposite side resists medial movements at the contralateral joint.) The ligament arises

as a fan-shaped sheet from the **articular tubercle** of the zygomatic arch (which is badly named, as DuBrul (1980) notes, because no articulation takes place at all here) and a ridge of bone that runs backwards from this point along the zygomatic process of the

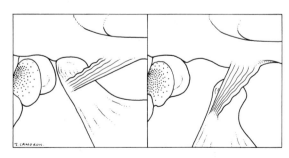

FIGURE 6·18 *The outer oblique band of the temporomandibular ligament. Forward excursion of the condyle is resisted by the posterior fibres of the ligament and backward movement by the anterior fibres. After Rees (1954).*

temporal bone. The fibres of the temporo-mandibular ligament then converge onto an area at the back of the neck of the condyle. The taut temporomandibular ligament helps keep the condyle, disc and temporal bone firmly opposed. Forward excursion of the condyle is resisted by the posterior fibres of the ligament and backward movement by the anterior fibres of the ligament (Rees, 1954). Hence this outer **oblique band** of constant radius, constrains the condyle such that it travels in a smooth arc as it slides over the disc as well as down and around the posterior slope of the articular eminence. At the same time it also maintains tight contact between the condylar head, articular disc and glenoid fossa and helps to maintain them in a **close-packed** position.

A deeper **horizontal band** of fibres within the temporomandibular ligament that resembles a small strap has been described by DuBrul (1980). This band also arises from the articular tubercle of the zygoma and passes backwards to the lateral pole of the condyle and to the back of the disc (Fig. 6.19). One function of this deep band is to resist firmly any retrusive movements of the mandible into the soft vascular tissues behind the condyle; it also anchors the lateral pole of the condyle to the articular tubercle of the zygoma. The presence of this deeper band ensures that the condyle never comes into

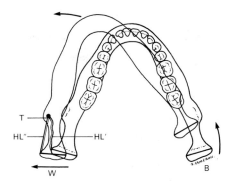

FIGURE 6·20 *The role of the left horizontal band of fibres (HL' and HL") at the chewing or working side temporomandibular joint during left lateral movement of the jaw. (W) and (B) are the working and balancing side condyles respectively and (T) is the attachment of the horizontal band of fibres to the articular tubercle of the zygomatic process of the temporal bone. The balancing side condyle translates forwards, down the articular eminence and medially. At rest the horizontal band is slack on the working side but as this condyle rotates about its vertical axis, the horizontal band becomes taught and prevents the lateral pole from moving any further posteriorly. The working side condyle is then forced to swing slightly laterally, constrained by the taught band fixed at the zygomatic tubercle. After DuBrul (1980b).*

contact with the post-glenoid process. As a result, it follows that in rotatory movements of the jaw towards the same side as the condyle under consideration, the lateral pole is constrained to orbit about this fixed radius so that it may swing outwards but cannot turn backwards into the tissues behind the joint (Fig. 6.20). The slight lateral shift that occurs as the condyles are forced to swing to one side rather than backwards into the tissues behind the joint capsule is known as **Bennett movement** or **Bennett shift**.

The temporomandibular joint in great apes

The general anatomical arrangement of the temporomandibular joint in great apes is similar to that in modern humans (Fig. 6.21), there being an intervening fibrous articular disc between the head of the mandibular

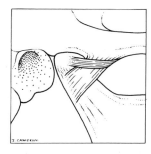

FIGURE 6·19 *The deeper horizontal band of fibres that make up the temporomandibular ligament has been highlighted here even though the oblique fibres actually conceal them.*

FIGURE 6.21 *Anatomical landmarks in the region of the glenoid fossa of a male gorilla.*

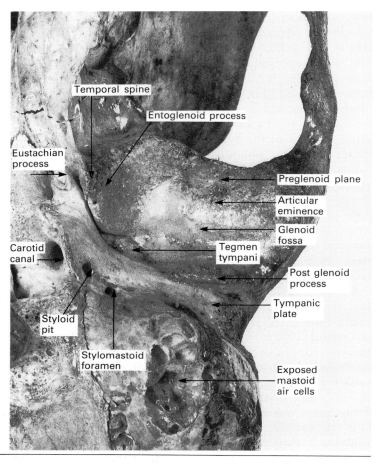

Temporal spine

Entoglenoid process

Eustachian process

Preglenoid plane

Articular eminence

Glenoid fossa

Carotid canal

Tegmen tympani

Post glenoid process

Tympanic plate

Styloid pit

Stylomastoid foramen

Exposed mastoid air cells

condyle and the glenoid fossa. The glenoid fossa of the great apes is more shallow than that of modern humans (see Fig. 6.35) but nevertheless is not completely without contour (Ashton and Zuckerman, 1954). With the teeth of a great ape skull occluded, the condylar head can be seen to rest, as it does in modern humans, against the posterior aspect of the articular eminence and not in the deepest part of the glenoid fossa. The thickened anterior band of the articular disc lies between the head of the condyle and the articular eminence anteriorly in great apes and not the thin intermediate zone as in modern humans (Sonntag, 1923, 1924). It seems that the whole of the disc is rotated backwards with respect to the head of the condyle in apes in the resting position so that

the thickened posterior band that lies over the top of the head of the condyle in modern humans, is positioned posteriorly behind the apex of the condyle in apes (Baycroft, 1988).

Bluntschli (1929) has suggested that the disc in great apes is of a more even thickness as compared with that of modern humans, but both anterior and posterior thickened bands can be identified in dissected specimens, although the maximum thickness of the disc in great apes may only be between 1 and 3 mm. It is important to note that the entire large entoglenoid process forms part of the upper compartment of the joint in great apes and has articular disc intervening between it and the head of the condyle over all its lateral surface (Fig. 6.22).

An important feature of the great ape

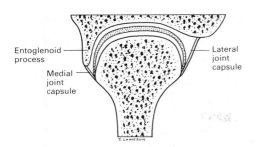

Entoglenoid process

Medial joint capsule

Lateral joint capsule

FIGURE 6·22 *Coronal section through the temporo-mandibular joint of a great ape illustrating how the large entoglenoid process forms the medial wall of the upper part of the joint compartment.*

temporomandibular joint is how closely the head of the mandibular condyle follows the contours of the entoglenoid process on the medial aspect of the joint. Little medial or lateral movement is possible here in many ape crania although some appear to be less constrained than others. The mandibles of modern humans are not so tightly constrained mediolaterally and the entoglenoid process is said to have migrated medially on each side allowing more freedom of movement here. Translatory movements of the head of the mandibular condyle in apes are also bound to course around and maintain close contact with the entoglenoid process. Smith (1984a) points out that Sicher (1937) claimed that chimpanzees do not exhibit condylar translation. However, as condylar translation has been clearly documented in macaques (Carlson, 1977) and as there is an extensive preglenoid plane in apes, this would seem unlikely but no details about condylar movements in living great apes are documented. Dissections of the temporomandibular joints of an orang-utan and two chimpanzees (Baycroft, 1988) revealed that, like the gorilla (Raven, 1950), the orang-utan has a large temporomandibular ligament on the lateral aspect of the joint. However, no such ligament could be identified in this study in either specimen of *Pan*, although Sonntag (1923) reported one in his dissection of a chimpanzee. Another significant feature is

that no deep horizontal portion of the temporomandibular ligament could be found in the dissected orang-utan, so that in this respect the ligament differed from the form of the modern human temporomandibular ligament described by DuBrul (1980).

The form of the temporomandibular joint of the great apes is dominated by a massive entoglenoid process and temporal spine as well as by a long post-glenoid process that is usually even longer than the most prominent part of the articular eminence (Figs 6.21 and 6.22). The head of the mandibular condyle is massive in the gorilla and the orang-utan, but closer to the dimensions of the human condyle in the chimpanzee. Often there is a clear impression of the post-glenoid process on the back of the mandibular condyle in great apes which may indicate that, unlike the modern human post-glenoid process, the process in great apes may help to prevent the condyle being displaced backwards. As great apes have large canine teeth that also prevent the mandible from being displaced backwards, it can be reasoned that wide opening movements of the mandible are those most likely to impinge the condyle against the post-glenoid process. During chewing the condyle may well still be quite free of contact with the post-glenoid process.

The contours of the articular surface of the mandibular condyle in apes conform very closely to the contours of the glenoid fossa. During lateral excursive movements of the mandible (that are also limited and guided by the canine teeth) the medial articular aspect of the balancing-side condyle rotates strongly against its entoglenoid process. Therefore, the function of the temporomandibular ligament in resisting lateral movement of the condyle is aided, if not wholly assumed, in apes by the entoglenoid process of the contralateral side (and this may explain why the temporomandibular ligament of some great apes is reduced or absent). However, the fact that there is a tough and extensive oblique external ligament in some specimens suggests that the primary function of this ligament is not to resist lateral movements but to maintain the condylar head, disc and articular eminence in tight contact during mandibular opening and closing movements. The axis of the condylar heads in apes, as in modern humans, form a right angle with the line of the posterior tooth row. These are near-parallel sided so that the medially extended axes of the condyles intersect more anteriorly than in modern humans. Simple hinge movements of the mandible in apes, therefore, have an axis of rotation that more nearly coincides with the transverse axes of the head of the condyles.

Movements of the condyle and mandible during chewing

Mammals are unique in their ability to chew unilaterally with maximal force focused on one side of the jaw only. Other jawed vertebrates apply maximal chewing forces in the sagittal plane and to both sides of the jaw simultaneously. Primates share an ability to chew mediolaterally with many other omnivores and herbivores. Simple bilateral chewing movements do, nevertheless, often occur in primates during swallowing and while 'chomping' soft and sloppy foodstuffs. However, forceful chewing typically occurs alternately on one or other side of the jaw. While condylar translation must accompany unilateral chewing, some other advantages for mandibular translation have also been proposed (Smith, 1985). Condylar translation around a centre of rotation in the region of the mandibular foramen is said to minimize displacement of the inferior alveolar nerves and vessels during simple opening and closing movements of the mandible. Condylar translation during opening also reduces the amount of stretch in the masseter and medial pterygoid muscles allowing these muscles to function within an efficient portion of the muscle fibre length–tension curve (see Chapter 2). In addition, such translation prevents compromise by the tongue and mandible of the airway and great vessels of the neck during wide opening of the mouth (Smith,

1985). Even so, most wide-opening movements of the mandible, such as yawning, are accompanied by an extension of the neck, to reduce further the chance of compressing and occluding vital structures behind the mandible. In great apes the only part of the skull that lies posterior and inferior to the mandibular condyle is the post-glenoid process. The tympanic plate and the mastoid process are at a higher level and even the vertebral column extends posteriorly at a greater angle, so the mandible is able to swing downwards and backwards with less danger of compressing the great vessels and airway. In bipedal hominids, the foramen magnum and the occipital condyles are positioned further forwards and further inferiorly than in great apes. As a result, the airway and other ascending and descending structures of the neck travel in a restricted space between the vertebral column and the retruded face.

Simple bilateral opening movements of the mandible

With the mandible at rest the head of the condyle in modern humans lies with the thin intermediate zone of the disc between it and the posterior aspect of the articular eminence. The thickened posterior band of the disc lies at the apex or highest point of the condyle. During simple bilateral opening movements of the mandible the condyles rotate across both the posterior and anterior bands in the lower joint compartment. As the condyle begins to pass the posterior band, however, the whole of the disc and the attached condyle slide together down the articular eminence while the head of the condyle is rotating against the disc in the lower compartment of the joint (Fig. 6.23). In fact the amount of opening at the front teeth that occurs initially when there is only simple hinge movement in the lower joint compartment and no translation at all is probably less than 1 mm. The sliding or **translatory** movement of the mandible relative to the cranium occurs only in the upper joint compartment. Because

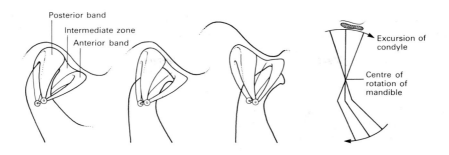

FIGURE 6·23 *Opening movements of the jaw bring about rotation of the head of the condyle in the lower joint compartment and translation of both the disc and condyle down the articular eminence. During* *simple opening movements, the centre of rotation of the mandible is said to be in the region of the mandibular foramen. After Rees (1954).*

during wider opening the condyle continues to rotate at the same time that the disc is sliding forwards, the total excursion of the condyle is greater than that of the disc alone. The presence of the thickened bands and thinner intermediate zones of the disc means that the condyle is not constrained to follow the sinuous contours of the bony glenoid fossa, but can move in a smooth arc of a circle. The moving thick and thin parts of the disc convert the irregular surface of the glenoid fossa into a smooth curve as the condyle rotates against it. The centre of this rotation, we have already said, is supposed to be near the mandibular foramen during simple opening and closing movements but this must vary greatly during more complicated chewing movements. Opening and closing movements of the mandible occur in a smooth and harmonious manner. During closing movements of the mandible the events occurring in the temporomandibular joint are a simple reversal of the opening sequence.

Movements of the working-side condyle

The side of the mouth where forceful biting contact takes place in unilateral chewing is referred to as the **working side**. The contralateral side where normally no tooth contact occurs is referred to as the **balancing side**. During powerful unilateral chewing the condyle and disc of the working side do not move far from their starting point on the posterior part of the articular eminence. This condyle simply orbits very slightly downwards and outwards in a vertical axis around a centre of rotation just behind the condylar head. In fact, we noted this movement previously when describing the constraining action of the horizontal band of the temporomandibular ligament (Figs 6.19 and 6.20). The point of contact between the condylar head and the posterior band of the disc moves across the condyle from lateral to medial as this happens. The horizontal band of the temporomandibular ligament of this condyle prevents its posterior displacement into the tissues behind the joint but allows slight lateral movement to compensate for the rotation (this is the Bennett shift and is usually only about 1.5 mm as measured at the head of the condyle). The lateral pterygoid muscle, which is attached to the anterior aspect of the neck of this condyle and which holds it firmly forwards against the articular eminence, opposes the backwards pull of the temporalis muscle which is attached to the more laterally positioned coronoid process; together they act as powerful stabilizers at this near-stationary condyle, but assist the rotation of this condyle about its vertical axis.

Movements of the balancing-side condyle

The condyle and disc of the balancing side rotate about the same common axis as the near-stationary working-side condyle during unilateral translation. However, in addition, both the condyle and the disc move forwards and medially and down the articular eminence together. The extent of movement, then, is much greater at this condyle and is essentially similar to that which occurs at both condyles during simple bilateral opening movements of the jaw and which we described earlier (Fig. 6.23). The digastric muscle of this side (the balancing side) also acts to lower the mandible during this period.

The chewing cycle

Hiiemae (1978) (see also Kay and Hiiemae, 1974) has divided the chewing cycle into three strokes that form a complete cycle. From a position of maximum opening there is a fast **closing stroke**. This is followed by a **power stroke** during which food is crushed and sheared between the teeth. Then an **opening stroke** completes the cycle. Usually the entire chewing cycle is one continuous movement. The power stroke is the most variable part of the human chewing cycle and usually consists of two phases. In **phase 1**, or the **buccal phase**, the lower teeth move medially and upwards from their first contact with the upper dentition to the point of maximal intercuspation. In the shorter **phase 2**, or **lingual phase**, the lower teeth continue to move medially but are displaced downwards as their buccal cusps slide against the palatal cusps of the upper teeth (Fig. 6.24). A summary of the opening and closing movements of the mandible together with the electrical activity of the muscles of mastication during the chewing cycle is presented in Fig. 6.25 (from Walker, 1978). One chewing cycle takes between 0.5 and 1 s but, interestingly, there are differences between men and women. Howell (1987) has demonstrated that, during unilateral chewing, men open their mouths wider, move more

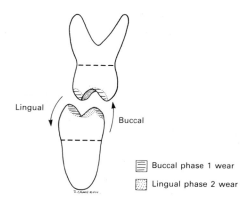

FIGURE 6·24 *Diagramatic representation of the buccal and lingual phases of the power stroke indicating the phase 1 and phase 2 wear facets on the upper and lower molar teeth.*

from side to side and have faster jaw movements than do women. Women, on the other hand, take longer for each chew and pause between chews in the intercuspal position for a greater period of time than do men. The slower chewing rate and velocity in women may relate to the reduced muscle mass (Fig. 6.26). It is likely that differences in the size of the jaws and the size of the muscles that move the mandible underlie this and that sexual dimorphism in hominid bite force can be linked with similar differences hypothesized for other hominoids (see later).

Forces acting on the mandible during mastication

Studies of human masticatory muscle activity have demonstrated that there is equal activity in the temporalis muscles of the balancing-side and working-side muscles, but that there is more activity in the masseter and medial pterygoid muscles of the working side. Hylander (1983) has nevertheless noted that there are large differences between masticatory muscle activity on the two sides in macaques. Fusion of the mandibular symphysis in hominoids (or the development of very strong ligaments in other animals), in a way that efficiently resists the different forces that are acting to separate the two halves of the

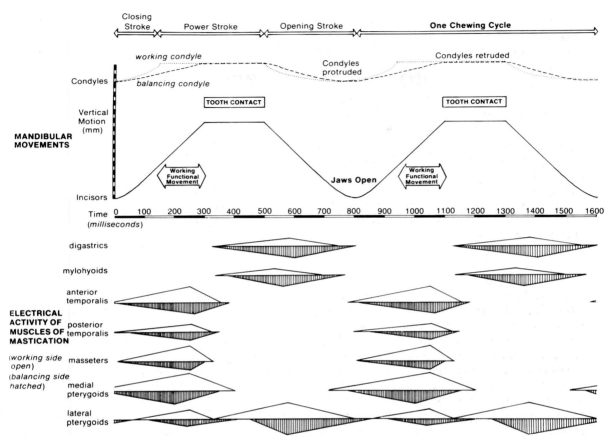

FIGURE 6·25 *Diagram of two idealized chewing cycles to demonstrate the correlation between movements of the incisors and mandibular condyles and the electrical activity in the muscles of mastication. After Walker (1978). Courtesy of Alan Walker and W.B. Saunders.*

mandible during unilateral chewing, is a way of transferring balancing-side muscle force to the chewing side. In this way extra muscle force can be employed on the working side and this makes unilateral chewing more efficient than bilateral chewing. The direction of pull of the various muscles that act on the mandible, together with the twisting, bending and shearing forces that occur in the mandible during muscle contraction and during chewing, have an important influence upon the bony morphology of the mandible. Some of these important forces that occur within the mandible during chewing are reviewed here. A great number of them have been demon-

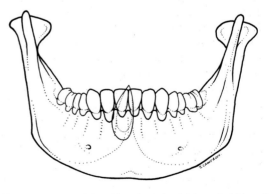

FIGURE 6·26 *Movements at the incisors measured during unilateral chewing in men (solid line) and women (dotted line). After Howell (1987).*

97

strated through the experimental work of Hylander.

During jaw opening and simple bilateral protrusion of the mandible the lateral pterygoid muscles squeeze the mandibular rami together (Fig. 6.9). This can be measured between the mandibular third molar teeth at the back of the dental arch where the distance between them can decrease by up to 0.5 mm (DuBrul and Sicher, 1954). It is usually claimed that the crowding of the tongue within the smaller retruded jaws and between the vertebral column posteriorly has resulted in an external brace or chin in humans designed to resist this 'squeezing force' in the same way that the internal brace or simian shelf (inferior transverse torus) is designed to resist it in great apes. When measured, however, this 'squeezing force' is much less than others that occur during the power stroke of mastication (Hylander, 1984).

Hylander (1975) has shown that, during unilateral mastication, forces acting on the balancing-side condyle are greater than those acting on the working-side condyle and Smith (1978) has calculated that the total condylar reaction force on both sides may be over 75% of the bite force. Smith has further noted that 70–80% of this total condylar reaction force may be borne by the balancing-side condyle. Indeed, this explains why after trauma to one temporomandibular joint or condyle, most pain is experienced when people attempt to chew on the opposite side to the injury. Similarly, biting edge to edge on upper and lower incisors is extremely painful when there is damage to the temporomandibular joint or the mandibular condyles as this is when reaction forces acting on the condyles are maximal (Hylander, 1975, 1979a).

Hylander (1984) has demonstrated that maximum stress and strain in the mandibular symphysis of the macaque occurs during the latter part of the power stroke. At this time the downwards force of the bite on the teeth of the working side is opposed by the upwards force of the balancing-side muscle

force and, as a result, the mandibular symphysis experiences a shearing force in the sagittal plane at the front, (**dorsoventral shear**, see Fig. 6.27). In addition, at the same time, the bite force on the working side is also pushing laterally as the lower teeth move across the upper teeth but the balancing-side muscles

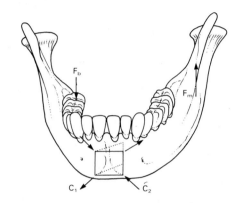

FIGURE 6·27 *During dorsoventral shear the solid square may be expected to deform into the shape represented by the dotted parallelogram. Fm represents the direction of the balancing side adductor muscle force. Fb represents the direction of the vertical component of the bite force. After Hylander (1984).*

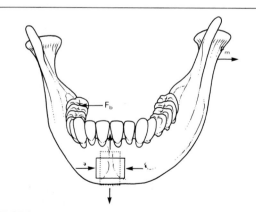

FIGURE 6·28 *During lateral transverse bending the solid square may be expected to deform into the shape represented by the dotted parallelogram. Fm represents the laterally directed component of the balancing side adductor muscle force. Fb represents the laterally directed component of the bite force. After Hylander (1984).*

(especially the deep head of masseter), whose insertion is now much further medially than its origin on the cranium, is pulling sideways in the opposite direction. The result is that the two halves of the mandible are bent in opposite directions in the transverse plane. This tends to pull it apart at the symphysis as happens to the 'wishbones' of chickens. This 'wishbone effect' is called **lateral transverse bending** and is much more important than dorsoventral shear (Fig. 6.28).

Other forces result from the strong pull of the masseter muscles everting the lower borders of the mandible and tending to turn the posterior tooth row of the working side inwards towards the tongue. These forces result in tensile stress at the lower border of the symphysis but compressive stress along its upper border. This is described as a twisting of the mandibular bodies about their long axes (Fig. 6.29). Hylander (1984) has proposed that in order to resist the resulting mandibular shearing stresses, that occur at the upper and lower borders of the symphysis in the macaque, the vertical height of the

symphysis has been increased, the thickness of the symphysis has been increased by a superior transverse torus, and the bony interface between the two mandibular bodies has been extended in the form of a simian shelf.

The combined effect of the condylar reaction force on the balancing side acting downwards, with the force transmitted through the symphysis from the balancing side, which is also acting downwards, together with the muscle force on the same side acting to pull upwards, results in a different pattern of stresses on the balancing side. The mandibular body is bent in the sagittal plane (Fig. 6.30) and a tensile stress results in the bone at the alveolar margin and a compressive stress in the bone at the lower margin of the mandibular body. Experimental work on human subjects that recorded tooth movement between teeth, while out of occlusion on the balancing side of the mandible, may also have measured this tensile stress in the alveolar bone (Picton, 1962).

There has been considerable debate about

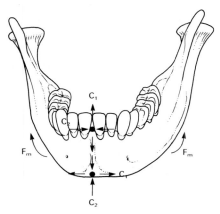

FIGURE 6·29 *The small arrows represent the expected patterns of strain along the labial aspect of the symphysis during symphyseal bending associated with twisting of the mandibular corpora about their long axes. The large arrows labelled Fm represent the direction of the adductor muscle force that tends to act to evert the lower border of the mandibular corpus and invert the post-canine alveolar process. After Hylander (1984).*

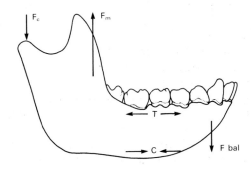

FIGURE 6·30 *Bending of the balancing side mandibular body in the sagittal plane. Fc represents the condylar reaction force on the balancing side which together with the force transmitted from the balancing side through the symphysis to the working side (F bal) are opposed by the adductor muscle force (Fm) on the balancing side. The result is tensile stress (T) in the alveolar bone and compressive stress (C) at the lower border of the mandible. After Hylander (1979b).*

whether the mandible functions as a lever with the condyle acting as a fulcrum, or whether the mandible is simply slung between muscles which 'link' it to the teeth during mastication. Clearly, during incisal biting and when the balancing-side condyle is in contact, via the articular disc, with the articular eminence, or during unilateral chewing, the mandible acts as a lever, and we have seen that condylar reaction forces are maximal in these situations. It may be, however, that the mandible sometimes acts as a link, when for example a bolus of food between the teeth is gently 'chomped on' during bilateral chewing and when while this is happening neither condyle is acting as a fulcrum. Nevertheless, during powerful unilateral chewing, the mandible is clearly functioning as a lever.

Some knowledge of the movements and forces acting during chewing, together with an understanding of the functional morphology of the muscles that move the mandible and of the temporomandibular joint, allows us to reconstruct fossil hominid masticatory systems more objectively.

The masticatory system of fossil hominids

Changes in the relative size of the masticatory system in early hominids have had an effect upon the morphology of the skull as have changes in the relative size of the brain and changes in posture. However, the confusing interplay between these different morphological adaptations makes it difficult to predict exactly which specific morphological features of the skull are solely related to shifts in masticatory strategy. Nonetheless, some broad conclusions can be drawn about differing hominid masticatory systems from a variety of observations about the overall shape of hominid skulls, about details of the attachments of the muscles of mastication and from comparisons of mandibular and temporomandibular joint morphology (as well as from comparative studies of hominid teeth). We have already described the man-

dibles of fossil hominids in this chapter and two other chapters of this book are devoted to the comparative anatomy of teeth. Elsewhere in this book the influence of the masticatory system on the facial skeleton is also described in some detail. In this section we will concentrate only on the more general features of the early hominid cranium that relate directly to the masticatory system and then outline the comparative morphology of the glenoid fossa in fossil hominids.

General features of the cranium that relate to the masticatory system of early hominids

We noted previously that many great ape crania have crests for the attachment of their temporalis muscles and that these sagittal crests (or keels) converge posteriorly, commonly fusing with the nuchal crest to form compound temporal–nuchal crests. The degree of emphasis on the muscle fibres of the posterior temporalis seems to be similar in *Australopithecus afarensis* and great apes as far as can be judged from the crests and markings raised on the cranium by the temporalis muscle. There is, however, a shift away from this and towards a greater emphasis on the anterior fibres of temporalis in specimens attributed to *A. africanus* and *Paranthropus*. There is evidence of even greater development of the anterior fibres of the temporalis muscle in *Paranthropus boisei*. Viewed from above, the cranial outline of later specimens of *P. boisei* (OH 5, KNM-ER 406) is extremely wide with flared zygomatic arches that reflect the expanded anterior temporalis muscle fibres. These widely flared zygomatics also give the masseter–medial pterygoid complex an advantage in bringing about transverse movements of the mandible and DuBrul (1977) has suggested that some muscle fibres (known collectively as the zygomaticomandibularis muscle in some animals) may have been especially well developed in *Paranthropus*. The mechanical advantage of the masseter muscle was also improved in *Paranthropus* by greatly increasing the facial height and by extending the insertion of the muscle anteriorly towards the premolars rather than opposite the molars as in great apes and *Homo*. These cranial adaptations are associated with the generation and dissipation of very high bite forces and occur through convergent evolution in many animals including, for example, *Hadropithecus*, *Theropithecus* (and probably *Gigantopithecus*) among primates and also in the giant panda, *Ailuropoda*, which along with the bear, *Ursus*, has been used as a comparative model for the masticatory system in early hominids (DuBrul, 1977, 1979, 1980).

The type of articulation that occurs between the temporal, occipital and parietal bones in large *Paranthropus* crania may well reflect these large masticatory forces generated around the cranium. In addition to the overlap that occurs on all hominoid crania where the squamous temporal bone overlaps the parietal bone, there is also an overlap of the parietal bone onto the mastoid portion of the temporal bone and of the parietal bone onto the occipital bone in *Paranthropus* (Fig. 6.31). Rak (1978) and Kimbel and Rak (1985) link this with resistance to extreme masticatory stresses in the vault bones of this region. The condition of overlapping vault bones in *Paranthropus* appears to be a unique adaptation among hominids. The well-developed and projecting mastoid angle of the parietal bone in great apes creates a distinct mastoid notch in the temporal bone. This, in turn, leads to the mastoid portion of the temporal bone articulating into a distinct **asterionic notch** formed by the posterior edge of the parietal bone and the lateral edge of the occipital bone (Fig. 6.32). The asterionic notch articulation is found in great apes, *Australopithecus afarensis* and probably KNM-ER 1805, a specimen which is usually attributed to early *Homo* (Kimbel and Rak, 1985).

Demes and Creel (1988) have estimated the bite forces generated in apes, fossil hominids and modern *Homo sapiens* crania by summing estimates of the total contributions made by

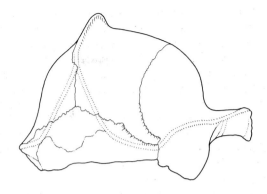

FIGURE 6·31 *The vault bones and the cresting patterns of a cranium of* Paranthropus boisei *illustrating the unique overlap of the temporal squama onto the parietal bone together with the overlap of the parietal onto the occipital squama.*

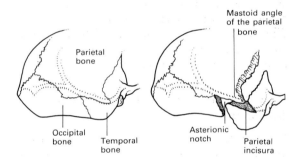

FIGURE 6·32 *Articulation of the parietal, temporal and occipital bones in the asterionic region of a great ape. After Kimbel and Rak (1985).*

each of the muscles of mastication. All extant species studied (except the pygmy chimpanzee) were sexually dimorphic with respect to their estimated bite force equivalents. Demes and Creel (1988) hypothesize from these estimates that the larger males might process a given amount of food effectively with fewer strokes of a massive jaw, while the smaller females achieve the same results with more strokes of a lighter jaw. We noted an alternative interpretation previously: Howell (1987) has shown that women pause for longer in the intercuspal position during the chewing cycle and so may apply force for longer periods of time to make up for the weaker musculature and smaller bite forces.

Nonetheless, when these data for bite-force equivalents are plotted against data for the molar-crown area in the same primate species (Fig. 6.33), it becomes clear that there is a consistent relationship between the two variables. Strong bite force is coupled with large molar teeth and weak bite force with small molar teeth. Force per unit area exerted on the molar teeth in all species must, therefore, be very similar, regardless of tooth size. The large molar surface areas of *Paranthropus* must have offered a large surface area for triturating a great quantity of food at any one time and at the same time increased the likelihood that small hard objects in the diet would fracture under the force of the bite (Demes and Creel, 1988). Evidence from microwear studies (Grine, 1986) supports the hypothesis that

Paranthropus (at least in southern Africa) did indeed feed on large quantities of small, hard objects. Microwear studies also support the conclusions of DuBrul (1977, 1979) that there was an emphasis on transverse mandibular movements during chewing in *Paranthropus*, and Hylander (1979b, 1988) has argued that the form of the body of the mandible in *Paranthropus* probably reflects an adaptation to counter large repetitive bending, twisting and shearing loads during unilateral mastication.

In contrast to this trend to increase the size of the masticatory system in *Australopithecus africanus* and *Paranthropus*, specimens attributed to early *Homo* demonstrate a reduction in the size of the jaws. Further details about the interrelationships of the facial skeleton and the masticatory system in early *Homo* and other hominids are presented in Chapter 11.

The glenoid fossa of fossil hominids

Comparative studies of fossil hominid temporomandibular joints have concentrated upon differences that exist in the morphology of the glenoid fossa. Noteworthy features in this region include the length and vertical orientation of the tympanic plate posterior to the temporomandibular joint, differences in the breadth and length of the glenoid fossa, and differences in the depth of the fossa and the associated development of the articular eminence.

Great apes have long bony tympanic plates that contribute to the tube that runs between the external auditory meatus and the tympanic membrane or ear drum (Fig. 6.21). The outer one-third of this tube is cartilaginous, but Weidenreich (1943) drew attention to the fact that it is the size of the masticatory system (the bizygomatic breadth for example) that determines how long this tube must be in order to reach the side of the head. In the fossil hominids also, the length of the tympanic plate reflects the size of the masticatory superstructures on the side of the cranium. The bitympanic widths and the lengths of the tympanic plates in *Paranthropus* are,

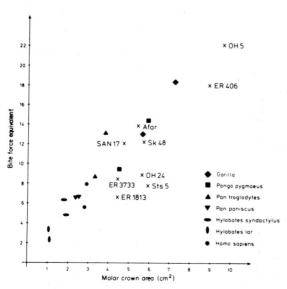

FIGURE 6·33 *Bite-force estimate plotted against molar crown size area. After Demes and Creel (1988).*

therefore, larger than in other hominids. Likewise, the bitympanic widths and lengths of the tympanic plates in small specimens of early *Homo* are the smallest among hominids. The orientation of the anterior aspect of the tympanic plate is almost flat and parallel with the Frankfurt Horizontal in great apes and *Australopithecus afarensis*. Increasingly, there is a trend for the tympanic plate to become more vertical in later hominids, but this may relate to the swinging down and expansion of the occipital and nuchal region rather than to any function of the temporomandibular joint.

The glenoid fossa in *A. afarensis* is broad and shallow, there being little of an articular eminence anteriorly (White et al., 1981; Kimbel et al., 1984) (Fig. 6.34). Specimens of *A. africanus* also have broad glenoid fossae but have a more pronounced articular eminence anteriorly so that the fossa is deeper when viewed from the side (Fig. 6.35). DuBrul (1977, 1979) noted that the glenoid fossa in Sts 5 resembles that of great apes in two respects; firstly, in that there is a long

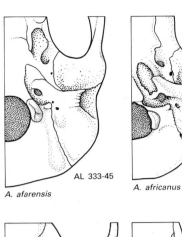

A. afarensis

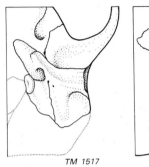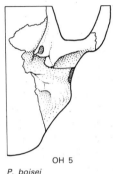

P. robustus TM 1517   OH 5 *P. boisei*

FIGURE 6·34 *(Above) The morphology of the glenoid fossa in* Australopithecus afarensis, Australopithecus africanus, Paranthropus robustus *and* Paranthropus boisei. *See text for description.*

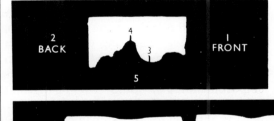

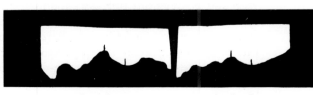

Explanatory figure

1. Anterior.
2. Posterior.
3. Articular eminence.
4. Articular fossa.
5. Temporal surface of mandibular joint.

FIGURE 6·35 *(Left) Sagittal section through the glenoid fossa and articular eminence of a modern human (explanatory figure) together with a male and female* Gorilla, *SK 48 (*Paranthropus*) and Sts 5 (*Australopithecus*). After Warick-James (1960).*

post-glenoid process that reaches well below the level of the articular eminence and which lies in front of the tympanic plate; and secondly, in that there is a well-developed preglenoid plane. Some specimens of *A. africanus*, however, (Sts 19) are quite contrasting and resemble the modern human condition more closely.

The glenoid fossa in *Paranthropus robustus* is also broad but differs from that of *Australopithecus* in that it is shorter from front to back. There is more depth to the fossa in *P. robustus* with a more pronounced articular eminence (Fig. 6.35). The post-glenoid process is reduced and merges with the superior margin of the tympanic plate. The glenoid fossa in *Paranthropus boisei* is especially foreshortened and derived, even with respect to the condition in *P. robustus*. For example, it is only 10 mm from the post-glenoid process to the articular eminence in KNM-ER 13750 (Leakey and Walker, 1988). DuBrul (1977, 1979) has emphasized the small postglenoid process in *P. boisei* as well as the foreshortened fossa, marked articular eminence anteriorly and the absence of a preglenoid plane. In addition, DuBrul notes the entoglenoid process in *P. boisei* that turns posteriorly at the medial part of the articular eminence. It even comes to overlap the medial end of the tympanic plate where DuBrul calls

it the **medial glenoid plane**. This combination of morphological features in *P. boisei* suggests a temporomandibular joint with strong transverse movements of the condyle on the posterior aspect of the articular eminence, with the working-joint surface primarily in the coronal or transverse plane. The jaw cannot rotate about a vertical axis at the working-side condyle because there is no preglenoid plane for the balancing-side condyle to ride forward on (DuBrul, 1977, 1979, 1980): DuBrul (1979, 1980) suggests that it would simply dislocate up into the temporal fossa. Instead, the extensive medial glenoid plane does not wall off the glenoid fossa medially like the entoglenoid process in large apes but rather allows the head of the condyle to slide medially and downwards (Fig. 6.36).

One specimen presently attributed to *P. boisei* (KNM-WT 17000) does not have a glenoid fossa that resembles those just described. This specimen has a huge glenoid fossa anteroposteriorly and a massive entoglenoid process medially that projects downwards, in a manner reminiscent of large apes (Leakey and Walker, 1988). Interestingly, however, the tympanic plates in this specimen are orientated at 50° to the Frankfurt Horizontal (see Chapter 11) and are not flat along the cranial base, suggesting once again that this reflects the expanded and 'swung down' occipital region rather than the form of the temporomandibular joint. The postglenoid process of KNM-WT 17000, however, remains anterior to the tympanic plate as in *A. afarensis*.

Specimens of both early and modern *Homo* share some of the features that characterize specimens of *Paranthropus*. These include nearly vertical tympanic plates that merge with the postglenoid process superiorly and deeper glenoid fossae that have steep articular eminences anteriorly (although there is more of a preglenoid plane in *Homo*). As we noted earlier, there is evidence that the age of an individual may affect the observed morphology of the glenoid fossa and that besides this there is great variation in glenoid fossa

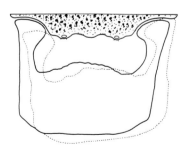

FIGURE 6·36 *The medial glenoid plane of the glenoid fossa in* Paranthropus boisei *allows the condyle to slide downwards and medially, primarily in the transverse plane. After DuBrul (1979).*

morphology within modern human populations. It is also clear that attrition and tooth loss may influence the height of the articular eminence below the level of the glenoid fossa. While it is possible to make general comments about glenoid fossa morphology in some groups of fossil or modern *Homo* (for example, it is commonly stated that the glenoid fossa in Neanderthals or Eskimos is shallow (Knowles, 1915; Stewart, 1933; Warick-James, 1960) there are insufficient data to attribute these features securely to either a particular hominid group or a specified cause.

CHAPTER SEVEN

THE MICROANATOMY AND DEVELOPMENT OF TEETH

The structure of teeth, especially the microanatomy of enamel, has come to figure prominently in studies of hominoid taxonomy and more recently also in studies of hominoid ontogeny. The aim of this chapter is to highlight important points about the structure and development of dental tissues that are becoming increasingly relevant to physical anthropologists.

The bulk of all tooth tissue is formed of **dentine** (Fig. 7.1) and the anatomical crowns of teeth are covered with a layer of **enamel** which is the hardest biologically formed substance known. The dentine of the root and usually a small part of the adjacent enamel at the neck or **cervix** of the tooth are covered with **cementum**. Fibres of the **peridontal ligament** attach into the cementum and suspend the tooth in its socket of **alveolar bone**. A chamber within the tooth known as the **pulp chamber** contains nerves, blood vessels and fibrous connective tissue that communicate with the nerves and vessels in the jaws through the **apical foramen** at

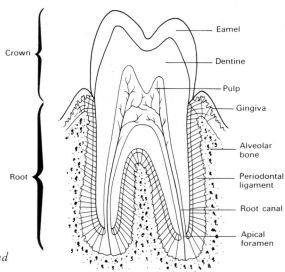

Crown

Root

Eamel

Dentine

Pulp

Gingiva

Alveolar bone

Periodontal ligament

Root canal

Apical foramen

FIGURE 7·1 *The tooth and its supporting tissues.*

the end of the root. Teeth are extremely sensitive to pain and temperature. This, together with feedback information about the position of the jaws in space (from **proprioceptive** nerve fibres in the periodontal ligament, see Chapter 3), protect teeth from excessive occlusal trauma and potentially damaging external stimuli (Orchardson and MacFarlane, 1980).

By the time teeth emerge into the mouth and come into occlusion the whole of the morphology of the crown has been established and the enamel that covers the crown is fully formed and matured. The cells that form enamel become shed from the surface of the tooth during tooth eruption and enamel cannot subsequently undergo repair during the lifetime of a tooth. After emergence into the mouth, enamel undergoes a post-eruptive maturation phase and can only be physically altered by ionic substitution, for example, by fluoride ions in the saliva, or else worn away by attrition or abrasion, or decayed away. Dentine and cementum differ from enamel in that the cells that form them remain alive and able to add small amounts of tissue to the mature tooth throughout the lifetime of the individual.

The embryological development of teeth

Teeth develop in the jaws between the first few months of foetal life and about 18 to 25 years of age in modern humans (or 12 to 15 years in the case of great apes). During the whole of this time active embryological tooth development is taking place. While contained within the jaws, teeth are protected from any functional influences over their morphological development, although diseases, especially high fevers, may disturb tooth germ development and thus affect their normal morphology thereafter. Following emergence into occlusion there is some functional influence over the last part of root development (Kovacs, 1971), but this is minimal. Because of this and the durability of teeth, great significance has been attached to studies of tooth morphology and microstructure.

The first stage of tooth development involves a downgrowth of epithelial cells, that line the primitive mouth cavity, into the jaws and an interaction between these epithelial cells and a proliferating mass of **mesenchyme** beneath it (Fig. 7.2) ((mesos (Gk) = middle; enchyma (Gk) = infusion),

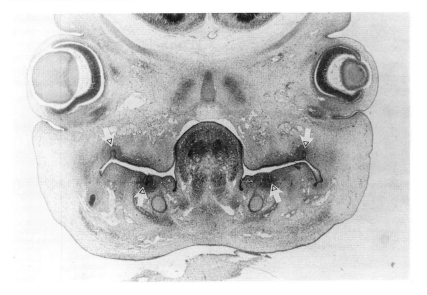

FIGURE 7·2 _Histological section through the head of an embryo made in the coronal plane. Epithelial downgrowths (arrowed) from the mouth develop into tooth buds which in turn mature in size and shape and develop into tooth germs and eventually teeth._

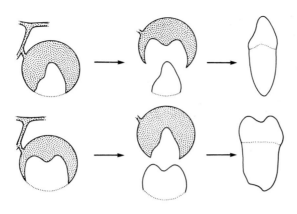

FIGURE 7·3 *Mature tooth buds comprise an epithelial component and a mass of mesenchyme beneath this. These components in a molar tooth bud or an incisor tooth bud can be separated and then recombined. It has been shown in experimental animals that it is the mesenchymal component of the tooth bud that determines the morphology of the tooth which develops. After Ten Cate (1985).*

thus mesenchyme is an embryonic connective tissue between the layers of ectoderm and endoderm in the embryo). It is the mesenchyme that determines the morphology of a tooth (Fig. 7.3). When molar mesenchyme is transplanted beneath incisor epithelium then molar teeth develop, and when incisor mesenchyme is transplanted beneath molar epithelium then, in time, incisor teeth develop (Kollar and Baird, 1969).

Gradually, the cells of the **inner enamel epithelium** (Fig. 7.4) of the tooth germ divide and expand at different rates, causing this sheet of epithelium to buckle and bulge at the sites of the future cusps and fissures of the future enamel–dentine junction. As the adult outline of the dentine horns is approached, the cells of the inner enamel epithelium induce a layer of cells directly opposite them to mature into **odontoblasts** or dentine-forming cells. These cells then begin to produce dentine under the cusp tips along the junction between the enamel and dentine of the tooth crown. Almost immedi-

ately the odontoblasts have begun to form dentine, the **ameloblasts**, or enamel-forming cells, that develop from the inner enamel epithelium directly opposite the odontoblasts, begin to form enamel on the outer aspect of the enamel–dentine junction (Fig. 7.5). In this way two hard tissues, enamel and dentine, begin to develop along the enamel–dentine junction of the growing tooth germ. The ameloblasts travel outwards away from the enamel–dentine junction and secrete enamel until the full thickness of enamel over the dentine is completed; they then become reduced in size and eventually die. The odontoblasts travel inwards from the enamel–dentine junction towards the pulp chamber of the tooth and secrete dentine until the tooth is fully formed. Odontoblasts, however, do not die but continue to secrete dentine very slowly (secondary dentine) throughout the life of the tooth.

At first, the cusps of teeth begin to calcify close together but then move apart from one another as the inner enamel epithelium expands between them, continuing to map out the final contours and proportions of the enamel–dentine junction. Eventually, as the mineralizing cusps coalesce the enamel– den-

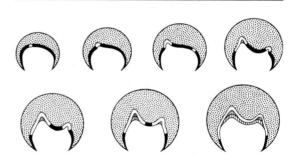

FIGURE 7·4 *As tooth buds mature the inner enamel epithelium (dark region) grows and buckles as a result of cell division. Cell division ceases first over the cusps of the future tooth (white regions). Eventually the whole of the future enamel–dentine junction is mapped out by the inner enamel epithelium and dentinogenesis proceeds towards the pulp and amelogenesis proceeds outwards from the enamel dentine junction. After Ten Cate (1985).*

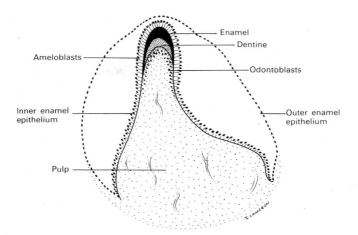

FIGURE 7·5   _Diagram of a fully-formed tooth germ that has just begun to form enamel and dentine._

tine junction occlusally assumes its final mature dimensions. Butler (1986) has drawn attention to the fact that the dentine horns or cusps along the enamel–dentine junction are taller in modern humans and the gorilla than they are in the orang-utan or _Sivapithecus_ (Fig. 7.6). This kind of variation results from a greater rate of cell division along the inner enamel epithelium under the future cusps in _Gorilla_ and modern humans and results in tall and pointed dentine horns. The thickness of enamel formed over the cusps has a lot to do with the final form of the tooth so that thin-enamelled gorilla teeth, for example, still have tall pointed cusps whereas the much thicker enamel of modern human teeth reduces the pointed contours of the dentine horns to a more rounded, bulbous, shape at the occlusal enamel surface.

Root development

In the same way that epithelial cells of the internal enamel epithelium induce odonto-blasts to mature and form dentine in the crown of the tooth, epithelial cells proliferate along the future outline of the root surface. This layer of epithelial cells that covers the future root is known as **Hertwig's epithelial root sheath** and its cells induce odontoblasts

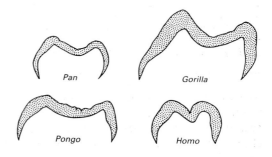

FIGURE 7·6   _The enamel caps of_ Pan, Gorilla, Pongo _and_ Homo _illustrating the more pointed dentine horns of those in_ Homo _and_ Gorilla. _After Grine and Martin (1988)._

to mature and form dentine in the root of the tooth (Fig. 7.7). Odontoblasts proliferate firstly in the cervical region of the root and then gradually begin to mature further towards the apex of the tooth root along the future cementum–dentine junction. As with dentine in the crown of a tooth, these cells then move inwards towards the pulp chamber of the tooth forming dentine behind them. There is no sharp distinction between dentine formation in the crown of a tooth and dentine formation in the root of a tooth. In fact dentine formation in a tooth is a continuous

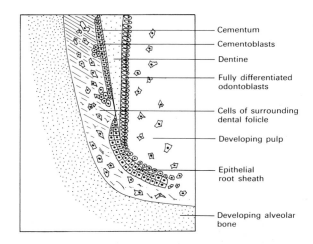

FIGURE 7·7 *Diagram of the proliferating epithelial root sheath of a tooth and the surrounding tissues. After Berkovitz (1978).*

process and dentine in the root of a tooth actually begins to form before the innermost dentine of the crown is complete.

Teeth commonly have either one, two or three roots. These develop as cells around the rim of the proliferating root diaphragm invaginate inwards across the apical opening and divide it into two or three giving two or three roots, respectively (Fig. 7.8). When there is late invagination of the root processes, so that the separate roots do not divide until late into root development, teeth are said to be '**taurodont**' because Keith (1913) considered that such teeth resemble those of the ox, (hence: taurus (Gk) = bull; odous (Gk) = tooth). The converse condition of very early division of the tooth root into two or three was considered by Keith to resemble the teeth of carnivora and, consequently, he named this process **cynodontism** (kyon (Gk) = dog). Taurodont molar teeth are known to be common in Neanderthals but they are also found in some modern human populations. The incidence is as high as up to 5.6% in the permanent dentitions of, for example, Israeli adults (Jaspers and Witkop, 1980). Taurodont teeth are even more common in people with various X-chromosomal aneuploidies or with syndromes that include

an ectodermal defect (Ogden, 1988). Slowing down of the mitotic cycle of epithelial cells in the root sheath and hence late bi- or tri-furcation development are thought to underly taurodontism in modern humans with chro-

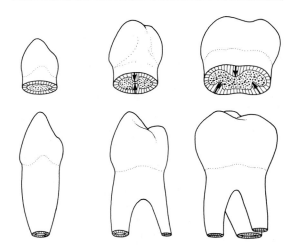

FIGURE 7·8 *In single rooted teeth the root diaphragm continues to proliferate towards the apex of the tooth. Two processes grow across the developing root diaphragm in two rooted teeth and three processes grow across the developing root diaphragm in three rooted teeth, forming a bifurcation and a trifurcation respectively. After Berkovitz et al. (1978).*

mosomal abnormalities (Jaspers and Witkop, 1980). In terms of its expression among hominids, taurodontism appears to be a continuous trait and may generally be ascribed to a polygenic system involving complex expression. Some suggestions have been put forward regarding the adaptive significance of taurodontism. These include adaptation to a high attrition environment, or more likely, delayed exposure of the bi- or tri-furcation to the oral environment as wear is accompanied by further eruption of the tooth (Blumberg et al., 1971). This, it is argued, might have delayed the onset of periodontal disease and tooth loss so commonly associated with bone loss around the molar root bi- or tri-furcation areas.

Enamel

The structure of enamel

Enamel forms in a regular incremental manner that reflects the rhythmic slowing down and

FIGURE 7·9 *Scanning electron micrograph of enamel prisms in a specimen of* Paranthropus boisei. *Each prism head is about 6 µm in diameter. This view illustrates how prisms undulate in their course from the enamel–dentine junction to the surface. In places, alternating varicosities and constrictions can be seen along the lengths of the prisms.*

speeding up of enamel secretion by the ameloblasts. Two types of incremental markings are present within enamel, cross striations and brown striae of Retzius. The first occur along the **rods** or **prisms** of enamel that pass from the enamel–dentine junction to the surface of the enamel (Fig. 7.9). (Prisms are tightly packed bundles of apatite crystals and can be seen across fractured surfaces of enamel or in sections cut from teeth.) Cross striations occur as fine dark lines in polarized light microscopy (PLM) but are also manifest as alternating varicosities and constrictions along the lengths of prisms when seen using scanning electron microscopy (SEM). Cross striations or varicosities are generally accepted as representing daily or **circadian** markings along the lengths of prisms (Okada, 1943; Boyde, 1964, 1976, 1989). The alternate slowing down and speeding up of enamel matrix formation results in the varicosities and constrictions in the same way that squeezing a tube of toothpaste faster and slower leaves an uneven trail of paste behind. There are, however, differences in chemical composition along the prism that may underly the cross striations seen with light microscopy that correspond to the varicosities and constrictions. The second form of incremental marking in enamel, the brown striae of Retzius, are coarser and more widely spaced than cross striations. These markings reflect regular but more marked disturbances that result in a slowing of matrix secretion and which occur at the same time in all the cells forming enamel; they occur with a similar periodicity throughout the whole of enamel formation.

The first enamel formed over the cusps of a tooth is laid down in an appositional manner (Fig. 7.10). The striae of Retzius can be seen to pass over the cusps in a way that reflects the concentric layers of enamel formation, which are laid down as the whole ameloblast sheet moves occlusally. When the cuspal enamel is fully formed, ameloblasts stop secreting enamel over the cusps. Other ameloblasts, however, begin to mature and then

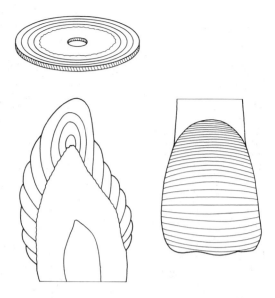

FIGURE 7·10 *Striae of Retzius look like tree rings in transversly-sectioned teeth. In longitudinal sections it can be seen that appositional growth under the cusps results in striae that do not reach the surface. Imbricational growth of subsequently formed striae are also manifest at the surface of the tooth in the form of perikymata. After Ten Cate (1985).*

continue to secrete enamel over the sides of the teeth. This enamel is laid down in an **imbricational** manner as each layer does not cover the whole of the previous layer but overlaps it at the surface of the tooth rather like roof-tiles. It follows that in this more lateral enamel the striae of Retzius come to the surface. They do this at outcroppings known as **perikymata**. Different teeth have different proportions of enamel increments buried beneath the surface of cusps or visible as perikymata on the lateral aspects of the tooth. Generally speaking, anterior teeth have the least proportion of hidden increments because the enamel is thinner over the incisal edge than it is in posterior teeth. There is an increase in the proportion of hidden increments towards the more posterior teeth of the mouth so that more of the total number of growth increments are expressed as perikymata at the surface in anterior teeth.

Both cross striations and striae of Retzius can usually be seen in ground sections of teeth along the lengths of the enamel rods or prisms as they pass from the enamel–dentine junction to the enamel surface. Retzius lines look like exaggerated cross striations with marked constrictions of the prisms and can also be associated with a deviation of the whole prism towards the neck of the tooth, especially in the cervical region. This reflects the extreme slowing of the ameloblast's secretion of enamel matrix at the time the Retzius line is formed.

Counts of seven, eight, or nine cross striations are most common between adjacent Retzius lines, suggesting a regular rhythmic process as an underlying cause (Dean, 1987a). The teeth of any one individual seem to have the same number of cross striations between adjacent Retzius lines suggesting a regular systemic rhythm (Fig. 7.11). While periodicities of six and 10 cross striations have been recorded in modern human teeth, striae of Retzius appear to occur in what has been called a **circaseptan** (or around seven day) rhythmic manner and are a phenomenon with no defined underlying cause at the present time.

Because teeth record these regular rhythmic disturbances during their formation and because teeth are forming from before birth to adulthood, it is possible to count the incremental markings in tooth enamel and use them to estimate the period of crown formation of an individual tooth (see later). Other events, such as birth or high fevers, cause disturbances which look like exaggerated brown striae of Retzius so that the true periodicity of cross striations between adjacent Retzius lines can only be counted when the Retzius lines associated with adjacent perikymata can be clearly seen near to the surface of the tooth.

Enamel prism packing patterns

Enamel is a composite material, i.e. it is composed of two phases: a mineral phase and an organic phase (Boyde and Martin

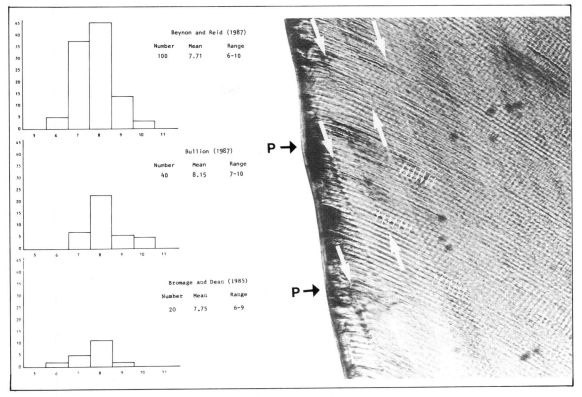

FIGURE 7·11 *The results of three independent studies suggest that the range of cross striations between adjacent striae of Retzius in the outer cervical enamel is 6–10 with a mean value of about 8. Courtesy of David Beynon, Tim Bromage and Simone Bullion. A modern human tooth with a cross-striation repeat interval of 7 is illustrated (light micrograph × 200). Adjacent striae of Retzius (large arrows) can be seen coming to the surface at perikymata (P). The small arrows point to daily varicosities and cross-striations along the length of prisms.*

1984). The mineral phase is an apatitic calcium phosphate (approximately $Ca_{10} (PO_4)_6, (OH)_2$) and accounts for the hardness of the tissue. Mature enamel is over 96% mineralized by weight but it resists brittle fracture far better than crystalline apatite alone because fine long **crystals** of apatite within the prisms (about $1/20$ μm diameter) are 'cemented' together by the organic phase, which is a complex of proteins. Because enamel is so highly mineralized it undergoes little change during fossilization. Enamel crystals tend to grow perpendicular to the surface of the ameloblasts that secrete them. Each ameloblast ends in a point at its secretory end known as the **Tomes' process** which, during enamel formation, fits into a **Tomes' process pit** in the developing enamel surface (Fig. 7.12). Marked discontinuities exist in enamel at what are called **prism sheaths** or **boundaries**. This is because crystals grow at right angles to the surface of the Tomes' process and where two surfaces of the Tomes' process come to a point at the tip of the cell the crystals must grow at right angles to each other and so are discontinuous here, forming a potential fault line. These boundaries define large bundles of crystals which are what we call the **prisms** or **rods** of enamel. Prisms are not completely separated from one another, as for example a bundle of pencils would be one from another, but are only partially

113

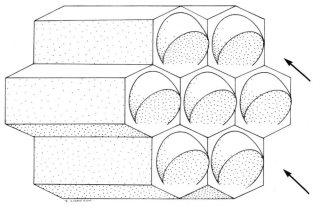
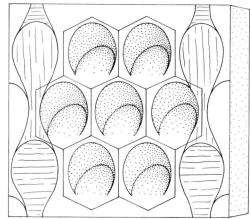

FIGURE 7·12 *A group of six ameloblasts (each hexagonal in cross section) with their projecting Tomes' processes are shown diagramatically on the left. Each Tomes' process fits into a Tomes' process pit in the developing enamel surface shown on the right. Pattern 3 prisms (illustrated by the shaded keyhole-shaped outlines on the right and left of the* enamel *block) are each formed by four ameloblasts. Their relationship to the ameloblasts and to the Tomes' process pits in forming enamel surfaces can be established from this diagram and Figure 7.14. Note that the prisms move away from the enamel surface they are forming at an angle (arrowed). Diagram courtesy of David Beynon.*

discontinuous at the prism sheaths or boundaries. Only here are there marked discontinuities in the orientation of the crystals. This has important consequences, in that this form of enamel is able to resist fractures better than if it were composed of completely separate bundles of prisms or rods.

Cross striations are manifest as varicosities along the enamel prisms or rods and these alternate with constrictions along their lengths. When many rods or prisms are packed together tightly, each varicosity coincides with the constrictions of its neighbouring prisms. It has been hypothesized that the varicosities form when enamel secretion is fastest and that the constrictions form when enamel secretion slows down during the daily cycle (Boyde, 1964, 1976). In order that constrictions and varicosities, which form at different times of day, can abut each other tightly, the forming enamel surface must be irregular. This explains the presence of the Tomes' process pits. The heads of prisms form in the depths of the pit but the tails,

while formed at the same time, lie in a plane superficial to the heads. In cross-sections of mature enamel that are polished flat and perpendicular to the prisms, adjacent heads and tails are then not those that formed on the same day.

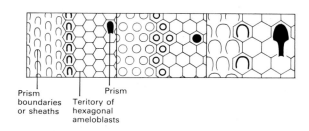

Prism boundaries or sheaths   Prism   Teritory of hexagonal ameloblasts

FIGURE 7·13 *Prism boundaries, the hexagonal outline of the ameloblasts and the outline shapes of the prisms are illustrated for pattern 2 (left), pattern 1 (middle) and pattern 3 enamel (right); note that the size of the hexagonal ameloblasts are smallest in pattern 2 enamel, intermediate in pattern 1 enamel and largest in pattern 3 enamel. After Boyde (1976).*

Three major types of prism packing pattern have been described by Boyde (1964, 1976, 1989): pattern 1, pattern 2 and pattern 3 (Fig. 7.13). These packing patterns describe the cross-sectional appearance of many prisms or rods seen end on as they approach the tooth surface. More correctly, they describe the appearance of the prism sheaths or boundaries. Boyde (1969) has shown that pattern 2 enamel is formed by small ameloblasts, pattern 1 by medium-sized ameloblasts, and pattern 3 by ameloblasts with the largest cross-section (20, 30, and 40 μm in cross-sectional area, respectively). Pattern 1 enamel is the commonest type found in Sirenia, Cetacea, and Chiroptera but is also found in other orders in the very deep enamel and the enamel near the surface of the tooth. Pattern 2 is found in Lagomorpha, Artiodactyla and the Perissodactyla. Pattern 3 enamel is found in Carnivora, Pinnipedia and Proboscidea. All three patterns are found in primates but patterns 1 and 3 predominate in the hominoids, pattern 2 being common in Old World monkeys.

In pattern-1 enamel the prism boundaries are round closed circles. The ameloblasts are oriented perpendicular to the prisms and the Tomes' process is shaped like a short cylinder projecting from the end of the ameloblast (as can be judged from the shape of its pit). In pattern-2 enamel, prisms are aligned in alternate rows but the ameloblasts pass each other in different directions to varying degrees (they **decussate**), but are not as in pattern 1 aligned perpendicular to the axis of the forming prisms. Pattern-3 enamel prism boundaries are keyhole shaped so that the prism appears to have a **head** and a **tail**. The ameloblasts that form pattern-3 enamel are also tilted relative to the axis of the prisms they form. The shape of the Tomes' process resembles more the tip of a finger than the short cylindrical form in pattern-1 enamel. The deepest part of the Tomes' process pit in the forming enamel surface of pattern-3 enamel corresponds to the tip of the Tomes' process and the gradual slope out of the pit

to the level of the rest of the forming enamel corresponds to the contour of the undersurface of the Tomes' process. The result is a keyhole-shaped prism boundary with a deep pit floor that slopes gradually to the surface of the rest of the forming enamel surface. The tail of the keyhole always points towards the cervix of the tooth. Because of this arrangement and the tilt of the ameloblasts relative to the axis of the forming prisms, more than one ameloblast (in fact four) contribute to each prism in pattern-3 enamel (Figs 7.12 and 7.14). The 'tails' and 'heads' of the keyhole-shaped prisms are, however, able to form varicosities and constrictions at the same time but at different levels in the forming surface. When the enamel is fully formed, each tail interlocks with each head perfectly and each varicosity is beside a constriction in its neighbouring prisms.

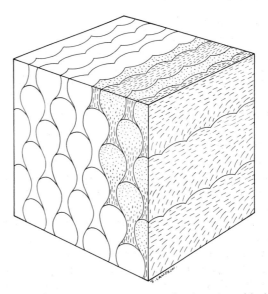

FIGURE 7·14 *Pattern 3 enamel prisms in a block of enamel illustrating the alternating varicosities and constrictions along the lengths of the prisms and the orientation of the enamel crystalites within each of the prisms. It is abrupt changes in the orientation of these crystalites that create the prism sheaths or boundaries.*

Enamel thickness in hominoids

The thickness of the enamel of cut or fractured teeth can be measured either directly or with various forms of microscopy. Because tooth enamel may fracture obliquely, a better measure of enamel thickness is the total area of enamel divided by the length of the enamel–dentine junction. Both of these can be measured from a section of the tooth carefully cut to pass perpendicularly through the tips of the dentine horns under the cusps (Martin, 1985). This measure of enamel thickness approximates the ideal measure (i.e. one of enamel volume divided by the surface area of the enamel–dentine junction). One interpretation of measurements of enamel thickness made in this way from sections of hominoid teeth has revealed that *Homo* and *Sivapithecus* retain the primitive condition of thick enamel which is characteristic of the great ape human clade (Martin, 1985). This conclusion is important in that thick enamel in *Ramapithecus* and *Sivapithecus* had previously been considered to be a derived condition, indicative of their hominid status. Measurements of enamel thickness in the modern great apes reveal that the African apes possess thin enamel that, it has been argued, is secondarily reduced relative to other hominoids (Martin, 1985). The orangutan has thicker enamel than the African great apes but thinner enamel than *Homo* and so has what has been called 'intermediate thick enamel' (Martin, 1985).

Accompanying these differences in enamel thickness there are differences in the enamel prism packing patterns at various depths within the enamel of these taxa. All hominoids studied are said to have a very thin layer (20–30 μm) of pattern-1 enamel both close to the enamel–dentine junction and at the surface of the tooth, the last formed enamel. *Homo* and *Sivapithecus* have pattern-3 prism packing patterns throughout the whole of the rest of the thickness of their enamel. Early accounts (Martin, 1985) suggested that the outer 20% of the enamel in *Pongo* and the outer 40% of the enamel in *Gorilla* and *Pan* showed a pattern-1 prism packing pattern with only the inner enamel of these taxa being pattern 3. More recently, Boyde and Martin (1987) have noted that only about

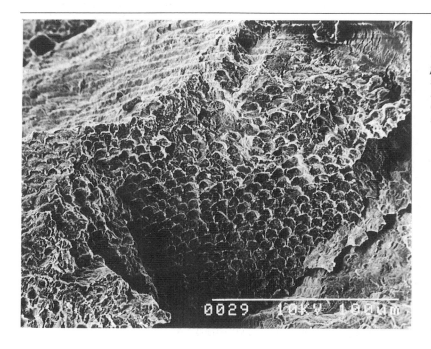

FIGURE 7·15 *Scanning electron micrograph of enamel prisms with a pattern 3 packing pattern near to the surface of the tooth in LH-21 (Australopithecus afarensis).*

100 μm of pattern-1 enamel can be identified in the cuspal regions of *Gorilla* and *Pan* but that this reduces in thickness cervically where pattern-3 enamel grades through to pattern 1 'in the same way as in Hominidae'. Similarly, in *Pongo* the pattern-1 enamel is apparently even more confined to the cuspal regions of teeth and grades into pattern-3 enamel towards the cervix sooner than it does in *Pan* and *Gorilla*. It seems that only a small amount of surface pattern-1 enamel (over and above that which always exists at the surface of hominoid teeth) is present in great apes and that it is pattern-3 enamel which predominates in the deeper layers and towards the cervix, as it does in other hominoids.

The thickness of enamel in early fossil hominids has also been investigated both by sectioning teeth (Grine and Martin, 1988) and by examining the naturally fractured surfaces of teeth (Beynon and Wood, 1986). The consensus of opinion is that all early hominids had thick enamel but that *Paranthropus boisei* stands apart from the other hominids in possessing extremely thick enamel, beyond that seen in any other primate. Pattern 3 enamel predominates in all early hominids studied so far (Fig. 7.15).

Enamel prism decussation

As enamel prisms pass from the enamel–dentine junction to the surface of the tooth they deviate from side to side (Fig. 7.16). In the cuspal regions of teeth, prisms 'worm about' very deviously and their appearance has given rise to the term **gnarled enamel**. Each prism in human lateral enamel makes approximately 2.5 turns before it reaches the surface of the enamel. Bands of prisms tend to deviate in the same direction together but there may be abrupt changes of direction between adjacent prisms. This is very marked in, for example, rodent enamel where one layer of prisms consistently travel at right angles to the next layer. When sections of teeth are prepared, some prisms are cut along their length and others are cut across their ends. The alternating pattern of what are

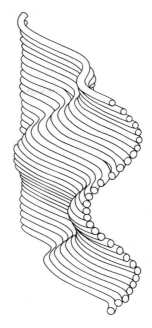

FIGURE 7·16   *Enamel prisms decussate as they pass from the enamel–dentine junction to the surface. After Ten Cate (1985).*

called **diazones** and **parazones** create a series of bands called **Hunter–Schreger bands** in the enamel (Fig. 7.17). These are also visible in fossil hominid teeth that have been naturally fractured, especially when they are viewed under alcohol with a light microscope. The pitch or angle of the Hunter–Schreger bands in early *Homo* are identical to those of modern humans but those of *Paranthropus boisei* are angled more obliquely to the enamel–dentine junction (Beynon and Wood, 1986). The enamel prisms of *P. boisei* apparently decussate a lot less than do those in other hominids, in fact the outer two thirds of the enamel is composed of rods that deviate very little from a straight course. Strong decussation may be an adaptation to resist enamel fracture. However, in rodents it is an adaptation to shear off the enamel of continuously growing incisor teeth cleanly, leaving an extremely sharp edge. The very straight enamel prisms of *P. boisei* may have

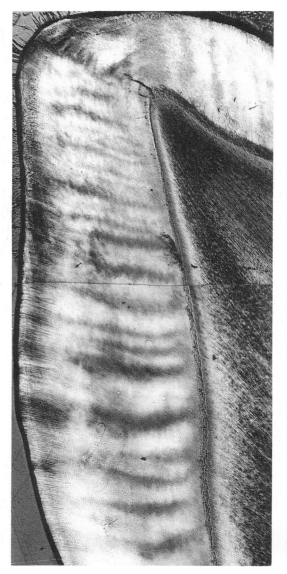 

FIGURE 7·17 *Sections made through enamel prisms cut some prisms longitudinally and some in cross section because of the fact that the prisms are decussating. This creates a banding effect visible in both ground sections of teeth and in scanning electron micrographs of cut surfaces of enamel. The bands are known as Hunter–Schreger bands and consist of parazones (para (GK) = besides) where prisms run in the plane of section and diazones (dia (GK) = through) where prisms are cut end on. The light micrograph is of a Neanderthal deciduous second molar. The higher power SEM of Chimpanzee enamel reveals the orientation of the prisms within the bands. Examples courtesy of Yoel Rak and Lawrence Martin.*

influenced the way in which the surface enamel wore and fractured during normal mastication and this needs to be considered in studies of early hominid tooth microwear.

Dentine

Dentine differs from enamel in that it is not as highly mineralized (about 70% by weight) and that it is composed of collagen as well as hydroxyapatite. Odontoblasts retreat from the enamel–dentine junction during dentinogenesis and move towards the pulp chamber of the tooth. They secrete behind them a dense feltwork of collagen in a matrix (**predentine**) that slowly mineralizes (Fig. 7.18). The odontoblasts have long cell processes that remain in tubules that run through the dentine. The path of each tubule represents the path travelled by each odontoblast as it formed dentine during crown and root formation. After completion

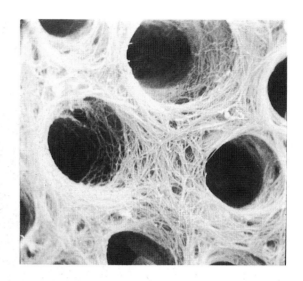

FIGURE 7.18 *Scanning electron micrograph of a predentine surface showing the feltwork of collagen fibres within the dentine. Dentine tubules are approximately 2 μm in diameter. Courtesy of Alan Boyde and Sheila Jones.*

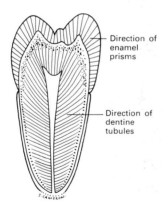

FIGURE 7.19 *Dentine tubules run from the enamel–dentine junction or the cement–dentine junction to the pulp chamber in the direction illustrated in (a). Tubules that become sealed off quickly (while they are still patent) with irregular secondary dentine in response to trauma or attrition form what are called dead tracts in the dentine. Peritubular dentine forms within the dentine tubules throughout life until the tubules are completely sclerosed (b). Physiological secondary dentine also forms slowly throughout life until sometimes the pulp chamber of the tooth is obliterated.*

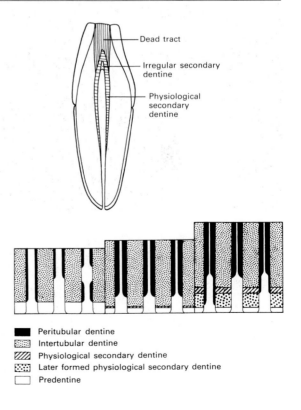

- ■ Peritubular dentine
- ▦ Intertubular dentine
- ▨ Physiological secondary dentine
- ▦ Later formed physiological secondary dentine
- ☐ Predentine

of tooth growth, odontoblasts remain alive and are capable of producing **secondary dentine** (Fig. 7.19). Small amounts of secondary dentine are produced throughout life but it can be laid down more quickly in response to attrition or trauma (**irregular secondary dentine**). Slowly, during life, a hypermineralized type of dentine that contains no collagenous matrix is laid down around the odontoblast processes in the dentine tubules. This is known as **peritubular dentine** and, as it forms, teeth become increasingly sclerosed and transparent such that an idea of the age of an individual can be gained from the appearance of the dentine in ground sections of teeth (Figs 7.20 and 7.21) (Miles, 1963).

Dentine formation is also influenced by the same systemic rhythms that affect enamel as it forms. Wider spaced incremental lines in dentine occur as small changes in the direction of the dentine tubules along the forming front. These are called **Owen's lines** (Fig. 7.22) and possibly correspond to the striae of Retzius in enamel. Smaller incremental lines between Owen's lines are known as **von Ebners lines**. These lines probably correspond to the cross striations of enamel, and are known to be daily incremental lines and may represent changes in the composition of the organic matrix of the dentine during the daily cycle (Okada, 1943; Yilmaz et al., 1977).

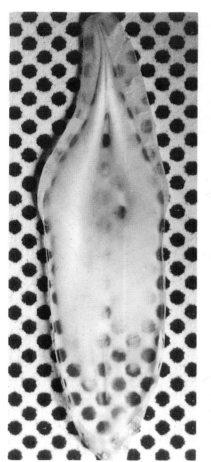

FIGURE 7·20 *Photograph of a tooth root with transparent sclerosed dentine in the apical region.*

FIGURE 7·21 *There is a good relationship between root sclerosis and the chronological age of an individual such that quite reliable estimates of age at death can be made from observations of root translucency. The graph shows the length of translucent root in 118 incisor teeth plotted against the age of their owners. Solid line = calculated regression line. Broken line = 95% confidence limits. After Miles (1963).*

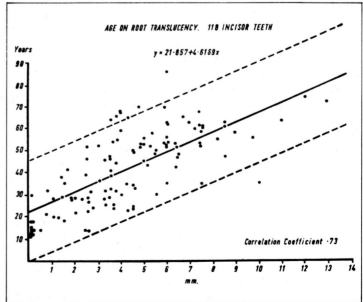

AGE ON ROOT TRANSLUCENCY. 118 INCISOR TEETH

$y = 21 \cdot 857 + 4 \cdot 6189 x$

Years

Correlation Coefficient ·73

m m.

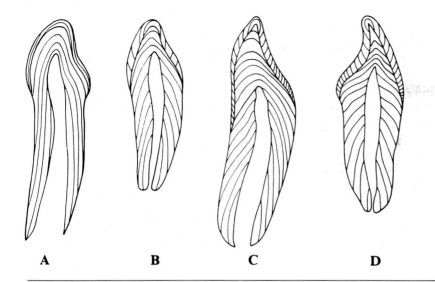

A B C D

FIGURE 7·22 *Drawings made from longitudinal ground sections through (A) a great ape deciduous canine, (B) a modern human deciduous canine, (C) a great ape permanent canine and (D) a modern human permanent canine. Owen's lines in the dentine of each of these teeth represent previous positions of the forming dentine surface during crown and root dentine development and can be used to provide information about the rate of root elongation.*

Cementum

Cementum forms over the root–dentine surface and usually also over a small area of the enamel at the cervix of the crown. Initially, cementum is formed by cementoblasts that retreat from the surface of the tooth and which do not therefore become incorporated into the tissue as it is laid down. This type of cementum is known as **acellular cementum**. Cementum that is laid down later, especially that in the region of the apex of the tooth root, tends increasingly, like bone, to incorporate cells into its substance. This type of cementum is therefore called **cellular cementum**. Cementum is laid down much more slowly than enamel or dentine and so records more widely spaced rhythms such as seasonal changes and annular rhythms in the form of slowed growth rings. Cementum rings or lines are commonly used to estimate the age at death of many animals and are also known as **lines of Salter** in modern human teeth.

Patterns of tooth growth in early hominids

The incremental nature of tooth growth has made it possible to study the way in which early fossil hominid teeth grew and to compare patterns of tooth growth among early and modern hominids. Fossil hominid material is precious but, ideally, the best way of documenting tooth-growth patterns is to cut sections of teeth and make direct comparisons with modern material prepared in the same way. Nevertheless, some naturally fractured surfaces of fossil hominid teeth preserve the enamel prisms well enough to be able to identify daily varicosities along their length. This has made it possible to calculate the total time of crown formation in these teeth by adding up the total number of days of growth within the tooth crown. In the case of molars (Beynon and Wood, 1987) crown-formation times appear to be within the range known for modern humans and great apes and in the case of premolars of *P. boisei* (Beynon and Dean, 1987) crown-formation time appears to be somewhat less than expected given the modern human range (although this premolar data is so far only for one tooth). Striae of Retzius are also visible in many of the same fossil teeth when the fractured surfaces are viewed under alcohol with a light microscope. The pattern of these lines has been used in another way, i.e. to reconstruct the way in which fossil hominid teeth grew.

When only a few ameloblasts are actively forming enamel at one time it takes a long time to complete the enamel of a tooth. However, when very many ameloblasts are all active together the enamel of a tooth can be formed very much more quickly (Fig. 7.23). Despite differences in tooth size, teeth can then be formed either quickly by employing many forming cells or slowly by employing fewer forming cells to suit the time available during the growth period. The amount of enamel secreted in one day by an individual ameloblast need not be different in the two cases to achieve this.

Deciduous teeth are required quickly and so accordingly form in a short period of time. Permanent modern human teeth have more time to grow and so form less quickly. Likewise, large gorilla canines can form in the same amount of time as smaller modern human canines but with no difference in the daily rate of enamel or dentine secretion by individual ameloblasts or odontoblasts. It seems that all the molar teeth of hominoids take about 2.5–3.0 years to form the enamel on their crowns (Beynon and Wood, 1987). The extremely thick enamel on the molars and premolars of *P. boisei* also appear to have formed within the same time as other hominoids but they seem to have achieved this by employing very many more active ameloblasts at one time.

The angle of the Retzius lines at the enamel–dentine junction (Fig. 7.23) reflects the **extension rate** of the ameloblasts as they become active towards the cervix of the tooth (Shellis, 1984): small angles (Df) indicate that large numbers of ameloblasts were becoming active in enamel formation; large angles (Ds) indicate that fewer ameloblasts were involved in enamel formation at any one time. The pattern of incremental lines in the teeth of *Paranthropus* indicate that, despite a similar daily rate of enamel formation along the prisms, large numbers of ameloblasts were able to complete their large thick-enamelled crowns in the same time as (or even somewhat quicker than) other hominoids.

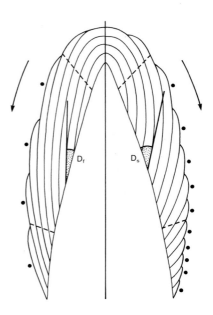

FIGURE 7·23 *The angulation of the striae of Retzius in the enamel of teeth gives some indication of the relative number of ameloblasts that are active during crown formation. On the left angle Df is small, the extension rate of the enamel is fast and there are fewer perikymata on the surface of the tooth (marked by dots). On the right angle Ds is greater than angle Df. The extension rate of the enamel is slower and there are more perikymata on the surface of the tooth. Crown formation on the left is faster than crown formation on the right even though each ameloblast may secrete enamel at the same daily rate. After Boyde (1964).*

Dentine formation, as already noted, precedes enamel formation, so that a fast extension rate of the dentine is clearly a prerequisite for a fast enamel extension rate. The same principle of ameloblast extension rates in the crown of a tooth applies to the odontoblasts in the root and some idea about the speed of root elongation can be gleaned from observing the angle that the forming dentine front makes relative to the long axis of the tooth. In this way the angle of the Owen's lines in sections of teeth indicates slow or fast root-extension rates. In partially formed roots, seen either on radiographs or on individual specimens, the **root cone angle** (Fig. 7.24)

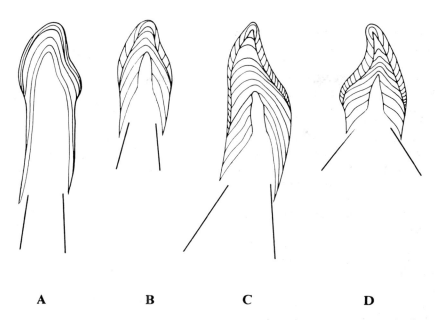

A   **B**   **C**   **D**

FIGURE 7·24 *The same teeth as in Figure 7.22 are illustrated incompletely formed. The angle formed by the mineralizing dentine front (the root cone angle) reflects the relative number of odontoblasts actively forming dentine and the rate of root extension. The deciduous teeth have a smaller angle than the permanent teeth and so form their roots more quickly. (Teeth not drawn to scale.)*

can be used as a rough indication of the speed of root elongation. Preliminary studies of Owen's lines in great ape teeth suggest a very fast rate of root elongation in ape teeth relative to the rates known for modern humans (Dean and Wood, 1981a). Root cone angles in some fossil hominid teeth appear to be much smaller than angles in samples of modern human teeth, suggesting a fast rate of root elongation in some early hominids relative to those in modern humans (Dean, 1985b), but these data need to be supported by histological studies of early fossil hominid roots.

Where it is not possible to section teeth to observe the internal structure of early fossil hominid tooth crowns, the surface manifestations of incremental lines can be used to provide information about tooth crown growth. Perikymata can sometimes be seen on the surfaces of well-preserved hominid incisors and canines (Fig. 7.25). If careful estimates are made of the hidden increments

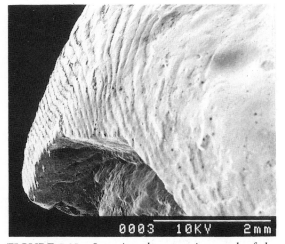

FIGURE 7·25 *Scanning electron micrograph of the buccal surface of a fossil hominid premolar tooth (Omo 33–507) from the Omo, Ethiopia. Perikymata on the surface of the tooth illustrate the layered or imbricational manner in which the enamel was formed. At the fractured portion of cervical enamel (bottom left of field) the direction of the striae of Retzius in the enamel can be seen. Replica courtesy of Yves Coppens and Gen Suwa.*

123

beneath the incisal edge of these teeth, good estimates of both the total crown formation time and of the extension rates of cervical enamel can be made. It seems that there are marked differences in the time it takes to form the incisors of *Paranthropus* and *Australopithecus* (Fig. 7.26). Those of *Paranthropus* formed quickly (in 18 months to two years) and have perikymata that are evenly spaced towards the cervix, indicating a fast extension rate. Incisors of *Australopithecus* took longer to form (approximately three years) and have perikymata that become very closely packed at the cervix, indicating a very slow enamel extension rate (Bromage and Dean, 1985; Dean, 1987b; Beynon and Dean, 1988).

Sequences of tooth development in hominoids

All teeth pass through three recognizable and homologous stages during their development: the initiation of calcification of the tooth; the completion of enamel formation (which is a definition of completion of the anatomical crown of the tooth); and closure of the apical canal of the root (that defines the end of root growth). Each of these three stages of development is best recognized by direct observation of the developing tooth. However, this is clearly not always possible, as growing teeth are embedded within the jaws and either radiographs or histological sections are required to study them.

Histological sections of teeth can record initial calcification up to six months before radiographs can, as early stages of mineralization are not very radio-dense and are 'burned out' on X-ray images of developing teeth (Hess et al., 1932). Radiographic images of teeth in the jaws also fail to record much of the buccal and lingual aspects of the crowns of teeth in the jaws. It is only usually possible to image the dense superimposed mesial and distal aspects of tooth crowns and roots. As a result of this, radiographic definitions of crown completion are based upon the end of

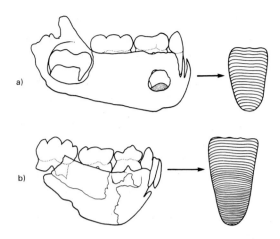

FIGURE 7·26 *(a) Represents a specimen of* Paranthropus *with evenly spaced perikymata on an incisor tooth which takes about 2.5 years to form its crown. The first permanent molar is completing its crown at the same time. (b) Represents a specimen of* Australopithecus *with perikymata that become increasingly close together at the cervix of the incisor. This specimen takes about 3.5 years to complete its incisor crown by which time the first permanent molar has not only formed its crown but has formed some root and emerged into functional occlusion. After Beynon and Dean (1988).*

enamel formation and the beginning of root formation as it is observed on the mesial and distal aspects of the tooth. Because in most teeth enamel formation continues for very much longer on the buccal surface, radiographic estimates of crown formation times are commonly shorter than the true times taken to complete enamel formation. Many of these difficulties do not detract from the clinical usefulness of radiographic studies in paedodontics or orthodontics. Nevertheless, observations made directly on isolated tooth germs for anthropological purposes cannot automatically be equated with observations made from either histological sections of teeth or with observations made from radiographs of developing teeth in the jaws of living (or dead) subjects.

Additional stages of tooth formation have been defined to describe fractions of crown completion and fractions of root growth

(Fig. 7.27). Crowns can be defined as one-quarter, one-half or three-quarters complete and, likewise, roots can be defined as one-quarter, one-half or three-quarters complete. These fractions of tooth growth correspond to proportions of the final crown height or the final adult root length and can only be objectively assessed in longitudinal studies where the mature proportions of the tooth are known (Demirjian, 1978). Because teeth do not grow, or extend, in a linear manner these fractions of crown and root growth do not represent equal amounts of time during the growth period of a tooth. The last quarter of crown formation of a modern human tooth may take three times as long to grow as the first quarter of crown formation. Likewise, the last quarter of root growth may take three or four times less time to grow than the first quarter of root growth. Failure to appreciate these important points has led to confusion when the same modern human stages of tooth formation have been adopted uncritically in comparative studies of non-human primate tooth development. We have already seen that molar and premolar tooth crowns attributed to *Paranthropus* have fast extension rates and so do not pass through homologous stages of tooth formation as defined for modern human teeth in the same proportions of time. The crowns of fossil hominid teeth are also of different sizes and proportions to modern human teeth. Radiographic images of these developing teeth can be very misleading when compared uncritically with modern human standards of tooth development. Similarly, the lengths of some early hominid tooth roots exceed those of modern humans (Fig. 7.28) and estimates from radiographs of the proportions of root formed are likely to be wrong if modern human standards are employed. It is worth emphasizing here by way of example that great ape premolars form their root bi- or tri-furcations before enamel formation is complete on the buccal aspects of the crown.

Swindler (1985), however, has observed that the whole period of dental development

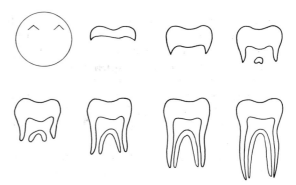

FIGURE 7·27 *Some commonly used stages of crown and root formation: initial mineralization of the crown, crown half-complete, crown complete, root bifurcation forming, root half-formed, root three-quarters formed and root completely formed. Many studies employ even more stages of tooth formation than are shown here.*

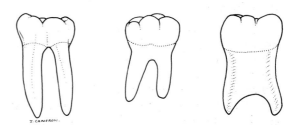

FIGURE 7·28 *Variation in the size and proportions of hominid molars make it difficult to employ modern human standards of tooth formation to all of them. Some have different proportions of root and crown height. Many have root bi- and trifurcations at different levels.*

from the initiation of calcification to the end of growth of the last tooth can be used as a comparative 'yardstick', and that this is useful for measuring the timing and sequence of other developmental events. Smith (1989) investigated the relationship between several life-history variables available in the literature and dental development in primates. For 21 species representative of 15 primate genera, correlations of age of M1 emergence with brain weight (Fig. 7.29) were among the highest attained, ($r = 0.98$ with adult brain

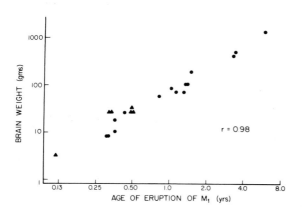

FIGURE 7·29 *Adult brain weight for 21 primate species plotted against age of eruption of the mandibular first permanent molar; circles = anthropoids; triangles; = prosimians. Note the logarithmic scale on both axes. From the rapid maturing* Cheirogaleus *(far left) to the slow maturing* Homo *(far right) dental maturation has a strong relationship with brain weight. After Smith (1989).*

FIGURE 7·30 *(Right) Summary charts of dental development (lower teeth only) in macaques, great apes and modern humans. Solid bars are crown formation times, dotted lines are root formation times. Thin vertical lines on the bars indicate the range of initiation of crown formation, the range of crown completion/root initiation and the range of root completion, where there is data for these in humans and macaques. E indicates the period during which a tooth is likely to emerge into occlusion. Data from Sunderland et al. (1987), Swindler (1985), Garn et al. (1959), Gustafson and Koch (1974), Dean and Wood (1981), Bowen and Koch (1970) and Beynon, Dean and Reid (unpub.).*

weight and $r = 0.99$ with neonatal brain weight). Smith also noted that other life-history variables (gestation length, age of weaning, interbirth interval, age of sexual maturity in females, age at first reproduction and lifespan) correlate with brain weight with an average of $r = 0.90$ (range 0.82–0.95).

The order of calcification of the deciduous and permanent dentitions is identical in Old World monkeys, apes and hominids. Even the order of molar cusp calcification is the same, although there are differences in the way the crowns coalesce that relate to the prominence of the ridges and crests of the occlusal surface of different teeth (Swindler, 1985). Despite the fact that the teeth of monkeys, apes and humans begin to form at equivalent times within the growth period, not all their crowns take the same relative times to form. For example, permanent great ape canines form over a longer proportion of the growth period than do permanent modern human canines (Fig. 7.30). It is also clear that the roots of monkey and great ape teeth form very much more quickly than do the

roots of modern human teeth, both absolutely and relative speaking (Dean and Wood, 1981a; Swindler, 1985). These facts are partly responsible for the differences in the sequence of emergence of teeth into occlusion that have been documented in great apes, modern humans and early hominids (see below).

While the overall sequence of initiation of tooth calcification is the same in hominoids, there is great variation in the timing of initiation of individual tooth calcification. For example, Fanning and Moorrees (1969) observed that calcification of third permanent molars in Australian Aborigines begins when second molar crowns are only half complete whereas in Caucasoids' third permanent molar calcification occurs later when the second permanent molar root is one-quarter formed. This kind of age range in the stages of tooth formation and sequence polymorphism exists in both great apes and modern humans, between both premolars and molars and has been carefully documented for most modern human posterior teeth by Garn and co-workers (1959, 1963) (Fig. 7.31). Similar variation in the coincident stages of tooth development can be found in the posterior dentitions of macaques, great apes, fossil hominids and modern humans and it is clear that little phylogenetic significance can be attributed to the 'pattern' of development in any single individual specimen. Pooled data for dental development in *populations* of monkeys, great apes and modern

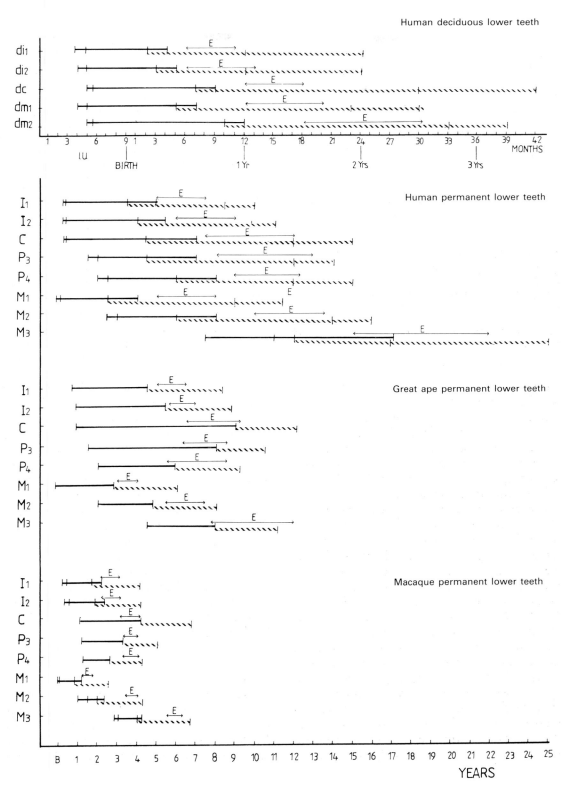

Human deciduous lower teeth

Human permanent lower teeth

Great ape permanent lower teeth

Macaque permanent lower teeth

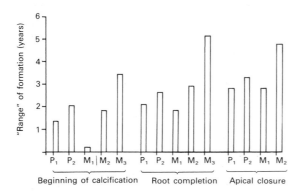

FIGURE 7·31 *A comparison of the range in chrono-logical age over which three stages of tooth formation (initial mineralization, completion of root length and apical closure) that have been recorded in one population of modern humans. After Garn et al. (1959).*

humans, however, do point to a gradual prolongation of dental-development events as the total length of the growth period increases from monkeys to modern humans, but this is not necessarily apparent from the developing teeth of any one individual.

Some data suggest that the emergence of the anterior teeth into occlusion (canines and incisors) in great apes occurs later in the growth period than in modern humans. Schultz (1935) noted that the permanent ape canine erupts very much later in the growth period than do modern human canines. Likewise, in great apes the permanent incisors sometimes even emerge into occlusion after the second permanent molars (Schultz, 1941) and always emerge closer in time to the M2 within the growth period than to the M1 (Fig. 7.30) (Nissen and Reisen, 1964; Willoughby, 1978; Smith and Garn, 1987). In modern humans, permanent incisors and first permanent molars both emerge into occlusion within a few months of each other during the growth period (Gustafson and Koch, 1974).

There is a consistent sequence of developmental events in juvenile specimens of *Paranthropus* where the first permanent molar completes its crown formation at the same time as the incisors and where the canines complete their crowns very shortly after this (Fig. 7.32). Although there are very few specimens where emergence of permanent fossil hominid teeth can be observed to be occurring with certainty, it also looks as if the incisors and the first permanent molars emerge into occlusion at a similar time within the growth period. The sequence of events in *Australopithecus* appears to be different (Fig. 7.32). By the time the slower growing incisor teeth have completed their crown formation, the first permanent molars have not only completed their crowns but have emerged into occlusion and have grown about one half of their root (see also Fig. 7.26).

The total morphological and developmental distinction in *Paranthropus* of small anterior teeth that grow quickly and emerge early relative to the first permanent molar, and the development in *Australopithecus* of larger anterior teeth that grow more slowly and emerge into occlusion later relative to the first permanent molar, seems to be a consistent feature in these two groups of juvenile early hominids (Dean, 1985d; Beynon and Dean, 1988; Smith, 1986a). This observation adds more weight to the generic distinction originally proposed for these hominids by Broom (1950) and by Robinson (1954). These differences in coinciding calcification stages and emergence sequence of the teeth, reflect the similar differences in sequences that exist between modern humans and modern great apes. However, it is also clear that early hominids were unique in their own right (both morphologically and developmentally) and that emphasizing the 'human-like' or 'ape-like' affinities of either group detracts from the more significant differences that can now be documented to exist between them (Beynon and Dean, 1988). These seem to be greater than has previously been appreciated.

It is likely that a relationship exists between the growth of the jaws, the growth of the teeth, the size of the tooth crowns and the

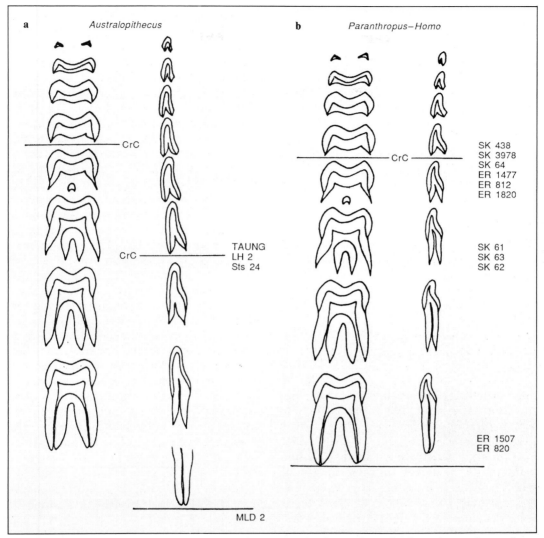

a   *Australopithecus*   **b**   *Paranthropus–Homo*

CrC

CrC

TAUNG
LH 2
Sts 24

SK 438
SK 3978
SK 64
ER 1477
ER 812
ER 1820

SK 61
SK 63
SK 62

ER 1507
ER 820

MLD 2

FIGURE 7·32 *Stages of first permanent molar formation are shown on the left. Stages of incisor formation that correspond with those of the first permanent molar are shown for* Paranthropus *early* Homo *and for* Australopithecus. *Individual specimens appear at the stages of formation of these two teeth observed at death and are shown for each genus. Maxillary specimens attributed to either genus follow the same pattern. OH 30, SK 839/852, SKX 162 (*Paranthropus*): Sts 2, LH-21, LH-3, LH-6 (*Australopithecus*): KNM-ER 808, KNM-ER 1590, SKX 21204 (early* Homo).*

age at which space is available in the jaws for the teeth to emerge into occlusion. The anterior teeth of great apes are very large and the anterior portion of the great ape mandible remains very narrow well into growth and development. There is little or no resorption of bone at the chin or mid-face during growth in hominoids that are prognathic (Enlow, 1968). The result is that only when the tooth row has elongated sufficiently posteriorly is there room in the jaws for the incisors and canines to emerge into occlusion. Modern human anterior teeth are much smaller than their great ape counterparts. In addition, there is continual resorption of the chin and mid-face during growth and development in modern humans and this contributes to the gradual widening and flattening of the face and anterior dental arch (Enlow, 1966, 1968). As a result, more space is available for the smaller teeth earlier in the growth period in modern humans and this may be one factor that underlies the relatively early emergence of the incisors and canines. The combined evidence of known crown formation times, known sequences of crown initiation and of known eruption times in both great apes and modern humans suggests that, in contrast to the posterior dentition, differences in comparative developmental sequences within the anterior dentition, may be sufficiently distinctive as to be phylogenetically useful.

Estimating age from the developing dentition

It has become well established that the calcification stages of teeth are a better indicator of chronological age than is the eruption status of teeth or even the ossification of the skeleton or hand (Gleiser and Hunt, 1955; Lewis and Garn, 1960) (Fig. 7.33). There is a higher correlation between dental and chronological age than between dental age and skeletal age (Demirjian, 1978). The skeletal system, as well as height and the onset of puberty, develop largely independently of the dental system. Two methods are widely employed to estimate chronological age from the developing dentition. The first and most commonly used is the 'atlas method', whereby the status of all the developing teeth are compared with a chart of tooth development and an individual is assigned a dental developmental age from the chart. The second method was developed by Demirjian et al. (1973) and is a more objective method that employs an index of development for several teeth which

are then summed together. The total score of an individual gives a good estimate of chronological age but the method requires that a large number of teeth are present in the jaws of each individual.

Enamel-formation times of teeth can be estimated histologically. Boyde (1963, 1964, 1989) has pointed out that this method can be used to estimate an age at death and is likely to be more accurate than methods that depend upon the chronology of other individuals. Obviously, however, this technique cannot be used for individuals who are still alive. Total counts of the number of daily cross striations in the crowns of teeth give good estimates of crown-formation times. Crown-formation times estimated from counts of perikymata on the incisor teeth of early hominids can also be used to estimate an age at death (Bromage and Dean, 1985; Dean, 1987b). When these teeth are associated with other developing teeth from the same individual, it is possible to calibrate events such as the emergence into occlusion of the first permanent molars (for example, Taung or LH2) or the completion of root formation of the first permanent molar or canine crown completion (LH6). Results from these studies indicate that dental-devel-

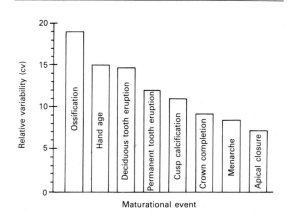

FIGURE 7.33 *Relative variability (coefficient of variation) of different maturational events. After Lewis and Garn (1960).*

TABLE 7·1 Published counts of perikymata on fossil hominid anterior teeth (Bromage and Dean, 1985; Beynon and Dean, 1988) together with the published range of ages at death for the specimens described in Bromage and Dean (1985) and Dean (1987)

Australopithecus			*Paranthropus*			Early *Homo*		
Specimen	Tooth	No. of perikymata	Specimen	Tooth	No. of perikymata	Specimen	Tooth	No. of perikymata
Sts 24a	I^1	135	SK 62	I_1	64	SK 74b	I_1	110
LH2	I_1	130	SK 63	I_1	86	KNM-ER 820	I_2	105
LH3	I^1	170	SK 73	I_1	>79	OH 6	I^2	95
LH3	I^2	180	SK 71	I_1	>79	KNM-ER 808	I^2	123
LH3	$\underline{C}$	168	OH 30	I_1	101			
LH6	I^2	116	OH 30	$\bar{C}$	>103			
LH6	$\underline{C}$	158	KNM-ER 812	I_1	86			
			KNM-ER 1477	I_1	92			
			KNM-ER 1477	$\bar{C}$	>80			
			KNM-ER 1820	I_1	82			
			KNM-ER 816A	$\bar{C}$	109			

Specimen	Estimated age at death[a] (years)	Specimen	Estimated age at death[a] (years)	Specimen	Estimated age at death[a] (years)
Sts 24a	3.2–4	SK 62	3.4–3.75[b]	KNM-ER 820	5.3–6.0[b]
LH2	3.2–4	SK 63	3.2–3.9[b]		
		KNM-ER 1477	2.5–3.0		
		KNM-ER 812	2.5–3.0		
		KNM-ER 1820	2.5–3.1		
		OH 30	2.7–3.2		

[a]Estimated range using seven cross striations and nine cross striations between striae.
[b]Estimates include root formation time for these specimens given in Bromage and Dean (1985).

opment events in early hominids occurred at ages more similar to modern great apes than to modern humans. More reliable ages at death for a Neanderthal juvenile estimated in this way have made it possible to associate the remains of one individual more confidently and these techniques are beginning to provide data about growth and development of early hominids (Dean et al., 1986; Stringer et al., 1990).

Figure 7.34 presents the juvenile fossil hominids for which it has so far been possible to estimate an age at death. It seems reasonable to conclude from this evidence that the stage of human evolution at which prolongation of the growth period began had not yet occurred in *Australopithecus* or *Paranthropus*. As yet there remains insufficient data to comment on the growth of early *Homo*. A preliminary summary chart of dental development in *Australopithecus* and *Paranthropus* is presented in Fig. 7.35, although many details remain to be confirmed.

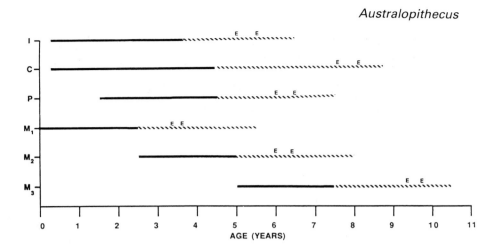

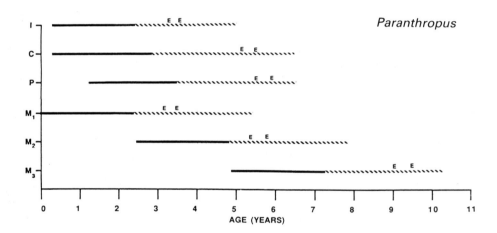

FIGURE 7·34   *Summary charts of dental developmental events as they are known at present for* Australopithecus *and* Paranthropus. *After Beynon and Dean (1988).*

CHAPTER EIGHT

HOMINOID TOOTH MORPHOLOGY

The majority of teeth comprise a crown and a root. However, some teeth, like the continuously growing anterior teeth of rodents (or *Daubentonia*), do not have separate crowns and roots. The anatomical crown of a tooth is that part of the tooth covered with enamel. This looks whiter than the root of the tooth which consists largely of dentine covered with cementum. The junction of the crown with the root is known as the **cervix** (cervix (L) = neck) or cervical region of the tooth and the junction of the enamel with the dentine or cementum of the root is known as the **cervical margin** (Fig. 8.1). The surface of a tooth that faces towards the mid-line of the jaws as it lies in the dental arch is known as the **mesial** aspect of the tooth (Fig. 8.2). The surface of the tooth furthest away from the mid-line is called the **distal** aspect of the tooth. The remaining surfaces of a tooth are named according to whether they face the palate, tongue, cheek or lips: these aspects are the palatal, lingual, buccal and labial surfaces of the teeth, respectively. Often, however, only lingual and buccal are used to refer to the 'inside' and 'outside' surfaces of teeth, irrespective of whether they are upper or lower teeth. The surface of a tooth that bites or **occludes** with the teeth opposing it is known as the **occlusal surface**.

There are four types of teeth known as **incisors, canines, premolars** and **molars** (Fig. 8.3). Incisors are blade-like teeth that cut and shear food at the front of the mouth.

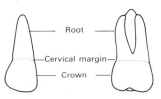

FIGURE 8·1 *The crown, root and cervical margin of an incisor and molar tooth.*

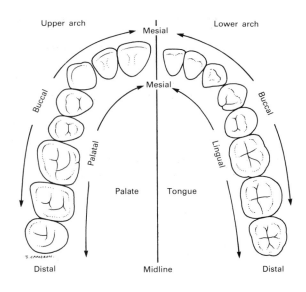

FIGURE 8·2 *The upper dental arch is illustrated on the left and the lower dental arch on the right. Each surface of each tooth is named either the occlusal, mesial (towards the midline), distal (away from the midline), buccal, lingual or palatal depending upon its orientation in the dental arch.*

133

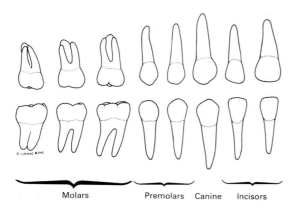

| Molars | Premolars | Canine | Incisors |

FIGURE 8·3   *Incisors, canines, premolars and molars from one upper quadrant and one lower quadrant of the mouth.*

All hominoids have eight permanent incisors, four upper and four lower, or two in each quadrant of the mouth. Canines are large teeth at the corners of the mouth, distal to the incisors, that can pierce food and whose relative size are also important to the social structure of many groups of animals, including primates. There is one permanent canine tooth in each quadrant of the mouth. Premolars are teeth intermediate in form between canines and molars and these commonly have two cusps (raised points on the crown) and are therefore sometimes called bicuspid teeth. There are two premolars in each quadrant of the mouth, usually called the first and second premolars. As the simian dental formula is derived from the general mammalian formula by loss of teeth and, in particular, by loss of the first two premolars which are nearest to the canine, the remaining premolars are also often referred to as the third and fourth premolars. Molar teeth have an expanded occlusal surface, with more cusps than premolars, for crushing and grinding food. There are three permanent molar teeth in each quadrant of the hominoid mouth making a total of 12. The **dental formula** for Old World primates summarizes the number of permanent-tooth types in one upper and one

lower quadrant of the mouth: $I\frac{2}{2} : C\frac{1}{1} : PM\frac{2}{2} : M\frac{3}{3}$. Molar teeth have the most complicated morphology of all the tooth types and their occlusal surface is characterized by cusps that project proud of this surface, which are either connected to each other by ridges of enamel or separated from each other by fissures that run around the base of the cusps.

The basic ground plan for mammalian molar teeth is a triangle, the triangular lower molar teeth fitting between the triangular upper molar teeth (Fig. 8.4). The upper triangles are called **trigons** (trigonum (L) = triangle) and their points, or apices, lie towards the palate (up and in). In the lower jaw the triangles are called **trigonids** and their apices lie towards the buccal (and so, like George Orwell, are 'down and out'). (The suffix 'id' denotes the lower teeth throughout.) Simple triangular molar teeth like these are called **tritubercular teeth** and are adapted to shearing, or slicing, through food. Much like sewing, or pinking shears, the sides of these triangular molar teeth slice past each other cutting the food trapped between them in the process.

No living mammals have teeth as simple as these tritubercular teeth. The molar teeth of living mammals are capable of puncturing and crushing food as well as of shearing food. They do this by adding a basin-like **talon** (**talonid** in the lower jaw), or **heel**, onto the rear (distal side) of the basic triangular tooth. These heels fill the spaces between the basic triangular teeth (trigons and trigonids) and provide enamel basins against which the cusps of the opposing teeth can puncture and crush food. Among primates, the tarsiers have basins (talonids) only on their lower teeth, retaining the primitive triangular-shaped molars in the upper jaw. Their molars are known as **tribosphenic** molars (tribein (Gk) = to rub; sphen (Gk) = a wedge). The remaining primates have **quadritubercular** molars with heels not only on their lower molars but also on their upper molars. In the small prosimian primates that are primarily insectivorous in diet there is a considerable

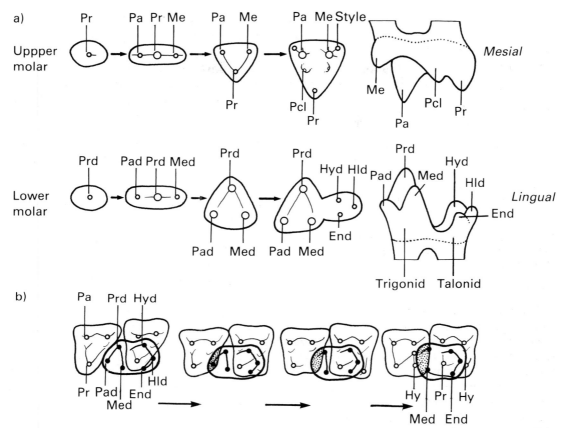

FIGURE 8·4 *(a) The theory of molar evolution according to Cope and Osborn. The tribosphenic molars illustrated in (b) gradually developed a hypocone while the paraconid reduced and was lost. After Bulter (1981). Abbreviations: Upper molar—* Pa, paracone; Me, metacone; Pr, protocone; Pcl, paraconule; Hy, hypocone. Lower molar—Prd, protoconid; Pad, paraconid; Med, metaconid; Hyd, hypoconid; End, entoconid; Hld, hypoconulid.

difference in height between the trigons and talons (and between the trigonids and talonids). This height difference retains the primitive shearing function of the trigons and trigonids. The larger bodied higher primates that are more herbivorous in diet have trigons and talons (and trigonids and talonids) that are more similar in height. They have sacrificed much of the primitive shearing capability of their molars for more efficient puncturing and crushing teeth.

There is a widely used nomenclature for describing cusps and ridges on the occlusal surfaces of mammalian molar teeth that was developed by Cope and Osborn nearly 100 years ago (Fig. 8.4). The basic triangular tooth, both the trigon and trigonid, have three cusps, one at each corner. The cusp in the apex of the triangle is the **protocone** or **protoconid** in the lower jaw (protos (Gk) = first). The cusp mesial (anterior) to the protocone is the **paracone** or **paraconid** in the lower jaw (para (Gk) = near or besides). This cusp is lost in the lower molars of higher primates so there is in fact no paraconid in monkeys, apes or humans. The last cusp on the triangle, the distal cusp, is called the **metacone** or **metaconid** in the lower jaw (meta (Gk) = after). There is only one cusp on the talon (heel) of the upper molars. This cusp is called the **hypocone** (hypo (Gk) = under). There can be up to three cusps on the talonid (heel) of the lower molars. The **hypoconid** sits next to (distal to) the protoconid on the buccal, or cheek, side of the talonid. Opposite the hypoconid, on the lingual side of the talonid, is the **entoconid** (entos (Gk) = within) and distally, between the hypoconid and entoconid is the smaller **hypoconulid** that completes a ring of cusps around the talonid. Minor cusps that occur on ridges linking major cusps are distinguished by the suffix **conule** (upper teeth) or **conulid** (lower teeth) and are attributed to the nearest major cusp. A collar of enamel around the base of the crown is known as the **cingulum** (cingulum (L) = a girdle) and other minor cusps that arise from cingula are given the suffix **-style** (upper teeth) or **-stylid** (lower teeth) and again are attributed to the nearest major cusp, for example the **protostylid** is situated next to the protoconid. The cusps of the human upper and lower molar teeth are shown and labelled in Fig. 8.5.

Several systems have been devised to annotate teeth to a position in the jaws. (Fig. 8.6). The Fédération Dentaire Internationale (FDI) two-digit system, used widely in the U.S.A. denotes each quadrant of the mouth by a number, beginning in the upper right (1) and passing clockwise as the mouth is viewed from the front to the upper left (2), lower left (3), lower right (4) and then likewise through the deciduous dentition (5)–(8) (see Fig. 8.6). Individual teeth are given a number from the midline of each quadrant: 1–8 for

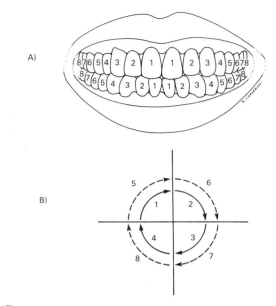

A)

B)

C)
F.D.I. System

Permanent dentition:

Maxillary right								Maxillary left						
18	17	16	15	14	13	12	11	21	22	23	24	25	26	27	28
48	47	46	45	44	43	42	41	31	32	33	34	35	36	37	38
Mandibular right								Mandibular left						

D)

Deciduous dentition

			55	54	53	52	51	61	62	63	64	65
			85	84	83	82	81	71	72	73	74	75

E)
Zsigmond system

Permanent dentitions

8	7	6	5	4	3	2	1	1	2	3	4	5	6	7	8
8	7	6	5	4	3	2	1	1	2	3	4	5	6	7	8

F)

Deciduous Dentition

	E	D	C	B	A	A	B	C	D	E
	E	D	C	B	A	A	B	C	D	E

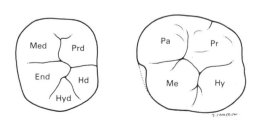

FIGURE 8·5 *Typical occlusal anatomy of upper and lower modern human molars.*

FIGURE 8·6 *Each of the teeth in one quadrant of the mouth are numbered 1–8. In the FDI system each quadrant is also given a number (1–4 for the permanent teeth and 5–8 for the deciduous teeth). Each tooth in the mouth can then be denoted as indicated in (c) and (d). In the Zsigmond system the quadrants are not numbered but the deciduous teeth are lettered A–E in each quadrant to distinguish them from permanent teeth. Examples of teeth denoted by this system are given in the text.*

the permanent teeth and 1–5 for the deciduous teeth. Two figures then denote the quadrant and the tooth in question. For example, 32 refers to the third quadrant (lower left of the permanent dentition) and the second tooth (the lower lateral incisor).

A simpler system is the **Zsigmond System** which is used widely in the U.K. The permanent teeth are numbered 1 to 8 in each permanent quadrant as before, but are lettered A–E in each deciduous quadrant. Two imaginary lines separate the jaws into left and right (the vertical line) and into upper and lower (the horizontal line) as the mouth is viewed from the front. Any tooth can then be easily notated in shorthand as $\underline{2|}$ or $\underline{|1}$ or $\overline{6|}$ or $\overline{|E}$ (see also fig. 8.6).

The easiest and perhaps the most widely accepted shorthand way of denoting tooth type and number is to abbreviate the type to a letter (upper case for permanent teeth and lower case for deciduous teeth), and the specific tooth to a number. For example: I1, I2, C, PM3, PM4, M1, M2, M3 or i1, i2, c, dm1, dm2. Upper and lower can be signified by placing the number as a subscript or a superscript (I_1 or I^1), or alternatively by using a line underneath to denote upper and a line above to denote lower ($\underline{C}$ or $\overline{PM}$). Finally, the left- and right-hand sides of the mouth can be indicated by an L or R in front of the full abbreviation. For example, the upper-right second molar would be abbreviated to RM^2, or the lower-right second deciduous molar to Rdm_2.

Modern human tooth morphology

Deciduous teeth

Primates, like the majority of mammals, have two generations of teeth in a lifetime and are therefore **diphyodont**. Small immature jaws are not large enough to accommodate permanent teeth for some time and so deciduous teeth come into functional occlusion first, while permanent teeth are still developing in the growing jaws. A subsidiary function of deciduous teeth is to act as a guide to align

developing permanent teeth beneath them and another is to maintain space in the jaws for later erupting **successional** teeth. Permanent teeth always begin to develop on the lingual aspect of deciduous teeth in the jaws or, in the case of premolars, between the widely splayed roots of deciduous molars. Small holes in the alveolar bone on the lingual aspect of deciduous teeth are known as **gubernacular canals**. These mark the path of a fibrous band that connects the permanent tooth germ to the lingual oral epithelium which possibly assists in guiding the permanent tooth during the eruptive process (hence: gubernaculum (L) = a helm or rudder). Besides being smaller and fewer in number than their permanent successors, several morphological features distinguish deciduous teeth from permanent teeth (Fig. 8.7).

(1) Deciduous tooth enamel is whiter, less hard and thinner than permanent tooth enamel.
(2) Deciduous tooth crowns are bulbous and have prominent cervical margins which tend to run around the tooth in the same horizontal plane (are less sinuous). They usually also have more pointed cusps than permanent teeth when unworn.
(3) The roots of deciduous teeth are lighter in colour than the roots of permanent teeth and are absolutely shorter than their successional teeth.
(4) Deciduous anterior tooth roots are proportionately longer than anterior perma-

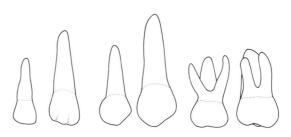

FIGURE 8·7 *A deciduous modern human incisor, canine and molar compared with their permanent modern human counterparts.*

nent tooth roots relative to their crown heights.

(5) Deciduous molar tooth roots are widely splayed and each arises directly from the cervix with little common root trunk.

(6) The pulp chambers of deciduous teeth are larger than those of permanent teeth relative to the size of the crown.

(7) Overall, deciduous teeth show much less morphological variation than do permanent teeth.

DECIDUOUS INCISORS

All incisor crowns are rounded at their distal incisal margin but more squared and sharper at their mesial margin (Fig. 8.8). Surprisingly, unworn upper central deciduous incisor crowns are wider than they are tall; upper lateral deciduous incisor crowns are about equally as wide as they are tall but have a markedly sloping incisal edge, less high at the distal than the mesial incisal margin. Lower deciduous incisors are all smaller than upper deciduous incisors. Lower central deciduous incisor crowns are the smallest of all, so that lower lateral deciduous incisors can be distinguished from them by: (i) their larger crown size; (ii) a longer root, and (iii) by a crown that is twisted distally on the root to follow the curve of the dental arch. The roots of deciduous incisors are more rounded in cross-section than permanent incisors, especially as compared with the lower permanent incisors.

DECIDUOUS CANINES

Deciduous canine crowns are bulbous and especially 'full' at the cervical margin (Fig. 8.8). They have pointed crowns and roots that are long in proportion to their crown heights. The apex of the cusp is symmetrically positioned midway between the mesial and distal aspects of the crown in the upper deciduous canines. Often, however, there is a tendency for the upper deciduous canine to have a longer mesial incisal slope away from the point of the crown and so to be somewhat asymmetrical in this respect.

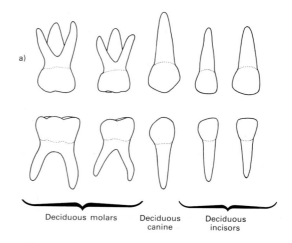

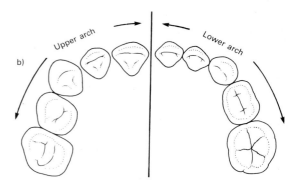

FIGURE 8·8 *Modern human deciduous incisors, canines and molars viewed from the buccal (a) and occlusal (b) view of the upper and lower deciduous dental arches.*

The lower deciduous canine crown is slimmer and smaller than the upper deciduous canine. It has an elongated *distal* slope away from the cusp tip and so can be distinguished from the upper deciduous canine in this way.

DECIDUOUS MOLARS

The upper first deciduous molar has two buccal cusps separated from two palatal cusps by a deep mid-line fissure (Fig. 8.8). There are usually four cusps in total but the distal cusps may be diminutive and the distopalatal cusp may occasionally be absent. The mesiobuccal aspect of the crown of both upper and lower first deciduous molars is characterized by a pronounced swelling known as the **molar tubercle of Zuckerkandl**. All upper molars, permanent and deciduous, have three roots (although some or all of them may be fused). Upper first deciduous molars have three roots that diverge markedly, directly from the cervix of the tooth. It is common to see areas of resorption on the inner aspects of completed deciduous molar tooth roots that are the result of pressure from the developing premolar teeth within the **trifurcation** (or **bifurcation** of lower deciduous molars). The upper first deciduous molar is wide buccolingually but the lower first deciduous molar is narrow buccolingually and elongated mesiodistally. The two buccal cusps lie in the mid-sagittal axis of the lower deciduous first molar so that the buccal surface slopes steeply inwards from the cervix to the cusp tips. Two smaller cusps lie on the lingual side of the tooth and the mesial of these, when unworn, can be extremely pointed and almost 'styloid like' and occasionally even taller than the mesiobuccal cusp which is usually the tallest cusp. Again, like the upper first deciduous molar, there is a pronounced molar tubercle mesiobuccally. Two widely divergent roots, one mesially and one distally, arise directly from the cervix of this tooth.

Upper second deciduous molars resemble upper first permanent molars very closely in their crown morphology but are inclined to be more bulbous, have a small mesiobuccal swelling and have three widely divergent roots. Interestingly, they also have a higher incidence of an extra cusp on the mesiopalatal aspect of the protocone, known as the **cusp of Carabelli**, than do permanent first molars (where the incidence is 50–70% in European populations). Lower second deciduous molars also closely resemble lower first permanent molars. However, while there are five cusps in the former, three buccally and two lingually, the buccal cusps are all similar in size, unlike the condition in permanent molars, and the lingual cusps of the crowns of deciduous teeth are more conical than those of permanent first molars. The mesial and distal aspects of the crown flare markedly from the narrow cervix and the buccal aspect is more bulbous than it is in permanent molars. The widely divergent mesial and distal roots of the lower second deciduous molar are another feature that distinguish them from permanent molars.

Permanent teeth

PERMANENT INCISORS
Maxillary permanent incisors are larger than mandibular incisors (Fig. 8.9). They have conical tapering roots, whereas lower incisors have roots that are flattened mesiodistally and are more 'blade-like'. Unlike deciduous incisors, permanent central incisors have three mamelons (small cusps), along the incisal edge when they first erupt. These mamelons often quickly wear away as do the regular surface incremental markings known as perikymata that are also easily visible on newly erupted modern human permanent anterior teeth. The cervical margin of permanent incisors (and canines) is very sinuous and rises towards the incisive edge on the mesial and distal aspects of the tooth. The mesial

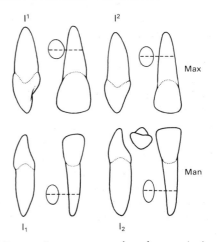

FIGURE 8·9 _Permanent modern human incisors._

cervical margin usually rises more than the distal and this is a way of distinguishing a left tooth from a right tooth. The marginal ridges of permanent incisors may be weakly or strongly developed. Prominent ridges are common in some modern populations (for example in Mongoloids), especially in the maxillary incisors and are known as 'shovel-shaped' incisors. The upper lateral, or second, incisor is smaller and less tall than the central incisor but like it and all other incisors, it has a sharp angular mesial corner and a rounded distal incisal corner. The cingulum of the upper lateral incisor is less pronounced than that of the central incisor but because of a more deeply concave palatal aspect to the tooth there is commonly a pit or palatal fossa that is bounded laterally by the marked marginal ridges that are continuous with the cingulum.

Mandibular permanent central incisors are smaller than mandibular lateral incisors, indeed they are the smallest of all the permanent teeth. The cingula and marginal ridges of lower permanent incisors are less well developed than those of maxillary incisors. The crowns of lower lateral incisors are wider and more 'fan shaped' than lower central incisors but do not slope downwards to the distal in the way that deciduous lower lateral incisors, or permanent great ape lateral incisors do. They tend also to have less-distinct mammelons when they first erupt. The crowns of lower lateral incisors are twisted distally to follow the curve of the dental arch and this is another feature that can be used to distinguish left from right as well as central from lateral incisors.

PERMANENT CANINES

Permanent canines are stout teeth and have longer roots than incisors (Fig. 8.10). The upper canine has a crown equal in height to the central incisors but the lower canine has the tallest crown of all permanent teeth, exceeding all the incisors and the upper canine in crown height. The upper canine is pointed when unworn and has a longer *distal* sloping

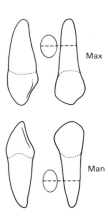

FIGURE 8·10   *Permanent modern human canines.*

edge (not mesial as in the deciduous upper canine) even though the point of the cusp is in the mid-axis of the tooth. This reflects the more expanded distal margin of the tooth. Two strong marginal ridges and a mid-line elevation on the palatal aspect of the upper canine define two shallow hollows, the mesial and distal **palatal fossae**.

The lower canine is narrower mesiodistally than the upper canine, but like the latter it has a shorter mesial slope from the cusp tip. In fact, the mesial aspect of the tooth crown is almost in a direct straight line with the mesial root face. In comparison with the upper canine, the lower has a poorly defined cingulum and weaker marginal ridges so that the lingual fossae are not so well demarcated. Very occasionally the lower permanent canine has a double root.

PERMANENT PREMOLARS

Premolars replace the deciduous molars and so, like permanent incisors and canines, are successional teeth (Fig. 8.11). More often than not premolars have two cusps and are therefore sometimes referred to as **bicuspid** teeth. All the premolars have a single root with the exception of the upper first premolar which has two roots, a buccal root and a

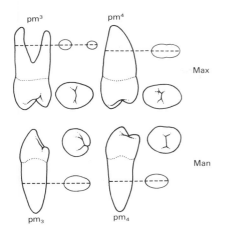

FIGURE 8·11 *Permanent modern human premolars.*

palatal root. The crowns of mandibular premolars have a marked lingual inclination, whereas the crowns of upper premolars are placed centrally on the root. In the mandible the lower first premolar is *smaller* than the second, but in the maxilla the upper first premolar is *larger* than the second. Besides having two roots the upper first premolar is distinguished by a depression on the mesial aspect of the root face and cervix. This is called the **canine fossa** because, it is alleged, the developing canine presses against the first premolar tooth germ during its development. Of the two cusps the buccal cusp is larger and the palatal cusp is positioned just mesial to the mid-bucco–palatal line. At the mesial marginal ridge of the upper first premolar, the fissure that divides the two cusps passes over the ridge to merge in the concave mesial face of the tooth. This, its two roots and the canine fossa are diagnostic of the upper first premolar. The upper second premolar is smaller than the first and has only one root. Both cusps are of near-equal size and are centrally placed with respect to each other so that the outline of the occlusal surface is oval when viewed from above. There is no fissure interrupting the mesial marginal ridge and no canine fossa. The mesial slopes of the

maxillary canine, first premolar and second premolar are, short, long and short, respectively.

The lower first premolar is the smallest of the human premolars and has only a small lingual cusp (occasionally it is scarcely bicuspid, closely resembling the canine). Occlusally, the tooth is almost circular in outline and a ridge from the buccal cusp (which is tilted lingually) divides the occlusal surface into two fossae mesially and distally. In general, the mesial fossa is smaller than the distal fossa. Commonly a mesiolingual groove (the **canine groove**) passes from the mesial occlusal fossa towards the cervix of the tooth. The second mandibular premolar is a larger tooth with the lingual cusp (of which there are sometimes even two) being more equal in height to the buccal cusp. The outline of the tooth is more square when viewed occlusally. A median fissure divides the buccal and lingual cusps and curves around the large base of the buccal cusp. In maxillary premolars the mid-line fissure always passes in a straight line from mesial to distal. When there are three cusps, three fissures radiate out from a central pit and the lingual of these fissures may then continue to create a groove over the lingual aspect of the tooth surface. Sometimes a prominent ridge that runs from the buccal to the lingual cusp in the bicuspid lower second premolars divides the central fissure creating two fossae occlusally as in the first lower premolar. When this occurs the mesial fossa is again commonly the smaller of the two.

PERMANENT MOLARS

Maxillary molars usually have four cusps (Fig. 8.12). Molars are the only teeth to have more than one buccal cusp but they often have an additional cusp associated with the mesiopalatal aspect of the protocone, the cusp of Carabelli. Moving distally along the molar tooth row there is a tendency for the hypocone (the distopalatal cusp) to become reduced in size and occasionally this cusp may be completely absent in which case the

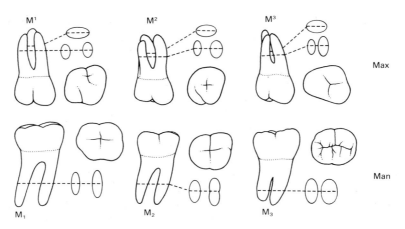

FIGURE 8·12 *Permanent modern human molars.*

third molar is tricuspid. The occlusal outline of the maxillary molars reflects this reduction of the hypocone through the molar row distally, with the upper first molar being quite square in outline but with the upper third molar being nearer triangular in occlusal outline (Fig. 8.13). The ridge between the protocone (mesiopalatal cusp) and the metacone (distobuccal cusp) is prominent in maxillary molars and separates the hypocone from the rest of the occlusal basin; this is known as the **oblique ridge** and separates the trigon anteriorly from the talon posteriorly. All maxillary molars have three roots, the largest of which is the palatal root, although two or even all three of these roots may be fused. Permanent molar tooth roots arise from a root trunk that is variable in length but which is greater than in any of the deciduous molar teeth and and which in extreme cases may be taurodont. Taurodont teeth (see Chapter 8) have a greatly extended common root trunk (and pulp chamber within), prior to the bi- or tri-furcation. Upper first permanent molar teeth have more widely divergent roots than upper second molars, which in turn have more divergent roots than upper third molar teeth. In fact, commonly, the three roots of upper third molars are fused together and are much

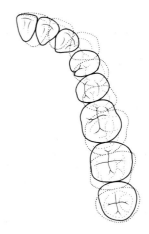

FIGURE 8·13 *The lower permanent modern human teeth (solid outlines) are shown in occlusion with their upper counterparts (dotted outlines).*

shorter than those of the first and second upper molars.

Mandibular molars have a primitive five-cusp morphology that may sometimes be reduced to four (Fig. 8.12). Only two cusps of the original primitive trigonid remain in lower molars, the protoconid (mesiobuccal cusp) and the metaconid (mesiolingual cusp), whereas all three (including the paracone) are still present in upper molars. The distal part of lower molars, the talonid, comprises the hypoconid buccally, the entoconid lingually and the hypoconulid distally. This pattern of two cusps buccal and one distal or distobuccal and two lingual creates a fissure pattern at the base of the cusps that resembles the letter 'Y' with the two short limbs pointing buccally and the single limb pointing lingually. As this pattern of fissures is identical to that in the molars of the Miocene ape *Dryopithecus*, it has become known as the *Dryopithecus* Y pattern (see later and Fig. 8.30).

Unlike lower second deciduous molars, the buccal cusps of lower permanent molars are not all the same size. The protoconid (mesiobuccal cusp) being the largest buccally and the hypoconulid the smallest. Overall, the lower first molar is longer mesiodistally than it is broad buccolingually. It has only two roots, one mesial and one distal, which arise from a short root trunk. There is a tendency for the two roots to become closer through the molar row distally along the tooth row with fusion of the third molar roots being very common. The lower second molar typically only has four cusps, the hypoconulid being lost, and this gives the tooth a squarer outline occlusally than the first lower molar. The fissure pattern of this tooth is usually cruciform when there are four cusps and only shows the *Dryopithecus* Y pattern when five cusps are present. The lower third molar is particularly variable in its occlusal morphology and also in its root form. It may have four or five cusps, separate or fused roots but these tend to be shorter than the first two lower molar tooth roots. In European populations the third molar crown is usually smaller than the second molar crown but in many African populations and Australo–Melanesian populations the third molar crown is commonly the largest of the molar series.

Comparative hominoid dental morphology

The shape of the dental arch in hominoids

Many general factors influence the form of the mandibular and maxillary dental arches, including: the width of the cranium at the glenoid fossa; the relative length of the jaws; the width of the mandible in the symphyseal region; the length of the posterior tooth rows; the size of individual teeth; and the degree and proclination of the teeth and the associated alveolar bone that supports them. Most important also are the balance of soft-tissue forces from the tongue, lips and cheeks that act to maintain the teeth in a position of equilibrium that is established during growth.

The form of the dental arcade in modern man has been described as parabolic or rounded as compared with the U-shaped or straight-sided dental arch of the great apes (Le Gros Clark, 1950). The large incisors of adult great apes are placed relatively further forwards than in adult hominids so that the posterior margins of the incisor sockets lie well in advance of a transverse line joining the anterior margins of the canine sockets (Fig. 8.14). This is not usually the case in the majority of fossil hominids although some specimens of *Australopithecus afarensis* tend towards the ape condition. Juvenile apes and humans all have more rounded parabolic dental arches than adult great apes and it is during the growth period that the shape of the dental arch of apes deviates from this form as the mandible and posterior tooth row lengthen and the permanent incisors and molars come into occlusion.

The mandibular dental arches of *A. afarensis* are narrow anteriorly but the molars and premolars form straight rows that give the arcade an overall V-shaped appearance (Johanson and White, 1979). The mandibular and maxillary dental arcades in *Paranthropus* and *Australopithecus africanus* differ little from one another and are more parabolic

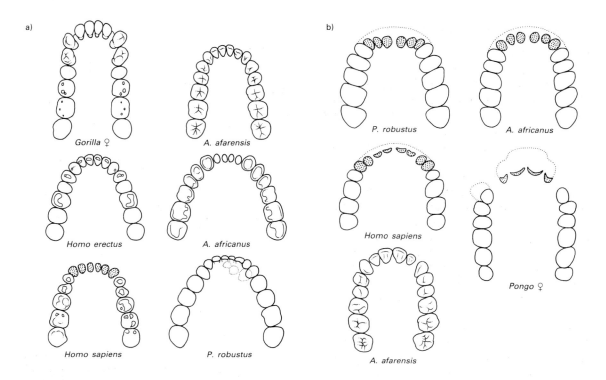

FIGURE 8·14 *In (a) the lower dental arcade of a female gorilla can be compared with several early fossil hominids and a modern human. In (b) the upper dental arcade of a female orang-utan is compared with that of three early hominids and a modern human dental arcade.*

than those of *A. afarensis*. The curve of the incisors and canines is more flattened across the front of the arch in *Paranthropus* and this is more marked across the alveolar bone which lies more or less on a straight line between the canines (Robinson, 1956). The dental arch of *Homo* is wider between the premolars than it is in *Australopithecus* giving the outline a more rounded contour anteriorly. A reduction in the width of the body of the mandible has also occurred in *Homo*. Leakey et al. (1964) noted that the width across the floor of the mouth between the premolars and first permanent molars was greater in *Homo habilis* than in *Australopithecus*. Robinson (1966) re-emphasized that he had described the width across the floor of the mouth in earlier hominids as less than

the width of the mandibular body measured at the first permanent molar or the second premolar.

The shape of the lower border of the corpus or body of the mandible, however, is determined by the contour of the basal bone of the mandible and not by the contour of the dental arch and alveolar bone. As such, it is somewhat independent morphologically. Scott (1957) demonstrated that while there is great variation in the shape of the dental arcade, the lower border of the mandible in almost all mammals conforms closely to the shape of a catenary curve (i.e. the curve formed by a fine chain suspended from each end). The length between the suspended ends of the chain is then a way of recording the shape of this curve.

Comparative hominoid tooth morphology

Hominoid tooth morphology is very complicated. This review begins with a comparative account of the deciduous teeth of hominoids and highlights some of the characteristics that distinguish apes from hominids, as well as some of the characters that distinguish each of the major fossil hominid taxa from each other. The emphasis in this account on deciduous tooth morphology is on how similar teeth compare across groups of hominoids and not so much on those aspects of morphology that typify a particular taxon. The account of the permanent teeth that follows is arranged by tooth type. Incisors, canines, premolars and then molars are discussed here in a comparative context.

Deciduous teeth

Four important morphological complexes distinguish the deciduous dentition of the great apes from those of the hominids.

First, the deciduous incisors of great apes (like the permanent incisors) have bigger, especially taller, crowns that are proclined more anteriorly in the alveolar bone than are deciduous hominid incisors (Fig. 8.15). Too few early hominid deciduous incisors are completely preserved to make meaningful comments about variation between groups of hominoids, but by and large it is difficult to distinguish deciduous incisors of early hominids from those of modern *Homo sapiens*.

Second, deciduous great ape canines (both upper and lower) are morphologically distinct from those of early and modern hominids (Bronowski and Long, 1952). Great ape deciduous canines (in both males and females), are very much taller and more pointed than hominid deciduous canines and project well above the level of the occlusal plane of the other teeth. Great ape deciduous canines have a *concave* posterior border extending down to the distal cingulum which extends backwards to form a talon or talonid

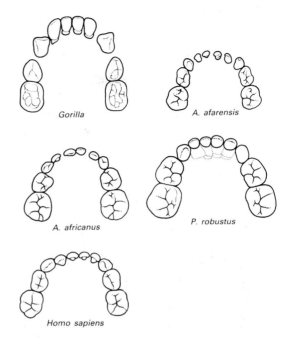

FIGURE 8·15 *Occlusal view of the deciduous dentition of an infant gorilla compared with those of three early hominids and a modern human infant.*

that projects beyond the level of the rest of the crown (Le Gros Clark, 1950). Hominid deciduous canine crowns, on the other hand, are small and spatulate and broaden out from the cervix towards the middle of the crown where they are widest. They are therefore rounded or *convex* above the cingulum distally and are usually in contact with the first deciduous molars posteriorly. Hominid deciduous canines do not project beyond the occlusal plane in the way that great ape deciduous canines do. Grine (1985b) has drawn attention to the variation in deciduous canine morphology among early hominids. Deciduous canines attributed to *Australopithecus*, from Hadar, Laetoli, Taung and Sterkfontein, are, like their permanent counterparts, larger than those from Swartkrans, Kromdraai and those of *Paranthropus boisei*. In addition, Grine (1985b) notes several other differences in deciduous canine

145

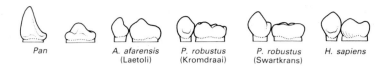

Pan   A. afarensis (Laetoli)   P. robustus (Kromdraai)   P. robustus (Swartkrans)   H. sapiens

FIGURE 8·16   *Lateral view of the deciduous canine and deciduous first molar of an infant chimpanzee compared with three early hominids and an infant modern human. The chimpanzee and the specimen from Kromdraai are seen from the lingual side, the other three specimens are viewed buccally. After Le Gros Clark (1950).*

morphology including the asymmetrically positioned occlusal apex of the lower deciduous canine crown in specimens from Swartkrans and of *P. boisei*. In these specimens, the highest point of the crown is towards the mesial aspect of the tooth and not midway as in other hominids from Laetoli, Hadar, Sterkfontein, Taung and Kromdraai.

A third major morphological distinction between the great ape deciduous dentition and that of hominids relates to the size of the deciduous canines. Marked diastemata, pre-canine in the maxilla and post-canine in the mandible, accommodate the projecting deciduous canines of great apes. No hominids are known that share this feature with juvenile great apes in the mandibular deciduous dentition although (see below) the permanent canines of *Australopithecus afarensis* are associated with diastemata in some cases and more of a pre-canine space occurs in the maxilla of some hominids such as Taung and Sangiran 4 than does a post-canine space in the mandible. Le Gros Clark (1950) notes that, like modern humans, the mandibular deciduous canines of the majority of juvenile early hominids are in direct contact with both the deciduous incisors anteriorly and the deciduous first molars posteriorly.

The fourth important morphological difference between the deciduous teeth of great apes and those of hominids is the form of the first deciduous molars, especially the lower dm$_1$ (Fig. 8.16). In great apes this is a sectorial tooth (occluding with the distal aspect of the upper deciduous canine) in which the crown is dominated by a conical

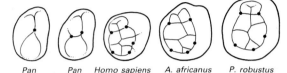

Pan   Pan   Homo sapiens   A. africanus   P. robustus

FIGURE 8·17   *Occlusal view of the deciduous lower first permanent molars of two infant chimpanzees compared with a modern human deciduous first molar and those of two early hominids. After Robinson (1956).*

protoconid. Behind this, but on a much lower level, is a shallow sloping talonid basin. There is sometimes a small rudimentary metaconid on the lingual slope of the protoconid and along the margins of the talonid there may be small tubercles that represent the entoconid and hypoconid (Le Gros Clark, 1950). No hominid early or modern shares this ape-like form of the lower deciduous first molar, as all are multicuspid or molariform (Fig. 8.17). Usually, hominid lower first deciduous molars have four cusps (but there can be five). All these cusps are at a similar occlusal level and there is always a well-marked anterior fovea. Besides the clear-cut distinction between great ape and hominid lower first deciduous molars, there is considerable variation in deciduous molar tooth form between different groups of early hominids. The following account of this variation is for the most part abbreviated to a description of trends that underlie deciduous molarization.

MAXILLARY DECIDUOUS MOLARS

The upper deciduous first molar of great apes is essentially bicuspid with a larger cusp buccally and a smaller one palatally, which lies at a lower level. Upper second deciduous molars in great apes are morphologically much more similar to their hominid counterparts than are the upper first deciduous molars. Grine (1985a) has documented the differences that exist in deciduous tooth morphology among early hominids. There is a trend in the palatal bevel (or slope) of the protocone of the upper first and second deciduous molars of early hominids that reflects an increase in size (or the 'degree of inflation') of this cusp (Fig. 8.18). In hominid specimens from Hadar and Laetoli, the palatal aspect of the protocone is strongly bevelled. Deciduous molars from Sterkfontein and Taung are, however, moderately bevelled but those from Swartkrans and those of _Paranthropus boisei_ are more rounded in outline reflecting an increase in the relative and absolute size of this cusp over and above that in other hominids. The paracone in upper deciduous molars from Laetoli, Hadar, Taung and Sterkfontein is larger than the metacone, but in robust australopithecines from Swartkrans and East Africa the metacone is enlarged relative to the paracone such that these two cusps are nearly equal in size. Robinson (1956) also notes that specimens of _Australopithecus_ can be distinguished from those of _Paranthropus_ by a well-developed buccal limb of the anterior fovea of the dm¹, which has a well-developed accessory cusp mesial to it, whereas those of _Paranthropus_ do not (Fig. 8.19). The overall occlusal outline of the maxillary deciduous second molars is also distinct in _P. boisei_ (Fig. 8.20). Here, the mesiodistal diameter of the crown exceeds the buccopalatal diameter. Other hominids from Laetoli, Hadar, Taung, Sterkfontein and Swartkrans have more quadrangular dm² crown outlines where the buccopalatal diameter exceeds the mesiodistal diameter. Grine (1985b) has also re-affirmed Robinson's observation that the hypocone in the maxil-

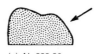

(a) AL 333-86   (b) OH 30

FIGURE 8·18 _Cross-sections through the mesial cusps of two deciduous upper second molars (a) AL 333-86 and (b) OH 30. Besides the difference in the absolute size between these teeth, the mesiopalatal cusp or protocone (arrowed) of OH 30 is relatively larger, more inflated and has a more rounded outline (is less bevelled) than the smaller protocone of AL 333-86._

A. afarensis   A. africanus   P. robustus

FIGURE 8·19 _Upper first deciduous molars from Hadar (AL 333-86), Sterkfontein (Sts 24) and Swartkrans._

AL 333-86   OH 30

FIGURE 8·20 _Two upper second deciduous molars from Hadar (AL 333-86) and Olduvai (OH 30)._

lary deciduous molars (as well as the permanent M¹s) of the Taung child are reduced in size and that this gives these teeth a more curved distopalatal outline than in other early hominids.

MANDIBULAR DECIDUOUS MOLARS

Among the mandibular deciduous molars of early hominids there is variation in the relative disparity of cuspal heights, in the relative size of cusps, and in the positions of certain cusps relative to others on the occlusal aspect of the teeth (Grine, 1985a).

First deciduous molars from Hadar and Laetoli have protoconids and metaconids that are considerably higher than the hypoconid and entoconid (Fig. 8.16). Deciduous first molars from Taung and Sterkfontein also show some disparity in cusp height but not to this degree. This difference in relative cusp height is reduced somewhat in the first deciduous molars from Kromdraai, but in those from Swartkrans and in the first deciduous molars of *Paranthropus boisei* the heights of both the talonid and trigonid are nearly equal.

As tooth cusp heights become more equal, so the size of the metaconid increases through the same series of early hominids. In all specimens attributed to *Australopithecus* (from Laetoli, Hadar, Taung and Sterkfontein) the protoconid remains the largest cusp of the lower deciduous first molar. In the hominids from Kromdraai, however, the metaconid is equal in size to the protoconid and in specimens from Swartkrans and of *P. boisei* the metaconid is the largest cusp on the dm_1. In the most molarized specimens the hypoconid is also larger than the protoconid and a buccal fissure that seems to be unique to *P. boisei* passes between the protoconid and hypoconid and runs across the cervical enamel margin.

Whereas the protoconid on the dm_1 is positioned mesial to the metaconid in *Australopithecus*, it is aligned transversely with the metaconid in *Paranthropus* (because of the expansion of the metaconid and hypoconid). Similarly, the second deciduous molars in specimens attributed to *Australopithecus* from Laetoli and Hadar show a protoconid mesial to the metaconid, while in the second deciduous molars from Swartkrans and those of *P. boisei* (the hominid with the most molarized deciduous teeth) these cusps lie side by side in the buccolingual plane.

These trends in relative cusp size and position result in the trigonid of the lower dm_1 in *Australopithecus*, being dominant over the talonid in mesiodistal length. In *Paranthropus*, however, the talonid of the lower dm_1 is relatively long, being dominant or at least equivalent to the length of the trigonid (Grine, 1985a).

Clearly, great variation exists in the morphology of the upper and lower deciduous dentitions among different groups of early hominids. Nevertheless, their overall general morphological similarity to those of later hominids, for example the Neanderthal child from Devil's Tower (Robinson, 1956), and to modern human deciduous teeth, serves to emphasize that the distinction between hominids and great apes is greater than any that exist between early hominid taxa.

Permanent teeth
PERMANENT INCISORS
Great ape incisors are renowned for their large size relative to the smaller posterior teeth, but this is especially pronounced in *Pan* and *Pongo* where the post-canine teeth are smaller than those of *Gorilla* (Fig.8.21). Permanent upper central incisors in great apes exceed the permanent upper lateral incisors both in mesiodistal width and in crown height. Permanent lower central incisors in great apes are slightly smaller in mesiodistal width than are the permanent lower lateral incisors. While all permanent hominid incisors are essentially similar to those of modern humans, those of *Australopithecus afarensis* are larger relative to the size of their posterior teeth than in any other early fossil hominids (Johanson, 1985 (Fig. 8.22). It is notable, however, that the anterior teeth of some specimens of *Homo* (KNM-WT 15000 and KNM-ER 1590) are also quite large relative to their posterior tooth size.

Both the upper central and lateral permanent incisors of *Australopithecus* and *Paranthropus* (indeed most early hominids) tend to be shovel-shaped with well-defined marginal ridges, but not to the same degree as those of *Sinanthropus* (Robinson, 1956). In comparison with other hominids, the

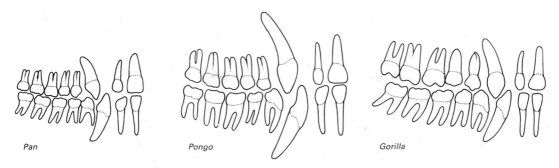

FIGURE 8·21 *The teeth of* Pan *(female),* Pongo *(male) and* Gorilla *(female) from one upper and one lower quadrant of the mouth viewed from the buccal.* *Note the large anterior teeth, sexually dimorphic canines and sectorial lower first premolars.*

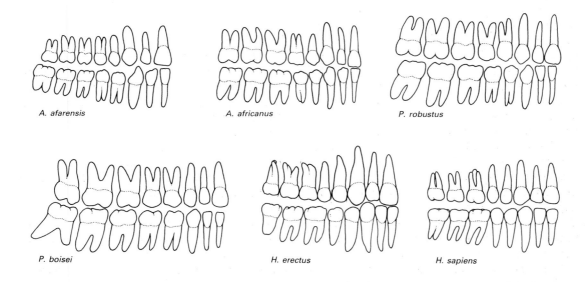

FIGURE 8·22 *The teeth of modern* Homo sapiens *compared with the teeth from five early hominid taxa. Composite dentitions have been assembled for each hominid taxa for one upper and the occluding lower quadrant of the mouth. Note the relatively* *large anterior teeth in* Australopithecus afarensis *and* Homo erectus, *the relatively small anterior teeth in* Paranthropus *as well as the three rooted upper premolars and large molars in* Paranthropus.

upper lateral incisors of *Paranthropus* tend to be greatly reduced in size with a markedly rounded and reduced distal incisal margin. However, this is also true of great apes, especially *Pongo*. It is most marked among the fossil hominids in *Paranthropus boisei* where permanent upper lateral incisors

may even resemble deciduous teeth and yet also be crowded in the arch (KNM-WT 17400). The lower permanent incisors of *Paranthropus* are most like those of modern humans in size and form but are distinguished from the latter in that the central mamelon is commonly at a lower level than the other two. Lower central incisors of *Australopithecus* are somewhat taller teeth than those of *Paranthropus* and, although only a few unworn specimens are preserved, these tend to have more mamelons (five or six) along their incisal edge. The lower lateral incisors of *Australopithecus* (e.g. Sts 24) are more pongid-like than are those of other hominids in that the distal aspect of the incisal edge slopes strongly downwards. This is reminiscent of the form of modern human deciduous lower lateral incisors and of great ape permanent upper (and often lower) lateral incisors but not of human permanent lower lateral incisors.

PERMANENT CANINES

The permanent canines of great apes are large and sexually dimorphic, most strongly so in the gorilla. Canine size in primates also scales in a positive allometric manner with body size so that with an increase in body size there is an even greater accompanying increase in canine size (Corruccini and Henderson, 1978; Wood, 1979). Unworn ape canines are tall and sharp and project beyond the occlusal plane such that there are wide diastemata in the mandibular and maxillary dental arches to accommodate them. Ape canines wear in a distinct way with elongated facets appearing along the distal aspect of the lower canine and on the mesial aspect of the upper canine. The upper great ape canine functions with the lower canine and sectorial premolar. The widest part of ape canines occurs at the base of the tooth crown and, as in deciduous ape canines, there is a concave posterior border and a distinct talonid extension distally, best developed in the lower canine. Johanson (1985) noted that the permanent canines and first premolars of *Australopithecus afarensis* project above the occlusal plane of worn posterior teeth and that there are pongid-like wear facets on these canines (Fig. 8.23). In addition, nine out of 20 specimens attributed to *A. afarensis* show evidence of a lower-canine/first-premolar diastema. These features are reported to contrast with the conditions observed in *A. africanus* where only one specimen out of 12 shows evidence of a diastema and where all the permanent teeth wear flat. There is then good evidence that the canine/first-premolar dental complex retains some ape-like affinities in *A. afarensis*.

In general, hominid permanent canines are spatulate in form and are larger mesiodistally in the mid-crown region rather than across the base. They have a rounded basal tubercle that forms a cingulum rather than an extended distal talonid and wear down flat at the tip to expose a diamond-shaped island of dentine occlusally. (Incisor teeth wear down to expose a rectangular-shaped area of dentine occlusally and this is a useful way of distinguishing some worn anterior teeth.) Upper canine teeth are quite symmetrical, with the apex of the cusp positioned midway between the mesial and distal aspects. The upper canines of *Australopithecus* are more pointed than those of *Paranthropus* and tend to have a raised mid-line ridge palatally, as well as

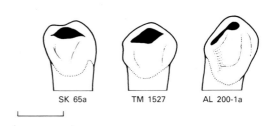

SK 65a   TM 1527   AL 200-1a

FIGURE 8·23 *Patterns of wear on three early hominid permanent canines. After Johanson and White (1979).*

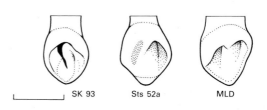

SK 93   Sts 52a   MLD

FIGURE 8·24 *Permanent maxillary canines from Swartkrans, Sterkfontien and Makapanskat viewed palatally. After Robinson (1956).*

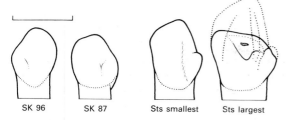

FIGURE 8·25 *Large and small permanent mandibular canines from Swartkrans compared with large and small permanent mandibular canines from Sterkfontein, viewed buccally. After Robinson (1956).*

two parallel palatal grooves (Fig. 8.24). The upper permanent canines of *Paranthropus*, however, have strong marginal ridges that recall the shovel-shaped incisor morphology. Lower permanent canines are strongly asymmetrical, and those of *Australopithecus* develop a cusp on the distal edge that slopes cervically more markedly than those of *Homo* or *Paranthropus* (Fig. 8.25). The lower canines of *Paranthropus*, like the upper canines, are smaller than those of *Australopithecus* (and also smaller than those of *Homo erectus*) (Weidenreich, 1937; Robinson, 1956; Wood and Stack, 1980) and generally closely resemble modern human canines in morphology.

Wolpoff (1978) claimed that canine size is inversely proportional to molar size in hominids, so that it would be 'normal' for canines to become smaller as body size and molar size increased in *Paranthropus*. However, Wood and Stack (1980) demonstrated that canine size in *Paranthropus* departed significantly from the positive allometric canine trend seen in other primates and so argued that it is probably a dietary specialization that links large cheek teeth with small anterior teeth.

PERMANENT PREMOLARS

The form of the lower first permanent premolar tooth in pongids is distinctly different from the majority of hominids. It is a sectorial tooth that occludes on its anterola-

teral surface with the upper permanent canine. The crown consists mostly of a greatly hypertrophied protoconid which is conical and pointed so that the whole tooth resembles a canine. Occasionally, a small metaconid is present on the lingual slope of the protoconid. There can then be no well-defined anterior or posterior fovea on the occlusal aspect of this tooth. Great ape lower first premolars have three roots; a strong mesiobuccal root that is continuous with the mesiobuccal aspect of the crown and two distally, one lingual and one buccal that may in fact be partially or completely fused. Lower first premolars of *Australopithecus afarensis* stand apart from other hominids in that they are commonly unicuspid (six out of 15 lack metaconids) (Johanson, 1985), and are oval in occlusal outline (Fig. 8.26). Wood et al. (1988) also noted that the root system of lower first premolars attributed to *A. afarensis* is relatively unchanged from the primitive condition seen in apes.

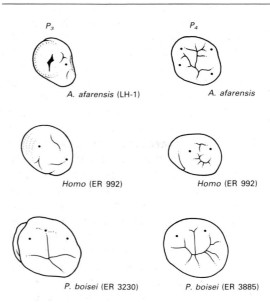

FIGURE 8·26 *(a) occlusal view of three lower permanent first premolars belonging to LH-1, KNM ER 992 and KNM ER 3230. (b) occlusal view of three lower second permanent premolars belonging to AL-400-1, KNM ER 992 and KNM ER 3885.*

Lower premolars of other early hominids are very variable in form. Robinson (1956) noted that in buccal view the premolar crowns of *Paranthropus* are near equal in width cervically and occlusally, whereas those of *Australopithecus* constrict more cervically. In addition, buccal grooves are more pronounced than lingual grooves in hominid premolars, although the mesiobuccal grooves in *Australopithecus* are deeper than those in *Paranthropus*. Robinson (1956) also noted that the lower first premolars of *Australopithecus africanus* and *Paranthropus* are asymmetric in occlusal outline with a large buccal cusp and a smaller lingual cusp. This lingual cusp lies on the same buccolingual axis or is shifted *mesial* to it as it is in modern human lower premolars. Wood and Uytterschaut (1987) have confirmed this together with the fact that lower first premolars attributed to early *Homo* from East Africa have lingual cusps either alongside the buccal cusp or *distal* to it. The lower first premolar teeth of early *Homo* also have the narrowest lower first premolar crowns of all early hominids (the broadest, i.e. the most buccolingually expanded relative to their length, being those of *A. africanus*). This was one fact that contributed to the description which led

Leakey et al. (1964) to propose *Homo habilis* as a new species distinct from *A. africanus*.

In hominids, the occlusal outline of the second lower premolar is not so markedly asymmetrical as the first since it is expanded in size. Expansion of the talonid (i.e. molarization) is most exaggerated in both the first and second lower premolars of *Paranthropus boisei* such that, overall, premolar tooth size is greater in this taxon than in other early hominids (Fig. 8.27). The talonid areas of *Paranthropus* in southern Africa are, in turn, larger than those of *Australopithecus* and account for their larger crown areas. Interestingly, the talonid in lower second premolars attributed to *Paranthropus* from southern Africa is more elongated mesiodistally than in East African robust australopithecines, which appear to be elongated buccolingually (Wood and Uytterschaut, 1987). Wood and Uytterschaut (1987) argued from the results of their study that expansion of the talonid is not to be expected as a simple consequence of increase in body size and must be seen as a special adaptation in these early hominids.

The root system of lower premolars in *Australopithecus afarensis*, as noted above, is

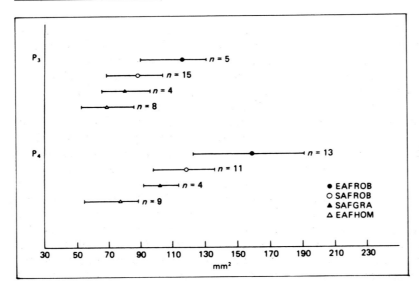

FIGURE 8·27 *Means and ranges of the measured crown base areas of the major taxonomic categories of early hominid mandibular premolars. After Wood and Uytterschaut (1987). EAFROB, East African* Paranthropus; *SAFROB, South African* Paranthropus; *SAFGRA, South African* Australopithecus; *EAFHOM, East African* Homo.

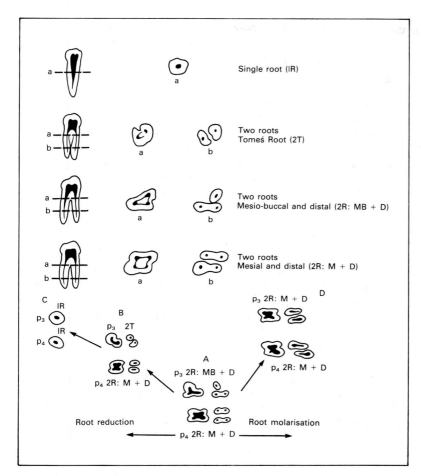

FIGURE 8·28 *Cross-sections through lower premolar roots reveal that roots may be single (1R), fused (2T), two rooted with a mesiobuccal and distal root (2R; MB + D) or two rooted with mesial and distal roots (2R; M + D). (A) The presumed primitive pattern of roots in the lower PM3 and lower PM4 that occurs frequently in great apes and Australopithecus afarensis. Premolar root reduction occurs frequently in specimens of Paranthropus robustus (B) and even further in specimens attributed to early Homo (C). On the other hand premolar root molarization is marked in specimens attributed to Paranthropus boisei (D). After Wood et al. (1988).*

similar to that of great apes, but the roots of other early hominid lower premolars are also interesting (Fig. 8.28). Both modern and early *Homo* lower first and second premolars are usually single rooted. The hominids from Swartkrans and Sterkfontein, attributed to *Paranthropus* and *Australopithecus*, respectively, have either double roots or, more commonly, a fused double root known as a **Tomes' root** in the majority of cases (nine out of 13 from Swartkrans) showing a trend to root reduction with respect to *A. afarensis*. However, lower first premolar roots of *P. boisei* show a trend towards two quite separate roots, one mesially and one distally, that are morphologically similar to molar tooth roots. Wood (1988) argued that this

molarized root form in the lower first premolars of *P. boisei* is not a likely continuation of a gradual trend to root reduction, that seems to occur in *Australopithecus* and *Paranthropus* from southern African sites.

The upper premolars of all hominids are bicuspid and are morphologically more similar than other tooth types. Robinson (1956) noted that the upper first premolars were more distinct from the upper second premolars in *Paranthropus* than they were in *Australopithecus*. Wood and Engleman (1988) have confirmed the larger crown areas of upper second premolars over upper first premolars in *Paranthropus*. Upper second premolars of *A. afarensis* are slightly smaller in size than first premolars from the same individual, as they are in modern humans (which as White et al. (1981) note is the primitive hominoid condition). The upper second premolars in *P. boisei* are, however, like the lower second premolars, consistently larger than the first but the size discrepancy between the two is less than for southern African specimens attributed to *Paranthropus* (10% and 22%, respectively) (Wood and Engleman, 1988). The tendency in early *Homo* is again for the second upper premolar to be equal or reduced in size relative to the first. Both palatal and buccal cusps lie either in the mid buccopalatal line or have a tendency for the palatal cusp to be positioned more mesial to this line in *Australopithecus* and *Paranthropus*. The buccal cusp is consistently slightly larger in hominid upper premolars but the buccal face of upper premolars in *Paranthropus* enlarge towards the cervical margin as they do in the lower premolars, whereas those of *Australopithecus* narrow somewhat cervically (Fig. 8.29). In *Paranthropus* the upper second premolar, being more molarized, has more of a distinct talon.

The root system of the upper premolars attributed to *Paranthropus* reflects their molarized crown form and, as such, the roots are robust and usually number three in total, with two buccal roots and one palatal root (Fig. 8.28). Upper premolars of *Australopithecus* on the other hand, usually only have one buccal root and one palatal root. It is notable that the specimen of early *Homo* from the Sterkfontein extension site (Stw 53) also has three-rooted upper first and second premolars (Hughes and Tobias, 1977), which is not a common finding in *Homo*.

PERMANENT MOLARS

Lower molars in early hominids are morphologically very similar. However, they may be distinguished from one another in several ways. One notable difference is their overall size (Fig. 8.30). Lower molars of *Paranthropus* are larger than those of *Australopithecus africanus*, which in turn are larger than those of *Australopithecus afarensis* and *Homo* (the latter two being similar in size). The lower first permanent molar is consistently smaller than the second molar in all early

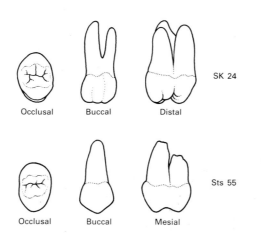

FIGURE 8·29 *(a) Occlusal, buccal and distal views of a maxillary first premolar from Swartkrans; note the three roots. (b) Occlusal, buccal and mesial views of a maxillary first premolar from Sterkfontein; note the double root only and the reduced width of the crown at the cervical margin. After Robinson (1956).*

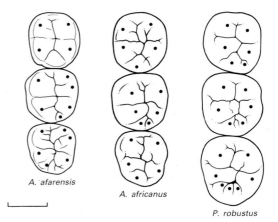

FIGURE 8·30 *The first, second and third permanent molars of specimens attributed to* Australopithecus afarensis *(AL 400-1),* Australopithecus africanus *(Sts 52) and* Paranthropus robustus *(SK 6) illustrating the comparative size of the teeth in the molar tooth row.*

hominids. On average, the third molar tends to increase in mesiodistal diameter relative to the second molar but reduces in average buccolingual diameter relative to the second molar in all early hominids (White et al., 1981). Within the same individual the third molar reduces in size relative to M2 in *A. africanus* but is equal or increases in size in *Paranthropus* and *Australopithecus afarensis*. Over and above relative molar size, however, it is the combination of very large anterior teeth combined with small posterior teeth in *Australopithecus afarensis*, for example, and the very small anterior teeth in *Paranthropus* combined with extremely large posterior teeth that are most usually held to distinguish these hominids (Fig. 8.22).

The lower molars of *A. afarensis* have prominent lingual cusps that appear to wear down less quickly than the buccal cusps. Compared with lower molar teeth attributed to *A. africanus* or *Paranthropus* they have expanded hypoconids that give the buccal aspect of the tooth a 'bilobate' appearance (White et al., 1981). The hypoconulid in *A. afarensis* is more mesially positioned than in other early hominids and this gives the lower first and second molar crowns a more square occlusal outline (Fig. 8.30). The third molar tooth in *A. afarensis* has a triangular occlusal outline that results from its narrow talonid. This tooth tends to have less well-demarcated cusps than its counterparts in other early hominids and is commonly extremely crenulated.

Robinson (1956) has described several features that distinguish lower permanent molars of *A. africanus* from those of *Paranthropus* in southern Africa. Lower molars of *A. africanus* are characterized by a shelf-like cingulum that runs around the protoconid (a similar feature occurs in the maxillary molars) or by a **protostylid** in this position (Fig. 8.31). (Small cusps associated with this protoconal cingulum are known as protostylids.) This is most marked in some specimens such as MLD 2, but is present in some form or other in the majority of

specimens of *A. africanus* with first permanent molars. These molars also have a well-defined anterior fovea separated from the central fossa by a trigonid crest that runs between the protoconid and the metaconid.

Lower first molar teeth of *Paranthropus* almost invariably have a **tuberculum sextum**, or sixth cusp, in the wall of the posterior fovea (Fig. 8.31). While this may occur in the second or lower third molars of other hominids it has only been described as a major feature in the first lower molars of the Taung child outside *Paranthropus*. Other features of the Taung M_1, however, resemble *Australopithecus*; for example, the well-defined anterior fovea and trigonid crest and the *Dryopithecus* Y pattern of fissures across the tooth where the anterobuccal limb of the Y and the lingual limb of the Y are directly in line with each other (Fig. 8.31). This pattern of sutures in *Paranthropus* first molars tends to follow the typical *Dryopithecus* Y pattern more closely, although in lower second molars it often resembles the pattern in *Australopithecus* lower molars. It is probable that the relatively increased area of the anterior cusps in *A. africanus* (and early

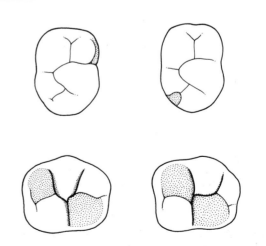

FIGURE 8·31 *Variation in the occlusal anatomy of early hominid lower molars. The protostylid, tuberculum sextum (or C6) and the Dryopithecus Y pattern are illustrated for hypothetical lower M1s.*

Homo) contribute to this difference in fissure pattern (Wood et al., 1983) (Figs 8.32 and 8.33).

It is clear that there is a trend to increase the size of the molar teeth through the series of hominids: *A. afarensis*, *A. africanus*, *Paranthropus robustus*, *Paranthropus boisei*. In addition, it is clear that some features of early *Homo* lower molars can be useful taxonomic indicators. For example, the buccolingual diameters of lower molars attributed to early *Homo* are the smallest of all early hominids, but the average mesiodistal diameter of these teeth actually exceeds those of *A. afarensis* (which are closest to them in overall size). It follows that the molar teeth of early *Homo* can be recognized as being long and narrow (Howell, 1978; White

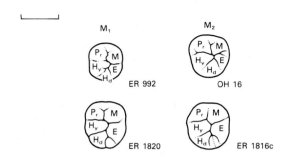

FIGURE 8·32 *Contrasting fissure patterns and cusp areas illustrated for two early hominid lower first permanent molars (KNM ER 992 and KNM ER 1820) and two early hominid lower second permanent molars (KNM ER 1816c and OH 16). After Wood et al. (1983).*

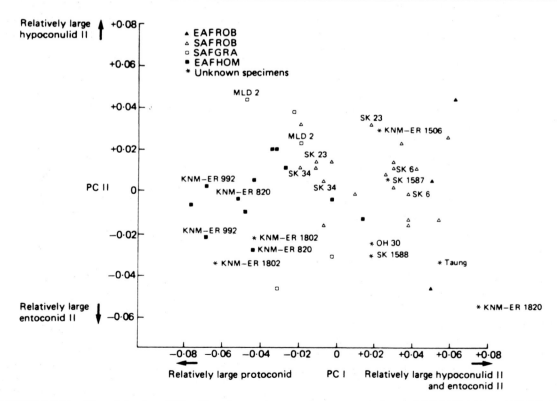

FIGURE 8·33 *Plot of the first (PCm I) and second (PCm II) principal components generated from relative cusp area data of hominid mandibular first molars; note specimens KNM ER 992 and KNM ER 1820 illustrated in Figure 8.32. After Wood et al. (1983).*

et al., 1981). Wood et al. (1983) have demonstrated that within *Paranthropus* there is a trend to have relatively smaller mesial cusps (protoconid and metaconid), and relatively large distal cusps (entoconid and hypoconulid). The trend is most marked in the M1s and least marked in the M3s. It is likely that the presence of a tuberculum sextum (C6) bears some relationship to the expanded talonid in *Paranthropus* and, interestingly, the relative cusp areas in the Taung M_1 fall nearer to the values for *Paranthropus* than to those for *Australopithecus* (Wood et al., 1983). Jungers and Grine (1986) have noted that within *Australopithecus* the buccolingual diameters of both the M1 trigonid and talonid increase faster than does the mesiodistal length, whereas these breadths scale isometrically with tooth lengths in *Paranthropus*. Furthermore, whereas the molars of *Paranthropus* from Swartkrans tend to broaden as length increases, *Paranthropus boisei* is notable for its comparative mesiodistal crown elongation.

The roots of hominid lower molar teeth follow the typical hominoid pattern, there being two: one mesial with a buccal and lingual root canal, and one distally, usually with only one root canal but again not uncommonly with two. Whereas modern human lower third molar tooth roots have a tendency to reduce and become fused, this is not necessarily the case in early hominids which may have markedly splayed distal roots, especially where there is an elongated talonid.

The upper molar tooth row length progressively reduces in the series *Paranthropus* (especially *P. boisei*), *A. africanus*, *A. afarensis* and early *Homo*. The molarized premolars in *Paranthropus* and *A. africanus* make for less discontinuity along the posterior tooth row in these groups of hominids, but there is a marked distinction between premolar and molar size in *A. afarensis* and *Homo* (Fig. 8.34). The majority of hominid upper permanent molars have four main cusps, but there is a tendency in the last

molar tooth (M³) for the metacone (and to a lesser extent the hypocone) to divide into two subequal cusps forming a distal occlusal rim. This occurs more frequently in *A. africanus* and *Paranthropus* than in *A. afarensis* (White et al., 1981). The buccal cusps of upper molar teeth from Hadar and Laetoli tend to be higher than those of other early hominids and remain unworn with respect to the palatal cusps which apparently wear down more quickly. This pattern is reported to be reversed in *Paranthropus* but intermediate in *A. africanus*, with more wear occurring buccally (Smith, 1986).

It is a feature of the upper molar teeth in *A. africanus* (especially the upper second molar), that the palatal aspect of the protocone is enlarged as a strong shelf-like projection resembling a Carabelli complex. While some specimens from Swartkrans (e.g. SK 13) have Carabelli cusps, there is usually only a trace of this complex in specimens attributed

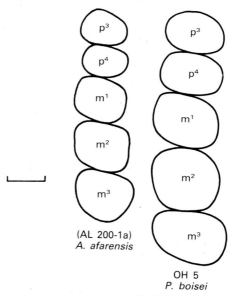

FIGURE 8·34 *Outlines of the permanent upper premolar and molar teeth in the tooth row of AL 200-1a (*Australopithecus afarensis*) and OH 5 (*Paranthropus boisei*); note the relatively small size of the premolars of AL 200-1a and that P4 is smaller than P3 in this specimen.*

to *Paranthropus*. Likewise, only one specimen from Hadar or Laetoli (LH 17) has a strong protoconal cingulum. White et al. (1981) claim that this complex may be indicative of early buccopalatal expansion of the crown in *A. africanus*.

Upper first molars of *Australopithecus* generally have better developed Carabelli traits and anterior trigonal ridges between the protocone and the metacone such that the anterior fovea is clear and distinct from the central fossa. This is also the case in the

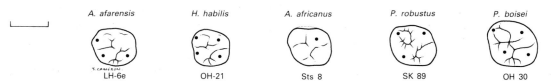

FIGURE 8·35 *Occlusal outlines of maxillary first permanent molars belonging to various early hominid taxa indicate that there is much less morphological variation among these teeth than there is among lower first permanent molars. After Wood and Engleman (1988).*

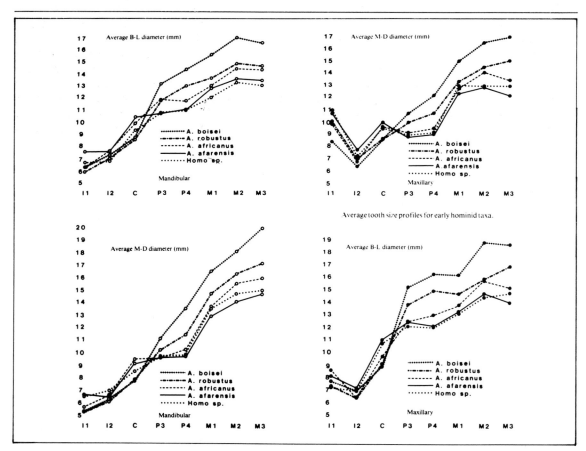

FIGURE 8·36 *Average bucco-lingual and mesiodistal diameters for permanent mandibular and permanent maxillary teeth of five early hominid taxa. After White et al. (1981).*

upper first permanent molars from Taung. Robinson (1956) noted that this is not the case in upper first permanent molars from Swartkrans but that teeth attributed to _Paranthropus_ are nevertheless characterized by well-scored buccal grooves (that end in deep pits) and a distal cuspule (Fig. 8.35). None of these features, however, necessarily hold true for the upper second molars of these hominids. Upper third molars of _Australopithecus_ are generally smaller than the second molars in the same individual (Fig. 8.36), but the upper third molars of _Paranthropus_ from southern African sites tend to be larger than the upper second molars in the same individual. The differential between M^2 and M^3 size is variable in _P. boisei_. The more distal upper molar teeth in early hominids tend to show reduction in the distobuccal angle of the crown outline as the relative size of the metacone decreases (Robinson, 1956: 91), whereas modern humans and modern great apes tend to reduce the hypocone through the upper molar series so that the disto_palatal_ angle of the crown reduces (Swindler, 1976), Johanson and White (1979) noted that the hypocones of the upper molars of _A. afarensis_ are fully developed and do not reduce through the upper molar series. Wood and Engleman (1988) noted that both southern African specimens attributed to _A. africanus_ and East African specimens attributed to early _Homo_ have buccolingually narrower crowns than _Paranthropus_. A summary chart of the average dimensions of the permanent teeth of early hominid taxa is reproduced from White et al. (1981) in Fig. 8.36.

Hominid upper molars have three roots; two buccally (one mesial and one distal) and one very robust root that is orientated palatally. Towards the distal end of the molar series there is a tendency for the roots to become less widely separated and for the third molar roots to be reduced in length. Wood et al. (1988) noted that the lengths of the roots of the upper posterior teeth of early hominids are longer than those of modern humans but that between groups of early hominids there are no significant differences in root length. (Specimens of _A. afarensis_, however, were not represented in this study.)

CHAPTER NINE

THE INTRACRANIAL REGION

The bones of the cranial vault together with the bones of the cranial base form the walls and roof of the **intracranial cavity**. The intracranial cavity is partially divided into compartments by tracts of fibrous **dura mater** which, together with the bones of the cranial cavity both support and protect the brain. The bones of the calvarium and the dural tracts also influence the way in which the brain may grow and expand.

From about the second year after birth in modern humans (Ohtsuki, 1977) or possibly even earlier (Boyde et al., 1990) the bones of the cranial vault develop inner and outer 'tables' of compact bone, between which is an intervening layer of cancellous bone known as the **diploë** (see Fig. 4.9). The diploë is an important site of **erythropoiesis** (red blood cell formation), others being the marrow spaces in the vertebral bodies, the sternum and iliac crests. Not surprisingly then, these bones have a rich nutrient blood supply from the **meningeal arteries** (poorly named, as in fact they supply little blood to the meninges). Some diseases such as anaemias affect the diploë, so that the vault bones may become thickened in response to long-term demand for red blood cell production. Other diseases (such as Paget's disease and syphilis) affect the inner and outer tables as well as the diploë and are easily revealed on radiographs of cranial vault bones. **Diploic veins** also exist within the diploë and drain blood away either to the **intracranial venous sinuses** (see below), or alternatively to the exterior of the vault. The dura mater, one of the meningeal linings that surrounds the brain, is intimately associated with the inner aspect of all the bones of the calvaria, so that the meningeal vessels which lie outside the dura (being then **extra-dural**) are compressed against the inner table of bone and so leave a pattern of grooves in the bone which marks their distribution (see Fig. 4.8). Between some folds of the dura mater, or in other places between the dura mater and the periosteum of the vault bones, a series of venous sinuses drain blood away from the brain, diploë (via diploic veins) as well as from the orbit and parts of the face via opthalmic and other veins. Other small veins pass all the way through the vault bones to communicate with the scalp or veins of the neck or the intracranial venous sinuses and these are known as **emissary veins**. Some emissary veins pass to the exterior of the cranium through their own **emissary foramina**, whereas others join with important nerves and arteries to pass with them through neurovascular foramina in the cranial base.

A great deal can be learned from studying the bony markings created by the various attachments of the fibrous dural tracts or from the grooves and impressions left by the brain, cranial venous sinuses and meningeal arteries on the internal aspect of the bones of the calvarium.

The interior of the modern human cranium

The interior of the cranial cavity is divided into three **cranial fossae** which can be seen most easily when the calotte is removed and the cranium is viewed from above (Fig. 9.1). These are named the anterior, middle and posterior cranial fossae and lie above, behind and below the level of the orbital cavities, respectively.

The anterior cranial fossa

The anterior cranial fossa is enclosed largely by the frontal bone and its orbital processes; these carry the weight of the frontal lobes of the brain and form the roof of the orbital cavities. In the mid-line of the anterior cranial fossa, interposed between the left and right orbital processes of the frontal bone, lies the cribriform plate of the ethmoid bone. The

cribriform plate is perforated with small holes through which olfactory nerves (which carry the sensation of smell to the brain), enter from the roof of the nose (Fig. 9.1). At the front of the cribriform plate is a raised mid-line crest, the **crista galli** of the ethmoid bone. In front of the crista galli there is also a raised **frontal crest** that rises a short distance along the mid-line of the frontal bone. Together, the crista galli and its anterior continuation the frontal crest form the anterior attachments of the **falx cerebri**, a tough sickle-shaped tract of dura that partially separates the right and left cerebral hemispheres of the brain. The posterior part of the floor and edge of the anterior cranial fossa is formed by the lesser wings of the sphenoid bone which extend across the floor of the cranial cavity from *pterion* to *pterion*. Occasionally, the pattern of sutures in the anterior cranial fossa varies and the frontal

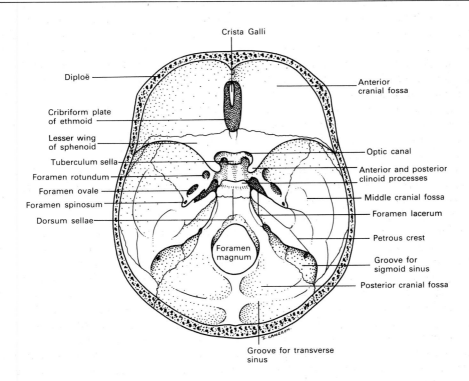

FIGURE 9·1　*The intracranial region of a modern human.*

bone completely encloses the cribriform plate of the ethmoid, so that the lesser wings of the sphenoid are prevented from articulating with the ethmoid in the anterior cranial fossa (see later).

The middle cranial fossa

The middle cranial fossa supports the temporal lobes of the brain and houses the pituitary gland in a small central depression, the **pituitary fossa** or **sella turcica**. This lies in a raised central portion of the middle cranial fossa (see Fig. 4.16). The pituitary fossa forms the most superior part of the body of the sphenoid bone, the centre of which (immediately beneath the pituitary gland) is completely hollowed out by a large **sphenoidal air sinus** in adults (see Chapter 4). The pituitary fossa is bounded in front by an elevated tubercle, the **tuberculum sellae**, and behind by a ridge of bone called the **dorsum sellae**. Two large foramina in the roots of the lesser wings of the sphenoid, the **optic canals**, are bounded laterally by the **anterior clinoid processes** that project backwards from the back of the anterior cranial fossa. Two more processes, the **posterior clinoid processes**, lie on either side of the dorsum sellae at the back of the pituitary fossa.

The most posterior aspect of the body of the sphenoid bone forms a primary cartilaginous joint with the **basilar process** of the occipital bone (known as the **spheno-occipital synchondrosis**). This joint is an important growth centre in the primate cranial base. The smooth slope posterior to the dorsum sellae and its continuation as the superior aspect of the basilar process of the occipital bone are known together as the **clivus**. Laterally, the middle cranial fossa lies at a lower level and is formed partly by the greater wing of the sphenoid bone and partly by the flat endocranial surface of the squamous temporal bone. A solid pyramid of bone, the petrous part of the temporal bone, projects inwards across the cranial base at an angle of 45° from the transverse axis of

the cranium and separates the middle cranial fossa from the posterior cranial fossa. The ridge or **petrous crest** on top of the petrous temporal bone forms the boundary between the middle and posterior cranial fossae. In the floor of the middle cranial fossa is a large oval foramen, the **foramen ovale**, that transmits the mandibular division of the fifth cranial nerve and a smaller foramen, the **foramen spinosum**, in the **spine** of the sphenoid bone (a part of the sphenoid bone that projects backwards into the junction of the petrous and squamous parts of the temporal bone). The foramen spinosum transmits the **middle meningeal artery** which grooves the floor and walls of the middle cranial fossa and supplies the diploë.

The posterior cranial fossa

The posterior cranial fossa supports the cerebellum and centrally the spinal cord is continuous with the brain stem through the **foramen magnum** of the occipital bone (Fig. 9.1). Most of the posterior cranial fossa is formed by the cup-shaped occipital bone, but anteriorly it is bounded by the clivus, the posterior surface of the petrous temporal bone and a small part of the squamous temporal bone that forms some of the lateral wall in the region of the sigmoid sinus (see below). The cerebellar hemispheres below and the occipital poles of the cerebral hemispheres above the transverse sinus, create four hollows on the internal surface of the occipital bone. In modern humans the impressions of the cerebellar fossae on the occipital bone are much larger than those of the cerebral fossae.

The supporting dural tracts of the brain

Within the cranial cavity several tough fibrous sheets of connective tissue known as **dura mater** divide the intracranial region into compartments, each of which houses a part of the brain. These compartments help to support the brain and prevent it 'sloshing

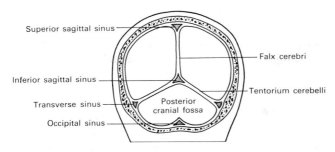

FIGURE 9·2 *A coronal section through the falx cerebri and tentorium cerebelli posterior to the foramen magnum.*

around' unduly when we rush about. Beneath and surrounded by the dura mater there are two other membranes that surround the brain, the **arachnoid** and the **pia mater**. All three membranes together are known as the **meninges** of the brain.

The posterior cranial fossa is mostly 'roofed over' by a tough sheet of dura known as the **tentorium cerebelli** that takes the weight of the overlying cerebral hemispheres (Fig. 9.2). This spreads downwards and outwards from the **falx cerebri** (the mid-line sickle-shaped fibrous sheet that partially divides the cerebral hemispheres into two halves). The falx and the tentorium are really continuous with each other over the roof of the posterior cranial fossa in the mid-line (Fig. 9.3). The tentorium then attaches at the internal occipital protuberance and around the rim of the posterior cranial fossa on both sides. It continues to sweep round the rim of the posterior cranial fossa and finally attaches to the petrous crests of both temporal bones and to the anterior clinoid processes. Both the falx and the tentorium have **free edges** that run respectively between the cerebral hemispheres and around the midbrain (see Chapter 10).

Bull (1969) has argued that the functional role of the dural tracts has changed during mammalian evolution. In birds and lower mammals the tentorium cerebelli forms a fibrous separating membrane, and the weight of the brain being trivial, is supported by the flat floor of the cranial base. In modern

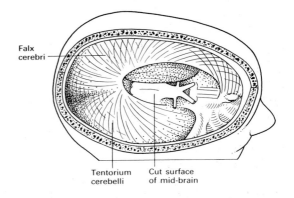

FIGURE 9·3 *The appearence of the falx and tentorium with the vault bones cut away laterally and the cerebral hemispheres removed from the cranial cavity.*

humans, following the massive increase in the size and weight of the brain and the rotation of the foramen magnum through 90°, the tentorium cerebelli acts as a support against gravity, to prevent the brain herniating through the foramen magnum. Thus, because of the greater sagittal angle of its inclination, by virtue of its increased area and because it is divided into two angulated planes, the tentorium cerebelli now transmits the weight of the occipital lobes of the brain outwards towards the rigid bony skull. This supportive role of the tentorium has become increasingly important during hominid evolution as the brain has undergone massive expansion.

163

Accordingly, the attachments of the dural tracts to the bone of the cranial cavity have become more pronounced.

Another important role of the dural tracts was emphasized by Moss and Young (1960) who hypothesized that the growing or expanding brain is 'tied down' by organized tracts of fibrous dura mater to the cranial base. The crista galli of the ethmoid and frontal crest in the anterior cranial fossa form the attachment of the falx cerebri and the petrous crests, clinoid processes and rim of the posterior cranial fossa form the bony attachments of the tentorium cerebelli. Moss (1963) investigated 'clubbing' of the crista galli in modern humans and concluded that it was the result of increased tension in the dura resulting from the backwards rotation of the neurocranium and falx cerebri. Moss also claimed that these changes were accompanied by a depression of the petrous crests to which the tentorium cerebelli attaches. Vertical artificial cranial deformation of American Indian skulls (Moss, 1958) also appears to result in depression of the clivoforaminal angles as if both *basion* (the most anterior point of the foramen magnum) and *opisthion* (the most posterior point of the foramen magnum) are rotating about a common centre in the posterior clinoid region of the dorsum sellae (Fig. 9.4). Bjork (1955) also noted that increased flexion of the cranial base angle is accompanied by depression of the foramen magnum and temporal bones. More recently, Anton (1989) has pointed out that induced changes in the cranial base following vault deformation are in fact more complicated than first suggested by Moss (1958).

The arrangement of the dural tracts then is not only of great importance for supporting the brain but also plays a great part in the development of the bony morphology of the vault and may affect the manner in which expansion of the brain (either phylogenetic or ontogenetic) takes place within the bony neurocranium. Figures 9.5 and 9.6 show how the tentorium cerebelli rotates posteriorly

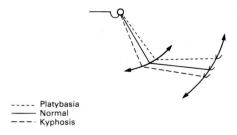

----- Platybasia
——— Normal
--- Kyphosis

FIGURE 9·4 *Rotation of the clivus and foramen magnum seems to occur around the posterior region of the sella turcica in artificially deformed crania. After Moss (1958).*

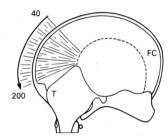

FIGURE 9·5 *Posterior rotation of the tentorium cerebelli (T) in prenatal humans between 40 mm and 200 mm crown rump lengths. FC is the falx cerebri. After Moss et al. (1956).*

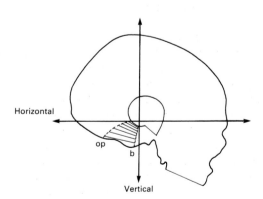

FIGURE 9·6 *Infant and adult modern human crania have been superimposed on a plane passing through the horizontal semicircular canal of the inner ear using the technique of Fenart (see Cousin and Fenart, 1971). The illustration demonstrates the downwards and forwards rotation of the modern human foramen magnum and posterior cranial fossa through some 30° during the postnatal growth period. OP = opisthion; B = basion.*

before birth as the cerebral hemispheres expand and how continued post-natal growth of the brain is accompanied by a downwards and forwards rotation of the foramen magnum.

The cranial fossae of the great apes

The morphologies of the three cranial fossae in great apes are largely determined by the size and shape of the brain. As a consequence of this (but also partly due to differences in expansion and disposition of the various regions of the brain), there are striking contrasts between the human and great ape intracranial regions. Biegert (1963) has noted the evolutionary expansion of the neopallium (see Chapter 12) that appears to have brought about flexion of the brain around an axis near to the pituitary fossa (Fig. 9.7). This, along with the expansion of the frontal and occipital lobes above the tentorium, as well as the cerebellum below the tentorium, has contributed to the elevation of the brain above the orbits and the downwards and forwards rotation of the posterior cranial fossa and foramen magnum in modern humans and some fossil hominids.

The anterior cranial fossa

Besides being smaller and markedly pinched towards the front, the anterior cranial fossa in great apes is characterized by a deep midline olfactory pit (Fig. 9.8). The cribriform plate of the ethmoid bone lies at the base of this deep pit. The rounded roofs of the orbits are raised above this level but slope downwards and backwards towards the middle cranial fossa, contrasting with the more nearly flat floor of the anterior cranial fossa in modern humans. The crista galli of the ethmoid is either diminutive or absent in apes, the falx cerebri hardly even raising a frontal crest of the frontal bone. The cribriform plate of the orang-utan is similar in width to that of *Pan* and *Gorilla* but not as long, and so is more rounded in outline (Dean, 1983). Wood-Jones (1948) drew atten-

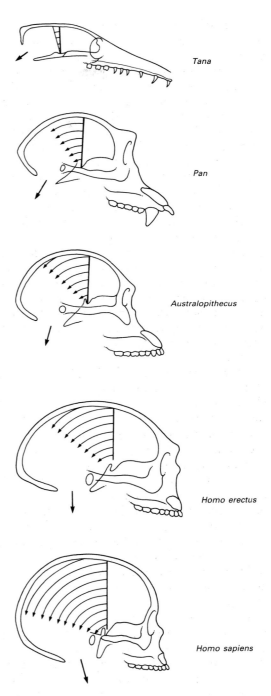

FIGURE 9·7 *Increasing expansion of the neopallium in a series of primates is accompanied by the reorientation of the foramen magnum. After Biegert (1963).*

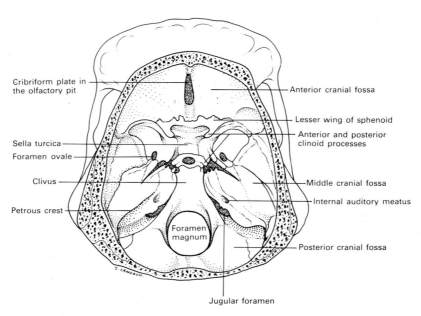

Cribriform plate in the olfactory pit

Anterior cranial fossa

Lesser wing of sphenoid

Anterior and posterior clinoid processes

Sella turcica

Foramen ovale

Clivus

Petrous crest

Middle cranial fossa

Internal auditory meatus

Foramen magnum

Posterior cranial fossa

Jugular foramen

FIGURE 9·8 *The intracranial region of a chimpanzee. Note that the dorsum sellae is fenestrated in this specimen.*

tion to the fact that there is also variation in the pattern of sutures between the bones of the anterior cranial fossa in apes. He considered the condition that predominates in the orang-utan (and modern humans) to be the primitive pattern. Ashley-Montagu (1943) recorded the incidence of a spheno-ethmoidal articulation in the anterior cranial fossa of *Pongo* as being 100%, in *Pan* as 85% and in *Gorilla* as 47%, the remaining animals in the study showing the frontal bone intervening posteriorly between the sphenoid and ethmoid bones behind the cribriform plate as in Fig. 9.8. Among modern human populations, Butler (1949) has recorded an incidence of this 'retro-ethmoid frontal suture' as 24% in 25 skulls from Bengal, and Murphy (1955), in a larger series of 455 Australian aborigine skulls, found only an 8% incidence of this particular pattern. Because of this variation, little taxonomic value can be attributed to these sutural patterns, but they are nevertheless an important reflection of differential growth rates at the spheno-

ethmoidal synchondrosis in hominoids and may have parallels in other parts of the cranium (see Chapter 11).

The middle cranial fossa

The middle cranial fossa of apes is also less expansive laterally than in modern humans. Whereas the human glenoid fossa lies below the thin squamous temporal bone of the middle cranial fossa, in great apes and in most fossil hominids the glenoid fossa lies more lateral to the brain under the root of the zygoma. The pituitary fossa is not so hollowed and is less well-defined in apes especially in the region of the dorsum sellae, where occasionally it can appear almost continuous with the clivus. There is little indication here of the upwards and backwards remodelling that occurs during growth in modern human skulls (Latham, 1972). Commonly, a patent **sphenoidal canal** (an embryological remnant marking the final site of fusion of the ossification centres of the body of the sphenoid bone) persists in most juvenile and some adult ape crania (Sprinz and Kaufman, 1987). This canal runs from the centre of the pituitary fossa to open on the external surface of the body of the sphenoid bone behind the articulation with the vomer (Fig. 9.9). Whereas the lesser wings of the sphenoid extend across the whole width of the junction between the anterior and middle cranial fossae in modern humans, they do not extend as far as *pterion* in great apes.

The posterior cranial fossa

The posterior cranial fossa in great apes is shallow and is characterized by a long flattened clivus and a less vertical posterior surface of the petrous temporal bone than in modern humans and some fossil hominids (Fig. 9.8). The foramen magnum is posteriorly positioned in adult ape crania and the attachment of the tentorium cerebelli around the rim of the fossa lies only a short distance above the posterior margin of the foramen magnum. The **cruciate eminence** or **internal occipital protuberance** then, lies a considerable distance below the position of the nuchal crest at *opisthocranion* externally.

The intracranial region of fossil hominids

Most early fossil hominids appear to have an anterior cranial fossa characterized by an olfactory pit (which is, however, not so deep as that in great apes) but with no prominent crista galli of the ethmoid bone (Fig. 9.10). This is true of *Australopithecus*, *Paranthropus* and early *Homo* from East Africa and is probably true of hominids from the Far East although the anterior cranial fossa is rarely well preserved in the *Sinanthropus* or Solo skulls. Full expansion of the frontal lobes occurred very late in hominid evolution and only in late Middle Pleistocene hominids is the anterior cranial fossa comparable in size to that of modern humans. Here, it may be argued, expansion of the forebrain has finally extended downwards so that the orbital cavities have been restrained from above

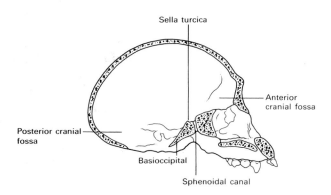

FIGURE 9·9 *Sagittal section through a juvenile chimpanzee cranium illustrating the levels of the cranial fossae and the sphenoidal canal that commonly runs from the pituitary fossa to the underside of the body of the sphenoid bone posterior to the vomer in apes.*

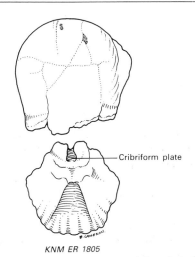

KNM ER 1805

FIGURE 9·10 *The cranium and facial skeleton of KNM ER 1805 (early* Homo*) seen from above and illustrating the cribriform plate in the olfactory pit of the anterior cranial fossa.*

during growth and remain at the level of the cribriform plate.

In contrast, the pituitary fossa is typically like that of modern humans even in the earliest hominids (OH 5 and Sts 5) with a well-developed dorsum sellae and prominent anterior clinoid processes. As previously mentioned, however, the lateral part of the middle cranial fossa is insufficiently expanded (even in *Homo erectus* crania) to overlie the glenoid fossae. Broom et al. (1950) noted the undercut anterior margin of the middle cranial fossae in Sts 5, which also exists in *A. afarensis* and which contrasts with the flatter morphology in apes.

The posterior cranial fossa of hominids is much more variable and is affected greatly by the relative position of the foramen magnum. Weidenreich (1943) noted in *Homo erectus* crania from Zhoukoudian like *Austra-*

lopithecus (Tobias, 1967) that the internal occipital protuberance is close to the foramen magnum and the impressions for the cerebellum are small in comparison with those for the occipital lobes of the cerebral hemispheres on the occipital bone. In *Paranthropus*, however, this is not so and the cerebellar fossae are expanded at the expense of cerebral fossae on the occipital bone (Fig. 9.11). A greater proportion of the occipital lobes of the cerebral hemispheres are presumably in contact with the parietal bones in these specimens. The posterior aspect of the petrous temporal bones are also near vertical or even

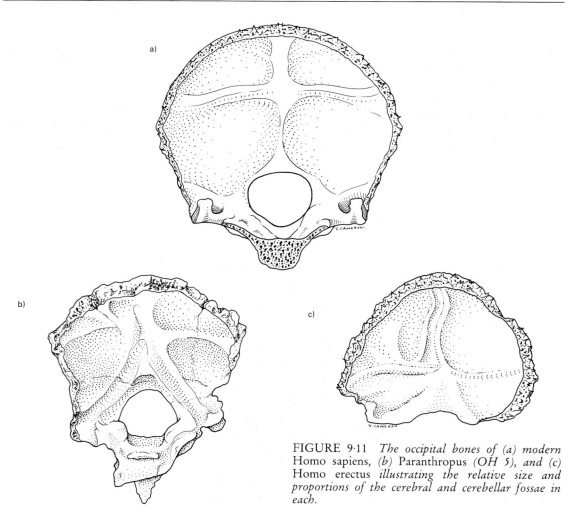

FIGURE 9·11 *The occipital bones of (a) modern* Homo sapiens, *(b)* Paranthropus *(OH 5), and (c)* Homo erectus *illustrating the relative size and proportions of the cerebral and cerebellar fossae in each.*

'undercut' in most specimens of *Paranthropus* much as they are in modern humans, *Homo habilis* (Stw 53, OH 24, KNM-ER 1813) and one specimen from Sterkfontein (Sts 19) (Broom et al., 1950; Clarke, 1977). It should be noted that Clarke has argued that Sts 19 might have come from Member 5 at Sterkfontein which implies that it may plausibly be attributed to *Homo*. Nevertheless, the vertical or undercut posterior petrous surface in these fossils contrasts yet again with the condition in *H. erectus* crania and other *Australopithecus* crania and KNM-WT 17000 where the posterior aspect of the petrous temporal bone slopes posteriorly and is thus much less vertical in its profile. As a consequence of this the petrous temporal bone is thicker front to back in these specimens. This 2.5-million-year-old specimen of *Paranthropus* (KNM-WT 17000) from West Turkana combines a long basioccipital (but still a relatively forwardly placed foramen magnum with respect to the bitympanic line) and a long occipital region. In this specimen, again like *H. erectus* crania from Zhoukou-

dian, there is a considerable slope to the posterior aspect of the petrous temporal bone which is consequently thicker front to back. There is then considerable variation among early hominid taxa in the morphology of the posterior cranial fossa and the confusing interplay between several functional systems make this fact difficult to interpret. Expansion of the cerebellum in the posterior cranial fossa beneath the tentorium cerebelli seems to be one factor that might underlie variation in the posterior cranial fossa (Dean, 1988a). This may be accompanied by an increase in endocranial base flexion, in a manner noted by Moss (1958) for modern humans. However, shortening of the basioccipital and the forward migration of the foramen magnum are other factors that may have compressed the cerebellum into the posterior aspect of the petrous temporal bones.

The cranial venous sinuses

The cranial venous sinuses (Fig. 9.12) drain blood from the brain as well as some blood

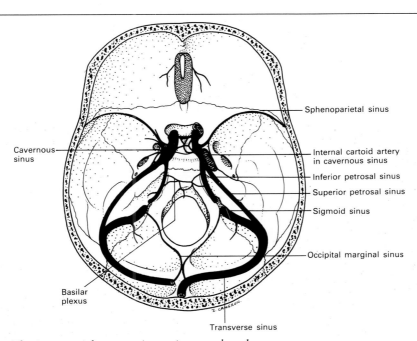

FIGURE 9·12 *The intracranial venous sinuses in a modern human.*

from the scalp, face, orbits and diploë. They also communicate with the **pterygoid plexuses** of veins (around the pterygoid muscles) with veins in the suboccipital region and with the **vertebral venous plexus**. The vertebral venous plexus has an **intraspinal** part that lies outside the dura (**extradurally**) and an **extraspinal** part that lies around the vertebrae. Both communicate with each other and extend for the full length of the vertebral column. The whole of the cranial venous sinus system and vertebral venous plexus is valveless which means that blood is potentially able to flow in any direction. Veins enter and leave the vertebral venous plexus along its entire length such that venous blood can potentially travel from the head to the abdomen bypassing the thoracic cavity. Eventually, however, venous blood must return to the heart, so that the vertebral venous plexus is only a temporary detour around the thorax.

Drainage of cool venous blood from the scalp and mid-face back into the **cavernous sinus** may be important for regulating the temperature of warmer blood in the internal carotid artery that supplies the brain. This mechanism is especially important and well developed in desert animals (Schmidt-Nielsen, 1979) but is also thought to operate in modern humans (and other primates) by regulating the temperature of blood to the brain (Hayward, 1967; Hayward and Baker, 1968; Cabanac and Caputa, 1979a, b; Baker, 1982; Critchley, 1985; Cabanac, 1986). It is an especially important mechanism during exercise when although body temperatures can rise to over 40°C brain temperatures may be regulated to 37°C in the cavernous sinus. Because blood is able to flow through this valveless venous sinus system in any direction, the cranial venous sinuses also play an important role in protecting the brain during normal transient physiological shifts in venous blood pressure. These shifts arise either from changes in intrathoracic pressure, e.g. during lifting heavy objects or when straining, or even from transient shifts in

blood pressure that occur on moving from a lying to a sitting or standing position.

The **superior sagittal sinus** begins in the region of the crista galli and runs backwards in a mid-sagittal groove on the inner aspect of the vault. At the internal occipital protuberance it usually deviates to the right to become continuous with the right transverse sinus, but may sometimes join the left one instead. At this point also, blood from the small **occipital sinus** flowing backwards from the **marginal sinuses** around the foramen magnum joins the confluence of sinuses at the internal occipital protuberance. The **inferior sagittal sinus** is smaller than the superior one and runs backwards in the lower free fold of the falx cerebri to end at the junction of the falx and the tentorium cerebelli. Here, together with the **great vein of Galen** from the brain, it becomes the **straight sinus** which runs backwards to the confluence. The straight sinus usually turns to the left to form the left transverse sinus just as the superior sagittal sinus usually turns to the right to form the right transverse sinus. These **transverse sinuses** run in the attached edge of the tentorium cerebelli around the rim of the posterior cranial fossa (Fig. 9.13). At the point where they meet the posterior aspect of the petrous temporal bone, the transverse sinuses are joined on each side by the **superior petrosal sinuses** that run on the petrous crests and then together become the **sigmoid sinuses**. These groove the junction of the squamous and petrous temporal bones in the posterior cranial fossa and run downwards and medially to leave the cranial cavity as the **internal jugular veins** via the **jugular foramina**. At the jugular foramen, the **inferior petrosal sinuses** join the internal jugular veins and so, together with the superior petrosal sinuses, drain blood from the cavernous sinuses into the jugular venous system.

The cavernous sinuses are very important and lie either side of the body of the sphenoid bone and pituitary fossa in the middle cranial fossa (Fig. 9.12). Left and right cavernous

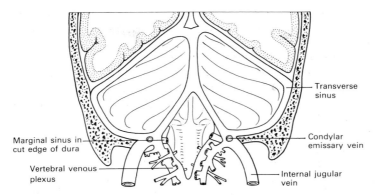

FIGURE 9·13 *Coronal dissection of the posterior cranial fossa of a modern human illustrating the many venous communications with the jugular bulb* *that lies in the jugular foramen. After Matsushima* et al. *(1983).*

sinuses communicate with each other via a meshwork of veins on the clivus and via the pituitary fossa. These sinuses also communicate with the orbital veins that drain part of the scalp and mid-face and with the pterygoid plexus of veins outside the cranial cavity. Blood from the cavernous sinuses usually passes via the superior and inferior petrosal sinuses into the jugular venous system but can also leave via the orbital veins to meet other veins of the face and scalp and via emissary veins to join the pterygoid plexus of veins. The **internal carotid artery** (together with many important nerves) passes through the cavernous sinus on each side of the cranium and so is completely bathed in venous blood for the last part of its course before supplying the brain.

When the marginal sinuses communicate anteriorly with the **basilar plexus** of veins on the clivus, the whole of the foramen magnum is encircled by a venous channel that communicates with the jugular bulb (the most superior part of the internal jugular vein that lies in the jugular foramen). Eckenhoff (1970) has noted that when pressure in the thoracic cavity is raised, as it is during heavy-weight lifting, venous blood is prevented from returning to the thorax via the internal jugular veins. Venous blood then leaves the jugular bulb and cranial cavity via other routes such as the vertebral venous plexus and by way of emissary veins until the intrathoracic pressure falls once again (Epstein et al., 1970). Once the pressure has returned to normal blood can then pass back into the thorax from the vertebral venous plexus where it joins the azygos vein and returns to the heart. Venous blood leaving the cranial cavity via emissary veins joins the *external* jugular veins which engorge during raised intrathoracic pressure as blood cannot enter the thorax until it reduces again. This is the reason we go red in the face when lifting heavy objects or when straining.

Occasionally, venous blood from the superior sagittal sinus and from the straight sinus that usually passes into the right and left transverse sinuses, passes instead into enlarged right and left marginal sinuses via the occipital sinus (Fig. 9.14). This is said to occur unilaterally in 4% or bilaterally in 2% of cases in modern humans (Woodhall, 1936, 1939). Kimbel (1984) concluded that out of a total of 581 humans investigated, the average incidence of grooves for enlarged occipital marginal sinuses in dried skulls is about 6%, but spatiotemporally distinct populations may present with increased frequencies. For example, 45% of 11 crania of late Pleistocene early modern *Homo sapiens*, from Predmost, Czechoslovakia, present with this pattern. It

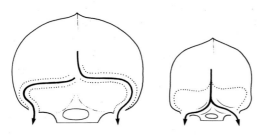

FIGURE 9·14 *Variation in the pattern of venous sinuses in the posterior cranial fossa of early fossil hominids and modern humans. In one the transverse–sigmoid sinus pattern drains venous blood to the jugular bulb and in the other the occipital–marginal sinus pattern drains venous blood from the brain by this alternative route. After Falk and Conroy (1984).*

is important to note that this pattern occurs most frequently unilaterally, co-existing with transverse-sigmoid drainage on the same and/or contralateral side of the head. Where this pattern of venous drainage exists, the usual direction of blood flow in the marginal sinuses is reversed and this system then drains blood to the jugular bulb and internal jugular veins by an alternative route to the more common transverse sinus–sigmoid sinus route. Blood that would have joined the transverse sinus from diploic veins and from the inferior cerebral and cerebellar veins presumably still does so. However, the reduced volume of blood may or may not run completely within the tentorium cerebelli and may or may not leave a groove in the occipital bone. More blood then joins the transverse sinus from the superior petrosal sinus at the start of the sigmoid sinus and there is always a groove in the bone here. In this situation, the occipital marginal and sigmoid sinuses both drain into the jugular bulb and internal jugular veins. It follows that modern humans or fossil hominids with enlarged marginal sinus systems still have jugular foramina, jugular bulbs and would have had or still have internal jugular veins. Other variations in the modern human pattern of cranial venous sinuses have been reviewed

and described by Browning (1953) and Hollingshead (1982).

Browning (1953) made the important observation that the width and depth of grooves created by venous sinuses on the intracranial aspect of the vault bones in modern humans, bear no relationship to the capacity of the sinus. Even weak grooves or grooves that are absent may be associated with sinuses that are equal in capacity to contralateral sinuses that create deep grooves in the vault bones. Shenkin et al. (1948) have also demonstrated another important point; a considerable proportion of blood in the internal carotid artery normally makes its way into the *external* jugular system and not into the internal jugular veins. As much as 22% of the blood in the external jugular veins is derived from the internal carotid artery and it must be concluded from their experiments that emissary veins communicating between the cranial venous sinuses and the extraspinal vertebral venous plexus and/or the external jugular veins extracranially are of considerable importance in returning cranial venous blood to the heart in modern humans. More recently, Zouaoui and Hidden (1989) have investigated the source of venous blood entering the vertebral venous plexus. These authors concluded that the venous plexus around the foramen magnum communicates with the extraspinal part of the vertebral venous plexus that runs along the superior oblique muscles of the neck and that this is an important route for drainage of the brain. This extraspinal part of the vertebral venous plexus has communications with the intraspinal part of the vertebral venous plexus and with the external jugular vein. However, direct communications from the intracranial venous sinuses in humans with the intraspinal part of the vertebral venous plexus seem to drain little or no blood from the brain (Zouaoui and Hidden, 1989). These studies demonstrate the variability of the intracranial venous system and also how little can be predicted about blood flow and function from studies of bony impressions alone.

The cranial venous sinuses in apes and fossil hominids

In an extensive survey of great apes, Kimbel (1984) has deduced that grooves in the posterior cranial fossa, indicating the presence of a significantly enlarged occipitomarginal sinus system, are very rare (only two out of 259 great ape specimens). Falk (1986b), however, noted a higher frequency but in a much smaller sample of great ape crania.

Of the fossil hominids, *Australopithecus afarensis*, *Paranthropus robustus* and *Paranthropus boisei* nearly all appear to have well-developed occipital marginal sinus systems (Fig. 9.14), although in several specimens (SK 1585, SK 859 and KNM-ER 407) it is well developed on only one side of the foramen magnum, there being a well-developed transverse sinus contralaterally (Kimbel, 1984). OH 5 is a clear example, however, of a specimen where enlarged occipital marginal sinuses exist unequivocally on both sides of the foramen magnum (Tobias, 1967; see also Fig. 9.11). Rak and Howell (1978) and Holloway (1981d) have both noted that a specimen of *P. boisei* from the Omo (Omo L333y–6) shows no evidence of occipital marginal sinuses. As noted above, however, these intracranial grooves of differing size are likely to reflect little of the true venous return pattern in early hominids. While no clear groove for an occipital marginal sinus can apparently be identified in KNM-WT 17000 (Walker and Leakey, 1988; Leakey and Walker, 1988) the grooves for the sigmoid sinuses in this specimen are easily as large as those in the larger brained Sts 5. It is hard to imagine where more blood could have drained from to produce an additional enlarged occipital marginal sinus in KNM-WT 17000, and greater variation in the pattern of cranial venous sinuses in *Paranthropus* than is presently recorded must remain a distinct probability (see also Holloway, 1988).

Few specimens of *Australopithecus africanus* or early *Homo* have been described with enlarged occipitomarginal drainage patterns but in many cases specimens are only partially preserved and the possibility of such a pattern existing cannot be ruled out. Exceptions are the Middle Pleistocene occipital bone from Vértesszöllös where there is evidence of a unilateral occipital marginal drainage system (Kimbel, 1984) together with those crania noted by Falk (1986b), which include Swanscombe, Guomde, Galley Hill, those from Predmost and Skhul 1 (the latter being the only one of these crania to have bilaterally enlarged occipitomarginal sinuses).

Falk and Conroy (1984) and Falk (1986b) attributed considerable taxonomic and phylogenetic significance to the high incidence of occipital-marginal sinus systems, as determined by intracranial bony markings, in *A. afarensis* and *Paranthropus*. According to these authors this is a shared derived character that supports a close phylogenetic relationship between these hominids. These authors supported their argument by proposing that the enlarged occipital marginal sinus pattern was a physiologically advantageous adaptation during the shift from quadrupedal to bipedal locomotion which diverted venous blood from the brain into the vertebral venous plexus. Drawing heavily on Eckenhoff (1970), they proposed that changes in intrathoracic pressure during bipedal locomotion result in blood being returned to the heart more easily and in greater quantities via the vertebral venous plexus in *A. afarensis* and *Paranthropus* than via the internal jugular veins. Falk (1986b) argued that later in hominid evolution the increased cranial venous outflow via emissary veins also entered the vertebral venous plexus and so was able largely to replace the occipital–marginal drainage into this plexus, so that the pattern of bony grooves in the posterior cranial fossa then presumably reverted to type. However, there is little evidence in modern humans of an increased communication with the vertebral venous plexus from enlarged occipital marginal sinuses that course around the foramen magnum. Matsushima et al. (1983,

173

p. 73) in a careful and extensive study of the veins of the posterior cranial fossa noted that a marginal sinus courses in the dura at the level of the foramen magnum and that the marginal, occipital and sigmoid sinuses and the condylar emissary veins all converge on the jugular bulbs and the internal jugular veins. The vertebral venous plexus, they add, also anastomoses with the internal jugular veins here (Fig. 9.13). If these various drainage routes all converge and communicate at the jugular bulbs, then it follows that nothing can be predicted about venous return to the thorax from the pattern of grooves in the posterior cranial fossa.

Eckenhoff (1970) noted that the function of a valveless venous return system is to allow *transient* shifts in the direction of flow during activities such as heavy lifting and childbirth. Sudden long-term obstructions to the jugular venous return system result in cerebral oedema and, as Eckenhoff pointed out, there is an acute increase in cerebrospinal pressure when the jugular veins are continuously compressed, due to blood being forced into the intraspinal vertebral venous plexus and thus encroaching upon the epidural space. Up to 1.5 l of blood may pass into the vertebral venous plexus during transient periods of raised intrathoracic pressure but this blood must eventually return to the thorax. It would be physiologically impossible then for the vertebral venous plexus to replace the jugular venous return system on a long-term basis in any animal. The fact that this does not occur in primates is clear from the small size of the azygos vein when compared with the size of the internal jugular veins.

Another explanation for the range of variation in the cranial venous sinus pattern has been proposed by Kimbel (1984) as simply being due to variation in a neutral trait within different populations of hominids. Alternatively, Tobias (1967) has suggested a link with early growth and development of the cerebellum in the posterior cranial fossa of some early hominids and it may well be that expansion of the cerebellum during foetal

development in these hominids squeezes blood in the transverse sinus to the region around the foramen magnum thus sending blood into the marginal sinus system which then becomes established as the dominant drainage system in adulthood.

Patterns of the middle meningeal arteries in hominids

It has been noted previously that the middle meningeal artery is predominantly a nutrient artery to the bone of the cranial vault and to the red marrow of the diploë. This artery enters the middle cranial fossa through the foramen spinosum of the sphenoid bone. Other meningeal arteries, both anterior and posterior are present but the pattern of grooves created by the middle meningeal artery are the most distinct and extensive. The middle meningeal artery soon divides into two branches, anterior and posterior

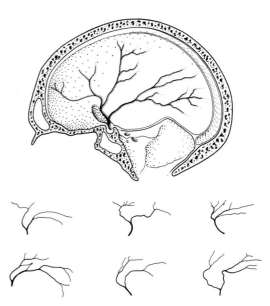

FIGURE 9·15 *The pattern of grooves on the lateral aspect of the vault bones created by the middle meningeal artery. Variation in this pattern is common and some are illustrated beneath. After Chandler and Derezinski (1935).*

(sometimes called frontal and parietal). The anterior branch passes upwards on the greater wing of the sphenoid bone, across *pterion* and onto the parietal bone where it distributes over the vault bones above the anterior and middle cranial fossae (Fig. 9.15). The posterior branch passes backwards to distribute over the remaining supratentorial vault bones above the posterior and middle cranial fossae.

Chandler and Derezinski (1935) have reviewed the literature and provide good data about the course and branching patterns of this vessel in modern humans. They concluded that the artery divided into anterior and posterior branches either near the foramen spinosum, or near the top of the greater wing of the sphenoid bone, or at a point between the two locations in about equal thirds of the 1200 crania examined. Double branches of both anterior and posterior divisions were noted in this series as was the complete absence of one or other division. The same authors also noted that the pattern on the right- and left-hand sides of the same cranium are likely to differ.

Schepers (1946) described in great detail both the pattern and course of the middle meningeal arteries in the great apes and certain fossil hominids including *Homo erectus*. These variations included many already noted above for modern humans and do not clearly distinguish any group from any other. Schepers noted additional variations such as the 'incomplete vascularisation of the occiput' and 'widely divergent anterior and posterior divisions' in some specimens. However, Tobias (1967) noted that the patterns of meningeal vessels even differ on each side of OH 5 and that 'little store can be set by these vascular patterns as indicators of taxonomic status'. Despite these findings, Saban (1977, 1986) suggested that some hominid genera can be distinguished by their meningeal branching patterns and, among other things, drew attention to the similarities between the meningeal artery patterns in *Homo* and *Paranthropus*. Saban holds that in *A. africanus* there are just two main branches, an anterior and a posterior one, but in *Paranthropus* there is a strong middle branch derived from the posterior branch. Leakey and Walker (1988) and Walker and Leakey (1988) noted that KNM-WT 17000 shows the pattern of meningeal arteries described by Saban for *Australopithecus*. However, these authors argue that smaller vault bones in specimens with small cranial capacities are likely to require a simpler nutrient meningeal artery supply and that this might underlie any differences that exist between fossil hominids.

CHAPTER TEN

THE ANATOMY OF THE BRAIN AND HOMINOID ENDOCASTS

Most of what is known about the brains of early hominids has been learned from studying natural endocranial casts of fossil brains or from studying the size and contours of the intracranial regions of hominid vault bones. For this reason the convolutions of the cerebral cortex and estimates of brain volume are topics that have dominated the field of palaeoanthropology. Information of this sort is, however, of limited value and can only be interpreted in the light of what is known about the comparative anatomy of the brain in extant primates. For these reasons the following account of the anatomy of the brain and the nervous system centres largely around the cerebral hemispheres and those topics that have figured prominently in studies of fossil hominids. Nevertheless, aspects of the anatomy of the limbic system, hindbrain and of some of the special sensory pathways are also included as these regions figure significantly in the evolution of the primate brain.

The cells that make up the nervous system

The nervous system is composed of nerve cells known as **neurons** and their supportive cells that are known collectively as the **neuroglia**. Neurons are electrically excitable

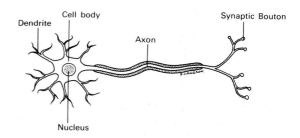

FIGURE 10·1 *A nerve cell or neuron and its component parts.*

cells specialized for the reception, integration and transmission of information. A typical neuron (Fig. 10.1) consists of a **cell body** (which has a nucleus within it) and a series of processes known as **dendrites** that convey information to the cell body. Dendrites normally branch profusely and make many connections with other neurons. A process known as the **axon** conveys information away from the cell body. Axons may be as short as 1 mm or as long as 1 m in length. The terminations at the distal end of the axon are known as **terminal boutons** or **synaptic boutons** and these may terminate on one or many other neurons, or on muscle cells or on the cells of glands. Intercellular communication in the nervous system occurs at

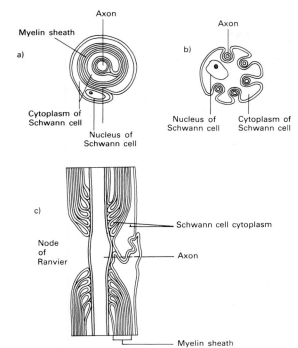

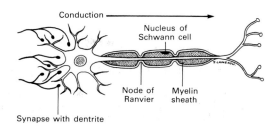

FIGURE 10·3 *Both inhibitory and excitatory nerve fibres synapse with the dendrites of a neuron. Conduction occurs along the axon in a saltatory manner between successive nodes of Ranvier.*

FIGURE 10·2 *Cross-sections through (a) a myelinated nerve axon and (b) a non-myelinated nerve axon. (c) A longitudinal section through the axon and Schwann cell of a myelinated nerve axon to show the structure of the insulating myelin sheath in the region of a node of Ranvier.*

synapses. Some axons are wound round with an insulating sheath known as **myelin** and so are called **myelinated nerves** (Fig. 10.2). Other axons are simply sunk into the cytoplasm of their supporting neuroglial cells and lack this extensive insulation and so are known as **unmyelinated nerves**. Small-diameter nerve fibres tend not to be myelinated but large ones (over 2 μm in diameter) are usually myelinated as this allows faster conduction of the nerve impulse. The impulse in myelinated nerves jumps in a **saltatory** manner from one break in the myelin sheath to the next along its length. This discontinuous type of conduction is faster than the simple continuous conduction in unmyelinated nerves (conduction velocities may reach

120 ms^{-1} in myelinated nerves but only 4 ms^{-1} in unmyelinated nerves). The breaks in the myelin sheath along the length of axons are known as **nodes of Ranvier** (Fig. 10.3).

The tissues of the central nervous system may be described as being composed of either **grey matter** or **white matter**. Grey matter contains cell bodies of neurons, dendrites and unmyelinated terminal portions of axons. Dense aggregations of neuronal cell bodies are known as **nuclei** within the brain or spinal cord, or alternatively, where they form swellings on nerves outside the central nervous system, they are known as **ganglia**. White matter contains primarily well-myelinated axons which, as a consequence, in cross-section appear white to the naked eye.

The general arrangement of the sensory and motor pathways of the central nervous system

The brain and spinal cord are known together as the **central nervous system**. Nerves that enter and leave the brain are known as **cranial nerves** and those that enter or leave the spinal cord are known as **spinal nerves**. Nerves carrying information about sensation, such as touch, pressure, temperature and pain, may enter either the brain or spinal cord as cranial or spinal nerves. Nerves that carry **special sensation** such as taste, vision, hearing or smell, enter the brain directly as cranial nerves. All these nerves, however, can be referred to as **sensory nerves**. The majority of cranial or spinal nerves are in fact **mixed nerves**, that is they contain bundles of both incoming sensory-nerve fibres and outgoing **motor-nerve** fibres. Motor nerves are nerves that leave the brain or spinal cord and eventually innervate secretory glands, or muscles that bring about movements. Together, sensory and motor nerves that are outside the central nervous system make up what is known as the **peripheral nervous system**. While no details of the peripheral nervous system will be described here, it is worth noting that there are 12 pairs of cranial nerves and 31 pairs of spinal nerves that leave or enter the central nervous system in a **segmental manner** in the same way that the vertebral column is segmental in its make-up (Fig. 10.4). Nonetheless, it must be emphasized that the internal structure of the spinal cord comprises columns of nuclei and tracts and is not segmented in the same way.

Once within the central nervous system, not all sensory nerves reach the cerebral cortex of the brain but many of them do. These nerves travel as great **tracts** (or bundles) of nerve fibres that reach the **thalamus**, a large central mass of grey matter that sorts and then relays incoming sensory impulses to the cerebral cortex or **pallium** (pallium (L) = a cloak) hence it covers the cerebral hemispheres like a cloak (Fig. 10.5). The majority of sensory nerve fibres cross to the other side of the body (or **decussate**) within the central nervous system so that their representation on the cerebral cortex is **contralateral**.

The function of the cerebral cortex is to analyse incoming sensory information by way of a cascade system of connections and interconnections that link both new nerve impulses from different regions of the brain and body with memories of past events that have been stored in the brain. Impulses that are the end result of this interaction of nervous information may end up in regions of

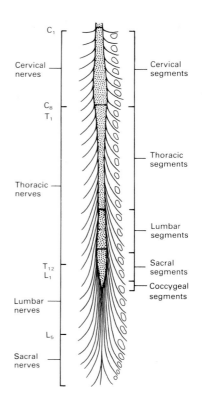

FIGURE 10·4 *Spinal nerves leave the spinal cord in a segmental manner. The level at which they pass out through the intervertebral foramina becomes increasingly drawn away inferiorly from their origin at the cord.*

the cortex that initiate motor-nerve activity. From these motor areas of the cortex, motor nerves pass downwards through bundles of nerves known as the **pyramidal tracts**. These end by synaptic contact with other nerve cells in the spinal cord which leave the central nervous system to end within muscles or glands (Fig. 10.6). The majority of motor nerves in the pyramidal tracts, like the sensory tracts, cross to the opposite side of the body. This means that the right motor cortex innervates the left side of the body and that

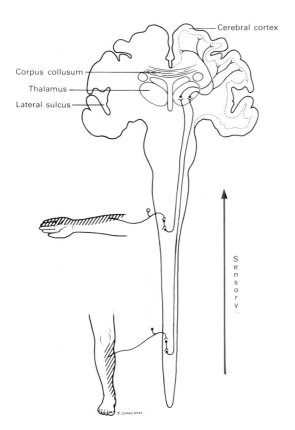

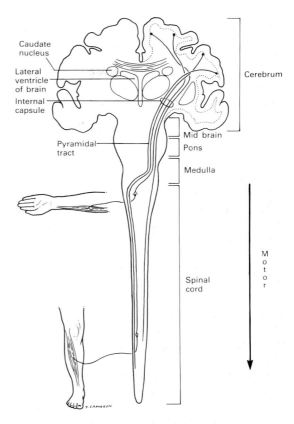

FIGURE 10.5 *One example of a sensory pathway from the dermatomes of the limbs to the cerebral cortex. Pain, touch and temperature are relayed to the cerebral cortex via the thalamus in the spinothalamic tracts. These nerve fibres decussate to the opposite side of the body in the central nervous system.*

FIGURE 10.6 *An example of a motor pathway from the cerebral cortex to the muscles of the limbs. Corticospinal tracts pass downwards through the internal capsule and form pyramidal tracts. The majority of fibres decussate, then all of them relay in the anterior horns of the spinal cord to innervate muscles of the limbs.*

the left motor cortex innervates the right side of the body.

By these pathways, incoming sensory information may bring about outgoing motor activity. For example, seeing and smelling food may result in your hands picking up something to eat or, alternatively, throwing it away depending on how all sorts of incoming (sensory) and stored information has been analysed within the brain. Generally speaking, an increase in the size of the cerebral cortex in primates has allowed a wider range of adjustments to environmental change by increasing the number of neural connections and interconnections necessary to bring about more delicately co-ordinated reactions.

The cerebral hemispheres

The two cerebral hemispheres are linked with each other through a large tract of **commissural** nerve fibres known as the **corpus callosum** (Fig. 10.7). The cerebral

179

cortex is only 2–5 mm thick on the surface of each hemisphere but each human cerebral hemisphere has an area of cortex of some 1200 cm². Beneath this thin layer, masses of white myelinated nerves form pathways to and from the cortex. Those that link regions within the same hemisphere are known as **association fibres**. As we have seen, ascending tracts of sensory fibres link the thalamus and also the **basal ganglia** (areas of grey matter within the cerebral hemispheres), with the cortex and descending nerves that form the pyramidal tracts pass from the cerebral cortex as far as the spinal cord and relay to innervate muscles and glands. These fibres that project up to the cortex and down from it are known as **projection fibres**. Non-mammalian vertebrates only have a small area of cerebral cortex and this is exclusively associated with the sense of smell or **olfaction** (Fig. 10.8). Because of its phylogenetically primitive nature this part of the cortex is known as the **archipallium**. The rest of the expanded mammalian cerebral cortex is associated with tactile, visual, and auditory sensations and is known as the **neopallium**. In primates the olfactory regions of the brain are very much reduced but the neopallium has become greatly expanded. It is said that about 20% of the cerebral cortex in modern humans can be precisely linked with primary sensory and motor areas for discrete regions of the body. The remaining 80% of the human brain is **association cortex**. The primary areas of the cortex are linked via commissural fibres in the corpus callosum to identical areas in the contralateral hemisphere and also by other association fibres to secondary and tertiary **association areas** in the same hemisphere (Fig. 10.9). In this way there is an inter-relationship through complex linkages between the activities of the sensory and motor projection areas. As the primate cerebral cortex has become more expanded and refined in its powers of sensory descrimination and its analytical functions, it has also become progressively structurally differentiated so that the extent and boundaries of

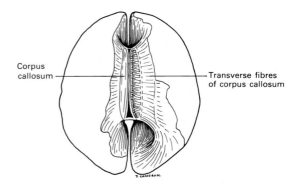

FIGURE 10·7 *The corpus callosum exposed from above with the right half dissected to show the transverse comissural fibres that pass between the two hemispheres of the brain.*

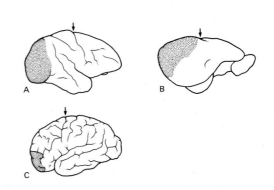

FIGURE 10·8 *Lateral views of (a) macaque, (b) tree shrew, and (c) modern human brains, illustrating the position of the central sulcus (arrowed) and the visual cortex (shaded).*

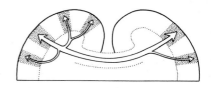

FIGURE 10·9 *Coronal section through both cerebral hemispheres with a diagrammatic representation of the comissural fibres that run between the hemispheres and the association fibres that link regions of cortex within one hemisphere.*

the various areas of the cortex can be clearly defined **histologically** (i.e. by examining the tissues of the brain microscopically). The visual cortex, for example, has a distinct structure and can be mapped along with other areas of the brain. In this way, **cytoarchitectonic charts** of the cortex can be built up and the degree and extent of differentiation of the brain in extant primate species compared. Brodmann (1868–1918) was responsible for mapping many of these cortical areas in modern humans and his numbers are still commonly used today. A rather bold and over ambitious attempt was also made to map the cortex of fossil hominid endocasts (Schepers, 1950) but this must really be discounted.

Nevertheless, some of the boundaries of these motor or sensory regions of the cortex also have a close topographical relationship to the convolutional patterns of the **gyri** and **sulci** on the surface of the brain. **Convolutions** are known as gyri. Sulci are the grooves between the convolutions, and **fissures** are especially deep sulci that separate large components of the brain. Convolutional patterns are especially well developed in higher primates because the local pressures and forces that occur during growth and development in these areas of the brain result in foldings along the boundaries of these functionally discrete regions.

The cerebral cortex can be divided into **frontal, parietal, temporal** and **occipital** lobes (Fig. 10.10). There are three particularly prominent sulci or fissures: the **sylvian** or **lateral sulcus** that delineates the temporal lobe; the **central sulcus** that divides the frontal lobe and primary motor cortex anteriorly from the parietal lobe and primary sensory cortex posteriorly; and the **lunate sulcus** that delineates the primary visual cortex (sometimes known as the **striate cortex**) and a small part of the occipital lobe at the back of the brain. The **interhemispheric fissure** is the most prominent of all and separates the right and left hemispheres of the brain.

The frontal lobes of the brain are associated with the primary motor areas. They are also involved in the ability to concentrate and focus attention on pursuing tasks to completion as well as with memory and recognition and the motor aspects of speech. The prefrontal cortex exerts a controlling influence over emotions, which are generated by the **limbic system** (see later) of the brain. Damage to the frontal lobes in human subjects may result in a loss of foresight and lack of purpose, with difficulty in stabilizing behaviour and with inability to speak or write coherently.

The parietal lobes are the site of the primary somaesthetic areas and the more posterior areas are concerned with the integration of visual stimuli with other sensory inputs. Damage to the motor or sensory areas affect the contralateral side of the body (because of the decussation of motor-nerve fibres in the brainstem or spinal cord). Other areas in the parietal region co-ordinate fine

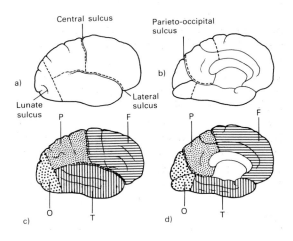

FIGURE 10·10 *(a), (b) Lateral and medial views of the human brain with the major sulci highlighted. (c), (d) The lobes of the brain on the lateral and medial aspects of one hemisphere that are defined by the major sulci; F frontal, P parietal, T temporal and O occipital lobes.*

181

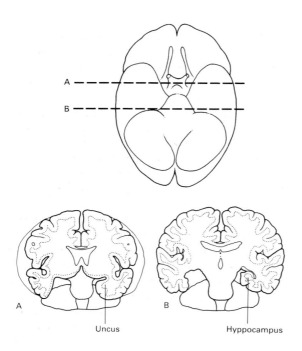

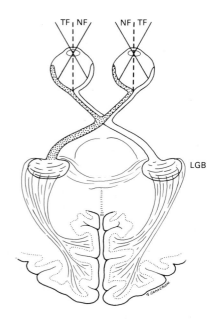

FIGURE 10.11   *(A), (B) Representation of the cut surfaces of the posterior part of the brain made in the coronal planes indicated. The positions of the uncus and the hyppocampus are illustrated.*

FIGURE 10·12   *Nerve impulses from the temporal field (TF) of one eye and the nasal field (NF) of the other eye travel in the optic tracts and relay in the lateral geniculate body (LGB). Fibres then project to the visual (or striate) cortex of the occipital lobes of the brain.*

movements and integrate gestures, for example with speech and the sensory aspects of language.

The temporal lobes are concerned with particular memory functions such as for example, of watching and hearing others speak or of hearing music. The medial pole or **uncus** of the temporal lobe is especially important in recalling long-term memories (Fig. 10.11). Damage to a large area at the caudal end of the temporal lobes results in various sensory **aphasias** (speech defects – see later), memory loss and a loss of musical appreciation or ability, depending upon the exact location of the lesion. The occipital lobes are the site of the visual cortex and damage to this region accordingly results in visual defects.

The visual system

Sight is arguably the most important of the special senses in man and large areas of the brain are involved in processing visual information. Light rays are focused on the retina of the eye which has photoreceptor cells within it called **rods** and **cones**. Within the retina the light rays are transformed into nerve impulses that leave the eyeball in the **optic nerve** (Fig. 10.12). Impulses from the nasal half of the visual field of one eye and of the temporal field of the other eye travel to the same side of the brain, but impulses from the nasal half of the visual field of each eye cross to the contralateral side of the brain in the **optic chiasma** (the structure which overlies the pituitary fossa). Because decussation in the optic chiasma is incomplete, impulses from the same field of view in each eye end up in the same hemisphere of the brain. This, together with the medial rotation of the eyes on the face and the consequent overlapping of the visual fields, accounts for the well-developed sense of stereoscopic vision in humans and higher primates. Lower vertebrate optic tracts decussate completely in the optic chiasma, so that these animals have no sense of stereoscopic vision.

From the optic chiasma nerve impulses continue in the **optic tracts** and relay in the **lateral geniculate body**. From here nerve fibres project in the **optic radiation** to the **primary visual cortex**. The primary visual cortex is found on each side of the **calcarine sulcus** on the medial aspect of the occipital lobe of the cerebral cortex and the small portion of the external aspect of the occipital lobe caudal to the lunate sulcus. Within this cortex, representation of the visual world is upside-down, with the central vision being near the occipital pole. **Visual-association cortex** occupies almost all of the rest of the occipital lobe in front of the lunate sulcus and even extends into the parietal and temporal lobes.

Cerebral dominance

Modern man appears to be unique in that each of the cerebral hemispheres is specialized in different ways. As we have seen, many areas of the cortex have connections via the corpus callosum to identical areas on the other side of the brain representing the other side of the body. Other areas are confined more discretely to one hemisphere. The cerebral hemispheres are commonly referred to as the **dominant** and **minor** hemispheres, but really they are simply complementary in their specializations. The speech centres are, by definition, situated in the dominant hemisphere, which as it happens is nearly always on the left (irrespective of whether the subject is right or left handed). Musical ability and appreciation is, by contrast, centred in the

temporal lobe of the minor hemisphere. The minor hemisphere is largely concerned with non-linguistic and non-mathematical functions, but is superior in pictorial, pattern, rhythm and pitch sense. It may be that the cerebral demands of language and speech are so great that it is advantageous to divide the roles of the hemispheres in this way. After injuries to the dominant hemisphere in childhood, it is still possible for the growing minor hemisphere to learn speech and language but these functions are crowded into this hemisphere at the expense of other cognitive abilities. This emphasizes the inadequate neuronal territories that exist for language under these circumstances and suggests perhaps that it is a disadvantage for language to be divided between the hemispheres when impulses would have to connect with each other via the long fibres of the corpus callosum.

Language

Whereas the muscles that are concerned with speech are represented bilaterally on the motor area of the cerebral cortex, the speech areas of the cortex are only found in one cerebral hemisphere (98% of the time on the left and 2% on the right). While this is established by four or five years of age, initially there is bilateral representation of speech and language but the hemisphere that regresses can no longer control the speech muscles in older individuals, even though it retains its ability to understand speech. In 1861 the French pathologist and anthropologist Pierre Paul Broca identified a damaged area of the cerebral cortex just in front of the primary motor areas controlling speech (in the posterior part of the third frontal convolution) in patients who had suffered from aphasia. In these patients speech was slow and laboured and articulation poor. In 1874 Carl Wernicke identified a different type of aphasia associated with damage to another part of the cerebral cortex: an area of cortex in the superior temporal convolution and a part of the parietal lobe of the dominant hemisphere more posteriorly positioned. Whereas patients with damage to Broca's area were unable to speak properly, they understood both written and spoken words. Damage to Wernicke's area, however, resulted in loss of comprehension of language. Superficially, speech sounds were nearly normal in patients with lesions of Wernicke's area but their sentences were more devoid of content and meaning.

Wernicke's area is connected to Broca's area by a tract of association fibres known as the **arcuate fasciculus** (Fig. 10.13). When words are heard and repeated, the auditory patterns of the words are relayed from Wernicke's area to Broca's area where the articulatory form is aroused and passed to the motor area that controls the muscles of speech. When words are read, connections from the visual cortex (via the angular gyrus) are first converted into an appropriate auditory form in Wernicke's area. Understanding written language as well as spoken language involves arousal of an auditory form of language in Wernicke's area. These speech areas of the human cerebral cortex are enlarged and give rise to asymmetries in 80% of brains (Geschwind, 1972) (Fig. 10.14). These areas are even present in infants and so are probably genetically determined, unlike the development of handedness which is under much more environmental influence. Nevertheless, many primates that have no language have asymmetric brains, e.g. the gorilla, but the degree of asymmetry and the pattern is different (LeMay, 1976; Holloway and De La Coste-Lareymondie, 1982). In great apes there is a **left-occipital petalia** or enlargement, but in *Australopithecus*, *Paranthropus* and *Homo* there is clear evidence of both distinct **left-occipital** and **right-frontal** petalial patterns in the same brain. This is even clearly present in the 2.5-million-year-old specimen KNM-WT 17000 attributed to *Paranthropus* (Holloway, 1988).

The fact that Wernicke's area involves part of the parietal lobe as well as a part of the

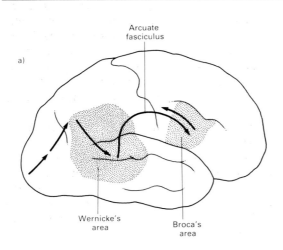
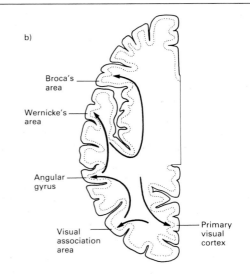

FIGURE 10.13 *Saying the name of an observed object involves the transfer of visual information to the angular gyrus where it is programmed to arouse the auditory pattern in Wernicke's area. From here the auditory form is transmitted to Broca's area via the arcuate fasciculus. In Broca's area the articulatory form is aroused and passed to the region of the motor cortex responsible for movements of the tongue, jaw and face. This region of the motor cortex lies immediately adjacent to Broca's area. After Geschwind (1972).*

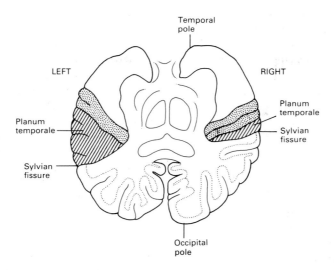

FIGURE 10·14 *Asymmetry of the human superior temporal lobes. The posterior margin of the planum temporale extends further backwards on the left so that the end of the lateral or sylvian fissure lies posterior to the corresponding point on the right.*

temporal lobe associated with hearing is important. A crucial aspect of language is the ability to label objects, so that something we see and touch and hear can be given a single name. In order that the brain can do this it is essential that sensory nerves from all these regions of the brain can associate with each other. Most animals are only able to make associations between a non-limbic stimulus, such as incoming visual, tactile or auditory information, and a limbic (or emotional) stimulus (Geschwind, 1965). Only modern humans are thought to be able to make associations between two non-limbic stimuli and it is this ability that makes it possible to learn names with which to represent objects. Holloway (1974) has cautioned against this simplistic hypothesis and noted that subcortical fibre systems are also likely to have been involved in any reorganization of the brain during hominid evolution and, particularly, a part of the thalamus known as the **pulvinar** that expands in concert with the parietal association cortex. The pulvinar (pulvinar (L) = a cushioned couch) is the prominent rounded portion of the posterior aspect of the thalamus that is characteristic of the brains of higher primates that have an expanded sensory association cortex. These sorts of sensory associations are likely to have been advantageous for many aspects of intelligence, from which language may in turn have been a secondary development. The area of the cortex where these sensory associations have evolved occupies a large part of the parietal lobe in the region of Wernicke's area in the human brain and is thought to have played a key role in the evolution of human language. These sensory association areas cannot be defined in the architectonic maps of the cortex of monkeys and are poorly developed in ape brains (Eccles, 1984).

Self-consciousness and conceptual thinking

Human beings have an inner picture of themselves, a self-awareness of their existence combined with a strong sense of feeling about

185

themselves in time. In other words self-awareness reflects 'a world that is tremendously extended over time so that an integrated picture built up from many sensory domains can be extended to retain events of the past and even projected into the future as images' (Jerison, 1973).

Because the concept of consciousness is an integrated association of sensory information, it is closely linked with acquisition of language and is an important aspect of hominid brain evolution. From a very early age, human infants can recognize an image of themselves in a mirror and even explore features they see on their faces with their hands. With the exception of great apes and some cetaceans, no other animals (as far as we can tell), appear to be able to recognize themselves in this way (Humphrey, 1986). This concept of 'self' is slowly built up during development as sensory information and experiences are stored and as the interconnections of the cerebral cortex are established. It is clear, none the less, that animals have feelings and respond to stimuli that come from within their own bodies and from the surrounding environment and that they also have a sense of awareness based upon memory. Much of this is manifest as either reflex or instinctive behaviour, but there is little or no evidence that animals can think in the sense that they can manipulate conscious symbols (Reynolds, 1976).

Definitions of human consciousness or self-awareness are necessarily vague but it is clear that hominid evolution is closely bound up with its progressive development. Important components include the ability to communicate using language and an uninterrupted memory of life events combined with a sense of curiosity about the environment and the future (Popper and Eccles, 1977). Representing the world with symbols or words which form the basis of language creates an abstract framework that becomes different from our inner selves. This framework is sufficiently distinct that we can manipulate ideas using language at a distance and so adopt a critical attitude towards ideas formulated in language. Like language, consciousness appears to be strongly linked with the dominant hemisphere which is further evidence of how closely associated the two are.

It has been argued that with the evolution of language comes the ability to imagine, to experience 'flights of the imagination', to be inventive (and even to deliberately deceive or to formulate excuses and false explanations). Recent evidence suggests that primates are skilled at deceiving and cheating others to their own advantage (Byrne and Whiten, 1987). Humphrey (1986) has attempted to place conceptual thinking into an evolutionary framework and has argued that consciousness is like having an 'inner eye' that makes us more aware of what is going on inside the brain. This provides a basis for being able to predict the feelings and thoughts going on in the brains of others. Humphrey argues that this capacity to 'mind read' others, was of enormous social advantage to early hominids, as they would then have been capable of a vastly increased range of more sensitive responses and interactions with their fellows. It is thought that caring for the dead is one form of evidence of conceptual awareness in early hominids as it reflects an awareness of the feelings, experiences and the fate of others (Hawkes, 1965; Dobzhansky, 1967). Other evidence of enlarged speech areas and asymmetries in the cerebral cortex (see below) are also occasionally taken as possible indirect evidence for language, cerebral dominance and the ability of early hominids to think conceptually.

Another hypothesis regarding the origins of language has recently been put forward by Dunbar (1990). Groups of primates devote a considerable amount of time to grooming, which is important for social bonding. Dunbar argues that group size among primates is a function of relative neocortical mass and also that the maintenance of social stability in large groups of hominids by grooming alone would have placed intolerable demands on the time budgets of individuals within these groups. The evolution of language is seen in this hypothesis as a more efficient method of social bonding that may be linked with the evolution of relative neocortical enlargement and larger numbers of individuals in groups of early hominids.

Limiting factors in comparative studies of the cerebral cortex of hominids

Various areas of the human cerebral cortex have expanded at the expense of other areas and this has had an influence upon the topography of the primary sulci of the cortex. It is especially notable that the expansion of the parietal association areas during hominid evolution has displaced the visual cortex of the occipital lobes posteriorly so that a great deal of the primary visual cortex now lies on the medial aspect of the cerebral hemispheres and not on the lateral aspect as in other primates. The lunate sulcus, which delineates the primary visual cortex, has shifted posteriorly in human brains relative to its position in the brains of apes and monkeys. The position of the lunate sulcus has been taken to indicate the degree of expansion of the parietal association areas in fossil hominid endocasts (when it can be identified). Holloway (1974) noted that only 10% of human endocranial casts have a lunate sulcus that can be clearly identified. (However, Falk (1986a) contested that 58% of 120 human hemispheres examined directly exhibited a lunate sulcus.) In general, the human brain develops many more secondary sulci than do the brains of monkeys or apes and these tend to obscure the pattern of the primary sulci, which in any case are very variable in their contours over the cortex.

While there is a good relationship between some sulci and the boundaries of discrete functional regions of the cortex, those on fossil endocasts obviously cannot be tested with architectonic maps or electrode stimulation and any conclusions reached are, therefore, of necessity speculative. Likewise,

it is an assumption that the swellings, which are presumed to be homologous with human speech areas of the cortex, are indeed so on early hominid endocasts.

Differences also exist in the number of cells per unit volume in the brains of higher primates. Some regions of the human brain have 50% more cells than similar regions of ape brains and greater degrees of connectivity (Le Gros Clark, 1971). Overall, however, it now seems to have been established that the number of brain cells remains remarkably constant throughout mammals. What increases is the number of interconnections, in other words, the cell density drops while the complexity of the neuronal network increases (Steele-Russell, 1979). These findings obviously complicate many of the comparisons that can be made between hominid brains and emphasize the fact that volume comparisons alone are misleading. Volume comparisons are, in any case, questionable to say the least in the absence of good data for body weights and these are particularly sparse and variable for early hominids (Chapter 14). Furthermore, the fact that no relationships can be drawn between human cranial capacity and, for example, measures of behaviour, intelligence or other skills, further underlines the inadequacy of estimates of brain volume as a measure of neural reorganization in hominids.

The limbic system

The idea of the limbic system arose from the **limbic lobe** originally described by Paul Broca as forming the lower medial edge or border of the cerebral cortex of the temporal lobes (limbus (L) = a border or hem). These days the limbic system refers to a loosely defined set of brain structures that include both cortical and subcortical elements (Fig. 10.11). **The hippocampus** is part of the limbic system associated with memory in modern humans, but it is generally recognized that the limbic system is especially closely associated with the emotions.

187

Many memories, such as smells, are linked with powerful emotions and the limbic system is closely bound up with sexual, fighting and feeding emotions among others. (Incidentally, the olfactory tracts are the only sensory tracts that do not relay in the thalamus and which project to the **ipsilateral** (same) side of the cerebral cortex or archipallium.) The close association of the olfactory and limbic systems underlies the affective and evocative aspects of the sense of smell. The **amygdala** is another part of the limbic system which may be involved in controlling rage responses and is more prominent in man than other primates, as are various other components of the limbic system.

Armstrong and Onge (1981), for example, have demonstrated that there are three times as many neurons in the human **anterior thalamic nuclei** of the limbic system as in the great apes. These authors emphasized that these results indicate that, in terms of numbers of neurons, humans diverged further from other hominoids in the limbic nuclei than in the **association nuclei**, which are traditionally related to 'higher' cortical functions. Vilensky et al. (1982) have drawn on these and other data that demonstrate significant neural expansion in the human limbic system and have proposed that modification to the limbic system preceded modifications to the neocortex during hominid evolution. Arguing that early hominids show little overall brain expansion relative to body weight (over and above great apes) and also that there appears to be little evidence for cortical reorganization in early hominids (*contra* Holloway, 1983), these authors have proposed that early expansion of the limbic system would correlate with the social and sexual changes believed to have accompanied early hominid evolution. While undetectable in the endocasts of early hominids, this would have promoted the memory-processing capacity and, in turn, the subsequent enlargement of the cerebral cortex and its interpretive and linguistic capabilities in these hominids. Yet again, these studies underline the fact that reorganization of the brain must have occurred during hominid evolution and that much of this is undetectable in fossil endocasts.

The brainstem

The brainstem comprises the **midbrain, cerebellum, pons** and **medulla** (Figs 10.15 and 10.16) and contains many of the vital centres that control respiration, heart beat and other

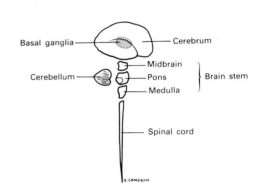

FIGURE 10·15 *The component parts of the brain stem are illustrated, together with the position of the basal ganglia within the cerebrum and the position of the spinal cord.*

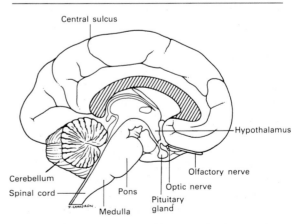

FIGURE 10·16 *A sagittal cut made through the interhemispheric fissure and corpus callosum continues inferiorly to expose the structures of the midbrain and brain stem in median section.*

autonomic functions that occur unconsciously. Parts of the brainstem contain the great tracts of fibres that pass to and from the cerebral cortex. As the motor fibres and the sensory fibres that pass to and from the cerebral cortex converge, they form a narrow zone of funnelling fibres in each cerebral hemisphere that are known as the **internal capsule** which converges upon the brainstem on each side from left and right hemispheres (Fig. 10.6). Expansions in various regions of the cortex require that parts of the brainstem that connect with these regions must also expand with them (such as the pulvinar of the thalamus mentioned above).

The cerebellum

The cerebellum is concerned with the unconscious control of balance, equilibrium and the synergism of muscle movement and so has important connections with the middle ear and with the motor and sensory pathways of the brain. It is likely then that bipedalism was associated with reorganization within the cerebellum during hominid evolution. Memories for skilled motor activities such as tool making, riding a bicycle or playing the piano are also stored in the cerebellum so that the acquisition of manual skills associated with opposability of the thumb are also likely to have been linked with reorganization or expansion of the cerebellum in early hominids.

A large part of the cerebellar hemispheres is concerned with the control of voluntary movements. Haines (1986) has drawn attention to the lateral hemispheres of the cerebellum and noted that an increase in the size of the cerebellum in primates relates almost exclusively to expansion of an area within it known as the **lateral** or **dentate nucleus**. Manato et al. (1985a, b), have noted that the number of cells in the lateral nucleus of the cerebellum increases in primates with the acquisition of planned skilled, diversified movements, especially of finely tuned hand and finger movements and opposability of

the thumb. It may also be that the width of the spectrum of motor abilities, rather than narrowly specialized motor patterns, underlies the lateral expansion of the cerebellum in primates. The ventral pons that lies against the clivus is a relay station between the cerebral cortex and the cerebellar hemispheres so that expansion of the ventral pons relates directly to the increase in size of the cerebral cortex and cerebellum. It is possible that changes in endocranial base flexion and petrous orientation (see Chapter 5) in certain fossil hominids relate to changes in the ventral pons and cerebellum (Dean, 1988a).

Patients with lesions of the cerebellum commonly experience slurring of their speech, tremor, and difficulties opposing their fingers and thumbs. The acquisition of tool making and speech that require precise control of the muscles of the hand and of the tongue are likely to have been associated not only with an expansion of various regions of the cerebral cortex but also with an accompanying expansion of the lateral aspects of the cerebellum and the ventral pons in hominids as well as with the expansion of the pulvinar of the thalamus that also has two-way connections to and from the cortex. It may be significant that new evidence for opposability of the thumb in _Paranthropus_ (see Chapter 18) occurs alongside changes in the morphology of the cerebellum (see later). Obviously these changes in the hominid brain are inadequately represented by the topography of the cerebral cortex alone.

Encephalization

Measuring the relative brain size of an individual requires a knowledge of the brain weight, the body weight and a clear idea about the standard of comparison. Jerison (1973) defined the encephalization quotient (EQ) as the ratio of actual brain size to expected brain size for living mammals using the formula: $0.12 \times (\text{body weight})^{0.667}$. A scaling coefficient of 0.667 has been claimed to fit the relationship between brain weight and body weight for a large number of mammals, but Martin (1982, 1983) has calculated an exponent of 0.75 on an even larger sample. Holloway and Post (1982), however, noted that depending on the data base chosen to calculate the EQ value, estimates for fossil hominids, with respect to chimpanzees and *Homo sapiens*, can vary by 20%. Holloway and Post (1982) suggested that the most useful measure of brain expansion is obtained by dividing the EQ for an individual by the EQ for modern humans (which they calculated as 2.87) to give a percentage figure describing the relative brain size of a specimen in comparison with modern humans. Modern humans thus have a relative EQ of 100 and *Proconsul africanus* of 48.8 (Walker et al., 1983). Figure 10.17 gives EQ values for several fossil hominids and chimpanzees as well as other data presented in Table 10.1 and illustrates the general trend of brain enlargement in later hominids.

As we have already seen there is clear evidence of neural reorganization among primates and a more comprehensive method of estimating the relative size of individual brain structures is to estimate the relative weight of different portions of the brain. Obviously, however, this is not a technique that can be applied to fossil endocasts. Stephan et al. (1970) estimated volumes of neural structures and body weights in many primates and developed a **progression index** (PI). The PI estimates how many times larger a given brain structure is over and above that expected in a basal insectivore. For example, the human neocortex is 156 times larger than expected and the amygdala is 3.0 times larger in simians and 3.5 times larger in *Homo sapiens*.

The endocranial casts of fossil hominids

While it is remarkable that any natural fossil hominid endocranial casts exist at all, there are relatively few of them and although data derived from them are particularly important, these data need to be interpreted within an objective neuroanatomical framework. Table 10.2 at the end of this chapter provides a list of estimated endocranial capacities for early fossil hominids that have been compiled from several sources. The following discussion of hominid endocranial casts sets out some of the morphological characteristics known for *Australopithecus* and *Paranthropus* as well as early and later *Homo* in the context of the foregoing account of the anatomy of the brain.

Crania of *Australopithecus afarensis* and *Australopithecus africanus* all appear to have endocranial capacities estimated at below 500 cm³ while specimens of *Paranthropus* all

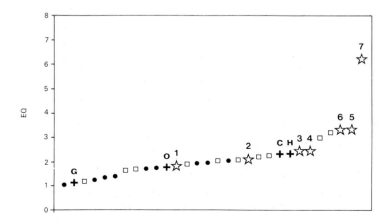

FIGURE 10·17 *Encephalization quotients for hominids (by species) and other higher primates (by genus). Closed circles = Old World monkeys; open squares = New World monkeys; + = greater and lesser apes, stars = hominids. G = Gorilla; O = orangutan; C = chimpanzee; H = gibbon. 1 = Australopithecus afarensis; 2 = Australopithecus africanus; 3 = Paranthropus boisei; 4 = Paranthropus robustus; 5 = Homo habilis; 6 = Homo erectus; 7 = modern Homo sapiens. Data from Table 10.1, Martin EQ.*

appear to have endocranial capacities of above 500 cm³ (with the exception of the older specimen KNM-WT 17000 (Walker et al., 1986) which has a small endocranial capacity of just 410 cm³ despite the large size of the cranium). Even this value, however, falls well within the most likely range of values (361–643 cm³) predicated by Tobias (1967) for these early hominids. The allometric effect of body size on these volumes still cannot be confidently assessed, but there is evidence for an increase in relative brain size from earlier to later hominids (Falk, 1985; Holloway, 1988). However, current estimates

TABLE 10·1 Body weights, brain weights and encephalization quotients.

Taxon	Body weight (g)	Brain weight (mg)	Martin[a] EQ	Jerison[b] EQ
Australopithecus afarensis	50 600	415 000	1.87	2.44
Australopithecus africanus	45 500	442 000	2.16	2.79
Paranthropus boisei	46 100	515 000	2.50	3.22
Paranthropus robustus	47 700	530 000	2.50	3.24
Homo erectus	58 600	826 000	3.34	4.40
Homo habilis	40 500	631 000	3.38	4.31
Homo sapiens	44 000	1 250 000	6.28	8.07
Gorilla gorilla	126 500	505 900	1.14	1.61
Hylobates	6 521	112 057	2.40	2.60
Pan troglodytes	36 350	410 300	2.38	3.01
Pongo pygmaeus	53 000	413 300	1.80	2.36
Cercocebus	7 433	107 800	2.09	2.29
Cercopithecus	4 245	66 133	1.96	2.05
Colobus	8 729	74 050	1.27	1.41
Erythrocebus	7 800	106 600	1.99	2.19
Macaca	7 280	90 330	1.78	1.95
Nasalis	15 100	94 200	1.07	1.24
Papio	17 043	168 357	1.74	2.05
Presbytis	8 861	83 400	1.42	1.58
Theropithecus	17 050	131 900	1.36	1.61
Alouatta	6 667	56 567	1.19	1.29
Aotus	960	18 200	1.67	1.52
Ateles	6 800	110 525	2.29	2.49
Brachyteles	9 500	120 100	1.93	2.17
Calicebus	1 088	20 700	1.73	1.59
Cebus	2 733	78 250	3.25	3.25
Chiropotes	3 000	58 200	2.25	2.27
Lagothrix	6 300	96 400	2.12	2.29
Pithecia	1 500	31 700	2.08	1.97
Saimiri	665	25 050	3.04	2.68

Brain weights (BrW) and body weights (BoW) for the fossil hominids are after McHenry (1988) and for the extant species are after Harvey et al. (1986). EQ = encephalization quotient (the ratio of the observed brain size and the expected brain size).
[a]The expected brain size is computed on the basis of: $Log_{10}(BrW) = 0.76Log_{10}(BoW) + 1.77$ (Martin, 1983; BrW in milligrams, BoW in grams).
[b]The expected brain size is computed on the basis of: $Log_{10}(BrW) = 0.67Log_{10}(BoW) + 2.08$ (Jerison, 1973; converted from the original formula to the units used here (BrW in milligrams and BoW in grams)).

Table 10.2 Endocranial capacities of fossil hominids

Specimen	Endo-cranial volume	Source
Australopithecus afarensis		
AL 333 45	500	Holloway and Post (1982)
AL 162 28	400	Holloway (1983)
AL 333 105 Juv	400	Holloway and Post (1982)
Australopithecus africanus		
MLD I	500	Holloway (1973a)
MLD 37/38	435	Holloway (1973a)
Sts 5	485	Holloway (1973a)
Sts 19/58	436	Holloway (1973a)
Sts 60	428	Holloway (1973a)
Sts 71	428	Holloway (1973a)
Taung Juv	405	Holloway (1970)
Paranthropus robustus/boisei		
L388y–6 Juv	448	Holloway (1981a)
SK 1585	530	Holloway (1973a)
KNM ER 406	510	Holloway (1973a, b)
KNM ER 732	500	Holloway (1973a, b)
OH 5	530	Holloway (1973a, b)
KNM WT 17000	410	Walker et al. (1986)
KNM WT 17400 Juv	400	Walker and Leakey (1988)
KNM ER 13750	475	Holloway (1988)
KNM ER 407	506	Falk and Kasinga (1983)
KNM ER 732	500	Holloway (1973)
Early *Homo*		
OH 7	674	Tobias (1987)
OH 13	673	Tobias (1987)
OH 16	638	Tobias (1987)
OH 24	594	Tobias (1987)
KNM ER 1470	752	Tobias (1987)
KNM ER 1590 Juv	810	Blumenberg (1985)
KNM ER 1805	582	Holloway (1978)
KNM ER 1813	509	Holloway (1978)
Homo erectus		
OH 9	1067	Holloway (1973)
OH 12	750	Holloway (1980a)
KNM ER 3733	850	Holloway (in Stringer, 1984)
KNM ER 3883	804	Holloway (in Stringer, 1984)
KNM WT 15000 Juv	900	Walker and Leakey (1986)
Sangirian 4	908	Holloway (1981c)
Sangirian 2	813	Holloway (1981c)
Sangirian 10	855	Holloway (1981c)
Sangirian 12	1059	Holloway (1981c)
Sangirian 17	1004	Holloway (1981c)

Specimen	Endo-cranial volume	Source
Solo I	1172	Holloway (1980b)
Solo IV	1251	Holloway (1980b)
Solo VI	1013	Holloway (1980b)
Solo IX	1135	Holloway (1980b)
Solo X	1231	Holloway (1980b)
Solo XI	1090	Holloway (1980b)
Lantian	780	Woo (1966)
Zhoukoudian II	1030	Weidenreich (1943)
Zhoukoudian III	915	Weidenreich (1943)
Zhoukoudian X	1225	Weidenreich (1943)
Zhoukoudian XI	1015	Weidenreich (1943)
Zhoukoudian XXI	1030	Weidenreich (1943)
Salé	880	Holloway (1981b)
Trinil 2	940	Holloway (1981c)
Archaic *Homo sapiens*		
Arago XXI, XLVII	1200	Day (1986)
Djebel Irhoud I	1305	Holloway (1981b)
Djebel Irhoud II	1450	Holloway (1985)
Petralona	1230	Protsch (in Stringer, 1984)
Kabwe/Broken Hill	1285	Holloway (1981a)
Elandsfontein	1225	Beaumont et al. (1978)
Dali	1120	Wu (1981)
Florisbad	1280	Beaumont et al. (1978)
Omo II	1430	Day (1972)
Laetoli LH–18	1367	Holloway (in Stringer, 1984)
Ndutu	1100	Rightmire (1983)
Steinheim	1100	Olivier and Tissier (1975)
Swanscombe	1325	Olivier and Tissier (1975)
Vertesszollos 2	1300	Thoma (1981)
Skhul IV	1555	Holloway (1985)
Skhul V	1520	Holloway (1985)
Skhul VI	1585	Holloway (1985)
Qafzeh VI	1570	Holloway (1985)
Neanderthal		
Neanderthal	1525	Holloway (1985)
Spy 1	1305	Holloway (1981b)
Spy 2	1553	Holloway (1981b)
La Quina 5	1350	Holloway (1981b)
La Quina 18 Juv	1200	Tillier (1984)
La Chapelle	1625	Boule (in Holloway, 1981b)
La Ferrassie 1	1689	Boule (in Holloway, 1981b)
Amud 1	1750	Ogawa et al. (1970)

Table 10.2   Continued

Specimen	Endo-cranial volume	Source
Gibraltar 1	1200	Keith (1915)
Gibraltar 2 Juv	1400	Dean et al. (1986)
Engis 2 Juv	1392	Fenart, Empereur-Buisson (1970)
Teshik-Tash Juv	1500	Tillier (1984)
Shanidar I	1600	Holloway (1985)
Le Moustier	1565	Holloway (1985)
Krapina B	1450	Holloway (1985)
Tabun I	1270	Holloway (1985)
Saccopastore I	1245	Holloway (1985)
Saccopastore II	1300	Holloway (1985)
Monte Circeo I	1550	Holloway (1985)
Modern *Homo sapiens*		
Cro-Magnon	1600	Day (1986)

	95% Limits		
	Lower	Upper	Mean
Modern humans	1 159	1 243	1 201
Chimpanzee	391	409	400
Gorilla	452	486	469
Orang-utan	385	409	397

Estimates quoted are in cm³. Where a range of values is given in the original source only the greatest value of the range has been quoted here. It should be noted, however, that only *one* source/estimate has been quoted for each hominid even though there may be others. This list is presented only as a broad overview and serious workers should consult the original literature rather than quote values given here. Data for modern humans and for great apes are quoted from Ashton and Spence (1958) but see also Table 10.1.

(McHenry, 1988) suggest that all australopithecines and paranthropines had relative brain sizes within the range of living monkeys and apes (Fig. 10.17) (however see Martin (1983) for an alternative interpretation).

Irrespective of endocranial volume, obvious morphological differences exist between hominid and great ape endocasts and between the endocasts of the hominid groups themselves; when compared with ape endocranial casts, the height of the brain of both *Australopithecus* and *Paranthropus* is greater (Fig. 10.18). In addition, the morphology of the frontal lobes overlying the orbits, as well as that of the temporal lobes (which are more expanded) is more reminiscent of later hominids and not of great ape brains (Holloway, 1974). The endocranial casts of *Australopithecus* brains may show more complicated convolutions than those of *Paranthropus*, but the parietal regions of SK 1585 and OH 5 appear to be expanded more laterally than those of *Australopithecus*. The cerebellum in SK 1585 and OH 5 is more rounded and human-like than the cerebellum of *Australopithecus*, and is also swung further anteriorly beneath the cerebral hemispheres in *Paranthropus* than in *Australopithecus* (Tobias, 1967; Holloway, 1972a, b). The smaller brain of the 2.5-million-year-old

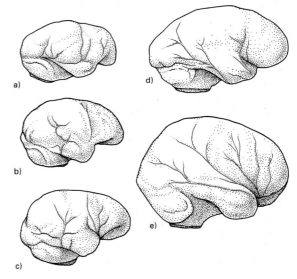

FIGURE 10·18   *Lateral view of five endocranial casts belonging to (a) chimpanzee, (b) Australopithecus africanus, (c) Paranthropus robustus, (d) Homo erectus and (e) modern Homo sapiens. After Holloway (1974).*

Paranthropus boisei specimen KNM-WT 17000 does not show the expanded parietal region or advanced cerebellar morphology of later *Paranthropus boisei* specimens which suggests these features evolved independently within this lineage and in parallel with early *Homo* specimens (Holloway, 1988).

Holloway (1972a, 1974) has also noted that the region corresponding to Broca's area in both *Australopithecus* and *Paranthropus* (SK 1585) appears more expanded than that in ape endocasts. These facts alone are enough to suggest that the visual cortex must have been displaced posteriorly in these hominids so that a greater proportion of it would be positioned on the medial aspect of the occipital lobes. Indeed, a more posterior position of the lunate sulcus has been described by Dart (1926), Schepers (1946, 1950), LeGros Clark (1947a, 1972) and Holloway (1974, 1981d) on several early hominid endocasts (including *A. afarensis*, Holloway (1983)). Le Gros Clark (1972), one of the most distinguished neuroanatomists ever to have examined the Taung endocranial cast concluded:

> It is really not possible to identify this sulcus with certainty from the impressions on the cast, but a careful examination of the original natural endo-cranial cast of the Taung skull does support the original contention of Dart that, in this particular individual at least, the simian (or lunate) sulcus must have been placed rather far back. For the sulcal impressions are sufficiently well marked to justify the negative con-clusion that it did not occupy the relatively forward position in which it is commonly found in gorillas and chimpanzees.

However, Falk (1980, 1983a, 1985) has challenged these findings and contests that in the Taung specimen, as well as several others from Sterkfontein and another from Hadar (AL 162–28) the position of the lunate sulcus is more anteriorly positioned and, therefore, more pongid-like. Tobias (1987) holds that the position of the lunate sulcus, on the Taung endocast at least, is 'indeterminate' and this is probably a wise conclusion. The position of the lunate sulcus is a significant indicator of cerebral reorganization but, as we have seen, there are other indications of brain reorganization in early hominids. These include the development of speech areas, the expansion of the cerebellum, the existence of human-like asymmetries and the possible reorganization of the limbic system which together underline the fact that wherever the lunate sulcus was positioned in these early hominids, they were likely to have progressed beyond apes in some aspects of their cerebral development. However, the position of the lunate sulcus in *A. afarensis* as described by Falk (1985), together with the small endocranial capacities noted for these few specimens, has prompted Falk (1985) to suggest that the trend towards brain enlarge-ment preceded that for cortical reorganization in early hominids.

Nevertheless, Holloway (1974), Falk (1983b) and Tobias (1987) all agree that clear signs of a human-like reorganization of the cerebral cortex is evident in specimens of early *Homo* (Fig. 10.18) and, in general, there is also evidence of an increase in cranial capacity in these specimens attributed to early *Homo* that exceeds the values known for *Australopithecus* and *Paranthropus*. The true significance of these cranial capacities can, once again, only be assessed in the light of reliable estimates for body weight, which it now appears may be quite contrasting among these specimens of early *Homo* (Chapter 14) (Lewin, 1987).

Other aspects of the morphology of the endocasts attributed to early *Homo* suggest that some specimens (e.g. KNM-ER 1813) may resemble *Australopithecus* more closely, whilst others (e.g. KNM-ER 1805) resemble *Paranthropus* and others still (e.g. KNM-ER 1470) are somewhere between *Paranthropus* and the *Homo erectus* specimens from Indone-sia (Holloway, 1976). Holloway added that the KNM-ER 1805 endocast is unique in its shape and that 'taken in conjunction with the dentition, could well represent an offshoot from either *A. robustus* or *A. africanus* lines'. Falk (1983b) again goes further and suggests that the KNM-ER 1805 endocast is 'clearly pongid-like' in the morphology of the frontal lobes and similar to earlier dated southern African australopithecines.

Endocranial casts of *H. erectus* from Indonesia (Holloway, 1980b) provides much clearer evidence of a predominance of left-occipital right-frontal petalia patterns. Holloway and De La Coste-Lareymondie (1982) speculated that these brain asymmetries were related to selection pressures operating on both symbolic and spatiovisual integration, which is consistent with the archaeological record. There is also a consistent expansion of Broca's area in these fossil brains. The mean capacity of 1151 cm³ in this sample is still below the value determined for later Neanderthals and the flattened *platycephalic* shape of the Solo brain endocasts are notable in that they are widest across the temporal lobes and not the parietal lobes as in Neanderthal and modern human endocranial casts.

Neanderthal cranial capacities are as large as, if not slightly larger than, those of modern humans. There is also a tendency for Neanderthal endocasts to be platycephalic and a suggestion that the visual cortex of the occipital lobes might have been somewhat increased, a fact that might go some way to explaining the bun or chignon of Neanderthal crania (Holloway, 1985). In all other respects, Neanderthal endocranial casts are essentially like those of modern humans. The large brow ridges of Neanderthals tend to give the impression that the frontal lobes of the brain may have been less well developed than in modern humans but this is not substantiated by the endocranial casts.

In any summary of the studies of early hominid endocasts it would be wise to take note of Falk (1986a) who stated that: 'in the final analysis, we must remember that despite their usefulness, the information that we glean from endocasts remains (literally) superficial'.

CHAPTER ELEVEN

THE FACIAL SKELETON OF HOMINOIDS

The facial skeleton is comprised of the supraorbital region, the orbital cavities, the nose and nasal cavity and the jaws. Broadly speaking, these can be designated as the upper facial, mid-facial and lower facial regions, respectively. While each of these three regions of the face have independent functions, they all influence the morphology of the facial skeleton and each region also has some effect upon the way in which the facial skeleton is joined to the neurocranium. Perhaps the most powerful influence on facial morphology is the masticatory system and it is this system in particular that dominates the morphology of the upper, middle and lower face of hominoids. The morphology of the hominoid mandible has been discussed in Chapter 5, so that this account of the facial skeleton centres largely upon the upper facial skeleton and the mid-facial skeleton of hominoids.

Facial prognathism in hominoids

The facial skeleton is separated from the neurocranium by the bones of the cranial base so that the ethmoid, sphenoid and temporal bones form a boundary between the brain and the face. The degree to which the face projects beyond the neurocranium has been the subject of many studies and usually it is the bones of the cranial base that are used as reference points from which to

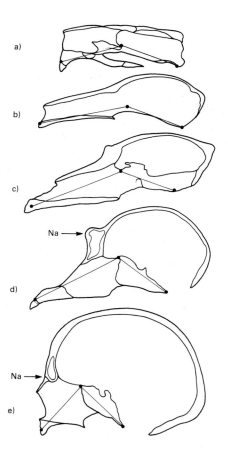

FIGURE 11·1 *The cranio-facial angle of Huxley (1863) described in the text is illustrated for (a) beaver, (b) lemur, (c) baboon, (d) chimpanzee, and (e) modern human. The position of nasion is indicated by the arrow on the chimpanzee and human crania.*

measure the degree of facial prognathism (Moore and Lavelle, 1974). Huxley (1863) defined the **craniofacial angle** as the angle between the most anterior point on the maxilla, the most anterior point on the sphenoid bone and the most anterior point of the foramen magnum (Fig. 11.1). (Another name for this angle is **sphenomaxillary angle**.) Using this angle, Huxley (1863, 1867) was able to identify major differences in the degree of prognathism within and among groups of animals including modern humans.

While endocranial reference points on the cranial base are relatively easy to locate on radiographs, or on sagitally sectioned skulls, this is not so on intact skulls. Some external (or exocranial) landmarks on the human cranium, however, have a constant relationship to endocranial landmarks. For example, _nasion_ (the junction between the nasal bones and the frontal bone in the mid-line) is always at the same level as the most anterior point on the cribriform plate of the ethmoid bone in human crania. _Nasion_ is, therefore, sometimes used to represent the foramen caecum of the ethmoid bone, at the anterior end of the cranial base, in craniometric studies of human skulls and in orthodontics. A problem arises in comparative studies of primate crania where endocranial and exocranial landmarks are not necessarily related in the same way as they are in modern human crania. Ashton (1957) was aware of this and (by way of example) defined the equivalent

of _nasion_ in great apes as 'the point where the line joining the upper limits of the frontal processes of the maxilla cross the mid-sagittal plane'. The reasoning behind this was that the anterior end of the ethmoid bone (marked by the foramen caecum – see Chapter 2) lies nearer to the level of the top of the maxilla than to the level of _nasion_, which is often much higher on the face of apes than in modern humans (Fig. 11.1). In the past, failures to recognize these sorts of anatomical differences in comparative studies, have led to misleading assertions about differences in facial prognathism between primate taxa.

Despite these problems, comparative studies of facial angles in primates, measured using carefully chosen landmarks, clearly demonstrate that infant great apes have flat or **orthognathic** faces like modern humans. However, the small facial angles of juvenile apes become progressively large with age as the degree of prognathism increases until skeletal maturity (Krogman, 1931a, b, c; Ashton, 1957) (Fig. 11.2). Some primates, _Papio_, for example, are extremely prognathic and a possible underlying cause may relate to the need for a large nose (see later). Other primates (e.g. _Theropithecus_) are still prognathic but have much taller faces (_nasion_ to mid-palatal line) that probably relate to an increasingly hard graminivorous diet. Clearly, complex interactions in facial morphology are unlikely to be measured using a single facial angle.

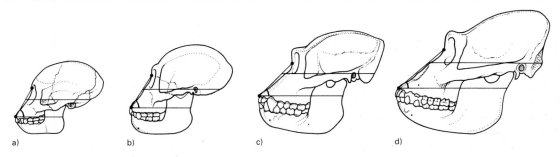

a) b) c) d)

FIGURE 11·2 _Lateral view of an ontogenetic series of_ Gorilla _crania (a) with deciduous dentition only, (b) with M1 emerged, (c) with M2 emerged, and_ (d) _an adult male. A facial angle,_ sellion–prosthion–Frankfurt horizontal, _shows how facial projection increases with age in apes._

Concentrating upon exocranial landmarks to measure facial morphology may also be misleading for other reasons. Ontogenetic expansion of the frontal air sinus or development of a pronounced *glabella* (see later) can move the exocranial landmarks of the upper facial skeleton further anteriorly in one comparative group than in another and so reduce the facial angle independent of facial changes related to the cranial base. Bilsborough and Wood (1988) have studied facial prognathism in fossil hominids and have avoided these problems by taking measurements from *porion* (the most superior point on the margin of the external auditory meatus) to a variety of landmarks on the face (Fig. 11.3). As expected, absolute measures of upper facial projection from *porion*, indicate that some larger fossil hominid specimens, e.g. specimens of *Paranthropus boisei*, as well as KNM-ER 1470 and KNM-ER 3733, have projecting upper faces (*porion–nasion*). They also indicate that smaller specimens attributed to *Australopithecus* or *Homo* have significantly smaller values with less projecting upper facial skeletons. Absolute measurements of lower facial projection (*porion–alveolare*) reveal that large specimens of *P. boisei* again have the largest values of all hominids, followed by specimens attributed to *Paranthropus robustus* and *Australopithecus*. Fossil hominid specimens attributed to *Homo* are notably much less prognathic in the lower face than are those of *Paranthropus* or *Australopithecus* and even the larger specimens such as KNM-ER 1470 or KNM-ER 3733 have absolute *porion–alveolare* measurements close to the mean for smaller *Australopithecus* specimens.

There is, nevertheless, still a problem with these absolute measurements, as Bilsborough and Wood (1988) point out. An elongated cranial base or an increased vertical or horizontal development of the face will also influence these measurements made from *porion* to the front of the face and tend to confuse comparisons between early hominids. In order to alleviate this problem these

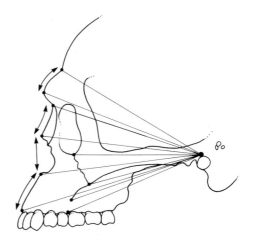

FIGURE 11·3 *Diagram from Bilsborough and Wood (1988) illustrating a variety of measurements of facial projection made from* porion *together with some of facial height.*

authors have 'collapsed' absolute measurements of facial projection onto the same horizontal plane by using an index of facial projection (Fig. 11.4). Presented in this way, measurements of facial prognathism reveal that *Paranthropus robustus* and *Australopithecus africanus* are equally prognathic in the lower face. There is also a very great range of variation in *A. africanus* such that Sts 71 and Sts 5 are at opposite extremes of the range for the entire hominid sample. Great variation also exists in lower facial prognathism within *P. boisei*, with KNM-ER 406 being much more prognathic than OH5 which has an extremely flat face. This is even more apparent between the 2.5-million-year-old specimen KNM-WT 17000 and the later specimen KNM-WT 17400. These two specimens could hardly be more extreme in their contrasting degrees of facial prognathism (Leakey and Walker, 1988), even though the latter, being subadult, may have become a little more prognathic had it lived.

Overall, a comparison of absolute measures of facial projection with indices of relative

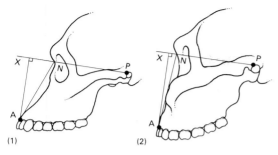

FIGURE 11·4   *Diagram showing the derivation of an index of facial projection taken from Bilsborough and Wood (1988). (1) Represents a hominid with a relatively prognathous face but which does not have a tall face superoinferiorly. (2) Represents a hominid that is relatively flat-faced or orthognathic but which has a taller face than the first example. A = alveolare; N = porion; N = nasion; X = a point on the extension of the nasion–porion line at which a perpendicular to that line joins it to alveolare. Relative facial prognathism = (XN / NP) × 100 and is derived as:*

$$\left(\frac{(Ap)^2 \times (NA)^2 \times (NP)^2}{2(NP)^2} \times 100 \right)$$

facial projection demonstrate that early *Homo* specimens are more flat faced than specimens attributed to either *Paranthropus* or *Australopithecus*. Nevertheless, Bilsborough and Wood (1988) note that the marked individual variation that exists within all early hominid taxa cautions against generalizations intended to diagnose any one group of hominids from another using these criteria. Rak (1988) has shown that when the position of the masseter origin relative to the tooth row is compared in *Australopithecus* and *Paranthropus*, *P. boisei* emerges as extreme with masseter origin shifted well forwards but KNM-ER 732 (a smaller specimen attributed to *P. boisei*) emerges as more like *P. robustus*. However, even the very prognathic KNM-WT 17000 specimen appears within the *P. robustus* range. This suggests that simple measures of facial prognathism can disguise more important functional relationships that exist within early hominids.

The morphology of the orbits and the upper facial skeleton

The orbital cavities protect the eyes and provide attachment for the muscles that move the eyes. The roof of the orbital cavity also provides support for the frontal lobes of the brain in modern humans (and, to varying degrees, in other primates). The average volume of the orbital cavity is broadly similar in modern humans, chimpanzees and orang-utans (the range of values being 16–31 cm^3), but the volume of the orbit of *Gorilla* is greater (32–37 cm^3). Schultz (1940) has demonstrated that the volume of the eye itself has little influence on the volume of the orbit (which relates more closely to body size among primates) and indeed the degree of protrusion of the eye from the orbit may also vary independently of the size of the orbit (Schultz, 1940). Values for *Gorilla* orbital volume are similar to those given for some fossil hominid specimens (Kabwe/Broken Hill, 42 cm^3; La Chapelle aux Saint, 40 cm^3; and Gibraltar, 35 cm^3) (Schultz, 1940).

The orbital cavity also demonstrates variation in the pattern of sutures between the bones of the medial wall in hominoids. The usual condition in modern humans, fossil hominids and the orang-utan is for the lacrimal bone to suture with the ethmoid bone on the medial wall of the orbit (Fig. 11.5). In the African apes, the more usual condition is for the frontal bone to extend downwards between these two bones and occasionally even completely prevent the lacrimal and ethmoid contacting each other (Wood-Jones, 1948; Sonntag, 1924). This distinct sutural pattern on the medial wall of the orbit of the African apes accompanies the similarly distinct sutural patterns that occur at *pterion* and in the anterior cranial fossa of African apes (see Chapters 4 and 10). All three regions involve the frontal bone and it is possible that they may be influenced by the more forward position of the orbits relative to the neurocranium in African apes.

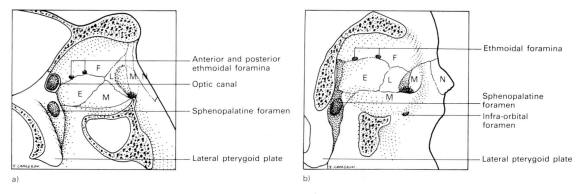

a)

b)

FIGURE 11·5   *The confluence of sutures on the medial wall of the orbit of (a)* Pan troglodytes *and (b) modern* Homo sapiens.

The supraorbital region of hominoids is variable and complicated. Besides the differing relationships that exist between the orbital cavities and the cranial cavity, the morphology of the supraorbital region is greatly influenced by the masticatory system (Fig. 11.6). *Sivapithecus* (e.g. GSP 15000) and the orang-utan appear to share several distinct supraorbital features and some of

these can be related to the face that is rotated backwards towards the neurocranium (**airorynchy**) (Shea, 1985). The **supraorbital rim** is a ridge of bone that begins in the region of *glabella* (the most anteriorly projecting point on the forehead in the midline) and arches smoothly over the orbit terminating at the frontozygomatic suture (Fig. 11.7). Clarke (1977) has described this

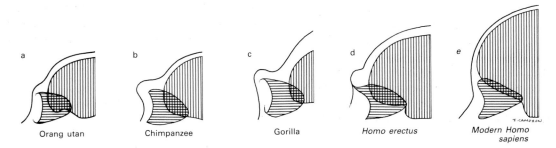

FIGURE 11·6   *The relationship of the cranial cavity to the orbital cavity in the three great apes compared with* Homo erectus *from Zhoukoudian and modern* Homo sapiens. *After Weidenreich (1943) and Shea (1985).*

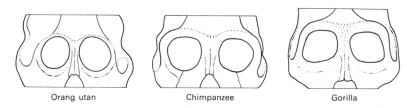

FIGURE 11·7   *The form of the orbital margin and supraorbital torus in the three great apes in frontal view.*

Orang utan   Chimpanzee   Gorilla

FIGURE 11·8   _The form of the supraorbital torus in the three great apes viewed laterally._

supraorbital rim or arch in the orang-utan (and _Paranthropus robustus_) as the **costa supraorbitalis** because the rims or bars of bone over the orbits 'resemble a pair of ribs emerging from a central backbone, represented by the glabellar region'. Because of the backwardly rotated face in _Sivapithecus_ and the orang-utan, the frontal bone rises up over the forehead immediately behind the costa supraorbitalis (Fig. 11.8). It follows that there is no depression or **supratoral sulcus** of the frontal bone behind the orbital rim as there is, for example, in the African apes.

In African apes the morphology of this region differs from that of the orang-utan and _Sivapithecus_ in that a complete bar of bone extends across the superior margins of both orbits, incorporating a prominent _glabella_ region at the mid-line (Figs 11.7 and 11.8). This is known as the **supraorbital torus** and it can be divided into three regions: a central _glabellar_ swelling that protrudes anteriorly; a medial portion of the arch between _glabella_ and the superior orbital notch, known as the **superciliary arch**; and a more lateral portion between the supraorbital notch and the zygomaticofrontal suture, known as the **supraorbital trigone** (Santa Luca, 1980) (Fig. 11.9). The orbital cavities of the chimpanzee and gorilla are more widely separated by the ethmoid bone than are those of the orang-utan and the orbital margins of the African apes lack the tall ovoid outline typical of the orang-utan and _Sivapithecus_.

The morphology of the supraorbital region in early hominids is influenced by the position of the orbits relative to the neurocranium, dependent for one upon the degree of expansion of the brain, but also upon the temporalis muscle. These factors affect the degree of postorbital constriction and the breadth across the zygomatic arches. The **temporal foramen** is bounded laterally by the zygomatic arches. The temporalis muscle passes through the foramen into the infratemporal fossa where it attaches to the coronoid process of the mandible (see Chapter 6). In African apes the temporal foramen is long and narrow (Fig. 5.29). It is shorter and wider in _Australopithecus africanus_ and in _Paranthropus_ (especially _Paranthropus boisei_) it is extremely short and very wide. This change in shape reflects the increasing development of the anterior part of the temporalis muscle. The ensuing degree of postorbital constriction of the cranium, expressed as a percentage of

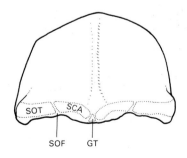

FIGURE 11·9   _The form of the supraorbital region of the frontal bone in the Ngandong_ Homo erectus _specimens, superimposed on the outline of Solo II. GT is the glabella_ torus; _SCA is the superciliary arch; SOF is the supraorbital fissure; and SOT is the supraorbital trigone. After Santa-Luca (1980)._

the biorbital breadth, clearly demonstrates this fact. In *A. africanus* the postorbital breadth is 73% of the biorbital breadth, compared with 78% and 64% in *Pan* and *Gorilla*, respectively. However, in specimens of early *Homo* (KNM-ER 1813 and 3733) values have risen to 81% and 86%, respectively, due primarily to expansion of the brain (Rak, 1983). *Paranthropus boisei* is a special case where the anterior temporalis muscle is more highly developed than in any other fossil hominids and the minimum postorbital width across the frontal bones is only 63% of the biorbital breadth (Rak, 1983). The temporalis muscle was so large and powerful in *P. boisei* that it 'squeezed' the neurocranium further posteriorly with respect to the face than in, for example, *A. africanus*.

It is easy to see that the greater the relative postorbital constriction and the wider the zygomatic arches, the longer the lateral portion of the supraorbital torus will extend as an isolated bar of bone over the orbits. A wide supraorbital torus is thus an especially prominent feature of *Paranthropus robustus* and *P. boisei*. However, expansion of the frontal bone due to increased cranial capacity in early and later *Homo*, also results in an absolutely wide supraorbital torus but its nature is different (see below).

In both *A. africanus* and *P. robustus* (as well as in the older 2.5-million-year-old specimen attributed to *P. boisei* (KNM-WT 17000) the *glabella* is prominent and the supraorbital torus arches over each orbit and descends to meet it medially. However, Rak (1983) noted that in more recent specimens of *P. boisei* the *glabella*, while still prominent, has shifted with the mid-supraorbital structures so that it is now positioned more superiorly such that the two supraorbital tori are no longer so highly arched but slope away downwards and laterally in more of a straight line towards the zygomaticofrontal sutures. As a result of this the orbital margins are less rounded in *P. boisei* than they are in *P. robustus*, because the margins now follow the sloping contours of the supraorbital torus

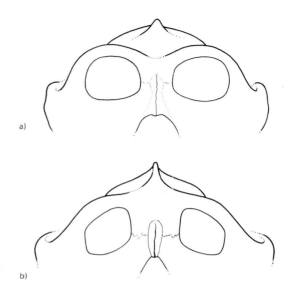

FIGURE 11·10   *The form of the supraorbital structures in (a)* Paranthropus robustus *and (b)* Paranthropus boisei. *After Rak (1983).*

(Fig. 11.10). The high position of *glabella* in *P. boisei* also reduces the depth of the supratoral sulcus, which is almost non-existent when compared with *P. robustus*.

The supraorbital tori of early *Homo* specimens tend to be expanded in width due to the increased size of the brain and its influence on the overall size of the frontal bone. This, combined with a reduced bimaxillary diameter, is typical of OH 24, KNM-ER 1813 and KNM-ER 3733 (Bilsborough and Wood, 1988). Specimens of *Homo erectus* from the Far East have a pronounced supraorbital torus that is straight, unarched and thickened towards the lateral aspect of the supraorbital trigone (in fact the trigone gets its name from this widened lateral part which forms a triangular mass of bone over the lateral aspect of the orbit) (Fig. 11.9). Occasionally, in these specimens *glabella* lies behind the plane of the supraciliary arches so that there is an anteriorly facing **glabellar concavity** in the mid-line of some specimens when viewed from above. This form of supraorbital torus contrasts with that of some

East African *H. erectus* specimens such as KNM-ER 3733 where the supraorbital tori are more rounded and less bar like across the orbits. Very few specimens are known from East Africa, however, and it is not yet clear what might be typical of this group. Specimens of *H. erectus* also often have a raised mid-line torus in the sagittal plane of the frontal bone, along the site of fusion of the metopic suture. This is known as the **metopic ridge** (Santa Luca, 1980).

The supraorbital tori of Neanderthal crania contrast with those of *H. erectus* in that they usually form a smooth convexity over each orbit (not angled at the superior orbital notch as in some *H. erectus* crania) and there is a tendency also for the torus to become thin laterally, rather than to thicken as it does in *H. erectus* crania. Neanderthal supraorbital tori are also extensively pneumatized by the frontal air sinuses (see below), whereas those of *Homo erectus* are less so, often being composed of near-solid bone.

The nasal bones and the mid-facial skeleton

Rak (1983) has noted that the nasal bones of *Paranthropus robustus* are wide at their base whereas those of *Paranthropus boisei* are narrow (Fig. 11.9). Both species, however, have nasal bones that are wide and flared superiorly, a feature which they share with most modern human crania and many great apes. Olson (1985) has drawn attention to the fact that the nasal bones of *Australopithecus africanus* are widest inferiorly but that in a specimen of *Australopithecus afarensis* (AL 333 105), like those of *P. boisei*, the nasal bones are narrow at their base (Fig. 11.11). Olson (1985) attributed considerable phylogenetic significance to the morphology at the base of the nasal bones. Eckhardt (1987), however, questioned the validity of these claims, and demonstrated that many of these morphological variations occur (albeit at low frequency) in the living great apes. Eckhardt (1987) showed that the wedge-shaped ('key-

stone') nasal bones, that are wide at the top, occasionally occur in apes but some weight must be given to the fact that they appear always to occur in *Paranthropus*, *A. afarensis* and *Homo*.

There is little about the morphology of the nasal bones in *Homo* that is remarkable except that in Neanderthals (Fig. 11.16) the inferior half of the bones have been swept markedly forwards to lie horizontally across the roof of the nasal aperture (Rak, 1987). While the Neanderthal nose is unique in several ways (Rak, 1987) it must be said that some other hominids, Cro-Magnon for example, have nasal bones (but not nasal cavities) that approach the Neanderthal condition.

The morphology of the mid-facial skeleton

The morphology of the mid-facial region is greatly influenced by the nose and by the masticatory system. Some of the functions of the nose are outlined elsewhere (see Chapter 9), but, besides the sense of smell, which is detected with the olfactory mucous membrane of the nose, these may include the warming and humidifying of inspired air (see Wolpoff, 1968) as well as the loss of body heat via the nasal mucous membrane (Scott, 1954) and possibly the cooling of arterial blood destined for the brain (see Chapter 9) (Cabanac, 1986; Dean, 1988b). It is likely that the area of the nasal mucous membrane that covers the nasal conchae (or turbinate bones) and the volume of the nasal cavity

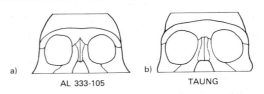

a) AL 333-105   b) TAUNG

FIGURE 11.11 *The morphology of the nasal bones in two juvenile fossil hominids (a) AL 333-105 (Australopithecus* afarensis *and (b) Taung (Australopithecus* africanus).

scale to body size in some consistent way as do all physiological variables. As, however, these variables have never been measured accurately in a representative series of primates, the precise way in which the physiology and the morphology of the nose and mid-face are interrelated can only be guessed at. Nonetheless, any changes in the height of the nasal cavity that result from masticatory adaptations are likely to be accompanied by changes in nasal breadth or depth. A good example of the interplay between respiratory and masticatory adaptations is the facial form of the Neanderthals (Rak, 1987).

The maxilla is primarily concerned with supporting the teeth and the zygomatic bones with providing attachment for the masseter muscles. Both the size and shape of the teeth and the shape of the dental arch, together with the forces generated during chewing, have a profound effect on the morphology of the mid-face. Rak (1983, 1985, 1987) has re-emphasized that masticatory forces generated during chewing are resisted in the facial skeleton. Two trends away from the general facial pattern of living hominoids can be recognized within the early hominids. The first accompanies molarization of the premolars and includes anterior migration of the origin of the masseter and an increase in the size of the anterior temporalis muscle (Rak, 1983, 1985). This trend can be identified as a morphocline through the series of fossil hominids: *Australopithecus afarensis*, *Australopithecus africanus*, *Paranthropus robustus* and *Paranthropus boisei*. The faces of hominid specimens attributed to early *Homo* demonstrate some parallels with *Australopithecus* and *Paranthropus* (see below) but, on the whole, they involve a reduction in the size of the masticatory system and so show none of the adaptations developed to resist increased chewing forces in the mid-face. The second trend that departs from the basic hominoid pattern is the unique morphology of the Neanderthal face where great emphasis on the use of the anterior dentition has resulted in a substantial mid-

facial prognathism (Rak, 1987).

Analysis of the teeth of *A. afarensis* suggests that there was little increase in the size of the premolars beyond the presumed ancestral condition, and also that there was less heavy wear on the teeth of *A. afarensis* than on those of *A. africanus* (Johanson and White, 1979; White, Johanson, and Kimbel, 1981). Accordingly, the face of *A. afarensis* is prognathic in the region of the premaxilla and is similar to that of the great apes and to specimens of early *Homo* in that the margins of the nasal aperture are thin and sharp and do not form **anterior pillars** (see later) for the support of occlusal forces. In short, these authors argue that there is nothing in the face of great apes or *A. afarensis* that show any adaptations to an increased masticatory load on the anterior post-canine dentition.

On the other hand, there is evidence to suggest that the premolars of *A. africanus* are more molarized than those of *A. afarensis* (Johanson and White, 1979) and it is clear that the premolars of *P. robustus* (and more so *P. boisei*) became increasingly molarized in their crown and root morphology. This trend to increase the surface area of the post-canine dentition was accompanied in the facial skeleton by a retraction of the palate such that the M3 is positioned closer to the articular eminence of the temporal bone and so that less of the palate protrudes anterior to *sellion*, the deepest point beneath *glabella* in the mid-line (Rak, 1983). Moreover, the origin of the masseter has shifted anteriorly in these specimens so that the force generated by this muscle lies directly lateral to the premolars and molars, rather than being located well behind them.

Increased chewing force in the premolar region, in the still quite prognathic face of *A. africanus*, was resisted by two rounded anterior pillars that run on either side of the nasal aperture (Fig. 11.12). The anterolateral aspect of the zygoma in *A. africanus* is flared outwards and forwards in the coronal plane, further lateral to the vertical plane of the

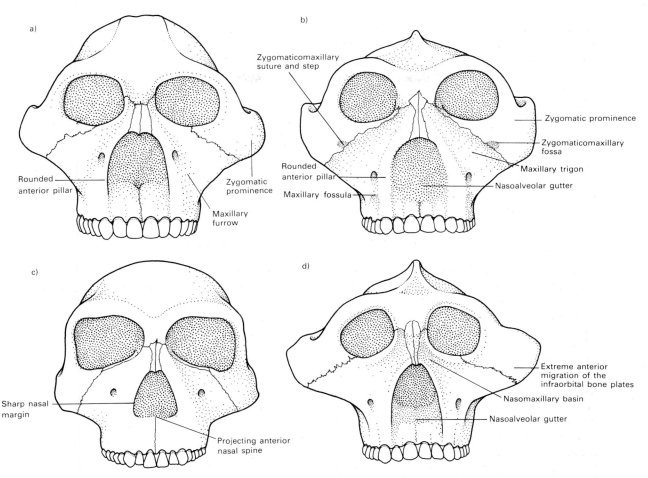

FIGURE 11·12 *Facial features characteristic of (a)* Australopithecus africanus, *(b)* Paranthropus robustus, *(c)* early Homo, *and (d)* Paranthropus boisei. *See text for details. After Rak (1983).*

lateral orbital margin than it is in the great apes. This arrangement acts to swing the origin of the masseter anteriorly and gives the mid-face a more 'diamond-shaped' anterior profile that contrasts with the 'square-shaped' great ape mid-face profile viewed anteriorly (Fig. 11.13). The maxillary process of the zygomatic bone of *A. africanus* meets the zygomatic process of the bone at a right angle, forming a feature known as the **zygomatic prominence**.

A good illustration of this morphological

shift within the mid-face is the comparison of growth changes between *A. africanus* and the chimpanzee (Rak, 1983). Superimposed tracings of the mid-face of juvenile and adult chimpanzees and juvenile and adult specimens of *A. africanus* seen in transverse section (Fig. 11.14) clearly demonstrate important differences. During growth of the face in the chimpanzee, most ontogenetic change (seen from above) occurs as the anterior face becomes prognathic during growth, whereas in *A. africanus* the lateral aspects of the face

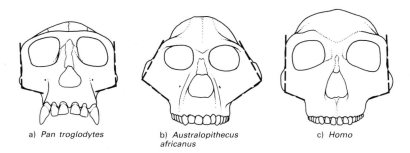

a) *Pan troglodytes* b) *Australopithecus africanus* c) *Homo*

FIGURE 11·13 *Outlines of the faces of (a)* Pan troglodytes, *(b)* Australopithecus africanus *and (c)* Homo. *The widest point on the* A. africanus *face is on the zygomatic bone but unlike the faces of* Pan *and* Homo *which are square in outline, there is more marked convergence of the facial outline towards the midline giving the* A. africanus *face a more 'diamond-shaped' profile. After Rak (1983).*

change most and swing markedly forwards, becoming much squarer in outline during growth when seen from above.

This adaptation to increased masticatory loads has progressed further in *Paranthropus*. The palate of *Paranthropus robustus* is retracted even further posteriorly than that of *A. africanus* and the premolars have become more molarized. Although the anterior dentition is reduced in size, two anterior pillars still support the maxilla either side of the nasal aperture and brace the maxilla against occlusal loading from the premolars. The rest of the anterior aspect of the maxilla is sunk deep to the periphery of the face. Two shallow depressions on the facial surface of the maxilla reflect the degree to which the bone has swung forwards at the **zygomaticomaxillary** suture. These depressions are called the **maxillary trigones** (Fig. 11.12). At the zygomaticomaxillary suture there is a raised ridge, the **zygomaticomaxillary step**, and laterally beyond this the zygomatic bone has become the most prominent aspect of the face. It is both swung more anteriorly and is more flared in *Paranthropus* than in *A. africanus*. Between the anterior pillars the alveolar bone forms a depressed **nasoalveolar gutter** that slopes gently into the floor of the nasal cavity.

Paranthropus boisei represents the extreme departure from the typical morphology of the hominoid face (Rak, 1983). The palate is

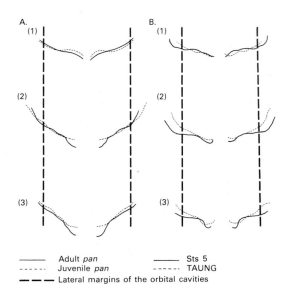

——— Adult *pan*
- - - - Juvenile *pan*
——— Sts 5
- - - - TAUNG
— — — Lateral margins of the orbital cavities

FIGURE 11·14 *Transverse sections of the face at three different levels demonstrating ontogenetic changes in (a)* Pan troglodytes *and (b)* Taung *and* Sts 5 *(*Australopithecus africanus*). Juveniles and adults are drawn to the same scale. In the chimpanzee the snout grows forwards most, whereas in* Australopithecus, *it is the periphery of the face that grows forwards most. After Rak (1983).*

markedly retracted. The zygomatic bones and the origin of the masseter muscle have migrated so far forwards and extended so much that they form 'visor-like' supports for the masticatory forces (Fig. 11.12). While there is still a nasoalveolar gutter (as there is in *P. robustus*), there are no longer any anterior pillars. Rak (1983) has argued that the extreme forward position of the infraorbital bone plates were able to support the occlusal load of the dentition in its entirety.

Compared with *Australopithecus* and *Paranthropus* the face of early *Homo* is somewhat less tall. As noted above, it also tends to be narrower across the maxilla, which together with the expanded supraorbital torus, presents a picture which contrasts with that of other early hominids (Fig. 11.12). The zygomatic bone in *Homo* is different to that of *Australopithecus* and *Paranthropus* in that it is either vertical or posteriorly sloping (see Fig. 11.15), although that of OH 24 (which, however, is quite distorted) is antero-inferiorly sloping but not to the degree seen in *Australopithecus* and *Paranthropus* (Bilsborough and Wood, 1988). The height of the zygomatic bone at its root on the maxilla is between 28% and 35% of the total facial height in *Australopithecus* and is greater than 40% in *Paranthropus*. The corresponding height for specimens of *Homo* fall within the range of those for *Australopithecus* but these are absolutely smaller and reflect only the smaller face of *Homo* (Bilsborough and Wood, 1988). While the mid-face of *Homo* is flattened, in comparison with other early hominids, the degree of prognathism at the alveolar bone of the lower face again overlaps with the range for *A. africanus*. Fransiscus and Trinkaus (1988) have drawn attention to the increased prominence of the nasal spine in early *Homo* and it is likely that this accompanies mid-facial retrusion while the volume of the nasal cavity is held constant relative to body size as the face reduces in size. One specimen of early *Homo* (KNM-ER 1470) appears superficially to resemble *Paranthropus*, at least in the form of the zygomatic bones. However, Bilsborough and Wood (1988) pointed out that it is the extreme retrusion of the mid-face rather than the true anterior migration of the root of the zygoma on the face of KNM-ER 1470 that underlies this apparent similarity. With the possible exception of KNM-ER 1470 and its appar-

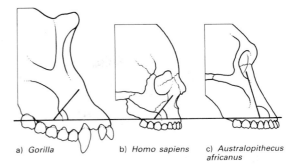

a) *Gorilla*   b) *Homo sapiens*   c) *Australopithecus africanus*

FIGURE 11·15 *Side view of the zygomatic region in (a)* Gorilla, *(b)* Homo sapiens, *and (c)* Australopithecus africanus. *Note the angular relationship of the root of the zygoma to the horizontal plane in these specimens. After Rak (1983).*

ently expanded zygomatics, there are no specimens of *Homo* that show any of the specialized morphological adaptations to increased masticatory stress, such as the anterior pillars, that the other early hominids do. Rather, the overall picture is one of facial reduction.

The Neanderthal mid-face

Evidence from several mandibles and complete skulls suggests that the anterior teeth of Neanderthals were large (especially anteroposteriorly) and, it is argued, were heavily loaded and became worn more quickly than the molars and premolars. Oddly though, Neanderthal crania do not show pronounced muscle markings at any of the attachments for the muscles of mastication. It is also now well established that the nasal cavity of Neanderthals was capacious, although as we have seen no clear function has been ascribed to this increase in size.

Maintaining a capacious nasal cavity (anteroposteriorly) and providing support for increased loads from the anterior dentition resulted in a unique facial morphology among Neanderthals that deviates from the generalized hominoid form (Rak, 1987). Rather than being aligned in the coronal plane, the

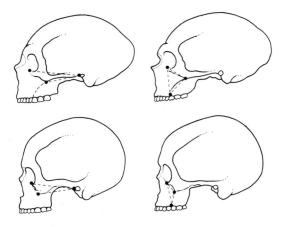

FIGURE 11·16 *The distance from the upper part of the zygomaticomaxillary suture to the ear and from the lower part of the same suture to the ear in the Neanderthal (upper left) and in modern* Homo sapiens *(lower left). The area of a triangle whose base is the upper and lower ends of the zygomaticomaxillary suture and whose apex is the alveolar margin above the upper first permanent molar in the Neanderthal (upper right) and* Homo sapiens *(lower right). Both these differences stem from the more sagittally orientated infraorbital plate in the Neanderthal. After Rak (1987).*

zygomatics and the maxilla have become sagittally orientated such that the whole of the lower mid-face has become drawn out anteriorly (Fig. 11.16). Rak (1987) and Demes (1987) have argued that this suite of characters reflects the fact that the Neanderthal face is adapted to resist bending and torsion in the sagittal plane that results from loading the anterior dentition. In the same way that patients who wear complete dentures have difficulty in preventing the upper plate falling down at the back when they bite only with their anterior teeth, the Neanderthal midface is adapted to resist this kind of torsion. Drawing the face out in this way probably also contributed to the large retromolar space behind the last molar seen in many Neanderthal mandibles and also to the remarkable nasal bones that have been mentioned above.

Sub-nasal morphology of the midface in hominoids

Ward and Kimbel (1983) have drawn attention to the morphology of the premaxillary region in hominoids. In *Sivapithecus* and *Pongo*, the premaxilla curves into the nasal cavity and joins the palatal process of the maxilla without a step in the floor of the nose. Both these genera also only have a single opening into the incisive canal on the anterior aspect of the palate (Schwartz, 1983). The African apes and early fossil hominids differ from this condition. In these specimens the nasoalveolar clivus curves back into the nasal cavity but joins the palatal processes of the maxilla at a marked vertical step in the floor of the nose (Fig. 11.17). African apes also differ from *Pongo* in having two openings in the incisive canal. Modern *Homo* seems to have been derived from the African ape pattern although the orthognathic face of modern humans may

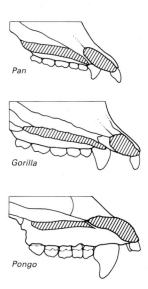

FIGURE 11·17 *Sagittal sections through the hard palate and premaxilla in* Pan, Gorilla *and* Pongo. *In the orang-utan there is no step between the premaxilla and the palatal process of the maxilla in the floor of the nose as the two are in smooth continuity. After Ward and Kimbel (1983).*

be responsible for superimposing some orang-utan-like features in the floor of the human nose.

The paranasal air sinuses

Some details of the paranasal air sinuses have been described in Chapter 4 but a comparative account of them is presented here. Many of the bones of the primate cranium are pneumatized and most of these **air sinuses** grow out from the nasal cavity. However, pneumatization of the temporal bones especially that of the mastoid process originates from the middle ear cavity via the **mastoid antrum**.

The functions of the paranasal air sinuses are unknown, but many theories have been proposed. These theories include: lightening the front of the skull to assist in its balance; increasing the area of olfactory mucous membrane in the nose (a role the frontal sinus does perform in cats); increasing the resonance of the voice; and the elimination of unnecessary bone between two functionally independent regions of the cranium. Of these, the last theory is usually held to be the most reasonable. However, among modern human populations there is considerable variation in the degree of pneumatization of, for example, the frontal bone, which is extensive in African populations but minimal in Australian aboriginals and Eskimos (see Blaney (1986) for a review of the literature). These observations favour the existence of a strong genetic influence for the diversity of frontal sinus development among modern human populations as opposed to or as well as a functional one.

The orang-utan has no ethmoidal air sinuses (or cells) and as the frontal sinus develops from the anterior ethmoidal air cells it can never develop a frontal sinus either (Cave and Haines, 1940; Cave, 1961). It could be argued that the reduced space between the orbits in the orang-utan and the backwardly rotated face that places the supraorbital rims very close to the neurocranium results in there being little room for these air spaces anyway. The African apes, however, like fossil hominids and modern humans have both ethmoidal and frontal air sinuses, although as mentioned previously, there is variation in the degree of pneumatization of the frontal bone between different groups of fossil hominids. Blaney (1986) has noted that relative to skull length (a reasonable indicator of body size) the chimpanzee has larger frontal sinus volumes than does the gorilla. Even between different sub-species of gorilla there may also be significant differences in the relative size of the frontal sinus. These facts underline the likely genetic influence on pneumatization.

The maxillary sinus (or **maxillary antrum**) is particularly extensive in great apes and may even excavate the palate and zygomatic regions. Conroy and Vannier (1987) have noted that even in the Taung juvenile the maxillary antrum has extended into the palate and zygomatic region, as it does in African apes. Previously, this was only documented in adult specimens of great apes and fossil hominid specimens attributed to *Paranthropus*. The body of the sphenoid bone is also extensively pneumatized in hominids and great apes. In mature great apes the sphenoidal air sinus commonly extends laterally, such that it separates the intracranial cavity from the greater wing of the sphenoid in the infratemporal fossa by up to 10–15 mm. In modern humans this air-filled cavity does not extend so far laterally but it may extend to and even invade the basioccipital posteriorly.

CHAPTER TWELVE

THE CERVICAL SPINE AND SUPPORT OF THE HEAD

All primates, as well as many other animals, habitually hold their heads so that their eyes look forward towards the horizon. Human subjects when asked to stare at their own eyes in a mirror, invariably orientate their heads so that the plane of the level floor is within 1° of the **Frankfurt Horizontal** (Downs, 1952). The Frankfurt Horizontal is a line drawn through *orbitale* (the lowest point on the inferior border of the orbit) and *porion* (the highest point on the bony external auditory meatus) when the skull is viewed in *norma lateralis*. Therefore, the Frankfurt Horizontal is, with good reason, often used as a reference plane in studies of primates that describe the spatial orientation or position of cranial landmarks relative to the presumed habitual posture of the head. Some anatomists have adopted another strategy and have made use of the fact that the horizontal semicircular canals of the inner ear, that are concerned with the sense of balance, always lie in a plane parallel with the ground in all animals and so can also be used to orientate skulls in their true habitual position (Cousin and Fenart, 1971; Fenart and Deblock, 1973).

The position of occipital condyles and foramen magnum

It has become established in the literature that there is a relationship between the posture of hominoids and the forward position of the foramen magnum and occipital condyles. Indeed even before the discovery of early hominid postcranial fossils a vertically orientated and anteriorly positioned foramen magnum was considered indicative of erect posture and bipedal locomotion (Dart, 1925; Broom, 1938a; Le Gros Clark, 1971). Bolk (1909) had earlier devised an index for humans and primates that expressed the pre-occipital part of the cranial base in terms of the whole cranium, and concluded that the modern human foramen magnum was anteriorly positioned relative to that of the great apes. He also concluded that the more posterior the position of the foramen magnum, the more vertical its orientation. Bolk believed these changes were due to changes in brain growth. Ashton and Zuckerman (1952, 1956) and Schultz (1955) confirmed the backward migration of the foramen magnum and occipital condyles during the growth of great apes, as well as the consistently forward position of these structures in modern humans. Ashton and Zuckerman used indices that excluded the effects of the facial skeleton but not the influence of the nuchal crests of large apes. Moore et al. (1973) and Adams and Moore (1975) have since also confirmed the changing orientation of the occipital condyles during growth in apes but noted that the orang-utan (a modified brachiator and so, it is argued, more habitually upright) exhibits the most vertically orientated occipital condyles. Biegert (1957, 1963) proposed that the negative

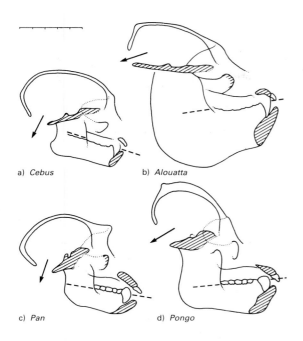

a) *Cebus*   b) *Alouatta*

c) *Pan*   d) *Pongo*

FIGURE 12·1 *Midsagittal sections through the skulls of* Cebus, Alouatta, Pan *and* Pongo. *The position and orientation (arrowed) of the foramen magnum is influenced by the relative size of the brain in primates. After Biegert (1963).*

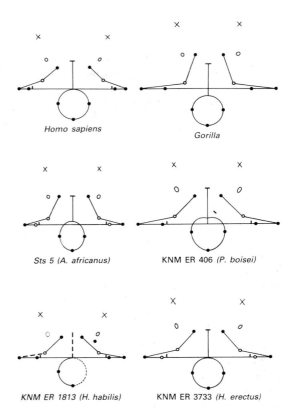

Homo sapiens   Gorilla

Sts 5 (A. africanus)   KNM ER 406 (P. boisei)

KNM ER 1813 (H. habilis)   KNM ER 3733 (H. erectus)

FIGURE 12·2 *Mean values of measurements taken on 30 specimens of modern* Homo sapiens *and* Gorilla *highlight major differences between the relative positions of landmarks defined on the cranial base. The same measurements are shown for four early fossil hominids;* Sts 5 *(*Australopithecus africanus*),* KNM ER 406 *(*Paranthropus boisei*),* KNM ER 1813 *(*Homo habilis*) and* KNM ER 3733 *(*Homo erectus*). Distinguishing features of each are described in the text. After Dean and Wood (1982).*

allometry of the brain together with the positive allometry of the masticatory system were a powerful influence on the position of the foramen magnum and occipital condyles and cited their relatively forward position in the small-bodied quadrupedal squirrel monkey (Fig. 12.1). This underlying allometric influence has also become clear in a comparative study of *Pan paniscus* and *Pan troglodytes* by Cramer (1977). The results of these studies suggest that it is difficult to predict a secure relationship between the forward position of the foramen magnum (or its orientation) and the habitual posture of hominoids. They also emphasize the potential role of the brain in influencing the position and orientation of the foramen magnum in primates.

The relative position of the foramen mag-

num has also to be assessed in basal view using the bitympanic line across the cranial base as a reference plane (Fig. 12.2). In this way the influence of the face and the nuchal regions on assessments of position can be overcome. However, other problems arise in that any differences in orientation of the tympanic plates that occur relative to the mid-line sagittal plane between comparative groups can still obscure the data. When studied in this way great apes all have a foramen magnum that lies well behind the bitympanic line even during infancy and so are quite distinct from early fossil hominids and modern *Homo sapiens*. The anterior margin of the foramen magnum in *Paranthropus* lies consistently in front of the bitympanic line; the foramen magnum of *Australopithecus africanus* also lies just in front of the bitympanic line but less markedly so compared with *Paranthropus*. Some specimens attributed to early *Homo* (SK 847, OH 24 and KNM-ER 1805) have a foramen magnum whose anterior margin, like that of *Paranthropus*, lies well in front of the bitympanic line. However, modern *Homo* as well as other fossils attributed to early *Homo* (KNM-ER 1813, 3733 and 3883) have a foramen magnum whose anterior margin lies on or slightly behind the bitympanic line (Dean and Wood, 1981b, 1982).

It is often suggested that the forward placement of the foramen magnum in hominids (relative to its position in great apes), is an essential adaptation to 'balance the head more efficiently' on an erect spine (e.g. DuBrul, 1977, 1979). This generalization, however, is again not sufficiently sound to make predictions about the position of the foramen magnum or the habitual posture of early fossil hominids. Schultz (1942) pointed out that many animals such as kangaroos, giraffes, camels, llamas, antelopes, birds (and even some dinosaurs) habitually support or supported a prognathic head on an erect spine for most of the time. It is all too easy to be misled by thinking of bones in isolation and not as an integral part of the musculoskeletal system which is always in equilibrium with itself.

Surprisingly, enlargement of the brain in hominids has not increased the relative weight of the head to be supported. Schultz (1942) demonstrated that in 35 primate species he studied, the wet weight of the head in all species equalled less than 10% of the body weight. In the chimpanzee the relative head weight is 5.8% and in modern humans 5.4% of the total body weight (the larger face of the former nearly equalling the heavier brain of the latter). Gibbons support much heavier heads relative to their body weight (7.3%) than do either modern humans or chimpanzees, but it is notable that there has been no trend to shift the foramen magnum forwards to balance the head more efficiently even though gibbons are true brachiators and are 'upright' during locomotion. Therefore, while it cannot be shown that the forward position of the foramen magnum and occipital condyles in early hominids and modern humans is 'essential', it clearly has had an effect upon the forces required to move and balance the head on the cervical spine and there have been accompanying changes to the muscles in the neck.

Schultz (1942) placed the occipital condyles of whole wet heads of primates on a transverse, horizontal sharp edge and measured the force needed to lift the face at its most forward point as well as the downwards force needed to pull the face up from its most nuchal point (Fig. 12.3). Modern human heads require only 15.9% of their weight to raise the face from the horizontal, whereas the average for all other primates is 37.3%. In all primates, however, the weight of the head is heavier in front of the occipital condyles than behind them. Very much more force is then needed to raise the face by pulling downwards from behind and this is the mechanism by which the nuchal muscles support the head in life; 22.3% of the total head weight in modern humans and an average of 120% in apes and monkeys are required to do this (Schultz, 1942). As might be predicted from their similar proportions, the heads of modern humans and foetal and newborn apes (where the occipital condyles and foramen magnum are more anteriorly positioned than in adults) balance in a near-identical way. During the development of an individual the weight of the face and head are always in equilibrium with the mass of the nuchal muscles that support them at any one time. In modern humans and fossil hominids, failure of the foramen magnum

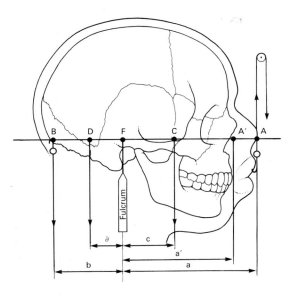

FIGURE 12·3 *Lateral view of a human head (with the skull outlined) mounted on a fulcrum (F) at the midpoint of the occipital condyles. Various lever systems about this axis are illustrated. C is the centre of gravity of the precondylar portion of the head and D is the centre of gravity of the postcondylar portion of the head. A is the most anterior point on the face, A' in the plane of prosthion and B inion. After Schultz (1942).*

The cervical vertebral column

There are seven cervical vertebrae in man. Of these the third, fourth, fifth, and sixth all share common features and are called **typical** cervical vertebrae. The **atlas** and **axis** (the first and second vertebrae), together with the seventh and last cervical vertebrae have features that distinguish them from the other cervical vertebrae and so are known as the **atypical** cervical vertebrae. Commonly, cervical, thoracic, lumbar and sacral vertebrae are notated in an abbreviated form as C, T, L and S together with the number in the series of vertebrae being denoted as a suffix. In this way a shorthand notation for any vertebra can be written quickly, e.g. C1, C2 or C6.

The cervical vertebral column is curved and has an anterior convexity to it like the lumbar region (Fig. 12.4). Each cervical vertebra (Fig. 12.5) has a foramen in it which when combined with all the other vertebral foramina forms the **vertebral canal** for the spinal cord. In the cervical region this canal is large, because it is nearest the brain, and triangular in cross-section.

and occipital condyles to migrate backwards (to varying degrees) during growth has resulted in a reduction in the mass of the nuchal musculature needed to support the head from behind in adults. There is evidence, however, from studies of rats, rabbits and humans, which suggests that changes in posture during the growth period are indeed accompanied by small changes in growth and remodelling within the cranial base. These changes have been shown to affect the degree of cranial base flexion and also the position of the foramen magnum (DuBrul, 1950; Bjork and Kuroda, 1968). Other factors, however, influence the form of the cranial base in hominids and on their own similar morphological observations in fossil hominids are again difficult to attribute solely to upright posture in any reliable way.

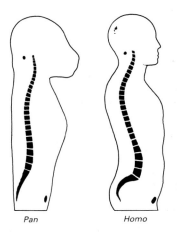

FIGURE 12·4 *The curvatures of the vertebral column. Median sagittal sections of the vertebral column of* Pan *and* Homo sapiens. *After Schultz (1961).*

213

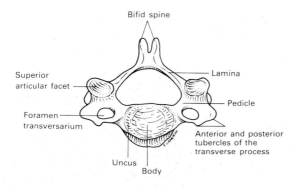

Bifid spine

Superior
articular facet

Lamina

Foramen
transversarium

Pedicle

Anterior and posterior
tubercles of the
transverse process

Uncus

Body

FIGURE 12·5 *The component parts of a typical cervical vertebra.*

A typical cervical vertebra consists of a mass of bone, the **body** at the front having an oval appearance when seen from above or below. Sometimes, the body of a vertebra is referred to as the **centrum**. Behind the body the **vertebral arch** encloses the vertebral foramen. Either side of the body at the front, the vertebral arch is formed by a short rounded **pedicle**. The posterior aspect of the vertebral arch is formed by two **laminae** which are flattened plates of bone that meet in the mid-line posteriorly. Two processes project sideways from the body and are known as the **transverse processes**.

Each transverse process of a cervical vertebra has a foramen in it, the **foramen transversarium** through which the vertebral artery travels cranially to supply the brain. This is a unique feature of all cervical vertebrae. The foramen nearly divides the transverse process into two, so that two tubercles project from the tips of the process at the back and at the front, rather than there being only one tubercle. These are known as the **anterior and posterior tubercles of the transverse process**.

Each typical cervical vertebra also has a **spinous process** that projects backwards from the mid-line where the two laminae join. The tips of the spinous processes of typical cervical

vertebrae are commonly bifid, at least in European populations (see later). At the sides, each typical cervical vertebra articulates with the vertebrae above and below it via **articular processes** that are angled upwards and backwards. The bone between the superior and inferior articular processes of the cervical vertebrae is known as the **lateral mass**.

The bodies of the vertebrae are joined together by fibrocartilaginous **intervertebral discs**, but in the cervical region the superior rim of the body projects upwards to meet the body above at the sides. This upwards extension of the body is called the **uncus** (uncus (L) = a hook) and is another special feature of the cervical vertebral column (Fig. 12.6). These **uncinate processes** form small atypical synovial joints with the body of the vertebra above; they are absent at birth but develop at about six years of age in humans and are called uncovertebral joints. These are probably adaptations to rotational movements of the cervical vertebral column in bipedal animals and are absent in other animals such as horses and cattle which swing their heads from side to side to look laterally rather than rotating their cervical vertebral column (Penning, 1988).

The first, second, and seventh cervical vertebrae are atypical. The first or atlas supports the skull and allows a nodding movement between the occipital condyles and two large oval concave superior articular facets. The atlas transmits the weight of the head to the second cervical vertebrae, the axis, via two rounded but flat inferior articular facets (Fig. 12.7). The atlas has no body to bear any of the weight of the skull and does not have a spinous process or flattened laminae, but simply has small tubercles in the mid-line of the thin ring-like anterior and posterior arches. These arches form a large vertebral foramen in the atlas, as big as the foramen magnum in the occipital bone.

The axis is unique because of the **dens**, a small peg of bone that protrudes upwards behind the anterior arch of the atlas. The

FIGURE 12·6 *Anteroposterior radiograph of disarticulated human cervical vertebrae. Courtesy of Theya Molleson and Helen Liversidge.*

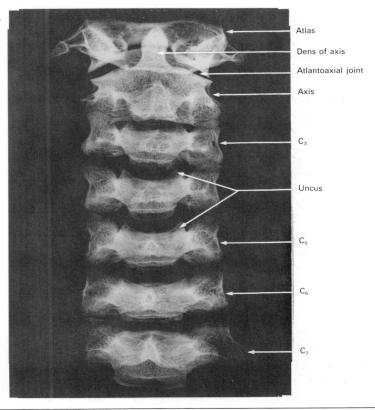

Atlas

Dens of axis

Atlantoaxial joint

Axis

C_3

Uncus

C_5

C_6

C_7

axis forms synovial joints between the flat horizontal inferior articular facets of the atlas and its own flat superior articular facets (Fig. 12.6). These joints allow a side-to-side movement. If these facets were not flat, rotation of the head would not be possible. This is a useful point to remember because the upper and lower surfaces of the atlas can be identified from the concave upper facets and the flat lower facets in modern humans.

The spinous process of the axis is large and rectangular as well as being bifid like the typical cervical vertebrae. Like all cervical vertebrae, both the atlas and the axis have a foramen transversarium in their transverse process. The last cervical vertebra, the seventh, is also an atypical cervical vertebra because the spinous process is especially long (but not bifid) and because of this it is known as the **vertebra prominens.**

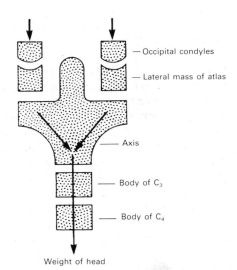

Occipital condyles

Lateral mass of atlas

Axis

Body of C_3

Body of C_4

Weight of head

FIGURE 12·7 *The weight of the head is transmitted through the occipital condyles to the lateral mass of the atlas and then to the axis. From here weight is transferred to the bodies of the cervical vertebrae.*

The cervical vertebrae of the great apes

The typical cervical vertebrae of the great apes are notable for their pronounced spinous processes (Figs 12.8–12.10). Unlike modern humans, however, where the seventh vertebra or vertebra prominens, bears the longest spinous process, in great apes it is more usually the fifth or sixth cervical vertebrae that does so. The axis is commonly the only bifid spinous process and the tips of this point markedly downwards. However, even this vertebra is not constantly bifid in apes (Sonntag, 1924) being more usually so in *Gorilla* and less often the case in *Pongo*. The remaining spines of the cervical vertebrae, with the exception of the atlas, end in an enlarged knob. Duckworth (1904) noted that the foramen transversarium is commonly

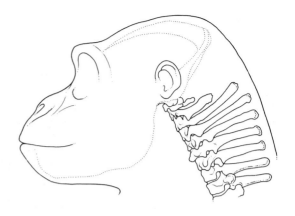

FIGURE 12·8   *The cervical vertebrae of a gorilla (C1 to C7) together with two thoracic vertebrae (T1 and T2) seen in lateral view. After Willoughby (1978).*

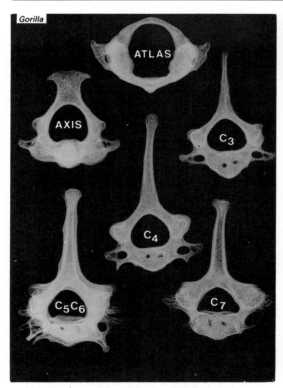

FIGURE 12·9   *Disarticulated cervical vertebrae from a modern human and a female gorilla. Note there is congenital fusion of the gorilla C5 and C6.*

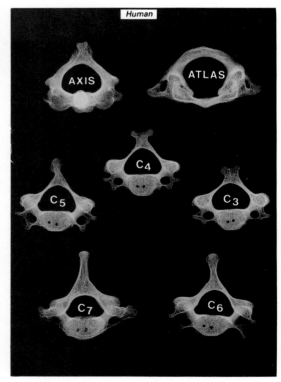

Radiograph courtesy of Theya Molleson and Helen Liversidge.

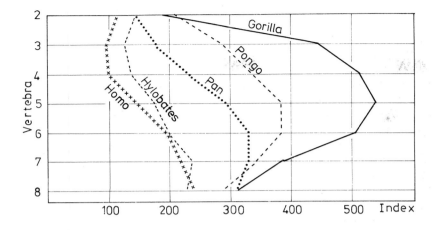

FIGURE 12·10 *The length of the spinous process of the cervical vertebrae (C1–C8) expressed as a percentage of the midsagittal diameter of the corresponding vertebral body for several hominoids. After Schultz (1961).*

absent in the seventh cervical vertebra of apes and even occasionally from the fifth or sixth. This is due to the underdevelopment of the **costal element** which forms the anterior element of the transverse process. It is not uncommon for the anterior tubercle of C7 to be reduced in modern humans also.

The superior facets of the atlas that bear the occipital condyles are more concave in apes than in modern humans and often an extra foramen for the vertebral artery is formed between the undersurface of these facets and the transverse process of the atlas. The mid-line anterior tubercle of the atlas is very prominent in large apes and points downwards rather than forwards as in humans. Another notable feature of great ape vertebrae is the well-developed uncinate processes of the bodies of the cervical vertebrae. When the axis is viewed anteriorly, the angulation of the plane of the superior articular facet to the mid-line is greater in monkeys and great apes than it is modern humans (Ankel, 1967). Seen in lateral view, the odontoid process of the axis is more vertical in modern humans and does not incline posteriorly as it does in other primates (Fig. 12.11).

The cervical vertebrae of fossil hominids

Few cervical vertebrae are preserved in the earlier fossil hominid record but there is a

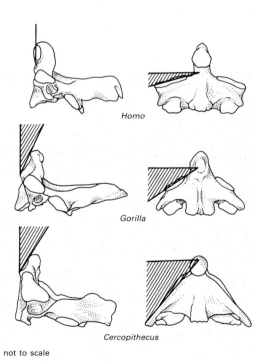

Homo

Gorilla

Cercopithecus

not to scale

FIGURE 12·11 *Three axis vertebrae belonging to* Homo, Gorilla *and* Cercopithecus *are illustrated in lateral and anterior view. The dens (or odontoid peg) of the axis is vertical when viewed laterally in the human but orientated more posteriorly in other primates. The superior articular facets of the axis are also more nearly horizontal when viewed anteriorly in modern humans than in apes and monkeys. After Ankel (1967).*

217

fragment of an atlas from Koobi Fora (KNM-ER 1825) and a near-complete axis from Swartkrans (SK 854) from which the spinous process has been fractured. Fragments of an atlas and axis exist from Hadar (AL 333-83 and 101) but are not well preserved. There is a well-preserved sixth or seventh cervical vertebra from Hadar (AL 333-106). Its substantial spinous process is not bifid (Fig. 12.12), but it is also notable for its body which is described as being unusually small relative to the remainder of the specimen (Lovejoy et al., 1982). The degree of curvature of the superior articular surface of the atlas, that forms a synovial joint with the occipital condyles is variable in hominids. Those attributed to *Australopithecus afarensis* and *Paranthropus* are more concave than those known for *Homo* and this may be an adaptation for extending the head further back in *A. afarensis* and *Paranthropus*.

Robinson (1972) has described an axis vertebra (SK 854) from Swartkrans that has been attributed to *Paranthropus*. While the odontoid process of this vertebra is fractured at its base it can be judged much thicker at the lower aspect of the body than at this site of fracture. The body of the axis has a raised ridge of bone (or keel) running superoinferiorly in the anterior mid-line and the bone of the neural arch (the pedicle and lamina) of SK 854 is quite rounded in cross-section. Robinson (1972) noted that the preserved right superior articular facet is large and has an appreciable degree of concavity, rather than being nearly flat as is usually the case in *Homo sapiens* but once again the specific significance of this feature is unknown. The preserved right inferior articular facet of this axis vertebra is also unusual. It appears to be inclined inferiorly more than in modern human counterparts and looks as if it is cut obliquely into the neural arch and not set out on a distinct protuberance as it is in modern humans. Together with what little is known of the atlas (KNM-ER 1825) described here, it seems that the relationship between the articular facets of the upper cervical

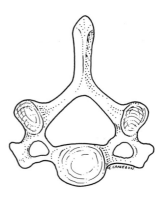

FIGURE 12·12   *The C6 or C7 vertebra (AL 333-106) from Hadar, Ethiopia seen from above and illustrating the long spinous process. After Lovejoy et al. (1982).*

vertebrae to each other and to the head in some early hominids may have been somewhat different to that found in modern humans.

Boule (1911–1913) and Boule and Vallois (1912) noted that the spinous processes of the cervical vertebrae of the Neanderthal specimen from La Chapelle-aux-Saints are not bifid. This is, however, not unusual in modern populations and there is really little to distinguish the cervical vertebrae of Neanderthals from those of modern humans, although the lengths of the spinous processes may be at the top end of the modern human range. Shore (1930) in a comparative study of modern human cervical vertebrae, noted that only 7% of the spines of Bantu lower cervical vertebrae are bifid whereas 71% are bifid in Europeans. Interestingly, all foetal cervical spinous processes are reported as bifid in both Bantu and European. Shore also described a variation in the inclination of the spinous processes in modern human populations and in the Neanderthal from La Chapelle-aux-Saints but other Neanderthal cervical vertebrae from Kebara and Shanidar suggest there are no real differences between Neanderthal and modern human cervical vertebral spinous processes.

The muscles that move and balance the head and neck in modern humans

The muscles of the neck are responsible for supporting and moving the head and the cervical vertebral column. These muscles can be divided into a group in front of the cervical vertebrae, the **prevertebral muscles**, and a group at the back of the neck, the **nuchal muscles** (Fig. 12.13). Generally speaking, the prevertebral group flexes the head and neck forwards and the nuchal group extends the head backwards. Other muscles, for example the **sternocleidomastoid** muscles, are also especially important for turning the head from side to side but are also involved in flexing or extending the head and neck. All these muscles, however, also constantly relay vital information to the brain about the orientation and position of the head and neck in space. This proprioceptive function may be more important for some muscles than their ability to contract, particularly for the smaller muscles that lie deep in the neck.

Many of the muscles of the neck and trunk take part of their name from the region in which they act, for example **capitis** when attached to the head, **cervicis** when in the neck, **thoracis** when in the thorax, **lumborum** when in the lumbar region. In the neck both deep muscles and superficial muscles act on the cervical vertebral column and head and these may originate from almost any part of the cervical vertebrae or even from the ribs, thoracic vertebrae, clavicle or scapula. Not all these muscles will be described here, as only those that create bony markings on the skull or the ribs, clavicle and scapula are of importance to anthropologists. The muscles of the neck in modern humans are described here simply as a deep group around the region of the foramen magnum, that are enveloped progressively by longer more superficial prevertebral and nuchal muscles. Their functions as prevertebral flexors, nuchal extensors and rotators of the head and neck will be considered in more

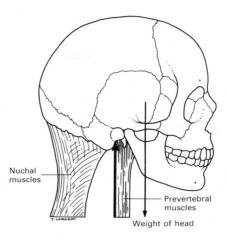

FIGURE 12·13 _The centre of gravity of the head falls in front of the occipital condyles (hence our heads fall forwards when we doze off in the upright position). Consequently, the mass of the nuchal muscles is greater than the mass of the prevertebral muscles which between them support the head in an upright position._

detail later in a comparative context along with a description of these same muscles in great apes.

The group of muscles which lie deepest in the neck, the **rectus capitis muscles** (Fig. 12.14), arise from the atlas and axis and attach to the cranium around the perimeter of the foramen magnum. They are named, from the front around to the back, anterior, lateral and posterior, but as there are two pairs of **rectus capitis posterior** muscles one is called **major** (because this pair is twice as long as the other), and the other is called the rectus capitis posterior **minor**.

All these muscles are short and can exert little leverage on the head but act to keep the synovial joints of the region closely packed together and, in addition, relay important information about head movement to the brain. Rectus capitis anterior originates from the lateral mass of the atlas, rectus capitis lateralis from the top of the transverse process of the atlas, rectus capitis posterior minor from the posterior tubercle of the atlas

219

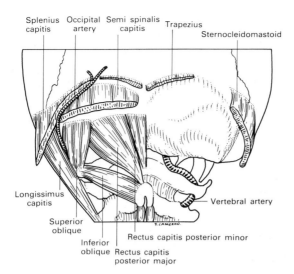

Splenius capitis Occipital artery Semi spinalis capitis Trapezius Sternocleidomastoid

Longissimus capitis

Superior oblique

Inferior oblique

Rectus capitis posterior major

Rectus capitis posterior minor

Vertebral artery

FIGURE 12·14 *The muscles of the suboccipital triangle in a modern human. Note the occipital artery which is claimed to groove the temporal bone between the attachment of the digastric muscle and the superior oblique muscle. Note also the vertebral artery as it winds out of the foramen transversarium of the atlas and around the lateral mass of the atlas to enter the foramen magnum.*

and rectus capitis posterior major from the spinous process of the axis. Posteriorly, there is another muscle on each side that passes obliquely inwards to the cranium from the transverse process of the atlas (this is the widest transverse process of all cervical vertebrae and is nearly as far lateral as the mastoid process of the temporal bone). This muscle is the **superior oblique** and it creates a characteristic 'comma-shaped' marking on the skull base, as well as a raised ridge of bone around its outer margin, the **occipitomastoid crest** (see later). This bony marking continues posteriorly around the outer margin of the rectus capitis posterior major to meet in the mid-line as the **inferior nuchal line** and so encircles all the deep muscles that attach to the squamous part of the occipital bone. Together, these five pairs of short deep muscles attach to the head in the same way the finger tips of both hands can support the base of a large ball when it is held high in the air.

The long prevertebral muscles
Several longer muscles (Fig. 12.15) also attach to the skull base and so usually take the name **longus** (longus (L) = long) or **longissimus**

(longissimus (L) = longest). In front of the foramen magnum the **longus capitis** attaches to much of the basioccipital and is an important flexor of the head (sabre-toothed cats with blade like stabbing teeth had especially well developed longus capitis muscles). This muscle shares its origin from the anterior tubercles of the transverse processes of C3 to C6 with the **scalenus anterior** muscle (see below). Arising from the anterior tubercle of the atlas is another long muscle, the **longus colli**, that does not run upwards to attach to the skull but rather spreads downwards from the bodies and transverse processes of C2 to T3. This muscle also flexes the neck and is responsible for the large downward-pointing anterior tubercle of the atlas in the great apes.

The long nuchal muscles
Behind the rectus capitis posterior muscles and the superior obliques, in the mid-line of the neck posteriorly is another muscle that marks the occipital bone, the **semi-spinalis capitis** (Fig. 12.16), where it fills in the space between the inferior and superior nuchal lines on the cranium. This muscle does the exact opposite of the longus capitis and is respon-

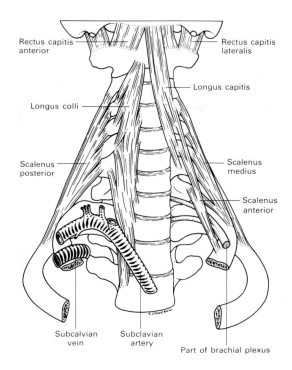

Rectus capitis anterior

Rectus capitis lateralis

Longus capitis

Longus colli

Scalenus posterior

Scalenus medius

Scalenus anterior

Subcalvian vein

Subclavian artery

Part of brachial plexus

FIGURE 12·15 _Anterior view of some of the prevertebral muscles together with the scalene muscles which attach to the first and second ribs. The subclavian vein and subclavian artery pass respectively in front of and behind scalenus anterior leaving grooves on the superior surface of the first rib._

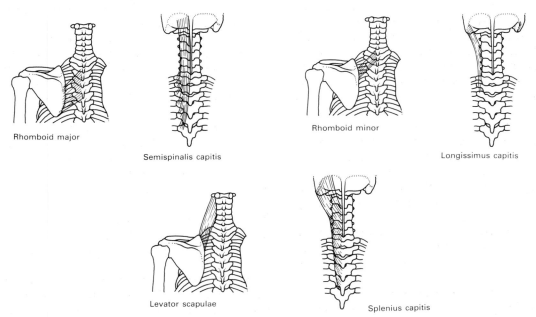

Rhomboid major

Semispinalis capitis

Rhomboid minor

Longissimus capitis

Levator scapulae

Splenius capitis

FIGURE 12·16 _Posterior view of the skull, vertebral column and thorax illustrating individual muscles described in the text._

221

sible for most of the support of the head in the upright position (which, as we have seen, in all primates falls forwards without support from behind). The semi-spinalis capitis arises from the transverse processes of the upper thoracic and lower cervical vertebrae C4, C5, C6 and C7. The longest muscle attached to the head, **longissimus capitis**, arises from the upper thoracic and cervical transverse processes and inserts into the temporal and occipital bones along a thin roughened line of attachment, deep to the mastoid process. This muscle is wrapped around by a more superficial muscle, the **splenius capitis**, once thought to resemble a bandage (splenion (Gk) = a bandage), that also attaches along a line deep to the mastoid process. Unlike longissimus capitis, splenius capitis continues backwards and almost meets in the mid-line posteriorly. The splenius arises from the spinous processes of the cervical and upper thoracic vertebrae as well as from a tough elastic ligament that runs from the external occipital protuberance to the prominent spine of C7. This ligament, the **ligamentum nuchae**, connects all the spines of the cervical vertebrae in man and is important in some animals for the support of the head. While present in man, the ligamentum nuchae is surprisingly absent in the great apes.

The posterior belly of the digastric muscle

While it is not a muscle that contributes to the movement or balance of the head, the attachment of the **posterior belly of the digastric** muscle is described in this section because it is intimately associated with the foregoing muscles and its bony markings have been described with them in some comparative studies. Immediately posterolateral to the rectus capitis lateralis (which attaches to the jugular process of the occipital bone) there is a narrow groove or notch in the temporal bone deep to the mastoid process in human skulls. This is the cranial origin of the posterior belly of the digastric muscle and is known as the **digastric fossa** or **groove**. Lateral to this groove are the lines of attachment of the longissimus capitis and splenius capitis as well as the sternocleidomastoid more superficially on the mastoid process. Taxman (1963) found that the posterior belly of the digastric muscle is also associated with the formation of a bony crest, the **juxtamastoid eminence** (see later), the most medial aspect of the digastric fossa. Another groove is also commonly present on human skulls just medial to the juxtamastoid eminence and is thought to be formed by the occipital artery as it passes on the inside of the posterior belly of the digastric muscle towards the nuchal muscles (but see later).

The scalene muscles

One group of muscles in the neck that originate from the anterior and posterior tubercles of the transverse processes of the cervical vertebrae do not attach to the skull. Instead they run downwards to the first and second ribs where they create important bony markings, especially on the first rib. Besides acting to support the vertebral column in the neck, they are also accessory muscles of respiration and raise the rib cage during sharp inspiration. These three muscles are called the **scalenus anterior**, **scalenus medius** and **scalenus posterior** (Fig. 12.17). As a group they look like a triangular mass of muscle in the neck and so like an asymmetrical (skalenos (Gk) = uneven) scalene triangle. Another muscle which is really part of the same series is so far back that it passes downwards to insert onto the scapula and so is called the **levator scapulae** muscle. Scalenus anterior inserts onto the **scalene tubercle** of the first rib. Anterior to this tubercle is a groove created by the subclavian vein and behind it is a groove for the artery (and nerves to the arm) and then behind this groove there is another roughened region on the superior surface of the first rib for the attachment of scalenus medius (Figs 12.18; 15.29).

a)

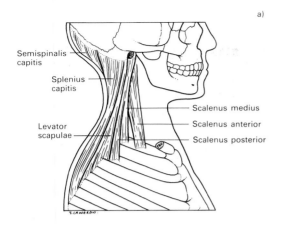

b)

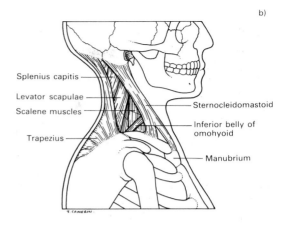

FIGURE 12·17 *The scalene muscles and levator scapulae in lateral view (a) (Figure 12.15 illustrates these muscles viewed anteriorly). When the trapezius and sternocleidomastoid muscles are added (b) the deeper muscles of the neck form the 'floor' of what*

is known as the posterior triangle of the neck. The borders of the posterior triangle are the anterior border of trapezius, the superior border of the clavicle and the posterior border of the sternocleidomastoid muscle.

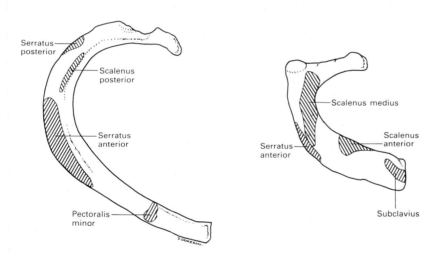

FIGURE 12·18 *Muscle attachments that create markings on the superior surfaces of the first and second ribs.*

The sternocleidomastoid and trapezius muscles

Superifical to all other muscles of the neck are the **sternocleidomastoid** (often incorrectly shortened to sternomastoid) and the **trapezius** muscles. Both these muscles attach to the superior nuchal line of the cranium, trapezius from behind forwards and sternocleidoma-

stoid from the mastoid process backwards (Fig. 12.19). Both muscles meet each other (nearer the mid-line posteriorly than to the mastoid process). Two parts of the sternocleidomastoid (the cleido-occipital and cleidomastoid parts) run downwards and forwards to attach to the medial aspect of the clavicle. The other two parts (the

223

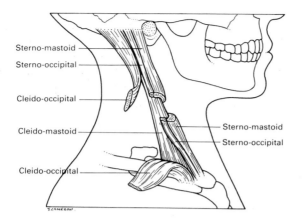

FIGURE 12·19 The sternocleidomastoid muscle can be divided into two parts that attach to the mastoid process (the sternomastoid and cleidomastoid parts) and two parts that attach to the occipital squama (the sterno-occipital and cleido-occipital parts). After Rouvier (1927).

sternomastoid and sterno-occipital parts) run downwards and more medially to attach to the manubrium of the sternum. The trapezius muscle fans outwards and downwards across the neck and back to the outer part of the clavicle and spine of the scapula and then inwards again (so forming a trapezoid shape in posterior view). It attaches in the mid-line posteriorly to the ligamentum nuchae in the cervical region and to all the thoracic spines below this.

The muscles of the neck in great apes

The neck of great apes is broader more squat and more muscular than that of modern humans and also often contains air sacs that protrude from the saccules of the larynx through deficiencies in the thyrohyoid membrane (see Chapter 13). Most of the muscles of the great ape neck are both absolutely bigger than those of man and more complicated, there being more 'bellies', 'bodies' and 'slips' to most of them. All apes and most other mammals have a muscle in the neck that has been lost in modern humans. It runs from the transverse process of the atlas to the outer aspect of the clavicle and is commonly named the **omocervacalis** or **atlanto-clavicularis** (Figs 12.20; 12.22; 12.23; see also Fig. 16.20). Other equally confusing names for this muscle appear in the compara-

tive literature such as **trachelo-acrominal** (trachelo (Gk) = neck), **levator-claviculae, omo-atlantic** (omo (Gk) = shoulder). Parsons (1898) considered that this muscle was functionally a part of the trapezius that was deeply attached to the atlas instead of the superior nuchal line. Owen (1831) considered it an extra member of the scalene, levator scapulae group that passes to the clavicle rather than to the ribs or scapula. He argued that it acts to fix the head of the scapula for the long head of the triceps to act from during extension of the elbow in quadrupedal locomotion.

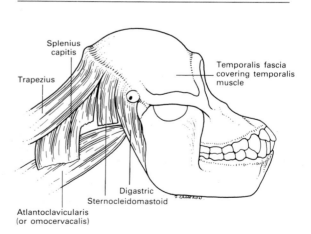

FIGURE 12·20 *Superficial muscles of the neck in great apes.*

Anteriorly, in the prevertebral region of apes, the rectus capitis anterior muscles lie side by side and are not spread apart by the longus capitis muscles being positioned between them as they are in man (Figs 12.21 and 12.22). The longus capitis in apes also contrasts with that of man in that it is markedly elongated and oval in cross-section at the basioccipital and its long axis runs from front to back rather than from side to

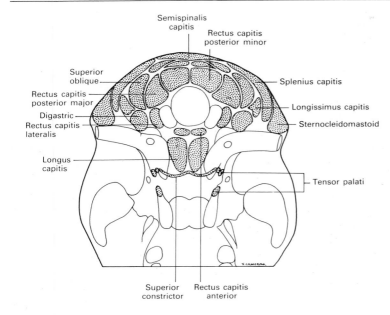

FIGURE 12·21 *Muscle attachments on the cranial base of a great ape. After Dean (1985a,c).*

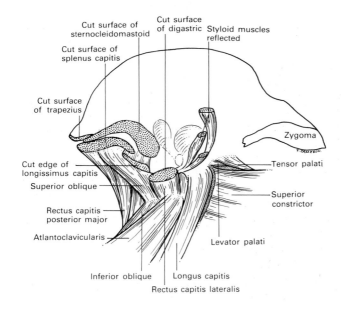

FIGURE 12·22 *A deeper dissection of the muscles of the great ape neck in lateral view.*

side across the basioccipital as in humans. This raises two bilateral prominences on the basioccipital. (A mid-line furrow created between these two large muscles can sometimes be seen running down the posterior wall of the pharynx in living great apes.) Because the basioccipital (and the longus capitis) are so elongated in this way in apes, the carotid artery and internal jugular vein lie beside these muscles and not so close to the wall of the pharynx as in modern humans.

The posterior belly of the digastric muscle in great apes is large and rounded in cross-section at its attachment to the cranial base. It leaves no bony marking or notch as in humans (Dean, 1985c) and is positioned over the temporal and occipital bones and so overlies the occipitomastoid suture. Nevertheless, the muscle still lies posterolateral to the rectus capitis lateralis in apes and deep to the muscles of the mastoid process. Because of this, a fossa for the longissimus capitis muscle in apes that resembles the digastic groove in modern human crania, can be mistaken for the origin of the posterior belly of the digastric muscle in great apes. No obvious groove for the occipital artery exists on the occipital bone of great apes.

Posteriorly, in the nuchal region, there are also important features in apes that contrast with those of modern humans. The most obvious is the bony nuchal crest (Fig. 12.20). The trapezius and sternocleidomastoid muscles rise off the edge of the nuchal crest posteriorly, but the temporalis muscle never reaches this posterior edge when there is a compound temporal nuchal crest. The free bony superior aspect of the crest has only one thin muscle attached along a part of it, the **occipitofrontalis** muscle which moves the scalp during facial expressions.

In the chimpanzee and gorilla, the splenius capitis gains attachment to the temporal and occipital bone beneath the lip of the nuchal crest and deep to the trapezius and sternocleidomastoid. In the great apes, extra support for the head comes from the **rhomboid muscles** (Owen, 1831; Winkler, 1988;

Miller, 1932) which usually only pass from the scapula to the thoracic spinous processes in man but here rise all the way up to the nuchal crest just deep to trapezius.

The semispinalis muscle (Fig. 12.23) is massive in all apes and is divided into a medial **biventer part**, so called because a tendinous intersection partially divides it horizontally into upper and lower bellies, and a more lateral **complexus part**. The biventer is thick and triangular in transverse cross-section at the cranial base and is continuous across the mid-line as there is no ligamentum nuchae (and so no **external occipital protuberance** in apes) to divide it. The lateral complexus part usually consists of a separate slip or 'strap-like' muscle that inserts more laterally to the mid-line beneath the splenius. Together the parts of the semispinalis capitis form a mass of muscle at the cranial base. Other muscles of the nuchal region in great apes differ only from those of modern humans in their size and internal divisions so that the remaining rectus muscles and longissimus capitis muscles are not described here but are simply illustrated in Fig. 12.23.

The nuchal and mastoid regions in early fossil hominids

In all fossil hominids the foramen magnum and occipital condyles are positioned further anteriorly than in great apes. The occipital region has also expanded (considerably in some hominids relative to great apes). However, despite an increased area for attachment of the nuchal musculature (Fig. 12.24), the muscle mass in this region has become reduced in size, both relatively and absolutely in comparison with great apes (Adams and Moore, 1975). This means that in hominids there is more bony space available for a reduced amount of nuchal musculature than there is in great apes.

Weidenreich (1951) considered that the length of the tympanic plates and width of the skull in the lateral region of the cranium was partly related to the degree of develop-

FIGURE 12·23 *The nuchal musculature and the muscles of the suboccipital triangle of great apes. Note that in great apes the rhomboid muscle can attach to the cranium and that the semispinalis capitis in large apes is divided into a medial biventer cervicis portion and lateral complexus parts.*

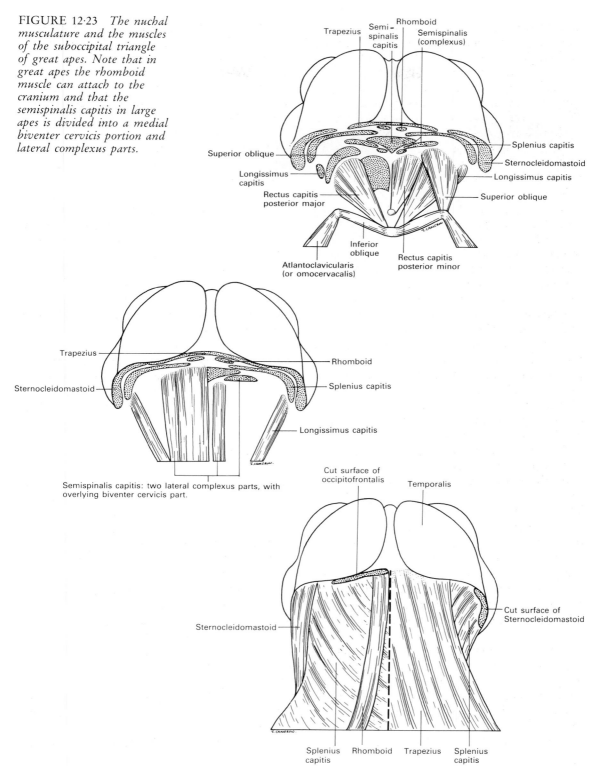

Semispinalis capitis: two lateral complexus parts, with overlying biventer cervicis part.

227

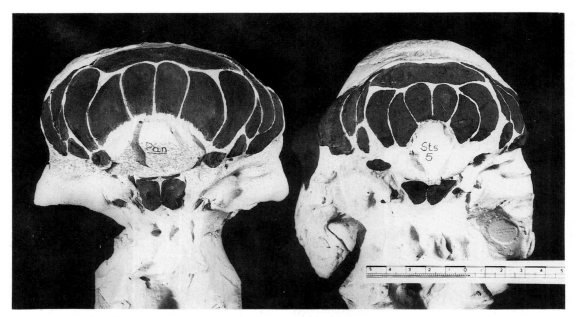

FIGURE 12·24 *Relative to great apes, all fossil hominids have a reduced area of attachment for the nuchal musculature on the cranial base. The muscle markings have been highlighted in a chimpanzee and are compared here with those derived from and highlighted on a cast of Sts 5* (Australopithecus africanus). *See Figure 12.21 for a key to the individual muscle markings.*

ment of the masticatory system. When this system is reduced in size, as it is in later hominids relative to the great apes, and there is an accompanying lateral migration of more medial structures on the cranial base (that are squeezed outwards by the forward movement of the occipital condyles), less space exists on the cranial base for muscle attachment. The result is the development of bony processes and grooves on the cranial base that provide a greater area of attachment to accommodate muscles. The mastoid process and the digastric fossa therefore seem to have developed at least partly in response to a decrease in the amount of bone available for their attachment to the cranial base. Overall then, muscles in the nuchal region of fossil hominids appear to be less crowded on the cranial base than is the case in the great apes but muscles attached to the cranial base more laterally seem to have become more crowded and are associated with grooves and crests that do not always exist on the cranial base of the great apes.

Larger fossil hominid specimens (AL 333–45, KNM-ER 406, KNM-ER 1805 and OH 5), demonstrate extensive bony cresting on the cranial base for the attachment of longissimus capitis, splenius capitis and sternocleidomastoid, on the mastoid process and nuchal crest. Smaller fossil hominid specimens (AL 288–1, AL 62–28, MLD 37/38, Sts 5, SK 47, OH 24 and KNM-ER 407) and also specimens of *Homo erectus* (KNM-ER 3733 and KNM-ER 3883), as well as specimens from Zhoukoudian and Java (that all have larger cranial capacities) show fewer pronounced muscle markings on the cranial base. However, in contrast to great ape crania, the majority of fossil crania do show some evidence of an external occipital protuberance for the attachment of a ligamentum nuchae.

All fossil hominids appear to have a digastric groove medial to the mastoid process and Broom et al. (1950), Tobias (1967), Ashton and Zuckerman (1956) and Weidenreich (1943) have all commented on the 'human-like' appearance of this fossa or groove in early hominids and later _H. erectus._

The mastoid process of early fossil hominids is a variable feature, in both its size and shape. It may be relatively large, as in specimens attributed to _Paranthropus_, or smaller, as in specimens attributed to _Australopithecus_, but it must be said that even great ape morphology varies considerably in this region of the cranium and little store can be set by small differences that may exist in mastoid morphology between the small numbers of specimens that have been attributed to early hominid taxa. Ashton and Zuckerman (1952) and Kimbel et al. (1985) have illustrated several apes and hominids in which the degree of variation in these features can be clearly seen.

Nonetheless, Olson (1985) has noted that the mastoid process of _Paranthropus_ is expanded relative to that of _Australopithecus africanus_, early _Homo_ and great apes and that the digastric muscle affords attachment to its inner aspect with no need to raise extra attachment area in the form of a juxtamastoid crest. Olson (1985) argued that the smaller mastoid process of _A. africanus_ and early _Homo_ required that the posterior belly of the digastric gain additional attachment from a juxtamastoid crest or eminence on the medial border of its area of attachment. Olson further suggested that the pattern in _Paranthropus_ is also present in a specimen from Hadar (AL 333–45), there being an inflated mastoid process, an absence of a juxtamastoid crest but an occipitomastoid crest incorporated into the medial aspect of the mastoid process. Drawing heavily on dental evidence (there being no comparable cranial evidence) Olson (1985), recognized a gracile group at Hadar (AL 128–23, 198–1, 288–1, 311–1 and 411–1) distinct from the group more similar to _Paranthropus_. He considered these specimens to be primitive members of the _Homo_ clade which would

predictably be expected to share a reduced mastoid process along with _A. africanus_, early _Homo_ and the great apes.

Kimbel et al. (1985) have refuted Olson's claims by once more pointing out the variable nature of the ape mastoid process (and the fact that Sts 5, Sts 25 and MLD 37/38, specimens attributed to _Australopithecus africanus_, are inflated well lateral to their supramastoid crests, and yet still form a part of Olson's gracile group). These authors also noted some confusion in the identification of the juxtamastoid eminence and the occipitomastoid crest in Olson's analysis and, furthermore, disagreed with the dental evidence put forward to support the existence of more than one group of hominids at Hadar.

Olson (1985), however, did provide new and important evidence from human and great ape dissections which demonstrated that the occipital artery never actually grooves the occipital bone medial to the digastric muscle and that it is the raised margins of the digastric and superior oblique muscles that create a depression between them in modern human crania.

The nuchal and mastoid regions of Neanderthals

Walensky (1964) studied the region of the mastoid process and digastric fossa in modern humans and Neanderthals (Figs 12.25 and 12.26). Neanderthal crania exhibit a very prominent occipitomastoid crest, a broad digastric fossa with no juxtamastoid eminence forming its medial border and a small mastoid process (even smaller than the occipitomastoid crest in most crania!). Modern human crania, on the other hand, have a greatly reduced occipitomastoid crest and have a narrow digastric fossa with a pronounced juxtamastoid eminence forming its medial border as well as a large projecting mastoid process which is usually bigger in males than females.

The mastoid process lies behind the axis of rotation at the occipital condyles in apes

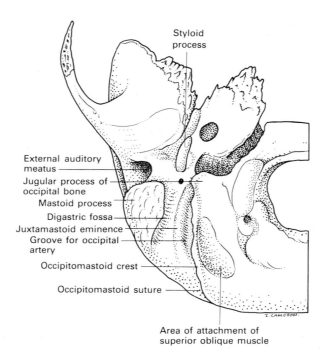

Styloid process

External auditory meatus

Jugular process of occipital bone

Mastoid process

Digastric fossa

Juxtamastoid eminence

Groove for occipital artery

Occipitomastoid crest

Occipitomastoid suture

T. CAMERON.

Area of attachment of superior oblique muscle

FIGURE 12·25 *Drawing of the cranial base of a modern human cranium. The groove for the attachment of the posterior belly of the digastric muscle is shown medial to the mastoid process, as well as the raised medial border of this marking which is called the juxtamastoid eminence. The 'comma-shaped' area of attachment of the superior oblique muscle has a lateral margin which is also raised and known as the occipitomastoid crest. The occipitomastoid suture and a groove for the occipital artery are positioned between the digastric and superior oblique muscles in modern humans.*

and in early hominids. As a result the sternocleidomastoid muscles that attach to the mastoid processes and superior nuchal line act with the nuchal muscles to support the head. In addition, these muscles also perform a vital role as rotators of the head from left to right in all hominoids. The size of the mastoid processes in early hominids, therefore, may well bear some relationship to the relative weight of the face to be supported.

In modern humans the sternocleidomastoid muscles have an additional function (Fig. 12.27). When the head is extended backwards, the long tips of the mastoid processes swing in front of the axis of rotation of the occipital condyles, so that the pull of the most anterior fibres of the sternocleidomastoid muscles (see Fig. 12.19) on the tips of the process act to

flex the head forwards and not to support the face as in great apes (Krantz, 1963). It is possible then, that large modern human mastoid processes provide mechanical advantage for the sternocleidomastoid to bring the head forwards when rising from a horizontal position. Perhaps the more important function of this muscle is to rotate the head from side to side. Large mastoid processes in early fossil hominids would certainly have rotated the head but may have had an additional and different function to those of modern humans. As such they might only have been able to act with the nuchal muscles to support the face as their most anterior fibres may not have been able to flex the head and neck forwards due to their position behind the axis of rotation of the occipital condyles.

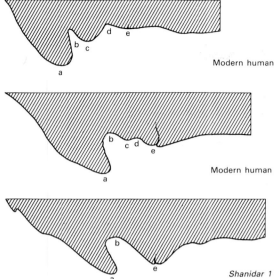

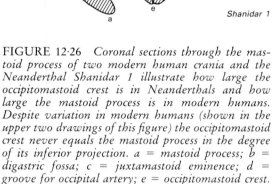

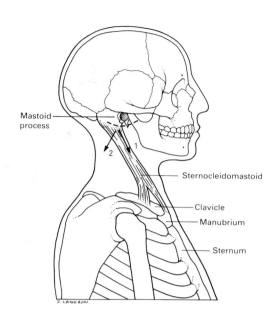

FIGURE 12·26   _Coronal sections through the mastoid process of two modern human crania and the Neanderthal Shanidar 1 illustrate how large the occipitomastoid crest is in Neanderthals and how large the mastoid process is in modern humans. Despite variation in modern humans (shown in the upper two drawings of this figure) the occipitomastoid crest never equals the mastoid process in the degree of its inferior projection. a = mastoid process; b = digastric fossa; c = juxtamastoid eminence; d = groove for occipital artery; e = occipitomastoid crest. After Walensky (1964)._

FIGURE 12·27   _Fibres of the cleidomastoid and sternomastoid portions of the sternocleidomastoid muscle (see Figure 12.19) act on the tip of the mastoid process. When the head is extended backwards (such that arrow 2 would be in line with the position of arrow 1 in the diagram) then the tips of the mastoid processes swing into a position in front of the axis of rotation of the occipital condyles. This allows these muscle fibres to act as flexors of the head. These muscle fibres become tense and can be palpated in the neck when rising from a horizontal position. After Krantz (1963)._

THE ANATOMY OF THE VOCAL TRACT

Bones associated with the pharynx, larynx and tongue are occasionally preserved in the fossil hominid record and have been used to reconstruct parts of the vocal tract and to make predictions about the mechanisms of speech, breathing and swallowing in early hominids. In this chapter we shall review the comparative anatomy of the vocal tract together with studies that have attempted to reconstruct the upper respiratory tracts of fossil hominids.

Vocalizations are produced within the upper respiratory tracts of animals but bear no relationship to the ability of an animal to communicate using language. (Parrots and budgerigars for example, are able to imitate human-speech sounds convincingly and yet clearly do not have language, as we understand it.) The ability to learn language and to communicate using either some form of sign language or speech cannot be predicted from attempts to reconstruct the architecture of the vocal tract of hominids. As such the anatomy of speech and of the vocal tract on the one hand and the origin of language on the other (see Chapter 10) remain quite separate issues.

The pharynx and larynx

Food passes downwards to the stomach via a muscular tube called the **oesophagus**. In fact, this muscular tube extends upwards from the stomach to the base of the skull (Fig. 13.1). At the top end, however, the front part of the tube is deficient and communicates with three other regions: the nose or **nasal cavity**, the mouth or **oral cavity**; and the airway via the **larynx**. This upper extension of the oesophagus with its communications anteriorly is known as the **pharynx** (pharanx (Gk) = a chasm). The lowermost deficiency in the anterior wall of the pharynx that opens opposite the inlet of the larynx, is known as the region of the **laryngopharynx**. Those regions of the pharynx opposite the oral cavity and nasal cavity are known respectively as the **oropharynx** and **nasopharynx**. The back of the muscular wall of the oesophagus (or the pharynx higher up), lies on the front of the vertebral column and its prevertebral musculature (see Chapter 12). The pharynx is able to slide up and down over these underlying structures during swallowing, speaking or singing.

The larynx is a valve-like structure that guards the opening into the **trachea**, or windpipe. During swallowing the larynx closes to prevent food entering the trachea. The larynx also closes during straining or while lifting heavy objects to retain air so that the thorax may act as a fixed rigid framework for muscles to take origin from. If it were not for the larynx it would be more difficult to make powerful arm movements and lift heavy objects. Despite these important functions, more often than

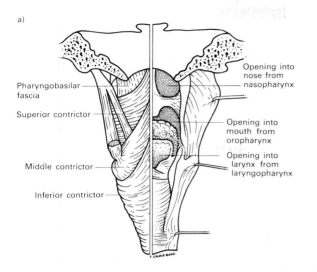

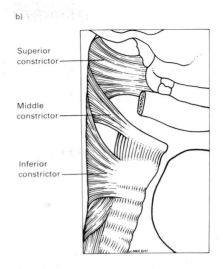

FIGURE 13·1 *(a) Posterior view of the pharynx with the constrictor muscles retracted to expose the nasopharynx, oropharynx and laryngopharynx on the right hand side. (b) Lateral view of the constrictor muscles and their respective attachments to the skull, hyoid bone and oblique line of the thyroid cartilage.*

not we associate the human larynx with the speech sounds that originate from vibrations of the **vocal folds** of the larynx, sometimes referred to colloquially as the vocal cords. These folds lie in the sagittal plane and are able to come together (or **adduct**) at the midline of the larynx and so close off the top of the trachea (see later and Fig. 13.6). They may also vibrate as air is forced between them during expiration and it is this vibration of the vocal folds that gives origin to speech sounds. In fact the larynx gets its name from the Greek word 'larungao' meaning to screech or scream. The quality of sound from the vibrating sound source is modified by the changing size and shape of the pharynx together with important interplay between the air-filled oral and nasal cavities. These spaces act as a resonator complex and have an effect on the volume and pitch of the sounds created at the vocal folds; this is called **phonation** (Fig. 13.2). For example, the vowel 'U' is produced when the tongue lies deep in the mouth and the pharynx is expanded in height and volume, so that both

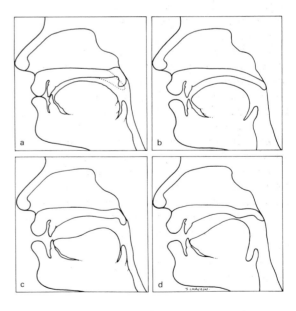

FIGURE 13·2 *(a) Tensor and levator palati together with a part of the superior constrictor act to close off the nasopharynx from the mouth during swallowing. (b) Forming the sound 'i'. (c) Forming the sound 'ah'. (d) Forming the sound 'u'. After DuBrul (1980).*

233

mouth and pharynx are acting as voluminous resonators. The vowel 'I' is produced when the pharynx remains expanded but the oral cavity is reduced in volume by raising the tongue. The reverse situation, with the pharynx constricted in both height and diameter to form a resonator of reduced volume but with the oral cavity expanded by a low-lying tongue, enables us to create the vowel sound 'A'. Hence we say 'aah' at the doctor when presenting a sore throat. Articulations between the lips, teeth, tongue, and hard and soft palate create **speech sounds** or consonants that result from momentary closures between any of these adjacent structures and the subsequent sudden or prolonged release of air flow through them. The muscles of the mouth and pharynx are continuously moving and changing the nature and quality of human speech sounds.

The nasopharynx and soft palate of modern humans and apes

The wall of the pharynx is composed of three circular muscles, wrapped around each other like three stacked plant pots (Fig. 13.1). These muscles are the **constrictor** muscles of the pharynx and are named the **superior**, **middle** and **inferior constrictors**. These muscles are covered on their outer surface by a layer of membranous fascia known as the **pharyngobasilar fascia**. This lining allows the pharyngeal musculature to contract and move independently of the other muscles in the head and neck and to slide over the prevertebral musculature. The pharyngobasilar fascia extends upwards and attaches to the base of the skull around the roof of the nasopharynx.

In great apes (Fig. 13.3) the superior constrictor muscle attaches to the periphery of the roof of the nasopharynx at the cranial base together with the pharyngobasilar fascia (Dean, 1985a). The superior constrictor muscle in modern humans, however, only attaches to the skull base at one small area, close to a raised tubercle on the basioccipital,

called the **pharyngeal tubercle**. The remainder of the superior margin of this muscle has been drawn downwards, away from the skull base, so that there is a greater space at the superior margins of the sides of the muscles of the nasopharynx, closed only on the outside by the pharyngobasilar fascia and on the inside by the mucous membrane that lines the roof the nasopharynx. The width of the modern human nasopharynx at the skull base is also expanded in comparison with that of great apes so that, overall, modern humans have a wider and taller nasopharynx than great apes do.

Two small muscles, destined to enter the top of the soft palate, take origin from the bones of the skull base just lateral to the periphery of the nasopharynx (Fig. 13.3). One of them, the **levator palati muscle**, squeezes through the pharyngobasilar fascia between the top of the superior constrictor and the skull base. It then runs downwards and inwards to enter the top of the soft palate. The function of this muscle is to pull

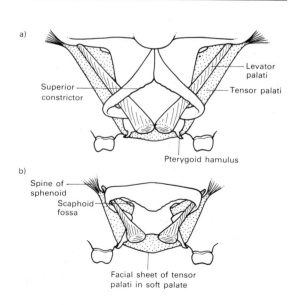

FIGURE 13·3 *The relations of the tensor and levator palati muscles to the superior constrictor muscle in (a) modern humans and (b) great apes.*

the soft palate upwards and so aid in closing the nose off from the mouth during swallowing. The other muscle is the **tensor palati**. This muscle attaches to the spine of the sphenoid and the scaphoid fossa on the cranial base in modern humans (see Chapter 4) and drops downwards on the *outside* surface of the superior constrictor but then converges to a tendon that hooks around the pterygoid hamulus of the medial pterygoid plate of the sphenoid bone (Fig. 13.3). The part of the tensor palati on the outside of the superior constrictor looks triangular in outline, when viewed from the side. The tendon of the tensor palati continues to travel around the pterygoid hamulus and spreads out into the roof of the soft palate as a sheet of tough fascia that joins the sheet of the other side. At rest, the soft palate is arched, but when the tensor palati tenses it pulls it taut across the roof of the mouth, the front part of the soft palate is depressed slightly towards the tongue (Fig. 13.2a). The levator palati muscle pulls the more posterior part of the soft palate upwards to close off the nasopharynx. The combined action of the tensor and levator palati muscles are responsible for closing the nose off from the mouth during swallowing and for many of the changes in speech sounds during speaking. Other muscles of the soft palate and pharynx assist in this sphincteric closure of the nasopharynx but these are not described here.

In great apes where the nasopharynx is narrower across the skull base than in humans, both tensor and levator palati take origin more medially than in modern humans. They both do this from the temporal bone as the tensor palati does not extend as far laterally as the sphenoid bone in apes. Both these muscles are associated with a large spine of bone that, at first glance, resembles the styloid process of the temporal bone. This spine of bone is called the **eustachian process**. Modern human crania do not have a eustachian process on the temporal bone, although a small raised portion of the tympanic plate may be associated with the attachment of the

levator palati muscle. As we have seen, the modern human nasopharynx is wider than that of apes and, as a result, the tensor palati has been displaced further laterally onto the small spine of the sphenoid bone (just medial to the foramen spinosum that transmits the middle meningeal artery, see Chapter 9). Modern human infants and great apes have a flat profile to the underside of the cranial base when it is viewed in the mid-sagittal plane, such that the palate lies much closer to the roof of the nasopharynx than it does in adult modern humans (see below). The levator and tensor palati muscles and the muscle fibres of the superior constrictor do not have to run so markedly downwards to reach the palate as they do in adult modern humans. Instead, they run almost parallel with the underside of the cranial base (see later and Figs 13.10 and 13.11).

The nasopharynx of fossil hominids

Some fossil hominid specimens attributed to *Australopithecus africanus* (Sts 5 and MLD 37/38) have prominent club-like eustachian processes that were presumably, like those of apes, associated with the muscles of the soft palate (Fig. 13.4). These fossil hominids are also narrow across the region of the cranial base associated with the nasopharynx and, in addition, have prominent elongated markings on the basioccipital. These markings on the basioccipital are for the attachment of the longus capitis muscles which form part of the prevertebral group of muscles that lie behind the pharyngeal constrictor muscles (see Chapter 12). All these features are typical great ape characters and are, in addition, associated with sagittally orientated petrous bones and with a cranial base profile that is flattened or 'unflexed' in lateral view (Dean, 1985a). Some other fossil hominid specimens also have club-like eustachian processes (notably TM 1517 from Kromdraai, and the *Homo erectus* crania described by Weidenreich (1943). The combination of characters found in apes and in *A. africanus* crania

FIGURE 13·4 *Photograph of MLD 37/38 (Australopithecus africanus) in basal view. The petrous pyramids are aligned at 60° to the coronal plane (angle cc–cc–pa). There are prominent muscle markings for the longus capitis muscle (Lc) and there are club-like eustachian processes (EP).*

differs from those typical of crania attributed to *Paranthropus* or early or modern *Homo*. Little functional significance can be attributed to the narrow nasopharynx of *Australopithecus* or to the bony markings of the cranial base in this region. Nevertheless, it is worth noting that specimens attributed to *Paranthropus* contrast those of *Australopithecus* and have a wider nasopharynx, a more flexed exocranial base angle and lack the prominent muscle markings present in the pharyngeal region of the cranial base of great apes and *Australopithecus*.

The larynx and hyoid bone in modern humans

The larynx is composed of a series of articulated cartilages that form the framework of what is essentially an inlet valve at the top of the trachea (Figs 13.5 and 13.6). The cartilaginous rings of the trachea are incomplete posteriorly, but the **cricoid cartilage** of the larynx that lies at the most superior aspect of the trachea is a complete ring of cartilage. This cartilage is taller behind than it is in front and so resembles a signet ring in shape. Surmounting the prominent posterior portion of the cricoid cartilage are a pair of **arytenoid cartilages**. (So called apparently because on looking into the laryngeal opening from above, the upper apices of the arytenoids appear to resemble the spout of a jug (arytaina (Gk) = a pitcher).) The arytenoid cartilages (most textbooks say) are able to swivel slightly about their own long axes and, in addition, are able to slide laterally away from one another, down the slopes or **articular facets** of the cricoid cartilage. The arytenoid cartilages resemble tall three-sided pyramids with sunken faces which have three especially prominent corners at the base. The prominent processes that point forwards from each arytenoid across the inlet of the larynx provide the posterior attachment for the vocal folds that run across the inlet. These processes are known as the **vocal processes**. The more lateral processes at the base of the arytenoid cartilage provide attachment for muscles that act on the arytenoids and are known as the **muscular processes**. Two very important muscles (one on each side of the larynx), the **posterior crico-arytenoid muscles**, insert onto the muscular processes and act to pull the arytenoids apart constantly (and **abduct** the vocal folds). In other words they hold the airway open to breathe through.

The raised posterior aspect of the cricoid cartilage and the two arytenoid cartilages cannot be seen anteriorly because the thyroid cartilage is positioned in front of them. The **thyroid cartilage** is the largest cartilage of the larynx and comprises two laminae which slope backwards either side of the mid-line. There is a marked **superior thyroid notch** below which is the **laryngeal prominence**.

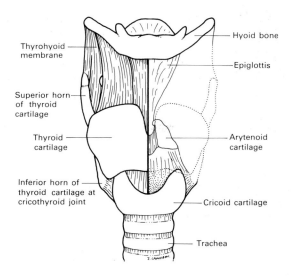

FIGURE 13·5 *The larynx viewed anteriorly with half of the thyroid cartilage removed to expose the arytenoid cartilage on the superior aspect of the cricoid cartilage at the back of the larynx.*

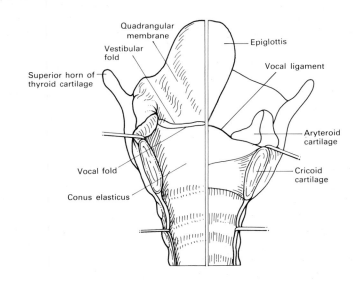

FIGURE 13·6 *The larynx sectioned posteriorly and retracted to show the cartilages and vocal ligament on the right hand side and the quadrangular membrane and free edge of the vestibular fold together with the conus elasticus and free edge of the vocal fold on the left hand side. The aryepiglottic muscle lies beneath the quadrangular membrane and runs between the arytenoid and epiglottic cartilages.*

The posterior margins of the laminae extend as the **inferior** and **superior horns** of the thyroid cartilage. These project towards the cricoid cartilage below and the hyoid bone above. The inferior horns of the thyroid cartilage form synovial joints with the cricoid cartilage and allow the thyroid cartilage to rock backwards or forwards with respect to the cricoid when the **cricothyroid muscle** contracts (Fig. 13.7). The lining of the trachea extends upwards into the cavity of the larynx within the thyroid and cricoid cartilages. The upper free margins of this elastic membrane (or cone called the **conus elasticus**), however, are strung between the vocal processes of the arytenoid cartilages at the back of the larynx and a point opposite them on the inner aspect of the thyroid cartilage in the mid-line

237

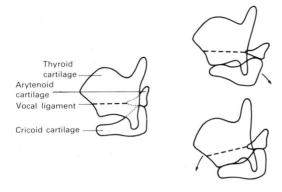

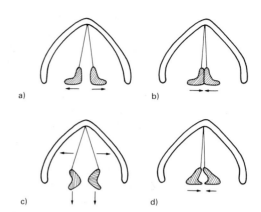

FIGURE 13·7 *Contraction of the cricothyroid muscle acts to tense the vocal ligament by rocking the thyroid cartilage relative to the cricoid cartilage (or vice versa if the thyroid cartilage is fixed in position).*

FIGURE 13·8 *Various muscles act on the arytenoid cartilages to adduct (a) and abduct (b) the vocal ligaments by bringing about medial and lateral movements of the arytenoids. Other muscle action brings about some rotation of the arytenoid cartilages about their vertical axes and also result in adduction (c) and abduction (d) of the vocal ligaments. (The medial aspects of the arytenoid cartilages in (d) have been drawn as they are in great apes, deeply excavated to form a hiatus intervocalis.)*

anteriorly. While the membrane or cone is circular in cross-section below, it comes to form a thin slit across the inlet of the larynx between these anterior and posterior points of attachment; this slit is known as the **glottis** and is formed by the vocal folds. The swivelling and sliding movements of the arytenoids (Fig. 13.8), together with the rocking movements at the cricothyroid joint (Fig. 13.7), are able to alter the tension in the vocal folds and the distance between them.

Above the attachment of the vocal folds on the inner aspect of the thyroid cartilage a leaf-shaped cartilage, the **epiglottis**, also gains attachment (Fig. 13.9). At rest the epiglottis lies against the back of the tongue, but during swallowing it is bent down over the inlet of the larynx so protecting the vocal folds from swallowed food. The epiglottis and the arytenoid cartilages are also covered by a membrane that stretches around the interior of the entrance (or vestibule) of the larynx, the **quadrangular membrane**. In the same way that the vocal fold is the free upper border of the conus elasticus, this membrane has a free lower border that hangs like a curtain from front to back of the laryngeal

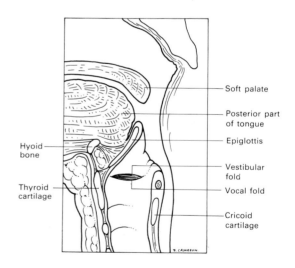

FIGURE 13·9 *The epiglottis lies beneath the back of the tongue. During swallowing it is raised into the tongue as the larynx rises in the neck. This, together with contraction of the aryepiglottic muscle, squeezes the epiglottis over the inlet into the airway.*

238

inlet. Because it looks so similar to the vocal fold it is occasionally called the false vocal fold, but more correctly it is known as the **vestibular fold** (Fig. 13.2). Between these two folds (the vocal and vestibular folds) that run across the laryngeal inlet from front to back there is a space that extends laterally towards the thyroid cartilage. This is called the laryngeal **ventricle**. It is from here, or more precisely from the **saccule** (the anterior most extension of the ventricle) that the **laryngeal air sacs** of the great apes originate before they pass out of the larynx through deficiencies in the **thyrohyoid membrane** and into the neck. In modern humans the saccule is a cul-de-sac that extends upwards from the anterior end of the ventricle between the thyroarytenoid and quadrangular membranes, but may occasionally extend above the thyroid cartilage and through the thyrohyoid membrane.

The modern human hyoid bone

The hyoid bone is a mobile bone which lies just above the larynx and just below the lower border of the mandible (Fig. 13.5). It is a U-shaped bone when seen from above or below and forms the skeleton of the

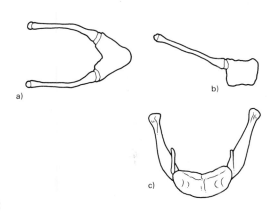

FIGURE 13·10 *(a), (b) Superior and lateral views of the hyoid bone of a chimpanzee. Note the body anteriorly is hollowed out on its posterior surface. (c) A modern human hyoid bone seen from the front for comparison.*

tongue. The anterior part of the 'U' shape is formed by the **body** of the hyoid bone and the sides of the 'U' are formed by its greater horns (Fig. 13.10). The hyoid bone is linked to the styloid process of the temporal bone by a ligament, the **stylohyoid ligament**, a variable length of which may be calcified. This ligament attaches to the lesser horn of the hyoid bone. The lesser horn of the hyoid bone is joined to the rest of the hyoid bone at the junction between the body and the greater horn. Besides providing attachment for one of the extrinsic muscles of the tongue the hyoid bone provides attachment for the strap muscles that run in the front of the neck (these are described in Chapter 6).

The larynx and hyoid bone of great apes

The morphology of the laryngeal cartilages in the great apes broadly parallels that of modern humans. However, the chimpanzee thyroid cartilage has deeper superior and inferior notches (that may almost divide the cartilage in young specimens). In the gorilla, the thyroid cartilage is smaller than in man but the cricoid is larger, whereas in the orang-utan the thyroid is larger and the cricoid smaller than in modern humans (Brandes, 1932; Sonntag, 1924). The posterior portion of the cricoid cartilage is taller in apes than in humans and reaches almost as high as the lower border of the hyoid bone and at its sides slants very abruptly downwards (Kelemen, 1969). In apes the hyoid bone has much more slender, greater horns, which are often not fused with the body of the hyoid (Fig. 13.10). The body of the hyoid itself is expanded anteriorly in the African apes such that there is a cavity or bulla at its posterior surface which houses an extension of the laryngeal air sac (Falk, 1975). In the orang-utan, however, the hyoid bone resembles that of modern humans more closely and is not expanded in this way anteriorly.

The whole of the great ape larynx lies higher in the neck than in human adults (see

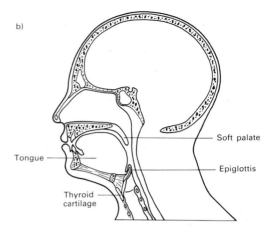

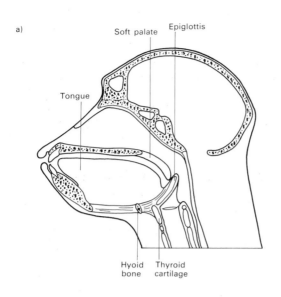

FIGURE 13·11 *Midsagittal section through the heads of (a) a chimpanzee and (b) a modern human to illustrate the differing positions of the larynx in the neck and the different relationship between the epiglottis and the soft palate. After Laitman (1977).*

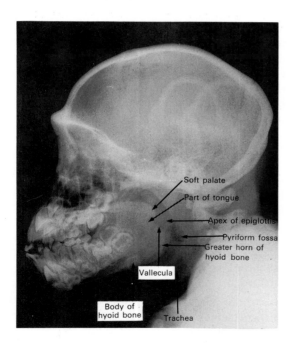

FIGURE 13·12 *Lateral skull radiograph of a young anaesthetized chimpanzee illustrating details of the pharynx, larynx and hyoid bone. Courtesy of Prof. G. DuBoulay.*

below and Figs 13.11 and 13.12) and the upper border of the thyroid cartilage is pressed tightly into the hyoid bone with only a small distance between these structures for the laryngeal air sacs to pierce the thyrohyoid membrane and enter the neck here. Linked with this, and with the closeness of the larynx in great apes to the base of the tongue, the epiglottis together with the entire laryngeal inlet slopes relatively further backwards than in adult humans, such that it faces the posterior pharyngeal wall. This arrangement assists in protecting the laryngeal inlet from swallowed food and liquids. The epiglottis is so high in apes that it remains in contact with the soft palate for much of the time. Because of the angulation of the entrance or vestibule of the larynx in apes, the impact of the inspiratory air current from the nose falls on the posterior part of the laryngopharynx behind the vocal folds (which remain under cover of the epiglottis).

The medial aspects of the arytenoid cartilages in great apes are deeply excavated at their base (Fig. 13.8d) such that even when they come together, an opening remains behind the vocal processes so that the airway cannot be closed off completely at any time (Brandes, 1932; Kelemen, 1969). This is called the **hiatus intervocalis** and is positioned behind the vocal folds and vocal processes of the arytenoids. In modern humans this hiatus only occurs rarely as a congenital abnormality. The posterior angulation of the laryngeal inlet and the high position of the laryngeal cartilages that are continuous with the soft palate in great apes severely limit the diversity of vocalizations among apes: most are shrieks and screeches of short duration. One great advantage that modern humans (and gibbons, which also lack a hiatus intervocalis) hold over great apes is the ability to close off the airway at the laryngeal inlet completely, because there is no hiatus intervocalis. This may mean that much more forceful coughing is possible which is especially important with upright posture. It also means that the thorax can be used as a more rigid framework for muscles of the forelimb to take origin from during energetic brachiation, in the case of gibbons, or while lifting heavy objects in the case of modern humans or possibly early hominids. The similar structure of the human and gibbon larynx is also interesting because of the complicated vocalizations of gibbons that are so important in their social structure and this too may be a factor that is related to the structure of the gibbon larynx.

The laryngeal air sacs of apes extend from the most anterior part of the ventricle of the larynx. This upwards extension is known as the saccule of the larynx, which lies lateral to the vestibular and vocal folds and pierces the thyrohyoid membrane and so enters the neck (Fig. 13.13). The air sacs form pouches of variable number and extent which lie in the deep cervical fascia of the neck beneath the platysma muscle. Some extend as far as the axillae laterally and others inferiorly to the sternum. They may be inflated during either inspiration or expiration and may act as support for the jaws, resonators of the voice or may even take a more active part in phonation. Kelemen (1969) reported that air forced out of the air sacs into the ventricle of the larynx may set up vibrations at the slit-like opening into the glottis between the vocal and vestibular folds. In this way apes may produce sounds during inspiration or expiration or, alternatively, sounds made up of double tones when combining vibrations from different regions of the larynx.

Vocalization and swallowing in modern humans, apes and fossil hominids

Several attempts have been made to reconstruct the vocal tracts of both early and later fossil hominids. Laitman and co-workers (1977, 1979, 1982) have noted the relationship between the degree of flexure of the external contours of the cranial base in the midsagittal plane and the position of the hyoid bone and larynx in the neck of primates. These authors argue that newborn human infants and primates with unflexed (or flattened) exocranial base angles have a hyoid bone and larynx positioned high in the neck such that the epiglottis is often, but not always, in contact with the soft palate. This arrangement of the soft tissues of the vocal tract and airway suggests that human infants under two years of age and other primates

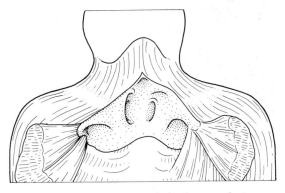

FIGURE 13·13 *Exposure of the laryngeal air sacs deep to platysma within the deep cervical fascia of the neck of a gorilla. After Raven (1950).*

with flattened cranial bases might be obligate nose breathers. (This, however, is known not to be so as Rodenstein et al. (1985) have convincingly demonstrated that human infants can easily breathe through their mouths by raising their soft palates and expanding their nasopharyngeal airways if the nose is obstructed.) A flattened cranial base does mean, however, that both modern human infants and other primates with a similar basicranial morphology, have a greatly reduced supralaryngeal region of the pharynx available for modifying speech sounds. During the swallowing of liquids (or suckling), the epiglottis in infants and great apes is close to the soft palate. Swallowed liquids, however, are deflected laterally to pass either side of the laryngeal inlet over the pyriform fossae and human infants are able at least to suckle and breathe at the same time.

Human newborn infants are unable to swallow solids, but adult apes, like adult humans, do so by closing off the nose from the oropharynx (with their soft palates) as the solid bolus of food passes over the laryngeal inlet. In modern humans the **aryepiglottic muscles** (see Fig. 13.6) squeeze the upper part of the arytenoid cartilages and epiglottis together (Falk, 1975) and the whole larynx is pulled up beneath the base of the tongue as the bolus passes overhead. Falk (1975) reported that in apes these aryepiglottic muscles are reduced or absent and the complete closure of the laryngeal inlet is effected by the downwards rotation of the epiglottis (which even at rest already lies obliquely over the inlet). This then is a more important mechanism for protecting the ape airway than it is for protecting the modern human airway. The reduced role of the epiglottis in protecting the modern human laryngeal inlet is emphasized by there being little problem swallowing either liquids or solids after its complete surgical removal.

Vocalizations through the mouth in human infants and primates where there is some contact between the soft palate and epiglottis requires that these be separated. Kelemen

(1969) suggests that this is often not possible for prolonged periods of time in great apes, as it is in adult humans during normal speech, so periods of vocalization are reduced. But this can hardly be said of human infants, some of whom are well practised at screaming for very long periods of time! Laitman and co-workers (1977, 1978, 1979, 1982) have hypothesized that many fossil hominids with flattened cranial base contours would necessarily have been obligate nose breathers and would have had reduced dimensions of the supralaryngeal pharynx, and that this in turn would have limited their capacity for sound modification or phonation. The form of the cranial base on its own, however, is not necessarily a reliable guide to the morphology of the upper respiratory tract and laryngopharynx in fossil hominids. These studies take no account of the position of the myohyoid line on the mandible, below which the hyoid bone of both infant and adult modern humans (and also apes and fossil hominids) must always be positioned. Were this not the case, the muscles attached to the hyoid and to the mandible could not function properly during swallowing (Falk, 1975). Some fossil hominid mandibles have extremely tall ascending rami that must have extended far down the neck. It is then not at all clear, for example, that *Paranthropus* (with a wide cranial base and wide nasopharynx superiorly (Dean and Wood, 1982; Dean, 1985a) flexed exocranial base angle and tall mandibular ramus (DuBrul, 1977)) was as distinct as the great apes are from modern humans in the anatomy of its vocal tract. It seems improbable that drawing parallels with adult early hominids and modern human infants is likely to provide a reliable basis from which to predict adult hominid vocalizations.

Other studies (Lieberman and Crelin, 1971) have attempted reconstructions of the vocal tract of Neanderthals using the angulation and orientation of the styloid process of the temporal bone, that extends downwards towards the lesser horns of the hyoid bone (Fig. 13.14). In this way it is argued that the

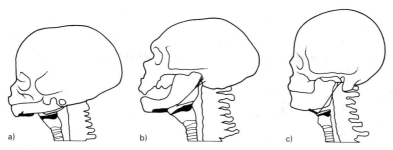

FIGURE 13·14   *The position of the larynx in the neck of (a) a neonatal infant, (b) a Neanderthal and (c) a modern human adult. After Lieberman and Crelin (1971).*

position of the hyoid bone can be estimated, the musculature of the floor of the mouth reconstructed and the volume and height of the vocal tract at least guessed at. The results of this study and others like it, suggest that Neanderthals would have a reduced supralaryngeal pharyngeal space and would have been unable to make the vowel sounds 'A', 'I', or 'U'. Unfortunately, once again these studies predict positions of the hyoid bone that are functionally improbable (Falk, 1975) and recent recovery (Arensburg et al., 1989) of a Neanderthal hyoid bone from Kebara Cave in Israel demonstrate that at least in its morphology the Neanderthal hyoid is entirely modern-human like. Duchin (1990) has emphasized that characteristic human speech sounds are produced in the oral cavity and so cannot be predicted from the morphology of the pharynx which is linked only with phonation. Interestingly, analysis of oral dimensions demonstrates that Neanderthals did not differ significantly from modern humans in this respect and so there is no evidence that their speech sounds would have differed from modern humans either (Duchin, 1990). In summary, it must be said that predictions about hominid vocalizations remain speculative and liable to revision in the light of future fossil discoveries.

CHAPTER FOURTEEN

BIPEDAL LOCOMOTION AND THE POSTCRANIAL SKELETON

The postcranial skeleton is that part of the skeleton which does not include the head, or cranium. In quadrupedal animals it is located posterior to the cranium, hence the name. However, in bipedal humans it lies beneath the cranium (Fig. 14.1). The complete post-cranial skeleton includes 177 separate bones, or 86% of the total 206 bones on the skeleton (Table 14.1). There are two main parts of the postcranial skeleton, the axial skeleton and the appendicular skeleton. The axial skeleton comprises the bones of the trunk and thorax including the vertebrae, sacrum, ribs and sternum. Although the number of vertebrae and ribs are variable both in humans and in apes, there are on average 51 bones in the axial skeleton, or about 25% of the total skeleton. The appendicular skeleton is made up of the bones of the upper and lower limbs and the respective limb girdles. There are 126 limb bones in both humans and African apes, or about 61% of the total skeleton, 64 of these are in the upper limbs (31%) and 62 in the lower limbs (30%).

The postcranial skeleton provides support for the body and at the same time acts as a lever system to facilitate muscle action. The major postcranial differences between humans and apes involve not only the shapes of the different postcranial bones but also the relative lengths of particularly the long bones and trunk (the bony levers) and the relative sizes of the joint surfaces (Fig. 14.2). The

TABLE 14·1   The number of bones in the human postcranial skeleton.

The axial skeleton		
Cervical vertebrae	7	
Thoracic vertebrae	12	
Lumbar vertebrae	5	
Total vertebrae		24
Sacrum	1[a]	
Coccyx	1[b]	
Ribs	24	
Sternum	1	
Total axial bones		51
The appendicular skeleton		
Clavicle	2	
Scapula	2	
Humerus	2	
Radius	2	
Ulna	2	
Carpals	16	
Metacarpals	10	
Phalanges	28	
Total upper limb bones		64
Pelvic bone (innominate)	2	
Femur	2	
Tibia	2	
Fibula	2	
Tarsals	14	
Metatarsals	10	
Phalanges	28	
Total lower limb bones		62
Total appendicular bones		126
Total postcranial bones		177

[a]Composed of five fused vertebrae.
[b]Composed of four fused vertebrae.

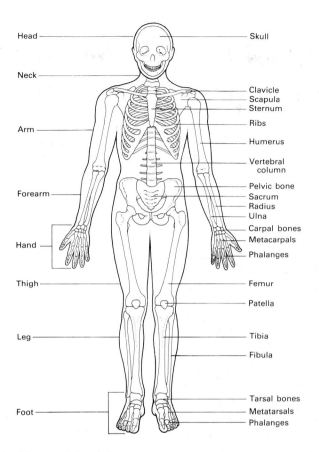

FIGURE 14·1 *The principal bones of the skeleton.*

structural requirements of bipedal locomotion provide an instructive introduction to these differences and to their mechanical significance.

The requirements of bipedal posture

In order to stand upright it is essential that the centre of gravity of the body remains directly over the rectangular area formed by

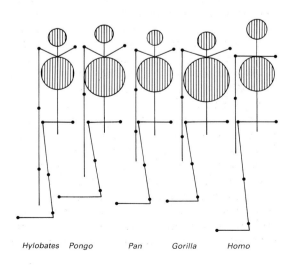

FIGURE 14·2 *Body proportions in humans and apes. Diagrams are drawn to the same trunk height. Upper circles represent relative head diameter and lower circles relative chest girth. After Schultz (1937).*

245

the supporting feet (Fig. 14.3). Quadrupedal (four-footed) animals have a relatively large supporting rectangle formed by the outer margins of their four feet and, therefore, have little difficulty in maintaining their balance. Bipedal humans, on the other hand, have a relatively small rectangle formed by the outer margins of the two feet. A small displacement of the centre of gravity can move it outside the supporting rectangle and cause the body to overbalance.

In adult humans the centre of gravity is located in the mid-line just anterior to the second sacral vertebra (MacConaill and Basmajian, 1969). While standing at rest, a line passing through this point and perpendicular to the ground also bisects the horizontal distance between the mastoid processes of the temporal bones as well as lines passing slightly anterior to the shoulder joints, between two points just behind the hip joints and between two points slightly in front of the knee and ankle joints. Even though this line does not pass directly through the centres of rotation of these joints (which would result in perfect equilibrium of the joints) only minimal muscle activity is needed to maintain this standing posture. The strong ligament binding the femur to the ilium and passing over the front of the hip joint (the **iliofemoral ligament**) helps to keep the trunk from falling backwards at the hip joint (see Chapter 19) and the strong ligaments in the knee (the **cruciates**) (see Chapter 22) help to keep the upper body and thigh from falling forwards at this joint.

Analyses of energy expenditure demonstrate the efficiency of human bipedal posture (Abitbol, 1988a). Oxygen consumption in erect standing is only 7% higher than when lying supine, or flat on the back. In contrast, dogs consume considerably more energy when standing in their normal quadrupedal stance than they do when lying down, most probably because they stand with their legs flexed, requiring continuous muscle activity. This suggests that in terms of energy expenditure humans at rest are much better adapted to

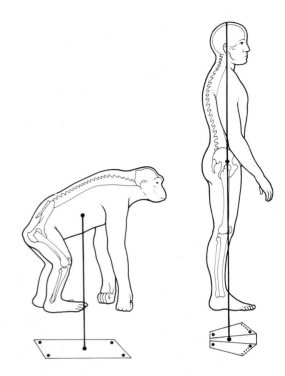

FIGURE 14·3 *The supporting rectangle and position of the centre of gravity in a bipedal human and a knuckle-walking chimpanzee. After Zihlman and Brunker (1979).*

bipedalism than dogs are to quadrupedalism (Abitbol, 1988a).

Moreover, there is also much more muscle activity in a chimpanzee standing upright on two feet than there is in a bipedal human. The bent-legged bipedal posture of the chimpanzee is energetically expensive because it requires continuous activity not only of the **hamstrings** but also of the **quadriceps femoris** and the **gluteal muscles** to keep the hip and knee joints from collapsing under the weight of the body (Fig. 14.4a).

There are a number of differences between humans and apes throughout the postcranial skeleton that allow humans to stand bipedally with their weight balanced over the supporting rectangle of their feet and with their hip and knee joints extended. When a chimpanzee stands upright, the pelvis rotates at the hip joint to raise the trunk over the lower limbs. This rotation of the pelvis seriously effects the lever advantages of the lower limb muscles, and particularly of the hamstring muscles, the prime extensors of the hip joint. This happens because as the chimpanzee pelvis rotates at the hip joint the posterior part of the pelvis, the ischium, moves closer to the femur. One reason why a chimpanzee walks bipedally with a flexed hip is to retain the maximum distance between the ischium and the femur and thereby retain maximum lever advantage for the hamstring muscles (Fig. 14.4a).

If a chimpanzee were to be able to stand upright efficiently with fully extended lower limbs and at the same time maintain the mechanical advantage of the hamstrings, modifications would have to be made to the chimpanzee pelvis. One of the main differences between a human pelvis and a chimpanzee pelvis is a greater angle between the superior part of the pelvis, the ilium, and the inferior part of the pelvis, the ischium. This greater angle allows the trunk to be held in an upright posture without effecting the relationship between the ischium and the femur. This angle represents, in essence, the point of inflection where the upper part of the body moved into the human vertical posture. The spinal curves (see Fig. 12.4), and particularly the lumbar curve, are also of utmost importance in bipedal posture in positioning the centre of gravity above the supporting rectangle formed by the feet.

There are a number of other differences in the postcranial skeleton between modern humans and apes that reflect bipedal posture in humans and its absence in apes. These include additional modifications to the human pelvis such as a reduction in the height of the pelvis and an anteriorly directed curvature of the iliac blades for efficient muscle action in bipedal walking. Furthermore, there are features that confer stability to the knee in an extended posture and to the ankle joint in terrestrial locomotion as well as a number of modifications in the human foot such as the unique longitudinal arch that are specific adaptations to bipedal locomotion. These features are discussed in detail in the subsequent chapters.

Some of the most obvious differences between the human and ape skeletons are, however, proportional differences, involving

FIGURE 14·4 (a) A bipedal chimpanzee with the major muscle groups that support the flexed hip and knee. (b) The major modifications required for perfect bipedalism in the chimpanzee are the angle between the ilium and the ischium (star) and the lumbar curve of the spine (arrow). After Kummer (1975).

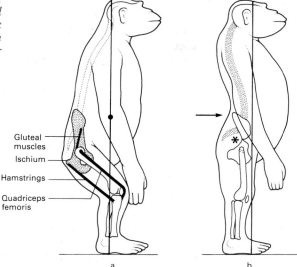

Gluteal muscles

Ischium

Hamstrings

Quadriceps femoris

a.

b.

not only the relative lengths of the long bones and trunk (the bony levers of the skeleton) but also bone girth and joint surface size directly reflecting the different requirements of weight transfer through the human and ape skeletons. These differences are discussed in detail in the following section.

Proportional differences in the long bones and trunk of humans and apes

Proportional differences in the long bones have traditionally been studied through indices. The most important of these are listed below.

(1) The intermembral index = [(humerus length + radius length) × 100]/(femur length + tibia length).
(2) The brachial index = (radius length × 100)/humerus length.
(3) The crural index = (tibia length × 100)/ femur length.
(4) The humerofemoral index = (humerus length × 100)/femur length.

The specific values of these indices vary according to the techniques used to measure the individual bones. Common measurements used to compute these indices are (after Schultz 1930, 1937) listed below (Fig. 14.5).

(1) Humerus length: the greatest distance between the head of the humerus and the capitulum parallel with the long axis of the bone.
(2) Radius length: the greatest distance between the radial head and the styloid process parallel with the long axis of the bone.
(3) Femur length: the greatest distance between the greater trochanter and the lateral condyle parallel with the long axis of the bone.
(4) Tibia length: the greatest distance between the middle of the edge of the medial condyle and the medial malleolus parallel with the long axis of the bone.

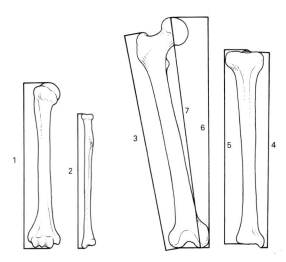

FIGURE 14.5 *Measurements of the length of the long bones. (1) Humerus length (Martin and Saller (1957 measurement no. 6.1), (2) radius length (no. 7.1), (3) femur length (Schultz, 1930), (4) tibia length (no. 13.1b), (5) tibia length (Aiello, 1981a), (6) bicondylar femur length (no. 11.2), (7) maximum femur length (no. 11.1).*

Alternative measurements of both femur length and tibial length are also used.

(5) Tibial length: the distance between the most superior point on the tibial spine and the centre of the inferior articular surface.
(6) Bicondylar femur length: the distance between the most superior point on the head and a plane joining the most inferior points on the lateral and medial condyles measured perpendicular to this plane.
(7) Maximum femur length: the distance between the most superior point on the head and the most inferior point on the medial condyle.

In practice, the difference in the alternative lengths of individual bones are relatively minor, there being a much greater difference in the indices characterizing different species.

The intermembral and humerofemoral indices demonstrate that among the higher primates (monkeys, apes and humans) modern

TABLE 14·2 Long bone indices by sex in humans and the large apes. After Aiello (1981a), [a]Coolridge (1933) and [b]Zihlman and Cramer (1978).

	N	Intermembral index		Humerofemoral index		Brachial index		Crural index	
		Mean	SD	Mean	SD	Mean	SD	Mean	SD
Homo sapiens (negro male)	20	69.7	(1.73)	71.4	(1.79)	77.9	(2.47)	82.4	(2.64)
Homo sapiens (negro female)	20	68.5	(2.11)	69.8	(2.45)	77.0	(2.36)	81.3	(2.55)
Homo sapiens (caucasian male)	10	70.3	(1.42)	72.8	(1.62)	73.4	(1.96)	79.8	(1.81)
Homo sapiens (caucasian female)	5	68.8	(2.68)	70.9	(3.41)	72.6	(2.07)	78.2	(0.84)
Gorilla (male)	9	120.2	(2.39)	118.0	(2.76)	81.1	(2.15)	77.8	(2.39)
Gorilla (female)	7	120.1	(3.58)	117.7	(3.00)	81.1	(2.67)	77.6	(2.07)
Chimpanzee (male)	12	108.0	(2.37)	101.1	(2.24)	91.9	(2.57)	79.8	(2.69)
Chimpanzee (female)	16	109.4	(3.03)	102.6	(3.71)	92.4	(3.76)	80.4	(1.97)
Pygmy chimpanzee (female)[a]	1	101.4	—	96.1	—	93.1	—	82.8	—
Pygmy chimpanzee (male and female)[b]	21 (15)*	102.2	—	98.0*	(2.00)	91.9	—	82.6	—
Orang-utan (male)	8	141.5	(3.07)	129.2	(3.42)	101.4	(4.03)	85.4	(2.79)
Orang-utan (female)	4	141.3	(2.36)	128.4	(1.70)	102.0	(1.41)	83.5	(0.58)

SD = standard deviation.

humans have unusually long lower limbs (and apes unusually short lower limbs) in relation to their upper limbs (Table 14.2). Of all of the apes, the pygmy chimpanzees approach the human condition most closely. Furthermore, the brachial index demonstrates that both humans and gorillas have unusually short radii in relation to their humeri and the crural index shows that modern humans as well as the apes have short tibiae in relation to their femora.

In ontogenetic terms, the intermembral index decreases in both the large apes and humans from its values in the neonates to its adult values (Table 14.3). The crural index either remains the same (in the chimpanzee) or shows a slight increase in humans and the remaining large apes, while the brachial index increases in the orang-utan and chimpanzee and decreases in both humans and gorillas. This decrease in humans and gorillas reflects the very low adult brachial indices in these two species in relation not only to the other large apes but also the monkeys (see also Chapter 17).

Further understanding of the significance of these proportional differences can be gleaned through comparison of the lengths

TABLE 14·3 Ontogenetic changes in body proportions in humans and the apes. Indices are determined on the basis of external measurements on cadavers and not on the basis of bone lengths. After Schultz (1956).

		N	Intermembral index[a]	Crural index[b]	Brachial index[c]
Hylobates	ad	80	165	85	108
Hylobates	nb	2	154	80	96
Symphalangus	ad	9	178	84	107
Symphalangus	nb	2	172	80	93
Pongo	ad	13	172	88	99
Pongo	nb	4	174	84	94
Pan	ad	30	136	83	91
Pan	nb	17	146	83	89
Gorilla	ad	5	138	83	80
Gorilla	nb	2	147	78	83
Homo	ad	25	88	83	76
Homo	nb	20	104	79	80

nb = new born, ad = adult.
[a]Intermembral index = [(total upper limb length × 100)/total lower limb length] where total upper limb length includes the length of the hand and total lower limb length includes the thickness of the sole of the foot.
[b]Crural index = [(leg length × 100)/thigh length] where leg length excludes the foot.
[c]Brachial index = [(forearm length × 100)/upper arm length] where forearm length excludes the length of the hand.

FIGURE 14·6 *The length of the lower limb in relation to the length of the upper limb in higher primates. Males and females of each species plotted separately. (■) = arboreal Old World monkeys; (▼) = baboons; (△) = macaques; (▽) = New World monkeys; (○) = gibbons; (◆) = great apes; (◇) = humans (caucasian and black). Po = orang-utan, P = chimpanzee, G = gorilla. The principal axis line is based on arboreal Old World monkeys. After Aiello (1981b).*

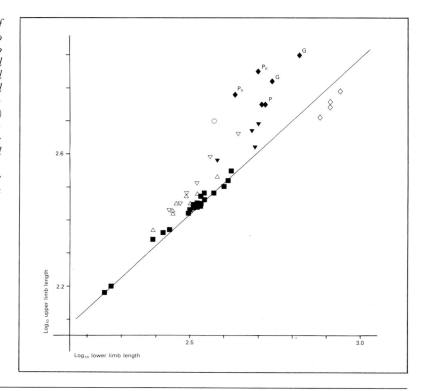

of the individual bones (or combinations of bones) to one another and to body weight. For example, the magnitude of the difference between apes, humans and the remaining higher primates in intermembral proportions is clearly reflected in the comparison of the length of the upper limb (humerus length + radius length) to the length of the lower limb (femur length + tibia length) (Fig. 14.6). In this case humans are much closer to the isometric relationship found in the arboreal Old World monkeys (colobines and cercopithecines) than are any of the large apes. These arboreal Old World monkeys resemble humans more closely than apes in having relatively long lower limbs in relation to their upper limb lengths. The more terrestrial Old World monkeys (macaques and baboons) together with the larger-bodied New World monkeys deviate from the arboreal Old World monkey pattern in the direction of the large apes and longer upper limbs for their lower limb lengths (or con-

versely shorter lower limbs for their upper limb lengths). The relationship between humerus length and femur length reflects this same general pattern (Fig. 14.7): modern humans fall on the principal axis characterizing the Old World monkeys, *Australopithecus afarensis* (AL 288–1) lies midway between modern humans and the apes and the orang-utan and gorilla are the most extreme.

The higher intermembral index in the large apes (and the low index in modern humans) results primarily from the alternation in the relationship between lower limb length and body weight rather than upper limb length and body weight (Fig. 14.8). Not only humans but also the large apes are consistent with the relationship between body weight and upper limb length that is found in the monkeys. However, in the relationship between lower limb length and body weight the large apes fall well below the line characterizing the monkey relationship. Comparisons between body weight and

FIGURE 14·7 *The length of the femur in relation to the length of the humerus in the higher primates. Symbols as in Figure 14.6. After Aiello (1981c).*

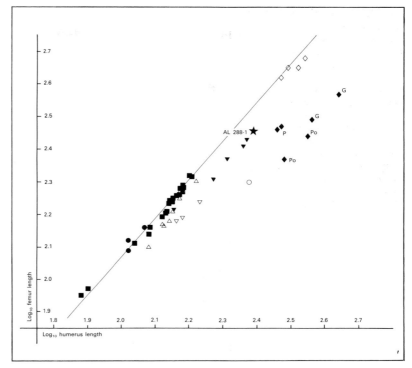

FIGURE 14·8 below *(a) The relationship between lower limb length and body weight in the higher primates. (b) The relationship between upper limb length and body weight in the higher primates. Symbols are as in Figure 14.6. After Aiello (1981b).*

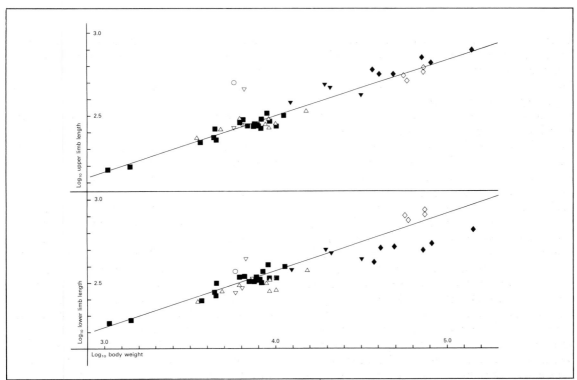

FIGURE 14·9 *The relation-ship between radius length and humerus length in the higher primates. Symbols are as in Figure 14.6. After Aiello (1981b).*

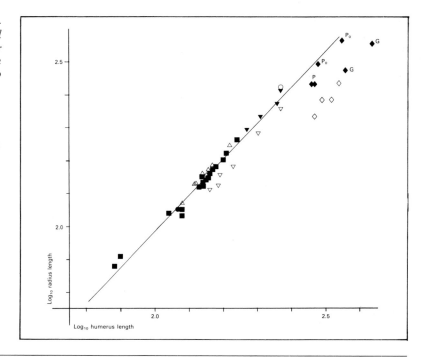

humerus length and body weight and femur length show the same general pattern, although the relationship between body weight and humerus length is even more consistent among humans and apes than is the relationship between body weight and upper limb length. This consistency results from the simple fact that radius length is much more variable in relation to body weight across primates than is humerus length (Fig. 14.9). Both humans and gorillas deviate in the direction of having short radii for their humerus lengths. Inclusion of the radius (and this variation) in the body weight to upper limb comparison introduces greater variation in this relationship than is found in the body weight to humerus length relationship.

The comparison between femur length and tibia length is the most consistent of all of the long bone relationships across the higher primates (Fig. 14.10). Tibia length has a high product-moment correlation with femur length ($r = 0.996$), however the relationship is negatively allometric. Tibia length does not increase as rapidly as femur length resulting in lower crural indices with increasing femur length.

Other indices characterizing further proportional differences in humans and apes are given in Table 14.4. The majority of these indices involve trunk length as a standard measure of body size. Trunk length, however, is not a consistent measure of body size across all primates (Fig. 14.11). It is longer relative to body weight in the arboreal Old World monkeys than it is in the remaining primates, including humans and apes. Indeed, across the higher primates indices involving trunk length have as good a chance of reflecting differences in trunk length as they do of reflecting differences in, say, chest girth or any other variable. However, because trunk length is consistent with body weight in humans and apes, indices involving this measure provide a fairly good basis for comparison when confined to these species.

In humans and apes, the relative differences in bone lengths are also reflected in the differences in the distribution of body weight according to its major segments (Zihlman, 1984). The relatively short-legged apes carry considerably less weight in their lower limbs

FIGURE 14·10 *The relationship between tibia length and femur length in the higher primates. Symbols are as in Figure 14.6. After Aiello (1981b).*

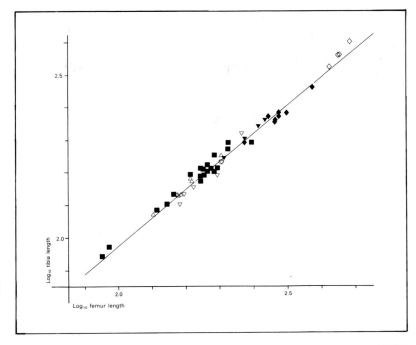

FIGURE 14·11 *The relationship between trunk length and body weight in the higher primates. Symbols are as in Figure 14.6. After Aiello (1981b).*

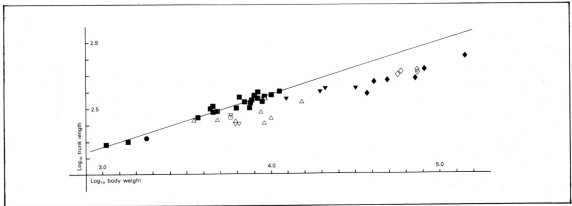

TABLE 14·4 Proportions of the foot, hand and head relative to trunk length in humans and the apes. After Schultz (1937).

	N	Relative foot length	Relative hand length	Relative hand breadth	Relative thumb length	Relative head circumference
Hylobates	80	52	59	20	52	26
Symphalangus	9	52	51	28	54	26
Pongo	13	62	53	34	43	26
Pan	30	50	49	34	47	26
Gorilla	5	47	40	51	54	29
Homo	25	48	37	43	67	30

FIGURE 14·12 *The weight of body segments as a percentage of total body weight in five female primates with the estimated condition in 'Lucy'. After Zihlman (1984).*

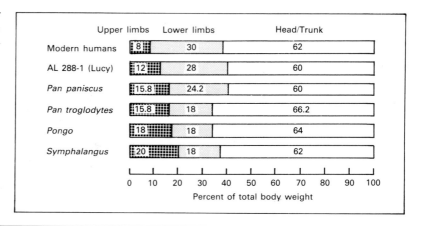

and more weight in their upper limbs than do relatively long-legged humans (Fig. 14.12). Furthermore, the pygmy chimpanzee (*Pan paniscus*) having a longer lower limb in relation to both body weight and upper limb length than do the other large apes carries a larger proportion of its body weight in its lower limb. In this sense it not only approaches the human condition more closely than do the other apes, but it has also been suggested to be most similar in its proportional relationships to the inferred condition for the early australopithecines (e.g. AL 288–1 'Lucy', Fig. 14.13; Zihlman, 1984).

Proportional differences in bone girth and joint surface size

Joint surfaces and long bone cross-sections are proportional in size to the amount of force that they transmit. For example, the mid-shaft circumference of the femur (FMC) in bipedal humans is larger in relation to the mid-shaft circumference of the humerus (HMC) than it is in any other higher primate (120% of expected size) (Fig. 14.14). The orang-utans, who rely to a large extent on their upper limbs for progression, lie at the opposite extreme and have the smallest FMC in relation to HMC of any higher primate (84% of expected size). The difference in joint surface size between species of different

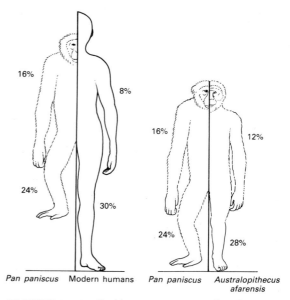

FIGURE 14·13 *Body proportions in modern humans and the pygmy chimpanzee* (Pan paniscus) *compared to the inferred proportions for 'Lucy'. After Zihlman (1984).*

locomotion patterns is also marked (Schultz, 1961; Robinson, 1972; Jungers 1988a). Indeed, the size of the upper surface of the body of the first sacral vertebra and of the auricular surface are only equivalent in humans and gorillas when the gorilla weighs approximately twice as much as the human (Fig. 14.15) (Schultz, 1961). Jungers (1988a)

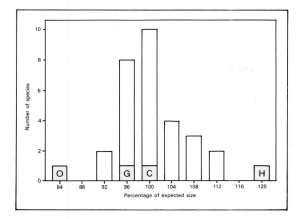

FIGURE 14·14 *The circumference of the femur expressed as a percentage of its expected size based on the actual circumference of the humerus in 30 species of higher primate. H = modern humans; C = chimpanzees; G = gorillas; O = orang-utans. Data from Aiello (1981a).*

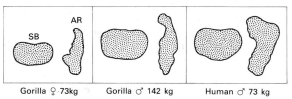

FIGURE 14·15 *The size of the upper surface of the sacral body (SB) and of the auricular surface (AR) of the same sacrum. The size of the joint surfaces of the bipedal human are equivalent to those of a male gorilla of approximately twice the body weight. After Schultz (1961).*

carries this line of analysis further and shows that bipedal humans have larger lower limb joints in relation to body size than do any of six species of greater or lesser apes (Fig. 14.16). In this analysis, *Australopithecus*

afarensis (AL 288–1) is midway between modern humans and apes. This might suggest that bipedalism in this species was either incompletely developed or not functionally equivalent to that seen in modern humans (Jungers, 1988a). Alternatively, new allometric data indicate that small bipedal humans have unusually small lower limb joints in relation to larger bipedal humans. The relatively small lower limb joints in Lucy might

FIGURE 14·16 *Male humans have unusually large hindlimb joints in relation to males of six other hominoid species (chimpanzees, pygmy chimpanzees, gorillas, orang-utans, siamangs and gibbons). Zero equals the expected joint size for body weight. The shaded area is the range of variation for non-human hominoid males. Deviations from zero are standardized by conversion to Z-scores. After Jungers (1988a).*

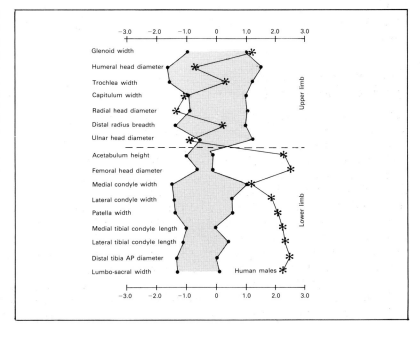

then just be a reflection of the small body size of this fossil and not of major differences in locomotor pattern (Ruff, 1988; Jungers, personal communication, 1988).

The above analyses could well lead to the belief that humans have very robust lower limb bones in relation to non-human primates. However, this robusticity is true only in relation to body weight and not in relation to bone length. The length–girth index of a bone reflects more the length of the long bone relative to body weight than it does any difference in girth. The short and stout femora of the apes in comparison with humans are a direct reflection of their relatively short femora in relation to their body weights. The increase in length of the human lower limb bones in relation to body weight outstrips any increase in girth resulting from bipedal locomotion and produces length–girth indices for the lower limb bones that lie below those seen in the large apes (Table 14.5). Indeed only those hominids with short lower limbs in relation to body weight (such

TABLE 14·5 Length–girth indices for the humerus and femur in humans and the large apes.

		N	Humerus	Femur
Homo	♂	20	21.6 (19.2–24.3)	20.5 (18.5–21.7)
	♀	12	21.1 (18.2–22.7)	20.3 (18.0–22.8)
Pan	♂	12	25.7 (21.7–28.4)	25.7 (21.3–29.2)
	♀	14	25.0 (22.4–28.0)	25.5 (22.7–27.5)
Gorilla	♂	12	25.6 (22.1–30.2)	31.7 (28.7–36.0)
	♀	12	22.5 (20.9–26.4)	29.6 (27.5–31.9)
Pongo	♂	12	19.9 (16.9–24.2)	25.1 (21.4–30.5)
	♀	12	19.4 (16.0–22.1)	24.3 (21.2–27.1)

[(Midshaft circumference × 100)/bone length]. Numbers in parentheses are ranges. After Schultz (1953).

as _Australopithecus afarensis_ AL 288) show the extreme length–girth relationships that reflect the relatively short femora of modern large apes (Fig. 14.17).

Modern humans with relatively short limbs also tend to have high length–girth indices, or stouter bones, than do modern humans

FIGURE 14·17 _The relationship between femur length and femur circumference in the higher primates. The principal axis is based on the arboreal Old World monkeys. Stars mark the position of selected fossil apes and hominids. Symbols are as in Figure 14.6. After Aiello (1981a)._

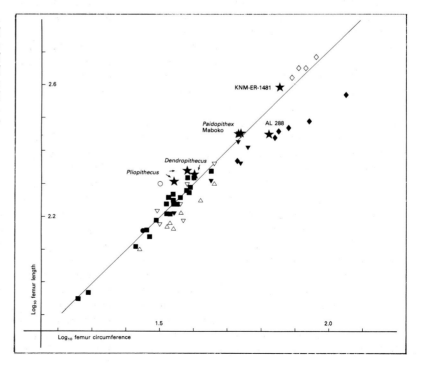

with relatively long limbs (Schultz, 1953). This suggests that both intraspecifically and interspecifically relative limb length has a greater effect on the length–girth index than does the relative magnitude of the bone cross-section.

Body proportions in the hominids

All available evidence points to the conclusion that hominids assigned to either *Australopithecus* or *Paranthropus* had body proportions that differ from those found in modern humans. In particular these hominids had short lower limbs for both upper limb length and inferred body weight. At least one fossil presently assigned to *Homo habilis* also shows evidence of non-human body proportions (Johanson et al., 1987). However, fossils assigned to *Homo erectus* and to later hominid taxa appear to have had body proportions that are within the range of variation found in modern human populations.

Australopithecus afarensis

The 40% complete AL 288–1 skeleton allows an accurate determination of not only the length of the lower limb relative to the length of the upper limb (as represented by the humerofemoral index) but also of limb length (as represented by either femur length or humerus length) relative to estimated body size. *Australopithecus afarensis* shows clear evidence of having had a short lower limb relative to the upper limb. The humerofemoral index of *A. afarensis* is 85% (Johanson and Taieb, 1976; Anon., 1976). This is midway between the indices for modern humans and the great apes (Table 14.2).

When the length of the humerus and the length of the femur are compared with estimates of body weight, it is apparent that the higher humerofemoral index of *A. afarensis* in comparison with that of modern humans results from a short lower limb in relation to body size rather than from a long upper limb in relation to body size (Jungers, 1982; Jungers and Stern, 1983; Wolpoff,

TABLE 14·6 Relative humerus length and relative femur length in human pygmies and *Australopithecus afarensis* (AL 288–1). After Jungers and Stern (1983).

	Humerus length/ pelvic height		Femur length/ pelvic height	
	Mean	SD	Mean	SD
Human pygmies (n = 9)	1.58	0.07	2.16	0.10
A. afarensis	1.51	—	1.77	—
Difference between human pygmies and A. afarensis in SD units	1.00		3.90	

SD = standard deviation.

1983a,b). For example, if the length of the pelvis (the summit of the iliac crest to the most inferior point on the ischial tuberosity) is taken as a reflection of body weight, the length of the AL 288–1 humerus relative to pelvic height is very similar to the length of the humeri of modern human pygmies relative to their pelvic heights. However, the lengths of the femora of AL 288–1 and modern human pygmies relative to pelvic height are quite different. AL 288–1 has a relatively much shorter femur than do the modern human pygmies (Table 14.6) (Jungers and Stern, 1983). This same relationship is also evident in the comparison between humerus (or femur) length and the length of the lumbar spine, although it is not as extreme (Wolpoff, 1983a,b). The AL 288–1 femur is also short when compared with the length of femora of modern humans of approximately the same body weight as inferred for AL 288–1, but it is the same absolute length as the femur of a pygmy chimpanzee of approximately the same body weight (Jungers and Stern, 1983).

Although the AL 288–1 skeleton provides the most complete evidence for australopithecine body proportions, other associated skeletons support the conclusion that the proportions not only of *Australopithecus*, but also of *Paranthropus* and (at least some

257

members of) *Homo habilis* differed from those of modern humans (McHenry 1978; Johanson et al., 1987; Grausz et al., 1988).

Other Plio-Pleistocene hominids

Olduvai Hominid 62 (discovered in 1986 at the FLK site in Bed I, Olduvai Gorge, Tanzania, dated to approximately 1.8 million years ago and assigned to the taxon *H. habilis* (Johanson et al., 1987)) includes radial, humeral, femoral and tibial fragments in addition to cranial and mandibular material. Although the postcranial material lacks epiphyseal ends (except for the proximal ulna), comparison with the AL 288–1 material suggests that the OH 62 humerus is longer and the OH 62 femur is shorter than (or of equal size to) the more complete *A. afarensis* material. Johanson et al. (1987) reconstructed the length of the humerus to be 264 mm and the femur to be 280 mm giving a humerofemoral index of 94%. This suggests that the intermembral proportions of OH 62 were less similar to those of modern humans than were those of *A. afarensis* (Table 14.2).

KNM-ER 1500 (discovered in Koobi Fora Area 130, dated to between 2 and 1.8 million years ago (Leakey, 1973; McHenry, 1978)) has been assigned to the genus *Australopithecus* (Leakey, 1973) but is considered here as *Paranthropus boisei* (see also Grausz et al., 1988). This skeleton includes tibial, femoral, radial, ulnar and fibular fragments together with some foot bones and part of a scapula. Various indices show that this fossil falls midway between humans and the great apes in its upper limb and lower limb proportions (Fig. 14.18) (McHenry, 1978) and in many ways is similar in these proportions to *A. afarensis* (Grausz et al., 1988).

TM 1517, the type specimen of *Paranthropus robustus*, (recovered from Kromdraai in 1938) includes a distal humerus and a proximal ulna together with some hand and foot bones and a cranium and mandible (Broom, 1938b, 1943; Broom and Schepers 1946; McHenry, 1974). Comparison of the

width of the articular surface of the distal humerus divided by the width of the trochlear surface of the talus places TM 1517 between humans and apes and again suggests that the upper limbs of some of the Plio-Pleistocene hominids were larger in relation to their lower limbs than is the case in modern humans.

The KNM-WT 15000 skeleton (*Homo cf. erectus*) is complete enough to determine skeletal proportions directly. The intermembral, brachial and crural proportions of this hominid have been reported to be within the modern human range of variation (Walker and Leakey, 1986). The only remaining associated skeleton known from this time period is KNM-ER 803 (McHenry, 1978). This fossil (discovered in 1971 in Ileret Area

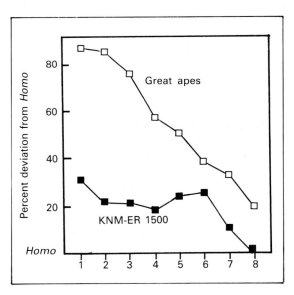

FIGURE 14·18 *Upper and lower limb proportions of KNM-ER 1500 in relation to modern great apes and humans. (1) Radius length/femur length, (2) radius length/tibia length, (3) radius head to bicipital tuberosity/femur lesser trochanter to superior neck, (4) ulna trochlear a–p diameter/bicondylar diameter of distal femur, (6) radial head diameter/average proximal tibial diameter, (7) average transverse and proximodistal diameter of ulnar trochlear notch/ average proximal tibia diameter, (8) average neck diameter of radius/average femur shaft diameter below lesser trochanter. After McHenry (1978).*

8A, dated to approximately 1.5 million years ago and assigned to *Homo* sp. indet. by Day and Leakey (1973)) consists of femoral, tibial, and ulnar shafts in addition to part of a radius shaft and parts of both fibulae some foot bones and other bits and pieces. The shaft cross-sections and reconstructed shaft lengths are all within the modern human ranges of variation (McHenry, 1978).

The Neanderthals

The intermembral, brachial and crural indices of the Neanderthals all fall within the modern human range of variation (Trinkaus, 1981, 1983a). However, there is a tendency for the Neanderthals to have short distal extremities. In modern humans both the crural and the brachial indices show a marked decrease in cold climates and the low Neanderthal indices are comparable with those of modern human cold-adapted peoples. This could reflect a cold-weather adaptation for the Neanderthals, decreasing the surface area of the extremities and reducing the heat loss from the appendages (Fig. 14.19) (Trinkaus, 1981). Alternatively, it could be related to an increase of power in the distal extremities. For example, a short tibia would shorten the load arm of **quadriceps femoris** relative to its power arm at the knee (Trinkaus, 1983a) thereby resulting in an increased power at the expense of speed. This would be consistent with other features of the knee which also act to increase the power of **quadriceps femoris** in the Neanderthals (see Chapter 22).

Prediction of stature

Prediction of stature rests on the assumption that there is a good correlation between long bone length and living stature. The best available data, the stature of U.S. military personnel measured before death and the lengths of their long bones measured after death, shows that this correlation varies between approximately 0.70 and 0.87 depending on population group and on the particular

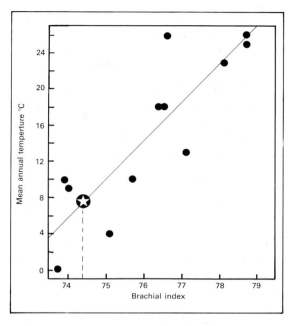

FIGURE 14·19 *The relationship between mean annual temperature and brachial index in 12 modern human populations. The Neanderthal average brachial index (star) is consistent with that of modern people living in cooler climates. After Trinkaus (1981).*

long bone used in the comparison (Trotter and Gleser, 1952, 1958). In all cases, the long bones of the lower limb have a higher correlation with living stature than do the long bones of the upper limb (Table 14.7).

This data set also shows that the relationship between a particular long bone length and stature is dependent on sex, race and generation. For example, for a given femur length U.S. caucasians are taller than U.S. blacks and within both population groups males are taller than females. And, again in relation to a given femur length, caucasian males who fought in the Korean War are taller than caucasion males that fought in the Second World War. The same relationship holds for U.S. black males, although the increase in stature is not as great as in U.S. caucasians (Trotter and Gleser, 1952, 1958). Because of this racial, sexual and generational

TABLE 14-7 Regression equations for estimation of maximum living stature (cm) of US Whites, Blacks, Mongoloids and Mexicans in order of preference according to standard errors of estimate. In all cases the bone lengths are maximum lengths. After Trotter and Gleser [a](1952: Tables 7 and 13) and [b](1958: Tables 3 and 12).

White males[a]	SE	R	White females[a]	SE	R	Black males[a]	SE	R	Black females[a]	SE	R
2.38 Fem + 61.41	3.27	0.87	2.93 Fib + 59.61	3.57	0.85	2.19 Tib + 86.02	3.78	0.80	2.28 Fem + 59.76	3.41	0.85
2.68 Fib + 71.78	3.29	0.87	2.90 Tib + 61.53	3.66	0.85	2.11 Fem + 70.35	3.94	0.77	2.45 Tib + 72.65	3.70	0.81
2.52 Tib + 78.62	3.37	0.86	2.47 Fem + 54.10	3.72	0.86	2.19 Fib + 85.65	4.08	0.77	2.49 Fib + 70.90	3.80	0.81
3.08 Hum + 70.4	4.05	0.79	4.74 Rad + 54.93	4.24	0.79	3.42 Rad + 81.56	4.30	0.71	3.08 Hum + 64.67	4.25	0.75
3.78 Rad + 79.01	4.32	0.76	4.27 Ulna + 57.76	4.30	0.76	3.26 Ulna + 79.29	4.42	0.71	3.31 Ulna + 75.38	4.83	0.65
3.70 Ulna + 74.05	4.32	0.76	3.36 Hum + 57.97	4.45	0.80	3.26 Hum + 62.10	4.43	0.72	2.75 Rad + 94.51	5.05	0.63

White males[b]	SE	R*	Black males[b]	SE	R*	Mongoloid males[b]	SE	R*	Mexican males[b]	SE	R*
2.60 Fib + 75.50	3.86	0.81	2.10 Fem + 72.22	3.91	0.81	2.40 Fib + 80.56	3.24	0.85	2.44 Fem + 58.67	2.99	0.87
2.32 Fem + 65.53	3.94	0.80	2.19 Tib + 85.36	3.96	0.80	2.39 Tib + 81.45	3.27	0.84	2.50 Fib + 75.44	3.52	0.86
2.42 Tib + 81.93	4.00	0.80	2.34 Fib + 80.07	4.02	0.80	2.15 Fem + 72.57	3.80	0.80	2.36 Tib + 80.62	3.73	0.83
2.89 Hum + 78.10	4.57	0.73	2.88 Hum + 75.48	4.23	0.76	2.68 Hum + 83.19	4.25	0.76	3.55 Rad + 80.71	4.04	0.76
3.79 Rad + 79.42	4.66	0.72	3.32 Rad + 85.43	4.57	0.72	3.54 Rad + 82.10	4.60	0.77	3.56 Ulna + 74.56	4.05	0.80
3.76 Ulna + 75.55	4.72	0.71	3.20 Ulna + 82.77	4.74	0.70	3.48 Ulna + 77.45	4.66	0.74	2.92 Hum + 73.94	4.24	0.74

R = product-moment correlation coefficient. * = correlation between the bones of the right side of the body only and stature. For individuals over 30 years of age reduce the estimated stature by 0.06 cm for each year over 30. SE = standard error.

variation, stature estimation from long bone length should ideally be based on formulae derived for the specific population from which the long bone comes. Considerable error can result from the use of inappropriate formulae for stature prediction. For example, stature estimates calculated from formulae derived on the basis of 29 different race/sex groups can vary by 13.3 cm from 164.1 to 177.4 cm (Geissmann, 1986b) for a given femur length (482 mm) (see also Feldesman and Lundy, 1988).

The lesson to be learned from this is that the accuracy of any stature prediction is dependent on close similarity in body proportions between the reference population and the skeleton for which stature is to be predicted. Following on from this, if the body proportions are not identical between the reference population and the unknown skeleton, stature predictions based on, say, the femur will not be the same as predictions based on other long bones. This lesson is particularly important for the early hominids that have body proportions demonstrably different from those of any living human population (see above). It also must be remembered that the majority of human fossils have smaller brains than do modern humans and, as a result, their crania are not as tall as would be expected in the reference population. This results in a slight over prediction of stature in these hominids based on modern human reference samples (Brown et al., 1985).

Furthermore, error can also result from the statistical technique used to generate an estimate. Traditionally, least-squares regression has been the favoured statistical technique for this type of prediction, however it has one major drawback when used to predict a stature from a bone that is particularly short (or long) in relation to the reference population (Oliver, 1976; Smith, 1984b). For very short bones this method overestimates stature and for very long bones it underestimates stature. This happens because the correlation between stature and

long bone length in the reference population has a direct effect on the slope of the regression line from which the predictions are derived (Fig. 14.20). The regression line always passes through the bivariate mean of the relationship (the mean of stature and of long bone length in the reference population), however the lower the correlation coefficient the lower will be the slope of this line in relation to the line that would seem visually to bisect the bivariate relationship. Given the correlation coefficients that are common between long bone length and stature, the slope of the regression line would be considerably lower than this bisecting line. This low slope elevates the line at the small end of the size distribution and depresses it at the large end. Any predictions based on this line would thereby inflate the stature predictions at the small end and deflate them at the large end.

Least-squares regression is most accurate when the length of the bone from which stature is to be predicted falls near the middle

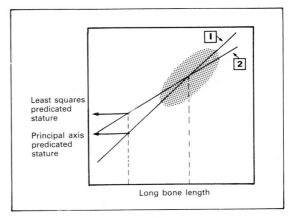

FIGURE 14·20 *The shaded area represents the ellipse of long bone lengths and statures in a reference population. The least squares regression line (2) will over-predict stature (relative to the relationship in the reference population) if the long bone upon which the prediction is based is much shorter than the long bones of the reference population. This is because the least squares regression line has a lower slope than the principal axis (or reduced major axis) line (1) which bisects the reference population ellipse. After Oliver (1976).*

of the length variation in the reference population. Where this is not the case, other statistical techniques that are largely independent of the correlation coefficient, such as principal axis or reduced major axis, produce more accurate predictions.

In addition to these sources of error in prediction, there is another and perhaps much more obvious source. This is variation in the reconstructed length of the long bone from which the prediction is to be made. Because few fossil long bones are found complete, a greater or lesser amount of reconstruction is frequently required and in certain cases length estimates derived by various authors on the basis of different techniques can vary considerably. A good example of this is the reconstructed length of the Sts 14 (*Australopithecus africanus*) femur where esti-

mates vary from 310 mm (Broom et al., 1950; Robinson, 1972) to 250 mm (Walker, 1973).

Stature prediction for the hominids

Formulae derived by Trotter and Gleser (1952, 1958) for males from the American military data set and for females from the Terry skeletal collection (Smithsonian Institution) are the most popular means of stature prediction for the fossil hominids (Table 14.7). Formulae are available for white, black, Mexican and mongoloid males and for white and black females. The choice of the appropriate formula to apply to a given case is dependent on the assessment of sex and closest reference population. For fossil hominids where sex is unknown it is arguably most prudent to base stature predictions on the average derived from the formulae for

TABLE 14·8 Principal axis formulae for the prediction of living stature from long bone length.

Reference population		Equation	Correlation coefficient
Human pygmies[a] (N=91)	Stature	= 4.99 humerus + 1.4	0.72
		= 5.69 radius + 19.8	0.71
		= 5.63 ulna + 10.9	0.71
		= 3.42 femur + 17.1	0.69
		= 3.29 tibia + 37.8	0.73
Human pygmies[b] (N=15)	Stature	= 3.8807 femur − 51.0	0.92
	ln (Stature)	= 0.9784 ln (femur) + 1.4480	0.93
Pygmy chimpanzees[b] (N=8)	Stature	= 5.8707 femur − 568.9	0.81
	ln (Stature)	= 1.2980 ln (femur) − 0.3246	0.82
South African Black[c] males (N=175)	Stature	= 3.422 femur + 0.002	0.90
		= 4.016 tibia + 0.222	0.90
		= 1.847 femur + tibia + 0.185	0.93
South African Black[c] females (N=122)	Stature	= 3.416 femur + 0.002	0.90
		= 4.051 tibia + 0.127	0.87
		= 1.853 femur + tibia + 0.072	0.92

[a]After Oliver (1976). Length of upper limb bones is maximum length, length of femur is bicondylar length and length of tibia is from the medial condyle to the medial malleolus. Bone length and stature are in centimetres.
[b]After Jungers (1988b). Length of femur is maximum femur length in millimetres. ln = natural logarithm.
[c]After Feldsman and Lundy (1988). Lengths of femur and tibia are bicondylar lengths. Bone length and stature are in centimetres. Equations give skeletal heights. To transfer these to living stature add 10 cm to skeletal heights under 153.5 cm, 10.5 cm for heights between 153.6 cm and 165.4 cm, and 11.5 cm for heights over 165.4 cm.

the two sexes and various population groups (McHenry, 1974). If the hominid limb bones are of the same general length as those of the modern human reference samples, estimates derived from these formulae most probably give the best currently available estimates of stature.

However, for the reasons stated above, these formulae can be expected to overpredict the stature of hominids with femora that are considerably shorter than the norm for modern human populations. Alternative formulae are, however, available (Table 14.8). Oliver (1976) provides principal-axis formulae for all six major long bones based on a reference sample of 91 modern human pygmies. Not only is this reference sample of small stature (approximating the small stature of some of the hominids) but also the principal-

axis formulae do not have the same predictive problems as do the least-squares regression formula. There are also available least-squares regression, principal-axis and reduced major-axis formulae for the femur based on two reference samples, modern human pygmies ($n = 15$) and pygmy chimpanzees ($n = 8$) (Jungers, 1988b). Moreover, formulae for the femur, tibia and femur+tibia are available based on a relatively small-bodied South African black population (Feldesman and Lundy 1988). For the 'short' hominids these formulae produce shorter statures than do the Trotter and Gleser formulae (Table 14.9, Fig. 14.21). Where, for example, the Trotter and Gleser formulae result in statures of between 126.1 cm and 128.4 cm (4ft 1in and 4ft 2in) for the four smallest bodied hominids (AL 288–1, Sts 14, OH 62 and KNM-ER

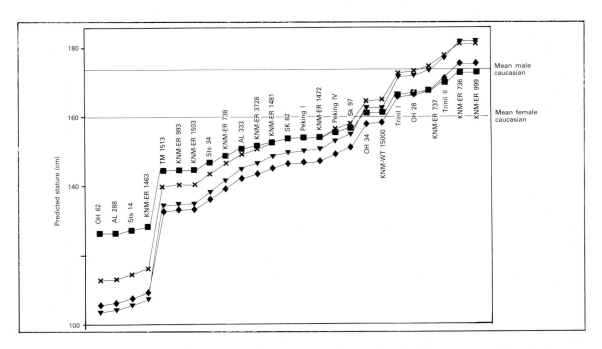

FIGURE 14·21 *Predicted statures for selected Plio-Pleistocene fossil hominids. The lower horizontal line represents the mean modern human female stature (North American caucasian) ($\bar{x} = 160.7$ cm; SD = 7.5) and the upper horizontal line is the mean male stature for the same population ($\bar{x} = 173.9$; SD =* 6.6) *(Trotter and Gleser, 1952). All predictions are based on femur length and the equations used are* ($\blacktriangledown$) = Jungers (1988b), ($\blacklozenge$) = Feldesman and Lundy (1988), ($\boldsymbol{X}$) = Oliver (1976), ($\blacksquare$) = Trotter and Gleser (1952). (See Tables 14.7, 14.8 and 14.9).

TABLE 14·9 Stature estimates (cm) for selected fossil hominids. Estimates based on femur length using four alternative predictive equations (see Tables 14·7 and 14·8). Taxonomic allocations are based on references cited in Geissmann (1988b), Feldsman and Lundy (1988) and McHenry (1988).

	Femur length (cm)	Trotter and Gleser 1952[a]	Oliver 1976	Jungers 1988b[b]	Feldsman and Lundy 1988[c]
Australopithecus afarensis					
AL-288-1	28.1 (2)	126.3	113.2	103.9	106.1
AL-333-3	38.6 (3)	150.6	149.1	144.7	142.0
Australopithecus africanus					
Sts 14	28.5 (4)	127.2	114.6	105.5	107.5
Sts 34	36.9 (1)*	146.6	143.3	138.1	136.2
TM 1513	35.9 (1)*	144.3	139.9	134.2	132.8
Paranthropus robustus					
SK 82	39.9 (1)*	153.6	153.6	149.7	146.5
SK 97	41.2 (1)*	156.6	158.0	154.8	150.9
Paranthropus boisei					
KNM-ER-0738	37.8 (1)	148.7	146.4	141.6	139.3
KNM-ER-0993	36.0 (6)	144.6	140.2	134.6	133.1
KNM-ER-1503	36.0 (10)*	144.6	140.2	134.6	133.1
KNM-ER-3728	39.0 (5)*	151.5	150.5	146.2	143.4
Paranthropus or *Homo?*					
KNM-ER-0736	48.2 (5)	172.7	181.9	181.9	175.3
KNM-ER-1463	29.0 (7)	128.4	116.3	107.4	109.2
Homo sp. indet.					
KNM-ER-0737	46.0 (1)	167.7	174.4	173.4	167.8
KNM-ER-0999	48.2 (5)	172.7	181.9	181.9	175.3
KNM-ER-1472	40.1 (8)	154.0	154.2	150.5	147.2
KNM-ER-1481b	39.5 (9)	152.7	152.2	148.2	145.1
KNM-ER-15000	43.2 (14)	161.2	164.8	162.5	157.8
Homo habilis					
OH 62	28.0 (15)	126.1	112.9	103.6	105.8
Homo erectus					
OH 28	45.6 (12)	166.7	173.1	171.9	166.5
OH 34	43.2 (11)	161.2	164.8	162.5	157.8
Pekin I	40.0 (13)	153.8	153.9	150.1	146.8
Pekin II	40.7 (13)	155.4	156.3	152.8	149.2
Trinil I	45.5 (12)	166.5	172.7	171.5	166.1
Trinil II	46.9 (12)	169.7	177.5	176.9	170.9

[a]Trotter and Gleser stature estimates are based on the average of the estimates for white males and females and black males and females using the 1952 equations.
[b]Jungers estimates are based on the human pygmy equation using raw, and not logged, data.
[c]Feldsman and Lundy estimates are based on the average of the estimates for South African black males and females. Reference numbers are given in parentheses.
* = length measurement which is the midpoint of a (sometimes considerable) range. This range is given after the following references as appropriate.
[1]McHenry (1974), Sts 34: 36.9 ± 6.1; TM 1513: 35.9 ± 7.2; SK 82: 39.9 ± 3.0; SK 97: 41.2 ± 3.0. [2]Jungers (1982). [3]Stern and Susuman (1983). [4]Jungers (1988). [5]Geissmann (1986b), KNM-ER 3782: 38.0 − 40.0. [6]Walker (1973). [7]Robinson (1978). [8]McHenry and Corruccini (1978). [9]Day et al. (1975). [10]McHenry (1978), KNM-ER 1503: 33.06 ± 39.98. [11]Day and Molleson (1976). [12]Day (1971). [13]Weidenreich (1941). [14]Feldsman and Lundy (1988). [15]Approximate length of AL 288-1 used as the upper limit for the length of the OH 62 femur (on the basis of assertion that the OH 62 femur is smaller and less robust than the AL 288-1 femur (Johanson et al., 1987)).

1463), the Jungers (1988b) human pygmy major axis formula results in statures between 103.6 cm and 107.4 cm (3ft 4in and 3ft 6in) for these same hominids.

These stature estimates suggest that at least some individuals belonging to *Australopithecus afarensis*, *Australopithecus africanus*, and the *Australopithecus/Paranthropus* group from Koobi Fora were well under 120 cm (4ft) in stature. The short individuals belonging to *A. afarensis* (AL 288–1) and *A. africanus* (Sts 14) are both undoubtedly adult, however it is not possible to determine whether the shortest hominid from Koobi Fori (KNM-ER 1463) is also adult (Day et al., 1976a). Furthermore, OH 62 (currently assigned to *Homo habilis*) may be as short, or even shorter, than these other small hominids. The fragmentary OH 62 femur is less robust than the AL 288–1 femur and Jungers (1988b) suggests that if the actual length of this specimen was only 1 cm less than the AL 288–1 femur, this small *H. habilis* hominid would have been as short as 100 cm (3ft 4in).

There are other larger individuals belonging to all these species. However, based on the equations derived from human reference populations, few of the analysed representatives of *A. afarensis*, *A. africanus*, *H. habilis*, or *Paranthropus* (southern Africa and Koobi Fora) seem to be over 160 cm (5ft 3in) in height and the great majority of them are under 150 cm (4ft 11in). Later occurring hominids (*Homo erectus*, Neanderthals and early modern humans) are with few exceptions taller than the earlier occurring group and have estimated statures that fall well within the modern human range of variation (Table 14.9, Fig. 14.21).

Prediction of weight

Body weight estimates for the hominids have been derived both from predicted stature and directly from the relationship between single (or multiple) skeletal measurements and body weight.

Weight estimates derived from predicted stature are based on the assumptions that the stature estimate is correct and that the fossil hominids have the same relationship between stature and body weight as the reference population used as the basis for prediction. Both these assumptions leave considerable room for doubt.

Stature-based weight predictions for the hominids have been based on one of two modern human reference populations. The first of these, used by McHenry (1974), are Jamaican school children (Ashcroft et al., 1966). Based on this population, he concluded that children of about the same height as his best estimates for the gracile australopithecines (145.1 cm) weigh approximately 35 kg (77 lb) and for the robust paranthropines (predicted height = 152.7 cm) approximately 42 kg (93 lb). The second reference population is composed of modern human pygmies. Wolpoff (1973) notes that the ratio between average height (cm) and average weight (kg) in this population is 3.7:1. Applying this to his stature estimates he arrives at an average weight of 37.3 kg (82 lb) for the gracile australopithecines. Although this ratio is correct for his pygmy reference population (average height = 140.5 cm, average weight = 37.7 kg) care should be taken in applying it particularly to larger and smaller individuals. The ratio decreases rapidly with increasing stature. Indeed for the average adult human (either caucasian or black) it underpredicts weight by approximately one quarter to one third.

Equations for the prediction of weight from height based on three reference populations, the Jamaican school children (data from Ashcroft et al. (1966)), Efé pygmies (Jungers and Stern, 1983) and New Guinea islanders (Aiello, unpublished results), are given in Table 14.10 and weight predictions for the hominids are given in Table 14.11. These equations can be expected to underpredict weight particularly for *Australopithecus* and *Paranthropus*. Because of their high intermembral indices, hominids in both

TABLE 14·10   Equations for the prediction of body weight (BW) from stature (STAT).

Sample	Equation	R
Kakar Islanders, New Guinea (N=362)[a]	$Log_{10}(BW) = 3.66 log_{10}(STAT) - 6.99$	0.80
Jamaican school children (N=42)[b]	$Log_{10}(BW) = 2.93 log_{10}(STAT) - 4.75$	0.99
Female human pygmies (N=81)[c]	$BW = 0.00011(STAT)^{2.592}$	0.71
Male human pygmies (N=94)[c]	$BW = 0.00062(STAT)^{2.241}$	0.67
Female human pygmies (N=80)[d]	$BW = 0.00013(STAT)^{2.554}$	0.68

[a]Principal axis equation where stature is in millimetres and body weight in grams. Data courtesy of Dr R. Harvey.

[b]Principal axis equation where stature is in centimetres and body weight in kilograms. The sample is made up of the average height and weight for each of seven age classes between the ages of 11 and 17 for male and female children of African, European and Mongoloid ancestry. Data from Ashcroft et al. (1966).

[c]Wolpoff (1983). Power curve regression equations for male and female Efé Pygmies.

[d]Jungers and Stern (1983). Power curve regression equation for the same sample of Efé Pygmies omitting one female with a questionable body weight.

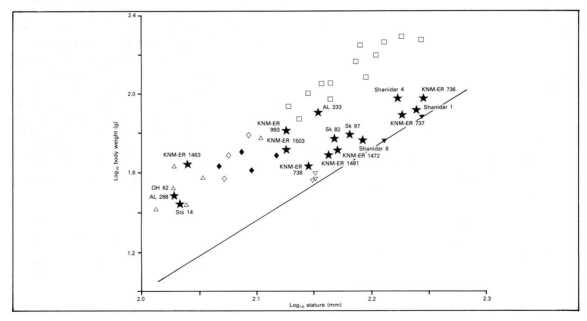

FIGURE 14·22 *The relationship between stature and weight in modern great apes, modern humans and fossil hominids. The principal axis line ($Log_{10}W = 3.66 Log_{10}S - 6.99$) is based on 362 Kakar Islanders (New Guinea). (★) = the predicted stature and weight of selected fossil hominids. The stature predictions are based on the Feldesman and Lundy (1988) equation (Table 14.8) and the weight predictions on the McHenry (1988) interspecific equation (Table 14.12). (◇) = Actual stature and weight of pygmy chimpanzees; (◆) = of chimpanzees; (□) = of adult gorillas; (△) = of juvenile gorillas; (▼) = of modern caucasian North Americans; (▽) of modern human pygmies.*

TABLE 14·11 Weight estimates for selected fossil hominids.

	Estimated stature[a]	Body weight (kg) estimates		
		Stature[b] based	Femur cross-section based	
			(2)	(1)
Australopithecus afarensis				
AL 288-1	106.1	12.3	38.9	29.9
AL 333-3	142.0	35.7	64.4	77.9
Australopithecus africanus				
Sts 14	107.5	12.9	37.2	27.4
Sts 34	136.2	30.9	—	—
TM 1513	132.8	27.6	—	—
Paranthropus robustus				
SK 82	146.5	40.0	55.1	57.8
SK 97	150.9	44.6	56.5	60.8
Paranthropus boisei				
KNM-ER 738	139.3	33.3	46.7	42.3
KNM-ER 993	133.1	28.2	57.8	63.5
KNM-ER 1503	133.1	28.2	51.6	51.1
KNM-ER 3728	143.4	36.6	45.5	40.2
Paranthropus or _Homo?_				
KNM-ER 736	175.3	77.1	70.7	93.3
KNM-ER 1463	109.2	13.7	47.4	43.5
Homo sp. indet.				
KNM-ER 737	167.8	65.7	64.8	77.0
KNM-ER 1472	147.2	40.6	51.6	51.1
KNM-ER 1481b	145.1	38.6	50.3	48.7
Homo habilis				
OH 62	105.8	12.3	39.3	30.3
Near Eastern Neanderthals				
Shanidar 1	173.4	73.3	65.8	81.1
Shanidar 4	167.0	63.9	70.7	92.9
Shanidar 6	155.7	49.4	54.7	57.0

Taxonomic allocations as in Table 14·9. Numbers in parentheses are the equation reference numbers from Table 14·12.
[a]Stature estimates based on the Feldsman and Lundy (1988) (see Table 14·9) equations except for the Near Eastern Neanderthals, based on the Trotter and Gleser (1952) male and female US White regression equations (Trinkaus, 1983a).
[b]Based on the principal axis equation for the Kakar Islanders (see Table 14·10).

these taxa would be expected to carry more body weight in their upper limbs than do modern humans. Zihlman (1984) suggests that _Australopithecus afarensis_ would carry 4% more weight in its upper limbs than do modern humans, however weight predictions would have to be increased by considerably more than 4% to match alternative estimates for these hominids based on other lines of reasoning (see below). Furthermore, these equations suggest that Neanderthals would weigh approximately the same as modern humans, contrary to other skeletal indicators which suggest that although of average stature, Neanderthals were considerably heavier than the average modern human.

Weight can also be predicted from either lower limb bone girth or joint surface size (Aiello, 1981c; McHenry, 1974; 1988; Jungers, 1988c). These variables might be expected to provide a more accurate reflection of weight than does stature because the joint surfaces and the cross-sections of the hindlimb bones must support the weight of the body. Figure 14.22 shows the relationship between estimated stature and weight predicted from the cross-sectional area of the femur just below the lesser trochanter for selected hominids (see also Tables 14.11, 14.12 and 10.1) (McHenry, 1988). The stature estimates are based on femur length and determined according to the Feldesman and Lundy equation (1988) (Tables 14.8 and 14.9). Wild-shot apes with known heights (heel-to-crown length)

TABLE 14·12 Least squares regression equations for the prediction of body weight (bw) from the cross-sectional area of the shaft of the femur.

Sample	Equation	R
African apes and humans (N=4 species)	(1) $\text{Log}_{10}(\text{bw}) = 1.189 \, \text{Log}_{10} (\text{femur area}) - 1.663$	0.94
Modern humans (N=30)[a]	(2) $\text{Log}_{10}(\text{bw}) = 0.624 \, \text{Log}_{10} (\text{femur area}) - 0.0562$	0.67

Numbers in parentheses are reference numbers for the equations.
After McHenry (1988). Bw in kilograms and femur area in millimetres. Femur area is the product of the transverse and anteroposterior diameters of the femoral shaft just below the lesser trochanter. Martin femoral measurements 9 and 10: Martin and Saller (1957).

and known weights are included for reference as is the stature–weight relationship for the New Guinea human population (Table 14.10). The stature–weight relationships for all of the fossil hominids exceed those expected on the basis of the New Guinea population. The smaller bodied hominids have predicted weights of between 2.73 (KNM-ER 1463) and 1.81 (Sts 14) times those expected, similar to the observed height–weight relationship of the chimpanzee reference sample (weights of between 2.56 and 1.83 times those expected). Some of the taller hominids also have these extreme height–weight relationships (AL 333–3, KNM-ER 993 and KNM-ER 1503), however the others are closer to the modern human reference sample with weights between 1.13 and 1.46 times the expected weights.

These results are based on only one of a variety of ways of predicting weight (McHenry, 1988) and on one of a variety of ways of predicting stature (Feldesman and Lundy, 1988). Other ways of predicting weight and stature would be expected to produce results that differ to a greater or lesser degree from these. For example, the Oliver (1976) stature equations produce taller statures for the hominids than do the Feldesman and Lundy (1988) equations (see Fig. 14.21 and Tables 14.8 and 14.9). These taller estimated statures result in stature–weight relationships for all of the hominids that are closer to those of the human reference population than do the shorter Feldesman and Lundy estimated statures. However, no matter which of a variety of techniques is used for prediction it is becoming clear that at least some of the hominids have predicted stature–weight relationships that are more similar to those observed in modern chimpanzees than in modern humans. Furthermore, because of the different height–weight relationships found in the hominids, the average body weights of hominid taxa are more similar to each other than might be expected from their differing femoral lengths or predicted statures.

Bipedal locomotion in humans and apes

Gait is the term used to describe the walking cycle. The primary unit of analysis in bipedal walking is the **stride**. A stride begins with the heel strike of the foot and lasts until that same foot contacts the ground again. During the **stance phase** of the stride the leg in question supports the weight of the body and during the **swing phase** it moves forward to begin the next stride. In human walking the stance phase for each leg takes up approximately 40% of the duration of the stride. The remaining 20% of the stride is divided into two periods when both feet are on the ground at the same time (Suzuki, 1985). In human running the two feet are never on the ground at the same time.

At the beginning of a walking stride (**heel strike**) the hip is flexed, the knee extended and the leg laterally rotated (Fig. 14.23). The adductor muscles then pull the weight of the body over the supporting limb. At the mid-stance phase of the stride the whole weight of the body is over the supporting foot and both the knee and hip are extended. As the stance phase progresses the ankle **dorsiflexes** and both the hip joint and the knee joint pass over and in front of the supporting foot. At **toe-off** (the final part of the stance phase) the body weight passes over the great toe, while the hip joint is now hyperextended, the knee extended and the foot dorsiflexed. At the beginning of the swing phase, the hip joint flexes as the leg moves forward. At the same time the knee also flexes and, as the swing leg passes the supporting leg, the knee starts to extend. Towards the end of the swing phase the leg laterally rotates in preparation for heel strike at the beginning of the next stride.

In bipedal walking the centre of gravity is at its highest during the mid-stance phase when the supporting leg is fully extended. It is at its lowest at the transition between stance and swing phases when both feet are in contact with the ground (Fig. 14.23). The centre of gravity also moves horizontally; during the stance phase it is positioned over the supporting leg and during the swing phase it moves laterally to a position over the alternate supporting leg (Fig. 14.24).

There are a number of features of human walking that minimize both the vertical and the horizontal movement of the centre of gravity and, thereby, minimize the amount of energy required (Saunders et al., 1953; Eberhart et al., 1954). Two of the most important of these features involve the pelvis

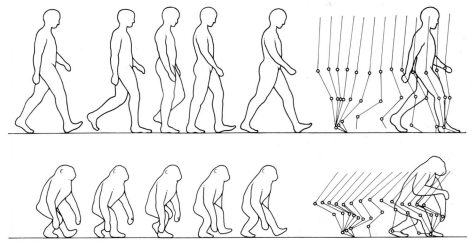

FIGURE 14.23 *Human and chimpanzee patterns of bipedal walking. After Okada (1985) and Yamazaki (1985).*

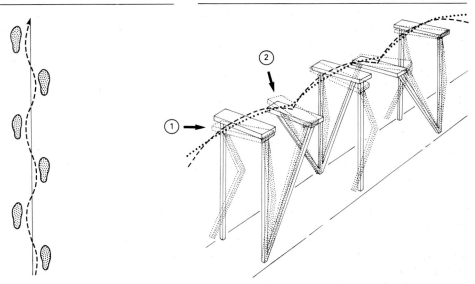

FIGURE 14.24 *Lateral movement of the centre of gravity (dotted line) during human bipedal walking. After Carlsoo (1972).*

FIGURE 14.25 *Pelvic tilt (1) and pelvic rotation (2) during human bipedal walking. After Saunders et al. (1953).*

and minimize the change in elevation of the centre of gravity during the stride (Fig. 14.25). Firstly, the pelvis (and the entire body) rotates around the stance leg carrying the swing leg (and the centre of gravity) laterally around the stance leg rather than up and over it. And secondly, during the mid-stance phase, the pelvis which is unsupported on the swing-leg side is held slightly below the horizontal. This has the effect of minimizing the height of the centre of gravity when it is at its highest point. In addition to these two features, slight flexion of the knee during the stance phase together with numerous other small movements of the knee, ankle and foot throughout the stride, also act to minimize the vertical displacement of the centre of gravity.

Horizontal displacement of the centre of gravity in human bipedal locomotion is minimized by the structure of the femur (Fig. 14.1). The human femur is angled towards the mid-line in a fashion which positions the knees (and the feet) close to the mid-line of the body. As a result, the centre of gravity moves a shorter lateral distance to lie above the alternating stance legs than if the knees (and feet) were positioned directly below the hip joints as they are in the apes.

During human bipedal locomotion the lower limb works as a compound pendulum making maximum use of kinetic energy. Electromyographic studies show that muscle activity is not intense (Suzuki, 1985) and there is surprisingly little activity in the hip and thigh muscles (Tuttle et al., 1979a,b). During the stance phase the pelvic rotators and abductors are active along with other muscles that either act to break the forward momentum or to propel the body forward (Fig. 14.26). During the main part of the swing phase the only muscles that are active are those that act to dorsiflex the foot (**tibialis anterior**) and extend the toes (Suzuki, 1985).

The average human walking speed is about 1.25 m s^{-1} which is equivalent to 4.5 km h^{-1} or 2.8 miles h^{-1} (Ralston, 1976). However, humans will choose to walk rather than run

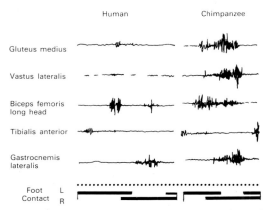

FIGURE 14·26 *Muscle activity (EMG) in the right lower limb during bipedal walking in a human and a chimpanzee. After Suzuki (1985).*

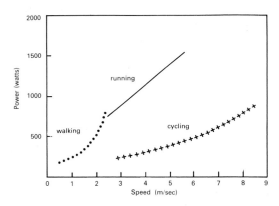

FIGURE 14·27 *The relationship in humans between energy use and speed of movement in walking, running and cycling. After Alexander (1984a).*

at speeds up to about 2.5 m s^{-1} (9 km h^{-1} or 5.6 miles h^{-1}) (Alexander, 1984a). Above this speed the energy required to walk exceeds that required to run (Fig. 14.27). Because humans, as well as other animals, tend naturally to adopt the most economical gait for a given speed, the transition will occur around this point.

The explanation for this lies with the magnitude of work done per cycle of walking

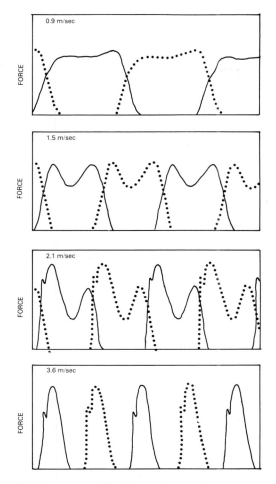

FIGURE 14·28 *The forces (vertical components) exerted by the feet in relation to time for human slow, moderate and fast walking speeds (top three graphs) and for human running (bottom graph). Solid lines represent the forces exerted by one foot and the dotted lines the other. After Alexander (1984a).*

step, the force peak for each step is higher and there are more steps per unit time. All these factors reflect the increase in work required to walk as speed increases.

The pattern for running differs from the pattern for walking in three ways: firstly, the time during which each foot is on the ground is shorter; secondly, the force pattern for each step shows far less fluctuation; and thirdly, the force peak for each step is higher. At about 2.5 m s^{-1} (the transition between walking and running) the work required to achieve the higher peak forces in running is exceeded by the work involved in sustaining the greater force fluctuation and longer step duration in walking and running becomes the energetically most economical gait (see Alexander (1980) for a more detailed analysis).

Bipedalism in the chimpanzee

Chimpanzee bipedal locomotion differs from human bipedal locomotion in the following ways (Jenkins, 1972). Firstly, neither the hip joint nor the knee joint are fully extended during bipedal progression (Fig. 14.23). Secondly, neither the knee joint nor the ankle joint pass behind the hip joint during the stance phase. As a result the propulsive force of the foot during the final stages of the stance phase is applied in front of the hip joint of the stance leg rather than behind. This results in a shorter bipedal stride length in the chimpanzee than either its quadrupedal stride length or human bipedal stride length (Reynolds, 1987). Thirdly, the chimpanzee femur lacks the human bicondylar angle which allows both the knees and the feet to be placed close to the mid-line of the body. This results in a greater lateral displacement of the centre of gravity during bipedal progression than is characteristic of humans. Fourthly, there is a reversed sequence of pelvic tilt during the stance phase (compare Fig. 14.25). In the chimpanzee the pelvis is elevated towards the swing leg during the stance phase rather than depressed as it is in

or running at a given speed (Alexander 1980, 1984a). This magnitude of work can be inferred from the pattern of forces exerted on the ground by the feet during the stride (Fig. 14.28). The forces are highest (and the work the greatest) while the leg is extending and thereby elevating the centre of gravity (body weight) in relation to the ground. As walking speed increases there is a greater fluctuation in the force pattern for a given humans. There are no pure abductor muscles of the hip in the chimpanzee and this tilt results from the shift of the entire trunk (and the centre of gravity) such that it comes to lie over the stance limb. These features produce both a greater vertical and a greater lateral displacement of the centre of gravity in chimpanzee bipedal locomotion. The centre of gravity is also lowest during the stance phase rather than during the swing phase as it is in humans (Fig. 14.23). The bent-legged bipedal progression involves a considerably larger muscle involvement than normally found in human bipedalism. Whereas in humans there is no long simultaneous activity of muscles throughout the stance phase, in chimpanzees (and other primates) there is notable activity in the extensors of the hip and knee which stabilize the bent-legged posture (Fig. 14.26) (Tuttle et al., 1979b; Ishida et al., 1985a, b).

There is no doubt that bipedal locomotion in chimpanzees and other primates differs considerably from human bipedalism. Kinematic studies (Prost, 1980) and electromyographical analyses (Fleagle et al., 1981) have shown that vertical climbing in primates is much more similar to human bipedal locomotion in joint excursion and in muscle usage and may have been the precursor form of locomotion to human bipedalism. This conclusion is supported by kinematic analyses (Okada, 1985), electroymyographic analyses (Kimura et al., 1979; Ishida et al., 1985b) and force plate studies (Kimura, 1985) which suggest that bipedalism in vertical climbers (chimpanzees, spider monkeys and orang-utans) is more similar to human bipedalism than it is in more terrestrial primates such as the Japanese macaques and baboons which do not habitually engage in such climbing behaviours.

Comparative energetics of bipedal and quadrupedal locomotion

The energy required in locomotion is normally measured in terms of the amount of oxygen consumed (m1 O_2) per unit body weight in grams (g) per unit distance in kilometers (km). Using this measure, the efficiency of human bipedalism in relation to quadrupedal mammals in general and to chimpanzees in particular is highly dependent on the speed of travel (Rodman and McHenry, 1980). At maximum running speed human bipedalism is twice as expensive energetically as estimated for a quadrupedal mammal of the same body size (Taylor et al., 1970; Fedak and Seeherman, 1979). However, at average walking speed (4.5 km h^{-1}) human bipedalism is slightly more efficient than is quadrupedalism in the average mammal (Rodman and McHenry, 1980), a conclusion which is consistent with the work of Fedak et al. (1974) demonstrating that human walking is 75% more efficient than human running. This work strongly suggests that the importance of bipedal locomotion in evolutionary terms lies with walking rather than running.

Energetic studies on chimpanzees and other primates provides tentative support for this idea (Rodman and McHenry, 1980). The costs of both bipedalism and quadrupedalism are identical in chimpanzees (Taylor and Rowntree, 1973) and at average walking speeds the chimpanzee consumes 150% more energy (per gram per kilometre) than does an equivalently-sized quadruped (Rodman and McHenry, 1980). Because bipedal modern humans are at least as efficient as the average mammalian quadruped at these speeds, there is no doubt that human walking is considerably more efficient than hominoid (or a least chimpanzee) quadrupedalism (Rodman and McHenry, 1980).

The other primates that have been studied so far help to put these conclusions in perspective. The capuchin monkey (*Cebus capucinus*) shows the same corresponding energy costs for bipedalism and quadrupedalism as the chimpanzee, but differs from it in being essentially as efficient in both modes of locomotion as an equivalently-sized mammalian quadruped (Taylor and Rowntree,

1973). Furthermore, the spider monkey (*Ateles geoffroyi*) which normally moves bipedally when on the ground, is as efficient in this mode of progression as is the average quadruped of its body size (Taylor and Rowntree, 1973).

Four tentative conclusions can be drawn from this work. Firstly, the relative efficiency of modern human walking in relation to human running is unique among primates where there is rather an identity in the energetic costs of these two modes of progression. Secondly, smaller-bodied primates are at least as energetically efficient in locomotion as are similarly sized quadrupeds. Thirdly, large body size and/or locomotor specializations in chimpanzees have reduced the efficiency of terrestrial locomotion in these primates. And fourthly, bipedalism in humans could well have evolved as a means of circumventing the inherent energetic inefficiency in terrestrial locomotion which may have been a characteristic of the hypothetical hominoid forbearer.

Walking speed and the Laetoli footprints

The Laetoli footprints provide a unique opportunity for the direct study of australopithecine bipedal locomotion. The central issues are firstly whether the stride lengths of the Laetoli footprint trails are unusually long (Leakey, 1979), unusually short (Charteris et al., 1981, 1982) or normal for a hominid of the inferred body size for the Laetoli hominids (Jungers, 1982; Reynolds, 1987). And secondly, following from this, whether the australopithecine gait was at all unusual in relation to the modern human bipedal gait.

The Laetoli footprints were discovered between 1977 and 1979 and comprise three trails, two from site G and one from site A (Leakey and Hay, 1979; White and Suwa, 1987). There is little doubt that the two site-G trails are hominid, however there is some doubt over the site-A trails (Alexander, 1984b; Tuttle 1984). Working on the assump-

tion that there is a direct correlation between stride length and speed, Charteris and co-workers (1981, 1982) conclude that the Laetoli hominids were walking at a very slow speed when the footprints were made. Modern humans walk at speeds of 0.8 m s^{-1} in small towns and about 1.7 m s^{-1} in large cities (Bornstein and Bornstein, 1976) and Alexander (1984b) noted that the walking speed of the Laetoli hominids would roughly correspond with modern small-town walking speeds.

Charteris and co-workers (1981, 1982) arrived at this result by comparing the relative stride length of the Laetoli trails (stride length (m)/inferred stature (m)) to the known

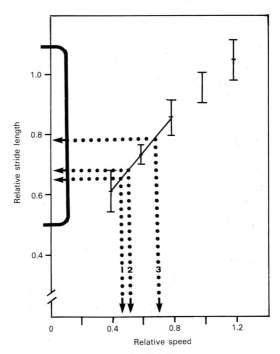

FIGURE 14·29 *Relative stride length (stride length/ stature) in relation to relative speed (speed/stature) for normal human adults. The bracket at the side represents normal human relative stride lengths. Lines 1 and 2 indicate the relative speeds for the Laetoli footprint trails (from site G) and line 3 is the relative speed for the larger trail when allowance is made for the short legs characteristic of the early australopithecines. After Charteris* et al. *(1981).*

273

relationship between relative stride length and relative speed (velocity (m s^{-1})/stature (m)) (Fig. 14.29). Their conclusion rests on the main assumption that their stature estimates for the hominids are correct. These estimates were derived directly from the footprints and the assumption that foot length makes up about 15.5% of stature as it does in certain pygmy populations (White, 1980). However, the australopithecines are known to have had relatively long foot phalanges which might be expected to increase foot length in relation to stature and also to have had relatively short lower limbs (see above). Either of these factors would be expected to result in shorter inferred statures for the australopithecines than would result from predictions based on a modern human sample. In the analysis of speed for the hominids, the shorter the stature, the higher the relative stride length and the higher the corresponding inferred relative speed.

In the case of the Laetoli australopithecines, it is likely that the inferred statures for the hominids are too tall and their corresponding inferred speeds are too slow. In fact Jungers (1982) suggested that relatively short lower limbs would require shorter stride lengths, while Reynolds (1987) demonstrated that, although short, the Laetoli stride lengths are not far different from the stride lengths observed for bipedal chimpanzees.

It is interesting to note that both stride length and speed may also be affected by the nature of the substrate. For example, another East African footprint trail from Koobi Fora and dating to approximately 1.5 million years ago also has very short stride lengths (Behrensmeyer and Laporte, 1981). The slow inferred relative speed (based on the analysis of Charteris and co-workers) for this trial may well be explained by slippery lake shore conditions where it is suggested that the hominids may have been walking through about 10 cm of water when the footprints were made.

CHAPTER FIFTEEN

THE HOMINOID THORAX AND VERTEBRAL COLUMN

The backbone has been described as the 'structural core of the vertebrate body' (Ankel, 1972). It is a segmented column that extends from the skull to the pelvis and (in animals with a tail) beyond the pelvis to form the bony core of the tail. In humans and other animals with varying degrees of upright posture it transmits the weight of the upper body to the legs. At the same time the vertebral column offers bony protection to the spinal cord, areas of attachment for the muscles that both support and move the trunk and articulation for the ribs the two most important functions of which are related to breathing and to protecting the organs of the thorax (thorax (Gk) = piece of armour to protect the chest and abdomen) and upper abdominal cavity (Fig. 15.1).

Each segment of the backbone is called a **vertebra**. The name vertebra (vertebra (L) = joint, from verto (L) = I turn) emphasizes the fact that the backbone is a flexible rather than a rigid column. Each vertebra is similar to a very short long bone in that the **body**, or weight supporting part of the vertebra, is constricted around its waist and expanded at either end (where it articulates with adjacent vertebra via an **intervertebral disc** (see below)). The epiphyses at each end of the body appear in humans around the time of puberty and fuse with the body by 24 years of age. In humans these vertebral epiphyses are ring like, while in the apes they are

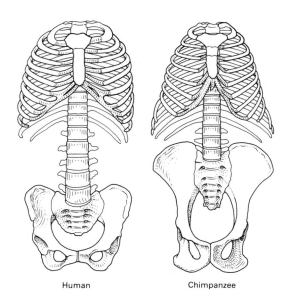

Human   Chimpanzee

FIGURE 15·1 *The trunk skeletons of an adult human and an adult chimpanzee. After Schultz (1950).*

complete, but very thin, discs that cover the superior and inferior surface of the body (Fig. 15.2).

In addition to the body, each vertebra has on its dorsal surface a **neural**, or **vertebral, arch** that encloses the **vertebral foramen** (foramen (L) = hole) through which the spinal cord passes (Fig. 15.3). The vertebral arch is made up of the **pedicles** at the sides of the arch and the **laminae** at the rear. Pedicles (pes (L) = foot) are literally the feet that support the rest of the vertebral arch and the structures projecting from it. Other structures are the **spinous process** that projects dorsally from the junction of the two laminae and the **transverse processes** that project laterally from the junction between the lamina and the pedicle on each side. These processes provide areas of attachment for the various muscles of the back. The relative sizes of the processes reflect the lever advantage of these muscles. Projecting both inferiorly (caudally) and superiorly (cranially) from the vertebral arch are paired **articular processes**, or **zygapophyses** (zygoma (Gk) = a yoke or bar connecting two parts;

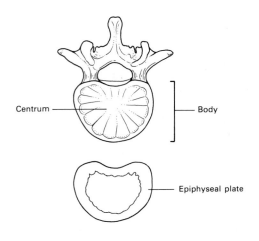

FIGURE 15·2 *An immature vertebra (a) with a separate epiphyseal plate (b).*

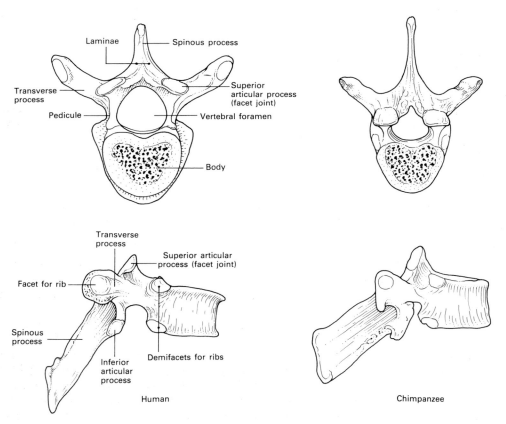

FIGURE 15·3 *Superior and lateral views of a human and a chimpanzee sixth thoracic (T6) vertebra.*

apophysis has been used since antiquity to mean any sharp or blunt process of bone). The cranial pair of articular processes face posteriorly (dorsally) and articulate with the immediately superior vertebra via its inferior anteriorly (ventrally) facing pair of articular processes. It is the orientation of these joints that largely determines the movement that can occur between adjacent vertebrae. Each vertebra, therefore, has three separate articulations with the vertebra lying immediately above or below it, the two pairs of zygapophyseal articulations, or facet joints, and the articulation between the bodies of the two adjacent vertebrae.

There are five different regions within the vertebral column and the vertebrae within each region are named accordingly (Fig. 15.4). The **cervical vertebrae** are the vertebrae of the neck, the **thoracic vertebrae** are the vertebrae of the chest and are the only vertebrae to articulate with the ribs, and the **lumbar vertebrae** are the vertebrae of the lower back. The **sacral vertebrae** are fused into a solid mass and articulate with the two **innominate** or **pelvic bones** and the **coccygeal vertebrae** are (at least in humans and apes) all that remains of the caudal or tail vertebrae found in many other animals.

The vertebrae themselves make up only about 75% of the total length of the presacral (not including the sacrum) backbone (Basmajian, 1975) in living humans. The remaining 25% is composed of the **intervertebral discs**. These are located between the bodies of each adjacent pair of vertebrae (Fig. 15.5). Each disc is composed of circularly arranged collagen fibres (**annulus fibrosus**) which enclose a softer gelatinous material called the **nucleus pulposus** (pulpa (L) = soft part of an animal's body) (Fig. 15.5). The discs act as shock absorbers and, in general, the thicker the disc the greater the movement possible between vertebrae. The discs can also be variably wedge shaped, so contributing to the various curves of the spine (see below). It is also interesting to note that astronauts returning from space are

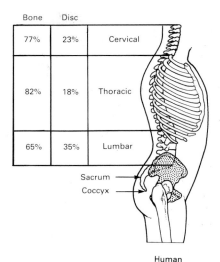

FIGURE 15·4 *Major regions of the presacral spine. The percentage of bone and disc for each region are given as percentages of the total height of that region. Data from Basmajian (1975).*

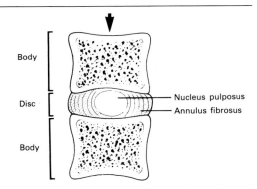

FIGURE 15·5 *The structure of a vertebral disc and its deformation under pressure.*

generally about an inch taller than before they left presumably due to the absence of gravity in space compressing the discs (Green and Silver, 1981).

In humans there are normally seven cervical vertebrae, 12 thoracic vertebrae, five lumbar vertebrae, five sacral vertebrae, and four coccygeal (or caudal) vertebrae making a total of 33 vertebrae (Table 15.1). The number of cervical vertebrae is stable at seven both within humans and across primate species, while the number of caudal (or tail) vertebrae

TABLE 15·1   The average number of vertebrae (thoracic, lumbar, sacral and caudal) in humans, apes and some monkeys. After Schultz (1961).

		Thoracic, lumbar, sacral vertebrae				Caudal vertebrae	
	N	Thoracic	Lumbar	Sacral*	TLS	*N*	Caudal
Homo	125 (116)*	12.0 (11–13)	5.0 (4–6)	5.2 (4–7)	22.2	68	4.0 (2–5)
Gorilla	81	13.0 (12–14)	3.6 (3–5)	5.7 (4–8)	22.3	64	3.0 (1–5)
Pan	162 (161)*	13.2 (12–14)	3.6 (3–4)	5.7 (4–8)	22.5	142	3.3 (2–5)
Pongo	127 (125)*	11.9 (11–13)	4.0 (3–5)	5.4 (4–7)	21.3	122	2.6 (1–5)
Symphalangus	29	12.8 (11–14)	4.4 (3–6)	4.7 (4–6)	21.9	25	2.2 (1–4)
Hylobates	319	13.1 (12–14)	5.1 (4–6)	4.6 (3–6)	22.8	312	2.7 (0–6)
Macaca	216 (214)*	12.1 (12–13)	6.9 (6–8)	3.0 (2–4)	22.0	162	17.0 (5–28)
Papio	66	12.5 (12–13)	6.4 (6–7)	3.2 (3–4)	22.1	34	19.4 (7–26)
Cerocopithecus	53 (52)*	12.2 (12–13)	6.9 (6–7)	3.0 (3–4)	22.1	20	26.6 (18–30)
Presbytis	122 (119)*	12.0 (11–13)	7.0 (6–9)	3.0 (2–4)	22.0	88	27.7 (18–31)
Nasalis	50	12.1 (12–13)	6.9 (6–7)	3.1 (3–4)	22.1	47	24.6 (23–26)
Alouatta	32	14.3 (13–16)	5.1 (5–6)	3.3 (3–4)	22.7	24	27.0 (25–28)
Cebus	32	13.8 (13–15)	5.7 (4–7)	3.0 (3–3)	22.5	20	23.5 (21–26)
Ateles	29	13.8 (13–15)	4.2 (4–5)	3.0 (2–4)	21.0	22	31.1 (28–35)

Sample sizes (*N*) in parentheses are the sample sizes for the sacral vertebrae. Numbers in parentheses are ranges. TLS = the total number of thoracic, lumbar and sacral vertebrae. Transitional (asymmetrical) vertebrae are not included in the ranges.

is highly variable both within species and across species (the latter depending on the presence/absence or relative length of the tail) (Table 15.1). The number of thoracic, lumbar and sacral vertebrae is also variable across species (Table 15.1), however the total number of thoracic, lumbar and sacral vertebrae is surprisingly stable at 22 in monkeys, apes and humans (Table 15.1) (Abitbol, 1987b). Variation in number of the various vertebral types in these primates therefore involves no increase or decrease in the total number of

vertebrae but a change in function or a shift in number from one region to another. This change in function occurs primarily at the transition between the thoracic and lumbar vertebrae or at the transition between the lumbar and sacral vertebrae (Abitbol, 1987b) (Fig. 15.6). For example, although there is considerable variation, apes normally divide their 22 TLS vertebrae into 13 thoracic, three to four lumbar and five to six sacral in contrast to the usual human count of 12 thoracic, five lumbar and five sacral (Table

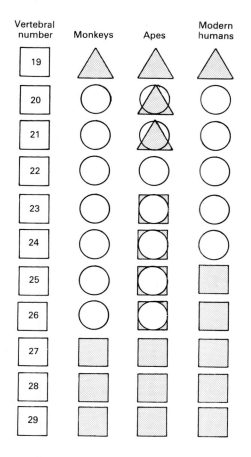

Vertebral number	Monkeys	Apes	Modern humans

FIGURE 15·6 *The 19th to the 29th vertebrae in monkeys, apes and humans. (△) = Thoracic vertebrae; (○) = lumbar vertebrae; (▦) = sacral vertebrae. Superimposed symbols indicate variation within the apes. After Abitbol (1987b).*

15.1) (Schultz, 1961).

Although vertebrae are made up of the same basic parts, each of the five vertebral types has specific distinguishing features. The structure and function of the **cervical vertebrae** have been discussed in detail in Chapter 12, however their distinguishing features can be usefully summarized once more.

(1) Each cervical vertebra has a **transverse foramen**, or hole, through both of its **transverse processes**.

(2) The bodies of the typical **cervical vertebrae** (C3–C7) are kidney shaped when viewed from above and saddle shaped when viewed from the side. The lateral sides of the superior (cranial) articular surface of one vertebra project upwards around the sides of body of the vertebra above it. These upward-projecting processes are called the **uncinate processes** (uncus (L) = hook) and form atypical synovial joints between the vertebral bodies called the **uncovertebral joint**.

(3) The superior (cranially directed) facet joints face posterosuperiorly while the inferior facet joints face anteroinferiorly.

(4) The **spinous processes** of typical cervical vertebrae are frequently **bifid**, or divided, near their tips.

Thoracic vertebrae can be recognized by the presence of articular facets for the ribs on the lateral sides of both the body and the transverse processes (Fig. 15.3). The facets on the lateral sides of the body are generally **demifacets**, i.e. they are half facets located near the superior (cranial) and inferior (caudal) borders on each side of the body. Each rib normally articulates with the demifacets of its own vertebra and the one above. Intervertebral discs separate these facets and the ribs are held in position by the **triradiate ligaments** (Fig. 15.7). Each rib also articulates with the tip of the transverse process of its own vertebra and is numbered according to this lower vertebra. For example, the seventh rib articulates with the demifacets of T6 and T7 and with the transverse process of T7. Exceptions to this rule are the first thoracic vertebra in humans and the last two thoracic vertebrae in both humans and apes. These vertebrae carry a single **costovertebral facet** rather than two demifacets and the corresponding ribs articulate only with the body of this vertebra. The facets on the transverse processes of the last two thoracic vertebrae are absent. The last two ribs are attached to these transverse processes by ligaments rather than via synovial joints as are the other ribs.

The thoracic vertebrae can also be distinguished by the orientation of their facet joints. The thoracic facet joints are vertically aligned, the superior pair facing posteriorly (dorsally) and the inferior pair anteriorly (ventrally) (Fig. 15.3). The only exceptions to this are the transitional first (T1) and the last thoracic (in humans T12 and in apes T13) vertebra. The superior pair of facet joints on T1 have the posterosuperior orientation of cervical facet joints while the inferior pair on the last thoracic vertebra have the sharp lateral orientation of lumbar facet joints (Figs 15.9 and 15.10).

If a complete set of thoracic vertebrae is present, they can be sequenced according to a number of criteria. Perhaps the quickest way to sequence not only the thoracic vertebrae but also the other presacral vertebrae is by general assessment of the size and volume of the vertebral body. This volume (as measured by the length, breadth and height of the body) increases steadily down the vertebral column from C3 to the second to last lumbar vertebra (Fig. 15.8). In addition, the cross-sectional shape of the thoracic vertebrae changes down the column (Fig. 15.9). The first few thoracic vertebrae have the sagitally narrow and transversely

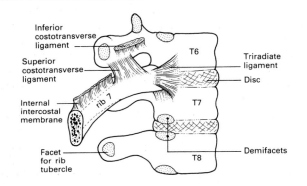

FIGURE 15·7 *The ligaments that connect the ribs to the vertebral column (the costovertebral ligaments).*

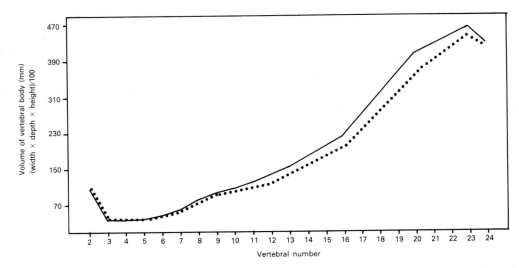

FIGURE 15·8 *The increase in volume (inferior transverse × inferior sagittal × middle vertical diameters) of the presacral vertebrae in North American whites (solid line) and blacks (dotted line). After Lanier (1939).*

broad 'kidney-shaped' cross-section of a cervical vertebral body. This compressed cross-sectional shape rapidly changes (by T4) to the typically heart-shaped cross-section of thoracic vertebral bodies. By the time the lower three thoracic vertebrae are reached this heart-shaped cross-section has broadened out again to a kidney shape that is also typical of lumbar vertebrae. However, the kidney-shaped cross-section of the lumbar vertebral bodies is far larger than that of either the thoracic or cervical vertebrae. In addition, the spinous processes of the thoracic vertebrae angle sharply downward (more so in humans than in the apes) completely covering and closing in the **vertebral canal**. The spinous processes are generally compressed from side to side at their tips; however the spinous processes of the upper thoracic vertebrae (generally T1 and T2) have bulbous tips while those of the lower three (T10, T11 and T12 in humans) are more stubby in appearance and project at a less acute angle from the body of the vertebra. In these features they approach the morphology of the superoinferiorly deep spinous processes of the lumbar vertebrae. There is also a transition in the morphology of the transverse processes in these lower thoracic vertebrae that is particularly evident in the lowermost vertebra (Fig. 15.9). There are two tubercles on the transverse process of the last thoracic vertebra that foreshadow the morphology of the transverse processes on lumbar vertebrae (see Fig. 15.13). These tubercles are the **upper (mammillary) tubercle**, and the **external (accessory) tubercle**. On L1 the mammillary tubercle fans out and forms the **mammillary process** which is the enlarged lateral side of the medially directed lumbar articular process (Fig. 15.10). The external, or accessory tubercle, seems to extend into a newly constructed lumbar 'transverse process'. An alternative view is that the lumbar transverse process is a rudimentary fused rib (Fig. 15.9).

The lumbar vertebrae are characterized by the presence of very distinctive facet joints (Fig. 15.10). The superior facet joints face towards the mid-line and are sharply concave while the inferior facet joints face away from the mid-line and are convex. These neatly interlocking joints restrict rotation and strongly prohibit both anterior and lateral slip of a superior lumbar vertebra in relation

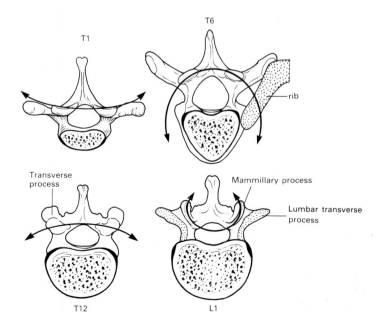

FIGURE 15·9 *The superior surfaces of three thoracic vertebrae (T1, T6 and T12) and the first lumbar vertebra (L1). The curved lines indicate the different orientations of the facet joints.*

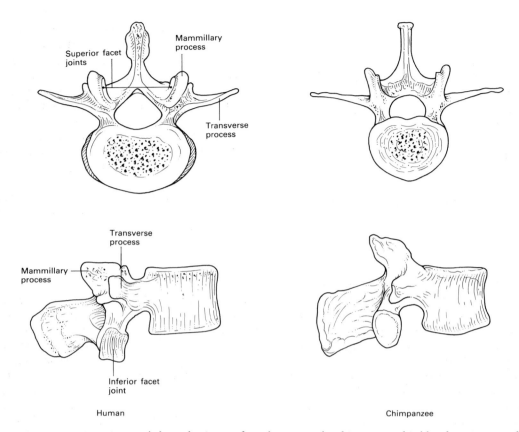

FIGURE 15·10 *Superior and lateral views of a human and a chimpanzee third lumbar (L3) vertebra.*

to its immediately inferior partner. It is interesting to note that this may well be the primitive form of vertebral articulation because it is found in all the vertebrae of both reptiles and birds.

In both humans and apes, the lumbar vertebrae can be sequenced by the size of their bodies (Fig. 15.8) and in humans by the distance separating the superior (or inferior) pair of facet joints (Fig. 15.11). In order to accommodate the wide human sacrum (Chapter 20) and the articulation between the last lumbar vertebrae (L5) and the first sacral vertebra (S1), the distance between the two superior (and/or the two inferior) facet joints in the majority of human backbones gradually becomes greater the lower the lumbar vertebra. At the same time the orientation of the facet joints in humans

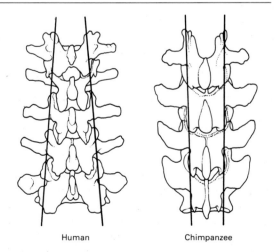

FIGURE 15·11 *Human and chimpanzee lumbar vertebrae from the posterior (dorsal) aspect. The solid lines show the difference in the distance between the facet joints in the two species.*

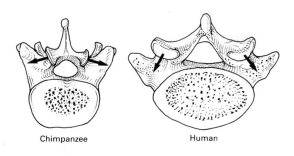

Chimpanzee   Human

FIGURE 15·12 *The last lumbar vertebra of a human and a chimpanzee from the inferior (caudal) aspect. Note the difference in the orientation of the facet joints on the vertebrae (arrows).*

becomes less mediolateral and more anteroposterior. At the lumbosacral articulation the anteriorly oriented inferior facet joints of the last lumbar vertebra serve a specific and very important function. They literally keep the last lumbar vertebra and the entire backbone from sliding anteriorly off the highly angled sacrum (Fig. 15.12). Apes which lack a highly angled sacrum also lack the change in orientation of and distance between the facet joints. Other features of the lumbar vertebrae are discussed below with the curves of the backbone.

The **sacral vertebrae** are traditionally defined as vertebrae that articulate with the **ilium** and whose bodies and vertebral arches (including spinous and transverse processes) are fused (Abitbol, 1987b). There are, however, many stages of partial sacralization from mere unilateral contact of one transverse process with the ilium to full fusion and compliance with the definition. In apes vertebrae beginning with number 23 (L4 in humans) can show some degree of sacralization while in humans it is comparatively rare (occurring in 10–15% of cases) for any more superior vertebra than number 25 (S1 in humans) to be sacralized (Figs 15.1 and 15.6) (Abitbol, 1987b; Wood, personal communication).

The **coccygeal vertebrae** (see Chapter 20) are rudimentary in humans and apes. Only the first coccygeal vertebra has any hint of either a transverse process or of facet joints. The remaining vertebrae are increasingly reduced vertebral bodies. In the foetus there are at least six coccygeal vertebrae present (in humans) and the last segment in adults represents the fusion of three or more of these foetal vertebrae. In adult humans there are normally four or less coccygeal vertebrae in women and four or more in men (Fraser, 1920).

Both the **sacral vertebrae** and the **coccygeal vertebrae** are discussed more fully in Chapter 20.

The backbone as a whole

The vertebrae are connected to each other via intervertebral discs that are interposed between the bodies of adjacent vertebrae and equally importantly by a system of strong ligaments (Fig. 15.13). The **anterior longitudinal ligament** runs the length of the backbone joining together the anterior (ventral) surfaces of the vertebrae from the sacrum to C2 (second cervical). This continues to the C1 as the **anterior atlanto-axial** and **atlanto-occipital membranes** (see Chapter 12). The **posterior longitudinal ligament** also runs the length of the backbone but is positioned inside the vertebral canal, joining together the posterior (dorsal) surfaces of the vertebral bodies. The **ligamentum flavum** (flavum (L) = yellow) also runs inside the vertebral canal connecting the inner surfaces of the laminae. The last ligaments, the **supraspinous** and **interspinous ligaments**, connect together adjacent spinous processes and in the neck extend to the occipital bone as the **ligamentum nuchae** which is very well developed in humans but rudimentary in apes (see Chapter 12).

These ligaments not only connect the vertebrae but also restrict movement of the spine. In so doing they economize on muscle involvement both in the maintenance of upright posture in humans and in various movements that involve the back. For exam-

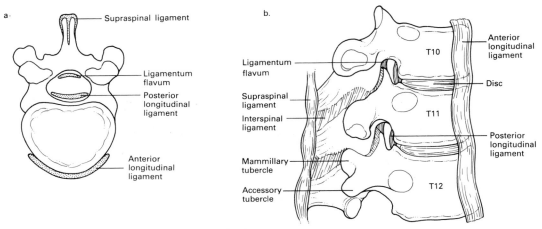

FIGURE 15·13 *The main ligaments of the vetebral column from the superior view (a) and the lateral view (b).*

ple, in extreme forward flexion of the trunk the muscles of the back are completely inactive and the ligaments alone take the strain of the upper body (MacConaill and Basmajian, 1969). There is some evidence that the **ligamentum flavum** is very much more elastic in the apes than it is in humans (Sonntag, 1924).

Movements of the backbone

Movements that we associate with the backbone such as flexion, extension and rotation involve not only the vertebral column but also the hip joints. For example, in toe-touching the majority of movement takes place at the hip joint while the remaining movement occurs primarily in the cervical and lumbar parts of the vertebral column. The rib cage restricts flexion or extension in the thoracic region (Green and Silver, 1981). Likewise, rotation, or twisting, of the trunk as in facing backward while the feet are still facing forward, involves all areas of the spine as well as the hip joints. Rotation in the spine brings the head around to 90° (Fig. 15.14). Any additional rotation occurs in the hip joints.

Of all the vertebrae the cervical vertebrae are the most mobile. In addition to

flexion–extension and rotation their joints also allow lateral flexion (abduction and adduction of the head) and circumduction (MacConaill and Basmajian, 1969). In apes the movements of the cervical spine are largely similar to humans; however the longer ape spinous processes may limit the range of extension in these species, the generally higher shoulders of these primates (Fig. 14.2) would also limit lateral flexion and the large jaws and enlarged air sacs in the neck would limit forward flexion (Schultz, 1961) (see Chapter 12).

Next to the cervical vertebrae, the lumbar vertebrae are the most mobile; however, their joints are basically capable only of flexion and extension and of lateral flexion. What might appear to be rotation in the lumbar spine is actually a 'conjunct rotation' or a combination of flexion and abduction or of extension and adduction (MacConaill and Basmajian, 1969). In the apes, flexibility in the lumbar region is severely restricted in relation to humans because of the reduced number of lumbar vertebrae and the close approximation of the ribs to the iliac blades (Fig. 15.1). This close approximation results from both the reduced number of lumbar vertebrae and the increased height of the iliac blades relative to the human condition.

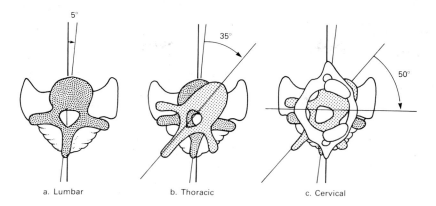

a. Lumbar   b. Thoracic   c. Cervical

FIGURE 15·14   *Axial rotation of the spine: (a) the lumbar region, (b) the thoracic region, and (c) the cervical region. After Kapandji (1974).*

Schultz (1961) mentions one male orang-utan whose ribs were less than 3 cm from the iliac blades permitting only the most minor lateral flexion in the lumbar region. The primary movement in the thoracic region is rotation.

Curvature of the backbone

When viewed from the side, the adult human backbone has four curves (Fig. 12.4). Both the cervical and lumbar regions are convex forward and the thoracic and sacral regions are concave forward (Fig. 15.4). The concave forward thoracic and sacral curvatures are the primary curvatures because these are the curvatures that are present in prenatal life. The cervical curvature appears at about three months of age when the infant begins to hold its head upright and the similar lumbar curvature at about 1.5 years of age when the child has mastered bipedalism. The secondary cervical and lumbar curves are the primary features enabling us to balance our weight efficiently over our feet in bipedal locomotion (see Chapter 14).

The convex forward curvature of the lumbar region is called **lordosis** ((Gk)=I bend). The concave forward curvature of the thoracic region is called **kyphosis** ((Gk)=bent or bowed forward). Extreme kyphosis results in a hump-back. In humans the degree of lordosis is generally more extreme in women than in men. This is related to the increased size of the birth canal in females and particularly to the more acute angle of the female sacrum (see Chapter 20). Humans are said to be shorter in the evenings than in the mornings and one of the reasons for this is that all the spinal curves increase throughout the day as a result of fatigue. Two other features that contribute to a decrease in stature through the day are the compression of the intervertebral discs and the reduction in the height of the arches of the feet.

The backbone of the large apes lacks the extreme curvatures of the human spine (Fig. 12.4). The cervical curve is missing in these primates although the straight neck may be extended (or angled) to some extent depending on the posture of the head in relation to the generally straight thoracic region (Schultz, 1961). Lumbar lordosis has been reported in non-human primates by both Cunningham (1886) and Schultz (1961), although it is not as extreme as it is in humans. It is unclear from these reports how much the reduced lumbar lordosis is a feature of the living animal or of the cadaver that has been stretched out on its back in an unnatural posture for analysis (see Cunningham, 1886).

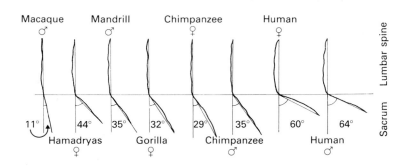

FIGURE 15·15 *The angular relationship between the lumbar and sacral regions of the spine in selected primates. After Schultz (1950).*

However, both apes and humans have an angle, or **promotorium** (Fig. 15.15) made by the last lumbar and the first sacral vertebra (also known as the **lumbosacral angle**). Not only does the promotorium become more pronounced in primates in concert with increasing upright posture (Abitbol, 1987a; Chapter 20), but it may also be related to the need to enlarge the birth canal in species with unusually large-headed infants. Schultz (1961) has reported unusually marked promotories for hamadryas baboons and mandrills that also have large-headed neonates in relation to other monkeys of their body sizes.

Proportional relationships

Proportional relationships involve the relative sizes of the individual vertebrae that make up the backbone as well as the relative lengths of major areas of the backbone. In comparison with monkeys and prosimians, both apes and humans are characterized by longer cervical columns and shorter lumbar columns in relation to total presacral column length (Table 15.2). The latter also contrast with the majority of monkeys and prosimians in the proportions of their individual vertebrae. In apes and humans the vertebrae are transversely broader than they are tall, while in monkeys and prosimians the reverse is the norm (Rose, 1975; Ankel, 1967, 1972).

In relation to apes, humans also have a relatively longer lumbar region in comparison to total presacral spine length (Table 15.2) (Schultz, 1961). Moreover, humans differ from apes in the size progression of their individual vertebrae down the spine. Vertebrae C5 and C6 are the shortest vertebrae in the human spine and beginning with C7 the vertebrae increase gradually in length through the series (Martin and Saller, 1959). In the gorilla, on the other hand, the shortest vertebrae are C7 and T1, the longest thoracic

TABLE 15·2 The lengths of the spinal regions expressed as a percentage of presacral spine length. After Schultz (1961).

	N	Cervical (%)	Thoracic (%)	Lumbar (%)
Homo	19	22.0	45.2	32.9
Gorilla	4	24.2	51.2	24.7
Pan	13	23.5	53.6	23.0
Pongo	8	24.6	50.9	24.6
Symphalangus	3	20.8	50.1	29.1
Hylobates	89	21.0	48.7	30.3
Old World monkeys	139	16.3	40.9	42.8
New World monkeys	22	16.8	46.0	37.2
Prosimians	26	18.4	46.6	35.0

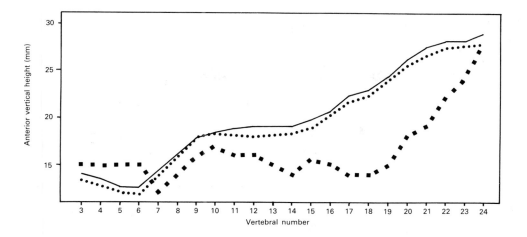

FIGURE 15·16 *The anterior vertical height of the presacral vertebral bodies in North American whites (solid line) and blacks (dotted line) and in gorillas (squares). Human data from Lanier (1939) and gorilla data from Martin and Saller (1959).*

vertebrae are T2–T3 and the last three vertebrae in the presacral spine increase more rapidly in length than is the case in the human spine (Fig. 15.16) (Martin and Saller, 1959).

Humans also have unusually wide lumbar vertebrae in relation to trunk length in comparison to other primates (Table 15.3) (Schultz, 1961). The absolute increase in transverse diameter betwen adjacent lumbar vertebrae is also greatest in humans (Rose, 1975). The weight of the lumbar vertebrae in relation to the total weight of all vertebrae also demonstrates the unusually robust human lumbar spine in comparison with the large apes (Table 15.4) (Schultz, 1961). However, the robusticity of individual lumbar vertebrae

TABLE 15·3 Relative breadths of thoracic and lumbar vertebrae. Figures represent the averages of the smallest breadth of the middle thoracic and the middle lumbar vertebrae expressed in percentage of trunk length. After Schultz (1961).

	N	Thoracic	Lumbar
Homo	30	5.4	8.0
Gorilla	7	5.2	6.9
Pan	20	4.4	6.5
Pongo	7	5.0	6.7
Symphalangus	3	4.4	6.6
Hylobates	48	4.0	6.5
Old World monkeys (7 genera)	78	3.3	5.3
New World monkeys (8 genera)	47	3.2	4.7
Prosimians (9 genera)	23	3.1	4.5

TABLE 15·4 Weight of the different types of vertebrae expressed as a percentage of the total weight of all vertebrae. After Schultz (1961).

	N	Cervical	Thoracic	Lumbar	Sacral	Caudal
Homo sapiens	5	12.7	35.7	30.6	20.3	0.6
Gorilla gorilla	4	18.5	41.6	22.1	17.4	0.4
Pan troglodytes	4	16.0	44.8	21.6	17.0	0.5
Pongo pygmaeus	3	20.4	40.9	23.6	14.4	0.7
Symphalangus syndactylus	2	14.0	40.3	30.7	14.2	0.8
Hylobates lar	4	12.0	37.8	34.9	14.0	1.3

TABLE 15·5   Robusticity of the lumbar vertebrae. After Rose (1975).

	N_s	N_v	Transverse diameter index	Sagittal diameter index
Homo	20	60	172.4 (149–203)	118.0 (101–139)
Pan	20	56	173.4 (131–217)	122.6 (97–169)
Gorilla	20	50	164.5 (120–219)	108.9 (82–139)
Pongo	20	57	167.6 (134–217)	117.8 (96–141)
Papio	20	60	121.6 (96–156)	76.5 (63–94)
Ateles	10	30	116.7 (76–170)	75.8 (51–104)
Colobus	20	60	97.6 (85–112)	59.6 (50–74)
Cercopithecus	20	60	83.7 (73–97)	49.1 (37–63)
A. afarensis				
AL-288-lak	1	1	154	89
AL-333-73	1	1	139	90
A. robustus[a]				
SK 853	1	1	169	114
A. africanus				
Sts 14	1	3 (4*)	127.0 (124–133)	97.8 (96–99)*

N_s = number of specimens. N_v = the number of vertebrae. Transverse diameter index = [(cranial transverse diameter × 100)/ventral craniocaudal length]. Sagittal diameter index = [(cranial sagittal diameter × 100)/ventral craniocaudal length].
[a]Classified as *Homo erectus* by Robinson (1972).

as measured by both the transverse and the sagittal diameter of the upper (cranial) surface of the vertebral body in relation to anterior (ventral) body height demonstrates little difference between humans and the large apes (Table 15.5). It can be argued, therefore, that the increase in the size of the human lumbar spine is due to a proportional increase in the length as well as breadth of the vertebral bodies. This can be viewed as a direct adaptation to the increased stress on this region resulting from upright posture and bipedal locomotion (Schultz, 1961; Rose, 1975).

The wedging of the lumbar vertebrae and lumbar lordosis

In humans the lumbar lordosis is normally, but not necessarily, reflected in the shape of the bodies of particularly the lower lumbar vertebrae. From L1 (first lumbar) to L5 in humans the posterior (dorsal) height of the vertebral body decreases in relation to the anterior (ventral) height of the same body producing an increasing dorsal wedging. The ratio of posterior height to anterior height is called the **lumbovertebral index** (Cunningham, 1886). Humans differ from the large apes in the marked dorsal wedging of the last lumbar vertebra (low lumbovertebral index) (Table 15.6). However, there is a considerable range of variation in the lumbovertebral index as well as a large degree of overlap between humans and apes in this index for each lumbar vertebra. Each of the modern human lumbar vertebrae may show dorsal wedging, while at least some of the lumbar vertebrae in the large apes can also show this feature (Rose, 1975).

Muscles that move the backbone

Except where indicated, the muscles that move the back are largely similar in humans and apes. The differences that do exist are most marked in the cervical region and these are discussed in detail in Chapter 12.

TABLE 15·6 The lumbovertebral index [(dorsal height of vertebral body × 100)/ventral height of body]. After [a]Rose (1975) and [b]Cunningham (1886).

		Lumbovertebral index[a]					Lumbovertebral index[b]
	N	L1	L2	L3	L4	*N*	Last lumbar
Homo	20	105.9 (98–118)	105.5 (95–114)	104.4 (94–114)	100.1 (89–109)	132	90.5 —
Pan	16	108.8 (100–125)	108.1 (92–124)	105.4 (84–118)	—	9	115.8 —
Gorilla	10	103.5 (101–107)	102.6 (96–107)	99.5 (91–106)	—	5	101.9 —
Pongo	17	109.9 (101–130)	109.4 (100–129)	106.7 (94–133)	—	4	103.5 —

Numbers in parentheses are ranges.

The main muscles that move the backbone fall into two general categories: the muscles of the spine, and the muscles of the abdominal wall. The muscles of the spine have both their origins and insertions on, or adjacent to, the spine, while the muscles of the abdominal wall form the sides and the front of the abdominal cavity.

The main muscles of the spine are a group of muscles known collectively as the **erector spinae** muscles (Fig. 15.17). (spinae (L) = spine, and derives from the rather fanciful similarity between the spine and a Roman racecourse. The central wall of the Roman racecourse was called a *spina* and this wall divided the racing circuit longitudinally in much the same fashion that the spine divides the back into two halves.) The name **erector spinae** literally means upright spine and describes the main function of these muscles in supporting the spine during upright posture.

The erector spinae muscles run bilaterally along the backbone, originating on the back of the sacrum and inserting into the vertebrae and the base of the skull. They originate as one muscle mass but as they extend upward they divide into three longitudinal columns

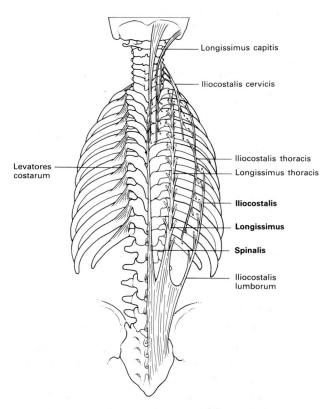

FIGURE 15·17 *Erector spinae and levatores costarum. After Romanes (1964).*

Longissimus capitis

Iliocostalis cervicis

Iliocostalis thoracis

Longissimus thoracis

Iliocostalis

Longissimus

Spinalis

Iliocostalis lumborum

Levatores costarum

on each side of the backbone. The most medial column, closest to the backbone, is called **spinalis** (meaning spine), the intermediate column is **longissimus** (= longest) and the lateral column is called **iliocostalis** (meaning from the ilium to the ribs). Each of these columns may be further subdivided (longitudinally) according to the region which they cover. Therefore, **iliocostalis lumborum** acts on the lumbar region, **spinalis thoracis**, **longissimus thoracis** and **iliocostalis thoracis** all act on the thorax, **spinalis cervicis**, **longissimus cervicis** and **iliocostalis cervicis** all act on the neck, and **longissimus capitis** acts on the head.

Because these muscles run longitudinally along the back, they act to extend the spinal column. They are also active in forward (ventral) flexion where they control and regulate the forward movement of the trunk (MacConaill and Basmajian, 1969). When the muscles on one side of the back contract they laterally flex and rotate the back. The main difference in **erector spinae** in the apes is the degree of fusion of the three columns (Sonntag, 1924). In particular, in chimpanzees the outer two columns can be fused over most of the back while the inner column separates only in the middle of the thoracic region.

Electromyographic studies show that at least two of the erector spinae muscles (longissimus thoracis and iliocostalis lumborum), have a similar function during bipedal locomotion in humans, chimpanzees and gibbons (Shapiro and Jungers, 1988). In the bipedal stride cycle, iliocostalis lumborum is active in all three species during the support phase of the limb on the opposite side of the body from the muscle, the contralateral limb. This suggests that the primary function of this muscle during bipedal locomotion is to counteract lateral movement of the trunk related to contralateral touchdown (Shapiro and Jungers, 1988). Longissimus thoracis is active during the support phases of both limbs (contralateral and ipsilateral (ipsi (Gk) = same)) and most probably functions to control both forward and lateral flexion of the trunk during bipedal walking (Shapiro and Jungers, 1988).

Although the erector spinae muscles are capable of producing some degrees of rotation, the main rotators of the back are the **transversospinalis** muscles (Fig. 15.18). These muscles are obliquely positioned and lie deep to the erector spinae muscles. They run cranially and medially from the transverse processes of one vertebra to the spinous process of another and are divided into three groups depending on the number of vertebrae they cross. The most superficial layer, crossing approximately five vertebrae, is **semispinalis** (divided into **thoracis**, **cervicis**, and **capitis**), the next layer, crossing approximately three vertebrae, is **multifidus** ((L) = much divided) and the deepest layer connecting adjoining vertebrae comprises the **rotatores**. This last group of muscles is most developed in the thoracic region where

FIGURE 15·18 *Transverso-spinalis and quadratus lumborum. After Green and Silver (1981).*

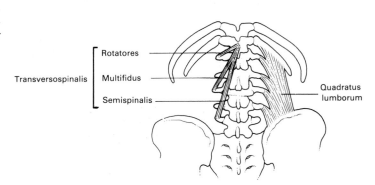

Transversospinalis
Rotatores
Multifidus
Semispinalis
Quadratus lumborum

rotation is the primary movement of the spine.

There are also a number of additional muscles of the back. The largest of these is **quadratus lumborum** (quadratus (L) = four sided) which forms part of the posterior abdominal wall (Fig. 15.18). This muscle originates on the iliac crest and iliolumbar ligament and inserts into the transverse processes of the lumbar vertebrae and into the last rib. It can flex the trunk laterally and aids breathing (see below). However, electromyographical studies show that its primary function during bipedal locomotion in humans as well as in chimpanzees and gibbons is to control movement in the sagittal plane (Shapiro and Jungers, 1988).

The remaining three spinal muscles are much smaller in size. **Levatores costarum** originates from the tip of each transverse process and inserts on the rib below (Fig. 15.17). Its main function is to raise the ribs in breathing (see below). **Interspinales** connect the spinous process of adjacent vertebrae in the cervical and lumbar regions. **Intertransversarii** are also found in the cervical and lumbar regions and connect the anterior tubercle of one vertebra to the posterior tubercle of adjacent vertebrae.

The main anterior (ventral) and lateral flexors of the trunk are the muscles of the abdominal wall. There are four muscles of the abdominal wall. Three of these, **internal oblique**, **external oblique** and **transversus abdominis**, pass anteriorly in three layers around the sides of the abdomen to enclose the fourth muscle, **rectus abdominis**, in a tendinous sheath formed from the anterior extensions of each of the three muscles (Figs 15.19 and 15.20). **Rectus abdominis**, runs a straight course (rectus (L) = straight) over the abdomen from the pubic bone inferiorly to the costal cartilages of the fifth, sixth and seventh ribs and the xiphoid process above. It is a major anterior flexor of the trunk. Rectus abdominis is most active when raising the head and shoulders or legs from a flat, supine posture, for example while doing sit-ups or when getting out of bed. Rectus abdominis is a paired muscle which is joined in the mid-line by the **linea alba** (albus (L) = white). The linea alba, or white line, is formed by the mid-line fusion of the sheath that surrounds each half of the paired rectus abdominis (Fig. 15.19). It extends down the centre of the trunk the full length of rectus abdominis from the xiphoid process to the pubis.

FIGURE 15·19 *Muscles of the trunk in cross section. After Green and Silver (1981).*

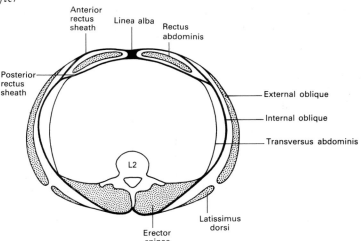

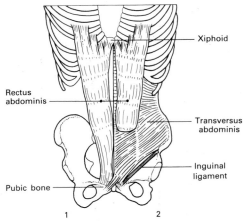
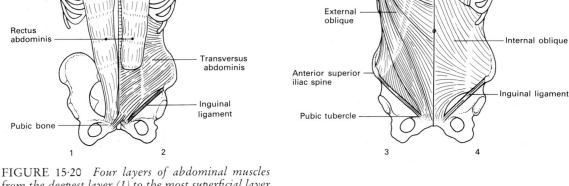

FIGURE 15·20 *Four layers of abdominal muscles from the deepest layer (1) to the most superficial layer (4). After Kapandji (1974).*

Both the **internal** and the **external oblique** muscles laterally flex and rotate the trunk (Fig. 15.20). In addition they are active during straining or lifting heavy objects. The external oblique forms the outer layer of abdominal wall muscles. Its fibres run downwards and forwards at right angles to those of the underlying internal oblique muscle. It originates on the fifth to twelfth ribs and inserts into the linea alba, iliac crest, pubic tubercle and **inguinal ligament**. The inguinal ligament (inguen (L) = groin) is a fibrous band that is, in effect, the lower border of the external oblique muscle that runs between the anterior superior iliac spine and the pubic tubercle and provides attachment not only for the internal oblique muscle but also for the transverse abdominis muscle. The internal oblique muscle lies deep to the external oblique muscle and makes up the middle layer of abdominal wall muscles. It originates on the iliac crest and thoracolumbar fascia posteriorly and on the inguinal ligament anteriorly. It passes forwards to insert into the lower three ribs, the linea alba and the pubic bone.

The third abdominal wall muscle, **transversus abdominis**, lies deep to both the internal and external oblique muscles and makes up the inner layer of abdominal wall muscles. It

originates on the lateral side of the trunk from the thoracolumbar fascia, the lower six ribs (costal cartilages), the iliac crest and the inguinal ligament and inserts into the pubic bone and linea alba. Its fibres run horizontally around the sides of the abdomen. It has little direct effect on the spine and its main function is to increase intra-abdominal pressure during breathing (Fig. 15.20 and see below).

The vertebral column in the Plio-Pleistocene

Fossil vertebrae are known for *Australopithecus afarensis, Australopithecus africanus, Paranthropus,* and *Homo* cf. *erectus.* The *A. afarensis* collection includes 24 vertebral elements, 15 from the AL 288–1 skeleton and nine isolated vertebrae from the AL 333 sample (Cook et al., 1983). *A. africanus* is represented by 17 vertebrae from the Sterkfontein (Member 1), 15 of these belong to the Sts 14 skeleton, one to the Sts 65 pelvic bone and the last is an isolated find (Sts 73) (Robinson, 1972). *Paranthropus* is represented by three vertebrae from Swartkrans and *Homo erectus* may be represented by an additional, but very fragmentary vertebra from this same site (Robinson, 1972). *Homo* cf. *erectus* is better represented by

14 presacral vertebrae from the KNM-WT 15000 skeleton from Nariokotome, West Turkana (Brown et al., 1985).

The vertebrae of all of these Plio-Pleistocene hominids have relatively long spinous and transverse processes (Fig. 15.21). In the _A. afarensis_ vertebrae the cervical spinous processes in AL 333 are long and there is some indication that the spinous processes of the upper thoracic vertebrae in AL 288 were also long (Cook et al., 1983). This has been interpreted by Cook et al. (1983) to indicate that erector spinae, the rhomboids, and trapezius were more massive in _A. afarensis_ than they are in modern humans. In _A. africanus_ the transverse processes of the lumbar vertebrae are particularly long in relation to the size of the vertebral bodies (Fig. 15.21). The transverse processes of L3 are the longest and these, as well as the processes of L4 curve distinctly upward (Robinson, 1972) (Fig. 15.21c). The length of these lumbar transverse processes would enhance not only the lever advantage of quadratus lumborum but also of psoas major. Both these muscles are important postural muscles in quadrupedal and bipedal locomotion.

The few vertebrae assigned to _Paranthropus_ that are complete enough also show long processes (Fig. 15.22). The last lumbar vertebra of _Paranthropus_ (SK 3981b) has an

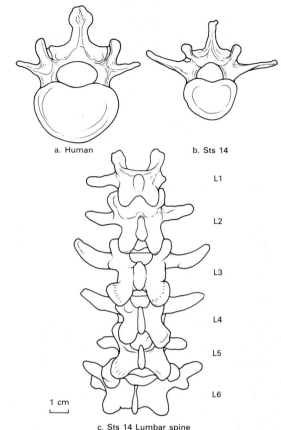

a. Human   b. Sts 14

c. Sts 14 Lumbar spine

FIGURE 15·21 _The second lumbar vertebra (L2) (a) in a human and (b) in Sts 14 from the superior (cranial) view. (c) The articulated Sts 14 lumbar spine from the posterior (dorsal) view. After Robinson (1972)._

FIGURE 15·22 _The last lumbar vertebra from the superior (cranial) view and the last thoracic vertebra from the lateral view in_ Australopithecus africanus _(Sts 14)_, Paranthropus robustus _(SK 3981) and modern humans. After Robinson (1972) (the thoracic vertebra have been reversed from the original)._

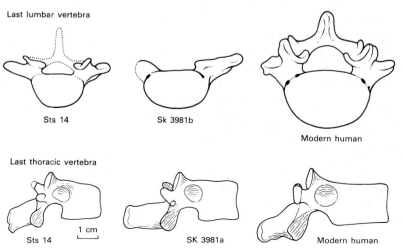

Last lumbar vertebra

Sts 14   Sk 3981b   Modern human

Last thoracic vertebra

Sts 14   SK 3981a   Modern human

upcurved transverse process (only the left side is preserved) that is reminiscent of *A. africanus*. However, this *Paranthropus* transverse process is longer (in relation to the sagittal diameter of the body but not the transverse diameter) than are the transverse processes of either *A. africanus* or modern humans. The last thoracic vertebra of *Paranthropus* (SK 3981a) has proportionately shorter transverse processes than that of *A. africanus*, but at the same time has a very much longer spinous process that projects more perpendicularly to the vertebral body than is the case either in *A. africanus* or in modern humans (Fig. 15.22). The length of the spinous process in *Paranthropus* is 107% the maximum anteroposterior length of the upper surface of the body, while it is 89% in the corresponding Sts 14 (*A. africanus*) vertebrae (Robinson, 1972).

In the KNM-WT 15000 *Homo*. cf. *erectus* the spinous processes of all of the vertebrae have been described as both relatively longer and less inferiorly inclined than is the case in modern humans (Brown et al., 1985).

In addition to the long spinous processes, the vertebrae of *A. afarensis* show other features that are not normally found in human vertebrae (Cook et al., 1983). Firstly, the epiphyses on the upper and lower faces of AL 288-1 vertebrae are more extensive than normally found in humans (Fig. 15.23). Rather than being ring like in form as they are in humans, these epiphyses are more extensive as are ape epiphyses. In AL 288-1 these epiphyses are fused to the vertebral body, but their edges are still distinct. In modern human terms this would indicate an age at death in the 20s or 30s (Cook et al., 1983).

Secondly, as is also the case for *A. africanus* vertebrae, the vertebral canals of the thoracic and lumbar vertebrae are large in relation to what appears to be a relatively small vertebral body (Figs 15.23 and 15.24). Cook et al. (1983) suggested that this might be an effect of body size as smaller animals tend to have proportionately larger canals. However, these

authors pointed out that the much larger AL 333x-12 (T10) (*A. afarensis*) vertebra also has the same large vertebral canal in relation to the size of the vertebral body.

In both absolute and relative measurements, the vertebral bodies of both *A. afarensis* and *A. africanus* appear to be quite small in relation to modern human vertebrae. Both cross-sectional and vertical absolute measurements of the AL 288-1 vertebrae are consistently below a sample of modern human vertebrae selected for small size (Cook et al., 1983). Robinson (1972) has clearly illustrated

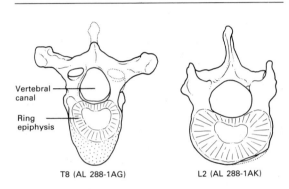

FIGURE 15·23 *Australopithecus afarensis vertebra. Note the large ring epiphyses, the relatively large vertebral canals and the pathology in T8 (stippled). After Cook* et al. *(1983).*

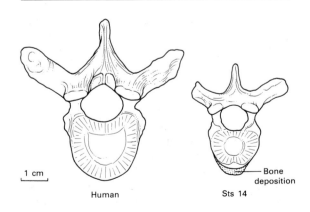

FIGURE 15·24 *The sixth thoracic vertebra (T6) in a modern human bushman and in Sts 14 from the superior (cranial) view. After Robinson (1972).*

this size difference between *A. africanus* and modern humans by demonstrating that the pelvic fragments of the Sts 14 skeleton are comparable in size to the pelvis of a small female Bushman. However, Bushman vertebrae are much larger than the Sts 14 vertebrae (Figs 15.21 and 15.24). In addition, in relation to the vertical height of the vertebral body, both the sagittal and the transverse diameters of *A. afarensis* and *A. africanus* lumbar vertebral bodies are small in relation to humans and apes (Table 15.5) (Rose, 1975). This indicates that the vertebrae in these fossils are less robust than is normally the case in either humans or apes.

Thirdly, in Sts 14 there are six lumbar vertebrae rather than the normal human five lumbar vertebrae (Fig. 15.21c) (Robinson, 1972). The first of these six lumbar vertebrae is transitional in form. It has a rib facet on the right side of the body but not on the left side (a feature of the thoracic vertebra) but has superior articular facets that are convex and face toward the mid-line (a feature of the lumbar vertebra). From a sample of one it is not possible to determine whether this is the norm for *A. africanus*. However, Cook et al. (1983) suggested that from the size progression of the preserved AL 288-1 (*A. afarensis*) vertebrae, five lumbar vertebrae would be a more probable number for this individual than six. The six Sts 14 lumbar vertebrae show a progressive dorsal wedging from L1 to L6 that is fully consistent with a well-developed lumbar lordosis at least in this species.

Finally, perhaps the most interesting feature of the vertebrae of these australopithecines is the deposition of bone, to a greater or lesser extent, along the anterior surface of the thoracic vertebral bodies in a position that would have been under the anterior longitudinal ligament in life. This is highly vascular bone that projects anterior to the fused epiphysis and seems to indicate a ventral, or anterior, growth of bone after the fusion of the ring epiphysis with the body of the vertebra (Cook et al., 1983). Such a

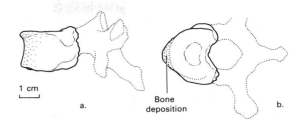

FIGURE 15·25 *The* Australopithecus afarensis *(AL 333-51) mid-thoracic vertebra from the lateral (a) and superior (b) view. After Cook* et al. *(1983).*

condition is present in the AL 333-51 adult mid-thoracic vertebra (*A. afarensis*) (Fig. 15.25) and in the lower thoracic vertebra of Sts 14 (*A. africanus*)(Fig. 15.24). It has also been observed in adult chimpanzees, orang-utans and macaques, although it is rare in modern humans. Cook et al. (1983) suggested that it might represent the normal condition in the small australopithecines and be related in some way to an early fusion of the more ape-like ring epiphyses of these hominids before the growth of the vertebral body was complete. The thoracic vertebrae of AL 288-1 (*A. afarensis*) show an extreme, and most probably, pathological development of this condition. The thoracic vertebrae have an extensive development of bone anterior to the ring epiphysis and this is most extreme on the eighth thoracic vertebra (Fig. 15.23). The bone had been laid down in concentric circles suggesting a gradual development to its maximum condition. Furthermore, the thoracic vertebrae show a considerable anterior wedging and approximation of the bony growths, or **osteophytes**, projecting from superior and inferior edges of adjacent vertebral bodies, suggesting that there had been considerable degeneration of the intervertebral discs (Cook et al., 1983). In spite of this rather extreme condition of the vertebral bodies (in the region of the anterior longitudinal ligament) there is little if any pathological change in either the facet joints or the intervertebral articular surfaces.

The pathologies present in the AL 288

thoracic vertebrae seem to correspond most closely to a human condition known as **juvenile kyphosis dorsalis**, or **Scheuermann's disease**. This disease is associated in humans with high activity levels and trauma and may also have a genetic component. Cook et al. (1983) suggested that for AL 288 this indicates pronounced anterior compression of the lower thoracic spine that could have resulted from a variety of activities including arboreal climbing, crutch walking or heavy lifting and carrying activities. These authors suggest that the heavy back and shoulder musculature indicated by the long spinous processes (and also robust vertebral arches) would be consistent with this interpretation.

Similar pathological conditions have not been reported for the few known *Paranthropus* vertebrae. However, in some ways these vertebrae are similar to the vertebrae of the small australopithecines. This is particularly apparent in the proportionately large vertebral canal (in the SK 3981a last thoracic vertebra) (Fig. 15.22). In addition to the already mentioned proportionately longer spinous processes, the *Paranthropus* vertebrae are also noticeably more robust (Table 15.5). Although the superoinferior (craniocaudal) lengths of the bodies of the corresponding *Paranthropus* and *A. africanus* vertebrae are similar, the transverse diameters of the bodies of the *Paranthropus* vertebrae are considerably larger than the *A. africanus* counterparts. For example, the transverse diameter of SK 3981a (*Paranthropus* last thoracic vertebra) is 123% that of the corresponding Sts 14 (*A. africanus*) vertebra and the transverse diameter of SK 3981b (last lumbar) is 128% of its Sts 14 counterpart (Robinson, 1972). If there was no difference in proportions between *Paranthropus* and the small australopithecines, Robinson (1972) suggested that on the basis of the similar superoinferior vertebral lengths, the statures of the individuals represented by these vertebrae also may have been similar. However, the larger diameters of the *Paranthropus*

vertebral bodies would suggest that this individual was considerably heavier than those individuals represented by the *A. africanus* (or *A. afarensis*) vertebrae.

To date little has been published about the virtually complete thoracic and lumbar series of KNM-WT 15000 (*Homo* cf. *erectus*). In addition to the previously mentioned long and less inferiorly inclined spinous processes, it has also been reported that the KNM-WT 15000 last cervical vertebra (C7) and thoracic vertebrae have small vertebral canals in relation to the size of their vertebral bodies (Brown et al., 1985). In this feature they contrast both with modern humans and with the small australopithecines and *Paranthropus*. The lumbar vertebral canals have been reported to be similar to modern humans in their relative sizes. This difference poses very interesting and, at present, unanswered questions about the relationship between body size, bipedal locomotion, vertebral body size, spinal cord size and brain size in the Plio-Pleistocene hominids.

The vertebral column in the Neanderthals

The Neanderthal cervical vertebrae have been discussed in Chapter 12. These, as well as the preserved thoracic and lumbar vertebrae, are within the range of modern human variation. However, as with many other features of the Neanderthal skeleton, all the vertebrae show a high degree of robusticity in comparison with the modern human sample (Trinkaus, 1983a) (Fig. 15.26). Neanderthal thoracic vertebrae have relatively robust bodies and vertebral arches (Trinkaus, 1983a). Although the European Neanderthals (La Chapelle-aux-Saints 1 and La Ferrassie 1) had transverse processes that were more laterally oriented than the human condition, Trinkaus (1983a) described the Shanidar transverse processes as dorsolaterally oriented as in modern humans.

Neanderthal lumbar vertebrae are robust in the size of their bodies, their vertebral

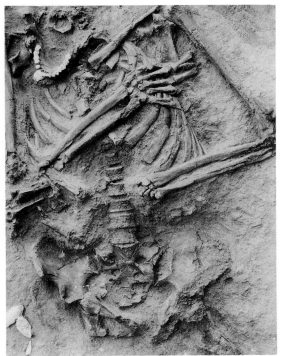

FIGURE 15·26 _The Neanderthal from Kebara, Israel. Courtesy of Y. Rak._

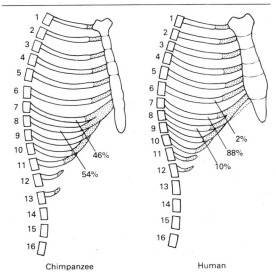

Chimpanzee   Human

FIGURE 15·27 _Sternal (or true) ribs in humans and chimpanzees. The numbers indicate the percentage of individuals in which the indicated rib is the last rib to articulate directly with the sternum. After Tredgold (1897)._

arches and their spinous processes (Trinkaus, 1983a). They also have relatively long transverse processes that taper laterally. As with the Plio-Pleistocene hominids, long vertebral processes suggest well-developed back musculature for stabilization and support of the trunk. The lower lumbar vertebrae show a marked degree of dorsal wedging indicating a lumbar lordosis similar to modern humans (Trinkaus, 1983a) in addition to both an absolute and a relative length of the lumbar vertebral column (relative to sacral length and bicondylar femur length) that is well within the range of modern humans.

The rib cage in humans and apes

The rib cage protects the organs of the thorax and upper abdominal cavity. It also provides areas of attachment for the muscles of the upper limb and abdominal wall and plays an important role in breathing. The rib cage in humans comprises 24 ribs (12 pairs) that are attached to the thoracic vertebrae in the back (dorsally) via synovial joints. Ventrally the majority of the ribs attach either directly or indirectly to the sternum (see below); however, the two lowermost ribs have no ventral attachment and are called floating ribs (Fig. 15.27).

The ribs: number

In mammals the number of ribs is highly variable, ranging from 24 pairs in the sloth (Edentata) to an average of 13 pairs in primates (Tredgold, 1897). Within primates, lemurs have the highest number of ribs, averaging about 14, and humans, orangutans, macaques and _Presbytis_ the lowest, averaging around 12 (Table 15.7). African apes average 13 ribs.

The ribs are generally divided into three different categories and the numbers of ribs within each of these categories is also variable. **True,** or **vertebrosternal, ribs** make up the first category. Ventrally these ribs are connected directly to the **sternum** via cartilaginous extensions from their anterior ends,

TABLE 15·7   The numbers of ribs in primates. After Tredgold (1897).

	N	Total number of ribs	Number of sternal ribs	Number of floating ribs	Cases with eight true ribs(%)
Lemurs and monkeys					
Lemuroidae	9	14.10	9.88	0.44	100
Platyrrhini	9	13.50	7.88	0.77	66
Cercopithecus	6	12.33	8.14	2.33	100
Macaca	27	12.04	7.96	1.45	96
Presbytis	30	12.03	7.09	2.00	17
Apes					
Gibbons	16	13.13	7.18	2.75	18
Pan	17	13.20	7.54	2.00	54
Gorilla	7	12.86	7.28	2.00	28
Pongo	12	12.15	6.92	2.25	8
Humans					
Negro	5	12.00	7.60	1.80	60
Caucasian	230	11.90	7.10	—	10

known as **costal cartilages** (costa (L) = a rib). In humans there are generally seven and occasionally eight true ribs. In the apes the number of true ribs varies from an average of 6.9 in orang-utans to 7.5 in chimpanzees (Table 15.7). In humans eight true ribs are found more frequently in males than in females and in both sexes on the right side rather than the left side of the body (Tredgold, 1897). A sexual correlation for the occurrence of the eighth vertebral sternal rib has not been reported for non-human primates. It was once thought that the lower incidence of the eighth true rib in human females was a direct consequence of the tight corsets worn by 19th century European ladies (see Tredgold, 1897). A more likely, but less colourful explanation, may simply be the fact that human females have a relatively shorter sternum than do human males (see below) offering a reduced area for rib attachment.

False ribs (also known as **asternal** or **costal ribs**) make up the second category. These ribs do not connect directly with the sternum but articulate with the next highest rib via their costal cartilages. In humans (and in orang-utans) the eighth to tenth ribs would

normally be false ribs while in African apes, depending on both the number of true ribs and the total number of ribs, either the eighth to eleventh or the ninth to eleventh would be false ribs. In human females the costal margins of the false ribs on one side of the rib cage make a more acute angle with those on the other side than they do in human males. A similar sexual dimorphism of lower thoracic shape has not been reported for the apes.

Floating ribs are the remaining category. These are the last two ribs, the eleventh and twelfth in humans and most orang-utans and the twelfth and thirteenth in the African apes. They are both very mobile and rudimentary. They have only one articulation with their corresponding vertebral bodies, lacking any articulation with the transverse processes of these bodies and lacking an anterior articulation. In both humans and apes the last floating rib can be rudimentary. In humans it can vary in length from 1–8 inches (Fraser, 1920).

In addition to the three categories of ribs found in all individuals, there are two types of supplementary ribs that may be found in

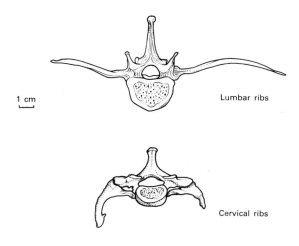

1 cm

Lumbar ribs

Cervical ribs

FIGURE 15·28 _Lumbar and cervical ribs. After Ankel (1967) and Basmajian (1975)._

both humans and apes. These may occur either bilaterally or unilaterally on the seventh cervical vertebra (**cervical rib**) or on the first lumbar vertebra (**lumbar rib**) (Fig. 15.28). Cervical ribs are much more common in humans than they are in the large apes while lumbar ribs are more common in the large apes (Tredgold, 1897). Where lumbar ribs occur in the large apes, they are firmly fused to the body of the first lumbar vertebrae and can be viewed as unusually long transverse processes (Ankel, 1967). Indeed both cervical and lumbar ribs are simply the exaggerated development of the part of that vertebra (the costal element) that normally forms the transverse processes. Both cervical and lumbar ribs have even been reported to occur in human foetuses and subsequently to disappear (Tredgold, 1897).

The ribs: morphology

A basic morphological pattern is shared by almost all of the ribs no matter to which category they belong (Fig. 15.29). The **head** of the rib is that part articulating with the vertebral body and intervertebral disc. The rib head is connected via the **neck** to the

tubercle which articulates with the transverse process of the vertebra. There is a sharp ridge on the upper surface of the neck called the **crista colli superior** (collis (L) = hill). This is for the attachment of the superior costotransverse ligament that binds the rib to the transverse process of the immediately superior vertebra (Fig. 15.7). The inner surface of the neck is smooth; however, there are roughened areas on the other surfaces of the neck for the attachment of additional ligaments. The rest of the rib is called the **body**, or **shaft**. A short distance lateral to the tubercle, the shaft turns rather sharply forward (ventrally). This is the **posterior angle** and also marks the attachment of **iliocostalis**. Anterior (ventral) to this angle the shaft curves inward and downward. In humans the seventh rib has the greatest downward curve as can be shown by placing the lower border of the rib on a flat surface and assessing the height the head then projects above the surface (Fraser, 1920).

The upper border of the rib shaft is rounder than the sharp lower border which has a groove, the **costal groove**, running along its inner surface. This groove is for the intercostal blood vessels and nerves and is a useful landmark for orienting the rib. The anterior end of the rib forms an oval hollow to receive the rounded end of the costal cartilage. This is even true for the floating ribs that have a short costal cartilage projecting from their tips.

Depending on the position of the rib some variation from this general pattern can occur and by far the most deviant rib is the first (Fig. 15.29). This rib can be best described as stubby and flat. It is tightly curved from its head to its neck to its anterior end with the long axis of its flat cross-section in the mediolateral rather than the craniocaudal plane. In some cases the head and neck may point downward in contrast to all of the other ribs where the head and neck point upward to a greater or lesser degree. This rib lacks both a posterior angle and a costal groove and its upper (cranial) surface shows

FIGURE 15·29 *The sixth rib from the posterior (dorsal) view (a) and the first, second and sixth rib from the superior (cranial) view (b).*

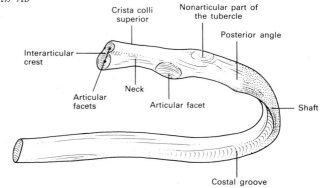

a.

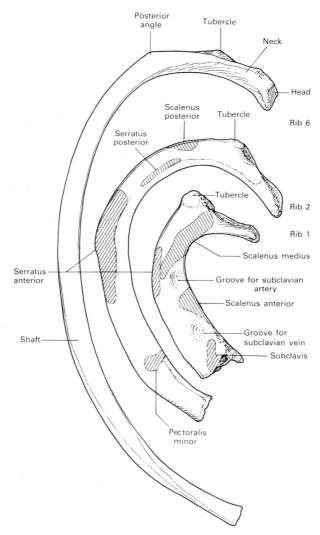

b.

a number of features which are absent in the other ribs. These include attachment areas of the **scalene** muscles as well as grooves for the **subclavian** vein and artery (see Fig. 15.29 and Chapter 12, Figs 12.7 and 12.18).

There is also variation in the form of the articular surface of the head. In the human first rib, as in the eleventh and twelfth, which normally articulate with the body of a single vertebra, the articular facet on the rib head is single and continuous. In the other ribs (and in the ape first rib) that articulate with the intervertebral disc and with demifacets on adjacent vertebra, the articular facet is separated in two by the **interarticular crest** (Fig. 15.29).

The second rib along with the tenth, eleventh and twelfth ribs can also show variations from the general pattern. The second rib can be viewed as a transition between the morphology of the first rib and the general morphological pattern of the majority of ribs (Fig. 15.29). The main similarity with the first rib lies in the mediolateral orientation of the long axis of its cross-section. In addition, the costal groove is not clearly defined and the posterior angle is either not present or in very close proximity to the tubercle.

The last three ribs show an increasingly simplified morphology. The human tenth rib may lack an articular facet on the tubercle and the human eleventh and twelfth ribs, which do not articulate with their respective transverse processes, are missing tubercles altogether. The eleventh rib has a mere trace of a costal groove while the often rudimentary twelfth rib is lacking both a costal groove and a posterior angle.

There is also variation in the position of the posterior angle that provides a potential aid in sequencing the ribs. The posterior angle lies most distant from the tubercle in the eighth rib where it is located about one-quarter to one-fifth along the length of the shaft. In both the more superior and more inferior ribs, the angle moves gradually closer to the tubercle. In the upper part of the thorax this decrease marks the constriction of erector spinae as it moves towards its attachment on the cervical vertebrae. On these more cranial ribs the length ratio gradually decreases to about one-eighth on the fifth rib and one-ninth on the second. On the lower (more caudal) ribs the ratio remains approximately the same, the reduction in its distance from the tubercle being proportionate to the reduction in total length of these lower ribs (Fraser, 1920).

The shape of the rib cage in humans and apes

In some features the human and ape rib cages are similar to each other and contrast with the rib cages of other primates, while in other features the human rib cage is quite distinct from the ape rib cage (Fig. 15.1). The greatest similarity in the rib cages of humans and apes is their general shape. In both groups it is transversely broad and dorsoventrally flat, contrasting with the transversely narrow and dorsoventrally deep rib cage normally found in monkeys (Fig. 15.30) (Schultz, 1961). Closely related to this difference in shape is the position of the vertebral column in relation to the rib cage. In both humans and apes it has migrated ventrally towards the

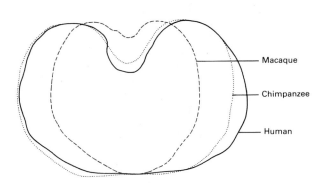

FIGURE 15·30 _The shape of the thorax in an adult macaque, a chimpanzee and a human. After Schultz (1956)._

centre of gravity in upright posture and lies more within the thorax rather than behind it as it does in other primates. Both the shape of the thorax and the position of the vertebral column develop during infancy in humans and apes. In the foetus the transverse and sagittal diameters of the chest are the same and the vertebral column lies at the back of the thorax. During subsequent growth the transverse diameter of the chest increases much more rapidly than does the sagittal

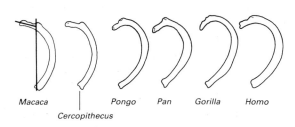

Macaca

Cercopithecus

Pongo   Pan   Gorilla   Homo

FIGURE 15·31 *The second rib of selected primates from the superior (cranial) view. After Schultz (1961).*

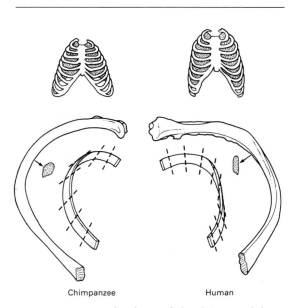

Chimpanzee   Human

FIGURE 15·32 *The shape of the thorax and form of the ribs in chimpanzees and humans. The dashed lines indicate the torsion of the ribs. After Schmid (1983).*

diameter in both humans and apes, but not in monkeys, while the vertebral column migrates to its more central position. Both ontogenetically and phylogenetically these features appear to be related to the development of a more upright posture. They are also associated with the migration of the scapula onto the back of the rib cage allowing greater mobility of the arm (Chapter 17) (Schultz, 1961).

These thoracic and vertebral similarities are also related to similarities in the structure of the individual ribs in these species. In particular the angle made by the neck of the rib (column) and the body of the rib is sharper in humans and apes than it is in the other primates (Fig. 15.31). As this angle 'points out the direction' to the body of the vertebra it can easily be seen that the vertebra would lie anterior, or ventral, to the most posterior part of the rib and thereby reflect the more anterior, or ventral, position of the vertebral column in these species.

Given these similarities between humans and apes, the differences in the shape of the thorax lie in its general contour. In humans (and gibbons) the thorax is barrel shaped when viewed from the front, while in the large apes it has a definite funnel-shaped appearance (Figs 15.1 and 15.32) (Schultz, 1961; Schmid, 1983). The flaring contour of the lower part of the ape thorax corresponds to the flaring ilia in these primates, while the more curved contour of the human thorax corresponds in a like manner to the curved human ilia. These differences in shape are also reflected in differences in the form and structure of the ribs.

(1) In the ape funnel-shaped thorax, the first few pairs of ribs increase more gradually in size than they do in humans where the increase is marked between the first and second pair of ribs.

(2) The last few human ribs are more curved than are the ape ribs corresponding to the curvature of the ilia. Furthermore, they have a decreasing radius of curvature

as the human thorax narrows toward the waist. In addition, the last rib in particular has a marked twist along its long axis bringing its medial surface into a superomedial orientation, cupping the bottom of the thorax.

(3) The cross-section of the ape rib is generally rounder and the human cross-section flatter mediolaterally.

(4) The long axis of the cross-section of the ribs follows the general curvature of the thorax. Corresponding to the funnel-shaped thorax of the apes, this axis is oriented uniformly in the superointernal to inferoexternal plane. In humans the axis changes from rib to rib according to the curvature of the thorax (Fig. 15.32). The axis also varies along the longitudinal axis of particularly the lower true ribs (Schmid, 1983). On the lateral side of the ribs, between the posterior and anterior angles of the ribs, the axis is vertical. The axis of the anterior part of the rib is in the typical orientation of the apes and gives the strong impression of being twisted medially.

The sternum

The name **sternum** (stereos (Gk) = solid or hard) was most probably given to this bone because of the solid feel of it through the skin of the upper chest. It provides the point of articulation for the ribs anteriorly and is divided into three sections which have been named in accordance with a fancied resemblance of the sternum to a gladiator's sword (Fig. 15.33) (Fraser, 1920). The most superior part is the **manubrium** ((L) = hilt or handle) and the next part used to be known as the **gladiolus** ((L) = little sword or blade) but is now generally referred to as the **body of the sternum**. The point of the sword, the most inferior part of the sternum, is the **xiphoid** ((Gk) = sword). The human sternum sits lower in females than it does in males. In females the upper border of the manubrium

is level with the body of the third thoracic vertebra, whereas in males it is level with the body of the second thoracic vertebra. In both sexes it is oriented at an angle of 20–25° to the vertebral column and the anteroposterior depth of the thorax is therefore wider at the lower end of the sternum than it is at the upper end.

The manubrium is the thickest and most robust part of the sternum. The large notch in its superior border is known as the **jugular**, or **suprasternal**, **notch**. To either side of the notch are the facets for articulation with the clavicles. These facets are concave mediolaterally and slightly convex from front to back. This shape, together with the fact that the joint has a disc, or meniscus, allows the clavicle to rotate as the arm is abducted, adducted, flexed or extended (Chapter 16). The robusticity of the manubrium is undoubtedly related to its role as the only contact between the shoulder girdle and the axial skeleton. Below the clavicular facets, on the upper part of its lateral margins, the costal

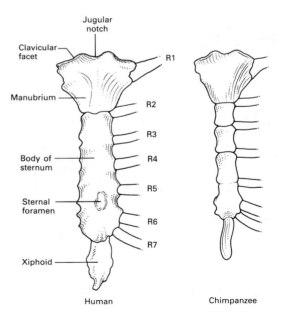

FIGURE 15·33 *The human and chimpanzee sternum. After Schultz (1961).*

cartilages of the first rib join the manubrium forming a synchrondrosis, or cartilaginous joint (see Chapter 2). The costal cartilages of the second pair of ribs join the manubrium at its junction with the body of the sternum and articulate with both the manubrium and the sternal body as a synovial joint.

The body of the sternum is made up of four separate parts, or **sternebrae**. These are completely fused in adult humans (by about 25 years of age) but are generally still visible as separate units. The third, fourth and fifth pairs of ribs articulate respectively at the junctions between the first and second, second and third, and third and fourth sternebrae. The sixth pair of ribs joins the lateral sides of the last (fourth) sternebra and the seventh pair (and eighth pair if present) articulate at the junction between the last sternebra and the xiphoid. In humans there may occasionally be a hole, or foramen, located in the mid-line in the lower third of the body of the sternum. This hole is called the **sternal foramen** and can range from 3 to 18 mm in diameter (McCormick, 1981). It occurs in between 7% and 8% of modern humans and is much more frequent in males (9.6%) than in females (4.3%) (McCormick, 1981).

The xiphoid is highly variable in both length and shape; however its lower end is normally bifid, or divided in two. It is set on a plane deeper than the sternal body, and its lower tip frequently projects forward. It fuses with the sternal body in middle life.

The sternum in humans and apes

The main differences in sternal morphology between humans and apes are proportional. Both humans and apes have relatively broad sterna in relation to prosimians and monkeys, the chimpanzees being closest to these other primates in the index between sternal length and breath (Fig. 15.34) (Schultz, 1930). Humans and apes also differ from these other primates in a greater degree of fusion of the sternebrae. In prosimians and monkeys the manubrium, sternebrae and xiphoid all remain separate during life, while in the apes there is normally fusion between the last two sternebrae. The chimpanzees most closely approach humans in the degree of sternal fusion.

Humans and chimpanzees have longer sternal bodies in relation to manubrium length than do either orang-utans or gorillas. There is also a sexual difference within species in the relative length of the sternal body where in human males the body of the sternum is generally twice as long as the length of the manubrium while in human females it is less than two times the manubrium length. This difference most probably relates to the narrower angle

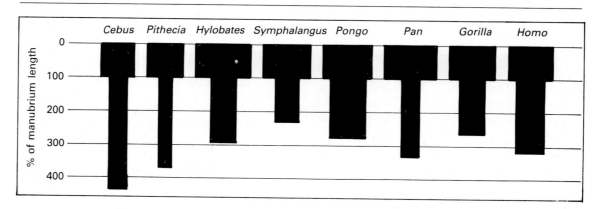

FIGURE 15·34 *Sternal proportions in selected primates. The diagrams are reduced to the length of the manubrium and the body of the sternum is presented as a percentage of sternal length. After Schultz (1930).*

between the costal margins of the false ribs in human females in comparison with that of human males (see above). In orang-utans there also seems to be a sexual difference in relative sternal length. Although both male and female orang-utans have sternal bodies that are at least twice the manubrium length, the sternum is relatively shorter in females than in males (manubrium length × 100/body length = 55.5 in 10 males and 58.1 in 17 females) (Schultz, 1930).

Breathing

Breathing in (inspiration) involves the increase in the capacity of the thorax that, in normal quiet breathing, results solely from contraction and flattening of the **diaphragm**. This lowers the air pressure in the thoracic cavity, resulting in an inrush of air down the **trachea** (wind pipe) and airway from the nose and mouth. During heavier breathing the rib cage itself expands further increasing the capacity of the thorax. Breathing out (expiration) is normally passive resulting from the elastic recoil of not only the lungs but also the wall of the thorax.

The diaphragm and breathing

The term **diaphragm** literally means wall across (dia (Gk) = across; phragma (Gk) = wall). The diaphragm is a sheet of muscle that separates the thoracic cavity from the abdomen. In humans it takes a circular origin from the base of the rib cage: from the back of the xiphoid process, from the lower six costal cartilages and from a series of tough ligaments spanning the muscles of the posterior abdominal wall (Fig. 15.35). These slips of muscle insert into a flattened **central tendon** which lies under the base of the heart. The diaphragm is dome shaped when at rest extending up behind the heart as far as the level of the fourth rib.

The chimpanzee diaphragm is, by and large, similar to the human diaphragm. However, the central tendon is much smaller. The muscular slips of the diaphragm originate

from the seventh to the thirteenth costal cartilages rather than the sixth to twelfth as in humans and from the first and second lumbar vertebrae in chimpanzees rather than the second and third as in humans (Sonntag, 1924). In the gorilla the central tendon is very large and contrasts with that of the chimpanzee.

When the diaphragm contracts it both flattens and descends (between a few centimeters in quiet breathing and up to 10 cm in forced inspiration). It thereby increases the volume of the thoracic cavity and at the same time compresses the contents of the abdomen. During forced expiration, contraction of the abdominal muscles increases the intra-abdominal pressure by reducing its volume. This squeezes the abdominal viscera upwards and thereby decreases the volume of the thoracic cavity.

Movement of the rib cage

The expansion of the rib cage during inspiration further increases the size of the thoracic cavity and thereby supplements the action of the diaphragm. In a normal human male the circumference of the chest at the level of the nipples can be increased by 7.5–10 cm (Green and Silver, 1981). Chest expansion results from three separate types of rib movement. These three movement types have been

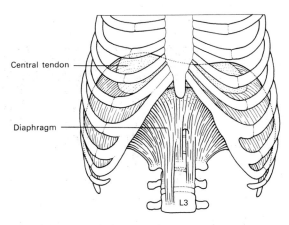

FIGURE 15·35 *The human diaphragm.*

likened to the motion of firstly a pump handle, secondly a bucket handle and thirdly a pair of spreading calipers (MacConaill and Basmajian, 1969) (Fig. 15.36). The pump-handle movement involves the anteroposterior expansion of the thorax. Here the upper ribs pivot around their vertebral articulations raising their anterior ends (and attached sternum). The result of this pump-handle motion is that the upper border of

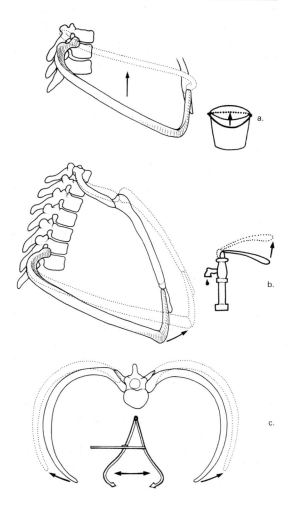

FIGURE 15·36 *Rib movements that expand the thorax and thereby draw air into the lungs. (a) The bucket handle movement, (b) the pump handle movement, and (c) the spreading caliper movement.*

the sternum (manubrium) moves upwards and outward and the body of the sternum (below the manubrium) pivots at its joint with the manubrium.

The bucket-handle movement involves the lateral expansion of the thorax. In this case the ribs that are attached to the body of the sternum (2–7) swing laterally like a bucket handle, rotating around both their vertebral and sternal articulations. The last movement, the spreading-caliper movement, involves the false ribs (8–10 in humans) and the floating ribs (11 and 12 in humans). When these ribs are elevated, the anterior ends (costal cartilages) of the ribs on one side of the thorax spread away from the anterior ends of the ribs on the other side in a fashion reminiscent of spreading calipers.

In addition to these three movements the ribs also actually change their shape during breathing, being straighter and, therefore, longer in inspiration. The costal cartilages of the true ribs (articulating with the sternum) also straighten during inspiration, giving a flatter contour to the front of the chest (MacConaill and Basmajian, 1969).

Muscles of rib movement and breathing

Several muscles act as accessory muscles in respiration. The **scalenes** help to bring about the pump-handle elevation of the first rib and sternum. The **sternocleidomastoid** may also aid in this function (MacConaill and Basmajian, 1969) (see Chapter 12 for a more complete discussion of these muscles). Two other groups of muscles, the **internal** and the **external intercostal muscles**, also play an important role in both inspiration and expiration (Fig. 15.37). As their name implies, the intercostals run between adjacent ribs. The internal intercostals run upward and medially from the upper margin of one rib to the lower margin of its partner. The external intercostals run superfical and at right angles to the internal intercostals, upwards and laterally between the edges of adjacent ribs. Opinions vary as to the specific function of these muscles in breathing. Some

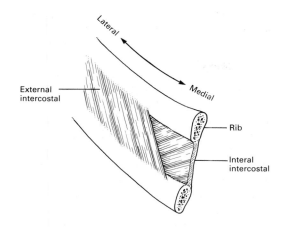

FIGURE 15·37 _The internal and external intercostal muscles._

believe that both groups of intercostals elevate the ribs during inspiration and thereby expand the rib cage. Others argue that it is only the external intercostals that are elevators while the internal intercostals depress the ribs during expiration and thereby reduce the volume of the rib cage. Another opinion is that the primary function of these muscles is to prevent the chest wall from imploding, or being sucked in by the low intrathoracic air pressure during inspiration. It is possible that the specific function of these muscles changes with the depth of respiration and that all these opinions contain some truth.

Many other muscles of the thorax and abdomen can aid the expansion and contraction of the rib cage during forced breathing. The **levatores costarum** and **serratus posterior superior** can assist in the elevation of the sternum while the **serratus posterior inferior** and **quadratus lumborum** can steady the lower ribs. In very laboured breathing the muscles of the upper limb such as **pectoralis minor, pectoralis major** and **serratus anterior** may also act to elevate the ribs. The **erector spinae** muscles together with other muscles of the back also assist by straightening the thoracic curvature causing the ribs to open out (Green and Silver, 1981).

The ribs in human evolution

Ribs, being relatively delicate in structure, are infrequently found preserved as fossils and when they are they are generally fragmentary, distorted and difficult to interpret. The most informative published samples of ribs are those of _Australopithecus afarensis_ and of Neanderthals.

There are over 30 rib fragments that belong to the AL 288-1 (_A. afarensis_) skeleton including one virtually complete first rib (Johanson et al., 1982). An additional nine fragments belong to the AL 333 sample (_A. afarensis_) (Lovejoy et al., 1982c). Schmid (1983) argued that the AL 288 rib fragments are more similar to ape ribs than to human ribs. Not only are they rounder in cross-section than are human ribs but they also show no indication of the extreme flattening in the middle section of the body of the rib. Moreover, they fit well onto the cast of a funnel-shaped chimpanzee thorax of the same trunk length as inferred for the Al 288 skeleton, while they make only a very imperfect fit with a similarly sized human barrel-shaped thorax (Schmid, 1983). The only human feature of the _A. afarensis_ ribs is the single articular facet found on the head of not only the AL 288-1ax right first rib but also on the AL 333-118 first rib. This suggests that these ribs articulated solely with the body of the first thoracic vertebra. This articular pattern is common in humans, while in the majority of apes the first rib has a double articular facet on its head and articulates with T1 and C7 and the intervertebral disc separating these vertebrae. Ohman (1986) believed that the human pattern is most probably related to the freeing of the upper limb from locomotion and the corresponding descent of the shoulder. However, Stern and Jungers (1987) argued that the univertebral articulation of the first rib need not reflect bipedal locomotion. Its presence in other large primates such as _Indri_ and _Propithecus_ among the prosimians and the siamangs among the gibbons suggested to them that the univertebral first rib articulation is related rather to a relatively large body size coupled with orthogrady (trunkal uprightness).

Neanderthal ribs have been reported as being very robust in relation to modern human ribs. The transverse diameter of the Shanidar ribs, for example, far exceeds the same diameter in modern human males while the superoinferior diameter of the same ribs falls within the modern human male range of variation (Trinkaus, 1983a). The robusticity index for 20 Shanidar ribs measured at the posterior angle is 53.8 (SD = 7.9) (data from Trinkaus, 1983a). This index is very close to that for 12 *A. afarensis* fragments identified as coming from the middle of the rib shaft

(53.7, SD = 5.9; AL 288 ribs; data from Johanson et al., 1982). The *A. afarensis* ribs are, however, proportionately smaller with a superoinferior diameter of 10.7 mm (SD = 0.952) as opposed to the Shanidar diameter of 14.7 mm (SD = 2.15). In addition to being unusually robust, Neanderthal ribs are also less curved than human ribs (Loth, 1938; Boule, 1911–13; see also Trinkaus, 1983a), the implication being that they possessed a larger thorax than is common in modern humans. Neanderthal ribs also show clearer evidence of muscle markings than do modern human ribs suggesting that the muscles of the back and shoulder region were very well developed in Neanderthals.

CHAPTER SIXTEEN

BONES, MUSCLES AND MOVEMENTS OF THE UPPER LIMB

The upper limb is made up of the pectoral girdle, the arm, the forearm and the hand (Fig. 16.1). The pectoral girdle (pectus (L) = breast) links the rest of the upper limb to the trunk and is composed of two bones, the **scapula** and the **clavicle**. The scapula is the large, flat, triangular bone that glides over the back of the rib cage as the arm moves up (abducts) or down (adducts), or backwards (extends) or forwards (flexes). The clavicle is a strut-like bone that articulates with the scapula above the shoulder joint at the **acromioclavicular** joint and with the **sternum** at the **sternoclavicular** joint at the centre of the chest (Fig. 16.2). The clavicle is the only part of the pectoral girdle that articulates directly with the trunk. The clavicle not only holds the shoulder joint at the side of the trunk, but also transmits forces from the arm to the sternum.

The clavicle articulates with the sternum via a disc which acts like a shock absorber and also helps keep the clavicle from disarticulating at the sternum. Excessive movement at the joint is also prevented by two ligaments and a small muscle (Fig. 16.2). The **sternoclavicular** ligament and the **costoclavicular** ligament secure the medial end of the clavicle to the sternum and the first costal cartilage. Lateral to the ligaments, the **subclavius** muscle which arises from the first costal

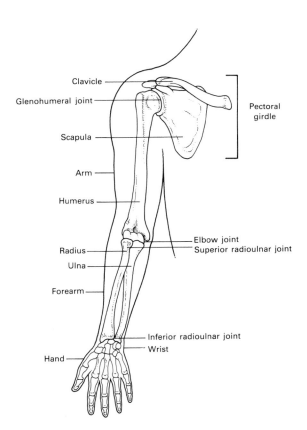

FIGURE 16·1 *The principal bones of the upper limb.*

cartilage and inserts into the inferior surface of the clavicle controls excessive movement of the clavicle.

The **acromioclavicular** joint is also strengthened by a series of ligaments that both bind the clavicle to the scapula and provide a protective cuff for the head of the humerus. The clavicle is secured to the scapula by the **acromioclavicular** ligament that binds it to the **acromion process** and by the **coracoclavicular** ligament that ties it to the **coracoid process**.

The upper arm articulates with the scapula at the **glenohumeral**, or shoulder, joint and is made up of a single bone called the **humerus** (Fig. 16.1). The forearm is made up of two bones, the **ulna** and the **radius**. These bones articulate with the humerus at the **elbow** joint and with the hand at the **wrist** joint. They also articulate with each other just below the elbow joint at the **superior**, or **proximal**, **radio-ulnar** joint and just above the wrist joint at the **inferior**, or **distal**, **radio-ulnar** joint.

The human hand is composed of 27 separate bones (Fig. 16.3). Eight (or nine in the orangutan) of these are **carpal bones** (karpos (Gk) = wrist), five are **metacarpals** (meta (Gk) = after), or bones of the palm, and 14 are **phalanges** (phalanx (Gk) = band or line of soldiers), or bones of the fingers and thumb.

The eight carpal, or wrist bones, are arranged in two rows. The **scaphoid** ((Gk) = a skiff or little boat), the **lunate** ((L) = moon), the **triquetral** ((L) = three-cornered) and the **pisiform** ((L) = pea shaped) form the proximal row and the **trapezium** ((Gk) = four cornered), the **trapezoid**, the **capitate** ((L) = head shaped), and the **hamate** ((L) = hook shaped) form the distal row. In the orang-utan there is a ninth carpal bone which is called the **centrale**. In humans and African apes this bone fuses with the scaphoid during foetal development (Virchow, 1929).

The joints between the bones are synovial joints and their capsules are strengthened by ligaments. Together the carpal bones form an arched structure (the **carpal arch**) with its

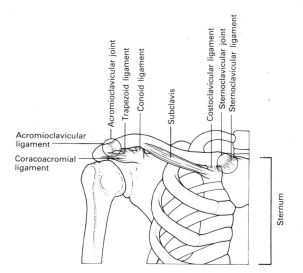

FIGURE 16·2 *The joints and ligaments of the pectoral girdle.*

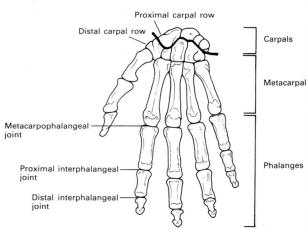

FIGURE 16·3 *The bones of the hand.*

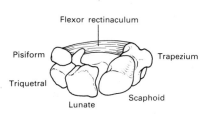

FIGURE 16·4 *The carpal tunnel from the proximal view.*

concave face on the palmar (anterior) side of the hand (Fig. 16.4). There is a strong fibrous band (the **flexor retinaculum**) that spans the arch on the palmar side and transforms it into a tunnel (the **carpal tunnel**). The long flexor tendons for the fingers pass through this tunnel and are thus prevented from bowstringing during wrist movement.

The main parts of the hand are the **palm**, the **fingers** and the **thumb**. The thumb is frequently called by its Latin name, **pollex**, and together the fingers and thumb are known as **digits**. For the purpose of anatomical description the hand is always oriented in the **anatomical position**, with the fingers

pointing downward and the palm facing forward (Fig. 16.5). The **wrist** (where your watch goes on the forearm) lies proximal to the hand and the fingers are the most distal part of the hand. The little (fifth) finger is on the medial side of the hand and the thumb is on the lateral side, while the **volar** ((L) = palm of the hand) surface of the hand faces anteriorly and the back of the hand, or **dorsal** surface, faces posteriorly.

Joints and movements of the upper limb

The arm can move at the shoulder joint in flexion, extension, abduction and adduction (Fig. 16.6). Flexion is the movement of the arm forward (anteriorly) in the sagittal plane and extension is its opposite backward movement. Abduction is the movement of the arm away from the body and adduction is the opposite movement towards the body.

The elevation of the arm at the shoulder in either flexion or abduction does not simply involve the movement of the humerus at the

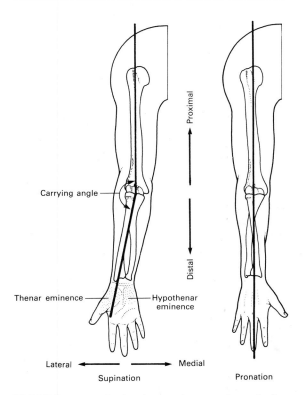

FIGURE 16·5 *Supination and pronation of the hand. Note that the bones of the forearm are parallel in supination and crossed in pronation. The carrying angle is present in supination and disappears in pronation when the axis of the forearm comes in line with the axis of the (upper) arm. After Kapandji (1982).*

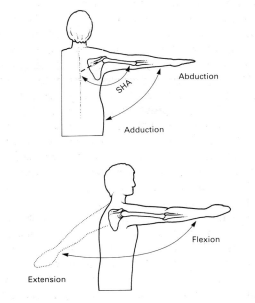

FIGURE 16·6 *Movements of the arm. SHA = Spinohumeral angle. After Inman* et al. *(1944).*

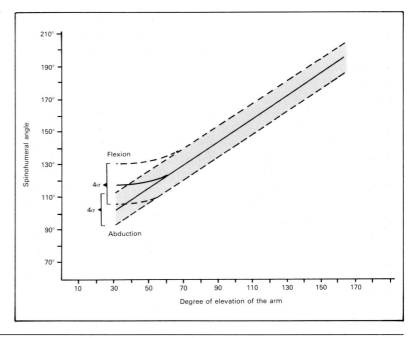

FIGURE 16·7 *The range of the spinohumeral angle in flexion and abduction of the human arm. After Inman* et al. *(1944).*

shoulder joint but also involves rotation of the scapula. In living humans, after the first 30° of abduction, or after the first 60° of flexion, there is a constant ratio of humeral movement to scapular movement (Fig. 16.7). Up until 170° of elevation, for every 15° of motion, 10° occurs at the shoulder joint and 5° occurs by rotation of the scapula on the thorax (Inman et al., 1944). Because the scapula is connected to the clavicle at the acromioclavicular joint and the clavicle is connected to the sternum at the sternoclavicular joint, rotation of the scapula also produces movements at each of these two clavicular joints. If the movements of the scapula and clavicle are artificially restrained the arm can be actively elevated 90° and passively elevated 120° at the shoulder joint alone; however, the lever advantage of the muscles is severely affected and muscle power may be reduced by approximately one third (Inman et al., 1944).

The elbow joint is a simple hinge joint that allows flexion and extension. Humans have a more restricted range of extension at this joint than do the great apes, however they

TABLE 16·1 Maximum angle of extension at the elbow joint in the higher primates. Angles measured on dry bones. Data from Knussmann (1967).

		Extension angle	
	N	Mean	SD
Humans	150	167.6	7.1
Gorillas	30	187.2	8.6
Chimpanzees	33	174.4	6.4
Orang-utans	14	181.3	9.2
Gibbons	48	167.0	5.8
Colobinae	11	153.0	8.4
Cercopithecinae	141	152.4	6.2
Callithricidae	7	146.9	6.6
Cebidae	47	152.5	7.9

SD = standard deviation.

have a greater range of extension than do the majority of other primates (Table 16.1). This is largely the result of differences in the size of the olecranon process of the ulna and in the depth and orientation of the trochlear, or sigmoid, notch. (see Chapter 17).

The elbow joint is closely associated with

the superior radio-ulnar joint which, together with the inferior radio-ulnar joint, allows pronation and supination of the forearm and hand. With the arm in the anatomical position the radius and the ulna lie parallel to each other with the ulna on the medial (little finger) side of the forearm and the radius on the lateral (thumb) side of the forearm (Fig. 16.5). The palm of the hand is in the supinated position. When the palm is in the pronated position with the palm facing posteriorly, the radius is twisted across the ulna. The superior radio-ulnar and the inferior radio-ulnar joints (together with the round head of the radius and the ball-shaped capitulum of the humerus) allow the radius to rotate around an axis that passes through the head of the radius and the distal end of the ulna and, thereby, to pronate the hand.

In humans, when the hand is in the supinated position and the forearm is extended, the forearm diverges laterally from the upper arm in what is known as the **carrying angle**, or the **vagus angle** (Fig. 16.5). The magnitude of the carrying angle is largely determined by the shape of the trochlea (part of the joint on the distal humerus). However, when the hand is pronated this angle disappears and the forearm lies in a straight line with the upper arm. This is also true for the other primates with a carrying angle similar to humans (Table 16.2). In these primates the carrying angle may be related to the use of the hand in a pronated position during locomotion and the mechanical necessity of maintaining the forearm in line with the upper arm for efficient weight transfer. In humans it may also have a mechanical explanation in the context of tool using. It is interesting to note that those primates with carrying angles of around 180° (forearm in line with the upper arm) can only pronate their forearms through approximately 90° and that this small degree of pronation would not be expected to disturb the alignment of the forearm significantly.

The major movements of the hand are described with reference to the anatomical position. Adduction occurs when the hand is angled towards the body, or ulnar deviated, and abduction occurs when it is angled away from the body, or radially deviated (Fig. 16.8a). When movement of the digits (the four fingers and the thumb) is described in isolation, adduction and abduction have different meanings because the reference axis is no longer the mid-line of the body but the middle digit of the hand. The digits are adducted when they are brought to lie adjacent to the middle (or third) finger and abducted when they are pulled away from the middle finger, or spread in a fan-like fashion (Fig. 16.8b). In the anatomical position, abduction and adduction of the

TABLE 16·2 The carrying angle and the normal range of radioulnar pronation and supination in several genera of higher primates. Data from [a]Knussmann (1967) and [b]O'Conner and Rarey (1979).	Carrying angle[a]			Range of pronation and supination[b]		
	N	Mean	SD	N	Mean	SD
Homo	150	166.7	4.8	8	156	8.5
Pongo	14	172.5	3.2	1	150	—
Hylobates	48	168.8	4.2	3	163	7.5
Cebus	17	176.6	4.5	6	118	4.2
Theropithecus	2	184.5	—	2	90	7.1
Cercopithecus	8	183.2	4.6	5	92	2.8
Papio	12	179.5	3.3	5	89	4.2
Erythrocebus	16	179.0	3.6	6	87	6.8
Macaca	102	182.3	4.3	33	79	6.1

SD = standard deviation. Angles and ranges are in degrees.

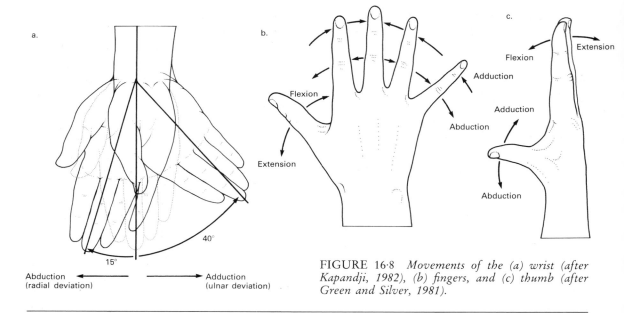

FIGURE 16·8 *Movements of the (a) wrist (after Kapandji, 1982), (b) fingers, and (c) thumb (after Green and Silver, 1981).*

hand as well as of the four fingers occur in the **coronal plane**. However, because the thumb lies at a right angle to four fingers when the hand is at rest, abduction and adduction of the thumb occurs in the **sagittal** rather than in the **coronal plane** (Fig. 16.8c). Flexion and extension of either the hand as a whole, or of the digits on their own, occurs

at right angles to the plane of abduction and adduction.

The complex movements of the wrist occur at two main joints, the **wrist joint** proper and the **midcarpal joint**. The wrist joint is the most proximal joint of the hand and is formed by the proximal row of carpal bones (the scaphoid, the lunate and the triquetral)

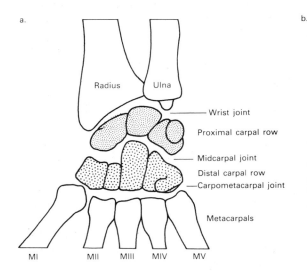

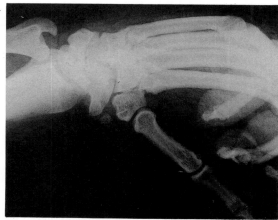

FIGURE 16·9 *(a) The wrist joint and midcarpal joint. (b) Radiograph of a chimpanzee wrist. Courtesy of Theya Molleson.*

and the distal end of the radius together with a disc covering the distal end of the ulna (Fig. 16.9). This joint is capable of flexion, extension, abduction and adduction. When the human hand is in the anatomical position the maximum range of abduction is about 15° and that of adduction about 45° (Kapandji, 1982). The range of abduction and adduction is much smaller when the wrist is fully flexed or extended. This is because the collateral ligaments of the wrist, the carpal ligaments, are tightened in these positions. From the anatomical position, the hand can be both flexed and extended by about 85°. However, only 50° of flexion and 35° of extension occurs at the wrist joint. The remainder of the movement occurs at the midcarpal joint.

The midcarpal joint is the joint between the proximal row of carpal bones (the scaphoid, the lunate, and the triquetrum) and the distal row of carpal bones (the trapezium, the trapezoid, the capitate, and the hamate). Although flexion and extension are the prime movements at this joint, it also has an additional movement. When the joint is extended the carpal bones are actually screwed together in a process known as **conjunct rotation**. The importance of this movement is discussed in Chapter 18.

With one exception, the carpometacarpal joints, the joints between the distal carpal row and the metacarpals, are relatively immobile (Fig. 16.9). The exception is the first carpometacarpal joint, the joint between the first metacarpal and the trapezium. This joint is responsible for a good percentage of the movement of the thumb in opposition to the fingers (see Figs 16.15 and 18.6). It is a saddle joint and, because of this shape, as the thumb is flexed the first metacarpal automatically rotates medially bringing the palmar surface of the thumb around to face the palmar surfaces of the fingers. Although the remaining carpometacarpal joints are relatively immobile they are capable of small amounts of movement and are highly important to the human prehensile (grasping) hand. The

detailed movements of these joints are discussed in Chapter 18.

The **metacarpophalangeal joints** (MCP) are the joints between the proximal phalanges of each digit and the distal ends of the five metacarpals (Fig. 16.3). These joints form the proximal row of hand knuckles. Because the metacarpal heads are condyloid, or rounded in shape, the metacarpophalangeal joints are capable not only of flexion and extension but also of abduction (movement away from the mid-line of the hand), adduction (movement towards the mid-line of the hand) and a combination movement of circumduction. The degree of flexion at these joints increases in humans from about 75–80° at MCP(1) (the thumb metacarpophalangeal joint) to 90° at MCP(2) to greater than 90° at MCP(5). MCP(1) does not allow active extension while the other MCP joints allow between 30° and 40° of active extension. The index finger has the greatest range of abduction and adduction at the MCP joint (about 30°) of any of the digits including the thumb (Kapandji, 1982). The index finger can also be easily moved independently of the other digits at the metacarpophalangeal joint. It is to this wide range of independent movement that it owes its name (index (L) = indicator) (Kapandji, 1982).

The **interphalangeal joints** (IP) are the joints between the phalanges of each digit (Fig. 16.3). There is only one interphalangeal joint between the two phalanges of the thumb and there are two in each of the remaining fingers, the proximal and distal interphalangeal joints. They are **hinge joints** and, therefore, only capable of simple flexion and extension. In humans the range of flexion of the proximal interphalangeal joints increases from about 90° for IP(2) to 135° for IP(5). It also increases in the distal interphalangeal joints from less than 90° for IP(1) to about 90° for IP(5). The proximal interphalangeal joints are not capable of any significant extension, while the distal interphalangeal joints show no more than 5° of active extension (Kapandji, 1982).

Osteology of the human pectoral girdle

The name **scapula** (skapto (Gk) = I dig) most probably refers to the resemblance of this bone to a spade (Fig. 16.10). It can be divided into two basic morphological areas, the **scapular blade** (or scapular body) which is the part of the bone resembling a spade and the **glenoid head**. One of the main features of the glenoid head is the **glenoid cavity** that articulates with the humeral head to form the shoulder joint. The two major areas of the scapula, the blade and the head, are separated by the surgical neck.

The blade of the scapula

Viewed from its dorsal surface the most notable feature of the blade of the scapula is the raised **spine** that runs horizontally from the vertebral border of the scapula (that edge which borders the vertebral column) and continues above and behind the glenoid cavity as the **acromion process** (Fig. 16.10). The areas above and below the spine give origin to all but one of the short scapular, or rotator cuff, muscles that function primarily to stabilize the humerus at the shoulder joint.

The area of the scapula above the spine is known as the **supraspinous fossa** and provides the area of origin for supraspinatus. The area below the spine is the **infraspinous fossa** and gives origin to infraspinatus. Teres major and teres minor also have their origins along the axial border (the border which faces laterally under the arm) of the infraspinous fossa. (Teres major is not a true rotator cuff muscle because it does not fuse with the capsule of the shoulder joint.)

The ventral, or anterior, surface of the blade of the scapula is concave and relatively featureless as would be expected for the surface of the scapula that glides over the rib cage when the arm is in motion (Fig. 16.10). The entire surface is known as the **subscapular fossa** and gives origin to the last of the rotator cuff muscles, the subscapularis.

With the exception of trapezius, the muscles that are responsible for both the upward and downward rotation of the scapula during arm movement are inserted into the edges of the blade of the scapula. Viewed from above, the U-shaped structure formed by the scapular spine, the acromion and the clavicle provides attachment for trapezius (an upward rotator of the shoulder) and deltoid (an abductor of the arm).

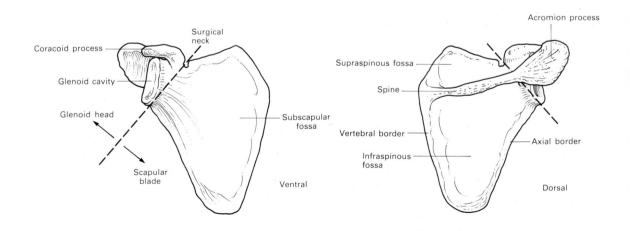

FIGURE 16·10 *The human scapula from the posterior (dorsal) view and the anterior (ventral) view.*

The glenoid head

The glenoid head is dominated by two major features. The first of these is the glenoid cavity that articulates with the head of the humerus to form the shoulder joint (Fig. 16.10). The glenoid cavity (glene (Gk) = a socket) is shallow and concave and is much smaller in size than the head of the humerus with which it articulates. A rim of fibrocartilage deepens the joint in life, but adds little to the stability of the joint.

The second major feature of the glenoid head is the **coracoid process** (korax (Gk) = crow), a hook-like process that projects ventral to, or in front of, the glenoid cavity. The coracoid process has at least three functions. Firstly, it is bound to the acromion process by the coracoacromial ligament and together with the acromion forms the coracoacromial arch. This structure is unique to humans and the greater and lesser apes and not only provides a protective cuff for the head of the humerus but also limits its superior movement (Fig. 16.2; and see Chapter 17) (Ciochon and Corruccini, 1977). Secondly, the coracoid process provides attachment for ligaments that bind the clavicle to the coracoid process and thereby strengthen the acromioclavicular joint. And thirdly, the coracoid process provides the origin for three muscles that are involved in movement of the arm, pectoralis minor, coracobrachialis and the short head of biceps brachii.

Osteology of the clavicle

The **clavicle** is the strut-like bone that articulates with the sternum at the **sternoclavicular** joint (Figs 16.1 and 16.2). It also articulates with the acromion process of the scapula at the **acromioclavicular** joint.

The name clavicle comes from the latin 'clavis' meaning key. Roman keys were S-shaped and 'clavicle' is thereby descriptive of the S-shaped form of the human bone. The medial part of the S-shaped curve is anteriorly convex and takes up approximately two-thirds of the length of the entire bone (Fig. 16.11). The cross-section of this part of the bone is roughly circular. The lateral one-third of the bone is posteriorly convex and the cross-section is considerably flattened so that the long axis of the cross-section is in the anteroposterior plane.

The superior surface of the clavicle is relatively smooth and featureless. The most notable feature on this surface is the **deltoid tuberosity**. This is a roughened and raised area of bone that lies on the anterior surface of the lateral curve of the clavicle. It marks the most medial point of attachment of the deltoid muscle.

The majority of the morphological landmarks of the clavicle are found on its inferior surface. The extreme medial end of the bone bears the small articulation of the sternoclavicular joint. Just lateral to this articular surface is a roughened area that marks the attachment of the costoclavicular ligament. This ligament ties the clavicle to the underlying costal cartilage and thereby acts to stabilize the sternoclavicular joint. On

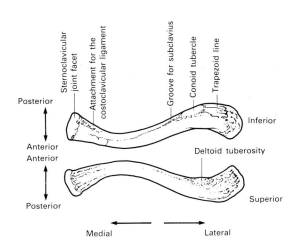

FIGURE 16·11 *The human clavicle from the inferior (caudal) view and the superior (cranial) view.*

317

the lateral, or shoulder, end of the bone there are additional roughened areas that mark the attachments of the two ligaments which tie the lateral end of the clavicle to the underlying coracoid process of the scapula and stabilize the acromioclavicular joint. The **conoid tubercle**, which is located on the posterior inferior surface of the lateral curve of the clavicle provides the attachment for the conoid ligament. The **trapezoid** line (or **oblique line**), which extends laterally from the conoid tubercle, marks the attachment of the trapezoid ligament. Medial to the conoid tubercle is a groove that extends over the central one-third of the inferior surface of the clavicle. This groove is the site of attachment of the subclavius muscle (a stabilizer of the clavicle).

Osteology of the humerus

The **humerus** (omos (Gk) = shoulder) is the longest bone in the upper limb (Fig. 16.1). The proximal end of the humerus articulates with the scapula at the shoulder joint. Its primary features are a head and two tubercles (Fig. 16.12). The head is hemispherical in shape, faces medially and is separated from the two tubercles by the **anatomical neck**. The **greater** and **lesser tubercles** (tuberosities) lie lateral and anterior to the head. These tubercles provide the area of insertion for the rotator cuff muscles. The subscapularis muscle inserts into the lesser tubercle and supraspinatus, infraspinatus and teres minor insert into the greater tubercle. The greater and lesser

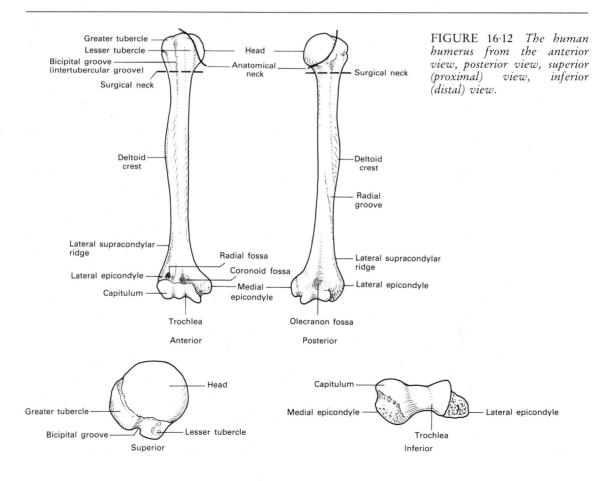

FIGURE 16·12 *The human humerus from the anterior view, posterior view, superior (proximal) view, inferior (distal) view.*

tubercles are separated from each other by the **intertubercular** (or **bicipital**) **groove** which continues down the anterior face of the shaft and, in life, houses the long tendon of biceps brachii. The proximal end of the humerus is separated from the shaft by the **surgical neck** which lies approximately 2 cm below the humeral head.

The proximal two-thirds of the shaft of the humerus are roughly circular in shape while the distal one-third is increasingly flattened in the anteroposterior plane. The main feature of the proximal third of the shaft is the **bicipital groove** which is bounded by ridges extending distally from the greater and lesser tubercles. The ridge extending from the greater tubercle is the better developed of the two and extends to the midpoint of the shaft. This ridge provides the insertion for pectoralis major. Lateral to this ridge, on the central portion of the shaft, is the raised and roughened **deltoid crest**, the insertion area for the deltoid muscle. The posterior surface of the midshaft may show the **radial** (or **spiral**) **groove** just distal to the deltoid crest. The most marked feature of the distal third of the shaft is the **lateral supracondylar ridge**. This is sometimes a very sharp ridge which extends proximally up the shaft from the **lateral epicondyle**.

The distal end of the shaft has two epicondyles (the lateral and the medial) and the articular surfaces for the radius and the ulna. The epicondyles are bony projections that lie slightly proximal to and at the sides of the articular surfaces. The **medial epicondyle** gives origin to the long flexor muscles of the hand and the **lateral epicondyle** to the long extensors. On the human humerus the medial epicondyle is always larger and more projecting than the lateral epicondyle. The distal articular surface, or **cubital surface** (cubis (L) = I recline), is divided into two major areas. The spool-shaped **trochlea** (trochilea (Gk) = pulley) is on the medial side of the articular surface and provides the articulation for the ulna. The round **capitulum** (capitulum (L) = little head)

on the lateral side of the articular surface articulates with the radius. The **olecranon fossa** (olene (Gk) = elbow; kranion (Gk) = head) is a large depression on the posterior surface of the distal humerus just above the trochlea. This fossa accommodates the **olecranon process** of the ulna when the arm is extended at the elbow. The **coronoid** and **radial fossae** are depressions on the anterior surface of the distal humerus just above the trochlea and the capitulum. They accommodate the **coronoid process** of the ulna and the **head** of the radius when the arm is flexed at the elbow.

Osteology of the ulna and radius

The **ulna** (ulna (L) = elbow or forearm) and the **radius** (radius (L) = a staff, rod or wheel spoke) are the two bones that make up the forearm. In the anatomical position, with the palm of the hand facing anteriorly, the ulna lies on the medial (little finger) side of the forearm and the radius on the lateral (thumb) side (Fig. 16.5). In this position the two bones lie parallel to each other.

The **trochlear notch** is the joint surface at the proximal end of the ulna that articulates with the trochlea of the humerus (Fig. 16.13). The surface of the trochlear notch is raised into a longitudinal keel and provides a good fit with the spool-shaped humeral trochlea. The **radial notch** is contiguous with the trochlear notch on the lateral side of the ulna and articulates with the **radial head** forming the proximal radio-ulnar joint. Just distal to the radial notch is a depression for the play of the **radial tuberosity** as the radius rotates over the ulna during pronation (see Chapter 17). The trochlear notch is bounded proximally by the **olecranon process** which provides the area of attachment for triceps brachii (the main extensor of the elbow-joint). The thickened and raised area of bone that forms the distal margin of the trochlear notch is the **coronoid process** (korōnē (Gk) = a crow). The roughened area of bone extending distally from the coronoid process is the site of

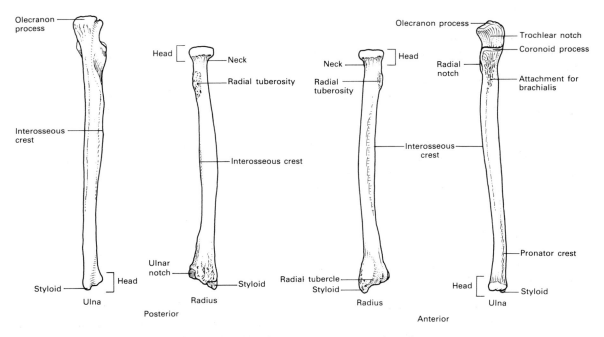

FIGURE 16·13 *The human ulna and radius from the posterior view and anterior view.*

attachment of brachialis (a flexor of the elbow joint). The proximal end of the ulna is the most robust part of the bone. The remainder of the bone tapers distally to a small **head** and **styloid** (stylos (Gk) = a stake) **process** that participate only indirectly in the articulation at the wrist joint (see Chapter 18). The human ulnar shaft is given a very angular appearance by the sharp **interosseous crest** that extends distally to the last one-quarter of the shaft. Here the roughened **pronator crest** on the anterior surface of the shaft becomes the most noticeable feature of shaft morphology.

The radius is slightly shorter than the ulna and, in contrast to the ulna, is broadest at its distal end. The proximal end of the radius is formed by the circular **head** that articulates with the ulna at the radial notch. Distal to the head is the relatively narrow **neck** of the radius and the **radial tuberosity**. The **interosseous crest** extends from the radial tuberosity down the medial side of the bone,

dividing at the distal end to enclose the **ulnar notch**. The ulnar notch articulates with the head of the ulna to form the distal radio-ulnar joint.

The distal surface of the radius articulates with two of the carpal bones. The lateral edge of the joint surface extends more distally than does the medial and forms the **styloid process** of the radius. The lateral, and more triangular, part of the joint surface articulates with the lunate and is separated by a weak ridge from the medial part of the joint surface which articulates with the scaphoid. On the posterior surface of the distal radius there are usually four grooves that carry the tendons of the muscles which extend the fingers.

Osteology of the hand

Over half the bones of the hand (14 out of 27) are phalanges, or bones of the fingers (Fig. 16.3). Of the 14 phalanges, five are proximal phalanges, articulating with the five metacarpals, four are middle phalanges, and five are distal phalanges. There are only four middle phalanges because the first digit, the thumb, only has two phalanges rather than three as in the remaining digits.

The phalanges belonging to each of the three types are very similar in their osteological features. All five proximal phalanges have oval concave facets at their proximal ends, or bases, for articulation with the spherical heads of the five metatarsals (Fig. 16.14). The distal ends, or heads, of these phalanges are pulley shaped, having two small condyles. The bases of the four middle phalanges have two shallow facets with a medial ridge that rides in the medial depression of the pulley-shaped head of the corresponding proximal phalanx. The heads of these middle phalanges have the same pulley shape as do the heads of the proximal phalanges. The five distal phalanges have bases that are similar to the bases of the middle phalanges; however, the distal phalanges can easily be distinguished by the flat expansions, or apical tufts, at their distal ends. These apical tufts support the finger tips.

The shafts of the proximal and middle phalanges are roughly semicircular in cross-section, being rounded on their dorsal surface and flat on their palmar surface with relatively sharp medial and lateral edges (Fig. 16.14). It is difficult from morphology alone to determine to which digit a particular phalanx belongs. If all the proximal, middle and/or distal phalanges are present the easiest way to determine this is by size. In humans the length formula for the proximal phalanges (PP) is commonly PPIII>PPIV>PPII>PPV>PPI and for the middle phalanges (MP) is MPIII>MPIV>MPII>MPV (Martin and Saller, 1959). For the distal phalanges (DP), the phalanx of the thumb is longer than the other distal phalanges and the length formula is commonly DPI>DPIV>DPIII>DPII>DPV (Martin and Saller, 1959).

Hand phalanges can also be relatively easily distinguished from foot phalanges, the former being generally broader, stronger and flatter

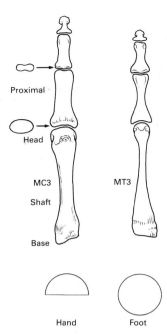

FIGURE 16·14 _Bones of the third digit of the hand compared to those of the foot._

(Fig. 16.14). Foot phalanges have a round cross-sectional shape, and a more lightly built shaft that expands out into heavy ends. The stout hand phalanges reflect their role in grasping while the foot phalanges, except for those of the great toe, are subjected to relatively little force.

The five metacarpals are also similar in morphology, each having a shaft, a proximal base and a distal head (Fig. 16.14). The metacarpal heads are condyloid, or rounded, in shape and the metacarpal shafts widen from the base toward the head. This feature distinguishes them clearly from the corresponding bones of the foot, the metatarsals. The metatarsals, with the exception of the first metatarsal, are broadest at the base and have a narrow shaft and head.

As with the phalanges, the digit to which the metacarpal belongs can be determined by relative size. In humans the common length formula for the metacarpals (M) is MII>MIII>MIV>MV>MI (Martin and Saller,

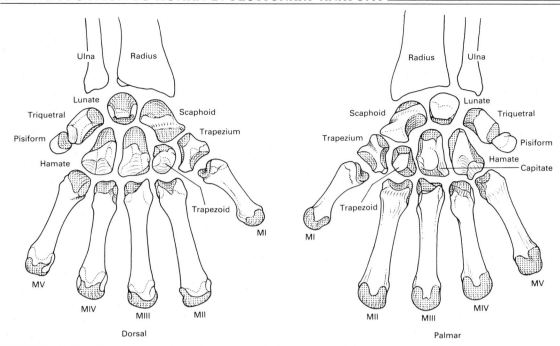

FIGURE 16·15 *Human carpal and metacarpal bones from the dorsal view and palmar view.*

1959; Susman, 1979). However, the metacarpals can also be distinguished by their morphology, and in particular by the morphology of their bases (Fig. 16.15).

MI   The proximal joint surface on the thumb metacarpal is saddle shaped or concavoconvex. There are no facets on the medial, or lateral, sides of the shaft at the base for articulation with adjacent metacarpals. Furthermore, there is frequently a flange running down the lateral side of MI for the attachment of opponens pollics, one of the short muscles of the hand (see Chapter 18).

MII   This metacarpal can be easily distinguished by a deep groove on its proximal joint surface, running in an anteroposterior (dorsopalmar) direction. This metacarpal has a facet on the medial side of its base for articulation with MIII but is lacking a similar facet on the lateral side of its base.

MIII   This metacarpal can also be easily distinguished by its styloid process which protrudes proximally from the dorsolateral corner (adjacent to MII) of the base. MIII also has facets on both sides of its base for articulation with the two adjacent metacarpals (MII and MIV).

MIV   This metacarpal has facets on both sides of its base (for articulation with MIII and MV) but lacks the styloid process of MIII.

MV   This metacarpal has only a single facet at its base (for articulation with MIV) but has a prominent tubercle near the base on its medial (nonarticular) side for the insertion of extensor ulnaris.

More subtle differences between the metacarpals in the asymmetry of their heads and the muscles markings on their shafts are discussed in Chapter 18.

The bones of the wrist, the carpal bones, are all irregular in shape and each has a distinctive morphology. The largest bone of the proximal carpal row is the scaphoid. This bone has a large and distinctive concave facet on its distal surface for articulation with the round head of the capitate (Fig. 16.15). If you place your thumb in this facet your index finger will rest on a large convex facet for articulation with the radius. The bone between these facets (and between your fingers) is the **body of the scaphoid** while the projecting bone is the **scaphoid tubercle**. The large facet on the tubercle articulates with the trapezium and trapezoid of the distal carpal row.

Medially, the scaphoid articulates with the lunate which is distinctive because of its half-moon shape. When the lunate is in its proper anatomical position, its distal concave facet is contiguous with the concave facet of the scaphoid and together they form the cup-shaped articulation for the head of the capitate. Likewise, proximally the large convex articular surface is contiguous with the corresponding scaphoid surface and together these surfaces articulate with the concave distal radial facet to form the wrist joint. There are three other articular surfaces on the lunate. Laterally there is a crescent-shaped facet for the scaphoid and medially there is an almost square and flat facet for articulation with the triquetral. This square triquetral facet is separated from the concave capitate facet by a narrow curved articular facet for the hamate.

The triquetral is the third major bone of the distal carpal row. It is irregular in shape with three articular surfaces. The flat, square lunate facet forms a right angle with the hamate facet which is larger than the lunate facet and is concavoconvex in shape. The third facet, for articulation with the pisiform, is small and oval in shape and separated from the others.

The last bone of the distal carpal row is the pisiform which articulates with the triquetral. This is the smallest of the carpal bones and is the only carpal bone with a single articular facet.

The capitate is the largest of all the carpal bones and occupies the central position in the distal carpal row. Its rounded proximal surface, or head, articulates with the lunate and scaphoid, resting in the concave cup formed by the distal surfaces of these bones. Additional articular surfaces extend distally from the head on both the medial and lateral sides of the bone. On the medial side there is a relatively flat facet for the hamate that tapers distally where part of the surface is non-articular. On the lateral side distal to the scaphoid facet and separated from it by a low ridge is the trapezoid facet. Distal to the trapezoid facet is a narrow bevelled facet for the base of the second metacarpal. This facet is contiguous with the relatively flat, triangular-shaped facet for the third metacarpal that faces distally. The apex of this triangular facet points towards the palmar side of the capitate and its base towards the broad non-articular dorsal side.

Medial to the capitate in the distal carpal row is the hamate, perhaps the most distinctive bone in the carpal skeleton. This triangular bone can be easily recognized by the large hook that projects from the palmar surface of the bone. This projection is called the **hook of the hamate**, or the **hamulus**. The base of the triangular hamate is the large distally directed facet for the fourth and fifth metacarpals. The articular surfaces for these two metacarpals are separated by a faint ridge. The articular surface on the lateral side of the bone is for the capitate and on the medial for the triquetral. These surfaces come together proximally to form the apex of the triangular hamate.

On the opposite side of the capitate are the two remaining bones of the distal carpal row, the trapezoid and the trapezium. The trapezoid is the smaller of the two and is very irregular in shape. Its non-articular dorsal surface is larger than its non-articular palmar surface and it has four articular facets that form a ring around its remaining surfaces. The distal facet is for the second metacarpal, the medial for the capitate, the proximal for the scaphoid and the lateral for the trapezium. Because this bone is so irregular it is often difficult to orientate correctly, however the distal second metacarpal facet can be easily recognized because it has the largest dorsopalmar diameter.

The last carpal bone is the trapezium. It has a large saddle-shaped facet on its distal surface for articulation with the first metacarpal. It can be recognized by this surface as well as by a tubercle and groove on its palmar surface. The groove is for the tendon of flexor carpi radialis. It has three other joint surfaces. On its medial side there is a concave surface for the trapezoid. Distal to this is a small linear articular facet for the second metacarpal. Proximal to the trapezoid facet and at an angle to it is the smaller proximally-facing facet for the scaphoid.

Muscles that move the arm and the pectoral girdle in humans

Movement of the arm requires not only muscles that move the humerus at the shoulder joint but also muscles that rotate the scapula.

Muscles that elevate the arms

UPWARD ROTATORS OF THE SCAPULA

Trapezius and **serratus anterior** are the two main muscles that act together to produce the upward rotation of the scapula (Fig. 16.16).

Trapezius (trapezium (Gk) = a small table) has a long, linear origin that extends from the base of the skull along the ligamentum nuchae to the thoracic spines and supraspin-

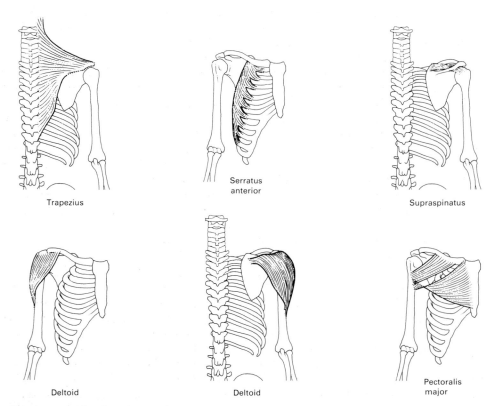

Trapezius

Serratus anterior

Supraspinatus

Deltoid

Deltoid

Pectoralis major

FIGURE 16·16 *Muscles that elevate the arm. After Green and Silver (1981).*

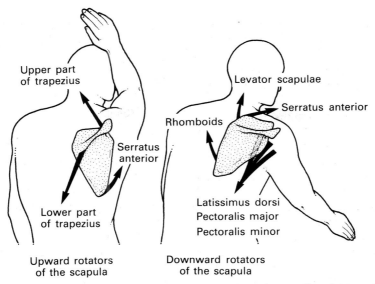

Upper part
of trapezius

Serratus
anterior

Lower part
of trapezius

Upward rotators
of the scapula

Levator scapulae

Serratus anterior

Rhomboids

Latissimus dorsi
Pectoralis major
Pectoralis minor

Downward rotators
of the scapula

FIGURE 16·17 *The force couples that rotate the scapula. After MacConaill and Basmajian (1969).*

ous ligaments. The upper fibres of trapezius insert into the lateral third of the clavicle and the lower fibres insert into the acromion and the spine of the scapula. The upper portion of this muscle makes up the upper portion of the force couple necessary for the upward rotation of the scapula (Fig. 16.17). The lower portion contributes to the lower portion of the force couple for scapular rotation. However, the main component of the lower portion of the force couple is serratus anterior. Serratus anterior (serratus (L) = saw) arises from the upper eight ribs, passes between the rib cage and the scapula and inserts along the entire length of the medial (vertebral) margin of the scapula. Contraction of the lower fibres of this muscle rotates the scapula upwards.

In humans, the electromyographic potentials, or activity, of both the upper portion of trapezius and the lower portion of serratus anterior (the force couple producing scapular rotation) increases linearly as the arm is elevated in both abduction and flexion. The greatest activity of the muscles is at 180° (the highest elevation of the arm) (Inman et al., 1944).

ABDUCTORS AND FLEXORS OF THE HUMERUS AT THE SHOULDER JOINT

At the same time that trapezius and serratus anterior are rotating the scapula, other muscles are either abducting or flexing the arm at the shoulder joint (Fig. 16.16). Muscles will abduct the arm (pull it up and away from the body in the lateral plane) if their origins and insertions lie in the transverse, or coronal, plane over the shoulder joint. Muscles will flex the arm (pull it up and in front of the body in the sagittal plane) if their origins and insertions position them across the anterior surface of the shoulder joint.

The **deltoid** (delta (Gk) = the triangular shaped fourth letter of the Greek alphabet) is the principal elevator of the arm. It has a U-shaped origin that wraps around the shoulder region, extending from the outer margins of the lateral part of the clavicle to the acromion and around to the spine of the scapula. It inserts into the deltoid tuberosity on the lateral side of the shaft of the humerus. In abduction, the electromyographic (EMG) potential, or activity, of this muscle rises linearly to 90° of arm elevation and is then constant in its activity for the remaining 90° of elevation. Contraction of the anterior fibres of this muscle (which pass anterior to the shoulder joint) flexes the humerus and the posterior fibres extend it. In flexion, the EMG potential, or activity, rises linearly to 90°, plateaus to 130° and then rises again linearly to full elevation (Inman et al., 1944); Basmajian and de Luca, 1985).

Supraspinatus, which arises from the supraspinous fossa of the scapula and inserts into the greater tuberosity of the humerus, is also active in abduction. It is one of four short scapular muscles, or **rotator cuff muscles**, which all have their origins on the scapula and their insertions around the head of the humerus. The other three rotator cuff muscles are infraspinatus, teres minor and subscapularis (see below). Together they provide stability to the shoulder joint during arm movement (but see Larson and Stern, 1986).

Pectoralis major (pectus (L) = the breast) is also active during flexion of the arm. It arises from the front of the chest (from the front of the medial half of the clavicle and the front of the sternum) and inserts into the crest of the greater tubercle of the humerus. Because it does not pass laterally over the shoulder joint it is completely silent during abduction. The upper, or clavicular, fibres of **pectoralis major** are active during flexion of the humerus.

Muscles that lower the arm

DOWNWARD ROTATORS OF THE SCAPULA

When the arm is lowered the scapula is rotated and the shoulder joint moves downward. At the same time, the humerus moves downward in relation to the scapula. This same movement also occurs during climbing when the upper limb is fixed and the body is hoisted up in relation to the fixed arm.

Levator scapulae, the **rhomboids** and **pectoralis minor** are the main muscles that act together to produce the downward rotation of the scapula (Fig. 16.18). Levator scapulae (levator (L) = a lifter) and pectoralis minor make up the upper portion of the force couple producing this rotation and the rhomboids make up the lower portion of the force couple (Fig. 16.17). Levator scapulae arises from the cervical vertebrae and inserts into the upper portion of the medial border of the scapula. Pectoralis minor arises from several ribs on the front of the chest and inserts into the coracoid process of the scapula. The rhomboids take their origin from the spines of the thoracic vertebrae and insert into the medial border of the scapula.

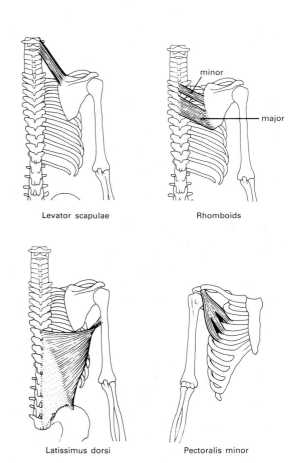

Levator scapulae

Rhomboids

minor

major

Latissimus dorsi

Pectoralis minor

FIGURE 16·18 *Muscles that lower the arm. After Green and Silver (1981).*

ADDUCTORS AND EXTENSORS OF THE HUMERUS AT THE SHOULDER JOINT

The principal muscle that lowers the arm is **latissimus dorsi** (latissimus (L) = broadest). This muscle arises from the back of the chest (from the lower thoracic spinous processes and the iliac crest) and inserts into the intertubercular groove of the humerus. When the arm is fixed, as in climbing, it can also draw the whole body upwards. It is a large muscle with a good lever advantage on the humerus and is active for the most part in adduction, although some fibres are also active in extension.

Pectoralis major is also an adductor of the humerus, while the posterior fibres of **deltoid** (passing across the posterior surface of the shoulder joint) act as an extensor of the humerus.

The muscles that rotate the scapula and elevate, or lower, the arm can also produce additional types of movement when they act together in different combinations. Two of these additional movements are **protraction** and **retraction**. Protraction is the forward movement of the scapula (and arm) as in pushing. Retraction is the backward movement of the scapula (and arm) as in standing to attention or pulling something toward the body. Protraction is produced by the simultaneous action of serratus anterior and pectoralis minor on the scapula and pectoralis major on the humerus. Retraction is produced by the simultaneous action of the middle fibres of trapezius and of the rhomboids on the scapula and latissimus dorsi on the humerus.

There are five additional muscles that act to rotate the humerus and also to stabilize the shoulder joint (Fig. 16.19). These are short muscles that arise from the scapula and insert into the proximal humerus. **Infraspinatus** arises from the infraspinous fossa of the scapula and inserts into the greater tubercle of the humerus and **teres minor** (teres (L) = rounded off smooth) arises from the axillary (lateral) border of the infraspinous fossa and also inserts into the greater tubercle (distal to the infraspinatus insertion). Both these muscles rotate the humerus laterally. **Subscapularis** arises from the subscapular fossa on the scapula and inserts into the lesser tubercle of the humerus and **teres major** arises from the lower part of the lateral border and inferior angle of the scapula and inserts into the medial edge of the intertubercular groove of the humerus. These muscles rotate the humerus medially and teres major also acts as an adductor. The last muscle, **coracobrachialis**, arises from the tip of the coracoid process of the scapula and

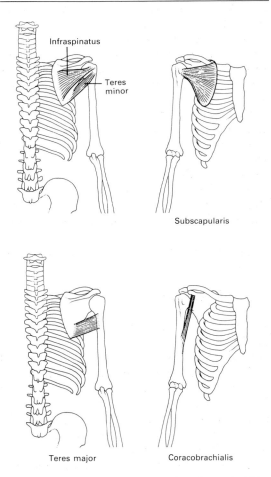

FIGURE 16·19 *The small muscles of the shoulder. After Green and Silver (1981).*

327

inserts into the middle of the humerus on the medial side. It both adducts and flexes the humerus and is analogous to the adductor muscles of the thigh.

Muscles that move the pectoral girdle and the arm in the apes

With the exception of two additional muscles, **atlanto-clavicularis** and **pectoralis abdominis**, the arm and shoulder muscles of apes are identical to those of humans (Fig. 16.20) (Miller, 1932). These two additional muscles are present not only in the apes, but also in all other primates; they have been lost in humans. Atlanto-clavicularis (omocervicalis of Miller (1932) see Chapter 12, Figs 12.20, 12.22 and 12.23) arises from the spinous process of the atlas vertebra and, in the apes, inserts into the lateral margin of the clavicle (see Chapter 12). Its action is to elevate the clavicle. In lower primates it inserts into the acromion of the scapula and is one of the upward rotators of the scapula. Pectoralis abdominis arises from the sheath of rectus abdominis on the front of the rib cage and inserts into the humerus. Like the lower fibres (sternal head) of pectoralis major, it serves as a flexor of the arm.

Other differences between the arm and shoulder muscles of humans and apes lie in the extent of their origins and insertions, in their electromyographic (EMG) potential, or activity, during arm movement and in the relative magnitude of the muscles of one functional group in relation to another.

Differences in arm and shoulder musculature between apes and humans

Origins and insertions

Humans and apes differ in the origins and insertions of the scapular rotator muscles. These differences result in corresponding differences in the lever advantage of these muscles. In both the gorilla and the chimpanzee, the occipital and cervical origins of **trapezius** are more extensive than they are in humans and the muscle is shortened and thickened in the cervical region (Fig. 16.20). In addition, the origin of this muscle does not extend to the last thoracic vertebra as it does in humans.

In comparison with humans, the origin of the **rhomboids** is different in the chimpanzees and gorillas (Figs 16.18 and 16.20). In both these primates, its origin extends from the occipital bone to the middle thoracic vertebra, while in humans the origin extends only from the last cervical or first thoracic vertebra to the middle thoracic vertebra. The cranial part of the origin of the rhomboids markedly reduces the angle of insertion into the medial border of the scapula in the apes and would provide a better lever advantage during scapular rotation than in humans.

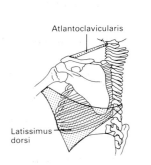 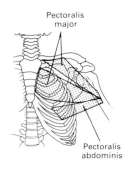 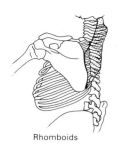 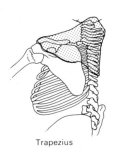

FIGURE 16·20 *Shoulder muscles in the gorilla. After Miller (1932).*

The origins and insertions in humans of both these muscles are more similar to those in the gibbons than they are to those in the large apes. This is not true of **latissimus dorsi**, however. In both humans, the African apes and the orang-utan, latissimus dorsi arises both from the lower thoracic vertebrae and ribs and from the iliac crest. Gibbons do not have the iliac origin. What differences there are in the origins of latissimus dorsi in humans and the large apes reflect the use of the ape upper limb in locomotion and, particularly, in climbing. In the large apes, the iliac origin of the muscle is much more extensive than the vertebral origin, while the reverse is true in humans (Figs 16.18 and 16.20). The iliac portion of latissimus dorsi is one of the major retractors of the arm. The large iliac origin of this muscle provides a direct transfer of weight from the lower body to the humerus during climbing and other types of locomotion when the arms are used above the head (Tuttle and Basmajian, 1977).

Electromyographic potential during arm movement

During elevation of the arm, the electromyographic (EMG) potentials of the major arm and shoulder muscles of the apes are essentially similar to those of the same human muscles (Inman et al., 1944; Tuttle and Basmajian 1977, 1978a). The exceptions to this are the upward rotators of the scapula, the cranial portion of trapezius and the lower, or caudal, portion of serratus anterior. During elevation of the arm, both these muscles show short bursts of high EMG activity early in the movement. Activity in these muscles then decreases to zero as the arm reaches its maximum elevation at 180° (Tuttle and Basmajian, 1977). In humans the activity in both of these muscles increases linearly to 180°. This difference in EMG activity may reflect a difference between humans and the apes in the rotation of the scapula in relation to the movement of the humerus during elevation of the arm. However, no direct

evidence is available for any of the apes. Tuttle and Basmajian (1977) have also suggested that it may be related to the more upward, or cranial, orientation of the glenoid cavity in the apes which contrasts with the lateral orientation in humans (Chapter 17).

EMG studies have also shown that in static quadrupedal posture in the large apes, pectoralis major and perhaps supraspinatus and subscapularis are active (Tuttle and Basmajian, 1978b). Furthermore, recent work on the rotator cuff muscles in chimpanzees has shown that teres minor is particularly active during the propulsive phase of arm swinging and that infraspinatus is an important stabilizer of the shoulder joint in pendant suspension (Larson and Stern, 1986).

Relative magnitude of functional groups

In their classic study of the shoulder musculature in primates, Ashton and Oxnard (1963) provided the weights for muscles that they considered to be used primarily in propulsion and in arm raising (Fig. 16.21). In humans the arm-raising muscles are larger than the propulsive muscles, while in the brachiators

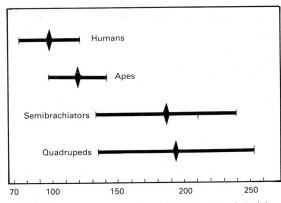

(Propulsive muscle weight × 100)/Arm raising muscle weight

FIGURE 16·21 *Locomotor contrasts in the relative weight of the arm-raising muscles (deltoid, trapezius, serratus magnus) in relation to the weight of the propulsive muscles (the pectoral muscles, latissimus dorsi, teres major) in the higher primates. After Ashton and Oxnard (1963).*

329

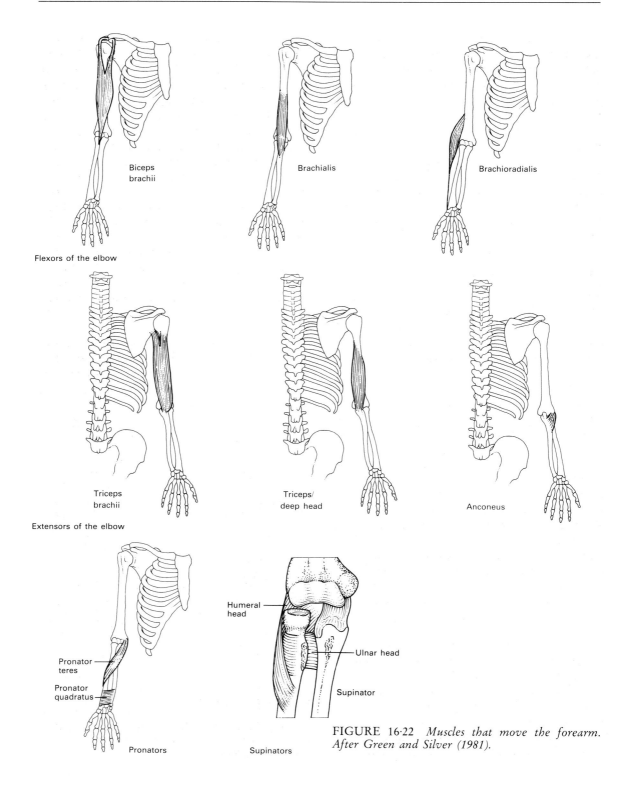

Biceps
brachii

Brachialis

Brachioradialis

Flexors of the elbow

Triceps
brachii

Triceps/
deep head

Anconeus

Extensors of the elbow

Pronator
teres

Pronator
quadratus

Humeral
head

Ulnar head

Supinator

Pronators

Supinators

FIGURE 16·22 *Muscles that move the forearm. After Green and Silver (1981).*

(which include the gibbon as well as the large apes) the propulsive muscles are larger than the arm-raising muscles. This difference most probably reflects the importance of the propulsive muscles in locomotion in the apes. The relative weights of the arm-raising and propulsive muscles in humans and the apes are part of a larger trend in the primates as a whole. In the quadrupeds, which do not use their arms above their heads, the propulsive muscles are almost twice as heavy as the arm-raising muscles and, in the semibrachiators, the ratio is intermediate, although closer to the quadrupeds.

Muscles that move the forearm in humans

The muscles that are involved with forearm movement are those that either flex or extend the elbow and those that either pronate or supinate the forearm (Fig. 16.22).

Muscles that flex the elbow

In flexion the elbow acts as a simple third class lever where the effort, or power, is applied between the axis of rotation (the joint) and the resistance (the hand). There are three muscles that flex the elbow, **biceps brachii**, **brachialis** and **brachioradialis**. Electromyographic (EMG) analysis shows that brachialis (brachium (L) = upper arm) is the all-around flexor of the elbow (Tuttle and Basmajian, 1974a; Basmajian and Latif, 1957; Pauly et al., 1967). This muscle passes directly over the anterior surface of the elbow joint, originating on the distal half of the anterior surface of the shaft of the humerus and the intermuscular septa and inserting on the coronoid process of the ulna.

The function of brachioradialis is to increase the speed and force of flexion (Tuttle and Basmajian, 1974a). It originates on the lateral side of the distal humerus and inserts into the distal end of the radius near the styloid process.

In contrast to both brachialis and brachioradialis, biceps brachii can function both as a flexor of the elbow and as a supinator of the forearm. As its name suggests (bi (Gk) = two; cep (Gk) = head), biceps brachii originates by two heads. The long head takes its origin from the scapula just above the glenoid cavity, passes over the head of the humerus and continues down the anterior surface of the humeral shaft in the **bicipital** (or **intertubercular**) **groove**. The short head originates on the coracoid process of the scapula. The primary insertion of biceps brachii is into the tuberosity of the radius, and its secondary insertion is by a thin aponeurosis which blends with the periosteum on the ulna. When the palm of the hand is facing upwards (supinated), biceps brachii acts as a simple flexor of the forearm and is most active during manipulation when the movement is not resisted (Fig. 16.23) (Tuttle and Basmajian, 1974a). However, when the palm of the hand is facing downward (pronated) biceps brachii acts as a supinator of the forearm. In a pronated posture, the

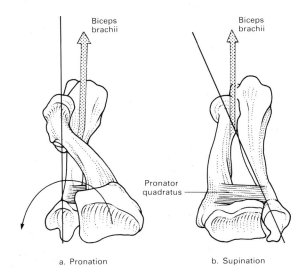

FIGURE 16·23 *The action of biceps brachii in supination. (a) Biceps brachii is wrapped around the posterior surface of the radius in pronation. As it contracts it pulls the radius over the ulna and into a position of supination. The axis of movement passes through the radial and ulnar heads (solid line). After MacConaill and Basmajian (1969).*

331

radius is twisted over the top of the ulna and the radial tuberosity faces posteriorly. This movement winds the tendon of biceps brachii around the pronated radius. When the muscle contracts it then acts to untwist (supinate) the radius by pulling the tuberosity back to its anteriorly facing position (see Chapter 17).

Muscles that extend the elbow

In extension, the elbow acts as a simple first-class lever where the axis (the joint) lies between the effort (or muscle force) and the resistance (or hand). The primary extensor of the elbow is **triceps brachii** (Fig. 16.22). This muscle originates in three heads. The first, or long, head originates from the scapula just below the glenoid and can serve as a support for the shoulder joint when the arm is abducted. The second, or lateral, head originates from the back of the humerus above the radial groove and the third, or medial, head originates from the back of the humerus below the radial groove. These three heads unite in a strong tendon to insert into the olecranon process of the ulna.

A second muscle, **anconeus** (agkon (Gk) = the bent arm or elbow), is also active during extension of the elbow. This is a small muscle that originates from the lateral epicondyle of the humerus and inserts into the olecranon. It serves as an all-around stabilizer of the elbow joint not only during extension, but also during pronation and supination (Tuttle and Basmajian, 1974a).

Muscles that pronate the forearm

There are two main muscles that pull the radius over the top of the ulna and thereby pronate the forearm and hand (Fig. 16.22). These muscles are **pronator teres** and **pronator quadratus**. Pronator teres originates from the anterior surface of the medial epicondyle of the humerus and inserts into the lateral side of the radius at the point of the maximum convexity of the shaft. Because it passes over the elbow joint it can also act as a weak flexor of this joint. Pronator quadratus is a simple pronator of the forearm. It arises from the lower quarter of the shaft of the ulna and wraps around the radial shaft to insert into the lower quarter of the shaft of the radius.

Muscles that supinate the forearm

The previously discussed **biceps brachii** (which also serves as a flexor of the elbow) and the appropriately named **supinator** are the two muscles responsible for supinating the forearm and hand. The supinator muscle originates from the lateral epicondyle of the humerus and from the ulna below the radial notch. It inserts into the front of the radius after wrapping around the back of this bone. The course of this muscle gives it the appropriate leverage to pull the radius from its pronated position (rotated diagonally across the top of the ulna) to its supinated position (parallel to the shaft of the ulna).

Muscles that move the forearm in the apes

As with the muscles of the shoulder and arm, the apes also have an additional muscle in the forearm, the **dorsoepitrochlearis**, that has been reduced to fascia in humans (Fig. 16.24) (Miller, 1932). In the apes, dorsoepitrochlearis originates from the tendon of latissimus dorsi, inserts into the medial epicondyle of the humerus and functions as

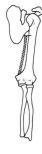

FIGURE 16·24 *Dorsoepitrochlearis in the gorilla. After Miller (1932).*

a tensor of the arm fascia. This muscle is also found in both the monkeys and the prosimians and, in these primates, inserts into the olecranon of the ulna and functions as an auxiliary flexor of the elbow.

In addition to the existence of this extra arm muscle in the apes, the major differences between the arm musculature of the apes and humans are the relative magnitude of the muscles of different functional groups, and the EMG potentials during elbow movement. The differences in the position of the insertions of these muscles into the bones of the arm are discussed in Chapter 17.

Differences in elbow musculature between apes and humans

Relative magnitude of functional groups

The main difference in the relative magnitude of the elbow musculature in humans and in apes is that the muscles that both flex and supinate the forearm are significantly more strongly developed in the apes in relation to the total arm musculature than they are in humans (Miller, 1932; Ashton et al., 1976) (Table 16.3). This increase in relative flexor magnitude in relation to both humans and other primates is related to the importance of elbow flexion in locomotion of the apes.

Electromyographic potential during movement

The main reference for the differences between humans and apes (gorilla) in electromyographic (EMG) potential for the elbow musculature is Tuttle and Basmajian (1974a). These authors reported that, during manipulation, brachialis and brachioradialis in humans and in the gorilla function in a basically similar fashion. However, there is some difference in the function of biceps brachii, particularly during quick flexion with the hand supine. Furthermore, during knuckle-walking locomotion in the gorilla, triceps is active and most probably acts to stabilize the joint while anconeus acts as a secondary extensor of the elbow and not so much as an all-around stabilizer of the joint as it does in humans. During pendant suspension when the animal is hanging by its hands from a support none of the arm muscles are active in the gorilla.

Muscles that move the hand in humans

The muscles that move the hand are divided into two major groups: the long muscles (sometimes called **extrinsic muscles**) and the short muscles (sometimes called **intrinsic muscles**). The long muscles of the hand originate from the epicondyles of the humerus or from the radius, ulna or interosseous

TABLE 16.3 Relative magnitude (weight) of arm muscles belonging to different functional groups expressed as a percentage of the total weight of the arm muscles. After Ashton et al. (1976).

	Elbow flexors		Elbow extensors		Forearm pronators		Forearm supinators	
	$\bar{X}$	SE	$\bar{X}$	SE	$\bar{X}$	SE	$\bar{X}$	SE
Humans	27.2	0.52	29.0	0.49	4.1	0.09	13.4	0.32
Apes	33.2	1.22	20.3	1.23	3.4	0.33	16.9	1.74
Semibrachiating monkeys	24.0	0.72	28.5	1.33	3.9	0.29	12.9	0.49
Quadrupedal monkeys	21.3	0.40	34.8	1.48	3.6	0.23	12.8	0.36

SE = standard error of the mean. $\bar{X}$ = mean relative weight.

membrane in the forearm. They insert into the carpal bones, the metacarpals or the phalanges and are responsible for flexion, extension, adduction, abduction and circumduction of the wrist and for flexion and extension of the fingers. The short muscles of the hand have both their origins and their insertions in the hand itself. These muscles are divided into three separate groups according to their location and function. The **thenar** (thenar (Gk) = hand) muscles are responsible for the fine movements of the thumb and are visible on the living hand as the **thenar**

eminence. The **hypothenar** (hypo (Gk) = under or less than) muscles are involved with movements of the little (fifth) finger and make up the **hypothenar eminence** on the ulnar side of the hand. The third group of short hand muscles are made up of the **interosseous** and **lumbrical** muscles and are involved with flexion and/or abduction and adduction of the fingers.

The long flexors of the hand

There are three main long flexors of the wrist, **flexor carpi radialis**, **flexor carpi**

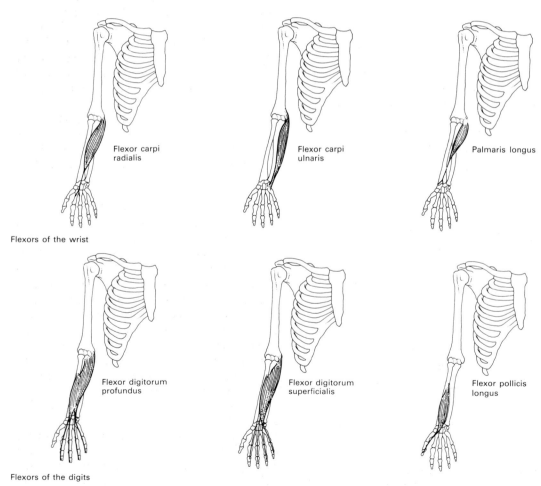

Flexor carpi radialis

Flexor carpi ulnaris

Palmaris longus

Flexors of the wrist

Flexor digitorum profundus

Flexor digitorum superficialis

Flexor pollicis longus

Flexors of the digits

FIGURE 16·25   *The long flexors of the hand. After Green and Silver (1981).*

ulnaris and **palmaris longus** (Fig. 16.25). Flexor carpi radialis arises from the common flexor origin on the medial epicondyle of the humerus and inserts in the radial side of the hand into the bases of the index and middle finger metacarpals. Its tendon passes through a marked groove on the palmar side of the trapezium. Flexor carpi ulnaris arises from the common flexor origin, the olecranon and the posterior border of the ulna and inserts into the hook of the hamate and the base of the fifth metacarpal. Its tendon contains the **pisiform**, a sesamoid bone. The third long flexor of the wrist, palmaris longus, is a weak muscle and is not always present in humans. It arises from the common flexor origin on the medial epicondyle of the humerus and inserts into the dense fascia in the palm of the hand.

There are also three long flexors of the digits: **flexor digitorum superficialis, flexor digitorum profundus,** and **flexor pollicis longus** (Fig. 16.25). Flexor digitorum superficialis arises from the common flexor origin, from the olecranon and from the anterior

border of the radius. It gives rise to four tendons which pass through the carpal tunnel (Fig. 16.4). As each tendon approaches its corresponding finger it splits in two and inserts on either side of the middle phalanx (Fig. 16.26). Flexor digitorum profundus originates from the anterior and medial surfaces of the ulna and interosseous membrane. In humans, it also gives rise to four tendons that pass through the carpal tunnel. Each slip passes through the split in the corresponding superficialis tendon at the level of the middle phalanx and inserts into the base of the terminal phalanx (Fig. 16.26). The third long flexor, flexor pollicis longus, arises from the front of the radius and interosseous membrane and inserts into the terminal phalanx of the thumb. This is the only flexor of the thumb phalanges. Flexor pollicis longus in its human form is not found in any of the apes (see below and Chapter 18).

The tendons of all three long digital flexors are held against their respective phalanges by fibrous **flexor sheaths** to prevent them from 'bow-stringing' as the fingers are flexed (Fig. 16.26).

The long extensors of the hand

There are nine long extensors of the hand and five of these have their origins on the posterior surface of the lateral epicondyle of the humerus (Fig. 16.27). Of these five, three are major extensors of the wrist: **extensor carpi radialis longus, extensor carpi radialis brevis,** and **extensor carpi ulnaris.** The long radial extensor inserts into the base of the second metacarpal, the short radial extensor into the bases of the second and third metacarpals and the ulnar extensor into the base of the fifth metacarpal.

The other two extensors of the hand that arise from the common extensor origin are extensors of the fingers. **Extensor digitorum** divides into four tendons that insert into the phalanges of the four fingers and **extensor digiti minimi** inserts by a tendon into the phalanges of the little (fifth) finger.

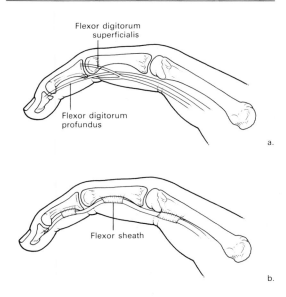

FIGURE 16·26 (a) The insertions of flexor digitorum profundus and flexor digitorum superficialis and (b) the flexor sheaths. After Kapandji (1982).

Extensors of the wrist

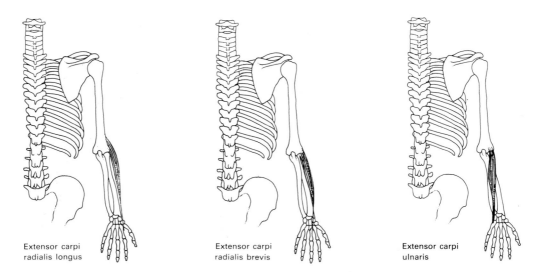

Extensor carpi radialis longus

Extensor carpi radialis brevis

Extensor carpi ulnaris

FIGURE 16·27 *The long extensors of the hand. After Green and Silver (1981).*

The remaining four long extensors of the hand originate from the posterior surface of the radius, ulna and interosseous membrane and function to extend either the index finger or the thumb. **Extensor indicis** inserts by a tendon onto the phalanges of the index finger and extends this digit. **Extensor pollicis longus**, inserts into the distal phalanx of the thumb and **extensor pollicis brevis** into the base of the proximal phalanx of the thumb. **Abductor pollicis longus** inserts into the base of the first (thumb) metacarpal and, as its name suggests, abducts (as well as extends) the thumb.

As the extensors pass over the posterior surface of the wrist they are held against the radius and ulna by the **extensor retinaculum** ((L) = a little network), a thickening in the fascia extending from the styloid region of the ulna and triquetral to the lower end of the radius. Ridges on the radius deep to the extensor retinaculum mark the passage of the extensor tendons. The most important of these ridges is the **radial tubercle** which deflects the tendon of extensor pollicis longus laterally towards the thumb (Fig. 16.13).

The short muscles of the hand

THE THENAR EMINENCE MUSCLES

There are three muscles, each with a separate function, that make up the thenar eminence and contribute to the prehensile movement of the thumb (Fig. 16.28). All three originate from the bones and ligaments of the radial side of the wrist. **Abductor pollicis brevis** inserts into the outer side of the base of the proximal phalanx of the thumb and functions as an abductor of the thumb. **Flexor pollicis brevis** inserts into the base of the proximal phalanx of the thumb and functions as flexor of the thumb. **Opponens pollicis** inserts into the lateral (radial) side of the thumb (first) metacarpal and rotates the thumb into opposition so that its palmar surface faces the palmar surface of the fingers.

HYPOTHENAR EMINENCE MUSCLES

There are four muscles that make up the hypothenar eminence (Fig. 16.28). Three of these muscles originate from the bones and ligaments of the ulnar side of the hand. **Abductor digiti minimi** inserts into the

336

Extensors of the digits and thumb

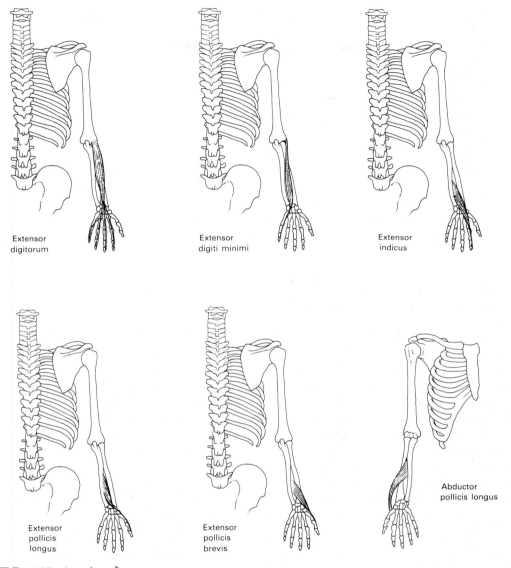

Extensor
digitorum

Extensor
digiti minimi

Extensor
indicus

Extensor
pollicis
longus

Extensor
pollicis
brevis

Abductor
pollicis longus

FIGURE 16·27 *(continued)*

medial side of the base of the proximal phalanx of the little (fifth) finger and abducts this finger. **Flexor digiti minimi** has a similar insertion but flexes the little (fifth) finger. **Opponens digiti minimi** inserts into the ulnar side of the fifth metacarpal and rotates the shaft of this metacarpal, thereby deepening the palm of the hand.

The fourth muscle, **palmaris brevis**, is a subcutaneous muscle that overlies the other muscles of the hypothenar eminence. It puckers the skin in this region and, thereby, further deepens the palm of the hand. This muscle originates on the flexor retinaculum and inserts into the skin on the ulnar border of the palm.

Thenar eminence muscles

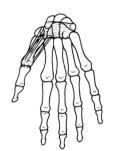
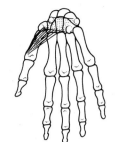

Abductor pollicis brevis   Flexor pollicis brevis   Opponens pollicis

Hypothenar eminence muscles

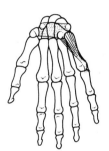
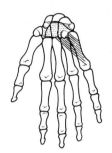
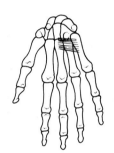

Abductor digiti minimi   Flexor digiti minimi   Opponens digiti minimi   Palmaris brevis

Interosseous muscles, adductor pollicis and the lumbricals

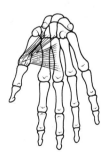
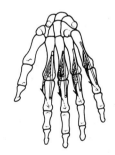

Dorsal interossei   Palmar interossei   Adductor pollicis   Lumbricals

FIGURE 16·28   *The short muscles of the hand. After Green and Silver (1981).*

INTEROSSEOUS MUSCLES, ADDUCTOR POLLICIS AND THE LUMBRICALS

There are two groups of interosseous muscles, the **dorsal** group and the **palmar** group (Fig. 16.28). The dorsal group is composed of four muscles. Each arises from the sides of adjacent metacarpals and inserts via an aponeurosis which merges with the extensor tendons of the middle three digits. This area of insertion is called an **extensor expansion**. The main function of the dorsal interosseous muscles is to abduct the middle three fingers away from the mid-line of the hand. Because they insert via the extensor expansions they are also capable of a secondary function (Fig. 16.29). Viewed from the side they span the metacarpophalangeal joint and, therefore, can flex this joint while at the same time pulling on the extensor expansion and straightening the interphalangeal joints. In other words, they can flex a straight finger.

The palmar interosseous muscles have the opposite function to the dorsal interossei, they adduct the fingers and thumb toward the mid-line of the hand (Fig. 16.28). The four palmar interosseous muscles arise from the palmar surface of the metacarpals of the thumb, index finger, ring finger and little finger and insert via the extensor expansion of the same digit. As with the dorsal interossei, the palmar interossei can also flex the straight finger.

The first palmar interosseous muscle that functions to adduct the thumb and is a relatively weak muscle. In fact this muscle is not recognized as a separate muscle by North American anatomists but is considered as a part of flexor pollicis brevis. **Adductor pollicis**, a much larger muscle, acts as the main adductor of the opposable thumb in both humans and apes. This larger muscle arises by two heads from the second and third metacarpals and inserts into the base of the proximal phalanx of the thumb.

The remaining short muscles of the hand are the four small **lumbricals** ((L) = worm). These muscles lie on the radial side of each

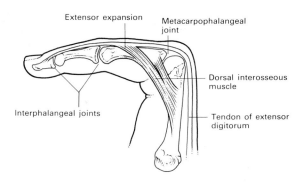

FIGURE 16·29 _The dorsal interosseus muscles allow us to 'flex a straight finger'. They span the metacarpophalangeal joint to flex the finger and at the same time extend the finger by pulling on the extension expansion._

of the four fingers and connect the tendons of flexor digitorum profundus to the extensor expansions. They integrate the actions of the digital flexors and extensors and, thereby, allow for fine co-ordination in the movements of the fingers. Because of their insertions into the extensor tendons, they can also function to flex the straight finger.

Muscles that move the hand in the apes

There are two major differences between the hand musculature of humans and that of the apes. The first of these concerns the relative magnitude of the muscles of different functional groups and the second includes the differences in morphology of individual muscles and tendons.

Tuttle (1967) has demonstrated that in relation to the total long hand musculature, the long flexors of the hand in the chimpanzee (_Pan troglodytes_) are more strongly developed than they are in humans (Table 16.4). In relation to the short hand muscles, Tuttle (1967) has also demonstrated that the muscles involved with the movement of the thumb (the thenar eminence muscles together with

TABLE 16·4 Percentages by dry weight of the major long and short muscles of the hand in chimpanzees (*Pan troglodytes*) and humans. After Tuttle (1967).

	Chimpanzees			Humans		
	N	Mean	Range	N	Mean	Range
Short muscles						
Thenar eminence	16	13.3	11.3–17.4	6	21.9	19.5–27.0
Hypothenar eminence	16	15.6	13.8–17.6	6	16.2	14.2–18.0
Adductor pollicus	16	10.6	7.9–13.0	6	17.3	14.9–22.6
Lumbricals, dorsal and palmar interossei	16	60.3	53.5–64.0	6	43.1	40.9–44.3
Long muscles						
Total flexors	7	58.9	57.3–60.7	3	50.7	48.0–53.1
Total extensors	7	26.1	24.8–27.5	3	35.2	33.5–37.8
Pronators	7	7.6	7.1– 9.2	3	9.3	7.9–10.3
Supinators	7	7.2	6.3– 8.2	3	4.7	3.9– 6.0

adductor pollicis) make up a larger percentage of the total short hand musculature in humans than they do in the chimpanzees.

The difference in the thumb musculature in the African apes and the orang-utan is also evident in the absence of flexor pollicis longus in its human form (Straus, 1942; Baldwin, 1989). It is true that a tendon corresponding to that of the human flexor pollicis longus tendon is present in 70% of chimpanzees (N=47), 69% of gorillas (N=16) and 11% of orang-utans (N=27) (Straus, 1942). How- ever, in many cases it is rudimentary and in the remaining it is attached to the radial belly of flexor digitorum profundus which also sends a tendon to the second digit (Straus, 1942). In these cases the phalanges of the thumb cannot be flexed independently of the phalanges of the second digit. Figure 16.30 illustrates some variations in flexor digitorum profundus reported by Straus (1942) for the greater and lesser apes. Furthermore, in some apes tendons from adductor pollicis insert into the distal phalanx of the thumb rather

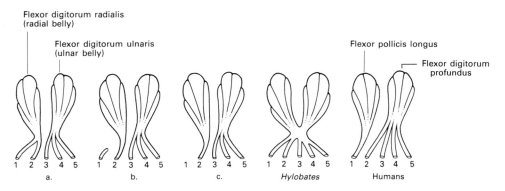

FIGURE 16·30 *The long flexors of the hand in humans and the apes. Humans alone have a separate flexor pollicis longus. (a), (b) and (c) are common variations that are found in the large apes. After Straus (1942).*

than into the proximal phalanx as they do in humans (Day and Napier, 1963; Tuttle, 1967). The first palmar interosseous muscle (or in North America the deep head of flexor pollicis brevis) is also commonly absent in the apes (Day and Napier, 1963; Tuttle, 1967). The presence of this muscle in humans is associated with opposability of the thumb (Day and Napier, 1961).

In the chimpanzees, the emphasis is on the lumbrical, dorsal interossei and palmar interossei muscle which make up a greater percentage of the total short hand musculature than they do in humans. These muscles may be important in limiting hyperextension at the metacarpophalangeal joints and resisting force on the middle phalanges during knuckle-walking (Chapter 18) (Tuttle, 1967). Another adaptation to knuckle-walking can be seen in the relatively short and heavily tendonized flexor digitorum profundus and flexor digitorum superficialis of the African apes (Tuttle, 1967, 1974b). Tuttle viewed this as an important adaptation in the support of the wrist during knuckle-walking. The short strong tendons passing over the palmar surface of the wrist would help prevent the wrist from buckling in the knuckle-walking posture (Chapter 18).

CHAPTER SEVENTEEN

THE HOMINOID ARM

The upper limbs of humans and apes are much more similar to each other than are the lower limbs (Fig. 17.1). This is largely because biped locomotion has resulted in greater anatomical changes in the lower limbs than in the upper limbs. Humans and apes share shoulder joints that are located on the back of the rib cage rather than at its sides and have a far greater range of upper limb mobility than is characteristic of all but the most arboreally acrobatic monkeys from the New World (Fig. 17.2). Humans and apes also share elbow joints that have a larger range of extension than the majority of the monkeys (Table 16.1) and wrist joints that are more mobile in the sense that the hand can be adducted (ulnar deviated) to a greater extent. In short, the entire upper limb in humans and apes is capable of a greater range of movement than is found in most monkeys,

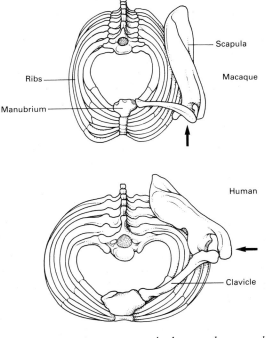

FIGURE 17·2 *A macaque and a human thorax and shoulder girdle from the superior (cranial) view. Both are drawn to the same anteroposterior depth. Note the different orientations of the shoulder joints (arrows) and the different shapes of the thorax. After Schultz (1950).*

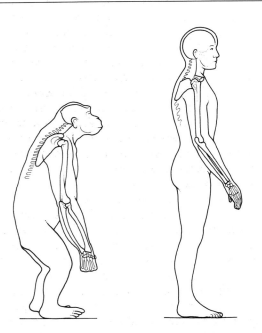

FIGURE 17·1 *The upper limb in a chimpanzee and a human.*

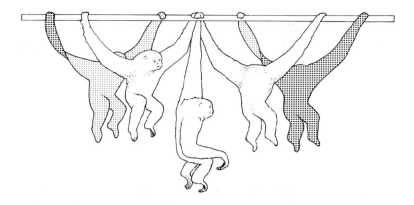

FIGURE 17·3   *A gibbon brachiating. The shadow figures emphasize the position of the trailing limb where the arm is laterally rotated and the forearm supinated. After Larson (1988).*

and particularly greater than is found in the monkeys of the Old World (cercopithecines and colobines).

Upper-limb mobility in the large apes and in the gibbons and siamangs has traditionally been associated with a type of arboreal locomotion described as **brachiation** (brachium (L) = upper arm) (Napier, 1963; Trevor, 1963; Le Gros Clark 1971; Tuttle, 1975). In this form of locomotion the arms are used fully extended above the head to suspend and propel the animal through the trees (Napier, 1963). The only primates that habitually use this form of locomotion as their primary means of movement however, are the gibbons and siamangs (Fig. 17.3). These primates are known as **true brachiators**. Gorillas, chimpanzees and orang-utans are **modified brachiators**. These primates are capable of using their upper limbs above their heads in locomotion, but they do use their lower limbs to a greater or lesser extent to provide support for the body from below. **Semi-brachiators** are monkeys, rather than apes, that are essentially quadrupeds but use their upper limbs above their heads to a greater extent than do purely quadrupedal monkeys. The **Old World semi-brachiators** are the colobine monkeys of Africa and Asia that use their forelimbs in association with the hindlimbs to check their momentum after leaping. The **New World semi-brachiators** are the spider monkeys (*Ateles*), woolly monkeys (*Brachyteles*, *Lagothrix*) and howler monkeys (*Alouatta*) that use their forelimbs in association with their prehensile tails to either suspend their bodies or propel them through space.

Similarities in morphology and mobility of the upper limb of humans and apes led to the idea that humans descended from an ancestor that brachiated (Le Gros Clark, 1971; Tuttle, 1974b; 1975). Indeed the term **pro-brachiation** was used to describe the hypothetical locomotor behaviour of primates inferred to be ancestral not only to the large apes and gibbons, but also to humans (Napier, 1963). These fossil pro-brachiators showed the early stages of adaptation leading to the more specialized locomotor patterns of living apes and humans.

Modern field studies of primate locomotion, together with modern comparative morphology, have challenged the importance of brachiation in the locomotor repertoire of living apes and in the ancestry of the human line (Fleagle, 1976). Rather, it is argued that the locomotor pattern that is actually common to the greater and lesser apes is **quadrumanous** (quadri (L) = four; manous (L) = hand) **climbing** where the upper limb is used above the head but not to suspend the body. Fleagle (1976) argued that it is most probable that the morphological features associated with upper limb mobility are related to forelimb-dominated climbing rather than to brachiation per se and that the true brachiation of the gibbons and siamangs has developed from this form of locomotion. Indeed, there is an accumulating body of literature that argues that the common morphological features of the upper limb skeleton in humans and apes are ancestral features inherited from a generalized climbing, rather than brachiating (or pro-brachiating), ancestor (Cartmill and Milton, 1977; Stern et al., 1977; Prost, 1980; Fleagle et al., 1981). In the apes, these ancestral features have been maintained in the modern populations by a continued selection for forelimb-dominated climbing and in humans they have been maintained in a slightly altered form by selection for upper-limb mobility associated with manipulation and tool use.

Variations in upper-limb morphology between apes and humans reflect the use of the upper limb in locomotion in the apes and its freedom from locomotion in humans. Furthermore, differences between the three large apes reflect the more aboreal locomotion of the orang-utans on the one hand and the more terrestrial locomotion of chimpanzees and gorillas on the other. For example, the human upper limb is shorter in relation to body size than is the case for any of the apes, and the more arboreal is the ape the longer is its upper limb (Fig. 17.4 and Chapter 14). This variation in upper-limb length results more from variation in the length of the radius and ulna in the forearm rather than from variation in the humerus in the upper arm. In relation to body weight, humans have very short radii while orang-utans have very long radii (Fig. 17.4). The **brachial index** [(radius length × 100)/humerus length] closely reflects this pattern and also emphasizes how short both the gorilla and human radii are in relation to humerus length (Table 14.2). The short gorilla radius is consistent with the well-known phenomenon where the distal parts of the limbs in very heavy (or **graviportal**) animals are reduced in length (Gregory, 1912). This significantly reduces the stress on the distal extremities during locomotion. The short human radius is consistent with the requirements of manipu-

FIGURE 17·4 *The length of the upper limb (■), humerus (◇) and radius (○) in humans, the large apes and gibbons expressed as a percentage of expected limb/bone length in relation to body weight (dotted line) based on 30 species of higher primates. Data from Aiello (1981a).*

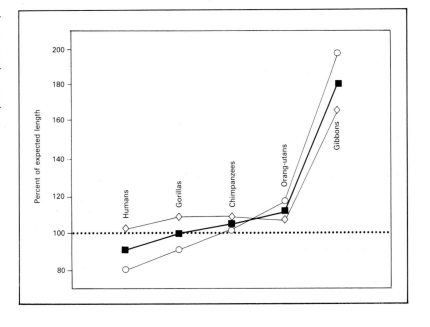

lation in that it allows the hands to be positioned comfortably in front of the body. Furthermore, Keith (1926) argued that elongation of the forearm would impair both the speed and precision of forearm movement (flexion, pronation and supination) during manipulation.

Other structural differences in upper-limb morphology between humans and apes concern the lever advantages of the muscles producing upper-limb movement as well as the morphology of the various joint surfaces. Such differences have major functional effects in: firstly, the ability to use the arm above the head; secondly, the strength of the forearm in pronation and supination; and thirdly the ability of the elbow joint to support weight during locomotion. These three areas are addressed in turn in the following discussions of the pectoral girdle and shoulder joint and of the elbow joint and forearm.

The pectoral girdle and shoulder joint in humans and apes

There are a number of osteological features of the scapula, clavicle and proximal humerus that are shared by humans and apes which show that the human arm can be raised as freely and fully as the ape arm. There are other features, however, that reflect the differences in the use of the forelimb in humans and the apes. In particular, there are differences that suggest that the human arm lacks strength in an elevated posture and is best adapted to use below the level of the shoulder.

Similarities in the human and the ape pectoral girdle and shoulder joint

The osteological features of the pectoral girdle and shoulder joint that indicate free mobility of the ape and human arm fall into two categories: those osteological features that enhance the mechanical advantage of the

muscles that raise the arm; and those features of the shoulder joint itself that enhance its mobility in a variety of planes of movement.

There are two features that increase the mechanical advantage of the deltoid, the principal muscle that abducts the arm at the shoulder joint (Chapter 16). Firstly, apes and humans have scapula with laterally projecting acromion processes (Inman et al. 1944; Ciochon and Corruccini, 1977) (Fig. 17.5). The strong lateral projection of the acromion process increases the lever advantage of deltoid by projecting the origin of the muscle laterally over the shoulder joint. Secondly, deltoid inserts low down on the shaft of the humerus in relation to other primates. In abduction the arm acts as a simple third-class lever and the low position of the deltoid insertion in humans and apes helps to increase the lever advantage of the deltoid (Fig. 17.6) (Miller, 1932; Evans and Krahl, 1945; Ashton and Oxnard, 1964).

In addition, humans and apes share features that enhance the action of the two main upward rotators of the scapula (trapezius and

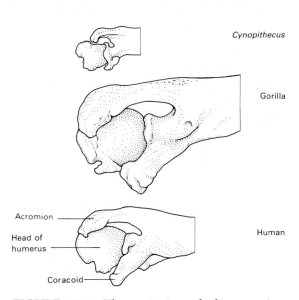

FIGURE 17.5 *The projection of the acromion process of the scapula in a monkey (*Cynopithecus*), gorilla and human. After Inman* et al. *(1944).*

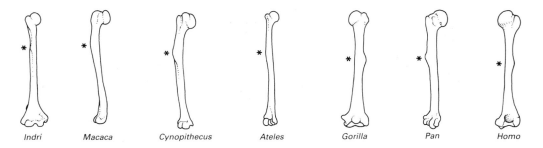

FIGURE 17·6 *The position of the deltoid insertion (star) in humans, the African apes and selected other primates. After Inman et al. (1944).*

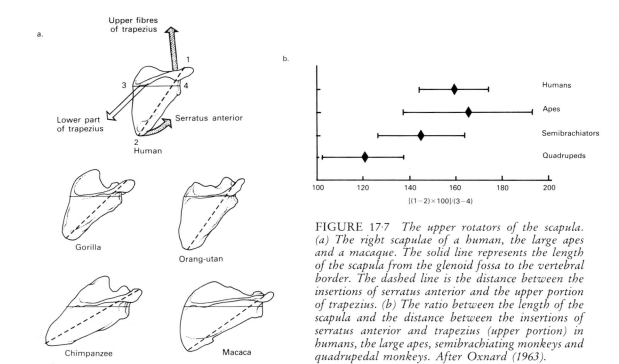

FIGURE 17·7 *The upper rotators of the scapula. (a) The right scapulae of a human, the large apes and a macaque. The solid line represents the length of the scapula from the glenoid fossa to the vertebral border. The dashed line is the distance between the insertions of serratus anterior and the upper portion of trapezius. (b) The ratio between the length of the scapula and the distance between the insertions of serratus anterior and trapezius (upper portion) in humans, the large apes, semibrachiating monkeys and quadrupedal monkeys. After Oxnard (1963).*

serratus anterior). Humans and apes have a high ratio between the length of the scapula and the distance between the insertions of these two muscles (Fig. 17.7) (Oxnard, 1967). This high ratio indicates a relatively great distance between the insertions of these muscles and, in relation to other primates, a correspondingly improved action of these muscles in scapular rotation.

Both the glenoid cavity and the humeral head have features that make the shoulder joint more mobile in a variety of planes and correspondingly less stable than that of the Old World monkeys (Fig. 17.8). The glenoid in both humans and apes is relatively wide, round and flat as opposed to narrow, elongated and more concave as in the Old World monkeys (Ciochon and Corruccini, 1976). In

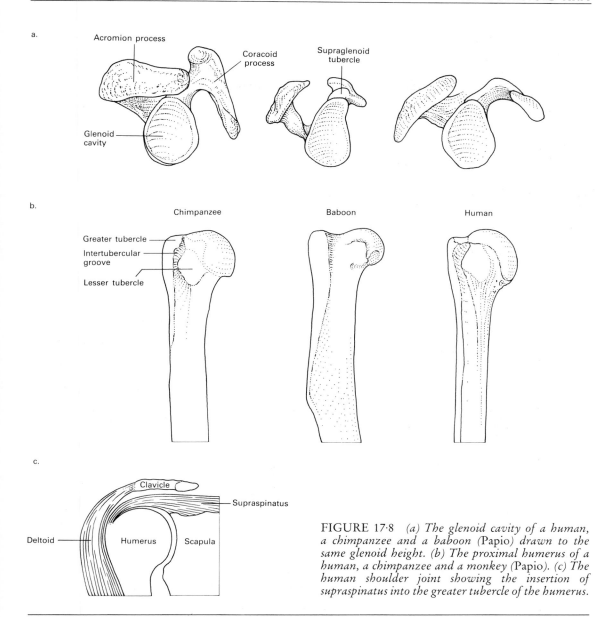

a.

Acromion process

Coracoid process

Supraglenoid tubercle

Glenoid cavity

b.

Chimpanzee   Baboon   Human

Greater tubercle

Intertubercular groove

Lesser tubercle

c.

Clavicle

Supraspinatus

Deltoid   Humerus   Scapula

FIGURE 17·8   *(a) The glenoid cavity of a human, a chimpanzee and a baboon (*Papio*) drawn to the same glenoid height. (b) The proximal humerus of a human, a chimpanzee and a monkey (*Papio*). (c) The human shoulder joint showing the insertion of supraspinatus into the greater tubercle of the humerus.*

particular, the ape and human joints lack a projecting supraglenoid tubercle. This is a well-developed knob of bone located immediately above the glenoid in other primates. It provides the origin for the long head of biceps brachii. In these primates this tubercle can limit the degree of upward movement of the arm by wedging into the intertubercular groove on the proximal humerus (Ciochon and Corruccini, 1976).

The head of the humerus is also characterized by a feature that increases the degree of upward movement possible at the shoulder joint. In both ape and human humeri, the greater and the lesser tubercles (and the corresponding intertubercular groove) are positioned well below the level of the humeral head (Fig. 17.8). This is particularly marked in the large apes, but also occurs not only in humans but also in the majority of the New World monkeys, the Old World colobine monkeys and the prosimians. It is only in the more quadrupedal cercopithecine monkeys of the Old World that the greater tubercle projects above the humeral head, a feature shared with many other terrestrial mammals (Zapfe, 1960). Low tubercles allow a greater degree of upward movement for the humerus at the shoulder joint (Larson and Stern, 1989), and also increase the leverage in rotation, abduction and adduction of the rotator cuff muscles that insert into the tubercles.

Perhaps the most important similarity between the proximal humerus in humans and in the apes is the medial rotation, or torsion, of the humeral head in relation to the plane of the distal (elbow) joint (Fig. 17.9). Torsion of the head of the humerus is defined as the angle made by the axis of the head of the humerus with the mediolateral axis of the elbow joint. Among the primates, humans show the greatest degree of torsion, followed by the African apes, the orangutan, the gibbons, the semi-brachiating New World monkeys and then by the Old World monkeys (Evans and Krahl, 1945; Knussmann, 1967).

The high degree of humeral torsion has generally been considered as a feature shared by humans and apes as the result of a common ancestral adaptation to either brachiation or forelimb-assisted climbing (Miller, 1932; Le Gros Clark, 1971). However, Larson (1988) has pointed out that the most specialized brachiator, the gibbon, has a small degree of torsion in comparison to humans and the large apes. Furthermore, the same author emphasized that torsion is only necessary, given a laterally facing shoulder joint, if it is important to position the elbow joint so that flexion and extension occur in the sagittal, or anteroposterior, plane. She explains the small degree of torsion in the gibbon as a specific adaptation to their highly specialized brachiating form of locomotion, and particu-

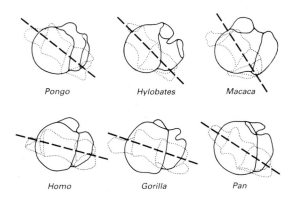

FIGURE 17.9 *The torsion of the head of the humerus in humans and selected primates (Martin and Saller, 1957: measurement no. 6.18). The dashed line represents the mediolateral axis of the elbow joint. The axis of the head of the humerus coincides with the horizontal plane. This axis is the line that bisects both the greater tubercle and the humeral head. After Evans and Krahl (1945).*

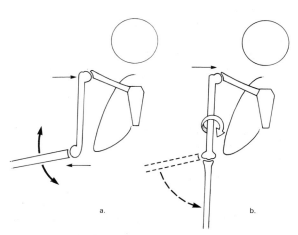

FIGURE 17.10 *Schematic diagram of the plane of action of the elbow joint. (a) The lateral set of the elbow joint in gibbons where the shoulder joint faces laterally, the humerus has a small amount of torsion and the action of the elbow joint is in the coronal plane (if the arm is not medially rotated at the shoulder joint). (b) The greater degree of torsion in the humerus of the large apes and humans positions the elbow joint so that the action is in the sagittal rather than the coronal plane.*

larly to the terminal support phase of the brachiating cycle just before the trailing arm releases the support in preparation for grasping a new support in advance of the body (Fig. 17.3). At this stage of the brachiating cycle, the trailing arm is laterally rotated at the shoulder joint and hypersupinated at the elbow. The small degree of humeral torsion in the gibbon helps to achieve this rather tortuous arm posture (Fig. 17.10). By orienting the plane of elbow movement more laterally (Larson's lateral set of the elbow joint), the gibbon's small degree of humeral torsion literally builds into the resting arm some of the lateral rotation necessary to their extreme locomotor pattern. Therefore, the stress on the elbow joint during locomotion is reduced.

Larson (1988) also suggested that the higher degree of humeral torsion in humans and the African apes is most probably not a shared feature inherited from a common ancestor, but a feature that has been achieved in parallel in response to different selection pressures. In the African apes, given a laterally positioned shoulder joint, a high degree of humeral torsion orients the elbow in the sagittal plane. Larson (1988) argued that this is essential to terrestrial locomotion and pointed out that the most terrestrial of the large apes, the gorilla, also has the highest degree of humeral torsion. The same author also argued that the equally high degree of humeral torsion in humans is a response, not to locomotion, but to manipulation. In this sense, given the laterally oriented shoulder joint which is so important to shoulder mobility, the anteriorly oriented elbow joint (resulting from the humeral torsion) would enable the forearm and hand to be used in front of the body for purposes of manipulation. Although it is also possible that humans inherited humeral torsion from a terrestrial, and perhaps knuckle-walking, ancestor, there is no skeletal evidence for humans having passed through a knuckle-walking stage of locomotion (see, for example: Tuttle, 1969a, 1974b, 1975; Stern and Susman, 1983) (Chapter 18). Fur-

thermore, although the evidence is confined to fragmentary proximal humeri, Larson also observed that the australopithecines (AL 288-1r, AL 333-107) may have had less humeral torsion than found in either living chimpanzees or humans. This also suggested to Larson that it is possible that this trait evolved in parallel in humans and apes.

It is interesting to note in this context that the human humerus acquires its high degree of torsion during ontogeny. From the second foetal month, as the scapula moves from its more lateral orientation at the side of the rib cage to its dorsal position on the back of the rib cage, the torsion of the head of the humerus increases to correspond with the more lateral orientation of the glenoid cavity (Martin and Saller, 1959).

Humeral torsion, whether or not it is achieved in parallel, is also associated with other features of the proximal humerus. In animals without humeral torsion, the bicipital (intertuberclar) groove is centrally located between the greater and lesser tubercles. Humeral torsion has the effect of displacing the bicipital groove medially (Fig. 17.9). It encroaches upon the lesser tubercle, which is significantly reduced in size in comparison with the primitive pattern that lacks humeral torsion (Inman et al., 1944). In the large apes the lesser tubercle is very small in relation to the greater tubercle, while in humans the lesser tubercle is larger, but is still far from equal in size to the greater tubercle (Zapfe, 1960). The bicipital groove also has a different form in those primates with a large degree of humeral torsion. In these primates, with the exception of the orang-utan, it is narrow and deep while in those with a smaller amount of humeral torsion it is wide and shallow. There is some variation, however, and the African apes do have a deeper and narrower groove than humans (Zapfe 1960). In these primates the bicipital groove undercuts the lesser tubercle, giving the lesser tubercle a distinctive flange-like morphology (Pickford et al., 1983; Hill and Ward, 1988).

A round, ball-shaped head of the humerus

has also frequently been mentioned as a shared feature between humans and the apes in the context of the greater range of arm movement common to these primates (Ciochon and Corruccini, 1976). However, a comparison of the index of the height and the breadth of the humeral head shows that humans stand midway between the ball-shaped humeral head of the great apes and the more elongated head of the monkeys (Knussmann, 1967). The human and ape humeri are more similar to each other in humeral torsion (although this may have arisen in parallel), in the low position of the tubercles of the humerus in relation to the head and in the projection of the head away from a proportionately slender humeral shaft (Fig. 17.8) (Ciochon and Corruccini, 1976).

Differences in the human and the ape pectoral girdle and shoulder joint

The osteological differences between the human and the ape pectoral girdle and shoulder joint are found primarily in the scapula and in the clavicle. These features suggest that the human arm is adapted for use in a lowered position and is less powerful in a raised position than is the ape arm.

Important features of the scapula include a relatively small supraspinous fossa coupled with a relatively large infraspinous fossa. This feature distinguishes humans not only from the apes but also from other non-human primates (Fig. 17.11). The relatively large fossae in the apes and other climbing primates have been interpreted by Roberts (1974) as indicating relatively large rotator cuff muscles which are needed to stabilize the shoulder joint when it is used under tension in climbing or suspensory locomotion (see also Larson and Stern, 1986).

In addition, measurements of the human scapula involving the insertions of the major upward rotators of the scapula resemble those of primate quadrupeds and not apes (Oxnard, 1967). This similarity suggests that the human

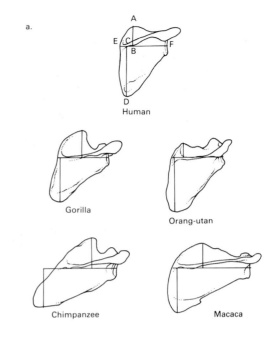

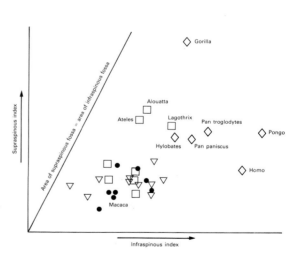

FIGURE 17·11 *The supraspinous and infraspinous fossae. (a) The supraspinous fossa and infraspinous fossa on the right scapula of a human, the large apes and a macaque. (b) Bivariate plot of the supraspinous index [(A-B)/E-F)] and the infraspinous index [(C-D)/E-F)] in selected primates. (◇) = humans and apes; (□) = New World monkeys; (●) = Old World Monkeys; (▽) = prosimians. After Roberts (1974).*

arm lacks power in elevation and is habitually used in a lowered position. These measurements include the angle of insertion of trapezius, the medial extent of the insertion of trapezius and the orientation of the components of the muscle couple made up of trapezius and serratus magnus (Fig. 17.12).

Two other features of the human pectoral girdle also bear a closer resemblance to primate quadrupeds than they do to apes and suggest that the human arm is habitually used in a lowered position. Firstly, the shoulder joint (glenoid cavity) faces laterally in humans and in the quadrupedal primates while it faces

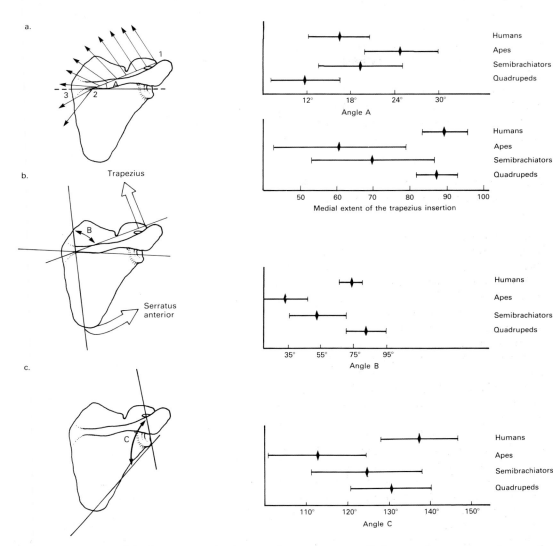

FIGURE 17·12 *Features of the human scapula that align it with primate quadrupeds rather than with the large apes. (a) The angle of insertion of trapezius (angle A) and the medial extent of its scapular insertion [(1–2)/(1–2 + 2–3)]. (b) The orientation of the components of the muscle couple made up of trapezius and serratus magnus (angle B). (c) The orientation of the glenoid cavity (angle C). After Oxnard (1963).*

351

more cranially (upwards) in the apes (Fig. 17.12). The high axilloglenoid angle in both humans and the quadrupedal primates confirms the difference in the orientation of the glenoid in these primates in comparison with the apes. Secondly, the lateral end of the clavicle lacks the pronounced cranial twist that characterizes it in the apes (Fig. 17.13). The degree of torsion shown by the flattened lateral end of the clavicle in relation to the anteroposterior plane of the medial end shows clear similarities between humans and the quadrupedal primates (Ashton and Oxnard, 1964).

From a comparative point of view the clavicle is one of the most poorly studied bones in the body. However, Schultz (1930) reported a considerable variation in the nature and degree of clavicular curvature in humans and in the different species of ape (Fig. 17.14). Gibbon clavicles have a single anteriorly convex curve while orang-utan clavicles are generally straight. Chimpanzee and human clavicles have S-shaped curves and gorilla clavicles are generally straight except for the lateral, acromial, end which is bent to various degrees. In view of the fact that the curvature of the human clavicle allows it to act as a crankshaft during elevation of the arm (Inman et al., 1944), differences in clavicular curvature would be expected to reflect major differences in shoulder orientation and function. In this context, Schultz (1930) noted that the clavicles of the orang-utans are not directed horizontally as in

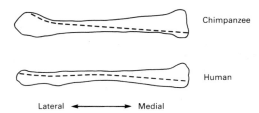

FIGURE 17·13 *A chimpanzee and human clavicle from the anterior (ventral) view. Note that the twist of the lateral end of the chimpanzee clavicle is absent in the human clavicle. After Oxnard (1984).*

humans, but are oriented at a steep angle consistent with the very high position of the orang-utan shoulder above the rib cage.

The pectoral girdle and shoulder joint in human evolution

The pectoral girdle and shoulder joint in the Australopithecines

The pectoral girdle and proximal humerus are very poorly known in the earlier stages of human evolution. The most discussed fossils from the Plio-Pleistocene period are AL 288-1l (the scapular fragment belonging to the Lucy skeleton *Australopithecus afarensis* (Johanson et al. (1982)) and Sts 7 (the *Australopithecus africanus* scapular fragment from Sterkfontein (Broom and Robinson, 1947; Broom et al., 1950; Vrba 1979)). In addition to these specimens, there are also descriptions in the literature of a scapular fragment from Koobi Fora (KNM-ER 1500

FIGURE 17·14 *Human and ape (left) clavicles from the superior view. Clavicles are drawn to the same length. After Schultz (1930).*

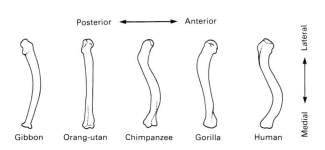

(Day et al., 1976a)) as well as a small number of proximal humeri (AL 288-1, Sts 7, KNM-ER 1473, Omo 119-73-2718) that are all robust and show evidence of well-developed shoulder and brachial muscles (Day 1978; McHenry and Temerin, 1979; Johanson et al., 1982).

THE PROXIMAL HUMERUS

The poor state of preservation of the australopithecine proximal humeri limits any comparative conclusions that can be drawn from their morphology. However, the humeral head (at least in *Australopithecus afarensis*) does not seem to be as spherical as it is in the apes. An index composed of its anteroposterior diameter divided by its superoinferior (vertical) diameter places the AL 288-1 humeral head close to the means of the monkeys and below that of humans (which is also less spherical than the apes) (data from Johanson et al. (1982)). Furthermore, the lesser tubercles in both the *A. afarensis* (AL 288-1) and *Australopithecus africanus* (Sts 7) humeri are larger in relation to the size of the greater tubercles than they are in the apes (Robinson, 1972; Johanson et al., 1982), and in this feature resemble modern humans (Fig. 17.15). However, the Sts 7 (*A. africanus*) humerus does have a prominent ridge separating the attachments of supraspinatus and infrapinatus on the greater tubercle that does not usually occur in modern humans (Robinson, 1972). Furthermore, the anatom-

ical neck of Sts 7 (*A. africanus*), separating the head from the lesser tubercle, is less well defined than is normally the case in modern humans (Robinson, 1972). There is also a marked hollow in Sts 7 between the lesser tubercle and the lower third of the articular surface of the head. As a result, the lower part of the tubercle forms a narrow and well-developed ridge unlike humans or any of the large apes (Broom et al., 1950).

Where these poorly preserved proximal humeri are ambiguous in their morphology, the fragmentary scapulae of both *A. afarensis* and *A. africanus* are more similar to modern ape scapulae than they are to the modern human condition.

THE SCAPULA OF AUSTRALOPITHECUS AFARENSIS (AL 288-1l)

The *Australopithecus afarensis* scapula is a small fragment which includes an intact glenoid cavity and a small part of the axillary (lateral) border of the blade. This specimen suggests that the glenoid cavity was cranially oriented in *A. afarensis* in a manner similar to that of the great apes and different from the lateral orientation found in humans (Stern and Susman, 1983; Susman et al., 1984). This conclusion is based on the analysis of the angle between the ventral scapular bar and the glenoid fossa (Fig. 17.16) indicating that the arm was habitually used in an elevated position that would be common during climbing behaviour.

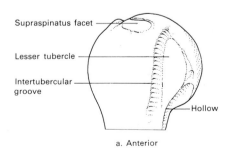
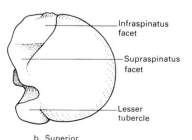

FIGURE 17·15 *The Sts 7 (*Australopithecus afarensis) *proximal humerus from the anterior* (ventral) view (a) and the superior (cranial) view (b). After Robinson (1972).

FIGURE 17·16 Australopi-
thecus *scapulae. The anterior
(ventral) view of the A. afarensis (AL 288-1l) scapula (a)
and a modern human scapula
(b). The low ape-like angle
between the ventral bar (VB)
and the glenoid cavity in A.*
afarensis *indicates that the glenoid is directed more cranially
than it is in modern humans.
After Stern and Susman (1983).
The posterior (dorsal) view (c)
and the lateral view (d) of the
A. africanus (Sts 7) scapula.
After Robinson (1972). Note
the well developed supra-
glenoid tubercle on the Sts 7
scapula, the origin of the coracoid process adjacent to the
supraglenoid tubercle and the
well developed dorsolateral
tubercle.*

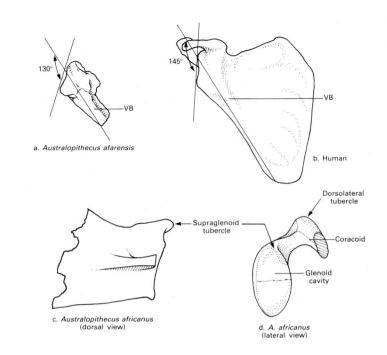

a. *Australopithecus afarensis*

b. Human

c. *Australopithecus africanus*
(dorsal view)

d. *A. africanus*
(lateral view)

FIGURE 17·17 *The coracoid
process and the dorsolateral
tubercle (arrows) on an orang-
utan, a chimpanzee, a human
and a baboon scapula.*

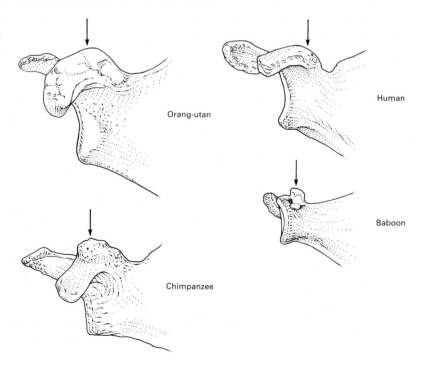

Orang-utan

Human

Chimpanzee

Baboon

THE SCAPULA OF
AUSTRALOPITHECUS AFRICANUS
(Sts 7)

The Sts 7 scapular fragment is more complete than the AL 288-1l fragment and, in addition to the glenoid cavity and part of the axillary border, preserves part of the coracoid and part of the spine (Broom et al., 1950; Campbell, 1966; Oxnard, 1968, 1969; Robinson, 1972; Roberts, 1974; Ciochon and Corruccini, 1976; Vrba, 1979).

The Sts 7 scapula is similar to the scapulae of AL 288-1l and the great apes and different from modern human scapulae in its more cranially oriented glenoid cavity (Vrba, 1979; Stern and Susman, 1983; Susman et al., 1984). Other features also underline its morphological similarity to the scapulae of the living great apes and particularly to the orang-utan (Fig. 17.17). As in the orang-utan, there is a more well-developed supraglenoid tubercle (the insertion for the tendon of the long head of biceps brachii) than is found in African apes or humans (Fig. 17.16) (Broom et al., 1950). Furthermore, the coracoid process arises practically from the supraglenoid tubercle as it does in the orang-utan. In humans it originates behind the plane of the glenoid and does not curve away from the glenoid as much as it does in either the fossil or the orang-utan (Broom et al., 1950).

There are two other features of Sts 7 scapula which suggest that the pectoral girdle of *Australopithecus africanus* was more similar in morphology to that of the living apes rather than to that of modern humans. Firstly, the dorsolateral tubercle and the adjacent area on the coracoid are both well developed. Vrba (1979) suggested that this area on the fossil coracoid was the attachment of the conoid ligament that transmits the weight of the upper limb to the clavicle. The size and development of this tubercle and adjacent area suggests that the upper limb was heavier and perhaps longer than would be expected in a similarly sized modern human. This suggested to Vrba (1979) that the forelimbs of *A. africanus* may have been longer relative

to their hindlimbs than is the condition in modern humans. Secondly, Vrba (1979) noted that the attachment of the conoid ligament on the coracoid process of the scapula of the apes is consistently nearer to the glenoid than it is in humans. Vrba suggested that this lateral placement of the attachment area in the living apes reflects a wider angle between the scapula and the clavicle than is found in humans. The lateral placement of the dorsolateral tubercle on the Sts 7 scapula would suggest a similar wide angle between the scapula and clavicle in *A. africanus*. The similarity between *A. africanus* and the living great apes on the one hand and the difference from modern humans on the other in the lateral position of the conoid ligament attachment would also imply a similar difference in the position of the pectoral girdle and shape of the thorax.

The pectoral girdle and shoulder joint in the Neanderthals

In comparison with the available fossils from earlier periods of human evolution, the Neanderthal pectoral girdle and shoulder joint are relatively well represented (Boule, 1911–13; Gorjanović-Kramberger, 1914; Vallois, 1928–46; McCown and Keith, 1939; Stewart, 1962; Endo and Kimura, 1970; Heim, 1974; Trinkaus 1983a). These fossils indicate that the Western European Neanderthals had very broad shoulders in relation not only to modern *Homo sapiens* but also to contemporary Neanderthal hominids from south western Asia (Shanidar/Tabun). This impression is supported by the absolute length of the western European Neanderthal clavicle as well as by the length of the clavicle in relation to the length of the humerus (claviculohumeral index).

The broad shoulders of the Western European Neanderthals in particular, may explain the observation that Neanderthal clavicles are very gracile in comparison to the normally very robust bones of the Neanderthal skeleton (McCown and Keith, 1939; Heim, 1974).

Trinkaus (1983a) has shown that the midshaft circumference of the Neanderthal clavicles are within the range of modern humans but the extreme length of the clavicles puts the clavicular robusticity index towards the bottom of the modern range.

Other aspects of the Neanderthal pectoral girdle and shoulder joint reflect the robusticity and muscularity of the Neanderthal skeleton.

THE NEANDERTHAL PROXIMAL HUMERUS

There are three obvious features of the Neanderthal proximal humerus that place it at the extremes of modern human variability (Trinkaus, 1983a). Firstly, there is an impressive development of the crests for the insertions of deltoid and pectoralis major reflecting the strength of these muscles in both the Western European and the South Western Asia Neanderthals. Secondly, the shaft is flat in the mediolateral direction. Thirdly, the shaft is also very robust in comparison with modern humans. The shaft is not only robust in its external measurements but also robust in its extreme cortical thickness (Ben-Itzhak et al., 1988). On average, both male and female Neanderthals have thicker humeral cortices than do modern humans; however, extreme cortical thickness is particularly evident in the right humerus of male Neanderthals. In other words the Neanderthals not only had thicker humeral cortices than do modern humans but also had a greater degree of sexual dimorphism in this feature. These findings suggest that Neanderthals had a greater muscle mass, and correspondingly more stress on their forearms, than do modern humans. In addition, they also suggest that there may have been a greater degree of difference between the sexes in the use of the right arm than is the case in the majority of modern humans (Ben-Itzhak et al., 1988). For example, male Neanderthals may have engaged in activities (such as spear throwing) that involved the use of considerable right-arm strength.

This pattern of extreme sexual dimorphism has not been reported for other areas of the Neanderthal postcranial skeleton. Patterns of sexual dimorphism in Neanderthal bone length and girth measurements fall within the modern human range of variation (Trinkaus, 1980; Heim, 1983). However, differences in stress on the postcranial skeleton resulting from behavioural differences would be expected to be more emphasized in bone mass (and cortical thickness) than in external bone measurements (Garn, 1970; Garn et al., 1972; Ben-Itzhak et al., 1988).

THE NEANDERTHAL SCAPULA

The morphology of Neanderthal scapulae implies that these hominids had very powerful shoulders that were, however, capable of the same type of movement found in modern humans (Stewart, 1962). The scapulae are large and robust but are generally within the extremes of the modern human range of morphological variation. For example, the Neanderthal glenoid cavities tend to be both narrower and more caudally orientated (as reflected by the low spinoglenoid angle) than is the norm in modern humans (Trinkaus, 1983a).

The Neanderthal scapulae do, however, show two features which suggest that the muscles that serve as lateral rotators of the humerus (infraspinatus and teres minor) were better developed in these hominids than is generally the case in modern humans.

The most important of these two features is the common occurrence on Neanderthal scapulae of a feature known as the **dorsal sulcus**, or **sulcus axillaris teretis** (Fig. 17.18) (Trinkaus, 1977). This is a depression on the dorsal (posterior) surface of the scapula adjacent to the axillary lateral border. It provides the area of insertion for teres minor. The majority of modern humans lack the dorsal sulcus and instead have a **ventral sulcus**, or **sulcus axillaris subscapularis**. This is a depression on the ventral (anterior) surface of the scapula adjacent to its axillary border. Its presence indicates that the inser-

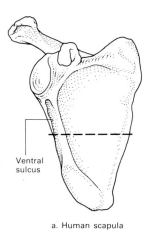

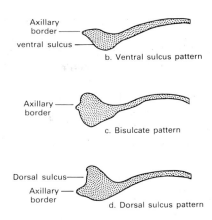

a. Human scapula

FIGURE 17·18 *(a) A modern human scapula. The dashed line indicates the position of the scapular cross-sections (b) showing the ventral sulcus pattern common in modern humans, (c) showing the bisulcate pattern and (d) showing the dorsal sulcus pattern common in the Neanderthals. After Trinkaus (1977).*

tion of teres minor is located on the ventral surface of the axillary border of the scapula in the majority of modern humans. A third, intermediate pattern is found in some Neanderthals, in some modern humans and in all known Palaeolithic anatomically modern humans. This pattern is the **bisulcate** pattern, or **facies axillaris bisulcata**. This pattern exhibits both the dorsal and the ventral sulci and indicates that **teres minor** inserted on both the ventral and the dorsal surface of the axillary border of the scapula.

In functional terms, the dorsal sulcus pattern that is common among Neanderthals, reflects a larger area of insertion for teres minor on the axillary border of the scapula than do either the bisulcate (intermediate area of the insertion) or the ventral pattern (more restricted area of insertion). This suggests that, on the whole, teres minor was well developed in the Neanderthals and that its function as a lateral rotator of the humerus was important in Neanderthal arm usage (Fig. 17.19).

The importance of lateral rotation of the humerus in Neanderthal arm usage is also suggested by the large infraspinous fossa in Neanderthal scapulae. This suggests that infraspinatus, another lateral rotator of the

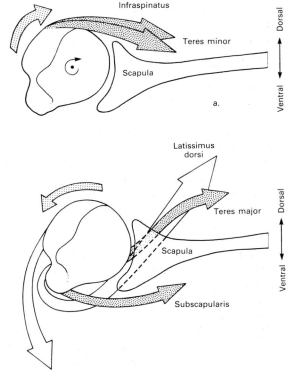

FIGURE 17·19 *Action of the shoulder muscles in (a) lateral rotation and (b) adduction and medial rotation. After Kapandji (1982).*

humerus, was also very well developed in the Neanderthals.

Trinkaus (1977) suggested that the Neanderthals had well-developed lateral rotators of the humerus to counteract the action of the equally well-developed adductors of the humerus (latissimus dorsi, pectoralis major and teres major). Because of the positions of their insertions on the humerus, these powerful muscles would medially rotate the Neanderthal humerus as they adducted the arm. The lateral rotators of the humerus (infraspinatus and teres minor) would also have to be well developed to counterbalance this medial rotation.

The elbow joint and forearm in humans and in the apes

The osteological differences between the human elbow joint and forearm and that of the great apes can be divided into two categories. The first category includes osteological features of the elbow joint itself that reflect its weight-supporting role in the large apes and the absence of this function in humans. The second category includes osteological features that reflect differences in both the strength and lever advantages of the brachial musculature.

Osteological differences in the elbow joint itself in humans and in the apes

In the past, people have thought that the human elbow joint is morphologically so similar to the ape elbow joint that it has limited value as a discriminator (Straus, 1948; Robinson, 1972). Indeed, the similarities between the human and ape elbow joints are particularly apparent in comparison with the morphology of the elbow joint in the Old World (Cercopithecine) monkeys (Fig. 17.20) (McHenry and Corruccini, 1975). For example, the hominoid distal humerus has a particularly wide trochlea while the narrow cercopithecine distal humerus has a medio-laterally narrow and anteroposteriorly deep trochlea. The hominoids have a lateral troch-

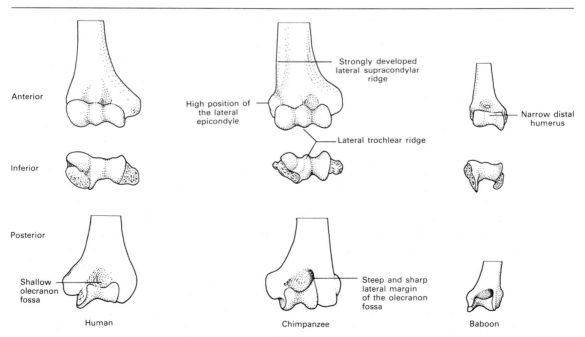

FIGURE 17·20 *Distal humeri of a human, a chimpanzee and a baboon (*Papio*) from the anterior, inferior and posterior aspects.*

lear ridge that separates the ulnar articulation (the trochlea) from the radial articulation (the capitulum) while the cercopithecines lack this ridge. The lateral trochlear ridge in the apes contributes to the stability of the articulation between the ulna and the humerus and frees the radius for pronation and supination in a variety of elbow postures. Associated with the wide range of pronation and supination is the round head of the hominoid radius. The elliptical head of the cercopithecine radius limits the range of forearm pronation and supination in these monkeys. In addition, the cercopithecine olecranon process is long and the trochlear notch has a more anterior orientation than that found in most hominoids. The long olecranon reflects the powerful extensors of the elbow in the cercopithecines and the orientation of the trochlear notch is associated with their inability to hyperextend the joint.

There are, however, more subtle differences between the human and ape elbow joints that involve particularly the morphology of the distal humerus and the proximal ulna (Patterson and Howells, 1967; Senut, 1981a, b). These differences reflect the weight supporting function of the joint in the apes and the absence of this function in humans.

THE LATERAL TROCHLEAR RIDGE
One of the most important of these differences is the morphology of the lateral trochlear ridge and the lateral margin of the olecranon fossa. This ridge is very strongly developed in the apes and gives the joint surface the appearance of two spools joined end to end. In distal (inferior) view the trochlear ridge runs diagonally across the joint in an anteromedial to posterolateral direction (Fig. 17.20). On the posterior surface of the ape humerus the trochlear ridge is contiguous with the steep and sharp lateral margin of the deep olecranon fossa.

The well-developed lateral trochlear ridge stabilizes the articulation between the ulna and the distal humerus throughout forelimb extension and flexion, while the steep lateral margin of the olecranon fossa continues to stabilize the joint in hyperextension and is well developed in the knuckle-walking chimpanzee and gorilla. Tuttle and Basmajian (1974a) suggested that hyperextension may be particularly important in knuckle-walking to stabilize the elbow joint against torque (or twisting) forces. In this context the steep lateral margin of the olecranon would prevent lateral dislocation of the ulna at the elbow during weight support.

There are a number of other features of the distal humerus and proximal ulna that are associated with the hyperextension of the ape joint and the ability to stabilize the joint in this posture.

DEPTH OF THE OLECRANON FOSSA
The olecranon fossa in the apes is considerably deeper in relation to its breadth than is the case for the human olecranon fossa (Knussmann, 1967). The ability to extend, or hyperextend, the elbow is largely dependent on the depth of the olecranon fossa because this fossa must accommodate the ulnar olecranon process as the elbow reaches its limit of extension. This limit will be reached more quickly with a shallow olecranon fossa (as in humans) than it will be with a deep fossa (as in gorilla).

It might seem that the length of the ulnar olecranon process would also effect the degree of possible elbow hyperextension; however this is not as important a factor in humans and apes as it is in monkeys and prosimians. The reason for this is that the ulnar olecranon process does not project as far proximally beyond the trochlear notch in humans and apes as it frequently does in the rest of the primates. As a result it does not contact the posteriodistal surface of the humerus as the elbow extends and thereby block further elbow extension. However, the orientation of the ulnar trochlear notch, together with the orientation of the humeral trochlea, is important to both the degree of possible hyperextension of the elbow and to its role in weight support.

ORIENTATION OF THE ULNAR TROCHLEAR NOTCH AND THE HUMERAL TROCHLEA

There is considerable variation in the orientation of the trochlear notch in humans, apes and monkeys (Fischer, 1906; Knussmann, 1967). In the apes the trochlear notch faces more proximally than it does in most monkeys where it has an anterior orientation. In humans the orientation is variable, in some individuals it faces more proximally and in others more anteriorly.

In primates, such as the African apes, that fully extend their forelimbs in locomotion, the ideal orientation for the trochlear notch would be fully proximal at a 90° angle to the long axis of the ulna (Fig. 17.21). It would then act as a cradle to support the humerus when the forelimb was fully extended during locomotion. However, this ideal orientation has one serious disadvantage; it seriously restricts the degree of elbow flexion (Fig. 17.21). The anterior border of the trochlear notch would abutt against the distal humerus at about 90° of flexion and the animal would have obvious problems in, say, getting food to its mouth. An angle of between about 20° and 30° to the long axis of the ulna as found in the African apes is an ideal compromise to the dual demands of elbow extension during locomotion and elbow flexion during manipulation or feeding (Fischer, 1906). In a fully extended forelimb posture there is still enough of the trochlear notch below the humerus to give it good support (Fig. 17.21). However, the trochlear notch also has a reduced anterior extent allowing full flexion of the elbow.

Monkeys do not fully extend their elbows during locomotion (Table 16.1) and therefore are not faced with the same biomechanical demands as the African apes. Their anteriorly oriented trochlear notches allow a full range of flexion although they limit the range of elbow extension (Fig. 17.21). The variable orientation of the trochlear notch in humans, ranging between the anterior monkey orientation and the proximoanterior ape orientation, is obviously unrelated to locomotion but may be related to differences in the habitual position of loading, for example, in tool-using or carrying postures (see below) (Trinkaus and Churchill, 1988).

CORONOID PROCESS ROBUSTICITY

Because the coronoid process supports the transmitted body weight during quadrupedal locomotion in the African apes, it is more robust than its human counterpart that lacks a weight-supporting function. Furthermore, the width of the trochlear notch (in relation

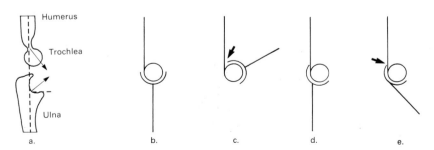

FIGURE 17·21 *Mechanical implications of the orientation of the trochlear notch. After Kapandji (1982). In both humans and apes the trochlea lies in front of the axis of the humerus (a). (b) The ideal proximal orientation of the trochlear notch in a weight-supporting upper limb. (c) The proximally oriented trochlear notch limits elbow flexion. (d) The anteroproximal orientation of the trochlear notch commonly found in apes. (e) The anterior orientation of the trochlear notch commonly found in monkeys limits elbow extension.*

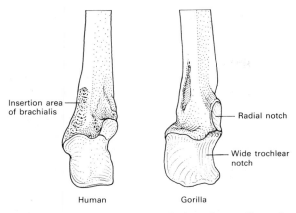

Insertion area of brachialis

Radial notch

Wide trochlear notch

Human   Gorilla

FIGURE 17·22 *The proximal ulna of a gorilla and a modern human drawn so that the lengths of their trochlear notches are the same.*

to its length) is frequently greater in the apes than it is in humans (Fig. 17.22) (Knussmann, 1967). This provides a greater joint surface area in these primates to transmit the increased weight through the elbow joint.

MORPHOLOGY OF THE CAPITULUM

Two features of the capitulum related to the greater degree of possible hyperextension in the ape elbow distinguish the majority of humans from the majority of apes. Firstly, in the modern humans the capitulum is

confined to the anterior surface of the humerus (Fig. 17.23), while in most apes the capitulum faces more distally and a proportion of the joint surface curves around and continues onto the posterior surface of the bone (Patterson and Howells, 1967). This gives the radius a greater arc of movement in the apes than in humans. Secondly, the actual shape of the articular surface on the capitulum distinguishes most humans from most apes. In humans the distolateral margin of the capitulum is generally a straight line that diverges anteriorly at an angle of approximately 45° from the transverse axis of the joint (Fig. 17.20). In apes this margin is generally more convex, curving out laterally before turning anteriorly (Patterson and Howells, 1967). This results in an increased area of articulation on the posteriodistal surface of the capitulum for the radius in hyperextension.

Osteological reflections of differences in the brachial musculature in humans and in the apes

These features reflect the differences in the leverage advantages of the brachial muscles in humans and apes as well as the greater strength or weakness of particular brachial muscles in these species.

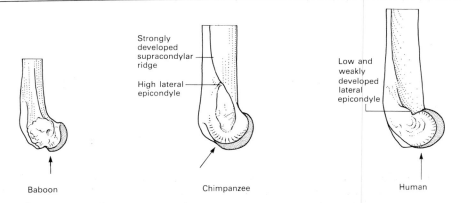

Strongly developed supracondylar ridge

High lateral epicondyle

Low and weakly developed lateral epicondyle

Baboon   Chimpanzee   Human

FIGURE 17·23 *The distal humerus from the lateral view in a baboon, chimpanzee and human. Note the difference in the extension of the articular surface of the capitulum onto the posterior surface of the bone (arrows).*

361

THE LENGTH OF THE OLECRANON PROCESS OF THE ULNA

When the elbow is extended it behaves as a first-order lever where the fulcrum (the joint) lies between the effort being applied to move the joint (triceps brachii) and the resistance (the hand). The olecranon represents the power arm of the lever system and, everything being equal, the length of the olecranon is proportional to the power exerted by the lever. Put another way, the greater the length of the olecranon in relation to the length of the forearm distal to the elbow joint the greater the power which can be exerted during extension of the elbow. In humans this ratio is greater than it is in any of the apes, indicating that the human forearm is adapted to slower and more powerful

movements in extension than are the forearms of particularly the orang-utan and the lesser apes (Fig. 17.24) (Knussmann, 1967). This adaptation in humans has been attributed to the power and control required by tool-using behaviour (Keith, 1926; Tuttle and Basmajian, 1974a; Aiello, 1981a).

INSERTION OF BICEPS BRACHII AND OF BRACHIALIS

In flexion of the elbow where the lower arm is the moveable lever (as in tool use) the arm behaves as a third-order lever. The power arm is the distance between the elbow joint (the middle of the ulnar trochlear notch or the head of the radius) and the insertion of brachialis (on the ulna) and of biceps brachii (on the radius). The lever arm is the physiological length of either bone: for the ulna it is the distance between the middle of the trochlear notch and the centre of the distal articular surface, the ulnar head, and for the radius it is the distance between the centres of the proximal and distal articular surfaces.

For the main flexor of the elbow, brachialis, there is little difference in the ratio between power and lever arms between humans and the large apes (Table 17.1) (Knussmann, 1967). However, all the large apes have marginally higher ratios and, thereby, stronger arms in flexion, than do humans. In

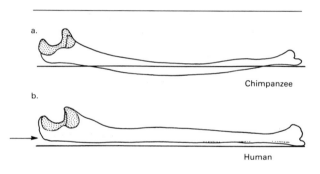

FIGURE 17·24 *A chimpanzee (a) and a human (b) ulna from the lateral view and a chimpanzee (c) and a human (d) radius from the anterior (ventral) view. Note the differences in the extent of the olecranon process (arrow) between the chimpanzee and human ulnae as well as the greater degree of curvature of the chimpanzee ulna and radius in comparison to the human bones.*

TABLE 17·1 The brachial flexors. Ratio between power and lever arms in humans and other primates. After Knussmann (1967).

	N	Brachialis		Biceps brachii	
		Mean	SD	Mean	SD
Homo	183	12.1	1.6	12.9	1.1
Gorilla	30	13.8	1.5	17.2	1.0
Pan	33	12.9	1.4	14.8	1.2
Pongo	14	13.3	1.7	11.5	1.0
Symphalangus	8	12.6	3.3	12.4	0.7
Hylobates	48	10.1	1.2	10.8	0.8
Colobinae	11	10.7	1.5	9.9	1.4
Cercopithecinae	145	9.5	1.2	11.0	1.2
Callithridae	19	13.3	6.5	12.8	1.1
Cebidae	49	11.8	1.4	10.8	1.5

this feature, humans are grouped with the apes in comparison with the monkeys. With the exception of the Callithricidae (marmosets and tamarins) humans and apes have more powerful forearms as measured by this index than do the remainder of the monkeys.

The situation is slightly different for biceps brachii which acts as both a flexor and a supinator of the forearm. African apes have long radial necks and as a result the ratio between the power and lever arms for biceps brachii (measured on the radius) in these apes exceeds the ratio in both humans and the orang-utan (Table 17.1). The degree by which the ratios of the African apes exceed the ratios of humans and the orang-utans surpasses any difference in the ratios between humans and apes in the previous brachialis comparison. One of the reasons for this difference may well be the role of the biceps brachii in supination (Chapter 16). However, the position of the radial tuberosity, which is a measure of the lever advantage of biceps brachii in supination, groups the orang-utan with the African apes.

THE ORIENTATION OF THE RADIAL TUBEROSITY

In the apes, and particularly in the orang-utan and chimpanzee, the radial tuberosity occupies a more medial position on the radius than it does in humans (Fig. 17.25). The more medial the radial tuberosity the greater is the mechanical advantage of biceps brachii as a supinator of the forearm and hand. The reason for this is that, as the forearm is pronated and the radius roles around the top of the ulna, the tendon of biceps brachii is wrapped around the upper part of the radial shaft (Chapter 16 and Fig. 16.23). This happens quite naturally as the radial tuberosity rotates (with the shaft) from a position facing medially toward the ulna to one facing posteriorly to one facing laterally away from the ulna. The more medial is the position of the radial tuberosity to begin with (when the forearm is supinated and the radius and ulna lie parallel to each other), the more lateral

will be its position when the forearm is in full pronation. This increases the power of biceps brachii by increasing the distance through which it pulls to supinate the arm.

Furthermore, the medial position of the radial tuberosity guarantees that the tuberosity lies medial to the axis of rotation of the forearm (the diagonal line through the proximal end of the radius and the distal end of the ulna) even when the forearm is nearing the final phases of supination. With the more anteriorly placed tuberosity common in modern humans, as the forearm reaches supination, the tuberosity moves laterally toward the axis of rotation of the forearm causing biceps brachii to lose its lever advantage in the final phases of supination. Therefore the medially placed radial tuberosity increases both the power and the effective range of action of biceps brachii in supination (Trinkaus and Churchill, 1988).

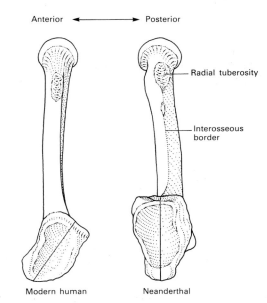

FIGURE 17·25 _In humans the radial tuberosity lies anterior to the mediolateral axis of the distal articular surface and to the interosseous border. In Neanderthals, as well as in the large apes, it occupies a more medial position in line with the mediolateral axis of the distal articular surface and the interosseous border. After Fischer (1906)._

The orientation of the radial tuberosity can be measured in relation to the transverse axis of the radial head (Ashton et al., 1976), in relation to the sagittal axis of the radial shaft (Knussmann, 1967) in relation to the axis of the distal articular surface (Fischer, 1906), or simply in relation to the position of the interosseous border (Napier and Davis, 1959; Trinkaus and Churchill, 1988). In humans with an anteriorly facing radial tuberosity, the posterior border of the tuberosity is in line with the interosseous border of the radius. In chimpanzees with a medially orientated tuberosity, the interosseous border bisects the tuberosity.

THE MORE PROXIMAL INSERTION OF THE BRACHIAL FLEXORS ON THE HUMERUS IN THE APES

In climbing and suspensory locomotor behaviours, where elbow flexion is an important component in progression, the forearm becomes the fixed and the upper arm the moveable lever (with the body weight). The forearm functions as a third-order lever system. The brachial flexors of the apes are inserted more proximally on the humerus than in humans and this enhances the lever advantage of these muscles (Miller, 1932).

FEATURES THAT ENHANCE THE POWER OF THE FOREARM AND HAND IN PRONATION

The most important feature enhancing the power of the forearm and the hand in pronation is the degree of bowing, or curvature, of both the radius and the ulna (Fig. 17.24). The more curved are these bones the greater the distance between them and the greater the distance between the origin and insertion of pronator quadratus. Furthermore, the more curved the radius is the greater the distance is the insertion of pronator teres from the axis of rotation of the forearm and, correspondingly, the greater is its lever advantage. The lever advantage of pronator teres is also increased by its more distal

insertion into the radius in the apes than in humans.

The radius is curved primarily in the transverse, or mediolateral, plane while the ulna is curved in the sagittal, or dorsoventral plane. The curvature of both these bones is greater in the African apes than it is in humans. In particular the transverse curvature of the radius (as measured by the greatest lateral distance of the centre of the radial shaft from the physiological long axis of the bone divided by the physiological length) clearly separates the straighter human radius from the more curved radii of the African apes and the orang-utan. The same relationship holds for the dorsal curvature of the ulna as measured by the greatest dorsal distance of the centre of the ulnar shaft from the physiological axis of the bone divided by the physiological length of the ulna (Knussmann, 1967). There are other ways of measuring the curvatures of the radius and the ulna (see for example, Martin and Saller, 1957), however all methods produce similar results demonstrating the bones of the forearm are less curved in humans than they are in the large apes.

THE SIZE AND POSITION OF THE LATERAL EPICONDYLE OF THE HUMERUS

The lateral epicondyle is well developed in the apes and projects well above the capitulum (Figs 17.20 and 17.23). In humans the capitulum is set on the same level, or only slightly below, the lateral epicondyle (Senut, 1981a, b). The larger lateral epicondyle in the apes has been suggested to increase the lever advantage of the long extensors of the wrist and fingers (Senut and Tardieu, 1985).

Other features of the distolateral part of the humerus have also been suggested to distinguish humans from apes and are, at least in part, related to differences in musculature. Firstly, the lateral supracondylar ridge in apes is, on average, more strongly developed than it is in humans (Figs 17.20 and 17.23) (Knussmann, 1967). This would suggest that

FIGURE 17·26 *The cross-sec-tion of the humerus at the level of the lateral epicondyle in a chimpanzee and a human. After Senut (1981b).*

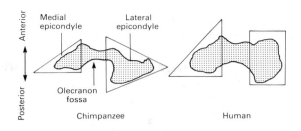

extensor carpi radialis longus and (to a lesser extent) brachioradialis are more strongly developed in the apes than they are in humans. Secondly, the pillars, or the bone on either side of the olecranon fossa, have different cross-sectional shapes. In the apes both the lateral and medial pillars are triangular in shape with the bases of the triangles adjacent to the olecranon fossa (Fig. 17.26). In humans, although the medial pillar is triangular in shape, the lateral pillar is quadrangular with a large anteroposterior diameter even near its lateral margin (Senut, 1981a, b). Thirdly, in humans the maximum anteroposterior breadth of the distal shaft is centrally located while in the apes, and particularly in the chimpanzee, it is medially located (Patterson and Howells, 1967). This suggests a lateral shift of the axis of maximum shaft strength in humans as compared with the chimpanzees. This is a direct result of the lateral flare of the supracondylar ridge in the chimpanzee.

The elbow joint and forearm in human evolution

The Plio-Pleistocene period (Australopithecines, Paranthropines and Early *Homo*)

The distal humerus is much better represented than is either the radius or the ulna in the fossil record for the earlier periods of human evolution (Day, 1978; McHenry and Temerin, 1979). Relatively well-preserved distal humeri are known from the southern African side of Kromdraai (Broom and Schepers, 1946; Le Gros Clark, 1947a,b; Straus, 1948) and from the East African sites of

Kanapoi (Patterson and Howells, 1967), Koobi Fora (Day, 1978), Hadar (Lovejoy et al., 1982b) and Melka Kunture (Chavaillon et al., 1977). There is only one published description of a well-preserved ulna which comes from the Omo valley (Howell and Wood, 1974) and no complete radii.

THE DISTAL HUMERUS IN THE PLIO-PLEISTOCENE PERIOD

All the known Plio-Pleistocene hominid distal humeri lack the steep lateral margin of the olecranon fossa that is so characteristic of the knuckle-walking chimpanzees and gorillas. However, they do show a considerable amount of morphological variation that is well illustrated by the comparison of the two best known specimens, TM 1517 (Kromdraai, southern Africa) and KNM-KP 271 (Kanapoi, Kenya) (Fig. 17.27). Patterson and Howells (1967) showed that the morphology of KNM-KP 271 is essentially similar to that of modern humans with the one exception of the chimpanzee-like medial placement of the greatest anteroposterior diameter of the distal shaft. In contrast, TM 1517 not only has more features that are similar to the distal humerus of modern chimpanzees, but also has features that are unique among living apes. Among the features that TM 1517 shares with the chimpanzees is the medially placed greatest anteroposterior diameter of the distal shaft (also shared with Kanapoi) as well as two features of the capitulum which are associated with hyperextension of the elbow. Firstly, the distal margin of the capitulum extends onto the posterior surface of the humerus as it does in the apes; and secondly, the distal margin of the capitulum

FIGURE 17·27 *The anterior (ventral), distal and posterior (dorsal) views of the TM 1517 distal humerus and the KNM-KP 271 distal humerus. The KNM-KP 271 (left) humerus is reversed for ease of comparison with the TM 1517 (right) humerus.*

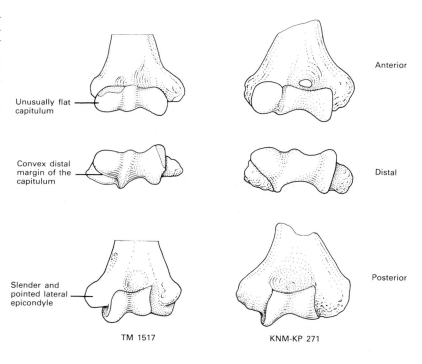

Unusually flat capitulum

Convex distal margin of the capitulum

Slender and pointed lateral epicondyle

Anterior

Distal

Posterior

TM 1517   KNM-KP 271

is convex in shape, turning outward laterally before curving anteriorly. TM 1517 is unique in relation to both apes and humans in having an unusually flat capitulum and a slender and pointed lateral epicondyle. Although there is some degree of overlap in the morphology of the distal humeri in humans and chimpanzees, Patterson and Howells (1967) suggested that the morphological difference between KNM-KP 271 and TM 1517 (*Paranthropus robustus*) is sufficient to place KNM-KP 271 in the separate taxon, *Australopithecus* sp. Other authors suggest that the Kanapoi humerus is so human-like that it should be classified in the genus *Homo* (Day, 1978).

Quantitative as well as qualitative assessments of these specimens, along with the analysis of more recently discovered specimens, have confirmed the difference in morphology between KNM-KP 271 and TM 1517 (McHenry and Corruccini, 1975). In addition, the results of these analyses suggest that there may be as many as three separate distal humerus morphologies represented in

the Plio-Pleistocene hominid collection (McHenry and Corruccini, 1975; Senut, 1981a,b; Senut and Tardieu, 1985) (Fig. 17.28). Largely on the basis of the degree of development and on the placement of the lateral epicondyle, Senut (1981a,b) has divided the known specimens from East Africa into two genera (*Homo* and *Australopithecus* including *Paranthropus*) and into three species groups (early *Homo*, *Australopithecus afarensis* and *Australopithecus boisei* which in our classification is *Paranthropus boisei*).

(1) Early *Homo* This group includes KNM-KP 271 (Kanapoi), Gombore IB 7594 (Melka Kunture) and AL 333W-29 (Hadar) and is characterized by: (1) a weak projection of the lateral epicondyle; (2) a low position of the lateral epicondyle relative to the capitulum; and (3) a moderate development of the lateral trochlear crest (Fig. 17.28).

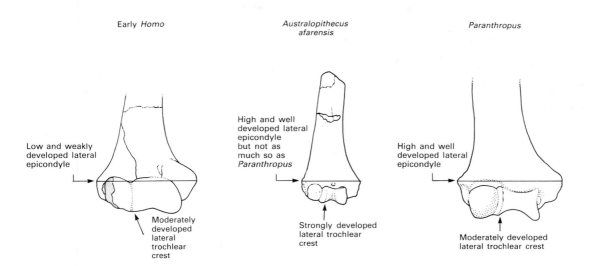

FIGURE 17·28 *Three different morphologies of the distal humerus found among Plio-Pleistocene hominids. The early* Homo *morphology is represented by Gombore IB 7594 (reversed for ease of comparison), the* Australopithecus afarensis *mor-* *phology by AL 288-1m, and the* Paranthropus *morphology by KNM-ER 739. Arrows highlight areas of difference between the humeri. See text for further explanation. After Senut (1981b).*

(2) *Australopithecus and Paranthropus* This group includes all remaining known specimens and is divided into two subgroups, *Australopithecus afarensis* and *Paranthropus* which both share: (1) a well-developed lateral epicondyle that is set high relative to the capitulum; and (2) an anteroposteriorly flattened shaft and lateral epicondyle (Fig. 17.28).

(3) *Australopithecus afarensis* This group includes AL 137-48A, AL 288-1m, AL 288-1s and AL 322-1 and is characterized by the two features definitive of the genus *Australopithecus* as well as by: (1) a strongly developed lateral trochlear crest that separates the chimpanzee-like trochlea into two parts; (2) a lateral epicondyle that is not as well developed as in the *Paranthropus* fossils; and (3) a lateral margin of the distal portion of the shaft that is rectangular in shape (Fig. 17.28).

(4) *Paranthropus* This group includes KNM-ER 739, KNM-ER 740 and KNM-ER 1504 and differs from the *A. afarensis* group in having a human-like poorly developed lateral trochlear crest and a shallow, centrally-placed olecranon fossa (Fig. 17.28).

The KNM-ER 739 specimen (which is the most completely preserved Plio-Pleistocene humerus) also has other features that distinguish it from modern humans (Fig. 17.29). The most obvious feature is its great robusticity. When its circumference is divided by its estimated length, the resulting robusticity index is larger than the means for not only humans but also the large apes and is almost outside the observed range of variation seen in the gorillas (that have the most robust humeri among the apes) (McHenry, 1973). The bone also has very powerful muscle markings (Day, 1978). In particular the markings for both deltoid and brachioradialis are massive, suggesting powerful shoulder abduction and elbow flexion. Furthermore, pronounced medial and lateral epicondyles suggest that the long muscles of the hand out that there is a considerable variation in size among the specimens from Koobi Fora that are included in the *Paranthropus* group and that this variation in size most surely represents sexual dimorphism within this one taxon.

The degree of morphological variation found among the distal humeri has led a number of authors to speculate that some of the Plio-Pleistocene hominids may not have been habitual bipeds (Leakey, 1971; Robinson, 1972; Kay, 1973; Day, 1978; McHenry and Temerin, 1979; Senut, 1981a,b; Senut and Tardieu, 1985). Climbing or suspension is the alternative locomotor pattern favoured at least for the gracile australopithecines (Senut and Tardieu, 1985), while the unique form of the *Paranthropus* humerus may also suggest some use of the forelimb in locomotion.

Perhaps the most surprising feature of the morphological variation in the Plio-Pleistocene hominid distal humeri is the fact that one of the most human-like specimens (KNM-KP 271) also dates from the earliest time period (ca. 4 million years ago; Patterson and Howells, 1967). In addition, if the early *Homo* group does represent a discrete functional or taxonomic unit and if the interpretation of Senut and Tardieu (1985) is correct, then there are two separate locomotor patterns represented in the Hadar *A. afarensis* collection: a more arboreal pattern represented by the majority of the specimens, and a more human-like pattern represented by AL 333-W29.

FIGURE 17·29   *KNM-ER 739 from the anterior (a) and the posterior (b) views. Courtesy of M.H. Day.*

(both the flexors and extensors) were also very powerful. Although these features suggest a functional similarity with the large apes, this humerus is unique in that it combines these features with a more human-like shallow olecranon fossa that lacks the steep lateral margin so characteristic in the weight-supporting ape elbow. In this context McHenry (1973) has suggested that the morphological uniqueness of this specimen may also imply a functional uniqueness.

The morphological variation found between all the Plio-Pleistocene distal humeri is considered by Senut and Tardieu (1985) to be too great to be explained by intraspecies variation (sexual dimorphism). They point

THE ULNA AND THE RADIUS IN THE PLIO-PLEISTOCENE PERIOD

The only well-preserved ulna from the Plio-Pleistocene period reported in the literature was discovered in 1971 in Member E of the Shungura Formation in the lower Omo basin of southern Ethiopia (Fig. 17.30) (Howell and Wood, 1974). This specimen (Omo Loc 40-19) shows both metrical and morphological differences from modern human ulnae and most probably should be referred to

FIGURE 17·30 *The Omo ulna (Loc 40–19) from the lateral view. Courtesy of M.H. Day.*

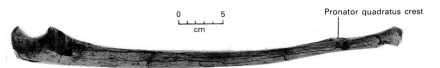

0   5
cm

Pronator quadratus crest

Paranthropus boisei (Howell and Wood, 1974; McHenry et al., 1976; Day, 1978; Feldesman, 1979; McHenry and Temerin, 1979). Together with the KNM-ER 739 humerus, this specimen has been a major piece of evidence for the suggestion that the forelimb morphology of the paranthropines differed significantly from that of *Homo* (Howell and Wood, 1974; Day, 1978; Feldesman, 1979).

The Omo 40-19 ulna differs from modern human ulnae in the length and morphology of the shaft, in the pattern of muscle attachments and in the morphology of the ulnar head. In particular, the length of the shaft coupled with its curvature and the morphology of its cross-section have been suggested to be reminiscent of knuckle-walking chimpanzees and gorillas (Howell and Wood, 1974). In absolute length the shaft is longer than would be expected for modern humans and shows the high degree of anteroposterior curvature that is characteristic of the chimpanzees and gorillas. The cross-section of the shaft lacks the prominent interosseous border characteristic of human ulnae. Howell and Wood (1974) suggested that this, together with the anteroposterior flattening of the shaft, reflects a different pattern of forearm stress than that found among modern humans. In addition, the head of the ulna is less convex than is the case in modern humans; but the functional significance of this is not clear (Howell and Wood, 1974).

The Omo 40-19 ulna differs from chimpanzee and gorilla ulnae in the morphology of the trochlear notch. In these apes the trochlear notch is oriented proximoanterially and the coronoid process is heavily buttressed. Both these features are adaptations to weight support during terrestrial (knuckle-walking)

locomotion (see above). The trochlear notch in the fossil ulna is anteriorly oriented and the coronoid process is not heavily buttressed. As a result, it is unlikely that the fossil forelimb was used in a fully extended posture during terrestrial locomotion as it is in the African apes (Day, 1978).

In addition, there are some differences in the pattern of muscle attachments. The hollow for the play of the radial tuberosity is only weakly expressed suggesting that biceps brachii may not have been as well developed as it is in the chimpanzee and gorilla. In addition, the supinator crest is also weakly expressed and the ulnar tuberosity is both flat and weakly buttressed. These two features reflect differences in development of supinator and brachialis in the fossil in comparison with modern humans (Howell and Wood, 1974). While there is indication that these muscles are relatively weakly developed in the fossil, the distal third of the shaft is marked by a prominent curved crest of bone for the attachment of a very well-developed pronator quadratus. Multivariate anlayses place Omo 40-19 closer to the chimpanzee and modern humans than to the orang-utan (McHenry et al., 1976; Feldesman, 1979). It differs from the orang-utan in being relatively shorter and more robust, in having a tuberosity that is positioned more proximally and in having a smaller distal end.

Other Plio-Pleistocene ulnae are either much more fragmentary than the Omo 40-19 ulna or have not been fully described (Day, 1978; McHenry and Temerin, 1979; Susman, 1988b). However, some ulnae, such as KNM-ER 1500b, which have been classified as *Australopithecus* (*Paranthropus* as used here) (Leakey, 1972; Day, 1978) show resemblances to Omo 40-19, while others

369

such as KNM-ER 803c (classified as *Homo*; Day and Leakey, 1974) are more similar to modern human ulna in having a prominent interosseous border, a supinator crest and a well-marked hollow for the play of the tuberosity of the radius (Day, 1978; see also Susman, 1988b).

There are no complete hominid radii known from the Plio-Pleistocene period. The fragments that are known do not show the differences in morphology between *Homo*, *Australopithecus* and *Paranthropus* that are evident in the ulnae and the distal humeri. Most of these fragments are generally similar to modern human radii (Day, 1978; McHenry and Temerin, 1979) in having rounded heads and necks, a strong interosseous border, and a marked pronator teres insertion (McHenry and Temerin, 1979). The possible exception is the SKX 3602, SKX 3699, and SKX 12814 composite *Paranthropus robustus* radius from Swartkrans (Susman, 1988b). If this composite radius actually belonged to a single individual as Susman suggested, it shows a unique combination of features. The distal part of the bone is human-like in form, lacking any indication of a knuckle-walking (weight supporting) adaptation of the distal (radiocarpal) joint surface (see Chapter 18) while the proximal part is very chimpanzee-like with a small head and long neck (Susman, 1988b).

The Neanderthal elbow joint and forearm

The morphology of the Neanderthal elbow joint, together with the morphology of the radial and ulnar shafts, suggests that the Neanderthals had more powerful forearms than does the average modern human (Trinkaus and Churchill, 1988). In particular, the Neanderthals had more power in pronation and supination through the full range of forearm rotation. Furthermore, the morphology of the proximal ulna suggests that they may have used their forearms with a

more flexed elbow joint than is the norm for modern humans.

The major evidence for these conclusions is in the morphology of the ulna and the radius. The Neanderthal distal humeri are not particularly different from their modern counterparts. In relation to the length of the humerus, the sizes of both the distal, or cubital, joint surface (mediolateral distance across the capitulum and trochlea) and the epicondyles (biepicondylar breadth) fall in the upper ranges of modern human variation (Trinkaus, 1983a). Furthermore, many of the Neanderthal distal humeri have well-developed lateral supracondylar crests suggesting that extensor carpi radialis longus was a powerful muscle (Trinkaus, 1983a). These features point to the obvious conclusion that Neanderthals had a more robust postcranial skeleton than is the norm for modern humans, a conclusion that is consistent with inferences drawn from virtually every area of the Neanderthal postcranium.

The major evidence for greater power in pronation and supination in the Neanderthal forearm comes from the medial position of the radial tuberosity (Fig. 17.25) (Trinkaus and Churchill, 1988). The more medial is the radial tuberosity the greater is the lever advantage of biceps brachii in supination (see above and Chapter 16). In this trait the Neanderthal radii are similar not only to the apes but also to the relatively rare hominid radii known from early periods. The hominid radii which show a medial tuberosity include *Australopithecus afarensis* (AL 288-1, AL 333-98), *Australopithecus* sp. (KNM ER 1500e) and *Homo habilis* (OH 62) (Trinkaus and Churchill, 1988). Furthermore, in many Neanderthals the radial neck is long, as it is in the African apes and *Paranthropus robustus*. This would increase the lever advantage of biceps brachii in elbow flexion.

Other features shared in common with living apes also suggest that Neanderthal forearms were powerful in pronation and supination. Firstly, the Neanderthal supinator crest (extending distally from the distola-

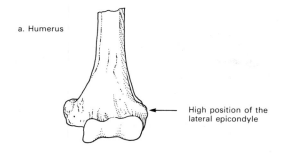

a. Humerus

High position of the
lateral epicondyle

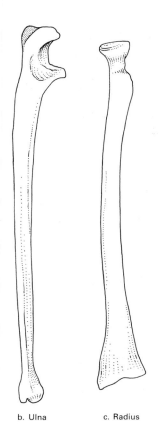

b. Ulna   c. Radius

FIGURE 17·31   *The Neanderthal (Shanidar I) distal humerus from the anterior (ventral) view (a), the ulna from the lateral view (b) and the radius from the anterior view (c). Note the high position of the lateral epicondyle (arrow) on the humerus and the anterior orientation of the trochlear notch on the ulna. After Trinkaus (1983a).*

teral margin of the radial tuberosity) is strongly developed indicating an equally well-developed supinator muscle. Secondly, Neanderthal radial shafts show a greater degree of lateral bowing than is normally found in modern humans (Fig. 17.31). This carries the insertion of pronator teres more lateral to the axis of rotation of the forearm and thereby increases its lever advantage relative to that seen in modern humans. Thirdly, Neanderthal ulnae have pronounced pronator quadratus crests that suggest that this muscle was also very well developed.

In addition to these features indicating that the Neanderthals had a more powerful arm in pronation and supination than is normally the case in modern humans, the Neanderthals also had a trochlear notch that is oriented more anteriorly than is the case in either modern humans or the living great apes (Fig. 17.31). Trinkaus and Churchill (1988) argued that the more anteriorly directed Neanderthal trochlear notch does not necessarily imply any difference in range of movement of the joint, but does imply a difference in the habitual position of loading. In particular, they suggested that the Neanderthal morphology would result in greater resistance to joint reaction forces when the elbow was partially flexed because in this posture the joint surface area perpendicular to the joint reaction force would be maximized.

The orientation of the trochlear notch, together with features of the ulnar and radial shafts, suggests then that the Neanderthal elbow and forearm were very powerful in pronation and supination and that the arm was used in a more flexed position than is common in modern humans (Trinkaus and Churchill, 1988). In addition to these features, the Neanderthal ulnar shaft was also relatively gracile and tended to be more rounded at midshaft than is the case in modern humans. This last feature is related to the poorly developed interosseous crest in the Neanderthals that is clearly present but not projecting (Trinkaus, 1983a).

CHAPTER EIGHTEEN

THE HOMINOID HAND

The human hand is a highly versatile organ the manipulative abilities of which fall into two general classes: *prehensile movement* and *non-prehensile movement* (Napier, 1980). Prehensile movements are movements where objects are grasped either between the digits or between the digits and the palm. Non-prehensile movements include pushing or lifting movements of the whole hand, or pushing, lifting, tapping and punching movements of the fingers.

The prehensile movements of the human hand have been divided into four different grips: the **hook grip**, the **scissor grip**, the **precision grip**, and the **power grip**, (Napier 1980). The first two of these grips are

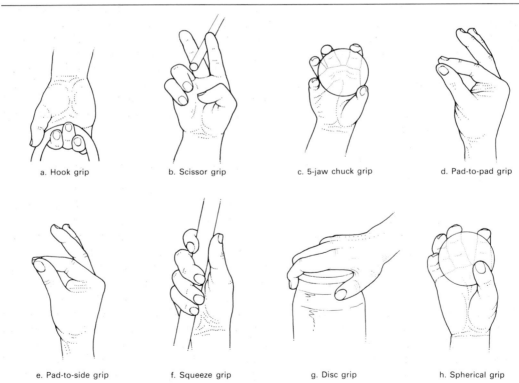

a. Hook grip   b. Scissor grip   c. 5-jaw chuck grip   d. Pad-to-pad grip

e. Pad-to-side grip   f. Squeeze grip   g. Disc grip   h. Spherical grip

FIGURE 18·1 *Prehensile capabilities of the human hand.*

relatively simple. The **hook grip** is used, for example, when carrying suitcases or briefcases (Fig. 18.1a). The thumb is not involved and the fingers are simply flexed to form a hook. The **scissor grip** is used to hold objects between the sides of the terminal phalanges of the index and the middle finger as in holding a cigarette. This grip is produced by the adduction of the index finger (movement of the index finger towards the middle finger) (Fig. 18.1b).

The remaining two grips require more complex movements of the hand. The **precision grip** is executed between the terminal digit pads (palmar surfaces) of the thumb and the pads of one, or more, of the remaining fingers (Fig. 18.1c–e). When a large object, such as a light bulb or jar lid, is held in this fashion all five fingers are used (five-jaw chuck). When smaller objects are held, either the thumb and the index and middle fingers are used (three-jaw chuck) or the thumb and index finger are used (pad-to-pad grip). A variant of the precision grip is the pinch grip, or pad-to-side grip, where the terminal digit pad of the thumb is opposed to the radial side of the index finger. This last grip is frequently used when inserting a key in a lock.

The **power grip** is executed between the fingers and the palm of the hand with the thumb acting as a buttress (Fig. 18.1f–h). There are three variants of the power grip. The **squeeze grip** is used to grasp cylindrical objects such as the handles of hammers or clubs. The handle is held diagonally across the palm. The fingers on the ulnar side of the hand (the little finger and the ring finger) are flexed around the handle to a greater extent than are the fingers on the radial side of the hand. The handle is braced against both the hypothenar area of the palm at the base of the little finger and the thumb, which is extended and rests along the axis of the handle. The second power grip is the **disc grip** which is used, for example, when tightening jar lids. In this grip, pressure is applied against the lid of the jar by the palm

of the hand while the slightly flexed thumb and the remaining fingers grasp the circumference of the lid. The third power grip is the **spherical grip** where large spherical objects, such as baseballs or cricket balls, are held against the palm by the flexed thumb and fingers.

The precision grips and the power grips require complex compound movements of the hand. These movements include not only the flexion of the fingers and the ability to spread the fingers (abduction) or move them together (adduction), but also two other movement capabilities. The first of these is **opposition**. This is defined by Napier (1980) as the ability to place the pulp surface of the thumb squarely in contact with, or diametrically opposite to, the terminal pads of one or all of the remaining digits (Fig. 18.1c). The second movement is the deepening or **cupping** of the palm of the hand. More specifically, this means that the hypothenar area (ulnar side of the palm) and the thenar area (radial side of the palm at the base of the thumb) are brought together. This is an essential movement in the squeeze grip, in the spherical grip and in the five-jaw chuck grip (where the tip of the little finger must be placed in fine adjustment to the size and shape of large objects manipulated by the fingertips).

Morphology of the human hand

The long human thumb relative to the length of the remaining digits (Fig. 18.2) is essential for both the precision and the power grips. Without it the tip (or pulp) of the thumb could not be opposed to the tips of the remaining digits. The various human grips are also dependent on the morphology of the joint surfaces between one bone and another. The shapes of these joint surfaces permit the simple movements of flexion, extension, abduction and adduction as well as the complex movements involved both in the cupping (or deepening of the palm) and in opposition of the thumb.

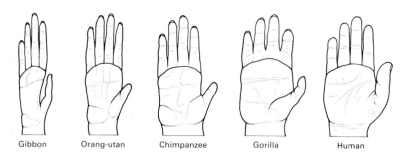

FIGURE 18·2   *Human and ape hands. After Straus (1942).*

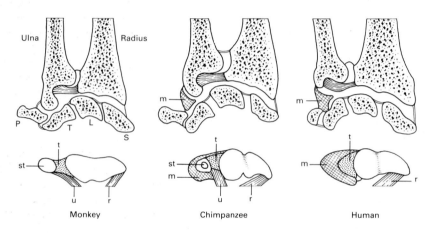

FIGURE 18·3   *The wrist joint in a monkey (Cercopithecus nictitans), a chimpanzee and a human. P = pisiform; T = triquetral; L = lunate; S = scaphoid; m = meniscus; st = styloid process of the ulna; t = triangular articular disc; u = palmar ulnocarpal ligament; r = palmar radiocarpal ligament. After Lewis (1972a).*

The wrist joint

The wrist joint in both humans and the large apes is a **radiocarpal** joint (Fig. 18.3). This means that the ulna does not articulate with any of the wrist bones, but is separated from the proximal carpal row by a disk-like pad of fibrocartilage called a **meniscus** ((L) = little crescent or half-moon). In other primates (with the exception of the gibbons) the styloid process of the ulna extends distally to articulate directly with the **triquetral** and the **pisiform**. The separation of the ulna from its articulation with the triquetral and pisiform in humans and apes permits a greater range of adduction of the hand (movement toward the ulnar side) and also may permit a greater range of supination at the wrist (Lewis, 1974).

In both humans and apes the radial articular surface is concave in shape and is extended medially by a triangular articular cartilage, the point of which is attached to the styloid process of the ulna (Fig. 18.3). The carpal articular surface, formed by the three closely connected carpal bones, is convex in shape. Adduction (deviation of the hand toward the ulnar side) and extension (bending the hand back on the wrist) at this joint are both

374

highly important to the distinctive human power grips and to the non-prehensile tapping or punching movements of the fingers. In the squeeze power grip used to grasp the cylindrical handles of tools, the hand is adducted or ulnar deviated. This brings the thumb, and the handle that it is buttressing, in line with the long axis of the arm and thereby increases the lever advantage in the use of that tool. In the power grips (as in individual flexion of the fingers, e.g. in tapping typewriter keys), the optimum position of the wrist is 30–40° degrees of extension (hyperextension or dorsiflexion). In this hyperextended position, it has been reported that the hand can exert four times the force that is possible when the wrist is in full flexion (Napier, 1980).

The midcarpal joint

In addition to its role in flexion and extension of the hand, the **conjunct rotation** characteristic of this joint is important in human manipulation. When the joint is extended, the triquetral (on the ulnar side of the proximal carpal row) is screwed against the lunate (the medial bone in the proximal carpal row) which is in turn screwed against the scaphoid (on the radial side of the proximal carpal row) (Fig. 18.4). This chain reaction is caused by the spiral form of the articulation between the triquetral and the hamate (on the ulnar side of the distal carpal row). This conjunct rotation not only stabilizes the otherwise relatively mobile carpal skeleton, but also is important to the human precision and power grips (Lewis, 1977). As the joint is extended and the carpal bones are screwed together, the scaphoid (on the radial side of the proximal carpal row) carries the trapezium (on the radial side of the distal carpal row) in a palmar direction. This brings the thumb into a roughly right-angled position in relation to the palm of the hand and to the four fingers, where it can be opposed in the attitude required by the precision and power grips. This may be one reason why the hand is commonly hyperextended when it is engaged in either of these grips.

The articulation between the capitate (the central bone in the distal carpal row) and the trapezoid (its neighbour on its radial side) is another feature of the distal carpal row that is important in the various human power grips. In the human hand these two carpal bones usually articulate with each other at the palmar or anterior half of their facing sides (Fig. 18.5). In non-human primates, the articulation between these bones is usually on the dorsal, or posterior, side of their facing surfaces. Lewis (1977) has suggested that human capitate/trapezoid articulation may be a direct adaptation to the compressive forces placed on the wrist by the thumb as

FIGURE 18·4 _Conjunct rotation of the midcarpal joint illustrated by the carpal bones of a chimpanzee in flexion (a) and extension (b). Bones are drawn from the dorsal aspect. See text for discussion. After Lewis (1977)._

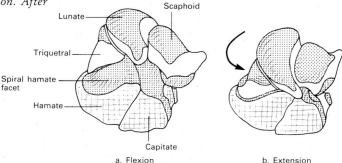

Lunate

Scaphoid

Triquetral

Spiral hamate facet

Hamate

Capitate

a. Flexion
b. Extension

FIGURE 18·5 *The distal carpal row of a chimpanzee and a human from the distal aspect. Arrows indicate the major points of difference between the human and ape distal carpal bones. (*) = An interosseous ligament which is present in the chimpanzee and absent in humans. After Lewis (1977).*

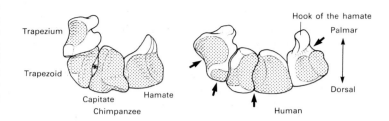

it is used as a buttress in the powergrips. These considerable compressive forces would be passed to the trapezium (at the base of the thumb) across the anterior part of the trapezoid to the capitate via the expanded anterior articulation between the two bones.

The bones and joints of the palm

The common length order of the **metacarpals** (M), the bones of the palm, is MII>MIII >MIV>MV>MI in humans (Martin and Saller, 1959; Susman, 1979). MII and MIII resemble each other and differ from MIV and MV in being both longer and more robust. This reflects the importance of these two metacarpals and of the radial side of the hand in the human power and precision grips (Figs 16.3 and 16.16).

The **carpometacarpal joints** (CMC) are highly important for human precision and power grips. The first carpometacarpal (CMC1) joint is the most mobile of the carpometacarpal joints and plays a major role in the opposition of the thumb to the remaining fingers. This joint, together with the other carpometacarpal joints, and particularly the second (index) and the fifth (little), produces the cupped human hand.

THE FIRST CARPOMETACARPAL JOINT (CMC1)

The joint between the first (thumb) metacarpal and the trapezium is one of the most important joints in the human hand. The great majority of thumb movements take place at this joint and not at the metacarpo-

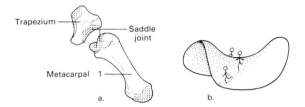

FIGURE 18·6 *(a) The human first carpometacarpal joint from the dorsal view. (b) A schematic illustration of the direction of movement determined by the shape of a saddle joint.*

phalangeal joint or the interphalangeal joint. Movements of these last two thumb joints are overwhelmingly confined to flexion, while movement at the CMC1 concerns rotation and opposition. The thumb CMC is shaped like a saddle and is usually described as a **saddle** or **sellar** joint (Fig. 18.6). The opposing joint surfaces of the trapezium and MI are indeed shaped like saddles and the axes of the depressed central troughs on these surfaces cause the metacarpal to rotate towards the palm of the hand when the joint is flexed. This, in combination with the position of the trapezium at an angle to its neighbouring carpal bones, automatically brings the palmar surface of the tip of the thumb into opposition with the palmar surfaces of the remaining fingers. This movement also brings the thenar eminence (the muscle mass lying over the first metacarpal) towards the centre of the palm, forming the radial edge of the cupped palm.

The joint surfaces on the human trapezium

(Fig. 18.5) and MI are both broader in a dorsopalmar direction and flatter than they are in the apes. These expanded surfaces help to resist the increased compressive forces that pass from the first metacarpal to the trapezium by the buttressing thumb in the power grip. The compressive forces in the human CMC joint average as much as 12.0 kg during simple pinching and up to 120 kg during strong grasping (Susman, 1988a). Together with the anterior (palmar) position of the joint between the trapezoid and the capitate (see above), these expanded surfaces offer a clear indication of the ability to execute human power grips.

THE SECOND CARPOMETACARPAL JOINT (CMC2)

The second carpometacarpal joint is the joint between the second metacarpal (MII) and (principally) the trapezoid (Fig. 18.7). This joint has much less mobility than the first carpometacarpal joint; however, it is capable of a small degree of flexion. Furthermore, the ulnar border of the MII also can be rotated away from the palm. This movement occurs when the fingers are spread (as they are in the five-jaw chuck precision grip) or clinched (as they are in the power grips) (Lewis, 1977) or during the pinch precision grip (pad-to-side) (Marzke, 1983). The flexion and rotation of the MII contributes to the cupped palm by helping to bring the radial side of the hand towards the centre of the palm.

The rotation of MII is made possible by a continuous convex articular facet on the ulnar side of its proximal articular surface (the side next to the third metacarpal) (Fig. 18.7b). This long facet articulates in a groove formed by the third metacarpal and the capitate. The shape of this groove causes the second metacarpal to rotate slightly around its long axis (sometimes referred to as pronation) during flexion. The rotation of the MII is also aided by the small joint between this metacarpal and the trapezium (on the thumb side of MII) (Fig. 18.5). This facet on the trapezium is angled in such a way that it causes the MII to rotate as it slides over it during flexion of the metacarpal.

THE THIRD CARPOMETACARPAL JOINT (CMC3)

This is the joint between the third metacarpal (MIII) and the capitate and is the least mobile of the carpometacarpal joints (Fig. 18.8). In the human hand it is capable of only a very small degree of flexion (Kapandji, 1982). The most notable feature of the otherwise relatively flat proximal articular surface of the MIII is a styloid process that protrudes well proximal to the plane of the joint surface at its dorso-radial edge. The opposing

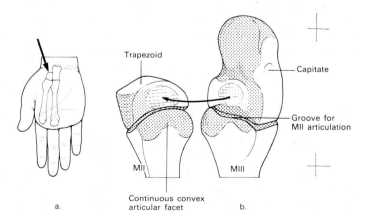

FIGURE 18·7 _The human second carpometacarpal joint. (a) Its location in the hand. (b) The disarticulated capitate (with M3) and trapezoid (with M2). The hatched area and arrow represent the severed massive capitate-trapezoid ligament. After Lewis (1977)._

Trapezoid

Capitate

Groove for MII articulation

MII

MIII

Continuous convex articular facet

a.

b.

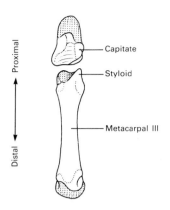

FIGURE 18·8 *The human third carpometacarpal joint from the dorsal view.*

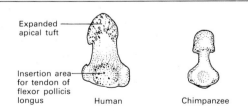

FIGURE 18·9 *(a) Cupping of the human hand. (b) The distal carpal row from the distal aspect showing the direction of flexion of the fourth and fifth carpometacarpal joints. Movement at these joints produces the cupping of the human hand. After Kapandji (1982).*

articular surface of the capitate is lipped to receive this styloid process. The styloid process stabilizes the joint which would otherwise rotate in response to the stresses on the MIII in the spherical and squeeze power grips (Marzke, 1983; Marzke and Marzke, 1987).

THE FOURTH AND FIFTH CARPOMETACARPAL JOINTS (CMC4 AND CMC5)

The fourth carpometacarpal joint is the joint between the fourth metacarpal (MIV) and the hamate and the fifth carpometacarpal joint is the joint between the fifth metacarpal (MV) and the hamate (Fig. 18.9). Both these joints are mobile and are important joints in relation to the cupping of the human palm.

CMC5 is a saddle joint; it permits about 25° of flexion. Because of the saddle shape of the joint, the ulnar border of MV rotates towards the centre of the palm during flexion (Dubosset, 1981; Kapandji, 1982; Marzke, 1983). The joint surface on the hamate for MIV is generally concave (Lewis, 1977) and permits about 10° of flexion. The flexion at both these joints, together with the slight flexion of CMC3 and the rotation during flexion of MV, brings the ulnar border of the palm (hypothenar area) towards the midline of the hand. This, together with the

position of the MI in opposition and the rotation (pronation) of MII in flexion, produces the distinctive cupping movement of the human palm.

The bones and joints of the digits

Human phalanges are both shorter and less curved than are ape-hand phalanges and lack the strongly developed flexor ridges that are common in these primates (see below). Furthermore, human distal phalanges are unique in comparison to those of other primates in having strongly developed apical tufts that are spade like in appearance (Fig. 18.10). In humans, the radio-ulnar diameter of the tip of the third distal phalanx is 69% of the radio-ulnar diameter of its base, while in chimpanzees this diameter is only 62% of

FIGURE 18·10 *Palmar view of the distal thumb phalanx of a human and a chimpanzee. Note the area of insertion for flexor pollicis longus on the human bone.*

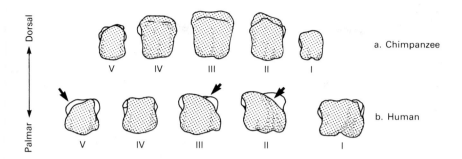

V   IV   III   II   I   a. Chimpanzee

V   IV   III   II   I   b. Human

FIGURE 18·11 *The morphology of the metacarpal heads in a chimpanzee (a) and a human (b). Arrows* *indicate the asymmetry of the human second, third and fifth metacarpal heads. After Lewis (1977).*

the radio-ulnar diameter (Susman, 1979). A further feature of the human distal thumb phalanx is a marked area of insertion for flexor pollicis longus (see below).

The metacarpophalangeal joints

The heads of all of the metacarpals are condyloid in shape; however, the heads of MII, MIII and (variably) MV are also asymmetrical (Susman, 1979) (Fig. 18.11). When the second and the third metacarpo-phalangeal joints are flexed the second (index) and third (middle) fingers naturally rotate and deviate towards the thumb. This is an important factor in the human ability to oppose the palmar surface of the tips of these two fingers to the palmar surface of the tip of the thumb. The asymmetry of the distal joint surface of the MV is a mirror image of the asymmetrical joint surfaces of MII and MIII and produces the opposite movement. When this joint is flexed the fifth (little) finger deviates toward the ulnar side of the hand. This is an important movement in the five-jaw chuck precision grip as well as in the various power grips.

The interphalangeal joints

Flexion of the interphalangeal joints of the four fingers produces the subsidiary **hook grip**. In addition, flexion of the interphalan-geal joint of the thumb, together with flexion of the interphalangeal joints of the four fingers, is an important component in all the various power and precision grips. There is little qualitative difference between the interphalangeal joints of humans and the apes.

Structure and function of the ape hand

Where the human hand is used solely for manipulation, the ape hand is used both for manipulation and for locomotion. This dual function is reflected not only in the proportions of the digits but also in the detailed morphology of the individual bones and their joint surfaces.

The most obvious feature that distinguishes the ape hand from the human hand is the length of the thumb in relation to the length of the remaining digits and particularly the index digit (Fig. 18.2). The ape thumb is much shorter in relation to the index digit than it is in humans and makes it difficult to oppose the pulp of the thumb to the tips of the remaining digits. This, together with the absence of asymmetry of the head of MIII (Fig. 18.11) and the inability to cup the palm of the hand, restricts the apes to primarily the hook grip and the pinch precision grip.

Using the hook grip, relatively large cylindrical objects, such as branches, are grasped horizontally across the palm. More slender objects are grasped using a variant of the hook grip which employs the double-locking capability of the long ape fingers (Fig. 18.12) (Napier, 1960). Double locking is the ability of these long-fingered apes to tuck the tips of their fingers into the creases at the bases of their fingers and then flex the folded fingers over themselves into the palm. Small objects are grasped in the pinch precision grip where the pulp of the thumb is opposed to the side of the index finger.

When the ape hand is used in locomotion, the short thumb has only a minimal role. In arboreal locomotion, the four fingers wrap around branches in a hook grip and in terrestrial locomotion they support the weight of the body (Fig. 18.13a). The African apes, the chimpanzee and the gorilla, use their hands in a **knuckle-walking** posture in terrestrial locomotion. The interphalangeal joints of the fingers are flexed and the weight is borne on the back, or dorsum, of the middle phalanges of primarily the third and fourth digits (Fig. 18.13b). The metacarpals are held in line with the wrist and the proximal phalanges are hyperextended at the metacarpophalangeal joints (Tuttle, 1967). The mainly arboreal orang-utans do not knuckle-walk when they are on the ground, but use their hands in a **fist-walking** posture. They form a fist by tightly flexing the terminal and middle phalanges and support their weight on the dorsum of their proximal phalanges (Tuttle, 1967).

There are four major areas of difference in the bones and joints of the ape hand in comparison with those of the human hand: (1) the features that confer stability to the hand in locomotion at the expense of the mobility necessary for the wide range of human precision and power grips; (2) the features that are associated with the strongly developed digital flexor muscles in the apes; (3) the features that are associated with knuckle-walking in the African apes; and (4)

FIGURE 18·12 *Double-locking facilitates the grip of slender branches or lianes. After Napier (1980).*

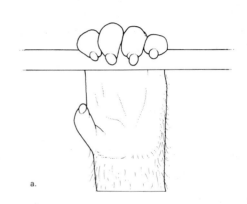

a.

Metacarpal III

Proximal phalanx

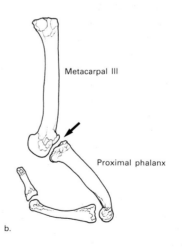

b.

FIGURE 18·13 *(a) An orang-utan employing the hook grip to hang from a horizontal support. After Napier (1980). (b) Lateral view of a chimpanzee third metacarpal and phalanges in the knuckle-walking posture. Arrow indicates the hyperextension of the carpometacarpal joint. After Tuttle (1969a).*

the features that are associated with climbing in the orang-utan.

Features that confer stability to the hand in locomotion

These features are found in two main areas of the hand of the African apes and the orang-utan, the midcarpal joint and the carpometacarpal joints.

THE MIDCARPAL JOINT

The major difference between the human midcarpal joint and that of the apes is the shape of the capitate (Fig. 18.14). In the apes, the capitate has a distinctly constricted wasp-waisted appearance. The head of the capitate is enlarged on the radial (lateral) side and the neck of the capitate, which connects the head to the main body of the bone, is narrow in the radio-ulnar dimension. The shape of the capitate enhances the conjunct rotation at the midcarpal joint (Lewis, 1977). When the joint is extended, the scaphoid fits onto the articular neck of the capitate (Fig. 18.14). Its anterior portion, which carries the trapezoid and the trapezium, becomes firmly wedged beneath the expanded capitate head. The resulting close-packed set of the bones not only enhances the stability of the midcarpal joint in extension, but also carries the trapezium towards the palm. This accentuates the carpal arch and increases the angle of the thumb in relation to the palm of the hand.

The functional explanation for this enhanced conjunct rotation is debated. Lewis (1972a,b, 1974, 1977) argued that it was originally an adaptation to meet the tensional forces on the wrist during forelimb suspensory movement. The conjunct rotation of the carpals would convert tensional forces to compressive forces at the joint surfaces. On the other hand, Jenkins and Fleagle (1975) argued that the constricted capitate neck (together with a spiral hamate facet for the triquetral) is an adaptation for stability in quadrupedal locomotion rather than suspensory locomotion. These authors supported their argument by noting that these features are most evident, not in genera that habitually exhibit suspensory behaviour, but in genera that move quadrupedally (including the macaques).

THE CARPOMETACARPAL JOINTS

The second to fifth carpometacarpal joints are much less mobile in the apes than they are in humans. In the second carpometacarpal joint (CMC2), the proximal end of the MII is deeply indented into the distal carpal row (Fig. 18.15). On the medial side (next to MIII) the MII has two separate facets that articulate with the capitate rather than the one continuous facet that is characteristic of the human joint (Figs 18.7b and 18.15b). On the lateral side (nearest the thumb) the joint between the MII and the trapezium is orientated in a plane parallel to the MII shaft (Fig. 18.5). This joint does not allow the rotation that is characteristic of the human joint.

There are three features of CMC3 that distinguish it from the human joint. The first

FIGURE 18·14 *A chimpanzee and a human capitate from the radial, distal and palmar aspects. After McHenry (1983). Note the wasting of the chimpanzee capitate (arrows) and the differences in the morphology of the joint surfaces for M2 and for the human M3 styloid process (M3S).*

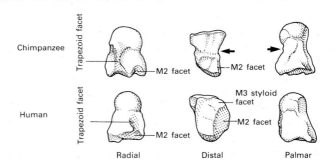

FIGURE 18·15 *The chimpanzee second carpometacarpal joint. (a) Its location in the hand. (b) The disarticulated capitate (with M3) and trapezoid (with M2). The articular surface on the second metacarpal is separated into a palmar and a dorsal part by a strong carpometacarpal ligament. Compare this with the human joint in Figure 18.7. After Lewis (1977).*

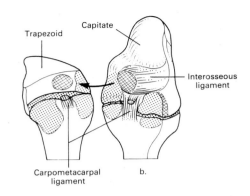

is the absence of a styloid process on the dorso-radial side of the MIII articular surface and the corresponding absence of the indentation on the opposing articular surface of the capitate to receive the styloid process (Figs 18.14). The second feature is the morphology of the joint surface itself. The articular surface on the capitate for the MIII is uneven (Fig. 18.14). The corresponding articular surface on the MIII matches the irregularities of the capitate surface resulting in a 'close packed' and relatively immobile joint between these bones. The third feature is the shape of the capitate articular surface for the MIII. In the apes, the capitate articular surface is constricted, or waisted, in the transverse, or radio-ulnar plane (Fig. 18.14). This constriction accommodates a strong ligament that runs in a distinct groove along the radial side of the capitate to the second and third metacarpals (Fig. 18.15b). The presence of this ligament in the apes, and its absence in humans, is one of the major distinguishing factors in the human and pongid carpal skeletons (Lewis, 1973).

In the apes, movement at CMC4 and CMC5 is severely limited by the hook of the hamate (Fig. 18.5). This is a bony protuberance that extends in a palmar direction from the body of the hamate. Contrary to the condition in humans, the hook of the hamate overhangs the bases of the fourth and fifth metacarpals and articulates with them.

The hook acts as a stop, or buttress, that prevents the flexion at these joints which in humans is a major factor in the ability to cup the palm.

Features that are associated with the strongly developed flexor muscles

In apes, the muscles that flex the hand and digits into a hook grip are relatively larger than they are in humans. The weight of the long flexor muscles expressed as a percentage of the weight of the total long muscles of the hand in apes significantly exceeds the condition found in humans (Table 16.4).

The development of the flexor muscles is reflected in a number of bony features in the hand and digits of the apes. In the apes the pisiform (the sesamoid in the tendon of **flexor carpi ulnaris**) is elongated in comparison with its human form and acts to enhance the leverage of this muscle (Marzke, 1983). Furthermore, the proximal phalanges show a variable degree of longitudinal curvature in the apes (Fig. 18.21) (Susman, 1979). The degree of longitudinal curvature corresponds roughly to the degree of arboreal behaviour in the apes. Next to humans, the gorillas, which are the least arboreal of the large apes, show the smallest amount of longitudinal curvature. The orang-utans, which are the most arboreal of the large apes, have the greatest amount of curvature. Longitudinal curvature is a remodelling response to the strong bending moments imposed on the proximal phalanges when the hand is used to grasp cylindrical branches during arboreal climbing (Preuschoft, 1973b; Susman, 1979).

A second feature of the proximal phalanges that reflects the well developed flexor muscles is the strongly developed flexor sheath ridges on either side of each phalanx (Fig. 18.16). These ridges mark the attachments of the flexor sheaths. (These sheaths are strong bands of fibrous tissue that hold the tendons of the digital flexors (flexor digitorum profundus and flexor digitorum superficialis) next to the palmar surface of the phalanges. The sheaths prevent the flexor tendons from pulling away from the phalanges (or bow-stringing) when the fingers are flexed (Chapter 16).

The middle phalanges also have well-developed flexor sheath ridges (Fig. 18.16). In addition, they show marked insertion areas for the tendons of flexor digitorum superficialis. These attachments are most marked on the third and fourth middle phalanges and reflect the relative strength of flexor digitorum superficialis in the apes.

Features associated with knuckle-walking in the African apes

In relation to the fist-walking orang-utans, the African apes have a limited degree of backward extension (dorsiflexion) both at the wrist joint and at the metacarpophalangeal joints of digits II to V. This limited degree of movement is associated with bony modifications of the wrist and of the metacarpophalangeal joints that prevent these joints from hyperextending, or collapsing, during knuckle-walking.

WRIST JOINT

In the chimpanzee and the gorilla, the distal articular surface of the radius is unique (Fig. 18.17): it is deeply concave and the dorsal rim of the distal edge of the radius projects downward to form a sharp ridge. During knuckle-walking, this ridge abuts against a well-developed ridge on the dorsum of the scaphoid. The scaphoid itself is firmly wedged

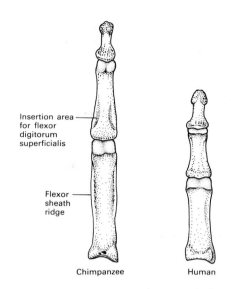

FIGURE 18·16 *Human and chimpanzee phalanges from the palmar view.*

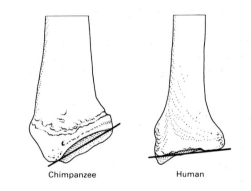

FIGURE 18·17 *A chimpanzee and a human distal radius from the lateral aspect. Note the difference in the orientation of the radiocarpal joint surface.*

beneath the expanded head of the capitate when the wrist is in the extended knuckle-walking posture (in conjunct rotation). These bony ridges on the radius and the scaphoid both limit the dorsiflexion of the wrist and increase its stability when the hand is supporting the weight of the body during knuckle-walking.

THE METACARPOPHALANGEAL JOINTS

During knuckle-walking there is considerable hyperextension at the metacarpophalangeal joints of the four fingers (Fig. 18.13). This is reflected in the extension of the articular surfaces of the heads of the metacarpals onto the dorsum, or backs, of the metacarpals (Fig. 18.18). Downward collapse of these joints is prevented by transverse ridges at the bases of the dorsal articular surfaces of each metacarpal head. In the hyperextended knuckle-walking posture, these ridges abut against the proximal articular surfaces of the proximal phalanges and prevent the collapse of these joints (Fig. 18.13) (Tuttle, 1967, 1969a). These ridges are most prominent on MIII and MIV which support the main weight during knuckle-walking; they may be absent in the pygmy chimpanzee (*Pan paniscus*).

In addition to features that limit backward extension (dorsiflexion) of the hand and fingers, there are two other features of the hand that are associated specifically with knuckle-walking. Firstly, the widest trans-verse (radio-ulnar) diameter of the heads of MIII and MIV are usually located dorsal in the African apes and palmar in the orang-utan (Susman, 1979). The dorsal expansion of the metacarpal heads in the African apes (Fig. 18.11) reflects the force that is transmitted through the dorsal part of this joint during knuckle-walking. Secondly, the proximal phalanges of the African apes, in relation to both humans and orang-utans, are short relative to the lengths of the metacarpals and relative to the dorsopalmar diameter of the metacarpal heads (Fig. 18.19). The short proximal phalanges reduce the ground reaction force on the hand during knuckle-walking (Susman, 1979).

Features that are associated with climbing in the orang-utan

The hand of the orang-utan lacks all the features that are specifically associated with knuckle-walking in the African apes. However, there are three features that can be interpreted as specific adaptations to arboreal locomotion in these primates (suspensory grasping or climbing).

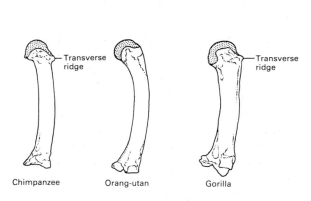

FIGURE 18·18 *The third metacarpal of a chimpanzee, orang-utan and gorilla from the medial aspect. Note that the metacarpal head extends further onto the dorsum of the bone in the knuckle-walking chimpanzee and gorilla than it does in the orang-utan. After Tuttle (1969a).*

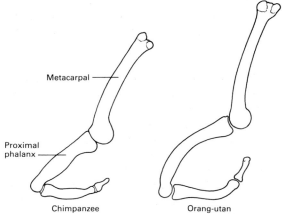

FIGURE 18·19 *The third metacarpal and phalanges of a chimpanzee and an orang-utan in the knuckle-walking posture. After Susman (1979). Note that the chimpanzee proximal phalanx is shorter relative to the length of the metacarpal than is the case in the orang-utan.*

(1) **The phalanges** The orang-utan phalanges are long, or attenuated, and generally show a greater degree of curvature than found in the phalanges of the African apes (Susman, 1979).

(2) **The metacarpals** The metacarpals MII to MV also show a greater degree of curvature than found in the metacarpals of the African apes. The cortex of the metacarpals is also thicker in relation to the dorsopalmar diameter of the metacarpal shafts (Susman, 1979).

(3) **The increased length of the fourth digit** The fourth digit of the orang-utan hand is increased in length in relation to the other digits. This is reflected in the length order of the proximal, middle and distal phalanges of the four fingers. In the African apes the length order for all three phalanges is normally III>IV> II>V. (However, there is no reliable size sequence for the distal phalanges in the gorilla.) A large percentage of orangutans deviate from this pattern. In 42% of the cases ($n=36$) the fourth proximal phalanx is longer than or equal to the third (Susman, 1979). This pattern is also true for the middle phalanges (68% of the cases studied deviated from the Africa ape pattern) and for the distal phalanges (54% of the cases deviated from the Africa ape pattern).

Phalangeal elongation increases the length of the hand and thereby enhances the effectiveness of the hook grip when grasping branches of a relatively large diameter during climbing. The marked longitudinal curvatures of the metacarpals and phalanges, together with the thick metacarpal cortex, resist the strong bending moments resulting from the elongated fingers (Preuschoft, 1973b; Susman, 1979).

The increased length of the fourth digit suggests that the axis of the hand has shifted toward the ulnar side. Susman (1979) suggested that this is a specific adaptation to vertical climbing where the hand is used to grasp vertical branches, etc. The ulnar digits (IV and V) would provide the major support resisting the downward force of the body weight. In this case the radial fingers (II, III and, minimally, I) would provide only secondary support to stabilize the ulnar digits.

It is interesting to note that with regard to many of the features of its hand the pygmy chimpanzee (*Pan paniscus*) is intermediate between the knuckle-walking chimpanzee and gorilla and the arboreal orang-utan. While maintaining closest morphological similarity to the other African apes, the pygmy chimpanzee resembles the orangutans in: (1) the cortical thickness of the MIII shaft; (2) the absence of well-defined ridges on the dorsal margin of the heads of the metacarpals; (3) the absence of dorsal expansion of the metacarpal heads; (4) the degree of curvature of the proximal phalanges; and (5) the attenuation of the middle and distal phalanges (Susman, 1979). These features might be explained by the more arboreal behaviour of the pygmy chimpanzee, together with its relatively small size, in relation to the other African apes.

The hand of Australopithecus afarensis

Carpal, metacarpal and phalangeal fossils excavated at Hadar, Ethiopia comprise the most complete set of early hominid hand bones from the Plio-Pleistocene period (Johanson et al., 1982; Lovejoy et al., 1982b). This collection includes five carpal bones (a pisiform, a trapezium, two capitates and a hamate), 18 identifiable metacarpals, 16 proximal phalanges, 10 middle phalanges and two distal phalanges. The hand of *Australopithecus afarensis* is discussed in the following references: Bush et al. (1982), Johanson et al. (1982), Marzke (1983), McHenry (1983), and Stern and Susman (1983).

These fossils lack any indication of a knuckle-walking adaptation (Tuttle, 1981; Stern and Susman, 1983). There are no transverse ridges at the bases of the dorsal articular surfaces of the metacarpal heads. The widest transverse diameter of the metacarpal heads is located anteriorly, and not dorsally as is common in the knuckle-walking chimpanzee and gorilla. The distal articular surface of the radius lacks the well-defined dorsal ridge that limits dorsiflexion of the wrist in a knuckle-walking posture.

The morphology of the hand shows a few similarities with modern humans; however, in the majority of its features it resembles a generalized hominoid hand (similar to the small end of the pygmy chimpanzee to common chimpanzee range) (Stern and Susman, 1983).

Morphological similarities to modern humans

There are only two features in which the *Australopithecus afarensis* hand appears to resemble the modern human hand, the ratio of the thumb length to finger length and the movement potential at the second carpometacarpal joint (Marzke, 1983).

THE RATIO OF THUMB LENGTH TO FINGER LENGTH

Because there is no complete hand in the Hadar collection, this ratio must be computed from bones of more than one hand, and without the distal phalanges. Using the only complete MI (AL 333w-39) and first proximal phalanx (AL 333-69) and the complete MIII (AL 333-16) together with the longest proximal phalanx (AL 333-63) and the longest middle phalanx (AL 333-88), Marzke (1983) determined that digit I was 50% of the length of digit III. Marzke noted that this is much closer to the mean of 53% for 10 modern humans than it is to the mean of 36% for seven chimpanzees. The tentative conclusion is that, in relation to digit III, the thumb of *Australopithecus afarensis* was closer in relative length to the modern human thumb than to the chimpanzee thumb.

THE SECOND CARPOMETACARPAL JOINT

The *Australopithecus afarensis* second metacarpals (*n*=3) all resemble those of modern humans, and differ from those of the apes, in having a continuous facet for articulation with the capitate. In addition, the joint on the trapezium for the MII is also similar to that in humans. It lies more in the coronal plane than in the sagittal plane and makes an angle of 121° with the facet for the trapezoid (Marzke, 1983). These two features together result in the distinctly human ability to rotate the second metacarpal.

However, the second carpometacarpal joint is not entirely like that of modern humans. The facet for the MII on the capitate is intermediate in orientation between the laterally facing facet in the chimpanzee and the more distally facing facet in modern humans (Fig. 18.20) (McHenry, 1983). In addition, the MII facet on the fossils extends all the way to the dorsal border of the capitate and is not terminated by the styloid process of the MIII, which is lacking in the fossil specimens (Marzke, 1983).

The shape of the capitate, itself, also shows some features that are similar to modern humans and other features that are ape-like in their morphology (Fig. 18.20) (McHenry, 1983). As in modern humans, the capitate is short in its proximo-distal length. The capitate facet for the MIII is also smoother, showing a reduced amount of 'cupping', than

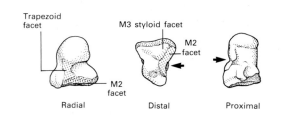

FIGURE 18·20 *The* Australopithecus afarensis *(AL 333-40) capitate from the radial, distal and palmar aspects. Compare with Figure 18.14. After McHenry (1983).*

is present in the living apes. However, as in the apes, the facet on the capitate for the trapezoid is dorsally placed, the capitate neck is waisted (facilitating the conjunct rotation) and the capitate facet for MIII is constricted in the mediolateral plane. This last feature might suggest the presence of the strong ligament connecting the capitate to the MII and MIII that is found in all apes. However, there is no groove on the radial side of the capitate to accommodate this ligament (McHenry, 1983) and the continuous articular facet on the MII for articulation with the capitate would argue against its presence.

Morphological similarities to living apes

The remaining morphological features of the *Australopithecus afarensis* hand bones are all most similar to those found in the living apes. The most important of these similarities are the following.

(1) **The first carpometacarpal joint** The articular surface on the trapezium for the MI is saddle-shaped. However, it is both narrower and more concavo-convex than is the broader and flatter joint surface of humans, *Paranthropus robustus* and *Homo habilis* (Stern and Susman, 1983; Susman, 1988a,b).

(2) **The fifth carpometacarpal joint** As in the apes, the joint surface on the hamate for MV is convex, rather than concave as it is in humans, and accommodates a concavity in the corresponding joint surface on the MV. In addition, as in the apes, the hamate joint surface continues on to the hook, or hamulus, of the hamate. These features suggest that the fifth carpometacarpal joint lacked the mobility found in humans which is necessary to cup the palm of the hand in both the power grips and the five-jaw chuck precision grip.

(3) **The distal phalanges** The distal phalanges are characterized by slender apical tufts (62% of the radio-ulnar diameter of the base as in apes), rather than the broad tufts (69% of the radio-ulnar diameter of the base) characteristic of human distal phalanges (Stern and Susman, 1983).

(4) **The strongly developed flexor apparatus** The fossil metacarpals and phalanges are characterized by all the features associated with a strongly developed flexor apparatus in the apes. These include: (1) an elongated rod-shaped pisiform; (2) longitudinally curved proximal phalanges (Fig. 18.21); (3) well-developed flexor sheath ridges on the proximal and middle phalanges; and (4) marked insertion areas for flexor digitorum superficialis on the middle phalanges (Marzke, 1983; Stern and Susman, 1983).

Based on these morphological features, Stern and Susman argued that the hand of *A. afarensis* would have been adapted to

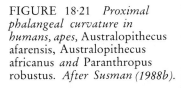

FIGURE 18·21 *Proximal phalangeal curvature in humans, apes,* Australopithecus afarensis, Australopithecus africanus *and* Paranthropus robustus. *After Susman (1988b).*

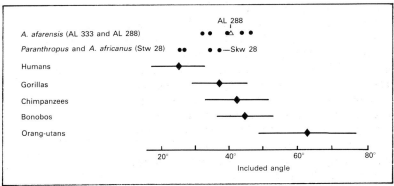

suspensory, or climbing, locomotion. In addition, Marzke (1983) noted that it would not have been suited to either the squeeze, disc, and spherical power grips or the five-jaw chuck precision grip. The morphology of the fifth carpometacarpal joint would prevent the cupping of the hand necessary to these grips. The absence of a styloid process on the MIII would suggest that this metacarpal was not subjected to the stresses of the spherical and squeeze power grips (Marzke and Marzke, 1987). In addition, the ape-like first carpometacarpal joint, together with the dorsally placed facet on the capitate for the trapezoid would not be suited to the high compressive forces exerted on the carpal skeleton by the thumb in the precision grips.

Marzke (1983) suggested that the human features of the *A. afarensis* hand would best facilitate the pad-to-pad and three-jaw chuck precision grips which may have been highly important in the incipient development of tool-using and making behaviour (see also Marzke and Shackley, 1986).

The hand of Paranthropus robustus

Recently reported hand bones from Member 1, Swartkrans (Susman, 1988a,b) have been assigned to *Paranthropus robustus* on two grounds. Firstly, over 95% of the cranial and dental remains from this member belong to

P. robustus and not to the alternative taxon, *Homo* cf. *erectus*. And secondly, the first metacarpal (SK 5020) is different from both another Swartkrans metacarpal (SK84) and a *Homo* metacarpal belonging to the KNM-WT 15000 skeleton from Koobi Fora (*Homo* cf. *erectus*). Both these latter metacarpals share a distinctive 'beaked' projection on the palmar surface of the metacarpal head which is not found in humans or in any other hominid taxa for which material is available and may prove to be distinctive of *H. erectus* (Fig. 18.22) (Susman, 1988a,b). The absence of this beak in SK 5020 suggests that this metacarpal belongs to *P. robustus* and not to *Homo* cf. *erectus*.

If this Swartkrans material is correctly assigned, it provides suggestive evidence that the *P. robustus* hand was more human-like than was the *Australopithecus afarensis* hand and may have been capable of modern human-thumb use in both power and precision grips (Susman, 1988a,b). A terminal thumb phalanx (SK 5016) provides two pieces of evidence that suggest this conclusion to Susman. Firstly, there is a clear area of insertion on the palmar surface (volar surface) of this bone for a tendon that has been interpreted by Susman (1988a,b) as indicating the presence of the uniquely human muscle flexor pollicis longus. This muscle is important to human-hand usage because it is the only muscle in humans that acts to flex the terminal phalange of the thumb. However, a tendon inserting into this thumb phalanx is present in monkeys and in a varying proportion of large apes (Straus, 1942; Baldwin, 1989; and Chapter 16). However, when it occurs in non-human primates it is connected to flexor digitorum profundus (Fig. 16.30). The presence of an insertion area for a tendon on the terminal thumb phalanx, therefore, does not necessarily indicate the presence of flexor pollicis longus and human thumb usage. Secondly, and perhaps more convincingly, the apical tuft of this terminal thumb phalanx is expanded in width well beyond the condition found in the apes (Fig. 18.23). Such wide apical tufts

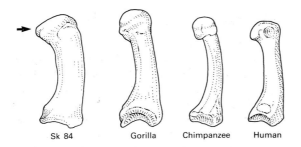

FIGURE 18·22 *The lateral aspect of the first metacarpal of* Homo erectus *(SK 84), a gorilla, a chimpanzee and a modern human. Note the beaked metacarpal head in* H. erectus *(arrow). After Napier (1959).*

FIGURE 18·23 *The relative size of the apical tuft in modern humans* (◆), *chimpanzees* (■), *gorillas* (▲), Paranthropus robustus *(SKX 5016), and* Homo habilis *(OH 7). After Susman (1988a).*

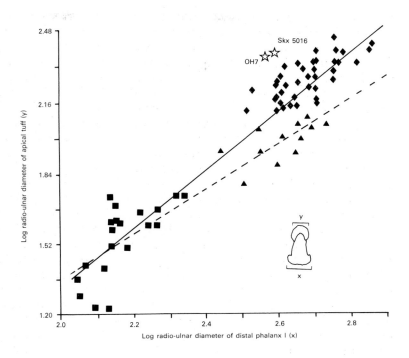

in humans support the fleshy and sensitive finger and thumb tips.

Furthermore, the *P. robustus* first metacarpal differs from ape first metacarpals in features that suggest it was able to withstand the considerable forces generated by human-like power grips. Not only is it a thicker and more robust bone than found in the apes, but it also has the distinctive human-like expanded proximal joint surface (Susman, 1988a,b). In addition, it is reported to also have had a well-developed opponens pollicis muscle as indicated by the presence of a human-like crest on the lateral margin of the shaft for insertion of this muscle. Human-like hand function is also suggested by the relatively short and straight proximal phalanges of the other digits.

Susman (1988a,b) suggested that the most likely explanation for the human-like morphology of the *Paranthropus* hand is tool-using behaviour. This suggestion rests on the assumption that such behaviour would generate the high levels of force for which the *Paranthropus* thumb appears to be adapted, while other non-cultural behaviours requiring precision manipulation, such as small-object feeding, would not. It is also supported by the presence of stone tools in Swartkrans Member 1 from which the fossils were also recovered.

The hand of Homo habilis

Fifteen hand bones were discovered at the FLK NN site (level 3) at Olduvai Gorge in 1960 (Fig. 18.24) (Napier, 1962). The hand bones were initially described by Napier (1962) and later included in the holotype of *Homo habilis* (Leakey et al., 1964). These bones are known as Olduvai Hominid 7 (together with other fossil material). There are three carpal bones (a right trapezium, a right scaphoid and a left capitate), four proximal phalanges, four middle phalanges, three distal phalanges and a single base of a

FIGURE 18·24 *The* Homo habilis *hand (Olduvai Hominid 7). Inset: Adult proximal phalanges. Courtesy of M.H. Day.*

second metacarpal. Some of the main references dealing with the Olduvai hand are: Napier (1962), Tuttle (1967), Lewis (1977), Susman and Creel (1979), and Susman and Stern (1979).

All but two of the proximal phalanges are generally believed to have belonged to a single juvenile individual (Day, 1976; Susman and Creel, 1979). The two adult proximal phalanges (FLK NN J and K) are different from the juvenile proximal phalanges and their taxonomic position is uncertain. They are smaller than the juvenile bones and a radiographic study has revealed that their medullary cavities are narrower distally and wider proximally than those of the juvenile bones (Susman and Creel, 1979). In addition,

the adult bones have disproportionately small heads, they are highly curved and have prominent flexor sheath ridges (Day, 1976). Day (1976) noted the clear morphological similarity between these bones and those of the black-and-white colobus monkey (*Colobus polykomos*). Day speculated on the possibility that these bones may not be hominid at all, but may have belonged to the giant fossil colobus (*Paracolobus chemeroni*) known from the East African Pleistocene.

The juvenile hand is similar to the *Paranthropus robustus* hand in the sense that it is more human-like than the *Australopithecus afarensis* hand. However, it still retains some primitive ape-like features.

Morphological similarities to living apes
The Olduvai juvenile hand is most similar to that of living apes in the morphology of the scaphoid and the proximal and distal phalanges (Susman and Creel, 1979).

(1) **The scaphoid** In virtually all aspects of its morphology, the scaphoid is more similar to the scaphoid of the apes than it is to that of humans (Fig. 18.25). Of particular importance is the morphology of the scaphoid tubercle. Although the tubercle is damaged in the fossil, its total volume would have been smaller than that of the body of the scaphoid, it is set at a sharp angle to the palmar margin of the bone and it very probably did not carry the articulation for the trapezium and the trapezoid (Susman and Creel, 1979). In apes these articulations are located on the body of the scaphoid and not on the tubercle. In the human wrist, where the distal carpal row is 'splayed apart', the scaphoid tubercle is equal in volume to the body of the bone, it lies in the plane of the palm and it carries the articulation for the trapezium and trapezoid (Lewis, 1977; Susman and Creel, 1979).

(2) **The proximal and middle phalanges**
The proximal and middle phalanges show no modern human features and resemble most closely those of the African apes. The most notable feature of these phalanges are the very well-marked insertion areas for the muscle flexor digitorum superficialis on the middle phalanges (Fig. 18.25). These insertions extend more distally than they do in any living apes (Susman and Creel, 1979).

Together with the marked curvature of the proximal phalanges, the ape-like structure of the scaphoid and the apparently well-developed flexor digitorum superficialis suggest that the Olduvai hand was well adapted to climbing (Susman and Creel, 1979). Electromyographic studies of flexor digitorum superficialis in living chimpanzees support this conclusion: this muscle is inactive during the support phase of knuckle-walking, but is particularly important in climbing and suspensory locomotion (Susman and Stern, 1979).

Morphological similarities to modern
Homo sapiens
Similarities to the hand of modern humans are

FIGURE 18·25 (a) The Homo habilis (OH 7) scaphoid compared to chimpanzee and human scaphoids. Note the chimpanzee-like small scaphoid tubercle on the OH 7 bone. (b) The OH 7 middle phalanges with very well developed insertion areas for flexor digitorum superficialis. (c) The OH 7 trapezium from the distal view compared to a chimpanzee and a human trapezium. Note human-like expanded first metacarpal joint surface on the OH 7 bone. After Susman and Creel (1979).

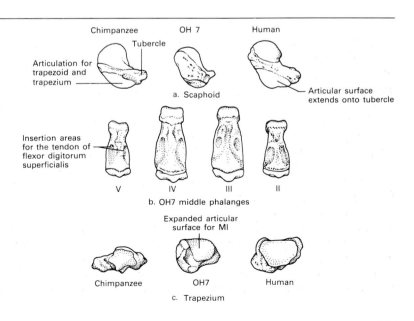

found in the first carpometacarpal articulation and in the distal phalanges.

(1) **The first carpometacarpal articulation** As in humans, the articulation for MI on the trapezium is broad in both the dorsopalmar plane and in the mediolateral plane and it also has a reduced convexity. This contrasts with the narrow and highly concave–convex joint found particularly in the chimpanzees (Figs 18.5 and 18.25) (Lewis, 1977; Trinkaus, 1989; Susman and Creel, 1979). The articular surface on the trapezium for the scaphoid is also broad as it is in humans and not small and narrow as it is in the apes (Susman and Creel, 1979).

(2) **The distal phalanges** The two distal finger phalanges of the Olduvai hand most probably belong to digits II and III (Susman and Creel, 1979). They are most similar to human distal phalanges, which, in relation to those of the apes, are short, very thin in the dorsopalmar dimension and have marked apical tufts. The major difference between the Olduvai distal phalanges and those of humans lies in the greater midshaft thickness of the fossil bones in both the dorsopalmar plane and the radio-ulnar plane.

The remaining distal phalanx of the Olduvai hand is generally believed to be a thumb, or pollical, distal phalanx (Napier, 1962; Day, 1976). However, Susman and Creel (1979) suggested that on morphological grounds it might well be a hallical (from the big toe of the foot) distal phalanx. In this context, Susman and Creel noted its robusticity, as well as the developed basal contour, the shape of the apical tuft, the bulging form of the lateral basal tubercles and the slight axial torsion of the apical tuft relative to the base. However, their metrical analyses contradict this morphological interpretation and suggest that this distal phalanx is most similar to a human pollical distal phalanx. If this interpretation is correct, the relationship between the length of this bone and the

length of the distal phalanx from digit III is more similar to the human condition than it is to the ape condition (Susman and Creel, 1979). This suggests that, in relation to the length of the fingers, the length of the thumb was more similar to that of humans than to that of the apes.

Both the morphology of the trapezium and the morphology of the distal phalanges, suggests that the *Homo habilis* hand was different from the *Australopithecus afarensis* hand and more similar to the *Paranthropus robustus* hand in its adaptation to the range of human grips. The expanded first carpometacarpal joint, characteristic of the *H. habilis* hand and of the *P. robustus* hand but not of the *A. afarensis* hand, could have resisted the strong compressive force transferred to it via the thumb used in the human power grip. The expanded apical tufts of the distal phalanges (Fig. 18.23), also found in the hands of *H. habilis* and *P. robustus* and not in the *A. afarensis* hand, could well be associated with enhanced precision manipulation characteristic of the various human precision grips.

The hand of the Neanderthals

The most complete hand skeleton known for the Neanderthals, as well as for any anatomically non-modern *Homo sapiens*, is the left hand of Shanidar 4 (Iraq) (Trinkaus, 1983a). It lacks only the lunate (which is known from the right hand) and one distal phalanx. The morphology of this hand agrees with the morphology of the other less-complete hand skeletons known from Shanidar (Shanidar 1, 3, 5 and 6) and with the published reports of other hand skeletons from Krapina, Yugoslavia (Gorjanovic-Kramberger, 1906), La Chapelle-aux-Saints, France (Boule, 1911–13), Kiik-Koba (Bonč-Osmolovsky, 1941), Amud, Israel (Endo and Kimura, 1970; Kimura, 1976), Mount Carmel, Israel (McCown and Keith, 1939), La Ferrassie, France (Heim, 1972, 1974) and the Crimea (Vlček, 1975). In addition to these

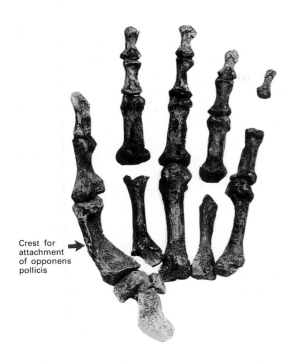

Crest for
attachment
of opponens
pollicis

FIGURE 18·26 *A Neanderthal hand from La Ferrassie, France. Courtesy of M.H. Day.*

sources, the morphology of the Neanderthal hand has been discussed by Sarasin (1932), Musgrave (1969, 1970, 1971, 1973) and Trinkaus (1983a).

In overall proportions and morphology, the Neanderthal hand (Fig. 18.26) is generally similar to that of anatomically modern humans. However, there are two areas of difference. The first of these is the morphology of the thumb, or pollex, and the second is a complex of features associated with the apparently greater degree of muscle development in the Neanderthal hand in comparison to that of modern humans.

The morphology of the thumb

There are two features of the morphology of the Neanderthal thumb that differ from that of modern humans: (1) the shape of the first carpometacarpal joint; and (2) the relative lengths of the proximal and distal thumb (pollical) phalanges.

THE SHAPE OF THE FIRST CARPOMETACARPAL JOINT

The joint surface for the trapezium on the MI is not uniformly convex in the dorsopalmar dimension and concave in the radio-ulnar dimension as it is in modern humans. In some Neanderthals, it is cylindrical in shape (convex in the radioulnar plane and straight in the dorsopalmar plane) (Shanidar 4, La Ferrassie 2) (Musgrave, 1971; Trinkaus, 1983a). In others it is condyloid in shape (convex in both the radioulnar and dorsopalmar dimensions) (La Chapelle-aux-Saints 1 and Kiik Kobi 1) (Trinkaus 1983a, Musgrave 1971).

The joint surface for the MI on the trapezium is saddle shaped in the Neanderthals; however, it is flatter than that in modern humans. Some authors have suggested that the morphology of the Neanderthal first carpometacarpal joint indicates the absence of a full precision grip (see especially Musgrave, 1971; Vlček, 1975). However, Trinkaus (1983a) suggested that it would offer more movement than the modern human joint.

RELATIVE LENGTHS OF THE PROXIMAL AND THE DISTAL POLLICAL PHALANGES

There is no evidence that the length of the Neanderthal thumb is different from the length of the modern human thumb in relation to the length of the index finger (Trinkaus, 1983a). However, there is a marked difference in the length of the pollical distal phalanx in relation to the pollical proximal phalanx. In modern humans, the distal phalanx is approximately two-thirds the length of the proximal phalanx. In Neanderthals the two phalanges are sub-equal in length (Trinkaus, 1983a). This difference results from an elongated distal phalanx and a shortened proximal phalanx (Trinkaus, 1983a).

From a mechanical point of view the Neanderthal proportions would decrease the load arm between the interphalangeal region and the metacarpophalangeal joint and thereby increase the effectiveness of the thenar eminence muscles (flexor pollicis brevis, abductor pollicis and adductor pollicis) when grasping large objects with the thumb (Trinkaus, 1983a). This would also increase the load arm between the interphalangeal joint and the finger tip and, thereby, decrease the effectiveness of flexor pollicis longus when grasping objects with the finger tip (Trinkaus, 1983a). Trinkaus (1983a) noted that all Neanderthals have large insertion areas for flexor pollicis longus and that the inferred greater size of this muscle may compensate for its relatively poor mechanical advantage.

Features associated with the apparently greater degree of muscle development in the Neanderthal hand

There are a number features of the thumb, the four ulnar digits and the carpal skeleton that are associated with the apparently strong muscles of the Neanderthal hand.

(1) **The thumb** The most notable feature of the thumb in this context is the development of the crest for the insertion of opponens pollicis (Fig. 18.27). This crest appears as a relatively large flange on the distal radial margin of MI. This muscle is an abductor of the thumb in a firm grip (Musgrave, 1971).

(2) **The four ulnar digits** The four ulnar metacarpals (MII, MIII, MIV, MV) are long, transversely narrow and have large heads (Fig. 18.26). The combination of narrow shafts and large heads increases the volume of the interosseous muscle spaces and would suggest large interosseous muscles (Musgrave, 1971). Large dorsal interosseous muscles are also suggested by the marked insertion areas on the ulnar side of MI and on the radial side of MII. In addition, there is a marked ridge, or **crista dorsalis**, that runs up the proximal quarter of the shaft on the

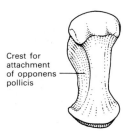

FIGURE 18·27 *The palmar aspect of a Neanderthal first metacarpal (after Musgrave, 1971). Note the well developed crest for the opponens pollicis muscle.*

back, or dorsum, of MII that marks the insertion of the first dorsal interosseous muscle.

Furthermore, the bases of the proximal phalanges, and particularly the base of PPII, are very large, making up approximately one-half of the length of the bone (Musgrave, 1971).

(3) **The carpal skeleton** The tubercles of the scaphoid and the trapezium as well as the hook (hamulus) of the hamate are very large in Neanderthals (Musgrave, 1971; Trinkaus, 1983a). These processes provide the attachments for the flexor retinaculum and suggest that the carpal tunnel in Neanderthals was large and that the flexors of the fingers were well developed. In addition, these processes provide the origins for the thenar and hypothenar muscles. The size of these processes also suggest that these muscles were well developed and that they had an increased moment-arm around the carpometacarpal joints.

All these features suggest that the Neanderthal hand was strong and powerful and, although not identical to modern humans in its morphology, the Neanderthal hand was capable of the wide range of complex manipulative functions that characterize the human hand.

CHAPTER NINETEEN

BONES, MUSCLES AND MOVEMENTS OF THE LOWER LIMB

The lower limb shares many similarities with the upper limb. Not only is it connected to the trunk by a girdle, the **pelvic girdle**, but also it is made up of three major units (Fig. 19.1). The proximal unit, the **thigh**, comprises one bone, the medial (the **lower leg**) two bones, and the distal (the **foot**) a number of small bones. However, the morphology of the bones themselves and the structure of the joints connecting them reflect the major difference in function between the two limbs. The forelimb in adult humans is free of locomotor function and is adapted for manipulation while the hindlimb supports the weight of the body and is the major locomotor organ.

The pelvic girdle is a basin shaped structure (pelvis (L) = a basin) that links the hindlimb to the trunk. It also transmits the weight of the upper body to the hindlimbs, supports the abdominal viscera, and provides the site of attachment for muscles that move the hindlimb and hold the upper trunk in a vertical position.

The single bone of the thigh is called the femur (femur (L) = thigh or possibly; ferendum (L) = bearing) and it articulates with the pelvis at the **hip joint**. The two bones of the lower leg are the **tibia** and the

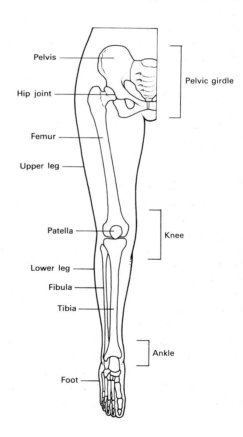

FIGURE 19·1 _The principal bones of the lower limb._

fibula. The tibia is the principal bone of the lower leg and the term tibia has been used since ancient times when it referred to an array of musical instruments of long tubular form made from the shinbones of animals and birds. The tibia alone articulates with the femur at the **knee joint**. The only other bone involved in this joint is the **patella**. The patella is a large sesamoid bone in the tendon of the quadriceps femoris muscle (which extends the leg at this joint). The fibula ((L) = a pin or skewer) is the second lower leg bone and it is long and thin as its name suggests. The fibula is positioned on the lateral side of the tibia and articulates with it at both the **superior** and the **inferior tibiofibular joints**. Together, the tibia and the fibula articulate with the **talus** to form the **ankle joint**. This joint resembles a carpenter's mortise joint where tibia and fibula form a U-shaped articular surface which encompasses the talus and in humans forms a stable hinge joint that restricts the range of movement of the foot at the ankle joint.

The foot is made up of 26 small bones including seven ankle (or **tarsal**) bones, five **metatarsal** bones and 14 **phalanges**. The human foot has sacrificed the prehensile, or grasping, capabilities of the non-human primate foot in favour of the stability that is essential for terrestrial bipedal locomotion. As a result, the joints connecting the tarsal bones to each other and to the metatarsals have very limited movement capabilities. Furthermore, the human foot is characterized by a non-opposable great toe as well as a longitudinal arch which functions as a spring in walking and effectively distributes the weight of the body over the sole of the foot in standing.

The foot is the most distal part of the body. However, because the foot is oriented perpendicular to the long axis of the body the terms **proximal** and **distal** can be confusing when used to describe particular orientations within the foot itself. Rather, the phalangeal end of the foot can also be termed

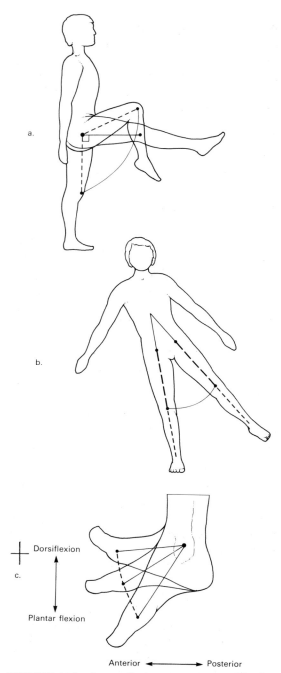

FIGURE 19·2 *Lower limb movements. (a) The hip is capable of a greater degree of flexion when the knee is bent. (b) Abduction of the lower limb always involves the simultaneous abduction of both hip joints. (c) Dorsiflexion and plantar flexion of the foot at the ankle joint. After Kapandji (1987).*

the **anterior** part of the foot, or forefoot, and the heel end the **posterior** part, or hindfoot (Fig 19.2c). The 'top' of the foot is its **dorsal** surface, while the sole of the foot is its **plantar** surface. The term **hallux** is used to refer to the great toe ('allex' = thumb or big toe; this word has been used since the 7th century A.D. but is unknown in either Latin or Greek and, as a result, is generally considered to be a barbarism).

Joints and movements of the lower limb

The hip joint is the most mobile joint of the lower limb. It is the classic ball-and-socket joint and allows the thigh to move in flexion and extension, in abduction and adduction, in rotation (around an axis passing through the head of the femur) and in circumduction. The degree of flexion and extension of the human hip joint is dependent on whether the knee is bent or not (Fig. 19.2a). If the knee is extended, tension in the hamstring muscles that pass down the back of the thigh from the pelvis to the tibia restrict flexion of the hip joint to about 90°. If the knee is bent these muscles are relaxed and the hip joint can be flexed to about 120°. In extension, the effect of the bent knee is reversed. If the knee is bent the hip joint can only be extended by about 10°, while if it is straight the joint can be extended by about 20°. This is because the hamstrings are more efficient hip extensors if they are not also functioning at the same time as knee flexors (Kapandji, 1987).

E? hip joint can be abducted by about 45°. However, abduction of a limb always involves the simultaneous abduction of both hip joints (Fig. 19.2b). Because of this a limb can be effectively abducted by about 90°. The maximum range of adduction in humans is 30°, of lateral rotation about 60° and of medial rotation beween 30° and 40° (Kapandji, 1987).

Because the knee joint is primarily a hinge joint its most marked movements are flexion

and extension. The human knee can be flexed by 120° if the hip is extended and by 140° if the hip is flexed. Because it is normally in full extension in human standing, extension of the knee joint refers to the movement which brings the knee back to full extension from any position of flexion. The human knee joint is also capable of about 30° of medial rotation and 40° of lateral rotation when the knee is flexed (Kapandji, 1987).

The ankle joint is also primarily a hinge joint; however, because the foot is at a right angle to the leg, movement of the ankle joint is described as either **dorsiflexion** or **plantarflexion** (Fig. 19.2c). The foot is dorsiflexed when the heel is pushed down and the toes are elevated and it is plantarflexed when the heel is pulled up and the toes are depressed. The normal human range of dorsiflexion is between 20° and 30° and of plantarflexion between 30° and 50°. Because of the shape of the ankle joint the foot also rotates laterally in dorsiflexion and medially in plantarflexion.

The complex movements of the foot in locomotion (discussed in Chapter 23) can be broken down into a series of simple movements that occur at the various joints in the foot.

(1) **Inversion and eversion** The foot is inverted when the sole of the foot is turned towards the mid-line of the body and it is everted when the sole is turned away from the mid-line (Fig. 19.3). Inversion and eversion are movements of the **foot plate** (or **lamina pedis**). This is the part of the foot which lies below the talus, the talus being the tarsal bone that articulates with the tibia and fibula at the ankle joint. The movement takes place at the **peritalar joint**. This is a compound joint comprising the articulations between the talus and two other tarsal bones, the **calcaneus** (heel bone) and the **navicular**.

(2) **Exorotation and endorotation** Exorotation and endorotation describe the move-

ment of the anterior part of the foot in relation to the stationary leg, talus and calcaneus or alternatively the movement of the leg, talus and calcaneus around the stationary anterior part of the foot (Fig. 19.3). The anterior part of the foot is exorotated when it is rotated laterally in relation to a stationary posterior part of the foot and endorotated when it is rotated medially. This movement takes

place at the **transverse tarsal joint** (Fig. 19.3). This is another compound joint. It runs laterally across the instep and comprises two single joints, the **talonavicular** joint and the **calcaneocuboid** joint.

(3) **Flexion and extension** As applied to the foot, flexion and extension occur only at the metatarsophalangeal joints or at the interphalangeal joints. The toes are flexed when they are curled and extended

FIGURE 19·3 *(a) The three major joints of the tarsal (ankle) region of the foot (the subtalar, talonavicular and calcaneocuboid) work together in two functional combinations (the peritalar joint and the transverse tarsal joint). After Shepard (1951). (b) Inversion and eversion take place at the peritalar joint (around axis A) and exorotation and endorotation at the transverse tarsal joint (around axis B).*

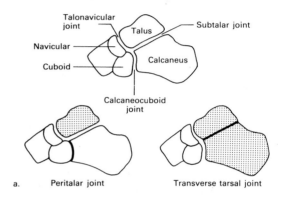

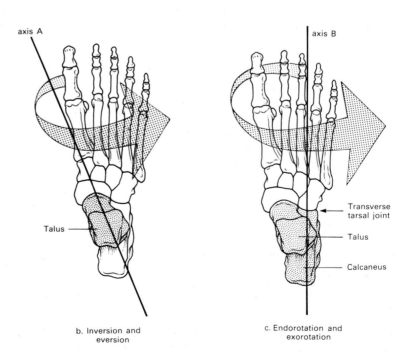

when the phalanges lie in line with the metatarsals.

(4) **Opposability** In non-human primates the great toe (hallux) can be opposed to the remaining digits. This movement involves flexion of the metatarsophalangeal and interphalangeal joints of the first digit together with movement at the joint between the first metatarsal and the medial cuneiform. This last mentioned joint is both flatter and more anterior facing in humans that lack an opposable great toe (see below).

Osteology of the human pelvic girdle

The pelvis is composed of two **pelvic bones** and the **sacrum** (Fig. 19.4a). The two pelvic bones are also known as **os coxae** (os (L) = bone; coxa (L) = the hip). These bones articulate with each other at the **pubic symphysis** and make up the front and sides of the **pelvic girdle**. The two pelvic bones articulate posteriorly with the sacrum at the **sacroiliac** joints to complete the basin-shaped pelvic structure. The sacrum is a continuation of the vertebral column and is composed of a number of vertebrae that are fused together to give strength and stability to this region. The name sacrum (sacer (L) = sacred) may refer to the dense, decay-resistant structure of this bone which in ancient times was believed to provide the basis for divine resurrection. Alternatively, the sacrum could have been termed divine because it helps protect the sacred organs of procreation.

Each pelvic bone is itself made up of three separate bones, the **ilium**, the **ischium** and the **pubis** (Fig. 19.4b). The ilium ((L) = flank) forms the upper part, or blade, of the pelvic bone. As one of the functions of the ilium is to support the gut, the name was most probably derived from 'ilia' ((L) = the small intestines). The ischium makes up the lower part of the pelvic bone and is the part of the pelvis which we sit on. The term ischium is

derived from the Greek 'ischion' which means the socket in which the head of the femur turns. Although all three bones are involved in the socket for the hip joint (**acetabulum**) the ischium makes the largest contribution. The pubis (pubes (L) = the hair of the genital region) forms the front of the pelvis. These three bones fuse into the solid pelvic bone around the time of puberty (between 13 and 15 years of age) (Krogman and Iscan, 1986).

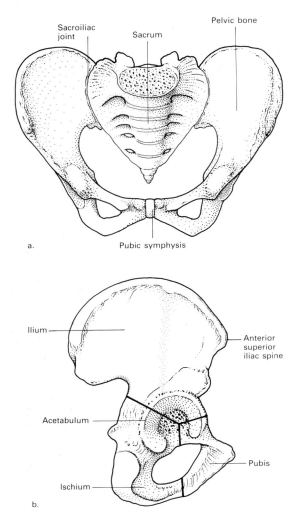

FIGURE 19·4 *The human pelvis from (a) the anterior (ventral) and (b) the lateral view.*

Prior to this age they are separated by the **triradiate cartilage** which functions like an epiphyseal cartilage in bone growth.

The complete pelvis is frequently described in two parts. The **greater pelvis**, or the **false pelvis**, refers to the upper part of the pelvis formed by the iliac blades (Fig. 19.5). The **lesser pelvis**, or **true pelvis**, is the remainder of the structure and is separated from the false pelvis by the **ilio–pectineal** line (**ilio–pubic line**), or **pelvic brim**. The true pelvis forms the birth canal in females.

In normal human bipedal standing, the human pelvis is oriented so that the most anterior part of the iliac blade (the **anterior superior iliac spine**) is vertically aligned over the most anterior part of the pubis (Fig. 19.4). In this position the true pelvis has an anterior–superior to posterior–inferior axis, the sacro-iliac joint is located posterior to the hip joint and the tip of the sacrum (the **coccyx**) is in line with the upper half of the body of the pubis (see Fig. 20.1).

Osteology of the human femur

The femur is the largest bone in the body and can be divided into three major parts, the upper (head and neck) end, the shaft and the lower (condylar) end (Fig. 19.6).

The upper, or proximal, end includes the **head**, **neck** and normally two **trochanters**. The head articulates with the acetabulum of the pelvis at the hip joint and makes up about two-thirds of a sphere. The **fovea capitatis** (fovea (L) = pit or grave) is a depression on the head for the attachment of the **ligament of the head** which has within it the blood vessels that feed the femoral head. The head is connected to the shaft of the femur by the neck. Lateral to the head and neck is the **greater trochanter** which is the bony protuberance that can be felt on the side of the thigh about a hands-breadth below the iliac blade of the pelvis. The term **trochanter** ((Gk) = a runner; tronchos (Gk) = wheel) may have been applied to this protuberance on the femur because of the way it can be

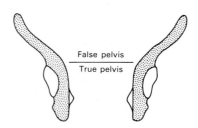

FIGURE 19·5 *Major divisions of the pelvis.*

seen to move in the act of running. The outer, or lateral surface, of the greater trochanter provides the area of insertion for gluteus medius, an important abductor of the pelvis in humans and extensor of the thigh in apes. The pit on the medial side of the greater trochanter (proximal to the femoral neck) that looks like a depression left by sticking a finger in clay is the **trochanteric**, or **digital**, **fossa** (digitus (L) = finger or toe). It provides the area of attachment for one of the lateral rotators of the thigh, obturator externus.

On the back of the proximal femur, the bump that forms the apex of the inverted triangle (whose other angles correspond to the head and greater trochanter) is the **lesser trochanter** which provides the area of insertion for the main flexor of the hip, iliopsoas. The lesser trochanter is connected to the greater trochanter by the **intertrochanteric crest**. The bump located on the proximal portion of this crest is the **quadrate tubercle** which is the area of insertion for a lateral rotator of the thigh, quadratus femoris.

On the front, or anterior, surface of the proximal femur, the **intertrochanteric line** marks the boundary between the neck and the shaft and is the line of attachment for the most important ligament of the hip joint, the **iliofemoral ligament** (Chapters 20 and 21; Fig. 20.9). The intertrochanteric line extends from a prominent projection on the anterior surface of the greater trochanter (called the **femoral tubercle**) obliquely down and across the front of the proximal femur and is

FIGURE 19·6 *The human femur from the anterior and posterior view.*

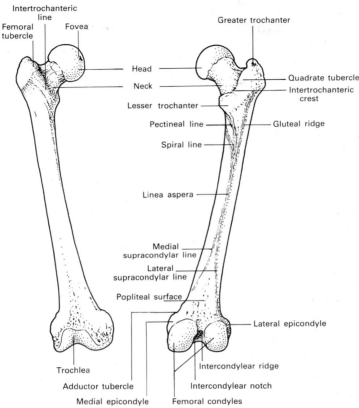

continuous below with the **spiral line**. The spiral line literally spirals around to the posterior surface of the femur to join the **linea aspera** and marks the attachment of the fascia covering one of the extensors of the knee, vastus medialis.

The shaft of the femur begins just below the lesser trochanter on the back of the proximal femur and at the level of the upper part of the spiral line on the front. The linea aspera (asper (L) = rough) is the major feature of the femoral shaft. It is a roughened ridge that extends down the back of the shaft from just below the lesser trochanter to a point in the lower third of the shaft where it separates into the two **supracondylar lines**. The **medial supracondylar line** ends at the

adductor tubercle on the **medial epicondyle** and the **lateral supracondylar line** ends on the **lateral epicondyle**. The supracondylar lines enclose a flat triangle of bone called the **popliteal surface** (poples (L) = horn or knee). The linea aspera and the supracondylar lines provide the area of insertion for many of the thigh muscles.

The **pectineal line** can frequently be recognized just below the lesser trochanter and immediately lateral to the spiral line. It is a roughened line about 3 cm long that marks the attachment of the pectineus muscle. The **gluteal ridge**, or **hypotrochanteric line**, is normally visible as a roughened ridge lateral to the pectineal line. This ridge marks the insertion of gluteus maximus. However, this lateral area of the femur can be exceedingly variable (see Chapter 21). Sometimes the gluteal ridge is combined with a linear depression on its outer border, the **hypotrochanteric fossa**, and sometimes this linear depression occurs on its own without the gluteal ridge. In other cases there is a distinct knob of bone called the **third trochanter** (or the **trochanter tertius** or the **gluteal tuberosity**), lateral to the lesser trochanter. This feature can occur by itself or in combination with either the gluteal ridge, the hypotrochanteric fossa or both and normally lies proximal to these landmarks.

The most obvious features of the lower end of the femur are the **condyles** (the articular surfaces) and the **epicondyles**, or bony expansions, lying proximal and lateral to the articular surfaces. There are actually three articular surfaces on the distal femur, the **medial** and the **lateral condyles** that articulate with the tibial condyles to form the knee joint and the **patellar trochlea**. The patellar trochlea (trochlea (L) = pulley) is the grooved surface at the front of the distal femur which carries the patella, the large sesamoid bone in the tendon or quadriceps femoris (Fig. 16.1). The trochlea is contiguous with the two condyles that make up the distal surface of the femur.

Viewed from the back the two femoral condyles are separated from each other by a large non-articular area called the **intercondylar fossa**, or **notch**, which in turn is separated from the femoral shaft by the **intercondylar ridge**. The **epicondyles** (epi (Gk) = upon), which are very obvious from this perspective, provide the area of attachment for the ligaments that help to stabilize the knee joint. There is a distinct groove on the lateral epicondyle that ends posteriorly in a pit which provides the origin for popliteus. When the knee is fully flexed the tendon of the popliteus lies in this groove.

Osteology of the tibia

The proximal end of the tibia is expanded into a broad and flat **tibial plateau** (Fig. 19.7). The plateau is divided into two articular

FIGURE 19·7 *The human tibia from the anterior and posterior view.*

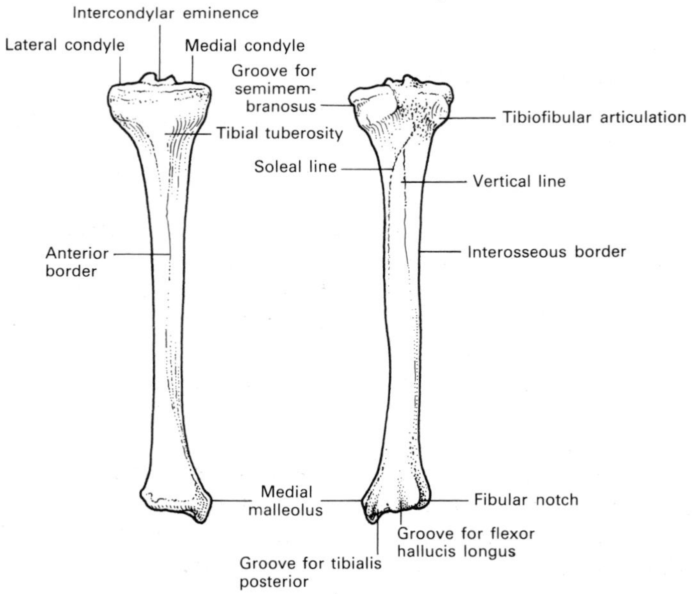

surfaces, the **articular condyles** (medial and lateral). These are separated from each other by a non-articular area which broadens anteriorly into a wide triangular surface. The **intercondylar eminence**, which is a raised area composed of two **tubercles**, or bumps, is located in the middle of this non-articular strip between the two articular condyles. These two tubercles are called the **tibial spines**.

The non-articular parts of the medial and lateral condyles support the articular surfaces from below. The articular facet for the superior **tibiofibular joint** is visible on the lateral condyle and a clearly marked groove for the insertion of semimembranosus on the medial condyle. The **tibial tuberosity** is located between and slightly below the medial and lateral condyles on the anterior surface of the tibia. It is a roughened knob of bone that provides the area of insertion for the **patellar ligament**.

Perhaps the most noticeable feature on the proximal half of the posterior surface of the shaft is the **soleal line**. This is a roughened line that curves from just below the lateral condyle to the medial side of the shaft just above the level of midshaft. Popliteus originates from the triangular area of bone above this line while the line itself marks the linear tibial origin of soleus. Lateral to the soleal line is the weakly developed **vertical line** and lateral to the vertical line the sharp and very obvious **interosseous border** runs the length of the lateral side of the bone. Tibialis posterior arises between the vertical line and the interosseous border, while flexor digitorum longus originates between the vertical line and the soleal line (which merges with the medial border of the shaft).

At midshaft the tibia is frequently triangular in shape (but see Chapter 22) with the three corners of the triangle made up by the interosseous border on the lateral side, the medial border on the medial side and the **subcutaneous**, or **anterior**, **border**, in the front. It is this subcutaneous (cutaneous (L) = pertaining to the skin) border that can be felt as a sharp ridge running down the front of the tibia in the living leg.

The shaft widens towards its distal end and the triangular shape of the cross-section becomes less distinct. The shaft ends with the articular surface for the **trochlea** of the **talus**. This surface faces distally and is anterioposteriorly concave and slightly convex from side to side. The **medial malleolus** projects downward on the medial side of this joint and the tibiotalar articular surface curves around onto the malleolus to articulate with the medial side of the talus. On its posterior surface the malleolus is marked by a clear groove for the tendon of tibilias posterior (as it passes into the foot) and lateral to this is a faint groove for the tendon of flexor hallucis longus. There is a concave notch called the **fibular notch** on the lateral side of the distal tibia immediately above the tibiotalar surface. Immediately above this is a roughened area for the **lower tibiofibular** ligaments. These ligaments are a direct continuation of the **interosseous membrane** and tightly bind the tibia and the fibula together. The anterior surface of the distal tibia is relatively featureless; however, occasionally it is marked by small anteriorly directed facets immediately above the tibiotalar surface. These facets are called **squatting facets** and are described more fully in Chapter 22.

Osteology of the fibula

The fibula is divided into a proximal end, or head, a shaft and a lower end (Fig. 19.8). The facet for the **superior tibiofibular** articulation is located on the broadened proximal fibular head. This facet is variably expressed, but in general faces anteriorly, proximally and medially. Lateral to the facet is the most proximal part of the bone, the apex of the head, or the **fibular styloid** process, which is the area of attachment for the **fibular collateral ligament** of the knee joint and for the tendon of biceps femoris. The head is connected to the fibular shaft by the slightly narrowed neck.

FIGURE 19·8 *The human fibula from (a) the lateral and (b) the medial view. (c) The posterior view of the fibular malleolus.*

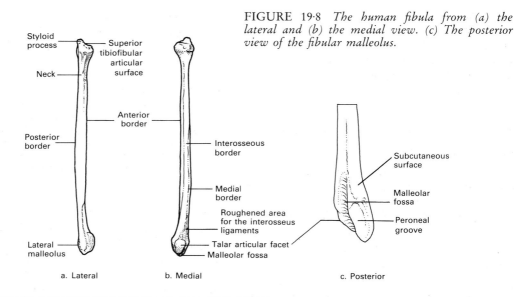

Styloid process
Neck
Superior tibiofibular articular surface
Anterior border
Posterior border
Interosseous border
Medial border
Roughened area for the interosseus ligaments
Lateral malleolus
Talar articular facet
Malleolar fossa
Subcutaneous surface
Malleolar fossa
Peroneal groove

a. Lateral   b. Medial   c. Posterior

The **shaft** of the fibula is long and narrow. Although it can be highly variable in the shape of its cross-section, four ridges can generally be recognized running down its length. The most anterior of these is aptly named the **anterior border**. Immediately medial to this is the equally sharp **interosseous border**. The narrow, and in some cases concave, surface separated by these margins gives origin to the extensor muscles of the foot.

Following around the shaft, the next and usually most obvious ridge is the **medial crest**. This crest is more rounded than the other edges of the fibula and merges with the interosseous border on the distal shaft. Tibialis posterior arises between these two ridges. The final ridge is the **posterior border**. The surface between the medial crest and the posterior border gives rise in its upper third to soleus and in its lower two-thirds to flexor hallucis longus. Completing the circuit around the fibular shaft, the peroneal muscles arise from the **external surface** of the shaft between the **posterior** and the **anterior borders**. It is easier to understand the relationships of the various fibular ridges when the fibula is properly articulated with the tibia.

The expanded lower end of the fibula forms the **lateral malleolus** which makes up the lateral side of the ankle joint. There is a triangular articular surface for the talus on the medial surface of the malleolus and immediately above this a small facet for articulation with the distal tibia. Above the tibial facet is a roughened area for the **interosseous ligaments** that bind the bones together. Posterior to the triangular fibulotalar articulation is the **malleolar fossa**, the area which provides the attachment for some of the important ligaments of the ankle joint. From the position of this fossa it is easy to tell whether the fibula is from the right leg or the left leg. With the fibulotalar joint surface facing you the malleolar fossa will be located on the side of the bone corresponding to the leg to which the fibula belongs.

The last feature of note on the malleolus is the shallow **peroneal groove** on the posterior surface of the malleolus. This groove is the distal extension of the **peroneal surface** of the fibula and carries the tendons of peroneus longus and peroneus brevis into the foot. These tendons follow a posteriorly directed course along the distal fibula and in so doing expose a roughly triangular surface of bone immediately under the skin. This **subcutaneous surface** is free of any muscles or ligaments. It lies lateral to the peroneal groove, extends from the malleolus up the shaft for about 5 cm and can be easily felt on the living leg.

Why there are two separate bones in the lower leg

Although the primitive tetrapod skeleton included two separate bones in the lower limb, it is unusual for modern mammals to have a fibula that is not at least partially fused to the tibia (Fig. 19.9) (Barnett and Napier, 1952, 1953). In fact only primates (except the tarsier) and carnivores (particularly cats and bears) have separate lower leg bones that allow a degree of 'give' at the ankle joint as the fibula moves in relation to the tibia. This morphology is particularly suited to the locomotor patterns of both these groups of mammals that involve a wide range of flexion and extension coupled with inversion and eversion strains at the ankle joint. In the carnivores this results from the need to move, often at high speeds, over irregular terrains and in the primates it results from arboreal locomotion (Barnett and Napier, 1953). In humans the mobile fibula has an added function during bipedal loco-motion (Weinert et al., 1973). During the initial heel-strike phase of the bipedal stride the fibula moves down in relation to the tibia. This deepens and stabilizes the ankle joint and also stretches the interosseous membrane and tibiofibular ligaments binding the tibia and fibula together, converting part of the compressive force at the ankle to tension in these soft tissues.

In other mammals the shaft of the fibula is either missing or is fused (to a greater or lesser degree) with the tibia. The shaft is missing in animals such as deer and giraffe that are both fast runners and digitigrade (running on their toes). In burrowing or swimming animals the fibula is robust and firmly fused (superiorly and inferiorly) to the

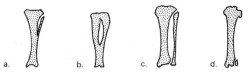

FIGURE 19·9 _Different degrees of tibiofibular fusion in (a) animals specialized for rapid jumping (e.g. Tarsier), (b) animals specialized for burrowing and swimming (e.g. Cape golden mole), (c) animals specialized for movement on uneven surfaces (e.g. most primates), and (d) animals specialized for fast running (e.g. deer). After Barnett and Napier (1953)._

tibia, while in rapid jumpers such as the tarsier and jerboa it is thin and flexible in its upper half but fused to the tibia along its entire lower half. The fibula articulates with the tibia at its upper end via a synovial joint and seems to function as a spring for these jumping animals (Barnett and Napier, 1953).

Osteology of the foot

There are 26 bones in the foot divided into three groups, the **tarsals** (seven bones), the **metatarsals** (five bones) and the **phalanges** (14 bones) (Fig. 19.10). The term tarsal (tarso

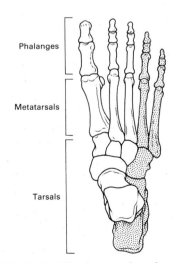

FIGURE 19·10 _The human foot from the dorsal aspect. The bones of the lateral arch are shaded and those of the medial arch are unshaded._

(Gk) = a wickerwork frame or flat basket) was used in ancient times to refer to a variety of flat or expanded objects. In modern terminology the tarsal bones are the ankle bones while the metatarsals are the bones of the instep and the phalanges are the bones of the toes.

The 14 pedal (pes (L) = foot) phalanges are arranged in the same fashion as are the hand phalanges. There are five **proximal phalanges**, four **middle phalanges** and five **distal phalanges**. The great toe has only two phalanges, the proximal phalanx and the distal phalanx. As with the hand phalanges, the easiest way to determine the digit to which a particular phalanx belongs is by size. In humans the size sequence for the proximal and distal phalanges (P) is normally PI>PII >PIII>PIV>PV and for the medial phalanges is PII>PIII>PIV>PV. The joint surfaces on the phalanges also correspond to those of the hand phalanges. The proximal facet on the proximal phalanges is an oval, concave hollow for the rounded metatarsal head. The distal facet on both the proximal phalanges and the middle phalanges is spool like in shape with a depressed central groove and raised condyles. The corresponding proximal facets of both the middle phalanges and the distal phalanges has a raised ridge that rides in the depressed groove of the anterior facets of the adjacent phalanges.

However, pedal phalanges can be easily distinguished from hand phalanges by their general morphology. This is particularly true for the middle and distal phalanges of the lateral four digits (see also Chapter 16). These bones are much smaller than the corresponding hand phalanges and give the appearance of being stunted. The proximal phalanges of these lateral four digits can also be easily distinguished. They are much more gracile than the corresponding hand phalanges, having a relatively narrow midshaft circumference and expanding or flaring outward at their proximal and distal ends. The proximal and distal phalanges of the great toe are much larger in size than the corresponding

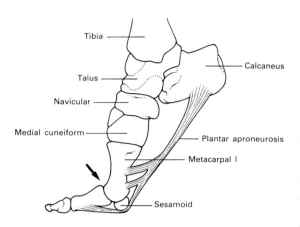

FIGURE 19·11 *The human foot from the medial aspect at toe-off. The articular surface of the first metatarsal head wraps around on the dorsal surface of the bone (arrow) thereby allowing a considerable degree of dorsiflexion at the metatarsophalangeal joint. After MacConaill and Basmajian (1969).*

phalanges of the other pedal digits and the proximal phalanx of the great toe is much more robust (midshaft girth/length) than any of the other hand or foot proximal phalanges. All the proximal pedal phalanges can also be distinguished from the corresponding hand phalanges by their round, rather than semicircular, cross-sectional shape (Fig. 16.14).

The five metatarsals can also be easily distinguished from the metacarpals. The first metatarsal (leading to the great toe) is particularly distinctive. It is a very robust bone (girth/length), and has a particularly large diameter in the dorsoplantar plane (Fig. 19.11). This allows it to resist the considerable dorsoplantar bending stress at the terminal, or toe-off, phase of the stride cycle when it is supporting the great majority of the weight of the body. It also has a mediolaterally expanded head with an articular surface that wraps around onto the dorsal surface of the metatarsal. This allows the considerable amount of dorsiflexion at this joint at toe-off.

The remaining four metatarsals also have heads with articular surfaces that extend onto the dorsal surface of the bone, allowing up to 90° of dorsiflexion at these joints (Fig. 19.13). Whereas at the corresponding metacarpophalangeal joints of the hand, the possible range of flexion exceeds the possible range of dorsiflexion (or extension), at the metatarsophalangeal joints of the foot the range of dorsiflexion greatly exceeds the possible range of flexion. On the dorsal surface of the anterior metatarsal shaft, the articular head of the metatarsal is separated from the epicondyles by a clearly defined groove running mediolaterally. The heads and anterior shafts are also mediolaterally compressed. This feature distinguishes them from the hand metacarpals with a shaft that expands anteriorly to a relatively broad head (Fig. 16.14). The metatarsals are also all longer than their corresponding metacarpals. Furthermore, the lateral four metatarsals, and particularly the second, third and fourth are very gracile bones with quite narrow shafts in relation to their shaft lengths (Chapter 23).

The digit to which a metatarsal belongs can be determined by either relative size or by morphology. As is the case with the hand metacarpals, the common human length sequence for the metatarsals (M) is MII>MIII>MIV>MV>MI (Martin and Saller, 1959). However, there are also marked distinctions in metatarsal morphology that permit easy recognition of at least the first and the fifth metatarsal.

MI The distinctive morphology of this metatarsal has already been described. However, it also has a unique base, or posterior joint surface (tarsometatarsal) that is kidney shaped.

MII This metatarsal has two articular facets on its lateral side adjacent to its base and variably a single facet on the medial side (Fig. 19.12).

MIII This metatarsal is shorter than MII and has an articular facet that wraps around its base from the medial side of the proximal shaft adjacent to the base to the lateral side of the shaft (Fig. 19.12). The facet on the lateral side of the shaft is larger than the facet on the medial side.

MIV The morphology of the base of MIV is very similar to MIII and it is sometimes difficult to distinguish the two metatarsals. However, MIV is generally shorter than MIII, the facets on the sides of the base are of more equal size and the medial facet is frequently not contiguous with the basal (proximal) facet (Fig. 19.12).

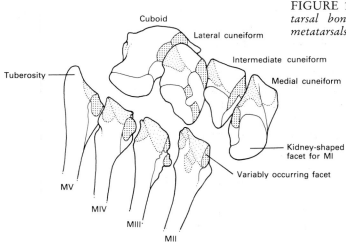

FIGURE 19·12 *A schematic diagram of the distal tarsal bones and the bases of the lateral four metatarsals of the human foot. After Lewis (1980b).*

Cuboid
Lateral cuneiform
Intermediate cuneiform
Medial cuneiform
Tuberosity
Kidney-shaped facet for MI
Variably occurring facet
MV
MIV
MIII
MII

MV Next to MI this is the most distinctive metatarsal. The basal articular facet is at a considerable angle to the shaft axis and lateral to this facet is a large non-articular tuberosity (Fig. 19.12). The posterior part of the shaft is also compressed in the dorsoplantar plane and the lateral margin of the bone traces a markedly curved path from the metatarsal head to the tuberosity (Fig. 19.13).

The seven tarsal bones are the **calcaneus,** the **talus,** the **navicular,** the **medial** (or first), the **intermediate** (or second), and the **lateral** (or third) **cuneiforms** and the **cuboid** (Fig. 19.13). The calcaneus (calx (L) = heel) is easily recognized because it is the largest of the tarsal bones (Fig. 19.13). On the superior surface of the calcaneus there are three articular facets for the talus, the convex **posterior talocalcaneal facet** and the concave **middle** and **anterior talocalcaneal facets**. The middle talocalcaneal facet rests on a shelf of bone, the **sustentaculum tali** (sustentaculum (L) = prop or support), that projects from the medial side of the calcaneus. In some individuals the middle and anterior joint surfaces can be fused into a long narrow concave surface. The remaining **calcaneocuboid joint** surface, is located anteriorly and the calcaneal tuberosity posteriorly. This is the part of the calcaneus that contacts the ground.

The talus provides the link between the leg and the rest of the bones of the foot. The talus is the only bone in the body that does not have a muscle attached to it. In other

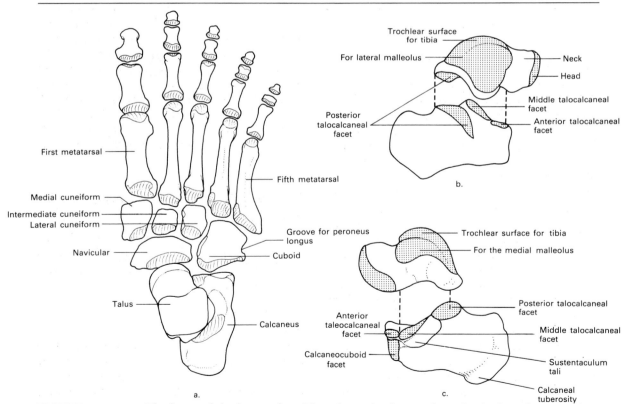

FIGURE 19·13   *(a) The bones of the human foot. The talus and calcaneus from (b) the lateral view and (c) the medial view.*

animals this bone is frequently called the **astragalus** (astragalus (Gk) = a die). These bones were used in playing dice because of their rather chunky six-sided shape. The large concave **posterior talocalcaneal joint** surface is on the inferior surface of the talus and on the superior surface is a even larger convex surface, the **talar trochlea**, for articulation with the tibia (Fig. 19.13). The **neck** of the talus projects anteriorly and supports the convex **head** that articulates anteriorly with the navicular. The inferior surface of the head and neck articulate with the calcaneus at the **anterior** and **middle talocalcaneal joints**.

The navicular (navicula (L) = a little ship) is located on the medial side of the foot and has a hollowed-out posterior surface and a convex anterior surface (Fig. 19.13). The hollowed-out posterior surface (for articulation with the rounded head of the talus) is frequently called the **acetabulum pedis** (pes (L) = foot; acetabulum (L) = cup). The convex anterior surface of the **navicular** has three articular facets for the three cuneiform bones. There is a projecting knob of bone, the navicular tuberosity on its medial surface.

The three cuneiform bones (cuneus (L) = a wedge) form the link between the navicular and the three medial metatarsals (Fig. 19.13). The wedge shapes of these bones contribute to the transverse arch of the foot (see below). The medial cuneiform is the largest of the three cuneiforms and the intermediate is the smallest (Fig. 19.12). The large medial cuneiform is also easy to recognize because it has an equally large kidney-shaped joint surface on its anterior surface for articulation with the first metatarsal.

The remaining tarsal bone is the **cuboid** (kuboeides (Gk) = cube shaped). It is located on the lateral side of the foot and articulates posteriorly with the calcaneus, anteriorly with the fourth and the fifth metatarsals and medially with the lateral cuneiform. Viewed from the side it forms the 'keystone' of the low lateral longitudinal arch of the human foot (Fig. 19.14). The calcaneus makes up the posterior arm of this arch and the fourth

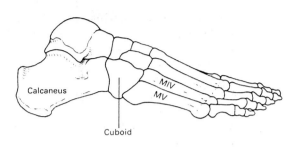

FIGURE 19·14 *The human foot from the lateral aspect.*

and fifth metatarsals the anterior arm.

When viewed from the top, the foot can be clearly divided into two halves (Fig. 19.10). The lateral half is the low arch formed by the calcaneus, cuboid and fourth and fifth metatarsals. The medial half is a higher arch formed by the talus (resting on the calcaneus), the navicular, the three cuneiforms and the medial three metatarsals. These two parts of the foot work together to achieve an efficient bipedal foot in humans. In apes, which lack the longitudinal arch, they produce a foot that is capable of weight support while retaining its primitive function as a prehensile organ (Chapter 23).

Muscles that move the pelvic girdle and hip joint

Because the pelvis is relatively immobile, the musculature in the hip region either moves the thigh in relation to a fixed torso or moves the torso in relation to a fixed thigh.

Muscles that flex the thigh at the hip joint

The main flexor muscle of the hip is **iliopsoas** (Fig. 19.15). This is a compound muscle made up of two muscles whose origins can be determined from their names. **Psoas major** (psoa (Gk) = the 'loin' region) arises from the vertebrae in the 'loin area' of the body

(T12–L5) and inserts onto the lesser trochanter of the femur. The other muscle comprising iliopsoas is **iliacus** which arises from the internal surface of the blade of the ilium (iliac fossa) and the lateral part of the upper surface of the sacrum and inserts onto the tendon of psoas major.

In the African apes the origins and insertions of these muscles are similar to the human condition. However, there is a third muscle that is closely associated with iliopsoas. This is **psoas minor** which arises from the last thoracic and first lumbar vertebrae and inserts via a long tendon onto the iliopectineal line of the pelvic bone. Because it inserts onto the pelvis and not onto the femur it cannot flex the thigh at the hip, but it does act to flex the trunk in the lumbar region. It is normally absent in humans.

An additional flexor of the hip is **rectus femoris** (rectus (L) = drawn in a straight line) (see Fig. 19.23). This is one of the four muscles making up **quadriceps femoris** but is the only one of this group that originates on the pelvis and so is able to flex the thigh at the hip. It arises from the anterior inferior iliac spine and the adjacent part of the acetabulum and inserts via a tendon into the patella which in turn is anchored to the tibial tuberosity by the patellar ligament. Rectus femoris acts simultaneously both to flex the thigh and to extend the knee and is exceptionally well developed in humans (and in leaping primates). The remaining three muscles that make up the quadriceps femoris (**vastus lateralis**, **vastus medialis**, **vastus intermedius**) arise from the tibia and insert in common with rectus femoris onto the patella. These muscles are discussed in a later section with the muscles that move the knee.

Tensor fascia latae, another flexor of the hip, is relatively insignificant in comparison with the previously mentioned flexor muscles, but has other important functions (Fig. 19.15). It arises from the anterior part of the iliac crest and the anterior superior iliac spine and inserts onto a strong band of tissue, the **iliotibial tract**, which itself inserts onto the lateral side of the tibia above the head of the fibula (see also Fig. 19.17). Tensor fascia latae, literally stretches this band on the side of the thigh (tendere (L) = to stretch; fascia (L) = band; latus (L) = side). It pulls it superiorly and anteriorly while the **gluteus maximus**, which also inserts onto the iliotibial tract (see below) pulls it superiorly and

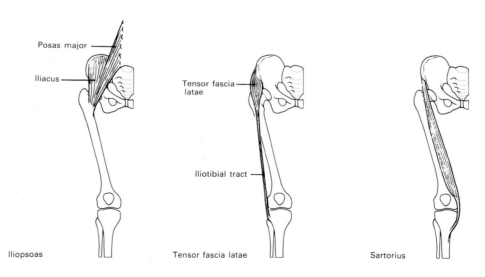

Posas major

Iliacus

Iliopsoas

Tensor fascia latae

Iliotibial tract

Tensor fascia latae

Sartorius

FIGURE 19·15 *Muscles that flex the thigh at the hip joint. After Green and Silver (1981).*

posteriorly. Together these muscles flex the hip and extend the knee. Furthermore, tensor fascia latae abducts and medially rotates the hip joint. Tensor fascia latae and the iliotibial tract also act as a strut to counteract the mediolateral bending stress on the femur and brace the knee when it is extended.

In the African apes tensor fascia latae arises from the fascia over the gluteal muscles and not from the iliac crest. Because of its different orientation, it flexes rather than extends the knee, although it also flexes, and abducts the hip joint as it does in humans. Tensor fascia latae is absent in the orangutans (Sigmond, 1975).

Sartorius is another muscle belonging to the flexor group that has an important additional function (Fig. 19.15). Sartorius (sartor (L) = tailor) has this name because it brings the leg into the 'cross-legged' position, the posture once habitually adopted by tailors. That is to say it flexes the knee and medially rotates the leg while simultaneously flexing, adducting and laterally rotating the thigh. It achieves this by running diagonally across the front of the thigh from its origin on the anterior superior iliac spine to its insertion on the upper part of the medial surface of the tibia. Sartorius in the African apes is similar to the human muscle in its origin, insertion and function.

The African apes have an additional small and insignificant flexor of the thigh that is absent in humans. This is **iliotrochantericus**. It arises in common with rectus femoris and/or from the iliofemoral ligament and, as its name implies, inserts onto the lesser trochanter of the femur.

Muscles that extend the thigh at the hip joint

The main extensors of the thigh at the hip joint are **biceps femoris**, **semimembranosus** and **semitendinosus** which together are known as the **hamstrings** (Fig. 19.16). The importance of these muscles to locomotion is reflected in the term 'hamstringing' which involves the intentional severing of the hamstring tendons of an animal in order to cripple it.

All three of these muscles arise on the ischial tuberosity where the recognizable facet for semimembranosus lies above (proximal to) that of the other two muscles and is separated from it by an oblique ridge (see Fig. 20.5). Biceps femoris also has a second

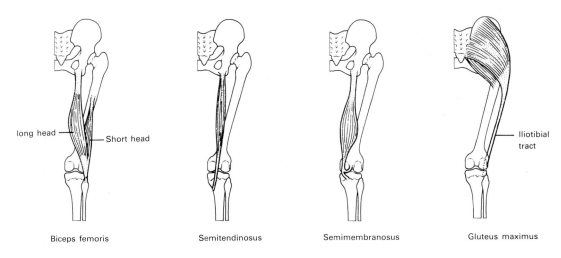

long head — Short head

Biceps femoris   Semitendinosus   Semimembranosus   Iliotibial tract   Gluteus maximus

FIGURE 19·16 *Muscles that extend the thigh at the hip joint. After Green and Silver (1981).*

411

origin (via its short head) from the lower part of the posterior surface of the femur (linea aspera) and the lateral supracondylar line. Biceps femoris inserts onto the head of the fibula and semimembranosus and semitendinosus superiorly onto the medial side of the tibia. As a result these muscles not only extend the hip joint but also flex the knee.

The origins and insertions of these muscles are similar in the African apes. However, because of the habitual flexed hip and knee posture of these primates, the hamstrings have a better lever advantage than is the case in humans. The hamstrings are about the same size (weight) as quadriceps femoris in chimpanzees. This reflects an equivalent functional importance of the muscles of the front and back of the thigh in chimpanzee locomotion (Zihlman and Brunker, 1979). In humans quadriceps femoris is about twice as heavy as the hamstrings.

Gluteus maximus (gloutos (Gk) = rump) is an additional important extensor of the hip (Fig. 19.16). As its name implies this is the largest muscle of the rump, or buttock. It arises in humans from the posterior part of the iliac crest, the sacrum, the upper part of the coccyx and the sacrotuberous ligament and inserts onto the gluteal ridge of the femur

and the iliotibial tract. It also acts as a lateral rotator of the thigh. This muscle is not active during normal walking but comes into action when more power is needed such as in climbing stairs, running, or rising from a sitting posture. Human buttocks are soft and 'flabby' in standing and walking because of the relaxation of this muscle. Gluteus maximus also has another important function in bipedal humans. It balances and controls the trunk during activities involving the upper limb such as carrying, digging, clubbing and throwing (Marzke et al., 1988).

Gluteus maximus has a very different form in the African apes (Fig. 19.17). Because of differences in the shape of the pelvis it has a much longer origin in the apes than it does in humans. It originates via an **aponeurosis** (apo (Gk) = from; neuron (Gk) = tendon), or fibrous sheet, from the distal half of the sacrum and upper part of the coccyx, from the sacrotuberous ligament and from the proximal part of the ischial tuberosity. It also has a much more extensive area of attachment in apes than it does in humans. The upper, or proximal, part of the muscle, the **gluteus maximus proprius** (proprius (L) = special, or one's own) is a thin muscle inserting on the iliotibial tract. The lower, or more caudal, portion, the **ischiofemoralis** (Sigmond 1974,

FIGURE 19·17 *Gluteus maximus in a human and a chimpanzee. After Zihlman and Brunker (1979).*

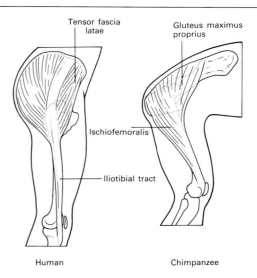

Tensor fascia latae

Gluteus maximus proprius

Ischiofemoralis

Iliotibial tract

Human

Chimpanzee

1975; Tuttle et al., 1975) is a much more robust muscle inserting along the entire length of the femur from the gluteal tuberosity to the lateral epicondyle. The orang-utan differs from the African apes in that gluteus maximus proprius and ischiofemoralis are two entirely separate muscles rather than the upper and lower parts of a single gluteus maximus muscle. The human gluteus maximus is equivalent to the ape gluteus maximus proprius, which, in these primate acts as an abductor and lateral rotator. The ischiofemoralis, which is absent in humans, acts as an extensor of the ape hip joint (Tuttle et al., 1975, 1978, 1979b).

Muscles that abduct the thigh at the hip joint

In humans the smaller gluteal muscles, **gluteus medius** and **gluteus minimus** are the main abductors of the thigh (Fig. 19.18). Gluteus medius also rotates the thigh, the anterior fibres rotating it medially and the posterior fibres laterally. In humans, gluteus medius covers gluteus minimus. The former arises from the lateral side of the ilium between the posterior and middle curved lines and inserts onto the greater trochanter

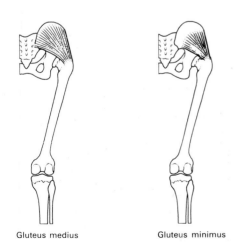

Gluteus medius   Gluteus minimus

FIGURE 19·18 *Muscles that abduct the thigh at the hip joint. After Green and Silver (1981).*

TABLE 19·1 The weight of the gluteal muscles expressed as a percentage of the total weight of the hip musculature in humans and the African apes. After Haughton (1873).

	Gorilla	Chimpanzee	Human
Gluteus maximus (%)	13.3	11.7	18.3
Gluteus medius (%)	14.0	14.6	12.3
Gluteus minimus (%)	1.8	4.2	3.5

of the femur while gluteus minimus arises 'under' it on the ilium between the middle and anterior curved lines and inserts medially to it on the greater trochanter (Fig. 19.18). The position of these muscles over the lateral side of the hip joint and their function as abductors of the thigh (with the hip joint extended) or alternatively abductors and stabilizers of the pelvis are unique features of human bipedal locomotion.

In humans, gluteus maximus accounts for a greater percentage of total hip musculature than do gluteus medius and gluteus minimus (Table 19.1) (Zihlman and Brunker, 1979). The reverse is true in the African apes. This reflects the importance to these primates of the small gluteal muscles as extensors of the thigh at the hip joint. With the hip flexed, the small gluteal muscles can also function as rotators of the thigh and abductors of the thigh (or pelvis) (Sigmon, 1974; Zihlman and Brunker, 1979; Stern and Susman, 1983). In the African apes, gluteus medius has a larger origin than in humans, whereas gluteus minimus has a smaller and much more linear origin (Fig. 19.19).

In some chimpanzees and gorillas the anterolateral (or more linear) portion of gluteus minimus is either completely or partially separate from the remainder of the muscle (as commonly occurs in the orang-utan) (Fig. 19.19). This separate muscle has been given the name **scansorius** which refers to its action in drawing the leg up towards

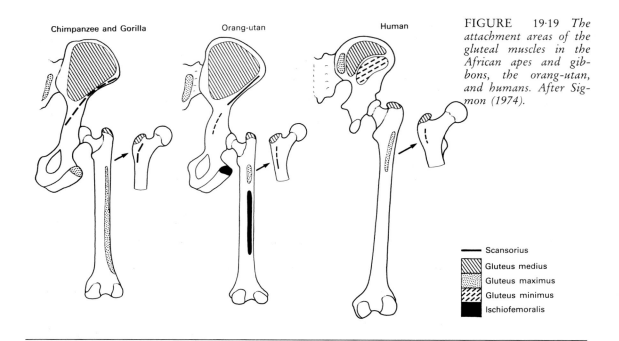

FIGURE 19·19 *The attachment areas of the gluteal muscles in the African apes and gibbons, the orang-utan, and humans. After Sigmon (1974).*

— Scansorius
▨ Gluteus medius
⋮ Gluteus maximus
⧄ Gluteus minimus
■ Ischiofemoralis

the body and its purported importance in climbing (Sigmon, 1974). Because of its anterior origin, this muscle (or the corresponding part of gluteus minimus) flexes rather than extends as well as abducts and medially rotates the thigh in the African apes (Sigmon, 1974).

Muscles that adduct the thigh at the hip joint

The adductors of the thigh are largely similar in humans and African apes. They comprise five separate muscles that fan out from the ischial tuberosity, ischiopubic ramus and pubis and insert along the entire length of the posterior part of the femur and onto the upper part of the medial surface of the tibia. (Fig. 19.20).

Adductor magnus is the largest of these muscles. It arises from the ischial tuberosity and ischiopubic ramus and has a long linear insertion extending from the greater trochanter along the linea aspera to a knob of bone known as the adductor tubercle on the medial epicondyle of the femur. Because part of this muscle arises posterior to the hip joint it can

also act to extend the flexed thigh. The remaining adductors originate anterior to the axis of the hip joint, and both medial and superior to the origin of adductor magnus. Because of their origin anterior to the axis of the hip joint they can also flex the extended thigh.

(1) **Adductor longus** arises from the pubis below the pubic crest and inserts along the middle third of the femoral shaft onto the linea aspera.

(2) **Adductor brevis** arises from the lower part of the pubis and inserts onto the upper third of the femur in a line running from the lesser trochanter to the upper part of the linea aspera.

(3) **Pectineus** (pecten (L) = comb or pubic region) is the most superior of the adductors, spanning the pubic region from its origin on the pubis to its insertion onto the upper half of the pectineal line, which connects the lesser trochanter and the linea aspera.

(4) **Gracilis** is the only one of the adductors that does not insert onto the femur. It

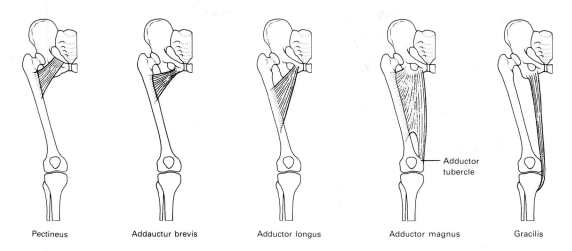

Pectineus   Addauctur brevis   Adductor longus   Adductor magnus   Gracilis

FIGURE 19·20   *Muscles that adduct the thigh at the hip joint. After Green and Silver (1981).*

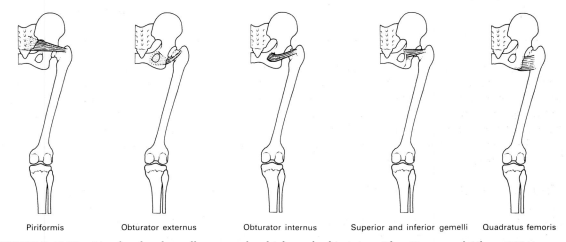

Piriformis   Obturator externus   Obturator internus   Superior and inferior gemelli   Quadratus femoris

FIGURE 19·21   *Muscles that laterally rotate the thigh at the hip joint. After Green and Silver (1981).*

originates on the body and inferior ramus of the pubis and inserts onto the upper part of the medial surface of the tibia. Because it traverses the knee it is also capable of flexing and medially rotating the leg. Its name derives from its long, slender and, therefore, gracile form in humans. However, in the African apes it is a wide, flat muscle considerably larger than that found in humans. The attachment of gracilis to the superficial fascia

of the leg may limit the amount of extension possible at the ape knee joint (Sonntag, 1924).

The lateral rotators of the thigh at the hip joint

Lateral rotation is an important movement in bipedal locomotion. This is reflected in the fact that in humans there are six separate muscles the function of which is lateral rotation (Fig. 19.21). In some ways these

415

muscles are analogous to the **rotator cuff** muscles of the shoulder. Not only are they relatively short muscles, their origins being very close to their insertions, but also their insertions are close to the joint itself, reducing their lever advantage. All these muscles pass around the back of the hip joint.

(1) **Piriformis** (pirium (L) = pear) arises from the sacrum and upper border of the greater sciatic notch and inserts onto the highest point of the greater trochanter of the femur.

(2) **Obturator externus** arises from the outer surface of the obturator membrane (covering the obturator foramen) and surrounding bone, runs outward and backward below the hip joint, wraps around back of the neck of the femur and inserts in the trochanteric fossa on the medial side of the greater trochanter. This tortuous route is the direct result of the habitually extended human thigh. When the human hip joint is flexed the obturator externus runs in a direct line from its origin to its insertion. It also takes a straight course in the African apes whose hip joint is habitually flexed.

(3) **Obturator internus** arises from the inner surface of the obturator membrane and runs in a direct line to its insertion onto the inner surface of the greater trochanter.

(4) **Superior gemellus** (gemellus (L) = twin or double) arises from the ischial spine and upper margin of the greater sciatic notch and inserts with obturator internus onto the inner surface of the greater trochanter.

(5) **Inferior gemellus** arises from the inferior margin of the sciatic notch and the lower margin of the lesser sciatic notch and inserts with obturator internus and superior gemellus onto the inner surface of the greater trochanter.

(6) **Quadratus femoris** arises from the lateral border of the ischial tuberosity and inserts onto the quadrate tubercle on the intertrochanteric crest.

Summary of the differences in the human and ape hip musculature

Sigmon (1974) noted that there are two major differences in the hip and thigh muscles in humans and apes.

(1) In humans the extensor musculature is designed for speed while in apes it is designed for power. In the African apes the ischial tuberosity is longer than it is in humans and the lower limb is shorter. This arrangement results in a long moment arm (ischial tuberosity) and a short lever arm (lower limb) and produces a more powerful force than is possible in humans. The reversed human condition results in a relatively weaker force, but one which produces a faster movement over a larger range of action than is possible for the apes (Sigmon, 1974).

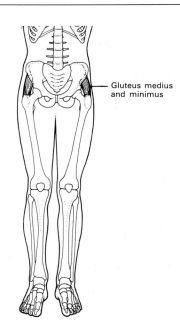

Gluteus medius and minimus

FIGURE 19·22 *The small gluteal muscles pass over the lateral side of the hip joint in humans and balance the pelvis during walking. When weight is taken off one leg, the small gluteal muscles on the opposite side contract to prevent the pelvis from collapsing towards the unsupported side of the body.*

(2) In humans there is an effective system of lateral balance control during erect bipedal walking. This results from the change in the shape of the iliac blades of the pelvis which carry the origins of the small gluteal muscles around to the lateral side of the body and allow them to work as effective abductors of the pelvis (on an extended thigh) (Fig. 19.22).

In addition to the above there are also muscles that are found in the African apes but are not commonly found in humans. These include scansorius and iliotrochantericus. Furthermore, gluteus maximus has an entirely different shape and two named parts (gluteus maximus proprius and ischiofemoralis) in the apes.

Electromyography of the hip and thigh muscles in humans and apes

There are marked differences between the activity of the hip and thigh muscles of humans and apes during bipedal locomotion. In humans there is surprisingly little activity in either hip or thigh muscles (Tuttle et al., 1979b) and high frequencies of biphasic and triphasic patterns (muscles fire, or are active,

two or three times during a stride) (Fig. 14.26). This contrasts with the chimpanzees where many of the hip and thigh muscles so far studied contract throughout the stance phase (Ishida et al., 1975; Tuttle et al., 1975, 1978, 1979a,b). There is also considerable gluteal and hamstring activity in bipedal standing in the African apes (Tuttle et al., 1979b) which is a requirement of their flexed hip and knee bipedal posture. In humans both the gluteal muscles and hamstrings are inactive during bipedal standing.

Muscles that move the leg at the knee joint

The majority of the muscles that act to flex, extend or rotate the leg at the knee joint also act on other joints of the lower limb and are discussed in detail either with the muscles that move the hip joint or the muscles that move the foot. There are only four muscles that act exclusively on the knee joint (Fig. 19.23). Three of these are extensors of the knee (**vastus lateralis**, **vastus medialis** and **vastus intermedius**) and the fourth (**popliteus**) both flexes the knee joint and laterally rotates the thigh (femur) on the

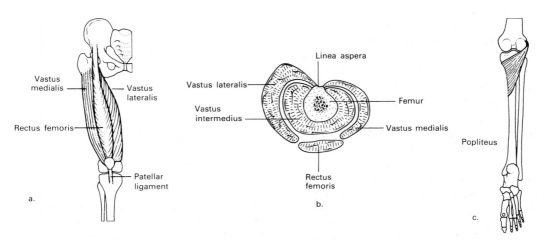

FIGURE 19·23 *Muscles that move the lower leg at the knee joint; (a) quadriceps femoris, (b) a cross-section of the thigh showing the position of the four quadriceps femoris muscles, (c) popliteus. After Green and Silver (1981).*

immobile leg or, alternatively, medially rotates the leg (tibia) in relation to a fixed thigh.

The three **vasti** muscles, together with a fourth muscle, **rectus femoris** (which arises from the pelvis and, therefore, also acts to flex the hip), insert onto the patella which in turn inserts onto the tibial tuberosity via a strong tendon called the **ligamentum patellae**, or **patellar ligament**. These four muscles together (known as the **quadriceps femoris**) make up the bulk of the anterior part of the thigh and are the primary extensors of the knee.

Vastus intermedius lies closest to the anterior surface of the femur and originates from most of its anterior and lateral surface. It also covers the medial surface of the femur but has no attachment on this surface. It lies both inferior to and between the other two vasti muscles which also arise from almost the total length of the femur. **Vastus medialis** has a linear origin extending from the lower part of the intertrochanteric line, along the medial lip of the **linea aspera** to the **medial supracondylar line** at the distal end of the bone. The equally linear origin of **vastus lateralis**, the largest of the vasti muscles, extends from the upper half of the intertrochanteric line and the lower part of the greater trochanter, along the lateral lip of the linea aspera to the **lateral supracondylar line**.

The origins, insertions and functions of these muscles are largely similar in humans and in the African apes. Furthermore, humans and the African apes are most similar to each other and to other primates that engage in a slow climbing locomotor behaviour in the extent of the origins of these muscles (and in the greater fusion of muscle groups) and contrast particularly with those primates that leap in the trees (Stern, 1971; Jungers et al., 1983).

Popliteus is a relatively small muscle that runs obliquely across the back of the knee joint from its origin on the lateral femoral condyle to its insertion on the posterior surface of the tibia. In humans one of its main functions is to 'unlock' the fully extended knee by laterally rotating the femur on the fixed tibia as the knee joint begins to flex. Popliteus is also very well developed in the African apes where its primary function is most probably the medial rotation of the tibia at the much more mobile ape knee joint.

In both humans and African apes, it is also connected at its origin to the lateral **meniscus**, a pad of fibrocartilage lying between the femur and tibia in the knee joint. Popliteus may also help to stabilize the joint by controlling the movement of this pad (Chapter 22).

Muscles that move the foot

As is the case with the muscles that move the hand, the muscles that move the foot are divided into two major groups, the long (or extrinsic) muscles and the short (or intrinsic) muscles. The long muscles originate outside the foot on the epicondyles of the femur or on the tibia, fibula or interosseous membrane of the leg. They insert onto the tarsal bones, the metatarsals or the phalanges and are responsible for dorsiflexion or plantarflexion of the foot at the ankle joint. They are also the major muscles responsible for inversion and eversion of the foot at the peritalar joint and flexion and extension of the phalanges at the metatarsophalangeal and interphalangeal joints.

The short muscles have both their origins and their insertions on the foot itself. In human bipedal locomotion these muscles are primarily responsible for 'fine-tuning' the position of the foot during the stance phase of the stride. In particular they position the anterior part of the foot in relation to the posterior part, maintain the mediolateral distribution of pressure on the foot, and increase the rigidity of the foot (Reeser et al., 1983). In the apes these short muscles of the foot are better developed than they are in the humans and play a prominent role in the use of the foot as a prehensile organ.

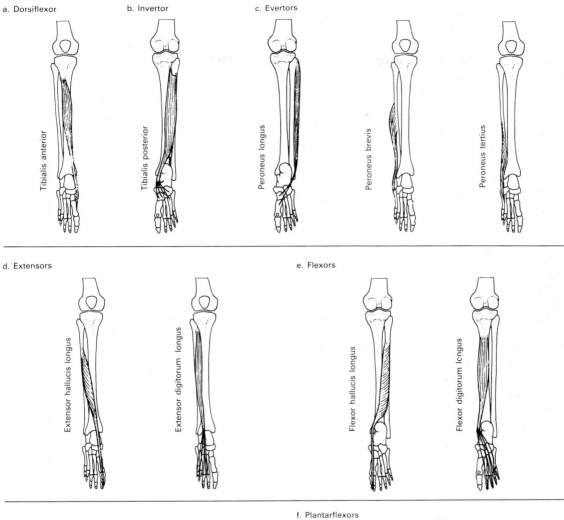

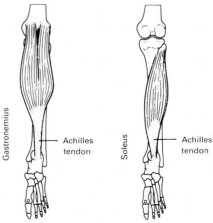

FIGURE 19·24 *The long muscles of the foot; (a) muscles that dorsiflex the foot, (b) muscles that invert the foot, (c) muscles that evert the foot, (d) muscles that extend the toes, (e) muscles that flex the toes, and (f) muscles that plantarflex the foot. After Green and Silver (1981).*

The long muscles of the foot

The action of the long muscles of the foot can (in general terms) be inferred from their position of origin. Those that take their origin from the anterior surface of the leg and pass over the anterior surface of the ankle joint are dorsiflexors of the ankle and (if their insertions extend to the phalanges) extensors of the toes. Those muscles that take their origin from the posterior surface of the leg and pass over the posterior surface of the ankle joint are plantarflexors of the ankle and (if their insertions extend to the phalanges) flexors of the toes. The more medially positioned these muscles are the more they tend to contribute to inversion of the foot and conversely, the more laterally positioned the more they contribute to eversion of the foot.

LONG MUSCLES THAT DORSIFLEX THE FOOT

The major dorsiflexor of the foot is **tibialis anterior** (Fig. 19.24). This muscle originates from the upper two-thirds of the lateral part of the tibia and from the lateral condyle of the tibia and inserts onto the dorsal surface of the base of the first metatarsal and onto the medial cuneiform. Because of its medial insertion it can also act as an invertor of the foot.

As tibialis anterior passes over the ankle joint it is held down and kept from 'bow-stringing' by the **extensor retinacula** (Fig. 19.25). There are generally two extensor retinaculae in the human leg. The most proximal (or superior) of these is a strong band of tissue extending from the fibula to the tibia. The distal (or inferior) retinaculum

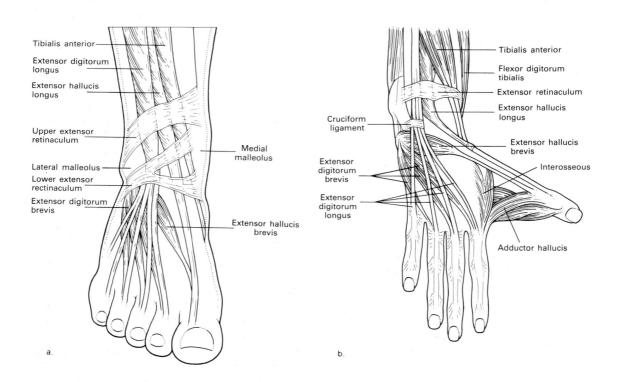

FIGURE 19·25 *Muscles on the dorsal aspect of the (a) human and (b) chimpanzee foot. After Sokoloff (1972).*

is generally Y shaped and more complicated. The base of the Y originates from the lateral side of the calcaneus, the upper limb of the Y attaches onto the medial malleolus and the lower limb sweeps around the instep of the foot, blending with the deep tissue (fascia) on the sole of the foot.

LONG MUSCLES THAT INVERT THE FOOT

Tibialis posterior is the prime invertor of the foot (Fig. 19.24). It arises from the proximal half of the posterior surface of the tibia (medial to and below the soleal line), on the posterior surface of the interosseous membrane and on the posterior fibular shaft. It becomes tendinous near the ankle joint, where it curls around the medial malleolus. The primary insertion is on the navicular tuberosity. However, small branches of the tendon fan out and insert onto all three cuneiforms, the cuboid, the tendon sheath of peroneus longus and the bases of the lateral four metatarsals. This broad insertion reflects the importance of inversion during the initial part of the stance phase in bipedal walking (heel strike). However, it differs only in minor details from the similarly broad insertion found in the chimpanzee foot. In chimpanzees the foot is normally inverted during arboreal locomotion (Lewis, 1980b). The broad chimpanzee insertion of tibialis posterior may also facilitate the grasping action of the foot by pulling the digital portion of the foot against the hallucial (great toe) portion (Lewis, 1964).

The human tibialis posterior insertion does differ from the chimpanzee insertion in that a branch of the tendon blends with the tendon of **flexor hallucis brevis** (one of the short muscles of the foot). Lewis (1964) argued that this arrangement is a unique adaption to bipedal locomotion. It stabilizes the human longitudinal arch by adding the power of tibialis posterior to that of the smaller flexor hallucis brevis. One of the main distinctions between the human and the ape foot is an increase in the musculature that flexes the

great toe in the human foot and the separation of this musculature from that of the lateral toes.

LONG MUSCLES THAT EVERT THE FOOT

Peroneus longus and **peroneus brevis** are the major evertors of the foot (Fig. 19.24). Both originate on the lateral part of the leg, peroneus longus from the upper part of the lateral surface of the fibula and peroneus brevis from lower on the dorsolateral surface of the fibula (peroneus (L) = fibula; (Gk) = pin). Both muscles become tendinous near the ankle and these tendons wrap around the lateral (fibular) malleolus in the **peroneal groove** on the base of this bone (Fig. 19.8). Peroneus brevis inserts onto the base of the fifth metatarsal, while peroneus longus inserts onto the plantar surface of the medial cuneiform and the base of the first metatarsal on the opposite side of the foot. To reach this insertion, its longer tendon curls around the cuboid and travels across its plantar surface in a well-developed groove.

A third peroneus muscle, **peroneus tertius**, also may variably occur in humans. When it is present it arises from the anterior surface of the distal fibula and inserts onto the shaft of the fifth metatarsal. As the other peroneal muscles it both everts and dorsiflexes the foot.

Electromyographic studies demonstrate that in humans both peroneus longus and peroneus brevis are active during the second half of the stance phase when weight is transferred onto the anterior part of the foot and medially onto the ball of the foot (Stern and Susman, 1983; Reeser et al., 1983). While both muscles are inactive in the chimpanzee during terrestrial locomotion they are active in arboreal locomotion (Stern and Susman, 1983). Stern and Susman suggested that eversion of the foot in chimpanzee arboreal locomotion is important in controlling the transfer of weight to the forefoot between the opposed hallux (great toe) and the second toe. Because peroneus longus inserts primarily onto the base of the first metatarsal in chimpanzees it has also been suggested that it might be important to the grasping, opposable great toe. However, Stern and Susman (1983) demonstrated that it is only active in chimpanzees (independent from peroneus brevis) during postural behaviours on supports of small diameters and on vertical trunks, and not during locomotion.

Long extensors of the toes

The two long extensors of the toes are the **extensor hallucis longus** and the **extensor digitorum longus** (Fig. 19.24). As the name implies extensor hallucis longus extends the great toe. It arises from the middle part of the anterior surface of the fibula and the interosseous membrane. Becoming tendinous near the ankle, it passes under the extensor retinaculum, expands and fuses with the joint capsule over the metatarsophalangeal joint and continues on to insert onto the base of the distal phalanx. Extensor digitorum longus is responsible for extending the remaining four toes. It takes its origin from the anterior shaft of the fibula and the interosseous membrane and, as does extensor hallucis longus, becomes tendinous near the ankle joint. However, as it passes under the extensor retinaculum it divides into four tendons which insert onto the four lateral toes. As each passes over its respective metatarsophalangeal joint it expands and fuses with the capsule (termed the **extensor expansion**) and then continues on to insert by a central slip onto the base of the middle phalanx and by two lateral slips onto the base of the terminal phalanx. In the chimpanzee these muscles are essentially similar to their human counterparts.

Long flexors of the toes

There are two long flexors of the toes, **flexor hallucis longus (flexor digitorum fibularis** in comparative anatomy) and **flexor digitorum longus (flexor digitorum tibialis** in comparative anatomy) (Fig. 19.24). Both these muscles also act as plantarflexors of the ankle.

As the name suggests, flexor hallucis longus flexes the great toe. It arises from the lower two-thirds of the posterior surface of the fibula, and becomes tendinous near the ankle. It curls around the talus in a well-defined groove on the posterior surface of this bone and continues under the sustentaculum tali in an equally well-defined groove to its insertion on the distal phalanx of the great toe. As with all the flexors of the toes it is held against the proximal phalanx by a well-developed flexor sheath.

It is often said that flexor hallucis longus does not exist in the apes. This is only true for the orang-utan where 21 out of 23 individuals lacked a long flexor tendon to the great toe (Straus, 1942). In the African apes flexor digitorum fibularis is analogous to the human flexor hallucis longus, having the same origin and entering the foot in the same manner. However, once in the foot the tendon divides into three branches which insert onto the distal phalanges of digits 1, 3 and 4 (Fig. 19.26). In these primates it functions not only to flex the great toe but also to flex digits 3 and 4.

In humans the second flexor of the toes, flexor digitorum longus, arises from the posteriomedial third of the tibial shaft,

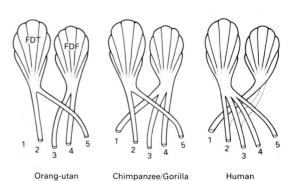

FIGURE 19·26 *The long flexors of the toes in the orang-utan, the African apes, and humans where flexor digitorum tibialis (FDT) is known as flexor digitorum longus and flexor digitorum fibularis (FDF) is known as flexor hallucis longus. After Straus (1942).*

becomes tendinous near the ankle and passes into the foot below the tendon of tibialis posterior and above the tendon of flexor hallucis longus. Once in the foot it divides into four tendons which insert onto the bases of distal phalanges 2 to 5.

In the chimpanzee this muscle is known as the flexor digitorum tibialis (Fig. 19.26). Its origin and passage into the foot is similar to the human condition. However, once in the foot its tendon divides and inserts onto the bases of only the second and the fifth distal phalanges (Keith, 1929; Straus, 1942; Lewis, 1964).

The major difference between the human and ape long digital flexors is that in humans the long flexor of the great toe is entirely separate from the long flexor of the lateral toes (although it may send very small and largely insignificant slips to the tendons of flexor digitorum longus leading to digits 2 and 3).

Long muscles that plantarflex the foot

Plantarflexion is the movement responsible for lifting the entire weight of the body over the foot as the heel 'lifts-off' during the stance phase of bipedal locomotion. Although the long flexors of the toes also act to plantarflex the foot, the **triceps surae** (sura (L) = calf of the leg) is the muscle primarily responsible for this movement and is the largest of the long muscles that move the foot (Fig. 19.24). It is a composite muscle made of two separate muscles (but arising from three heads) that in humans unite in a single tendon (the **tendo calcaneus** or **Achilles tendon**) before inserting on the heel (calcaneal tuberosity). In humans the tendon is long being about 65% of the total muscle length while in apes it is very short (Prejzner-Morawska and Urbanowicz, 1981). In both humans and apes the attachment is at mid-level on the calcaneal tuberosity and above the attachment the tendon is separated from the calcaneus by a bursa.

Gastrocnemius (gastro (Gk) = belly; kneme (Gk) = leg) is the largest of the two muscles comprising the triceps surae and arises by means of two heads, one from the lateral and one from the medial femoral condyle. In humans and gorillas the medial head is larger, and extends more distally, than the lateral head (Sigmon and Farslow, 1986), while in chimpanzees the heads are equally developed (Fig. 19.27). Sesamoid bones are found at the point of origin of both heads in the chimpanzees but are normally absent in gorillas and humans. Because this muscle crosses the knee joint it also functions as a flexor of the knee.

The second part of the triceps surae is **soleus** (solea (L) = sandal or sole) which derives its name from its fancied resemblance to a flat fish (*Pleuronectes solea*). The muscle is indeed flat and lies deep to gastrocnemius. In humans it has a horse-shoe shaped origin extending from the back of the fibula to the soleal line on the back of the tibia. In apes its origin is generally confined to the fibula; however, there may be an accessory origin on the tibia (Prejzner-Morawska and Urbanowicz, 1981).

An additional muscle, **plantaris**, is closely associated with the triceps surae. Plantaris ((L) = shoot or twig) is a long and delicate muscle. It arises from the lateral condyle of

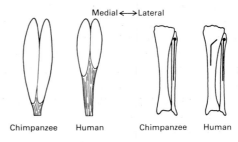

FIGURE 19·27 *Triceps surea in humans and chimpanzees. (a) The medial head of gastrocnemius is larger and extends more distally than the lateral head in humans. The two heads are of equal length in the chimpanzees. The tendinous part of the muscle is also much larger in humans. (b) the origin of soleus is confined to the fibula in the chimpanzee but extends from the fibula onto the tibia in humans. After Prejzner-Morawska and Urbanowicz (1981).*

423

the femur near the inner side of the lateral head of gastrocnemius and quickly forms a long tendon. While normally present in humans, plantaris is frequently absent in the chimpanzee (and orang-utan) and always absent in the gorilla (and gibbon). It is best developed in the leaping prosimians and in the monkeys (Sigmon and Farslow, 1986).

The triceps surae makes up the great bulk of the calf of the leg. The human calf muscles are twice as heavy as those of the chimpanzee calf relative to total body weight (Zihlman and Brunker, 1979). The well developed human triceps surae reflects the fact that these muscles lift the entire weight of the body over the foot during the second half of the stance phase in bipedal locomotion. The well-developed Achilles tendon is also a direct reflection of bipedal locomotion. Alexander (1984a) has suggested that the Achilles tendon behaves like a spring, storing energy at one stage in the locomotor cycle and releasing it at another, thereby allowing humans to conserve a considerable amount of energy particularly while running. The small Achilles tendon in the apes therefore implies a reduced efficiency of triceps surae in relation to humans.

LONG MUSCLES FOUND IN CHIMPANZEES BUT NOT IN HUMANS

There is one long muscle that is found in the chimpanzees but not in humans. This is the **abductor hallucis longus** which in the chimpanzee originates as part of the **tibialis anterior** and becomes a separate muscle that inserts onto the base of the first metatarsal. It functions to abduct the great toe in chimpanzees (and other primates) and its absence in humans reflects the absence of an opposable great toe.

The short muscles of the foot

There is only one short muscle on the dorsal surface of the foot (Fig. 19.28). This is **extensor digitorum brevis** the function of which, as the name implies, is to extend the

toes. In humans it originates on the anterior part of the dorsal surface of the calcaneus and is divided into four parts. The most medial part (sometimes called **extensor hallucis brevis**) inserts via a tendon onto the base of the proximal phalanx of the great toe. The tendons of the lateral three parts each insert onto a tendon of **extensor digitorum longus** running to the second, third or fourth digits. There is no tendon running to the fifth digit, although it may be vestigially represented by **peroneus tertius**, a muscle that is variably found in humans.

Extensor digitorum brevis in the chimpanzee and the gorilla is similar to the human muscle. It differs only in a more marked division between the parts of the muscle going to the great toe and to the lateral three digits (Fig. 19.25). This corresponds to the greater separation between the great toe and the lateral toes in these primates.

There are many more short muscles on the plantar side of the foot. In the human foot these muscles provide powerful flexion of the great toe, stabilize the foot during bipedal progression and reflect a more medial position of the long axis of the foot than is found in apes. In the apes these muscles reflect the

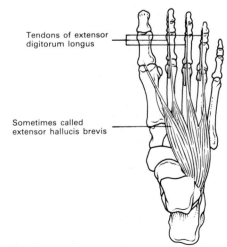

Tendons of extensor digitorum longus

Sometimes called extensor hallucis brevis

FIGURE 19·28 *Extensor digitorum brevis. After Green and Silver (1981).*

grasping power of the toes and, in particular, the opposability of the great toe.

The short muscles are organized into four layers (Fig. 19.29). By convention, the first layer is the most superficial and the fourth layer is the deepest, lying closest to the plantar surface of the bones.

THE FIRST LAYER OF SHORT MUSCLES

There are three muscles in the first, or most superficial, layer of short muscles (Fig. 19.29a). **Flexor digitorum brevis** occupies the central portion of this layer. It originates from the medial process of the calcaneus and

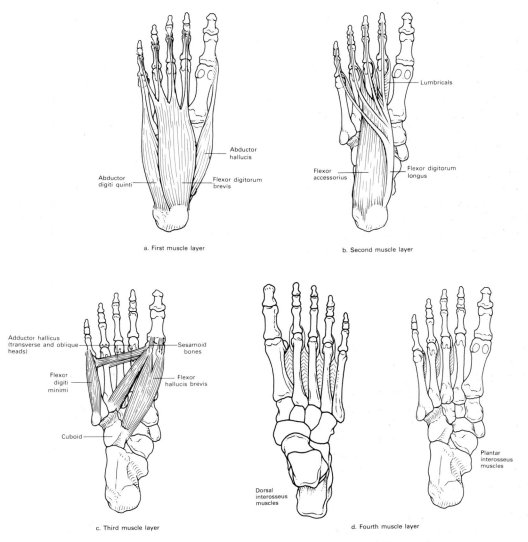

a. First muscle layer

b. Second muscle layer

c. Third muscle layer

d. Fourth muscle layer

FIGURE 19·29 *The short muscles of the plantar surface of the foot divided into (a) the first muscle layer, (b) the second muscle layer, (c) the third muscle layer, (d) the fourth muscle layer. After Green and Silver (1981).*

425

inserts by four tendons onto the middle phalanges of the lateral four toes. The other two muscles are located on either side of flexor digitorum brevis. **Abductor hallucis** originates medial to the origin of flexor digitorum brevis from the medial process of the calcaneus and inserts onto the medial side of the base of the proximal phalanx of the great toe. The third muscle, **abductor digiti quinti** (= abductor of the fifth toe), originates lateral to the origin of flexor digitorum brevis from both processes of the calcaneus and inserts onto the proximal phalanx of the little toe.

The surprising thing about the human foot is the relatively large size of the abductors and particularly of abductor hallucis. One might expect this muscle to have degenerated in a foot lacking an opposable great toe. However, electromyographic analyses have confirmed the importance of both these adductor muscles to the stabilization of the human foot during bipedal locomotion (Mann and Inman, 1964; Gray and Basmajian, 1968) and in particular to the positioning of the anterior part of the foot relative to the posterior part and to the leg (Reeser et al., 1983).

The main difference between humans and apes in the muscles of the first layer concerns flexor digitorum brevis (Keith, 1929). In humans the origin of this muscle on the calcaneus gives it a firm base to allow it to stabilize the uniquely human longitudinal arch during locomotion. Electromyographic analyses have demonstrated that it is active during locomotion when the force of the long flexors is insufficient or when more force is needed at the interphalangeal joints to enable the toes to grip the substrate (Reeser et al., 1983). In the African apes the parts of this muscle leading to the lateral digits (the fourth and fifth, and sometimes the third) originate from the tendon of flexor digitorum longus and not from the calcaneus as in humans (Fig. 19.30). This condition is also found in many monkeys and may result in the increased grasping power of the lateral digits in these primates (Keith, 1929).

THE SECOND LAYER OF SHORT MUSCLES

The four **lumbrical** muscles together with the **flexor accessorius** (or **quadratus plantae**) make up the second layer of short muscles (Fig. 19.29b). The four lumbricals originate from the medial sides of the four tendons of flexor digitorum longus and insert onto the base of each of the corresponding proximal phalanges and extensor expansions of the

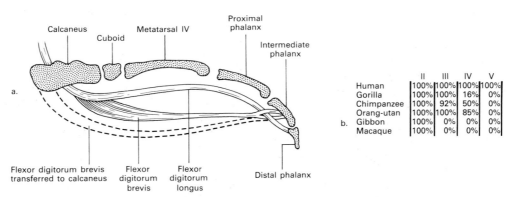

	II	III	IV	V
Human	100%	100%	100%	100%
Gorilla	100%	100%	16%	0%
Chimpanzee	100%	92%	50%	0%
Orang-utan	100%	100%	85%	0%
Gibbon	100%	0%	0%	0%
Macaque	100%	0%	0%	0%

FIGURE 19·30 *(a) A schematic section of the foot of a monkey showing the two origins of flexor digitorum brevis from the calcaneus and from the tendon of flexor digitorum longus. (b) The percentage* *of cases in which the parts of flexor digitorum brevis leading to the four lateral digits originate from the calcaneus rather than from the tendon of flexor digitorum longus. After Keith (1929).*

lateral four toes. As with the lumbricals of the hand they flex the metatarsophalangeal joints without flexing the interphalangeal joints. They are much more strongly developed in the African apes (with their long and prehensile toes) than they are in humans.

Flexor accessorius in humans originates by two heads from the concave medial surface of the calcaneus and from the area anterior to the lateral tubercle of the calcaneus. Flexor accessorius inserts onto the lateral margin of the tendon of flexor digitorum longus. It is commonly assumed that its function is to straighten out the pull of flexor digitorum longus, aligning its action along the long axes of the phalanges (Wood-Jones, 1949; Lewis, 1962). However, electromyography has shown that (together with flexor digitorum brevis) its primary function is to supplement the work of flexor digitorum longus when more flexing power is required during bipedal progression (Reeser et al., 1983).

In the African apes, flexor accessorius is frequently absent. In those individuals that do have it (and in many other primate species) it is smaller than the human muscle and originates by a single head from the lateral aspect of the calcaneus (Lewis, 1962). The large, double-headed flexor accessorius is unique to humans.

THE THIRD LAYER OF SHORT MUSCLES

Three muscles make up the third muscle layer, **flexor hallucis brevis, flexor digiti minimi brevis,** and **adductor hallucis** (Fig. 19.29c). These muscles are confined to the anterior part of the foot between the cuboid and the bases of the proximal phalanges.

Flexor hallucis brevis is a large muscle in humans, originating from the cuboid and inserting by two tendons onto the base of the proximal phalanx of the great toe. The medial tendon merges with the tendon of abductor hallucis and the lateral tendon with the tendon of adductor hallucis (see below). There is a large sesamoid bone in each of

these tendons. The size of this muscle reflects the importance of flexion of the great toe in bipedal locomotion. In the African apes this muscle is much smaller and has a single insertion on the medial side of the base of the proximal phalanx.

Of the muscles of this layer, flexor hallucis brevis shows the greatest difference between humans and the apes. However, there are also differences in adductor hallucis. In humans this muscle is divided into two largely separate parts. The oblique head originates from the bases of the second, third and fourth metatarsals and inserts (in common with the lateral tendon of flexor hallucis brevis) into the lateral side of the base of the proximal phalanx of the great toe. This head is powerful and assists flexor hallucis brevis to flex the great toe. The second head of adductor hallucis is the transverse head. It originates from the metatarsophalangeal ligaments of the second to the fifth toes and inserts with the oblique head. This head is often very weak and sometimes atrophied in humans, reflecting the absence of an opposable great toe.

In the African apes adductor hallucis takes a form intermediate between the human pattern (where the two heads are largely separate) and the powerful fan-shaped muscle found in the monkeys (and typical of the foetal condition in humans) (Keith, 1929). However, consistent with the ape opposable great toe, the transverse part of the muscle is always much better developed than is the case in adult humans.

The third muscle of the third layer is flexor digiti minimi brevis. In both humans and the African apes, this muscle originates from the base of the fifth metatarsal and inserts onto the lateral side of the base of the proximal phalanx of the little toe.

THE FOURTH LAYER OF SHORT MUSCLES

The **interosseous** muscles make up the fourth muscle layer (Fig. 19.29d). As in the hand, two groups of interosseous muscles are found in the foot. The **plantar interosseous** muscles function to adduct the toes and the **dorsal interosseous** muscles function to abduct the toes. In humans, the three **plantar interosseous** muscles arise from the plantar and medial sides of the third, fourth and fifth metatarsals and insert onto the medial side of the base of the proximal phalanx and the extensor expansion of the same toes. They adduct (or pull) the lateral toes towards the second toe. These muscles are small and relatively unimportant in the human foot.

The four dorsal interosseous muscles have a double origin from the sides of adjacent metatarsals. The first dorsal interosseous muscle arises from the opposing sides of the first and second metatarsals and inserts onto the medial side of the base of the proximal phalanx of the second digit. The second, third and fourth dorsal interosseous muscles arise from the opposing sides of the remaining metatarsals and insert onto the lateral sides of the second, third and fourth metatarsals. This arrangement abducts the digits away from the second toe.

The main difference between the human interosseous muscles and the African ape muscles is that in these apes the digits are adducted and abducted in reference to the third digit (described as the **mesaxonic** (mesus (Gk) = middle; axon (Gk) = axis) pattern) rather than in reference to the second digit (described as the **entaxonic** (endon (Gk) = within) pattern) (Fig. 19.31). This distinction can be easily seen in practice. In the apes it is the third digit that has a double dorsal interosseous insertion, while in humans it is the second digit.

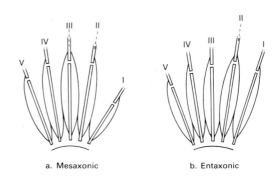

a. Mesaxonic   b. Entaxonic

FIGURE 19·31 *Schematic diagram of the dorsal interosseous muscles showing (a) the mesaxonic pattern (centring on the third digit) found in African apes and other higher primates and (b) the entaxonic pattern (centring on the second digit) common to humans. After Sigmon and Farslow (1986).*

CHAPTER TWENTY

THE HOMINOID PELVIS

Of all the bones in the postcranial skeleton, the human and ape pelves (= plural of pelvis) show perhaps the most obvious differences (Fig. 20.1). The human pelvis is short, squat and basin shaped, while the ape pelvis, especially the ilium, is elongated and two-dimensional in appearance. The major components of the pelvis also lie in completely different angular relationships to each other. Moreover, the ape pelvis occupies a different orientation in relation to the lower limb than it does in bipedal humans. In the lateral view the ape sacrum is positioned anterior to the hip joint and forms an approximate continuation of the line of the spine. In humans the sacrum lies posterior to the hip joint and is at a marked angle to the spine. The ape thigh is normally flexed at the hip during locomotion and is oriented perpendicularly to the long axis of the pelvis. This greatly enhances the lever advantages of the extensors of the hip that have their origin on the ischium and the flexors of the hip that have their origin on the elongated ilium. These features as well as others discussed in the following sections are direct reflections of the requirements of human bipedalism on the one hand and ape quadrupedalism on the other.

For purposes of discussion the different parts of the ape and human pelvis are described in turn, the ilium, the ischium the pubis, the acetabulum and the sacrum.

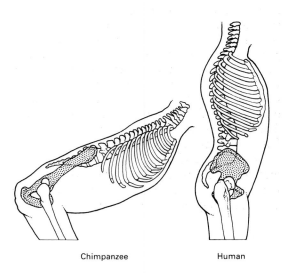

FIGURE 20·1 *The position of the pelvis in relation to the leg and the vertebral column in a chimpanzee and a human. After Schultz (1969a).*

The ilium in humans and apes

The ilium is divided into three surfaces and three crests or margins (Fig. 20.2). The **gluteal surface** is the posterolateral surface of the blade and provides the area of origin for the gluteal muscles while the **iliac fossa** is the anteromedial surface of the blade and gives origin to the iliacus muscle. The remaining surface is the **sacral surface**. This surface is directed medially and carries the

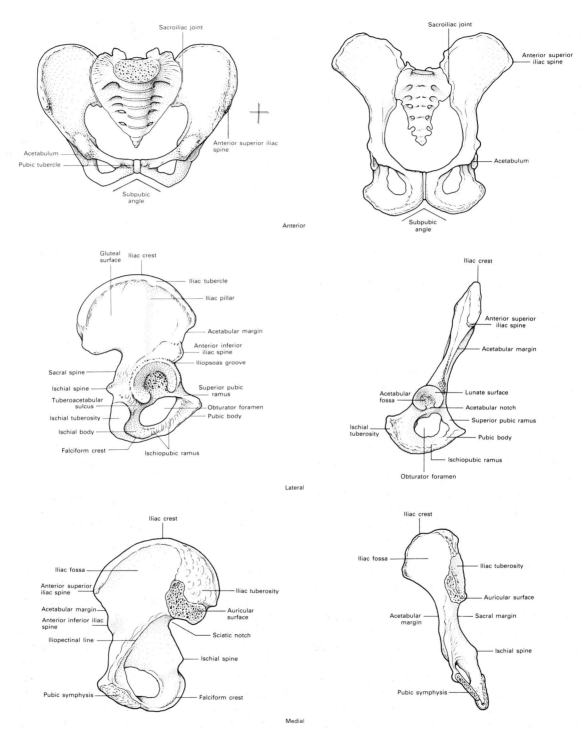

FIGURE 20·2 *A human pelvis and a chimpanzee pelvis from the anterior, lateral and medial aspects.*

articulation for the sacrum (the **auricular surface**) and the **iliac tuberosity** which lies above it. The **iliac crest** is the superior edge of the blade, the **acetabular margin** the anterior edge of the human ilium and the lateral margin of the ape ilium and the **sacral margin** the posterior edge of the human blade and medial edge of the ape blade.

Humans are the only primate in which the triangular-shaped ilium is wider than it is high (Straus, 1929) (Table 20.1). However, the unique feature of the human ilium is not its great width but its reduced height. Functionally this brings the sacroiliac joint close to the hip joint thereby reducing the stress on that part of the ilium that transmits the entire weight of the upper body from the backbone to the hip joint in bipedal posture.

There are a number of other important features of the human ilium. The most important of these features is the orientation of the blade. In humans it forms the side of the pelvis resulting in a convex gluteal surface, a concave iliac fossa and a very distinctive S-shaped iliac crest (Fig. 20.3). The upper part of the S-shaped curve is formed by the superior border of the iliac tuberosity which is long and directed posteriorly. The upper bend is the **spina limitans** which marks the transition between the border of the iliac tuberosity and the iliac fossa and the lower bend is the **iliac tubercle**. The iliac tubercle lies at the proximal end of the **iliac pillar** (a thickening of bone) which runs from the iliac crest down the blade to the acetabulum (Fig. 20.2).

These features form a complex that reflects the demands placed on the pelvis by bipedal locomotion. The curved, mediolaterally orientated iliac blade brings the small gluteal muscles (gluteus medius and gluteus minimis) into a position at the side of the pelvis where they can act as abductors of the pelvis and thereby support the upper body during bipedal locomotion (Chapter 19 and Fig. 19.22). The iliac pillar reinforces the anterior part of the blade against the pull of these muscles. The curved iliac blade also has an

TABLE 20·1 The relative width of the iliac blade [(ilium width × 100)/ilium length] in humans, apes and selected monkeys. After Straus (1929).

	N	Relative iliac width	Range
Humans	244	125.5	107.9–139.5
Gorilla	13	91.8	83.6–100.5
Chimpanzee	16	66.0	57.6– 73.7
Orang-utan	23	73.8	62.7– 87.4
Hylobates	18	48.7	41.1– 57.7

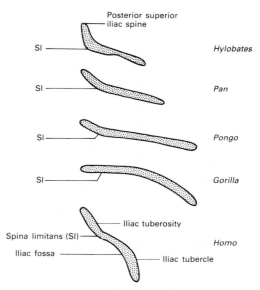

FIGURE 20·3 *The iliac crest from the superior (cranial) aspect in a gibbon, the large apes and a human. SI = spina limitans. After Waterman (1929).*

important effect on the abdominal muscles that attach to the ventral half of the iliac crest (external and internal oblique muscles) (Chapter 15). It brings these muscles into a position in relation to the ribs that they can help to balance the trunk above the pelvis during bipedal walking. Furthermore, the posterior extension of the iliac crest between the spina limitans and the **posterior–superior iliac spine** enhances the lever advantage of the main muscle that supports the spine above the pelvis, the erector spinae.

Features of the acetabular margin and the sacral margin of the ilium also reflect the demands of bipedal locomotion (Fig. 20.2). On the acetabular margin, the strongly developed **anterior–inferior iliac spine** provides the origin for one of the main muscles that extends the leg at the knee, rectus femoris. Large anterior–inferior iliac spines are not found in apes and, among non-human primates, only occur in the vertical-clinging and leaping prosimians (galagos and lemurs) where rectus femoris is an important and well-developed leaping muscle. The anterior inferior iliac spine in humans also provides the site of attachment for the iliofemoral ligament. This is a strong band of tissue that extends from the anterior–inferior spine across the front of the hip joint to the intertrochanteric line on the femur (see Fig. 20.9). When the hip joint is fully extended the iliofemoral ligament tightens and helps to maintain balance by preventing the joint from hyperextending.

The main feature on the sacral margin of the human ilium is the **greater sciatic notch** (Fig. 20.2). This large U-shaped indentation is functionally related to the orientation and position of the sacrum in its articulation with the ilium. The long axis of the **auricular surface** (auricula (L) = little ear), that part of the sacroiliac joint on the sacral surface of the ilium, is oriented in an anteroposterior direction and extends along the top of the greater sciatic notch from the **iliopectineal line** to the posterior–inferior iliac spine. Both the large posterior extension of the iliac blade and the greater sciatic notch accommodate this articulation. No other living primates have either of these features. The anteroposterior orientation of the sacrum, together with its unique morphology are specific adaptations to bipedal locomotion (see below).

In addition to length, the ilium of the chimpanzee and other apes differs from the human ilium in the orientation of the blade, in the size of the iliac tuberosity and the auricular surface and in the relatively smooth and featureless acetabular and sacral margins. The iliac crest of the ape lacks the human S-shaped curve and projects laterally from the mid-plane of the body (Fig. 20.3). The resulting anterior orientation of the iliac fossa and posterior orientation of the gluteal surface may be very important in maintaining the trunk in an upright posture during sitting or squatting (Waterman, 1929). The posteriorly directed gluteal muscles and the anteriorly directed iliacus would be in an ideal position to support the trunk in this position. This orientation of the iliac blade would also result in a different mechanism of pelvic balance during facultative bipedalism in the apes. The lesser gluteal muscles would stabilize the pelvis by acting as medial rotators on a flexed thigh rather than as abductors on an extended thigh as they do in humans (Stern and Susman, 1981, 1983).

Furthermore, the long iliac crest may well be an adaptation to climbing in the apes. One of the most important climbing muscles, latissimus dorsi, takes its origin from the iliac crest and the lumbar spine and inserts into the humerus (Chapter 19). During climbing it draws the entire weight of the lower body upwards on the flexed forelimb. The long iliac crest could enhance the function of this muscle by giving it a long iliac origin at a distance from the mid-line of the body.

Where the human and ape iliac blades are similar in their great breadth, the distance from the spina limitans to the posterior–superior iliac spine makes up a much smaller proportion of the total length of the crest in the apes than it does in humans (Fig. 20.3). This reflects the fact that the sacral surface of the ape iliac blade (comprising the iliac tuberosity and the auricular surface) is a much smaller proportion of total iliac width than it is in humans (Fig. 20.2). The small size of both the auricular surface and the iliac tuberosity is a direct functional consequence of the quadrupedal posture of the apes. Because the apes do not support the entire weight of their upper bodies on their pelves and hindlimbs, the sacrum and auricular surfaces are not as large as they are in humans nor is the sacrum oriented in an extreme anteroposterior plane. As a result the acetabular margin of the pelvis forms a smooth curve from the posterior–inferior iliac spine to the moderately developed ischial spine. The acetabular margin of the ape pelvis also forms a smooth curve from the projecting anterior–superior iliac spine to the acetabular margin. The anterior–inferior iliac spine which interrupts this curve in humans is absent in apes. The rectus femoris muscle that takes its origin from this area of the ilium is weakly developed in apes as is the iliofemoral ligament that in humans is so important in balancing the trunk in bipedal posture.

The ischium in humans and apes

The most obvious feature of the ischium is the large knob of bone known as the **ischial tuberosity** (Fig. 20.2). The ischial tuberosity provides the area of attachment for the hamstring muscles (biceps femoris, semimembranosus and semitendinosus) that extend the hip joint, for adductor magnus, one of the principal adductors of the thigh and for the sacrotuberous ligament. The tuberosity is connected to the **ischial body** which is joined to the ilium just superior to the ischial spine and to the pubis via the **ischio-pubic ramus** or **inferior pubic ramus**.

The human ischial tuberosity is also unique in that the origin for one of the hamstrings, semimembranosus, has been 'pulled-up' close to the rim of the acetabulum (Waterman, 1929; Stern, 1971; Stern and Susman, 1983). As a result the distance between the superior part of the tuberosity and the acetabular rim, the **tuberoacetabular sulcus** (Stern and Susman, 1983) or the **sulcus tuberoglenoidalis** (Weidenreich, 1913; Waterman, 1929) is very short. This elevated position of the tuberosity could be a direct result of the stresses placed on the sacrum and the ischial tuberosity by bipedal locomotion (Weidenreich, 1913; Waterman, 1929; Stern

and Susman, 1983). In bipedal locomotion the weight of the vertical trunk acts to depress the anterior part of the sacrum (Fig. 20.4) (Weisl, 1954). The sacrotuberous ligament, which connects the sacrum to the ischial tuberosity, prevents the posterior, or caudal, part of the sacrum from rising in response to this pressure. The resulting tension in this ligament could be directly related to the 'pulled-up' position of the ischial tuberosity. The **falciform crest** on the medial surface of the ischial body marks the ischial attachment of the sacrotuberous ligament.

The ischial spines which lie immediately below the greater sciatic notch on the two ischia (= plural of ischium) are another adaptation to bipedal locomotion. Abitbol (1988b) argued that the prominent human spines are the direct result of the stress placed on the ischia by the horizontally orientated pelvic floor, or diaphragm, that must support the weight of the abdominopelvic organs in bipedal posture. The pelvic floor inserts into the two ischia via the ischial spines either directly or indirectly through the sacrospinous ligament and the tendinous arch surrounding the muscles of the pelvic floor (Fig. 20.4) (see also Waterman, 1929).

The ape ischium differs from the human ischium in relative length as well as in the form of the ischial tuberosity. The ape ischium is long relative to a number of variables including acetabular diameter, minimum width of the iliac blade and diameter of the first sacral vertebra (Robinson, 1972; Lovejoy et al., 1973; McHenry, 1975a,e; Stern and Susman, 1983). In addition, McHenry (1975e) has shown that the ratio of the power and load-arms of the hamstrings (based on the length of the ischium and the distance from the hip joint to the ground) is more advantageous in apes than humans when the hip is flexed. This means that the long ape ischium is adapted to power in hip extension while the human ischium would be adapted for speed and range of movement (Smith and Savage, 1955; Sigmon and Farslow, 1986).

The form of the ischial tuberosity is also different in the apes (Fig. 20.5). It is absolutely wider than the human tuberosity, lacks the clearly demarcated facets for the hamstrings and adductor magnus and, together with the lower body of the ischium, is everted, or twisted outward. This eversion could, at the same time, result from the strong pull of the muscles that extend the hip and also increase their lever advantage in

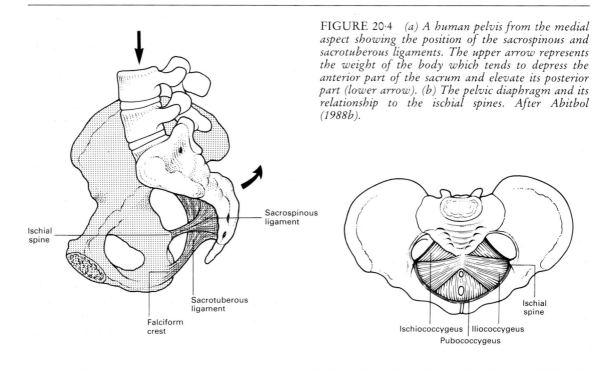

FIGURE 20·4 *(a) A human pelvis from the medial aspect showing the position of the sacrospinous and sacrotuberous ligaments. The upper arrow represents the weight of the body which tends to depress the anterior part of the sacrum and elevate its posterior part (lower arrow). (b) The pelvic diaphragm and its relationship to the ischial spines. After Abitbol (1988b).*

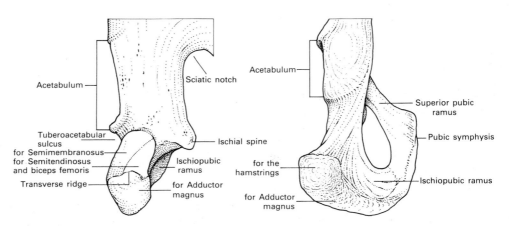

FIGURE 20·5 *The ischial tuberosities of a chimpanzee and a human from the posterior aspect. The bones are oriented so that the rims of the acetabulae are vertically aligned.*

hip extension (Sigmon and Farslow, 1986). The tuberoacetabular sulcus is also relatively wider in the apes than it is in humans and the ischial tuberosity lacks the 'pulled-up' appearance so characteristic of the human tuberosity.

The pubis in humans and apes

The main part of the pubic bone is the **pubic body** which carries the surface (the **pubic symphysis**) by which it articulates with its opposite number. The pubic symphysis is a fibrocartilage joint and the two bones are separated by a disc of fibrous tissue. Superiorly, the body of the pubis is connected to the rest of the pelvic bone by the **superior pubic ramus** and inferiorly it is connected to the ischium by the **inferior pubic ramus**.

The **pubic tubercle** is a bump just lateral to the symphysis on the superior pubic ramus and the **pubic crest** is that part of the ramus that lies between the pubic tubercle and the symphysis. The pubic crest is the site of attachment for one of the main muscles that supports the gut, the rectus abdominis, and the pubic tubercle is associated with the strongly developed inguinal ligament in humans (Chapter 15). This ligament gives added support to the abdomen in bipedal posture. Because of the differences in posture in apes, rectus abdominis does not exert the same force on the ape superior pubic ramus and the inguinal ligament is also correspondingly small. Both the pubic tubercle and a recognizable pubic crest are absent in these primates.

The **iliopubic**, or **iliopectineal**, **eminence** marks the junction of the ilium and the pubis. It is a raised area of bone that separates the superior pubic ramus from the anterior–inferior iliac spine. The **iliopsoas groove** between these two features conducts the iliopsoas muscle from its origins on the iliac fossa and the spine to its insertion on the proximal femur (Chapter 19). This muscle is one of the main flexors of the hip and in bipedal posture helps to support the trunk on the extended thigh. Because of the different orientation of the pelvis and thigh in the apes, the pelves of these primates lack both an iliopsoas groove and an iliopubic eminence.

The space enclosed by the superior pubic ramus, the inferior pubic ramus and the body of the ischium is the **obturator foramen**. This foramen is closed by a membrane, hence the name obturator (obturo (L) = I stop up). Kummer (1975) has argued that the area of the obturator foramen is under no mechanical stress in either quadrupedal or bipedal locomotion and that the superior pubic ramus is stressed in tension rather than in compression.

The length of the pubic bone in the apes does not differ materially from the length of the human pubic bone in relation to the length of the ischium (Schultz, 1930). However, there is some difference in the ischiopubic angle where the ape pubis forms a more acute angle with the ischium than does the human pubis (Fig. 20.6).

The body of the ape pubis is deeper than it is in humans and the subpubic angle is narrower even than the narrow human male subpubic angle (Fig. 20.2). The two pubic bones may fuse into a solid mass across the

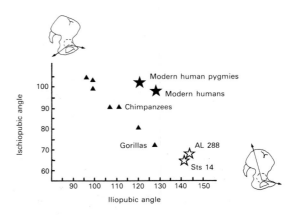

FIGURE 20.6 *The mean values of the iliopubic and ischiopubic angles among selected Old World primates and fossil hominids (n=270). After Berge* et al. *(1984).*

435

pubic symphysis in the apes as they do in many mammals and if this occurs it does so at approximately 17 years of age in the chimpanzee (Todd, 1923). In humans the two pubic bones do not normally fuse.

The acetabulum and hip joint in humans and apes

In humans and apes the ilium, ischium and pubis meet in the acetabulum, the socket for the hip joint. However, these bones do not meet in the centre of the acetabulum. The largest part of the acetabulum is made up by the ischium and, particularly in non-human primates, the centre of the acetabulum is invariably located considerably below the upper border of this bone (Fig. 20.7) (Schultz, 1969b). During foetal development, the ilium and ischium are the sole bones involved in the articulation with the femur and it is only later in development that the pubic bone becomes part of the acetabulum.

In humans the acetabulum faces inferiorly, laterally and anteriorly (Kapandji, 1987). In coronal section the superior border of the acetabulum overhangs the centre of the femoral head by about 30° (Fig. 20.8) (Kapandji, 1987). This superior border sustains the greatest pressure during bipedal locomotion and, as a result, the articular cartilage is thickest on both the acetabulum and the femoral head in this superior region. In horizontal cross-section, the human acetabulum faces anteriorly at 30–40° from the coronal plane (Kapandji, 1987). Because of this orientation, when the head of the femur is articulated with the acetabulum in upright posture the front (anterior) of the joint is open in the sense that the acetabulum does not fully encompass the cartilage-covered femoral head. This disposition of the acetabulum results at the same time in joint stability during upright posture and in maximum freedom of movement of the leg anterior to the body in flexion, adduction and medial rotation.

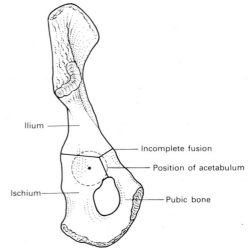

FIGURE 20·7 *A pelvic bone of a juvenile chimpanzee from the medial aspect showing the position of the acetabulum in relation to the ilium, ischium and pubis. After Schultz (1969b).*

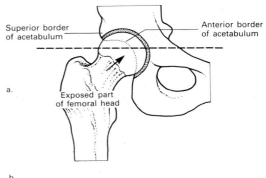

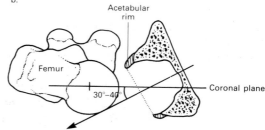

FIGURE 20·8 *The orientation of the head of the femur in relation to the acetabulum from (a) the anterior (ventral) view and (b) the superior (cranial) view. The dashed line in (a) indicates the approximate plane of (b). After Kapandji (1987).*

In neither humans nor apes is the acetabulum completely lined by articular cartilage (Fig. 20.2). Rather the articular surface is a horse-shoe shaped band of cartilage called the **lunate surface**, or **fascies lunata** (fascia (L) = a band or bandage; luna (L) = the moon or crescent shaped). The non-articular area making up the inferior circumference of the acetabulum between the two horns of the lunate surface is called the **acetabular notch** while the non-articular central part of the acetabulum is called the **acetabular fossa**. The acetabular fossa is deeper than the articular lunate surface and is separated from the inner surface of the pelvis by a very thin plate of bone. In both humans and apes the acetabulum is deepend by a fibrocartilaginous ring called the **acetabular rim**, or **laubrum acetabulare** (laubrum (L) = lip or rim). This ring has a slightly smaller diameter than does the bony acetabulum and, together with atmospheric pressure, helps to hold the head of the femur in the socket.

The hip joint is also strengthened, and the head of the femur held in the acetabulum, by powerful anterior and posterior ligaments (Fig. 20.9). In humans the anterior ligaments are Z-shaped. The upper arm of the Z is the superior band of the iliofemoral ligament that extends from the anterior–inferior iliac spine on the pelvis to the upper part of the trochanteric line on the femur. The crossbar of the Z is the inferior band of this same ligament which originates with the superior band on the anterior–inferior spine but inserts into the lower part of the femoral trochanteric line. The lower arm of the Z is the pubofemoral ligament which stretches between the lower part of the trochanteric line and the superior ramus of the pubic bone. The ischiofemoral ligament is the only posterior ligament of the hip joint. It spans the area between the ischium just below the acetabulum and the inner surface of the greater trochanter anterior to the trochanteric fossa. In humans both the anterior and the posterior ligaments are twisted around the femoral neck in the same direction. Flexion

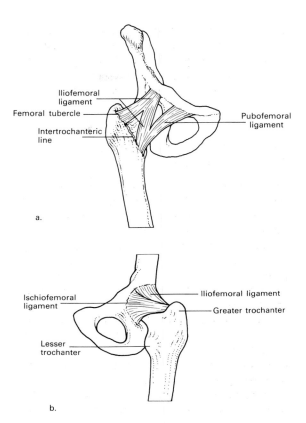

FIGURE 20·9 _The ligaments of the hip joint from (a) the anterior (ventral) view and (b) the posterior (dorsal) view._

of the hip joint untwists the ligaments and extension further twists the ligaments. In apes which move with a habitual flexed-leg posture, the ligaments are both weaker than the corresponding human ligaments and run in a straight line from their origins to their insertions.

In general, the deeper the acetabulum the more restricted is the movement of the hip joint. Schultz (1969b) has measured the maximum depth of the acetabular fossa in relation to maximum acetabular diameter. Among monkeys, apes and humans, the baboons have the deepest acetabulae, while orang-utans have the shallowest. In comparison with the apes, humans have on average a shallower acetabulum than do the African apes, but an acetabulum that is deeper than that of the orang-utan. Ruff (1988) reached similar conclusions. However, in his analysis the positions of humans and chimpanzees are reversed with chimpanzees having shallower acetabulae than humans. Citing support from work on carnivores (Jenkins and Camazin, 1977), Ruff (1988) suggested that acetabular depth is related to joint mobility and that the relatively shallow acetabulum of the orang-utans directly reflects greater mobility of the hip joint in this species. However, Schultz (1969b) warned that because the intraspecific variation in acetabular depth greatly exceeds any average interspecific difference between humans and apes, any locomotor inferences must be made with caution.

There is also considerable intraspecific variation in the width of the acetabular notch relative to maximum acetabular diameter. However, in relation to humans as well as to the other apes, the orang-utan has an exceptionally narrow acetabular notch (Schultz, 1969b). Schultz (1969b) suggested that this is the primary feature related to the exceptionally wide range of movement at the hip joint in the orang-utans, to the absence of a ligamentum teres, and to the corresponding need for an extensive articular surface in this species.

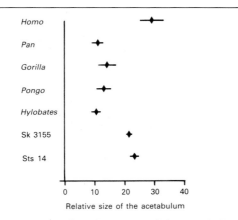

FIGURE 20·10 *The relative size of the acetabulum [(transverse acetabular width × 100)/iliac height] in humans, the large apes, gibbons and selected fossil hominids. After McHenry (1975c).*

As with the size of all the joint surfaces of the hindlimb, the size of the acetabulum in humans is large relative to body size (Jungers, 1988a). It is also large in relation to various skeletal measures such as the combined length of the ilium and ischium (Schultz, 1969b), trunk length (Schultz, 1969b), ilium height (Fig. 20.10) (McHenry, 1975c), and ilium width (Napier, 1964). Although the analyses are yet to be carried out, it is assumed that acetabular size will show the same pattern of positively allometric scaling within humans as does the head of the femur (Ruff, 1988). In this context, large humans would be expected to have acetabulae that are much bigger in relation to body size than would smaller individuals.

The sacrum and coccyx in humans and apes

The sacrum is a wedge-shaped bone that is at the same time part of the vertebral column and part of the pelvis (Fig. 20.11). The body of the human sacrum is comprised of five fused vertebra that not only make up the dorsal aspect of the pelvis but also provide a strong foundation that supports the weight of the upper body and anchors the trunk to the hindlimb. Attached to the apex of the sacrum, the coccyx comprises the last four rudimentary segments of the spinal column. The name coccyx (kokkyx (Gk) = cuckoo) may have been applied to this bone because of its resemblance to the beak of a cuckoo. Alternatively, an old common name for the coccyx, the whistle-bone, suggests that the name could have been derived from the similarity in sound between the breaking of wind from the anus and the sound of the cuckoo.

The vertebral origin of the sacrum (and coccyx) can be clearly recognized in the morphology of these bones. This is most obvious at the proximal end of the sacrum (sacral body) where there are two well-developed superior articular facets and where the body of the first sacral vertebra provides the articular surface for the fifth lumbar vertebra. The transverse processes of the sacral vertebra are fused to one another and expanded into wings, or **alae** (singular = ala) that provide the articulation between the sacrum and the ilium. Viewed from the anterior (or pelvic) surface the parallel rows of pelvic **sacral foramina** clearly define the bodies of the five sacral vertebrae that lie

between them. The margins of these foramina are rounded laterally indicating the course of the nerves that emerge from them.

The dorsal surface is divided into a series of crests. The **medial crest** corresponds to the fused spinous processes of the sacral vertebrae, the **intermediate crest** to the articular facets and the **lateral crest** to the transverse processes, or tubercles. The

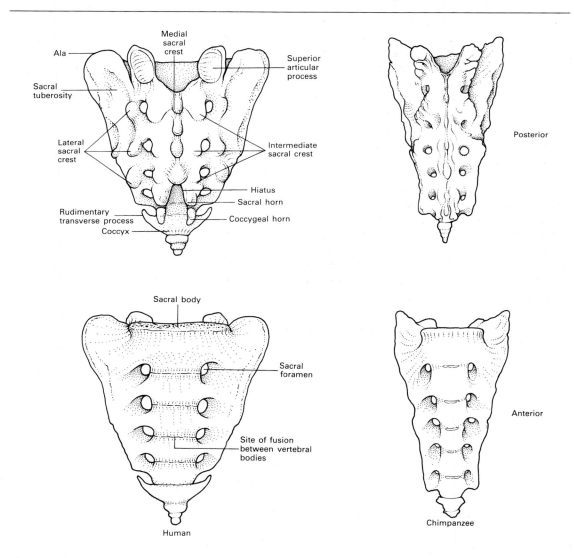

FIGURE 20·11   *A human sacrum and a chimpanzee sacrum from the posterior (dorsal) and anterior (ventral) aspects.*

439

absence of the fourth and fifth (or sometimes just the fifth) sacral spines and laminae produces a **sacral hiatus** that is bounded laterally by the **cornua** (or **horns**) of the sacrum that are contiguous with the intermediate crests above them. The parallel rows of the four dorsal sacral foramina lie between the intermediate and lateral crests. The **sacral tuberosity** provides the attachment for the strong sacroiliac ligaments that bind the sacrum to the iliac tuberosities and is located on the superior part of the sacrum between the lateral crest and the **auricular surface**. The auricular surface is the ear-shaped sacral part of the sacroiliac joint that corresponds to the auricular surface of the ilium. On the sacrum it extends from the level of the first sacral vertebra to the third sacral vertebra.

The human sacrum is wider than non-human primate sacra (Fig. 20.12) and is positioned at a more acute angle to the lumbar spinal column that it supports. Both these features are directly related to bipedal locomotion. The width of the sacrum increases the distance between the sacroiliac joints and positions them more vertically over the hip joints. This reduces the stress on the pubic symphysis by reducing the rotation of the pelvic bones about the sacroiliac joint (Leutenegger, 1977). The broad, horizontally oriented dorsal surface of the sacrum, together with the large iliac tuberosities also provide increased leverage for the muscles of the back that balance the spine over the pelvis (erector spinae).

The horizontal orientation of the human sacrum has been attributed to the demands of the enlarged foetal head at the time of birth (Schultz, 1969a; Lovejoy et al., 1973; Martin, 1983). However, the orientation of the sacrum together with its position distal to the acetabulae also may be explained by the demands of bipedal locomotion (Stern and Susman, 1983; Abitbol, 1987a). Stern and Susman (1983) have suggested that the posterior position of the sacrum relative to the acetabulae would have a desirable locomotor effect in bipedalism by bringing

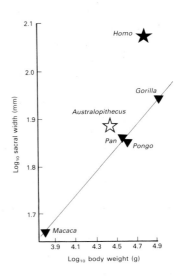

FIGURE 20·12 *The relationship between body weight and the width of the sacrum in humans, selected primates and Sts 14* (Australopithecus africanus). *After Leutenegger (1977).*

the line of gravity of the vertically oriented trunk closer to the femoral heads rather than far in front of them. This explanation is an alternative to that of Robinson (1972) who argued that the posterior position of the sacrum is necessary to increase the anteroposterior size of the human birth canal that would otherwise be unacceptably small as a result of the short human ilium.

In the apes there is generally a greater degree of variation in the number of vertebra that go to make up the sacrum than there is in humans (Abitbol, 1987b; but see Schultz, 1961). If sacral vertebrae are defined as those vertebra that have any type of articulation whatsoever with the ilium, all apes have at least three sacral vertebrae while some may have as many as seven (Abitbol, 1987b). The more cranial sacralized vertebrae articulate with the elongated blades of the ilium that extend cranially along the lateral margins of the spine. In humans there tends to be a smaller range of variation (between four and six sacral vertebrae) with the great majority of humans having five. If sacral vertebrae are defined as those vertebrae that are totally sacralized (from the most distal sacral vertebra that articulates with the coccyx to the most proximal that bears the promontorium (Chapter 15)) there is a smaller range of variation in both humans and apes. However, the pattern of this variation, with a greater amount in the apes, remains the same. The variation in the number of sacralized vertebrae in the apes also distinguishes them from the more quadrupedal monkeys and could reflect mild pressures to sacralize the lumbar region of the spine in response to the stress on this region resulting from the semierect posture of the apes (Abitbol, 1987b).

In addition, the ape sacrum is also less massive than the human sacrum. Specifically, it is narrower than the human sacrum in relation to body weight (Fig. 20.12), trunk length and the diameter of the sacral body (articular surface for L5) (Leutenegger, 1977; Abitbol, 1987b). It also tends to be oriented at less of an angle to the lumbar vertebral column above it (Chapter 15 and Fig. 15.15) and shows less longitudinal curvature than does the human (male or female) sacrum. Furthermore, it has a smaller auricular surface than does the human sacrum. All these features reflect the fact that it does not support the entire weight of the upper body in bipedal posture.

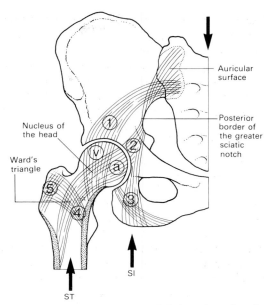

FIGURE 20·13 *Weight transfer and trabecular systems in the human pelvis and proximal femur. The upper arrow represents body weight and the lower arrows the resultant force in standing (ST) and sitting (SI). 1 = Pelvic trabeculae which are in line with the vertical (or supporting) bundle (V) of trabeculae in the proximal femur; 2 = pelvic trabeculae which are in line with the arcuate bundle of trabeculae in the proximal femur; 3 = the sacroischial trabeculae of the pelvis; 4 = the trochanteric bundle in the proximal femur; 5 = the trabecular bundle in the proximal femur that reinforces the greater trochanter. (☆) = the weak area of the femoral neck between the nucleus of the head and Ward's triangle. After Kapandji (1987).*

Weight transfer through the human pelvis

In human bipedal posture weight is transmitted from the sacrum and the sacroiliac joints through the pelvis to the acetabulum and on to the head and neck of the femur. In a sitting posture weight is transferred from the sacrum through the pelvis to the ischial tuberosities. In both the pelvis and in the head and neck of the femur the internal architecture of the bone, the trabecular orientation, corresponds to these lines of force (Fig. 20.13) (Kapandji, 1987).

In the pelvis there are three major trabecular systems (Kapandji, 1987). The first of these extends from the lower part of the auricular surface on the ilium in a gentle arc to the superior surface of the acetabulum. This trabecular system is in line with one of the main systems of the head and neck of the femur, the **vertical**, or **supporting, bundle**. This femoral system extends from the head of the femur vertically to the medial cortex of the shaft and transmits the main compressive forces through the femoral head and neck. The second pelvic trabecular system extends from the superior part of the auricular surface on the ilium to the border of the greater sciatic notch to the inferior border of the acetabulum. This system is in line with the second major femoral system, the **arcuate bundle**. This femoral system extends from the lower part of the femoral head to the lateral side of the femur below the greater trochanter and counteracts the shearing stresses in the head and neck of the femur. The third pelvic trabecular system, the **sacro-ischial trabeculae**, extends from the auricular surface to the ischium and supports the weight of the body while sitting.

In addition to these trabecular systems there are two additional accessory systems in the proximal femur (Fig. 20.13). The first of these, the **trochanteric bundle**, fans out from the medial side of the shaft to the greater trochanter. The second, and less important, extends from the base of the greater trochanter to its apex, and reinforces the greater trochanter against the pull of the small gluteal muscles. The area of intersection between the arcuate bundle and the trochanteric bundle just inferior to the neck of the femur is known as **Ward's triangle**, while the area of intersection between the arcuate bundle and

the vertical bundle in the head of the femur is called the **nucleus of the head**. The relatively weak area in the neck of the femur between the vertical bundle and the trochanteric bundle is the area of the neck of the femur which most frequently fractures.

These trabecular systems are specific to bipedal locomotion and to the human pelvis and proximal femur. The trabecular systems in the ape proximal femur are much more diffuse than they are in the human femur and reflect the different trajectory of weight transfer in these primates.

Obstetric pelvis in modern humans and apes

In addition to the functions of the pelvis in locomotion, posture and visceral support, the pelvis must also function in birth. The internal dimensions of the pelvis must be large enough for the foetus to pass through the **birth canal**, or **true pelvis**. In non-human primates there is an allometric relationship between the size of the head of the foetus and the size of the maternal birth canal. Smaller primates, such as the marmosets and the tamirins, tend to have infants with larger

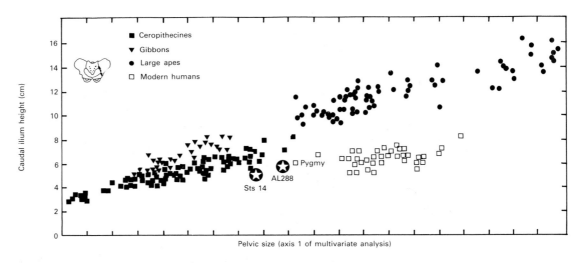

FIGURE 20·14   *The relationship between pelvic size and iliac height in humans, selected primates and*   *selected fossil hominids (n=370). After Berge et al. (1984).*

crania relative to the dimensions of the maternal birth canal than do larger primates such as the chimpanzees or gorillas (Leutenegger, 1982). Humans are unique among large-bodied primates in deviating from this trend and producing infants with large heads relative to both maternal weight and maternal birth-canal dimensions. There are two reasons for this deviation: firstly, relative to body size, human infants have larger heads than non-human primate infants; and secondly, the demands of bipedal loco-motion and bipedal posture in humans have severely restricted the size of the true pelvis and the birth canal in comparison to the size of these structures in the apes.

The short ilium characteristic of bipedal humans reduces the distance between the sacroiliac joint and the hip joint (Fig. 20.14). This shortens the anteroposterior diameter of the true pelvis. The short ilium also lowers the sacrum, bringing it down into the pelvic cavity. In bipedal humans the sacrum thereby becomes the dorsal wall of the birth canal. This reduces its anteroposterior diameter, particularly in the mid-plane and the outlet (Tague and Lovejoy, 1986). In apes, charac-terized by a higher position of the sacrum,

the dorsal wall of the pelvic midplane and outlet is composed of soft tissue (Fig. 20.15).

Humans, as well as non-human primates that produce infants with a large head relative to the dimensions of the maternal birth canal, are characterized by pelves that are sexually dimorphic (Leutenegger, 1982).

Sexual dimorphism in the human pelvis

The majority of the features that distinguish the female pelvis from the male pelvis are features that maximize the size of the female birth canal by increasing the size of the female true pelvis in relation to that of a similarly sized male.

In human females, the pelvic inlet is generally circular in shape and has a larger absolute circumference (measured by the length of the iliopectineal line) than the heart-shaped pelvic inlet of males (Fig. 20.16). The relatively large female pelvic inlet is also reflected in the ischio-pubic index (Schultz, 1949) and in the interiliac index (Straus, 1929) (Table 20.2).

Other features increase the anteroposterior diameter of the pelvic mid-plane and outlet

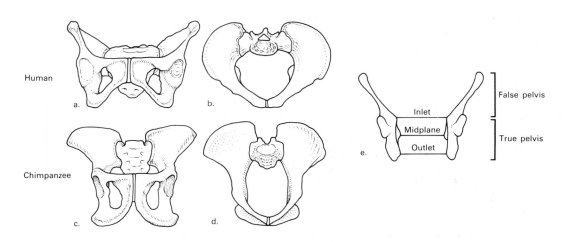

FIGURE 20·15 (a–d) A human pelvis and a chim-panzee pelvis from the anterior (ventral) and superior (cranial) aspects. After Tague and Lovejoy (1986).

(e) Cross-section of a human pelvis showing the levels of the pelvic inlet, midplane and outlet.

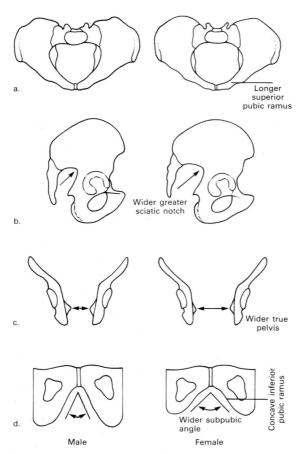

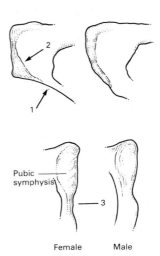

FIGURE 20·17 *A female pubic bone can be distinguished from a male pubic bone by (1) the concave inferior pubic ramus, (2) the presence of the ventral arc, and (3) a narrow, knife-like inferior pubic ramus just below the pubic symphysis. After Phenice (1969).*

FIGURE 20·16 *Sexual dimorphism in the human pelvis. In relation to the average male pelvic morphology the female pelvis has (a) a more circular pelvic inlet with a longer superior pubic ramus, (b) a wider and shallower greater sciatic notch and a more horizontally oriented sacrum, (c) a wider and less funnelled true pelvis, and (d) a shallower pelvic cavity and a wider sub-pubic angle.*

(Fig. 20.16). In females the greater sciatic notch forms a larger angle and is shallower than it is in males and the sacrum is more horizontally oriented as well as being shorter and less curved. Furthermore, the ischial bodies are everted (twisted outward) and thereby increase the transverse diameter of the pelvic outlet. This results in the characteristic female parallel-sided true pelvis and contrasts with the funnel-shaped true pelvis found in males.

In addition, the inferior border of the ischiopubic ramus is concave in females and straight or convex in males. This feature is directly related to the large subpubic angle characteristic of human females and to a unique feature of human birth. During birth the infant is normally positioned in the birth canal facing dorsally with its head flexed. The head is born underneath the pubic arch and ventral (or anterior) to the ischial tuberosities when the infant unflexes its head through the broad subpubic angle.

It is possible to visually sex a human pelvis with up to 95% accuracy by assessing the concavity of the ischiopubic ramus as well as the following two additional dimorphic features that do not relate directly to the size of the birth canal (Phenice, 1969; Lovell, 1989) (Fig. 20.17). Firstly, the ventral arc is absent in males. The ventral arc is a slightly elevated ridge that swings across the ventral (anterior) surface of the pubis and marks the site of attachment of the arcuate ligament (inferior pubic ligament). Secondly, the medial aspect of the inferior pubic ramus just below the pubic symphysis is narrow in

TABLE 20·2 Sexual dimorphism in the primate pelvis. After [a]Straus (1929) and [b]Schultz (1949).

		Ischium–pubis index[b]			Relative inlet breadth[b]			Interiliac index[a]		
		N	Mean	Range	N	Mean	Range	N	Mean	Range
Humans	♂	50	79.9	71.0– 88.0	50	44.1	40.6–48.4	50	157.0	106.5–195.5
	♀	50	95.0	84.0–106.0	50	49.5	47.2–52.2	44	121.9	82.9–171.4
Gorilla	♂	26	93.6	82.0–108.3	26	33.0	29.6–34.6	6	108.0	99.1–116.8
	♀	15	98.3	85.4–113.7	15	37.5	34.8–41.6	3	89.0	87.5– 91.7
Chimpanzee	♂	21	82.6	69.7– 95.2	21	36.9	31.1–41.4	7	98.6	84.0–120.9
	♀	30	87.0	78.9– 98.9	30	41.0	36.8–47.0	4	88.7	80.6–102.5
Orang-utan	♂	24	97.8	84.6–108.4	24	36.0	30.9–39.4	9	82.7	70.0– 95.9
	♀	26	107.0	96.2–118.1	26	42.9	39.2–46.1	14	74.8	57.9– 92.7
Gibbon	♂	11	95.0	85.4–102.6	11	49.6	42.7–56.8	9	85.0	69.6–109.3
	♀	7	108.1	95.0–118.9	7	52.0	48.1–58.1	8	78.1	70.0– 89.5

Ischium–pubis index = [(pubis length × 100)/ischium length]. Relative inlet breadth = [(maximum inlet breadth × 100)/maximum pelvic breadth] where the maximum pelvic breadth is measured between the iliac crests or between the anterior iliac spines, which ever is greater. Interiliac index [(height of the upper ilium × 100)/height of the lower ilium] where the upper ilium height is the false pelvic portion of the ilium above the iliopectineal line and the lower ilium height is the true pelvic portion below the iliopectineal line.

females and broad in males where it provides the site of attachment for the penis.

It is not possible to be completely accurate in sexing pelves either visually or metrically because a large number of female pelves do not correspond to the common female, or **gynecoid**, pelvic type. Some female pelves are **android** (andros (Gk) = a man), being like males with a heart-shaped inlet while others are like apes, or are **anthropoid**, where the sagittal diameter of the inlet is greater than the transverse diameter. A small minority are extremely broad and flat in shape and are termed **platypelloid** (platys (Gk) = flat or wide; pella (Gk) = bowl).

Sexual dimorphism in the ape pelvis

The non-human primate pelvis shows the greatest degree of sexual dimorphism in those species that bear the largest infants in relation to maternal pelvic size (Schultz, 1949; Straus, 1929; Leutenegger, 1974, 1982; Mobb and Wood, 1977; Wood and Chamberlain, 1986). Adult female primates have higher ischio-pubic indices (and therefore relatively longer pubic rami) than males (Schultz, 1949) (Table 20.2). It should be noted, however, that the apes are characterized by a greater range of overlap in this index than are humans and monkeys and, therefore, the index is of little practical value in sexing ape pelves. This is because apes have relatively much smaller neonates in relation to maternal pelvic size than do either humans or most monkeys. Sexual dimorphism (with a considerable degree of overlap between males and females) is also evident in the interiliac index in primates (Straus, 1929). Mobb and Wood (1977) and Wood and Chamberlain (1986) have demonstrated that the larger birth canal in non-human primate females is not related to any allometric differences between males and females but is a true reflection of the increased size of the pelvic inlet in females (contra Streudel, 1981).

Obstetrics in humans and apes

Both the shape of the human birth canal and the corresponding manner of human birth are unique among primates. The unique morphological feature of the human birth canal is the transversely oval shape of the pelvic inlet (Fig. 20.15). The birth canal in all other living primates is sagittally oval not only at the level of the pelvic inlet, but also at the level of the pelvic mid-plane and outlet. The human birth canal is transversely oval at the level of the pelvic inlet, sagittally oval at the midplane and nearly circular at the outlet.

The long axis of the head of the human foetus, as in all primates, is the sagittal, or occipitofrontal, axis. During human birth the foetus either aligns the sagittal axis of its head transversely across the pelvic inlet to correspond with the greatest diameter of the inlet, or aligns its head obliquely across the inlet (Fig. 20.18). At this stage the neck of

the foetus is generally flexed to present the shorter occipitobregmatic diameter of the head (as opposed to the occipitofrontal diameter) to the pelvic inlet. As the largest diameter of the birth canal changes from the transverse diameter of the inlet to the sagittal diameter of the mid-plane, the head of the infant rotates through 45–90°. At this stage of birth the foetus is normally facing dorsally in the birth canal with the occipital part of the foetal skull oriented toward the pubic, or anterior, edge of the birth canal. The shoulders of the infant are aligned transversely (or obliquely) across the pelvic inlet. As the birth progresses, the foetus extends its flexed head and it is born through the relatively wide female subpubic arch. The shoulders and rest of the body rotate after the head through the pelvic mid-plane and outlet.

Birth in apes is a relatively simple affair in comparison to human birth. The long sagittal axis of the head of the ape foetus aligns itself along the long anteroposterior diameter of the ape pelvic inlet and maintains this orientation through the birth canal, without the need to twist (Fig. 20.18). Because the longest axis of the ape foetal head is considerably smaller than the longest axes of the birth canal throughout its extent, ape mothers do not have difficulty in birth.

The australopithecine and paranthropine pelvis

The great majority of australopithecine and paranthropine pelves are fragmentary, grossly distorted or from juvenile individuals. By far the best specimen is AL 288-1 from Hadar, Ethiopia (*Australopithecus afarensis*) preserving a sacrum and a complete pelvic bone that is only seriously distorted in the region of the iliac tuberosity (Johanson et al., 1982). Following this in relative preservation is the Sts 14 pelvis from Sterkfontein, South Africa (*Australopithecus africanus*) including the complete right pelvic bone and most of the left together with a fragmentary sacrum.

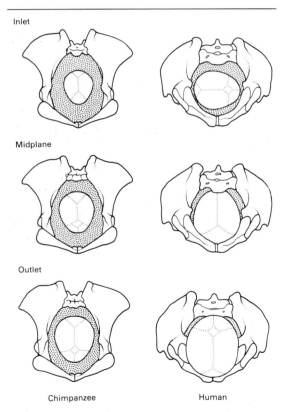

FIGURE 20·18 *Mechanisms of birth in a chimpanzee and a human. The pelvis and foetus are seen from the inferior view. After Tague and Lovejoy (1986).*

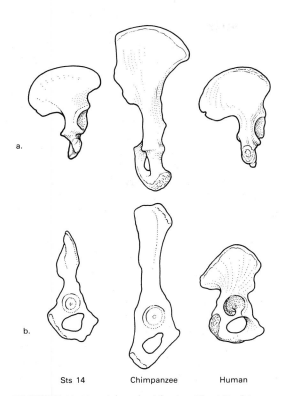

FIGURE 20·19   *Australopithecine (Sts 14), chimpanzee and human pelvic bones (a) depicted to show the widest aspect of the iliac blade, and (b) oriented with the acetabular rims parallel to the plane of the page. Note in (a) that both the australopithecine and modern human pelvic bones have short ilia and in (b) that only in the modern human pelvic bone is the wide iliac blade evident. After Oxnard (1975) and Broom and Robinson (1950).*

However, the right pelvic bone is badly distorted, particularly in the pubic region (Day, 1973).

These pelves, together with the more fragmentary australopithecine and paranthropine specimens, share a number of features with human pelves that are hallmarks of bipedal locomotion (Fig. 20.19). These include the short and wide iliac blade, a well-developed sciatic notch and an equally well-developed anterior inferior iliac spine. Furthermore, the sacrum is wide in relation to the diameter of the sacral body, to trunk

length and to inferred body weight (Fig. 20.12) and it is positioned behind the line drawn between the acetabulae. In addition, the ischial tuberosity is faceted for the attachment of the hamstring and adductor magnus muscles.

The australopithecine and paranthropine pelves also have a number of features that align them with the non-human primates as well as other features that are unique.

The australopithecine and paranthropine ilium

The most obvious non-human feature of the blade of the ilium is its orientation when articulated with the sacrum. The ilium lacks the strongly developed iliac fossa of the human blade and projects laterally from the mid-line of the body in a fashion similar to that of the chimpanzee (Fig. 20.19). Associated with this is the pronounced lateral flare of the blade (Fig. 20.20) and the well-developed (or beaked) anterior superior iliac spine that lies at a greater distance from the anterior inferior iliac spine than is the case in modern humans.

The extreme iliac flare effects the position of the iliac pillar which is located much more anteriorly than it is on the more vertical human iliac blade (Fig. 20.21). The auricular surface and the iliac tuberosity are both

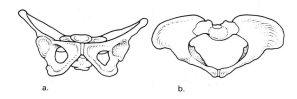

FIGURE 20·20   *The reconstructed AL 288-1 (Australopithecus afarensis) pelvis from (a) the anterior (ventral) and (b) the superior (cranial) aspects. After Tague and Lovejoy (1986). Compare with Figure 20.15.*

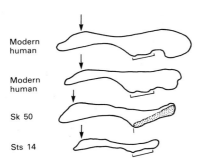

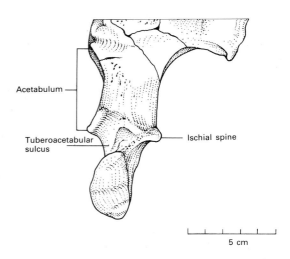

FIGURE 20·21 *Cross-sections of the iliac blade (from the posterior superior iliac spine to the deepest point in the notch below the anterior superior spine) in modern humans,* Paranthropus *(SK 50), and* Australopithecus africanus *(Sts 14). Note the different positions of the iliac pillar (arrows). The brackets represent the extent of the auricular surface. The shaded area in SK 50 is reconstructed. After McHenry (1975c).*

FIGURE 20·22 *The ischial tuberosity of* Australopithecus afarensis *(AL 288-1) from the posterior aspect. The bone is oriented so that the rim of the acetabulum is vertically aligned. Compare with Figure 20.5.*

small relative to the total width of the australopithecine iliac blade. The small iliac tuberosity together with the anterior position of the iliac pillar and the lateral flare of the blade result in an iliac crest that does not have the extreme S-shape that is characteristic of the human crest.

The australopithecine and paranthropine ischium

The ischium differs from that of modern humans in two ways. Firstly, the ischium is relatively longer than it is in most modern humans although it may not fall entirely outside the human range of variation. This has been demonstrated for *Australopithecus afarensis* relative to both minimum iliac width and acetabular diameter (Stern and Susman, 1983) and for the other australopithecines relative to acetabular diameter, femur length, width of the first sacral body, height of the ilium, iliac breadth and length of the lower limb (Robinson, 1972; Lovejoy et al., 1973; McHenry, 1975e). Secondly, the ischial tuberosity lacks the 'pulled-up' appearance of the human tuberosity which brings the origin of

the semimembranosus muscle close to the acetabular border. The facet for the origin of the hamstring muscles sits at a relatively sharp angle to the facet for the origin of adductor magnus in *Australopithecus afarensis* (Stern and Susman, 1983) (Fig. 20.22). In modern humans the surfaces are at less of an angle to each other and both surfaces appear to have been pulled-up toward the acetabular border resulting in a narrower tuberoacetabular sulcus (see above).

The australopithecine and paranthropine pubis

The superior public ramus in these early hominids is relatively longer than in modern humans and lies in a different angular relationship to both the ilium and the ischium (Fig. 20.6) (Berge and Kazmierczak, 1986). Moreover, the well-developed human pubic tubercle is absent from the superior surface of the ramus on the *Australopithecus afarensis* pubis. This surface is broad adjacent to the pubic symphysis and narrows laterally to a relatively sharp crest.

Well-developed ramparts (ridges) run down

FIGURE 20·23 *The* Australopithecus afarensis *(AL 288-1)*
pubic bone from (a) the anterior
aspect and (b) the supermedial
aspect.

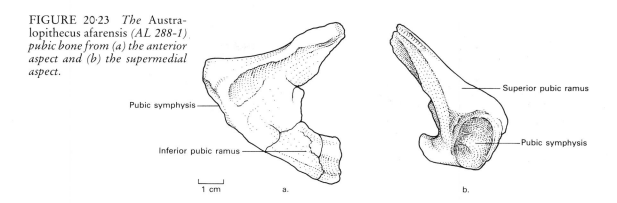

the ventral face of the pubic body in AL 288-
1 (*A. afarensis*) parallel to the pubic symphysis
(Fig. 20.23). The pubic body is wider than
it is in humans while the sub-pubic angle is
larger than it is in modern apes lying midway
between human male and female angles
(Tague and Lovejoy, 1986). The inferior
pubic ramus in *A. afarensis* lacks the concave
shape of modern human females (Fig. 20.16).
It has a straight inferior margin as it does in
modern human males. In modern apes it is
different still and has a convex lower margin
(Fig. 20.15).

The australopithecine and
paranthropine acetabulum

The best preserved australopithecine acetabulae are the AL 288-1oa acetabulum
(*Australopithecus afarensis*) and the STS 14
(*Australopitheous africanus*) acetabulum.
Stern and Susman (1983) reported that the
articular surface of the AL 288-1 acetabulum
forms less of a circle than does the acetabulum
of modern humans. In particular they suggested that the pubic contribution to the
articular surface is less in this hominid than
it is in modern humans. By dividing the
height of the acetabulum measured from a
line connecting to the ends, or horns, of the
articular surface by the maximum diameter
of the acetabulum they demonstrated that the
A. afarensis acetabulum is similar to at least
some African apes (Fig. 20.24). The average
value of this ratio for 98 modern humans is

0.906 (SD = 0.028), while *A. afarensis*
measures 0.78. Stern and Susman (1983)
noted that ratios below 0.80 occur in 5–10%
of chimpanzees and that the average in gorillas
is 0.82.

Human characteristics have, however, been
reported for the Sts 14 (*A. africanus*) acetabulum. Relative to acetabular diameter, the
depth, the notch width and the thickness of
the acetabular walls of the specimen fall into
the human range of variation (Schutz, 1969b).
In absolute size the Sts 14 acetabulum is
similar to the chimpanzee acetabulum and
much smaller than either gorilla or modern
human acetabulae. However, relative to both
iliac fossa width and to iliac height, the Sts
14 acetabulum is only small in relation to
modern human acetabulae (Napier, 1964;
McHenry 1975c). The SK 3155 acetabulum
has the same relative size as does the Sts 14
acetabulum (McHenry, 1975c).

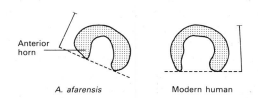

FIGURE 20·24 *The lunate surface of the acetabu-*
lum in Australopithecus afarensis *(AL 288-1) has a*
shorter anterior horn and is not as high in relation
to the line connecting its anterior and posterior horns
as is the case in the modern human acetabulum.
After Stern and Susman (1983).

The australopithecine and paranthropine sacrum

Sacra (= plural of sacrum) are known for both *Australopithecus afarensis* and *Australopithecus africanus*. (No paranthropine sacra are presently known.) In both australopithecine species the sacra are relatively wide as they are in modern humans (Robinson, 1972; Leutenegger, 1977) (Fig. 20.12). However, based on the more complete *A. afarensis* sacrum, they differ from those of modern humans in the following three ways: the sacrum is flat rather than curved in the longitudinal direction (Stern and Susman, 1983); it sits at less of an angle to the lumbar vertebral column than it does in humans (Abitbol, 1987a) and it lacks well-developed transverse processes on the upper segment (Fig. 20.25). The lateral angles on these transverse processes provide attachment areas for ligaments that stabilize the sacrum. Their absence in *A. afarensis* implies a less-developed system of sacral stabilization in erect posture than is the case modern humans (Stern and Susman, 1983).

Differences between the australopithecine and paranthropine pelves

On present evidence the ilia of *Australopithecus afarensis* and *Australopithecus africanus* are identical (McHenry, 1982, 1986; Berge and Kazmierczak 1986). *Paranthropus robustus* differs from these more gracile australopithecines primarily in its relatively wider iliac blade (McHenry, 1975c,d; Berge and Kazmierczak, 1986) that projects above the auricular surface to a greater extent than it does in the gracile forms (Berge and Kazmierczak, 1986). In addition, the auricular surface is relatively larger than it is in the gracile forms (Berge and Kazmierczak, 1986). These features are evident in SK 3155 (*P. robustus*) which is the same absolute size as the gracile fossils (Fig. 20.26). Therefore, the differences in these features are not merely

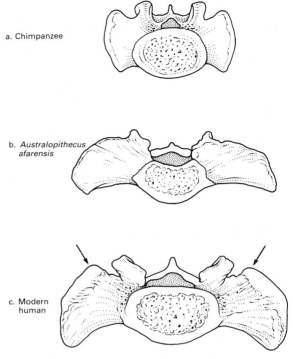

a. Chimpanzee

b. *Australopithecus afarensis*

c. Modern human

FIGURE 20·25 *Chimpanzee,* Australopithecus afarensis *(AL 288-1an), and human sacra from the superior (crania) aspect. Note the well developed transverse processes on the human sacrum (arrows). Views (b) and (c) after Stern and Susman (1983).*

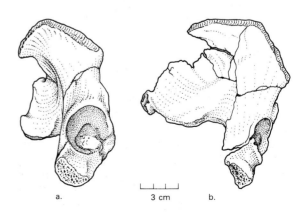

a.   3 cm   b.

FIGURE 20·26 *The SK 3155 pelvic bone viewed from (a) the plane of the acetabulum and (b) the orientation offering the greatest breadth of the iliac blade.*

the result of a difference in body size between *Paranthropus* and *Australopithecus*. (However, see Brain et al., (1974) for the alternative opinion that the SK 3155 pelvis is not *Paranthropus* but *Homo* and possibly even *Homo erectus*.) Robinson (1972) argued that the length of the ischium separated *Paranthropus* from *A. africanus*. However, more recent analyses have shown that there is no difference in the relative length of this bone in the two species (McHenry, 1975d,e; Lovejoy et al., 1973; Stern and Susman, 1983).

The australopithecine and paranthropine pelvis as a whole

Although these pelves look more like the pelves of modern humans than like the pelves of living apes, they are characterized by a unique morphology that has no exact parallel among living animals. They share with modern humans the following features: (1) a short, wide ilium; (2) a well-developed sciatic notch; (3) a well-developed anterior inferior iliac spine; (4) a wide sacrum; and (5) a short ischium. However, these pelves differ from those of modern humans in their extreme width. Not only are these pelves particularly wide at the level of the iliac crests (Berge and Kazmierczak, 1986), but they are also wide at the level of the acetabulae. This unusual width is reflected in the long pubic ramus, in the large iliopubic and subpubic angles and in the very broad, or platypelloid, pelvic inlet.

In addition, there is a suite of features that suggest that the australopithecine and paranthropine pelves may not have supported the body in the same manner that is characteristic of modern humans. Among these features are: (1) relatively small sacral and acetabular joint surfaces; (2) a small sacral angle; (3) a relatively flat sacrum; (4) an ischium that lacks the 'pulled-up' tuberosity; (5) a flared iliac blade; and (6) a relatively very small iliac tuberosity.

Although there is little doubt that the australopithecines and paranthropines were capable of bipedal locomotion, there is some disagreement as to whether the morphology of the pelves indicates habitual bipedalism. There are two current interpretations. On the one hand Lovejoy and co-workers (Lovejoy, 1974, 1975, 1978, 1988; Lovejoy et al., 1973) argued that the differences in pelvic morphology distinguishing the australopithecines from modern humans do not necessarily imply any difference in gait. On the other hand, Stern and Susman (1983) and Prost (1980) argued that the australopithecine morphology indicates an adaptation to climbing as well as to a less than human type of bipedal locomotion.

Lovejoy supported habitual bipedalism on the basis of his interpretation of the function of the small gluteal muscles as abductors of the hip. He argued that the wide lateral flare of the australopithecine iliac blade carried the small gluteal muscles over the lateral aspect of the hip joint allowing them to function as abductors of the pelvis on an extended thigh. The length of the iliac blade, together with the corresponding length of the femoral neck (Chapter 21), would give the muscles an even more favourable lever advantage in abduction than human iliac-femoral morphology and considerably reduce the force on the femoral head at the hip joint. This last point would be consistent with the relatively small acetabulum (and femoral head) characteristic of all known australopithecine and paranthropine hip joints. It is also claimed that the anterior position of the iliac pillar, coupled with the iliac flare, indicates that the function of the abductors was the same in both humans and the australopithecines.

Stern and Susman (1983) on the other hand argued that the anterior position of the iliac pillar, together with the more coronally orientated and posteriorly placed iliac blade, suggest that pelvic stabilization during bipedalism was closer to that in the apes rather than in humans. In the apes, the small gluteal muscles stabilize the trunk during bipedalism by acting as medial rotators on a flexed thigh

451

rather than abductors on the extended thigh. Their argument is also supported by the previously discussed characteristics which indicate that the pelvis did not support the weight of the upper body in the same fashion as it does in humans. In addition, the wide iliac flare could have served the same function in the australopithecines as it does in living apes giving an advantageous attachment to the important climbing muscle, latissimus dorsi (Schmid, 1983) (Chapters 16 and 19).

The obstetric pelvis in the australopithecines

There are surprisingly few pelves in the fossil record that are complete enough to determine the size of the birth canal. However, obstetric inferences are possible for *Australopithecus afarensis* and *Australopithecus africanus*.

Obstetric inferences for *A. afarensis* are based on the pelvis of AL 288-1. This pelvis is represented by a complete sacrum (AL 288-1an) and a left pelvic bone (AL 288-1ao). The condition of these specimens is discussed in Johanson et al. (1982). Reconstruction of a complete pelvis has been undertaken by both Berge et al. (1984) and Tague and Lovejoy (1986). The striking feature about both these reconstructions is that the pelvic inlet of *A. afarensis* is unusually wide in the transverse diameter (Fig. 20.20) (Berge et al., 1984; Tague and Lovejoy, 1986). In relation to overall pelvic size, the *A. afarensis* pelvis is far wider than those of either living humans or living apes (Berge et al., 1984). In addition, the sagittal diameter of the pelvic inlet is only 58% of the transverse diameter and this is considerably below the modern human average of 78% (Tague and Lovejoy, 1986). This extreme oval shape of the pelvic inlet has been likened to the platypelloid type of modern human pelvis (Tague and Lovejoy, 1986).

Also, in contrast to the human female pelvis, the *A. afarensis* birth canal is transversely oval not only at the level of the pelvic inlet, but also at the level of the pelvic mid-plane and the pelvic outlet. As a result, Tague and Lovejoy (1986) suggested that the *A. afarensis* foetus did not rotate in the birth canal at the level of the pelvic mid-plane as do human foetuses, but maintained a transverse orientation through the birth canal (Fig. 20.27). This inferred position of delivery is unique not only in relation to modern humans but also in relation to living apes.

Based on the assumption that the head of the *A. afarensis* foetus was the same size as the head of a newborn chimpanzee, Tague and Lovejoy (1986) argued that the size of the head of an *A. afarensis* neonate would approximate the dimensions of the birth canal. These authors therefore inferred that birth would have been more difficult in the early australopithecines than it is in the modern apes. However, it has recently been demonstrated that for a mother of the inferred body size of AL 288-1 it would be unlikely that the *A. afarensis* neonate would be as large as a chimpanzee infant. Therefore, birth would have been easier than suggested by Tague and Lovejoy (Leutenegger, 1987). Tague and Lovejoy (1986) also suggested that the pelvic sexual dimorphism characteristic

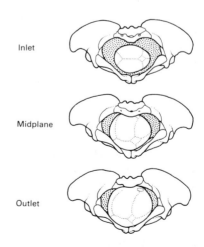

FIGURE 20·27 *The inferred mechanism of birth in* Australopithecus afarensis *(AL 288-1). The reconstructed pelvis (and fetus) are seen from the inferior view. After Tague and Lovejoy (1986).*

of modern humans had not yet developed in these early hominids. This inference is based on similarities between the *A. afarensis* inferred female pelvis and modern human male pelves in the angularity of the sacrum, in the index of pelvic funnelling and in the straight inferior border of the ischiopubic ramus.

Obstetric inferences for *A. africanus* have been based on the Sts 14 pelvis (Leutenegger, 1972, 1982). This is the most complete of the *A. africanus* pelves and is represented by both the right and the left pelvic bones and about half of the sacrum. Because of the distortion of the left pelvic bone in particular, Day (1973) has warned that the current reconstruction of this pelvis contains major anatomical errors.

The acetabulopubic length of Sts 14 (measured on the undistorted side) shows the same unusual width relative to overall pelvic size as does the *A. afarensis* pelvis (Berge et al., 1984). However, if the measurements of the size of the birth canal are considered to be correct in spite of the recognized distortion of the reconstruction, the pelvic inlet differs in shape from that of *A. afarensis*. The anteroposterior diameter is 8.5 cm and the transverse diameter is 9.9 cm giving an index of 85.9%. Although this still shows the transversely oval shape of the pelvic inlet it is not platypelloid as is the inlet in *A. afarensis* and is more rounded than is usually found in modern humans (Tague and Lovejoy, 1986).

Based on these pelvic dimensions and on the assumption that the head of the *A. africanus* neonate was the same size as the head of a modern chimpanzee neonate, Leutenegger (1972) has argued that birth would have been quick and easy in these australopithecines.

The Homo *sp. pelvis*

There are three *Homo* pelves described in the literature dating from between approximately 1.9 million years ago and the appearance of the Neanderthals. These are: (1) the KNM-ER 3228 pelvic bone from Koobi Fora, Kenya (Rose, 1984); (2) the OH 28 pelvic bone from Olduvai Gorge, Tanzania (Day, 1971, 1982, 1984); and (3) the Arago XLIV pelvic bone from Tautavel, France (Sigmon, 1982). These pelves show a morphology different from both the australopithecines and modern humans. A fourth pelvic bone from the Broken Hill mine, Kabwe, Zambia, shows a mixture of the *Homo* sp. and modern *Homo sapiens* morphology (Stringer, 1986).

Because all the known pelves lack both the sacrum and the pubis, our knowledge of the morphology of these hominids is based entirely on the ilium and the ischium. All these pelves are more similar to the morphology seen in modern *H. sapiens* than are the australopithecines (Fig. 20.28). Features shared in common with modern *H. sapiens* are: (1) the iliac proportions; (2) the relative length of the ischium and the 'pulled-up' appearance of the ischial tuberosity; (3) the size of the acetabulum; (4) the position of the sacroiliac articular area; and (5) the acetabulosacral buttress.

In addition, these hominids have certain features in common with the australopithecine pelvic morphology (Rose, 1984). In particular: (1) the iliac blade flares laterally in relation to the lower portion of the hip bone; (2) the iliac blade is set at a relatively wide angle to the sagittal plane; and (3) the anterior superior iliac spine is very protuberant.

There are also features that are unique to these hominids (Day, 1971, 1982, 1984; Sigmon, 1982; Rose, 1984). These include a well developed iliac pillar and iliac tubercle. The pillar is not only more developed than that found in modern humans but the cortical bone making up the pillar is very thick in relation to the cortical bone on the medial side of the iliac blade opposite the pillar (Stringer, 1986). In addition: (1) the gluteal fossa is very deep in the area of origin of gluteus medius and the iliac fossa is correspondingly shallow; (2) the sacral surface

453

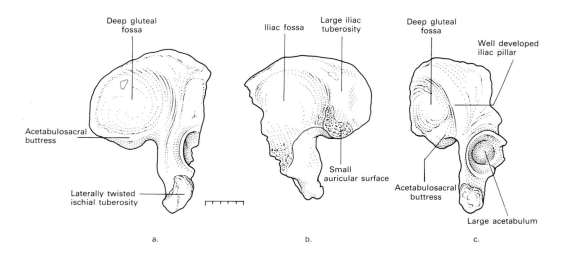

FIGURE 20·28 *The KNM-ER 3228 pelvic bone showing (a) the gluteal surface, (b) the iliac fossa and sacral surface, and (c) the iliac pillar. After Rose (1984).*

of the ilium is set at an acute angle to the remainder of the blade; (3) the auricular surface is small and set inferiorly on the sacral surface; (4) the iliac tuberosity is very large; and (5) the ischial tuberosity is twisted so that it faces not only posteriorly but also laterally.

The size of the greater sciatic notch appears to be an indicator of sex in these specimens as it is in modern *H. sapiens*. OH 28, Arago XLIV and E 719 have wide (female type) greater sciatic notches, while KNM-ER 3228 has a narrower notch reminiscent of modern human males.

In comparison to the other pelvic bones, the fragmentary Kabwe specimen (E 719) lacks the posterolateral orientation of the ischial tuberosity and also shows a clear ridge separating the true pelvis from the iliac fossa (Stringer, 1986). The other innominates show a smooth transition between these areas (Day, 1971). This pelvic bone appears to be transitional between archaic *H. sapiens* and modern humans; however, it may simply represent a primitive morphology shared by all non-modern hominids of its age (Stringer, 1986).

Function of the Homo sp. pelvis

The majority of the unique features of the *Homo* sp. pelvis can be explained by the lateral orientation and flare of the iliac blade (Rose, 1984) and there is general agreement that the *Homo* sp. pelvis as a whole shows a number of features that are consistent with human bipedalism (Day, 1971; Rose, 1984). These include: (1) a well-developed abduction mechanism indicated by the robust iliac pillar; and (2) the erect bearing of the spine as indicated by the large iliac tuberosity that gives rise to the muscles (e.g. erector spinae) and ligaments that support the spine on the pelvic base. In addition, the strong hip extensor mechanism is indicated by the impressive attachment areas for gluteus maximus and the hamstrings as well as by the well-developed acetabulosacral buttress. The well developed attachment areas for the iliofemoral ligament suggests that the body weight also passes behind the acetabulae.

The Neanderthal pelvis

Before the discovery of the Kebara 2 Neanderthal pelvis in 1983 (Bar-Yosef et al., 1986) there were no complete Neanderthal pelves (Trinkaus, 1983a). The Neanderthal pelvis was thought to be largely similar to that of modern humans with one exception. This exception was the unusual length and slenderness of the pubic ramus (Fig. 20.29) (Stewart, 1960; Trinkaus, 1976a, 1983a) which was taken to indicate an exceptionally large birth canal in Neanderthals in comparison to that of modern humans (Trinkaus, 1984).

The Kebara pelvis contradicts this interpretation in two areas. Firstly, the elongated slender pubic ramus is not associated with an enlarged birth canal. Secondly, the total morphology of the Neanderthal pelvis is different from that of modern humans.

In general, the morphology of the Kebara pelvis indicates that in Neanderthals the pelvic bones and sacrum lie in a different relationship to the hip joints (acetabulae) than they do in modern humans (Fig. 20.30) (Rak and Arensburg, 1987). It appears as if the iliac blades are twisted outward and the sacrum pushed forward. The outward rotation of the iliac blades also affects the orientation of the acetabulae and in Kebara the plane defined by their rims lies in a more lateral position than it does in modern humans. The outward rotation of the iliac blades requires the elongate pubis to span the distance from the more laterally facing acetabulae to the mid-line of the body. As the promontory of the sacrum lies closer to the line connecting the anterior margins of the acetabulae in Kebara, so the pubic symphysis lies further anterior to this line than it does in modern humans. The overall result is a pelvic inlet that differs little in the size from modern humans while the superior pubic ramus is 20% longer than the modern human norm (Fig. 20.31) (Rak and Arensburg, 1987).

Other features of the Kebara pelvis are most probably directly related to its general orientation. These features include: (1) an iliac blade that when viewed from the side of the pelvis is symmetrical, fanning out anteriorly to the same degree as it does

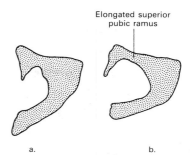

Elongated superior pubic ramus

a.   b.

FIGURE 20·29 *The pubic bone of (a) a modern human male and (b) a Neanderthal male. Note the elongated superior pubic ramus in the Neanderthal. After Trinkaus and Howells (1979).*

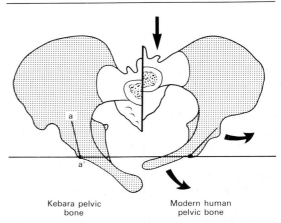

Kebara pelvic bone   Modern human pelvic bone

FIGURE 20·30 *The Kebara Neanderthal pelvic bone (left half) compared to a modern human pelvic bone (right half). The arrows indicate that in relation to a modern human pelvis, the Kebara sacrum lies closer to the plane of the anterior rims of the acetabulae (horizontal line), the Kebara acetabulae are oriented more laterally (line a–a'), and, as a result the Kebara superior pubic rami must be longer in order to meet in the midline. After Rak and Arensburg (1987).*

posteriorly; (2) the anterior superior and inferior iliac spines are protuberant and the notch between them is deep; (3) the massive iliac buttress lies close to the two spines and the portion of the gluteal surface posterior to this buttress faces more posteriorly than it does in modern humans; and (4) the body of the ilium, immediately above the acetabulum, is very narrow in comparison

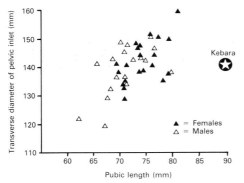

FIGURE 20·31 *The relationship between the length of the pubic bone and the transverse diameter of the pelvic inlet for modern human males and females and the Kebara Neanderthal. After Rak and Arensburg (1987).*

to modern humans. As the result of the symmetrical nature of the iliac blade and the anterior position of the sacrum, the greater sciatic notch is very shallow.

In addition, when viewed from the front, the body of the pubis appears to have been stretched anteromedially (Rak and Arensburg, 1987). Not only does this result in the elongated superior pubic ramus but also in an unusually wide subpubic angle (110°). In addition, the region anterior to the acetabulum appears extremely long anteromedially and low superoinferiorly.

The Kebara pelvis has been sexed as a male (Rak and Arensburg, 1987). From other Neanderthal pelves it appears as if the Neanderthals were characterized by an unusual sexual dimorphism where the female superior pubic ramus is shorter than that in the male (Rosenberg, 1986). If this is the case it is possible that the Kebara morphology represents a unique Neanderthal sexual dimorphism affecting the relative position of the pelvic inlet and the rotation of the iliac blades (Rak and Arensburg, 1987).

The obstetric pelvis in Neanderthal

Prior to the discovery of the complete Kebara 2 Neanderthal pelvis (Bar-Yosef et al., 1986) Trinkaus (1983a, 1984) suggested that the Neanderthal pelvis would have accommodated an infant with a head that was 15–25% larger in volume than the head of an average modern human infant. Furthermore, he suggested that it was possible that the gestation period of Neanderthal was 12–14 months, this extra time being the time needed to produce an infant brain size 15–25% larger than the average human neonate. Alternatively, it was also suggested that Neanderthal foetuses grew at different rates than modern human foetuses and, after a nine-month gestation period, were born with brains relatively larger than those of modern human infants (Dean et al., 1986).

These ideas were based on pubic bones from the then known Neanderthal pelves from Europe and south-western Asia (Trinkaus, 1984). In comparison with modern humans these pubic bones are mediolaterally elongated relative to available measures of stature (femoral length), body size (femoral head size) and brain size. The underlying assumption was that pubic ramus length correlated with the size of the female birth canal (particularly at the levels of the pelvic mid-plane and inlet). Not only has this purported correlation been questioned for humans (Greene and Sibley, 1986), but also the Kebara 2 pelvis throws serious doubt on this correlation for Neanderthals. Although this pelvis is male it clearly demonstrates that although the Neanderthal superior pubic ramus was indeed long the transverse diameter of the true pelvis (the birth canal in females) was within the range found in modern humans (Fig. 20.31). It remains true that some Neanderthal infants and juveniles had cranial capacities far in excess of modern humans of equivalent age (Dean et al., 1986). However, evidence from the pelvis no longer supports hypotheses of either a longer gestation period or a faster rate of foetal brain growth in order to produce these large brained young Neanderthals. Pelvic evidence suggests that any explanation for this phenomenon is most probably to be found in the postnatal growth patterns of these hominids.

CHAPTER TWENTY-ONE

THE HOMINOID FEMUR

The human femur is both absolutely and relatively longer than the ape femur (Fig. 21.1). The femur of an adult pygmy chimpanzee (*Pan paniscus*) is approximately 85% (84.1–87.1%) of the length of the femur of an adult human (pygmy) of about the same body weight (data from Jungers and Stern (1983)). Furthermore, the femur of an average male gorilla (body weight = 140 kg) is only about 82% of the length of the femur of an average male caucasian (body weight = 74 kg) (data from Aiello (1981a)). Allometric analyses have confirmed that the human femur is considerably longer than the ape femur not only in relation to body weight (Aiello, 1981b; Jungers, 1984, 1985) but also in relation to skeletal measures such as femur transverse diameter (Aiello, 1981b,c), humerus length (Aiello, 1981b), height of the iliac blade (Jungers, 1984) and skeletal trunk length (Biegert and Maurer, 1967; Aiello, 1981b).

The degree of obliquity of the femoral shaft, the **bicondylar angle**, also serves to distinguish human and ape femora. The bicondylar angle is the angle that the shaft of the femur makes with the vertical when the two femoral condyles are resting on a horizontal level surface (the bicondylar, or infracondylar, plane). Humans have a larger bicondylar angle, or a more valgus angle, than do the large apes, although the angles of some smaller primates that engage in climbing approach the human condition (Stern and Susman, 1983). A large bicondylar

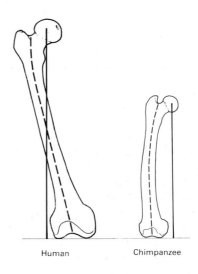

Human   Chimpanzee

FIGURE 21·1 *Human and chimpanzee femora from the anterior aspect. The solid line represents the load axis and the dashed line the axis of the shaft of the femur. After Walmsley (1933).*

angle is important to human bipedal locomotion because it positions the distal end of the femur, together with the knee joint, the lower leg and foot, close to the mid-line of the body. As a result, the centre of gravity of the body need move only a short distance laterally to lie above the stance leg during walking (Chapter 14). If the bicondylar angle were smaller and the knees and feet were thereby positioned directly below the hip-joints, as is the case with the apes, the distance through which the centre of gravity

457

would have to move to lie over the stance leg would be considerable. Under these circumstances bipedalism would be less efficient than it is in humans where the femora have a high bicondylar angle.

Specific values for the bicondylar angle are dependent on the definition of the axis of the shaft of the femur from which the angle is measured (Walmsley, 1933; Kern and Straus, 1949; Heiple and Lovejoy, 1971). However, modern humans have bicondylar angles that average about 9–10° while the African apes have much lower angles of about 1–2° (Table 21.1).

Because of the large human bicondylar angle, the weight transfer down the shaft of the femur is much different from that in the apes with their low bicondylar angles and their more perpendicular femora. In particular, Walmsley (1933) pointed out that in the apes the **load axis** (the perpendicular line from the femoral head to the bicondylar plane) never intersects the **axis of the shaft** of the femur (Fig. 21.1). In human adults, on the other hand, the load axis crosses the axis of the femoral shaft from medial to lateral at about midshaft. Human infants have femora that resemble those of apes to the extent that the axes are uncrossed and the load axis passes down the medial side of the

TABLE 21·1 The bicondylar angle in modern humans, apes and selected hominid fossils. After [a]Heiple and Lovejoy (1971). [b]Walker (1973) and [c]Stern and Susman (1983).

	N	Angle	SD
Modern human male[a]	168	9.43	1.93
Modern human female[a]	98	10.50	2.40
Gorilla[a]	16	1.72	3.80
Chimpanzee[a]	16	0.97	1.93
STS 34[a]		15	
TM 1513[a]		14	
KNM-ER-993[b]		14–15	
AL-333–4[c]		9	
AL-129[c]		15	

shaft. The typical human pattern of crossed axes develops with the bicondylar angle and the onset of efficient bipedal locomotion some time before three years of age (Walmsley, 1933).

There are a number of morphological differences between the human femur and the ape femur that directly reflect the difference in weight transfer. These include the following.

(1) Mediolateral curvature of the femoral shaft: where mediolateral curvature occurs in the human femoral shaft it corresponds closely to the characteristic pattern of crossed axes. The curvature is

FIGURE 21·2 *Relative area of the weight bearing surface of the lateral femoral condyle in relation to that of the medial condyle in humans and selected non-human primates. Data from Kern and Straus (1949).*

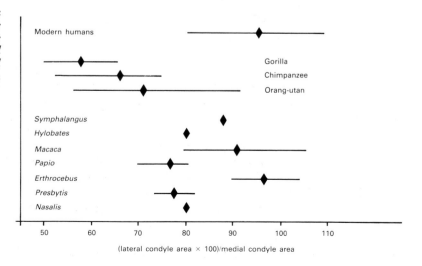

laterally convex in the upper part of the femur above the crossing point (Walmsley, 1933). In African apes, where the axes do not cross, the femora are laterally convex throughout their lengths (Fig. 21.1).

(2) Cortical thickness in the femoral shaft: in humans the thickness of the cortex on the lateral side generally exceeds the thickness of the cortex on the medial side at midshaft. This reflects the human pattern of weight transmission along the lateral aspect of the shaft beginning at, or slightly above, midshaft where the axes cross (Kennedy, 1983b).

(3) The relative size of the femoral condyles: the size of a joint surface can generally be taken as a rough reflection of the amount of force to which the joint is subjected (Kern and Straus, 1949; Jungers, 1988a). In humans the area of the lateral condyle of the femur approaches the area of the medial condyle, and transmits a similar load (Morrison, 1970; Walker and Hajek, 1972). In the African apes the medial condyle is considerably larger than the lateral condyle (Fig. 21.2) (Kern and Straus, 1949; Tardieu, 1981; Ruff, 1988) and most probably also transmits a larger proportion of the load.

Further morphological differences between human and ape femora

Other differences between the morphology of the human and ape femur can be divided into those that relate to (a) the structure of the head and neck of the femur, (b) the hip musculature, (c) the femoral shaft, and (d) the femoral condyles.

Differences between human and apes in the head and neck of the femur

THE SIZE OF THE HEAD
Relative to body size, the head of the human femur is larger than the head of the ape femur (Ruff, 1988; Jungers, 1988a). This is consistent with the other relatively large joint

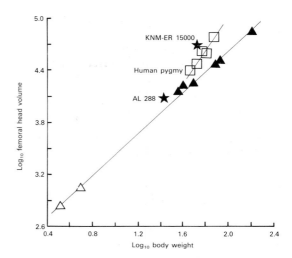

FIGURE 21·3 _The volume of the femoral head in relation to body weight in humans (□), the large apes (▲), macaques (△) and selected fossil hominids. After Ruff (1988)._

surfaces of the lower limb in the human skeleton (Chapter 14) and reflects the greater proportion of body weight carried by the lower limbs in bipedal locomotion than in quadrupedal locomotion. Although the head of the human femur is uniformly larger than the head of the ape femur, within humans the size of the head is strongly positively allometric (Fig. 21.3) (Ruff, 1988). This means that larger bodied modern humans have femoral heads that are much bigger relative to body weight than do smaller bodied modern humans. Why this should be so is not clear. Femoral head size is not similarly positively allometric in the apes and neither is the size of the cross-section of the femoral shaft in humans (Ruff, 1988). The only other feature recognized at present to be positively allometric is the cross-section of the neck of the femur. Ruff (1988) suggested that the positive allometry of both the femoral head and the femoral neck might indicate that an increase in body size in bipedal humans increases the loading magnitudes relatively faster than would be the case in quadrupedal apes.

THE SET OF THE HEAD ON THE NECK

The set of the head on the neck has also been considered an important distinction between humans and apes (Stern and Susman, 1983; Jenkins, 1972). In humans the articular surface of the femoral head extends onto the anterior surface of the neck, or sits squarely on the neck, while in chimpanzees it extends onto the posterior surface of the neck (Fig. 21.4). Jenkins (1972) argued that this difference is related to the mechanics of the hip joint and in particular to the deep and anterolaterally facing acetabulum in humans (Chapter 20) and the shallow and laterally facing acetabulum in chimpanzees. Recently, however, this distinction has been questioned. Asfaw (1985) found all three morphologies in a modern human skeletal sample and when the 'chimpanzee' pattern occurs it is found most frequently in females.

LENGTH OF THE FEMORAL NECK

There are a variety of ways to measure the relative femoral neck length and each of the three most popular methods gives different results for the relative neck length of humans in relation to either the apes or the fossil

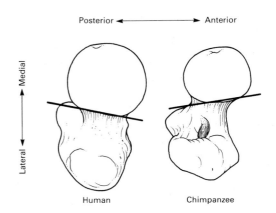

FIGURE 21·4 *The superior view of the femoral head in a human and a chimpanzee. In humans the edge of the articular cartilage typically angles from anterolateral to posteromedial across the neck of the femur. In chimpanzees the articular cartilage angles in the opposite direction. After Stern and Susman (1983).*

hominids. If neck length is measured from the intertrochanteric crest to the junction between the head and the neck and compared with the transverse diameter of the femoral shaft just below the lesser trochanter, the African apes have shorter femoral necks than

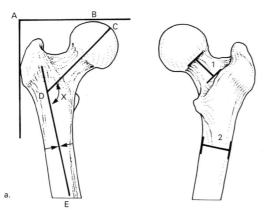

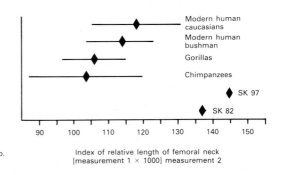

FIGURE 21·5 *(a) Measurements commonly taken on the proximal femur. 1 = Neck length from the intertrochanteric crest to the junction between the head and neck; 2 = the transverse diameter of the femoral shaft just below the lesser torchanter; (A–B)* = biomechanical neck length; (C–D) = neck length measured along the axis of the neck; (D–E) = axis of the shaft; angle X = the neck-shaft angle. (b) The index of the relative length of the femoral neck. After Napier (1964).*

do modern humans (Fig. 21.5) (Napier, 1964). However, if neck length is defined as biochemical neck length (the length of a line from the lateral most point on the greater trochanter to its tangential point of intersection with the most cranial part of the femoral head (Lovejoy, 1975) and compared with the length of the femur, the neck length in the chimpanzees is close to the human mean (Fig. 21.5). Furthermore, if neck length is measured along the axis of the neck from the top of the head to the junction of the neck axis with the shaft axis (Martin and Saller, 1957) and again compared with femoral length, femoral neck lengths of all the great apes exceed the average of the means of eleven modern human populations (Martin and Saller, 1959) (Fig. 21.5).

Some of the contradiction in these results may be explained by the use of femoral length in the last two comparisons as a standard of reference. As noted above, the length of the femur in all the large apes is relatively shorter than expected in relation to body weight, while in humans it is relatively longer. These latter two comparisons may just as well be measuring differences in relative femoral length as in relative neck length. In addition, any measure of neck length which also includes the head (e.g. Martin and Saller, 1957) would confuse differences in the size of the femoral head between species with differences in the actual length of the neck.

THE NECK–SHAFT ANGLE
The angle made by the axis of the femoral neck with the axis of the femoral shaft (Fig. 21.5) is variable in human populations and this variation within populations may be as great as 23° (Martin and Saller, 1959). The mean of the averages of 23 human populations for this angle is 126.9° (SD=2.93) (Martin and Saller, 1959), the mean for modern Europeans is 128.5° (SD=4.7, $n=50$) and for modern North American Indians 126.8° (SD=4.4, $n=50$) (Trinkaus, 1976b). However, the mean for Upper Palaeolithic modern humans is considerably lower at 117.4±1.6°

(SD=5.8, $n=14$) and for *Homo sapiens* from Skhūl, Israel, is higher at 130.0±4.0° (SD=7.0, $n=3$) (Trinkaus, 1976b). The neck-shaft angles for the African apes given by Zapfe (1960) fall near the modern human mean at 129° for both chimpanzees and gorillas.

In modern Europeans the neck-shaft angles decrease during ontogeny from an infant angle of about 140°. This decrease is also typical of other human groups and of the apes (Martin and Saller, 1959). In functional terms, a high neck-shaft angle in humans is related to a greater range of joint mobility than is possible when the neck-shaft angle is lower (Kapandji, 1987).

THE INTERTROCHANTERIC LINE
Chimpanzees also lack the anterior intertrochanteric line marking the attachment for the iliofemoral ligament (see Chapters 19 and 20). In humans this ligament is very strong and functions to maintain upright posture. Because the centre of gravity passes behind the hip joint, the strength of this ligament prevents the trunk from falling backwards at the hip joint (Fig. 20.9). The tension in the ligament produces the roughened anterior

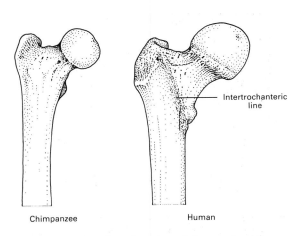

FIGURE 21·6 *Anterior view of the proximal femur in a chimpanzee and a human. Note the presence of the intertrochanteric line in the human.*

intertrochanteric line (Fig. 21.6). Because this ligament is shaped like a Z the intertrochanteric line is most strongly developed inferiorly and superiorly (where it terminates in a distinct femoral tubercle). The iliofemoral ligament does not have the same function in chimpanzees that move with an habitually flexed hip. It is therefore much weaker and does not leave an indication of its attachment in the form of either an intertrochanteric line or a femoral tubercle.

Differences between humans and apes that reflect hip musculature

CROSS-SECTION OF THE FEMORAL NECK

In humans the cortical bone in the neck of the femur is concentrated around the distal half of the neck circumference (Fig. 21.7) (Lovejoy, 1988). In the proximal half of the circumference the cortical bone is very thin. This is possible in humans because the smaller gluteal muscles (gluteus minimus and gluteus medius) function as abductors of the hip and pull in line with the neck of the femur (Chapter 19). They thereby considerably reduce the bending stress on the superior half of the femoral neck, reducing the need for thick cortical bone in this area. In the apes, on the other hand, no muscles pull in line with the femoral neck and the cortical bone is thick around the entire neck circumference.

THE GREATER TROCHANTER

The greater trochanter in the two species differs considerably in shape. The human trochanter is low in relation to the superior border of the neck in comparison with the apes (Fig. 21.8) (McHenry and Corruccini, 1976a) and the trochanter lacks the marked lateral projection, or flare, of the human trochanter. In addition, the trochanteric fossa on the medial surface of the trochanter is extremely deep in the chimpanzee giving a large area of insertion for the obturator externus muscle, one of the lateral rotators of the thigh at the hip (Chapter 19).

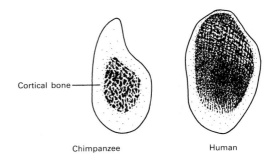

FIGURE 21·7 *Cross-sections through the neck of the femur showing the differences in internal structure between a chimpanzee and a human. After Lovejoy (1988).*

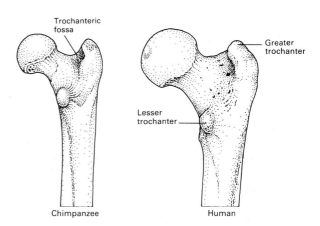

FIGURE 21·8 *Posterior view of the proximal femur in a chimpanzee and a human. Note the differences in the height of the greater trochanter in the two species.*

THE LESSER TROCHANTER

In chimpanzees the lesser trochanter is approximately the same absolute size as it is in humans (Figs 21.8 and 21.9). In relation to the overall size of the chimpanzee femur it is therefore much larger. When the chimpanzee femur is placed on a flat surface resting on its two condyles and its proximal end, it is the lesser trochanter that supports it proximally rather than the posterior surface of the greater trochanter as is normal in

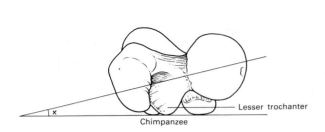

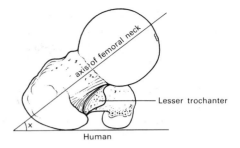

FIGURE 21·9 *Superior view of the proximal femur in a chimpanzee and a human. The angle of femoral torsion (X) is the angle made by the axis of the femoral neck with the tangent of the posterior surface of the femoral condyles. Note the differences in the size and position of the lesser trochanter.*

humans. The large size of the lesser trochanter reflects the relative development of the muscle that inserts into it, the psoas muscle (a flexor of the thigh at the hip joint) (Chapter 19).

ORIENTATION OF THE LESSER TROCHANTER

In both humans and the apes the position of the lesser trochanter is highly variable and, contrary to the opinion of some authors (Napier, 1964; Day, 1969a), does not directly reflect the locomotor differences that distinguish living humans from apes (or from earlier hominids) (Lovejoy and Heiple, 1972). In humans and apes the lesser trochanter can be located on the centre of the posterior surface of the proximal shaft or it can be more medially positioned. Not infrequently it is so medially positioned that it is clearly visible on the medial side of the shaft when the femur is viewed from its anterior side. The position of the lesser trochanter is strongly influenced by the torsion of the femur (Lovejoy and Heiple, 1972). The torsion of the femur is defined as the angle made by the axis of the femoral neck with the tangent of the posterior surface of the femoral condyles. The larger this angle, or the more anteverted is the femoral neck, the more the lesser trochanter is carried, or twisted, around to the medial margin of the femoral shaft.

There is an extremely large range of variation in the torsion of the human femur ($-25°$ to $+42°$) although negative values (retroversion of the femoral neck) are very rare (Martin and Saller, 1959). In gorillas, negative values occur in 17% of cases, in chimpanzees in 29% of cases and in orangutans in 80% of cases (Martin and Saller, 1959).

GLUTEAL RIDGE OR GLUTEAL TUBEROSITY

The **gluteal ridge**, or **gluteal tuberosity**, is that part of the upper lateral branch of the linea aspera that serves as the attachment of gluteus maximus (Figs 19.6 and 21.10) (Hrdlicka, 1937). There has been a considerable amount of confusion in the terminology applied to this feature. Some anatomy books (e.g. Gray's) use the term gluteal tuberosity to refer to this ridge. However, in the physical anthropology literature the term gluteal tuberosity has been traditionally reserved as an alternative name for the **third trochanter** (see below) (Hrdlicka, 1937). Furthermore, some authors consider the gluteal ridge and the **hypotrochanteric crest** to be one and the same thing (Pearson and Bell, 1919). However, Hrdlicka (1937) preferred to distinguish them, reserving the term gluteal ridge for the commonly occurring rugose ridge and the term hypotrochant-

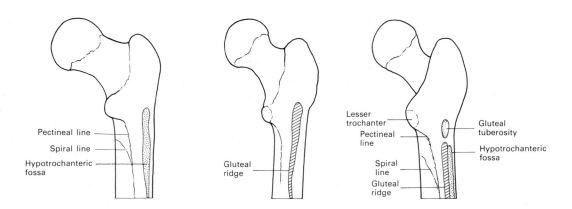

FIGURE 21·10 *The gluteal ridge, hypotrochanteric fossa and third trochanter in human femora. After Martin and Saller (1959).*

eric crest as an alternative name for the smoother linear area that corresponds to an oblong third trochanter (see below).

The gluteal ridge (as recognized by Hrdlicka) is very common in human femora being present, for example, in 98.5% of American Caucasian femora ($n=1000$) (Hrdlicka, 1937). In this sample a ridge exists without a corresponding third trochanter in 68.2% of cases. In other human populations a ridge can be totally absent in from 11% to 13% of cases (Hrdlicka, 1937). In gorillas, on the other hand, the ridge is absent in 92.2% of adults ($n=77$) although it is missing in only 63.6% ($n=22$) of young animals and in 86.7% ($n=45$) of juvenile animals (Hrdlicka, 1937). This suggests a decrease in the frequency of this trait in ontogeny. In chimpanzees the ridge is missing in only 15.5% of adults ($n=97$). In these apes it is absent in 10.7% of young animals ($n=28$) and 9.7% of juvenile animals ($n=31$), suggesting a decrease in frequency through the growth period (Hrdlicka, 1937). The orang-utans are midway between the chimpanzees and gorillas in the frequency of occurrence of the gluteal ridge. The ridge is absent in 54.3% of adults ($n=46$) and in 31.3% of juveniles ($n=32$). In human populations the ridge tends to increase in frequency during ontogeny rather than decrease in frequency although this pattern is variable (Hrdlicka, 1937).

THE HYPOTROCHANTERIC FOSSA

The hypotrochanteric fossa is an elongated groove that is variably found in primate femora but has not been observed to occur in other mammals (Fig. 21.10) (Hrdlicka, 1934b). It can occur in place of a gluteal ridge and when it does it occupies the position of the ridge on the lateral side of the proximal part of the posterior femur. When it occurs in combination with the gluteal ridge it is located lateral to the ridge. In modern humans the frequency of occurrence of the hypotrochanteric fossa varies from a high of 89.6% in Eskimos ($n=718$) to a low 61.2% in Cantonese Chinese (Hrdlicka, 1934b). However, 15–30% of those individuals that do have a fossa only have a trace of a fossa. In the great apes the fossa is normally well developed and is present in 73.8% of gorillas, 75.8% of chimpanzees and 80.4% of orangutans. In the African apes it is normally situated lower on the shaft than in humans, has a oblique orientation and is placed more laterally on the shaft. In humans it lies between the gluteal ridge and the lateral border of the shaft while in the apes it involves the lateral border of the shaft

(Hrdlicka, 1934b). In the African apes the fossa is also 'spacious and rough' rather than relatively narrow and linear as is normally the case in humans.

A curious thing about the fossa in humans is that its frequency of occurrence is greater in children and juveniles than it is in adults. The fossa begins to develop in the fifth foetal month and increases in frequency and size through the growing period (Hrdlicka, 1934b). In adulthood it declines in both size and frequency and appears to be obliterated by the encroaching gluteal ridge or by the formulation of a third trochanter. In other words remodelling occurs when bony deposits fill in the fossa during adult life (Hrdlicka, 1934b). In chimpanzees the situation is reversed and there is a much higher frequency of hypotrochanteric fossae in adults than in juveniles. The fossae in gorillas also appear to have a higher frequency in adults than juveniles, although the trend is not as marked as in chimpanzees. Orang-utans, among the great apes, are the most similar to the human condition where there is a greater frequency of hypotrochanteric fossae in adolescent femora.

THE THIRD TROCHANTER (OR GLUTEAL TUBEROSITY)

The **third trochanter** (or gluteal tuberosity) is an oblong, rounded, or conical area of bone that occurs variably on, instead of, or above the gluteal ridge (Fig. 21.10) (Hrdlicka, 1937). It provides attachment for part of the gluteus maximus muscle. The third trochanter is present in between 11.5% (Cantonese Chinese) and 43.4% (North American Indians) of modern human femora (Hrdlicka, 1937). In the great majority of cases the third trochanter takes an oblong form, while in fewer individuals it is rounded or conical. In a very few cases it is multiple (Hrdlicka, 1937). Femora with more pronounced third trochanters also appear to be slightly more **platymeric** than average (see below) (Hrdlicka, 1937). Among the great apes, the third trochanter is present in 9% of chimpanzees although it is never as well

developed as it can be in humans. It is not found in gorilla or orang-utan femora (Hrdlicka, 1937).

OBTURATOR EXTERNUS GROOVE

The **obturator externus groove** is a groove variably left on the posterior surface of the neck of the femur by the tendon of the obturator externus muscle (Fig. 21.11). This muscle originates from the body of the pubis, ischiopubic ramus and membrane that closes the obturator foramen (Fig. 19.21). In humans its fibres converge into a tendon which passes behind the neck of the femur to insert into the trochanteric fossa. As the tendon passes across the posterior surface of the femoral neck it can leave a noticeable rounded groove that traces a diagonal upwards course towards its insertion. Because of the habitual flexed-hip posture of the chimpanzee, the tendon passes in a straight line from its origin on the pelvis to its insertion in the trochanteric fossa. It does not wrap around the back of the femoral neck and as a result there is no obturator externus groove on the posterior surface of the chimpanzee femoral neck.

The locomotor significance of the presence of an obturator externus groove is, however, debated. Although many anatomists consider the groove to be a sign of bipedal posture (e.g. Day, 1969a; Lovejoy, 1975) this interpretation has been challenged. Stern and

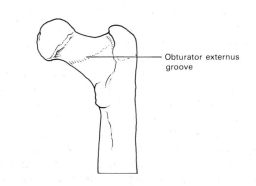

FIGURE 21·11 *Posterior view of the* Paranthropus robustus *(SK 97) femur showing a well defined obturator externus groove. After Robinson (1972).*

Susman (1983) argued that an obturator externus groove is simply a reflection of the close proximity of the acetabulum to the ischiopubic ramus. The shorter this distance the less room for the obturator externus to wrap around the femoral neck and the greater the likelihood of increased pressure on the neck. They suggested that the presence of the obturator externus groove reflects a reduced ischial length rather than a bipedal posture.

Difference between humans and apes in the morphology of the femoral shaft

The most noticeable difference between the femoral shaft of humans and apes is the presence in humans and the absence in apes of a well-developed **pilaster**. The pilaster is a beam-like structure which supports the linea aspera and extends down the back of the femoral shaft (Fig. 21.12). It is present in varying degrees of development in 71% of American Caucasian femora (Hrdlicka, 1934a) and is not necessarily correlated with the degree of expression of the linea aspera which is universally present in adult human femora. In other words the linea aspera may exist in all degrees of development without a pilaster and the pilaster may exist when the linea aspera is only minimally or moderately developed. In humans the linea aspera may appear as early as the middle of foetal life, while the pilaster first appears in late childhood or adolescence. Both structures increase in frequency and expression through adult life. In cases where the femur has a considerable anteroposterior curvature, the pilaster appears to compensate for the extent of curvature, straightening the back of the femoral shaft.

The pilaster is universally absent in the African apes, although it is present in some New and Old world monkeys (Hrdlicka, 1934a). In the African apes the linea aspera is present in a slight to moderate development.

The presence of a pilaster in humans and its absence in apes results in a very different shape of the midshaft cross-sections of human

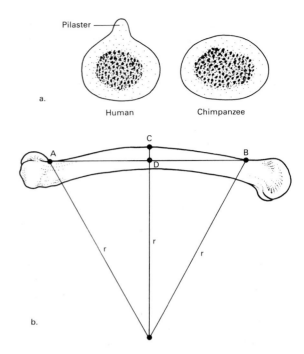

FIGURE 21.12 *(a) Cross-sections of a human and chimpanzee femur at midshaft showing the pilaster in the human cross-section. (b) The radius of curvature (r) of the shaft of the femur is computed from the length of the cord (h) connecting the two lowest points on the anterior border of the shaft (A–B) and from the distance between this cord and the most anteriorly projecting point on the anterior border of the shaft (C–D). See text for discussion.*

and ape femora. In humans the anteroposterior diameter of the cross-section is, on average, considerably larger in relation to the transverse diameter than in apes. This results in a generally higher **pilasteric index** (anteroposterior diameter × 100/mediolateral diameter) in humans than in apes. Although the range of variation in this index can be great (72.5–136.7 in one southern German population) the average of the means of 35 modern human populations is 109.7 (SD=5.64) (data from Martin and Saller, 1959). Among the large apes, gorillas have an average index of 80.0 (SD=5.9, $n=16$, range 69.5–89.6), chimpanzees an average index of 90.8 (SD=7.9, $n=10$, range

79–107.1) and orang-utans an average index of 80.9 (SD=3.6, n=11, range 73.8–86.4) (Aiello, 1981a).

Other proportional relationships in the human and ape femoral shafts include the following.

(1) **Meric index.** The meric index describes the anteroposterior flattening of the proximal shaft taken at 25% of shaft length (at about the level of the spiral line). Values for this index in humans range between 56 and 128 while the average of the means of 29 different human groups (Martin and Saller, 1959) is 76.9 (SD=5.25). In many cases in humans there is a correlation between a low meric index (a so-called platymeric shaft) and a weakly developed pilaster. However, this is not always the case (Martin and Saller, 1959). Among the great apes, the orang-utans have the flattest upper shaft with an index of 71.2 (range 67–75), gorillas have an average index of 81.0 (range 72–89) and chimpanzees an average index of 82.8 (range 72–96).

(2) **The point of minimum shaft breadth.** In humans the femoral shaft is noticeably narrow (in its transverse dimension) in the midshaft region while in apes the shaft is of more uniform proportions throughout its length (Fig. 20.1). The 'waisting' of the human femoral shaft corresponds to the crossing of the load and anatomical axes and occurs at between 54.7% and 59.8% of the shaft length as measured from the bicondylar plane (Kennedy, 1983b).

(3) **Robusticity.** The human femur is more gracile than the ape femur as measured by the robusticity index (midshaft circumference × 100/shaft length) and the shaft breadth index (transverse shaft breadth at midshaft × 100/shaft length) (Chapter 14 and Fig. 14.5). Allometric analyses suggest that this difference between humans and the apes results more from a reduction in the length of the femur in

relation to body weight in the apes than an increase in midshaft dimensions in humans (Aiello, 1981b).

(4) **Anteroposterior curvature of the shaft.** Martin and Saller (1957, measurement no. 27) defined the anteroposterior curvature of the femoral shaft as the radius of curvature of the anterior border of the shaft (Fig. 12.12b). The radius of curvature (r) is computed according to the following formula:

$$r = \frac{a^2 + 4h^2}{8h}$$

where h is the length of the cord connecting the two lowest points on the anterior border of the femoral shaft (Fig. 12.12b) and a is the distance between this cord and the most anteriorly projecting point on the anterior border of the shaft. The shorter the radius of curvature, the more curved is the shaft of the femur.

There is a great deal of variability in the radius of curvature of human femoral shafts. Martin and Saller (1959) reported a range between 81 and 127 (mean = 101.8) for a single population of North American Indians. The African ape femora reported by Martin and Saller (1959) have a more curved shaft than even the most extreme in this human population. The chimpanzee has a radius of curvature of 77.5 and the gorilla of 70.8.

Differences between humans and apes in the morphology of the distal femur

In addition to the previously described differences in the relative size of the femoral condyles the following features differentiate human and ape distal femora.

ELLIPTICAL PROFILE OF THE CONDYLES

Both the lateral and the medial condyles in humans have an elliptical profile when viewed from the side (Fig. 21.13). This maximizes the area of contact at the knee joint when the knee is in extension and thereby minimizes

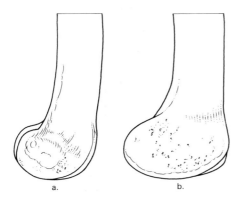

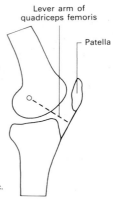

FIGURE 21·13 *Lateral view of the distal femur of (a) a chimpanzee and (b) a human. The elliptical shape of the human femoral condyles both maximizes the area of contact between the femur and tibia and*

(c) projects the patella anteriorly thereby increasing the lever arm of the main extensor of the knee, quadriceps femoris.

the load on the knee (Heiple and Lovejoy, 1971). In addition, the elliptical form projects the patella anteriorly thereby increasing the lever arm of the main extensor of the knee, quadriceps femoris (Lovejoy, 1975).

GENERAL PROPORTIONS OF THE CONDYLES

The distal surface of the human femur has a square outline rather than a rectangular outline which is characteristic of the apes (Fig. 21.14). This is directly related to bipedal locomotion, reflecting both the elliptical form of the condyles and the large size of the lateral condyle.

SYMMETRY OF THE CONDYLES

The distal surface of the human femur is also more symmetrical around the parasagittal plane passing through the middle of the patellar surface (the trochlea) than is the ape femur (Fig. 21.14). This most probably is a reflection of the large size of the human lateral condyle and the differences in weight transfer through the knee in humans as compared to the apes.

PROJECTION OF THE LATERAL LIP OF THE TROCHLEA (PATELLAR GROOVE)

In humans the **patellar groove**, or **trochlea**,

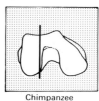

FIGURE 21·14 *Human and chimpanzee femoral condyles from the inferior view. The outline of the human condyles is square and that of the chimpanzee condyles rectangular. The human condyles are symmetrical around the patellar groove and the chimpanzee condyles are asymmetrical. The human patellar groove is deeper than the chimpanzee groove and has an anteriorly projecting lateral lip (arrow).*

is deeper than it is in the apes and the lateral lip of the groove projects anteriorly to a greater extent than does the medial lip (Fig. 21.14). In addition the distance from the lateral lip to the deepest part of the groove is longer than the distance from the medial lip. The angle formed by these two sides of the groove is relatively invariant at 147.93 ± 8.97° (Wanner, 1977).

A deep patellar groove with a high lateral lip is generally considered to be an important hominid feature reflecting bipedal locomotion (Le Gros Clark, 1947a; Preuschoft, 1970;

Heiple and Lovejoy, 1971). Heiple and Lovejoy (1971) argued that this groove is directly related to the high human bicondylar angle through the action of quadriceps femoris which has a tendency to pull the patella laterally (Fig. 21.15). These authors believe that the high lateral lip would help prevent this dislocation. However, Wanner (1977) has demonstrated that there is no correlation in humans between the height of the lateral lip and a high bicondylar angle as would be predicted by the Heiple and Lovejoy argument. Rather, Wanner (1977) suggested that patellar dislocation is avoided by the pull of the muscle vastus medialis. In humans this muscle inserts on the medial side of the patella and its pull would counteract the tendency of the other three quadriceps femoris muscles to dislocate the patella laterally (Fig. 21.15). Following on from this line of argument, Stern and Susman (1983) suggested that the main function of a deep patellar groove and high lateral lip in humans is to prevent dislocation of the patella on a flexed knee (at angles of flexion in excess of 20°). Rather than being a direct reflection of bipedal locomotion then, these features are viewed as a necessary structural correlate of a high bicondylar angle when the knee is subjected to high angles of flexion.

SHAPE OF THE SUPERIOR BORDER OF THE PATELLAR SURFACE

In humans the lateral side of the superior border of the patellar surface extends more proximally than does the medial border (Fig. 21.16). In the apes both the lateral and the medial sides of the patellar surface have the same proximal extent.

The australopithecine and paranthropine femur

The specimens which have been described in the literature and have formed the basis for the description of australopithecine and paranthropine femoral anatomy are for the most part fragmentary fossils. This makes

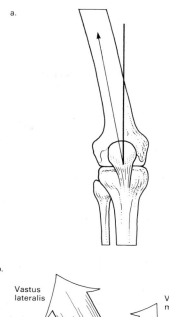
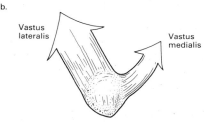

FIGURE 21·15 *(a) Anterior view of the human knee. The axis of the femur (solid arrowed line) is at an angle to the vertical. The resultant action of quadriceps femoris would tend to pull the patella laterally. (b) Note the medial insertion of vastus medialis. The action of this muscle counteracts the tendency of the other quadriceps muscles to laterally dislocate the patella.*

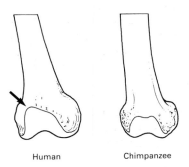

FIGURE 21·16 *Anterior view of the distal femur in a human and a chimpanzee. Note the proximal extension of the lateral portion of the patellar groove on the human distal femur (arrow).*

469

TABLE 21·2 Differences in the anatomy of the femur between modern humans, chimpanzees and australopithecines.

Feature	Humans	Australopithecines	Chimpanzees
PROXIMAL FEMUR			
Head size	large	small	small
Neck length	medium	long	short
Neck angle	large	small	large
Neck cross-section	round	compressed	round
Greater trochanter	flaring	non-flaring	non-flaring
Intertrochanteric line	variable	variable	absent
Obturator externus groove	variable	variable	absent
DISTAL FEMUR			
Bicondylar angle	high	very high	low
Lateral condyle profile	elliptical	elliptical	round
Epiphysis shape	square	variable	rectangular
Epiphysis symmetry	symmetrical	variable	asymmetrical
FEMUR SHAFT			
Robusticity index*	lower	higher	higher
Pilasteric index	higher	lower	lower

*Robusticity index = [(midshaft circumference × 100)/bicondylar length].

interpretation difficult. However, size does seem to be the primary difference between australopithecine and paranthropine femora. Table 21.2 summarizes the main differences between these hominid femora and those of humans and apes.

The proximal femur in australopithecines and paranthropines

The general proportions of the australopithecine and paranthropine proximal femur are very different from those in either humans or chimpanzees. The two most obvious features are the greater relative length of the femoral neck (particularly in the paranthropines) and the relatively smaller femoral head (Fig. 21.17) (Walker, 1973; Day, 1969a, 1978; McHenry and Temerin, 1979).

BIOCHEMICAL NECK LENGTH

Relative to a variety of different measures biomechanical neck length is long in comparison to that of modern humans. This is particularly true of the larger and temporally

later occurring paranthropines. The *Australopithecus afarensis* specimens show neck lengths that are closer to the modern human mean (and modern ape mean) in relation to femoral length (actual length for AL 288-1 and reconstructed length for AL 333-3) (Stern and Susman, 1983). Stern and Susman (1983) suggested that this might possibly reflect the fact that *A. afarensis* is less divergent from the pongid pattern than are the later australopithecines and paranthropines.

RELATIVE HEAD SIZE

There has been some debate in the literature over the relative size of the femoral head in the early hominids. In comparison to femoral length, the diameter of the head falls within the human range (Walker, 1973). However, in comparison with the transverse diameter of the femoral shaft just below the lesser trochanter, the diameter of the head falls well below the modern human mean (Fig. 21.17). This might, however, suggest that these fossils had relative head sizes similar to

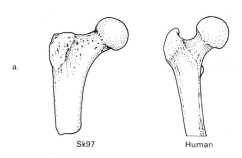
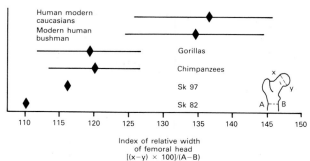

a.

FIGURE 21·17 *(a) Anterior view of the proximal femur of* Paranthropus robustus *(SK 97) and a modern human. (b) Index of the relative width of the femoral head in humans, the African apes and two* Paranthropus *hominids. After Napier (1964).*

modern humans but with much more robust femoral shafts (Walker, 1973). Other authors have argued that the cross section of the femur is highly correlated with body weight across primates (Aiello, 1981b,c; McHenry, 1988) and that this would also be the case in australopithecines and paranthropines. Therefore both the head size and the length of the femur would be relatively small in these hominids. This conclusion is confirmed by size-corrected multivariate analyses of the femur (McHenry and Corruccini, 1976a,b; Corruccini and McHenry, 1978) but disputed by univariate allometric analyses based on a modern human Amerindian population (Wolpoff, 1976).

Recent reassessment of the relative head size in the smaller australopithecines (e.g. AL 288-1) suggests that the head is small in relation to that of a modern human of average body weight (e.g. 50 kg), but not necessarily to a modern human of the same body weight as predicted for the fossil (e.g. 25–30 kg) (Jungers, 1988a; Ruff, 1988). This interpretation is based on the fact that within modern humans the size of the femoral head is strongly positively allometric in relation to body size where larger individuals have much larger femoral heads relative to body size than do smaller individuals (Fig. 21.3).

HEAD MORPHOLOGY

The femoral head in the early hominids is

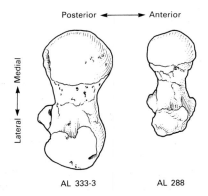

FIGURE 21·18 *The superior view of the femoral head in two* Australopithecus afarensis *specimens. In AL 333-3 the edge of the articular cartilage is square in relation to the neck of the femur; In AL 288-1ap the articular cartilage angles from the anteromedial to the posterolateral in relation to the axis of the femoral neck. Compare with Figure 21.4. After Stern and Susman (1983).*

hemispherical in shape rather than making up about two-thirds of a sphere as it does in modern humans (Fig. 21.17) (Walker, 1973). The paranthropines (SK 97, SK 82, OH 20) have femoral heads that sit squarely on the femoral necks as do the larger *Australopithecinus afarensis* fossils (Fig. 21.18) (e.g. AL 333-3) (Stern and Susman, 1983). The smaller *A. afarensis* fossils (e.g. Al 288-1ap) resemble the chimpanzee condition where the articular surface of the femoral head extends onto the

471

posterior surface of the neck (compare Fig. 21.4). This has been interpreted to suggest a marked locomotor difference within the australopithecines (Jenkins, 1972). However, it may simply reflect sexual dimorphism (Asfaw, 1985).

NECK-SHAFT ANGLE

The majority of the australopithecines and paranthropines have a lower neck-shaft angle than do modern humans, however both the *Australopithecinus afarensis* specimens for which measures are available (AL 288-1ap) (Johanson et al., 1982; Stern and Susman, 1983) and AL 333-3 (Stern and Susman, 1983) have angles that are closer to the means of modern human samples.

Bellugue (1962) and Kapandji (1987) suggested that in modern humans a small neck-shaft angle is related to other features that are also characteristic of the australopithecine femur and pelvis. These features include a femoral head that just exceeds a hemisphere rather than being more than two-thirds of a sphere, a transversely wide femoral shaft and a broad pelvis. However, the interrelationship of these features and their mechanical significance have yet to be established for either modern humans or for the fossil hominids.

LATERAL FLARE OF THE GREATER TROCHANTER

The lateral margin of the greater trochanter in most australopithecines and paranthropines is flush with the lateral margin of the shaft rather than projecting laterally from it as is the case in modern humans (but see AL 333-3). This difference may reflect a difference in the function of the pelvic abductors. Alternatively, Lovejoy et al. (1973) considered the modern human lateral flare as a means of reducing the bending stress on a long femoral neck. They argued that the medial shift of the femoral shaft relative to the greater trochanter would reduce the apparent neck length without the simultaneous reduction in the lever arm of the abductor muscles (Fig. 21.19).

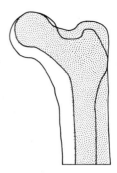

FIGURE 21·19 *A* Paranthropus *proximal femur (SK 82) (stippled) superimposed on the outline of a modern human proximal femur. Note the equivalent positions of the most lateral aspect of the greater trochanter in the two specimens in spite of the differences between them in the length of the femoral neck. After Lovejoy (1974).*

The absence of the lateral flare of the greater trochanter in these early hominids most probably also explains its apparent small size (Day, 1969a). Furthermore, the australopithecine and paranthropine greater trochanter barely projects above the neck of the femur (Fig. 21.17). In modern humans the trochanter is low in relation to the superior border of the neck in comparison with the much higher ape trochanter (Fig. 21.6) (McHenry and Corruccini, 1976a,b).

SHAPE OF THE CROSS-SECTION OF THE FEMORAL NECK

The cross-section of the femoral neck in the australopithecines and paranthropines is uniformly ellipsoid in shape, while the necks of both humans and apes are more circular in shape. The cross-sectional shape of the australopithecine femoral neck is undoubtedly related to the increased bending moments imposed on the neck by its unusual length. The greatest diameter of the neck is in the superoinferior plane where the bending moment would be the greatest.

The internal morphology of the neck as revealed through X-rays is also related to the ellipsoid shape of the cross-section (Walker, 1973). Preliminary analysis suggests that the cortex on the inferior part of the neck is thickened for a considerable distance along the neck. In addition, the trabeculae that extend into the femoral head begin where the cortex thins about midway along the length of the neck and not at the junction of the shaft and neck as in humans (Walker, 1973). However, the early hominids share with humans an absence of a thick cortex in the superior half of the circumference of the neck (Fig. 21.7) (Lovejoy, 1988). This may be related to the similar pull of the small gluteal muscles in line with the femoral neck in these hominids and modern humans.

OBTURATOR EXTERNUS GROOVE

The obturator externus groove is variably present in the australopithecines and paranthropines. For example, the *Paranthropus* proximal femora from South Africa and Olduvai (SK 97, SK 82, OH 20) and KNM-ER 1503 (cf. *Paranthropus*) have clearly defined grooves (Fig. 21.11) (Day, 1969a; Walker, 1973; Day et al. 1976a,b; Stern and Susman, 1983), while KNM-ER 738 and KNM-ER 814 (cf. *Paranthropus*) do not (Stern and Susman, 1983). There is some controversy whether the *Australopithecus afarensis* fossils have this groove. Lovejoy et al. (1982a) believed the groove to exist in AL 288-1 and AL 333-95. However, Stern and Susman (1983) are of the strong opinion that the grooves on these fossils are not true obturator externus grooves. They suggested that the presence of a small bump, or tubercle, at the base of the femoral head gives the illusion of a short obturator externus groove and point to the variable occurrence of similar tubercles and grooves in prehensile-tailed cebids, Old World monkeys and chimpanzees.

INTERTROCHANTERIC LINE

The intertrochanteric line is variably expressed in humans and therefore its absence does not necessarily indicate the absence of bipedal locomotion (Lovejoy and Heiple, 1972). It is also variably present in the early hominids.

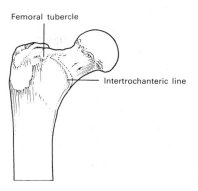

FIGURE 21·20 *Anterior view of the* Paranthropus robustus *(SK 97) proximal femur. After Robinson (1972).*

It is most strongly developed in AL 333-3 (*Australopithecus afarensis*) (Stern and Susman, 1983) and only weakly developed, if present at all, in the others (*Australopithecus* and *Paranthropus*) (Day, 1969a; Walker, 1973). If the line exists in these other hominids it curves medially on its way towards the lesser trochanter (Fig. 21.20). This medial curvature may be a result of the unusually long femoral neck in the australopithecines and paranthropines (Walker, 1973). Day (1978) emphasized that even though the line is not well developed in many of these hominids, the femoral tubercle is, suggesting that only the transverse band of the iliofemoral ligament (which attaches to this tubercle) was well developed (see Fig. 20.9).

The australopithecine and paranthropine femoral shaft

Because of the fragmentary nature of the majority of australopithecine and paranthropine femora, the total morphology of the shaft is not well known. However, there is sufficient evidence to suggest that the femoral shaft is round to sub-round both at its proximal end and at the midshaft. This rounded morphology provides a sharp contrast with the anteroposteriorly flattened shafts characteristic of the archaic *Homo* group hominids (Kennedy, 1983a,b) and of the African apes (Fig. 21.21). In addition, the best preserved shafts from Koobi Fora (KNM-ER 736, *Homo*; KNM-ER 993, *Australopithecus* (cf. *Paranthropus*)) have mediolateral dimensions that are the same throughout their length (Kennedy, 1983a,b). In this feature they are similar to the African apes and different from modern humans with a midshaft waisting and also from the archaic group of hominids with a distal waisting (see below). Furthermore, the cortical thickness of these femora is not unusually large in relation to that of modern humans (Kennedy, 1983a,b). Prior statements to the contrary (Walker, 1973; Lovejoy et al., 1982a) were made in the absence of good comparative data (Kennedy, 1983a,b).

Moreover, it has been suggested that the shafts of australopithecine and paranthropine femora are considerably more robust than is the norm in either modern human populations or the African apes (Table 21.3) (Kern and Straus, 1949). However, the conclusion is based on the comparison of lower shaft width to bicondylar width and may be a better reflection of reduced bicondylar width in the australopithecines and paranthropines rather than unusual shaft robusticity. Walker (1973) has also suggested (on the basis of a composite reconstructed femur) that these early hominids not only had shafts that were more robust than the norm for modern humans but also that the shafts were characterized by well-developed pilasters.

The australopithecine and paranthropine distal femur

The known australopithecine and paranthropine distal femora share some, but not all, of the features that are found in modern human distal femora (Tardieu, 1981; Stern and Susman, 1983). They all have: (1) a high bicondylar angle, (2) a deep patellar groove, and (3) an elliptical profile of the lateral condyle. However, there is some variation in the magnitude of the bicondylar angle

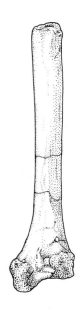

FIGURE 21.21 *Posterior view of the KNM-ER 993 femur. After Leakey and Leakey (1978).*

TABLE 21.3 Robusticity of the femur shaft in modern humans, apes and selected hominid fossils. After Kern and Straus (1949).

	N	Mean	Range
Humans	58	44.6	35.2–58.8
A. africanus		56.6	—
Gorilla	31	45.4	39.6–51.4
Chimpanzee	14	40.2	37.1–44.6
Orang-utan	28	42.0	32.5–49.3
Symphalangus	3	44.4	43.8–45.2
Hylobates	3	41.9	40.7–44.4
Macaca	17	52.8	46.8–60.0
Papio	6	51.1	45.7–58.7
Theropithecus	1	48.0	—
Cercocebus	3	51.2	50.0–52.0
Cercopithecus	5	52.3	46.2–57.1
Erythrocebus	2	58.0	53.1–62.9
Semnopithecus	2	55.9	53.8–57.9
Nasalis	1	52.8	—

Robusticity = [(shaft width × 100)/bicondylar width] where shaft width is measured at a distance 1.15 times the bicondylar dimension above the bicondylar plane.

among the australopithecines in addition to marked variation in other features that distinguish human from ape distal femora. These additional features include: (1) the proportions of the distal condyles, (2) the symmetry of the condyles, and (3) the morphology of the patellar groove.

THE HIGH BICONDYLAR ANGLE

The bicondylar angles of the currently known specimens are equal to, or greater than, the mean angles found in modern humans and far in excess of those found in modern apes (Table 21.1). The most complete femur for which measurements are available is KNM-ER 993 (Fig. 21.21) (cf. _Paranthropus_) (Walker, 1973). The large bicondylar angle of 15° confirms the equally high estimates for the more fragmentary _Australopithecus africanus_ specimens (Sts 34 and TM 1513) (Lovejoy and Heiple, 1970) and for the _Australopithecus afarensis_ specimen (AL 129-1a) (Stern and Susman, 1983). However the larger _A. afarensis_ specimen (AL 333-4) has an estimated angle of 9° which is consistent with the mean for modern human males.

Earlier estimates of a lower bicondylar angle for _A. africanus_ (a minimum of 7° (Le Gros Clark, 1947b) and 7° (Kern and Straus, 1949)) are based on a measuring technique that underestimates the true bicondylar angle as measured on complete femora (Heiple and Lovejoy, 1971). Kern and Straus defined the axis for the femoral shaft (for fragmentary femora) as that line connecting the midpoint between the two condyles with the midpoint of the shaft at a distance of 1.15 times the bicondylar width measured from the bicondylar plane. The femoral axis defined as such differs considerably from the femoral axis defined for complete femora as either (1) the line connecting the midpoint of the shaft of the femur just below the lesser trochanter with the midpoint of the shaft at a position 25% of the maximum length of the femur from its distal end (Heiple and Lovejoy, 1971), or (2) the line passing through as many of the midpoints of the shaft as possible

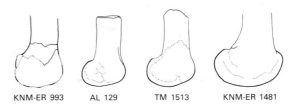

KNM-ER 993   AL 129   TM 1513   KNM-ER 1481

FIGURE 21·22 _Lateral view of the distal femur of selected fossil hominids. Afte Tardieu (1983). Compare with Figure 21.13._

and distally through the midcondylar point (Walmsley, 1933).

THE ELLIPTICAL PROFILE OF THE LATERAL CONDYLE

The lateral condyle in the australopithecines and paranthropines has the elliptical profile of the human lateral femoral condyle (Fig. 21.22) (Heiple and Lovejoy, 1971). However, australopithecines and paranthropines differ from modern humans in a slightly smaller posterior extension of the ellipse. Heiple and Lovejoy believe that the larger body size of modern humans requires a greater posterior extension. By presenting a larger contact surface in knee extension, this feature reduces the load in the bipedal knee.

PROPORTIONS OF THE DISTAL CONDYLES

The distal surface of femoral condyles of some of the early hominids has a square outline and overlaps the range of variation in this feature found in modern _Homo sapiens_ (Fig. 21.23). Others have rectangular distal outlines which fall within the range of variation seen in modern apes. This feature tends to separate the smaller _Australopithecus afarensis_ fossils (AL 288 and AL 129) from the larger individuals of this species (AL 333) and it also separates KNM-ER 993 (cf. _Paranthropus_) from Sts 34 and TM 1513 (_Australopithecus africanus_) and from other specimens that have been assigned to the genus _Homo_ (Tardieu, 1983).

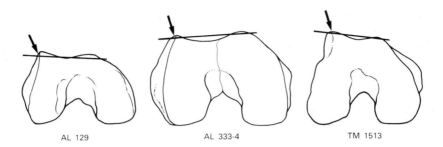

AL 129   AL 333-4   TM 1513

FIGURE 21·23 *Inferior view of the femoral condyles of (a) AL 129 and (b) AL 333-4 (both* Australopithecus afarensis) *and (c) TM 1513* (Australopithecus africanus: *reversed for ease of comparison). Arrows indicate the lateral lips of the patellar grooves. Compare with Figure 21.14. After Tardieu (1983).*

SYMMETRY OF THE CONDYLES

The same distinction between specimens as seen in the distal outline of the condyles is also evident in the symmetry of the condyles. Those specimens that have rectangular distal outlines also tend to have asymmetrical condyles while those that have squarer distal outlines have more symmetrical condyles.

MORPHOLOGY OF THE PATELLAR GROOVE

The australopithecines and paranthropines resembled modern humans in having a deep patellar groove (Fig. 21.23). The patellar groove in the apes approaches a flat surface. Furthermore, the great majority of the australopithecine and paranthropine distal femora have the high lateral lip that is characteristic of modern humans. However, the degree of projection of the lateral lip and the detailed morphology of the patellar groove varies at least among the *Australopithecus afarensis* fossils that have been studied.

The small *A. afarensis* fossils such as AL 129-1a have lateral lips that project further than do those in the larger specimens. In these smaller specimens the tangent to the lips of the patellar groove slopes in the same direction as it does in humans while in the larger specimens it slopes in the opposite direction.

THE DISTAL FEMUR AS A WHOLE

The features of the distal femur shared by humans and the early hominids suggest that the knee may have been used in extension. However, the differences between the early hominid and modern human distal femur also suggest that movement at the knee was unlikely to have been identical (Stern and Susman, 1983). This is particularly true when the variation in the early hominid distal femur is taken into consideration.

The small *Australopithecus afarensis* specimens (AL 288 and AL 129) share very few features with modern humans other than the high bicondylar angle (Tardieu, 1981; Stern and Susman, 1983). The overall functional morphology of these distal femora, together with complementary analyses of the tibial plateau (Chapter 22), is not incompatible with arboreal locomotion. The similarity in the morphology between these small *A. afarensis* specimens and the larger and later occurring KNM ER 933 (cf. *Paranthropus*) deserves further consideration.

The larger *A. afarensis* specimen (e.g. AL 333) is more similar to *Australopethicus africanus* (Sts 34 and TM 1513) as well as early members of the genus *Homo*, and modern humans in epiphyseal proportions and symmetry. However, Stern and Susman (1983) emphasized the difference in the morphology of the patellar groove and projection of the lateral lip of the trochlea in AL 333. They suggested that the knee in these larger *A. afarensis* individuals may not have been used in the same fashion as it is in modern humans.

The femur in early members of the genus Homo

For the purposes of this discussion, early members of the genus *Homo* (archaic *Homo*) include fossils belonging to *Homo habilis* and *Homo* sp. indet. from the African Plio-Pleistocene, *Homo erectus* (excluding the material from Trinil, Java: see below) and European and south-western Asian Neanderthals. The remarkable thing about the femora of these 'archaic group' fossils is the consistency in morphology that spans almost 2 million years and extends from Africa to Asia and to Europe (Fig. 21.24). This morphology includes: (1) anteroposterior flattening of the shaft reflected in the absence, or virtual absence, of a pilaster; (2) medial convexity of the shaft; (3) very low point of minimum shaft breadth; and (4) a cortex that is thicker on the medial side of the shaft at mid-shaft rather than on the lateral side of the shaft as in modern humans.

This morphological pattern has been reported for Neanderthal 1 (Walmsley, 1933), Tabun E1 and Ehringsdorf (McCown and Keith, 1939), the Shanidar Neanderthals (Trinkaus, 1983a), La Ferrassie 1 and 2 (Heim, 1974), *Homo erectus* from Zhoukoudian (Weidenreich, 1941), Olduvai Hominid 28 (Day, 1971), and in KNM-ER 1481, 1472, 737 and 803 (Kennedy, 1983a,b). The pattern for the Lower and Middle Pleistocene hominids has been summarized by Kennedy (1983b) and for the Upper Pleistocene hominids by Trinkaus (1976b).

Complete femora belonging to these archaic group hominids are rare. However, where they exist and are relatively undistorted (KNM-ER 1481 (*Homo* sp.) and some Neanderthals) they show a pattern of weight

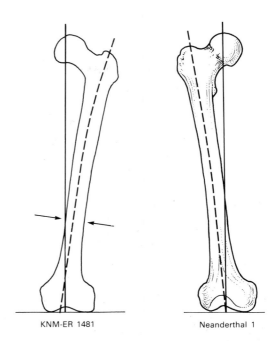

KNM-ER 1481   Neanderthal 1

FIGURE 21·24   *Anterior view of two 'archaic group' femora: (a) KNM-ER 1481 and (b) Neanderthal I. Note the medial curve of the shaft axis (dashed line), the low point of minimum shaft breadth (arrows), and the low point at which the load axis (solid line) intersects the shaft axis. Femora are drawn to the same bicondylar length.*

transfer that is distinct from both modern humans and the living great apes (Fig. 21.24). The load axes (the vertical line from the femoral head to the bicondylar plane) in these femora intersects the anatomical axis at the distal end of the femur rather than higher up the shaft as is the case in humans. Furthermore, the load axis falls midway between the medial and the lateral condyles rather than through (or even lateral to) the lateral condyle. The anatomical axes of these femora also curve medially and thereby reflect the distinct medial convexity of the shaft. This pattern of curvature contrasts markedly with the two patterns of lateral convexity found in humans and in the African apes (Fig. 21.1).

Both the low position of minimum shaft breadth and a thick medial cortex at midshaft (in relation to the lateral cortex) could well reflect this different pattern of weight transmission in these femora (Kennedy, 1983a,b) and it is possible that this different pattern reflects a higher activity level than is common in modern human populations (Trinkaus, 1976b). Trinkaus (1976b) argued specifically for Neanderthals for a relationship between a high level of activity, a rather low neck-shaft angle and a resulting high level of stress on the proximal femur. High levels of activity have been correlated with lower neck-shaft angles which in turn are correlated with higher levels of stress on the proximal part of the femoral shaft. On the medial side of the proximal shaft, the higher stress levels are counteracted by the marked medial convexity of the shaft. This convexity (or surface swelling) is the manifestation of the increased cortical thickness resisting this bending stress and resulting in an increase in the transverse diameter of the femur.

In modern humans a low neck angle also correlates with a high transverse diameter of the femur at midshaft (Van Gerven, 1972;

Trinkaus, 1976b). The Neanderthals as well as the other hominids in the archaic *Homo* group have relatively wide and anteroposteriorly flattened midshafts (very low pilasteric index) together with an absence of a pilaster.

Additional features of the earlier archaic group femora

Despite the similarities in femoral morphology found in the archaic group hominids there are a number of features that distinguish the earlier members of this group from the later Neanderthals and are reminiscent of the morphology found in *Paranthropus* and *Australopithecus* (Fig. 21.25). These are particularly evident in KNM-ER 1481 (*Homo* sp.) and include the narrow anteroposterior diameter of its femoral neck, its long neck length, the large transverse diameter of its shaft (also shared with Neanderthals), and the minimal projection of its greater trochanter (McHenry and Corruccini, 1976a,b). In fact KNM-ER 1481 only differs from *Paranthropus* (SK 97 from Swartkrans) in its large relative head size (in which it is more

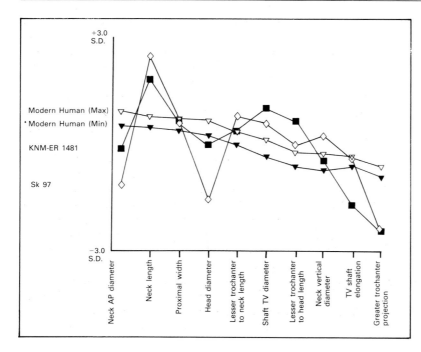

FIGURE 21·25 *Comparison of shape variables between the proximal femora of modern humans, KNM-ER 1481 and SK 97. After McHenry and Corruccini (1976a).*

reminiscent of Neanderthals and modern humans) and in the anteroposterior flattening of its shaft (McHenry and Corruccini, 1976a,

Many of the features found in KNM-ER 1481 are also found in the KNM-WT 15000 femur (*Homo* cf. *erectus*). In particular, the biomechanical neck length of KNM-WT 15000 (85 mm) is over 3 SD. from the modern human mean and it is relatively as long as *Paranthropus* (Brown et al., 1985). In addition the neck angle in this specimen (110°) is 5 SD. below the average neck angle reported by Lovejoy et al. (1973) for a modern human population. But as with KNM-ER 1481, the head size of this femur is absolutely larger than the femoral head found in *Australopithecus* and *Paranthropus*, falling within the range of variation of modern humans (Kennedy, 1983a; Brown et al., 1985). However, where the information is available for the fossil hominids, the head size in *Paranthropus*, in KNM-ER 1481 and in KNM-WT 15000 scales with the length of the femur in the same fashion as it does in modern humans (Walker, 1973; Kennedy, 1983a; Brown et al., 1985). This means that relative to the length of the femur there is no difference in head size in these hominids, although there may be considerable difference in absolute head size, or in head size relative to the transverse diameter of the upper part of the femoral shaft.

While much more fragmentary in nature, the femur belonging to Olduvai Hominid 62 (*Homo habilis*) (Johanson et al., 1987) is not inconsistent with this morphology. Although the neck is broken it has been reported to have been flattened in the anteroposterior plane, and has been reconstructed to have been the same length as found in *Australopethicus afarensis* and to have a neck/shaft angle similar to this australopithecine but higher than that found in KNM-WT 15000. However, this specimen differs from all the australopithecines and is similar to modern humans in that it has a clear obturator externus groove on the back of the femoral neck.

Femora belonging to the early *Homo* group also differ from the Neanderthals and modern humans in the proportions of their shafts. They have both a lower meric index (at the level just below the lesser trochanter) and a lower pilasteric index (at midshaft) than the later hominids, indicating that at both levels the shaft is narrower in the anteroposterior plane in relation to the transverse diameter.

Furthermore, these early femora have a distinct internal structure as revealed through X-rays. Weidenreich (1941) first noticed for the *Homo erectus* femora from Choukoutein that the walls of both the shaft and the neck were extraordinarily thick and that this was combined with a very reduced medullary canal (the marrow and fat-filled central cavity of the bone). He also reported that the arrangement of the trabeculae (fine struts of bone reinforcing the interior of the shaft) in the proximal end of the bone was different in these fossils in comparison to modern humans.

Kennedy (1983b) has elaborated on Weidenreich's work and established (in addition to the greater thickness of the medial cortex in relation to the lateral cortex in the archaic group femora) three further points in relation to the internal architecture of the femur. Firstly, in the Zhoukoudian femora as well as in KNM-ER 737 and in OH 28 (Day, 1971), the cortex on the inferior aspect of the neck remains thick further proximally so that it may even reach to the base of the head. In modern humans the cortex in this region thins at the base of the neck. Secondly, in the Zhoukoudian femora, and also in the Neanderthals (Weidenreich, 1941), the internal architecture of the neck of the femur is different from that in modern humans. The two main compressive trabecular struts are not separate as they are in humans (Chapter 20) but are blended together and give the appearance of being diffuse. Thirdly, the femoral cortex is thicker, both absolutely and relative to total shaft diameter, than it is in modern humans. This is particularly true, for example, in KNM-ER 1481 where at midshaft in the mediolateral plane the cortex occupies 73.9% of the shaft diameter where in modern human males it occupies only 57.6% (Kennedy, 1983b) and in apes it occupies only 40–43% (Aiello, 1981a).

It has also been suggested that a convex, rather than a flat, popliteal surface may distinguish at least the Trinil 1 femur (possible *H. erectus*, but see below) from modern femora (Dubois, 1926, 1927). However, a convex popliteal surface has been found in 49.5% of a sample of 900 modern human femora (Pearson and Bell, 1919) and in 51% of an additional sample of 100 Romano-British femora (Day and Molleson, 1973). Furthermore, the marginally high popliteal index in some of the archaic group of hominids (e.g. KNM-ER 1481) is a better reflection of a narrow transverse diameter of the femur in this region than any unusual bowing of the popliteal region. This narrow transverse diameter is a direct result of the low position of minimum breadth of the shaft which is characteristic of all the archaic group hominids (Kennedy, 1983b).

The anomalous position of the Trinil femora

In total there are six femora in the Trinil collection that have been assigned to *Homo erectus* (Fig. 21.26). Trinil I is a complete, but pathological, left femur that was discovered by Eugene Dubois at Trinil in 1892, the year following his discovery of the first *H. erectus* (*Pithecanthropus erectus*) skull cap from the same site (Dubois, 1894). The well-known pathology of this femur involves a large bony growth (exostosis) protruding from the medial side of the upper third of the bone that affects both the external and internal morphology of the shaft throughout its length (Day and Molleson, 1973; Kennedy, 1983b). The bony outgrowth appears to have been associated with the muscle adductor brevis and is usually described as myositis ossificans although it could also be diaphyseal aclasia (Keith's disease) (Day and Molleson, 1973). There is no evidence that it results

from fluorosis as was once suggested by Soriano (1970) (Day and Molleson, 1973).

The importance of this femur lies in its seemingly modern morphology and, therefore, in its contrast to the other archaic group femora (Weinert, 1928; Weidenreich, 1941; Le Gros Clark, 1964; Day and Molleson, 1973; Day, 1984). It is largely on the early discovery and interpretation of this specimen that the generally held assumption that *H. erectus* had a postcranial skeleton identical to modern humans rests (Le Gros Clark, 1964; Pilbeam, 1972). However, there are problems associated with its acceptance as *H. erectus* which include questions over its association with the Trinil *H. erectus* skullcap, its morphological similarity to the other femoral fragments from Trinil and the effect of the pathology on its morphology.

The other Trinil femora (II–VI) are problems in themselves. Femora II–V were disco-

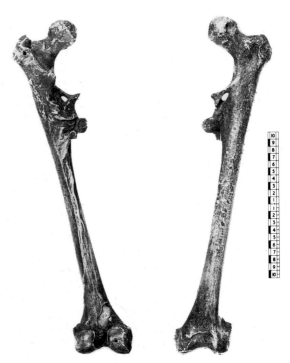

FIGURE 21·26 *The Trinil I femur. Courtesy of M.H. Day.*

vered in 1932 in the Leiden Museum in Holland among material that was said to have been excavated at Trinil in 1900. There is no definite evidence for the location of original discovery of femur VI which came to light (again) in the museum in 1934. This femur may have originated from the site of Kedung Brubus and in any case may not be hominid (Day and Molleson, 1973).

Day and Molleson (Day and Molleson, 1973; Day, 1984) are of the opinion that the five hominid femora from Trinil are all of modern morphology and may not be contemporaneous with the *H. erectus* skull cap. Although relative dating techniques are equivocal on the contemporaneity of the fossil material from Trinil (Day and Molleson, 1973; Day, 1984) suspicions about their age and association are based on three pieces of evidence. Firstly, more recent Upper Pleistocene deposits are now known to have existed at the site above the Middle Pleistocene Trinil deposits (Bartstra, 1982). Secondly, none of the femora were directly associated with the *H. erectus* skull cap. There is no recorded location of discovery for femora II–V and even femur I is known to have been found 10–15 m from the site of discovery of the skull cap. Thirdly, the morphology of the femora is out of character with that of other proved archaic femora (Zhoukoudian, OH28).

Although Kennedy (1983b) agrees that there are problems with the Trinil femora in relation to association and dating, she does not agree with Day and Molleson that all the femora are completely modern in form. Reserving final judgement on the pathological Trinil I, she demonstrated that the other Trinil hominid femora (II–V) are similar to known *H. erectus* femora in the form of their distal shafts and particularly in their large cortical diameters, high cortical indices and external distal shaft dimensions. However, these same femora differ from *H. erectus* and are similar to modern humans in the morphology of their proximal and midshaft regions (Fig. 21.27). At this stage in time

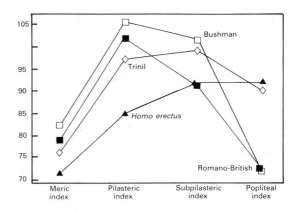

FIGURE 21·27 *Comparison of femoral shaft indices between modern humans (Romano-British and Bushman), Trinil hominids, and* Homo erectus. *Indices represent the anteroposterior diameter of the shaft divided by the mediolateral diameter multiplied by 100 at 25% of the length of the shaft from the proximal end (Meric index), at midshaft (Pilasteric index), at 75% of the length of the shaft (Subpilastric index) and at the level of the popliteal surface (Popliteal index). After Kennedy (1983b).*

further judgement must be reserved on this material until dating and association problems are resolved.

The Neanderthal femora

The main difference between Neanderthal femora and the femora of earlier members of the archaic group is in the general level of robusticity (Fig. 21.24). In particular, Neanderthal femora not only have consistently large gluteal ridges but also have a high frequency of hypotrochanteric fossae (Trinkaus, 1976b). The robust gluteal ridges suggest that gluteus maximus (which inserts onto the ridge) was also larger and more powerful in the Neanderthals than it is in modern humans. Hypotrochanteric fossae were found in 66.7% of European Neanderthals ($n=12$) as well as in some of the south-western Asian Neanderthals (Trinkaus, 1976b; 1983a) and have also been associated with increased strength of gluteus maximus. The Neanderthal hypotrochanteric fossae

481

have a different form than is generally found in modern humans. Rather than lying lateral to the gluteal ridge (between it and the lateral shaft border) the Neanderthal fossae are distinct depressions within the gluteal ridges (Trinkaus, 1976b). Such hypotrochanteric fossae are also found in Neanderthal juveniles (Roc de Marsal 1 aged 2–2.5 years, La Ferrassie 6, aged 3 years) (Trinkaus, 1976b).

Possibly associated with the well-developed gluteal ridge is the general rounded contour of the proximal shaft. Rather than being anteroposteriorly flattened as is the case in earlier members of the archaic group, the meric index indicates that in Neanderthals this region is more rounded as it is in modern humans. Other metrical parameters such as head size relative to shaft length and midshaft circumference relative to shaft length also show an overlap with modern humans. However, in these indices (which reflect general robusticity) the Neanderthals do lie consistently above the modern human mean (Trinkaus, 1976b, 1983a).

The earliest occurrence of the modern human femoral pattern

In view of the long duration of the archaic group femoral morphology (from the Lower Pleistocene through to the middle of the Upper Pleistocene) it is important to note the earliest occurrence of the modern human morphology. Kennedy (1984) recognized the modern femoral morphology (including relatively and absolutely thinner cortical bone, a wider medullary canal, and a rounder shaft with a higher point of minimum shaft breadth) in two African hominids that may well date prior to 100 000 years B.P. (Fig. 21.28). The first of these is Omo 1 from Member 1, Kibish Formation in the Omo River Valley, Ethiopia. Although this femur is very fragmentary, it shows no archaic features. This

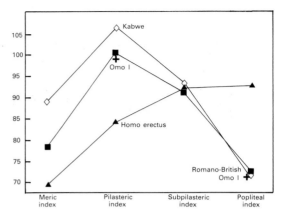

FIGURE 21·28 *Comparison of femoral shaft indices between modern humans (Romano-British),* Homo erectus, *Kabwe and Omo 1. After Kennedy (1984).*

is perhaps not surprising in view of the modern morphology of the associated cranium (Day, 1969b, 1972; Day and Stringer, 1982). This material may be as old as 130 000 years B.P. (Butzer et al., 1969). However, the uranium–thorium date is problematical (Butzer et al., 1969; Broecker and Bender, 1972). All that can be definitely said about the age of this material is that it was found in situ in deposits with at least one extinct species (*Elephas recki*).

The second African hominid that appears to have anatomically modern femora is Kabwe (formerly Broken Hill) from Zambia. This material is associated with the African Middle Stone age and is likely to be well in excess of 125 000 years old (Gentry and Gentry, 1978; Butzer, 1979; Vrba, 1982). The modern nature of the five femoral fragments (from a minimum of two individuals) is surprising in view of the fact that the Kabwe cranium and pelvis with which this material is purported to be associated have many archaic features. The main concern with this material is one of confident association of the femora with the cranial and pelvic material.

CHAPTER TWENTY-TWO

THE HOMINOID KNEE JOINT AND LOWER LEG

As with the pelvis and the femur, the morphology of the human knee joint and lower leg reflects a predominantly bipedal locomotor pattern and differs considerably from that of the ape knee joint and lower leg. The human knee joint has a number of features that stabilize it in a fully extended posture, while the ape joint lacks these features and permits a greater range of lower-leg rotation than does the human joint. The human tibia and fibula, the two bones of the lower leg, also reflect the use of the fully extended human leg for support. These bones are straighter than are the corresponding ape bones, and the proximal and distal joint surfaces of the main bone of the lower leg, the tibia, are more perpendicular to the mechanical axis of the shaft and to the line of force transmitted through the bone in upright posture. Furthermore, the ankle joint is a more stable joint in humans than it is in apes; restricting effective movement of the leg to the sagittal (anteroposterior) plane.

Structure and function of the knee joint in humans and apes

Many of the differences in the proximal tibia (and distal femur) that distinguish humans from apes are directly related to the differences in the mechanics of the knee joint. Because of this the morphology of the proximal tibia and distal femur are important

indicators of locomotor type in the fossil record.

In both humans and apes, the knee joint relies almost entirely on soft tissue (ligaments, muscles, menisci) for its structural stability. The two main bones involved in the joint, the tibia and the femur, have largely incongruous joint surfaces, the tibial condyles being flat and the femoral condyles being more rounded (Fig. 22.1). There are no bony reinforcements

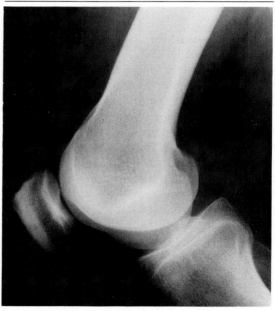

FIGURE 22·1 *Lateral radiograph of a human knee. Courtesy of Peter Abrahams.*

that keep these differently shaped surfaces from disarticulating. The only other bone involved in the knee joint is the patella ((L) = small pan, dish or plate). The patella is the largest sesamoid bone in the body. It is located at the front of the knee in the tendon of quadriceps femoris and its primary function is to increase the lever advantage of this muscle by moving its insertion anterior to the centre of rotation of the joint. The patella is particularly effective in this role when the knee is approaching full extension (150–180°) (Haxton, 1945).

The differences in structure and function of the knee joint in humans and apes involve: (1) patellar morphology, (2) the ligaments of the knee, (3) the menisci, (4) the locking mechanism of the human knee, (5) movement and the shape of the condyles, and (6) joint congruity and rotational ability. These features are discussed in turn along with their implications for the fossil record.

The patella in humans and apes

The human patella is variable in shape, sometimes being circular or elliptical, but most commonly being rounded on its proximal margin and pointed on its distal margin, or roughly heart-shaped (Fig. 22.2) (Martin and Saller, 1959). Its average diameter in a variety of human populations is 4–4.5 cm which is slightly more than 50% of the distance between the two femoral epicondyles.

Quadriceps femoris inserts on its proximal surface. Fibres continue over the anterior surface of the patella and are continued as the patellar ligament. In humans vastus medialis inserts on the medial, rather than the proximal, margin of the bone. In this position, it keeps the patella from laterally dislocating as quadriceps femoris (including vastus medialis) extends the knee (Chapter 21 and Fig. 21.15). The danger of dislocation results from the bicondylar angle of the human femur which alters the otherwise straight line of action of the quadriceps muscles, so that the patella tends to be pulled laterally as the knee extends.

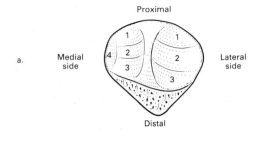

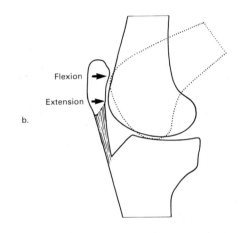

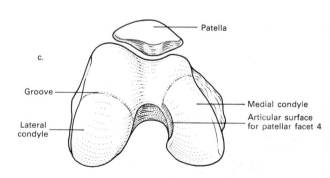

FIGURE 22·2 *The human patella. (a) The posterior surface of the patella. Each of the paired facets 1–3 articulate in turn with the patellar groove (trochlea) as the knee moves from full extension to full flexion. Facet 4 articulates with the medial side of the patellar groove in full flexion. (b) Lateral view of the knee in extension (solid line) and flexion (dotted line). Arrows indicate the point of articulation between the patella and the patellar groove in extension and flexion. (c) Inferior view of the femoral condyles and patella.*

The posterior, or articular, surface of the human patella is divided in two by a ridge running the proximodistal length of the bone (Fig. 22.2). This ridge is displaced medially to the extent that the lateral side of the articular surface is noticeably wider than the medial side. The longitudinal ridge rides in the patellar groove on the femoral trochlea as the knee flexes and extends. The wider lateral facet reflects a more extensive articulation with the lateral part of the trochlea than with the medial. Each side of the ridge may be divided into three facets (Fig. 22.2). As the knee flexes, or extends, the three paired facets (or areas corresponding to the facets if they are not immediately visible) articulate with the trochlea one after another with the lowermost pair (the extension facets) articulating in full extension and the uppermost pair (the flexion facets) articulating in flexion (Fig. 22.2). There is an additional facet on the extreme medial edge of the patellar articular surface that articulates with the medial side of the patellar groove in full flexion (Fig. 22.2). The size of the patella gives a general indication of the size of the quadriceps femoris muscle and its thickness can be taken as a reflection of the magnitude of the moment arm of that muscle around the centre of rotation of the knee joint (Trinkaus, 1983a).

The patella in the African apes differs from the human patella in three ways. Firstly, it is both absolutely and relatively smaller than the human patella. This reflects the relatively smaller quadriceps muscle that is characteristic of the apes in relation to humans. Secondly, vastus medialis does not insert onto its extreme medial edge. Thirdly, the posterior articular surface is flatter than is the case in humans, reflecting the flatter trochlear surface (patellar groove) of the ape femur.

The ligaments of the knee joint

The stability of the knee joint in humans and apes comes primarily from the ligaments that bind the joint together, the **collateral**

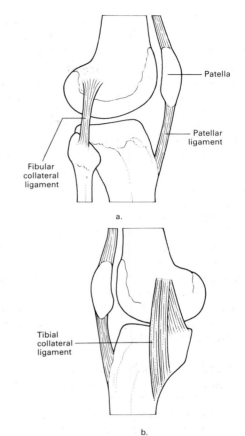

FIGURE 22·3 *The ligaments of the knee. (a) The fibular collateral ligament, (b) the tibial collateral ligament.*

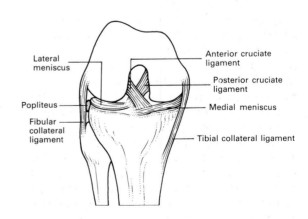

FIGURE 22·4 *The cruciate ligaments.*

485

ligaments on the medial and lateral sides of the joint and the **cruciate ligaments** within the joint capsule (Figs 22.3 and 22.4). Muscle and tendinous expansions also help to stabilize the joint capsule. In particular, the joint is reinforced on its medial side by vastus medialis, on its lateral side by the iliotibial tract, and posteriorly by the complex tendinous insertion of semimembranosus. The front part of the capsule has been replaced by the patellar ligament, the patella and the lower part of quadriceps femoris.

The collateral ligaments

There are two collateral ligaments, the **tibial collateral ligament** on the medial side of the joint and the **fibular collateral ligament** on its lateral side. They tighten in full extension and in this posture hold the joint together and limit rotation and hyperextension. In flexion they are loose and permit more rotary movement at the joint. The tibial collateral ligament arises from the medial epicondyle of the femur and attaches in two places to the tibia (Fig. 22.3). The deep part of the ligament attaches along the medial border of the tibial plateau while the superficial part attaches well down on the medial surface of the shaft where there is a distinct roughened area. The fibular collateral ligament is cord like in form and stands away from the joint (Fig. 22.3). It arises from the lateral femoral epicondyle and attaches on the head of the fibula. The tendon of popliteus passes under this ligament. Both these ligaments are similar in humans and in the African apes.

The cruciate ligaments

There are also two cruciate ligaments, the **anterior cruciate ligament** and the **posterior cruciate ligament** (Fig. 22.4). These ligaments lie inside the knee joint and their name comes from the fact that they cross each other when viewed both anteriorly and laterally (crux (L) = a cross). The anterior cruciate arises from the anterior intercondylar space on the tibial plateau, runs upwards and posteriorly and attaches on the inside of the

lateral condyle of the femur. The posterior cruciate arises from well back on the posterior intercondylar space, runs upwards and anteriorly and attaches on the inside of the medial condyle.

There is an easy way of remembering the origins, insertions and function of these ligaments. Pegington (1985) suggested that if the palm of the right hand represents the right tibial plateau the index and middle fingers of the left hand, when placed on the right palm, represent the two cruciate ligaments. If the index finger is placed forward and the middle finger back in scissor fashion, the index finger represents the anterior cruciate and its knuckle (on the lateral side of the knuckle for the middle finger) represents the attachment of this cruciate on the lateral femoral condyle. In a similar fashion the middle finger represents the posterior cruciate and its knuckle its attachment on the medial condyle.

One of the main functions of the cruciates is to keep the femur from sliding off the tibia when the joint is either flexed or extended (Fig. 22.5). The posterior cruciate tenses when the joint is flexed and keeps the femur from sliding anteriorly off the tibial plateau and the anterior cruciate tenses when the joint is extended keeping the femur from sliding posteriorly off the plateau. Particularly in hyperflexion of the knee, the taut posterior cruciate may produce a groove on the medial side of the intercondylar notch that partially obscures the intercondylar line (Fig. 22.5c) (Martin, 1932; Singh, 1959; Trinkaus, 1975). This groove is particularly prevalent in human populations that engage in habitual squatting and has a high correlation with other indications of such a posture (Martin, 1932; Trinkaus, 1975). When the knee is fully extended, the taut anterior cruciate sits in another indentation (or secondary notch) on the lateral side of the intercondylar notch (Fig. 22.5d) (Siddiqi, 1934). Such secondary notches have been taken to indicate a fully extended and, therefore, bipedal knee for *Australopithecus africanus* (TM 1513) (Le

FIGURE 22·5 *(a) The posterior cruciate ligament limits forward displacement of the femur in flexion and (b) the anterior cruciate ligament limits backward displacement of the femur in extension. (c) Posterior view of a human femur showing the groove that can result from the pressure of the taut posterior cruciate ligament in hyperflexion. After Martin (1932). (d) Inferior view of the femoral condyles showing the secondary notch that can result from the pressure of the tense anterior cruciate ligament in extension. After Siddiqi (1934).*

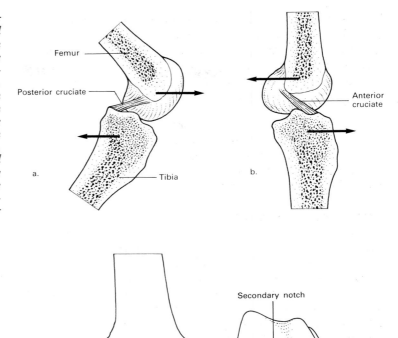

Gros Clark, 1947b). However, Kern and Straus (1949) warned that such secondary notches are occasionally found in monkeys and apes and therefore may not necessarily reflect a bipedal knee.

A second function of the cruciates is to limit medial rotation of the tibia in relation to the femur (Fig. 22.6). Returning to the palm-of-the-hand analogy, if the right palm (corresponding to the right tibial plateau) is rotated medially the fingers of the left hand (the cruciates) cross and limit further medial rotation. However, they have no effect on lateral rotation. If the palm (the tibial plateau) is rotated laterally the fingers (cruciates) untwist and have no limiting ability. A final function of the cruciates is in relation to the locking mechanism of the knee (see below).

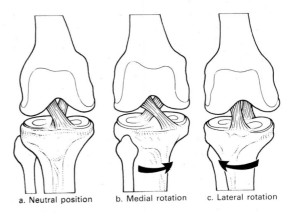

FIGURE 22·6 *(a) The cruciate ligaments in the neutral position. (b) In medial rotation they twist around each other and limit further rotation. (c) In lateral rotation they untwist and have no limiting ability. After Kapandji (1987).*

The tibial attachments of both cruciate ligaments are similar in chimpanzees and humans as is the femoral attachment of the anterior cruciate ligament (Fig. 22.7). There is, however, a considerable difference in the femoral attachment of the posterior cruciate ligament (Fig. 22.8). In chimpanzees it attaches to the medial femoral condyle in a position that is both more anteriorly placed and nearer to the mid-line of the intercondylar notch than is the case in humans. The cruciates in the chimpanzee control both the

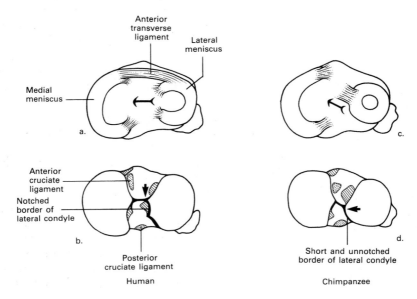

FIGURE 22·7 *Superior view of the tibial plateau in (a–b) a human and (c–d) a chimpanzee. Humans (a) differ from chimpanzees (c) in the presence of the anterior transverse ligament and in the semicircular shape of the lateral meniscus. The posterior border of the human lateral tibial condyle (b) is notched to receive the attachment of the posterior horn of the semicircular meniscus and the tibial spines are horizontally oriented across the plateau (arrow). The posterior border of the chimpanzee lateral condyle is short and unnotched and the tibial spines form an angle to the mediolateral plane of the plateau. In (b) and (d) the hatched areas represent the attachments of the menisci and the dotted areas the attachments of the anterior and posterior cruciate ligaments. After Senut and Tardieu (1985).*

FIGURE 22·8 *Posterior view of the knee in a human and a chimpanzee. Note the difference in the femoral attachment of the posterior cruciate ligament.*

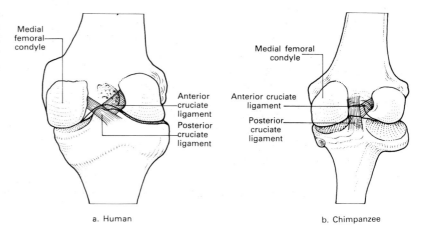

slide of the femur and the medial rotation of the tibia as they do in humans. However, because of the different attachment of the posterior cruciate on the medial femoral condyle, the chimpanzee cruciates allow more medial rotation of the tibia than is the case in the human knee (Corbridge, 1987). In addition, the attachments of the cruciates in the chimpanzee are not consistent with the locking mechanism that is characteristic of the human knee (see below).

The menisci and their function in humans and apes

The term **meniscus** (menis (L) = little half-moon or crescent) describes the shape of the two pads of fibrocartilage that separate the femur and the tibia in the knee joint (Fig. 22.7). These menisci increase the congruency (or fit) between the rounded femoral condyles and the flat tibial condyles and are also important for joint function.

In humans both the medial and lateral menisci are crescent shaped. The medial meniscus is firmly attached to the joint capsule and to the tibial collateral ligament on its medial side and to the tibia on its lateral side via its anterior and posterior ends (**horns**). The lateral meniscus is also firmly attached to the tibia via its horns and because it is the smaller of the two menisci it attaches between the horns of the medial meniscus. The lateral meniscus is also attached posteriorly to the popliteus muscle. The two menisci are connected to each other by the **anterior transverse ligament**.

Pure extension or flexion of the knee involves movement of the femur on the static menisci (and tibia) and rotation or gliding at the knee involves movement of the femur and menisci on the static tibia (Barnett, 1953). In addition to facilitating movement of the knee joint, the menisci also help to stabilize the knee in full extension when their anterior borders wedge into shallow grooves on the articular surfaces of the femoral condyles (Fig. 22.2).

There are a number of important differ-ences between the menisci of humans and of African apes. Firstly, in the African apes (and other primates) the lateral meniscus is completely circular rather than crescent shaped (Fig. 22.7). Secondly, the lateral meniscus has only a single point of attachment to the tibia rather than two as in humans. Thirdly, the lateral meniscus is connected on its posterior surface to the **ligament of Wrisberg** which attaches on the inside of the medial femoral condyle (Fig. 22.9). This ligament is infrequently found in humans but when it is it pulls the lateral meniscus backward in flexion and prevents the lateral meniscus from sliding too far forward in hyperextension (Girgis et al., 1975). Fourthly, the anterior transverse ligament is missing. In humans this ligament keeps the menisci from separating, or pulling apart from each other, during full extension of the knee. Finally, the medial meniscus is not attached to the tibial collateral ligament but is separated from it by a bursa.

The differences in the menisci relate to the functional differences in the knee joint in bipedal humans on the one hand and quad-rupedal African apes on the other (Corbridge,

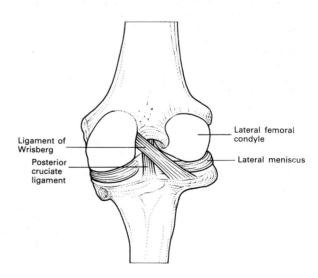

FIGURE 22·9 _Posterior view of a chimpanzee knee showing the Ligament of Wrisberg._

1987). They are particularly related to a greater capacity for rotation in the ape knee in comparison to the human knee. Furthermore, these differences can be detected from dry bone and, therefore, from fossil specimens (Tardieu, 1981; Senut and Tardieu, 1985). A bipedal tibial plateau is distinguished by two areas of insertion for the lateral meniscus that can normally be recognized on the dry bone by the shape of the posterior border of the lateral condyle (Fig. 22.7). Specifically it is long and notched by the posterior insertion of the meniscus. The African ape plateau has a single area of insertion for the lateral meniscus anterior to the tibial spine. The posterior border of the lateral plateau is therefore not notched but is relatively short and continuous (Fig. 22.7). Although the presence of a notched posterior border indicates a human-like bipedal knee, not all human tibial plateaus show this feature (Spencer, 1989). This feature tends to be absent in smaller bodied humans with retroverted tibial plateaus (see below).

The locking mechanism of the human knee

The human knee can be locked in an extended posture. The locking mechanism of the human knee refers to the sequence of movements that places the human knee in its most congruent, close-packed position and tenses the ligaments that stabilize the joint in this posture. The basic feature of the human locking mechanism is a noticeable medial rotation of the femur in relation to the tibia that begins approximately 30° from full extension and is greatest during the last 10° (Barnett, 1953).

The lateral condyle of the femur is shorter than the medial condyle by approximately 1 cm (Fig. 22.10). As the joint extends the lateral condyle reaches the extent of its articular surface before the medial condyle and achieves maximal congruence when the shallow groove outlining the anterior border of the surface wedges against the raised border of the lateral meniscus and the anterior

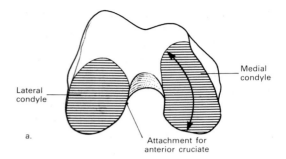

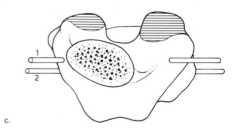

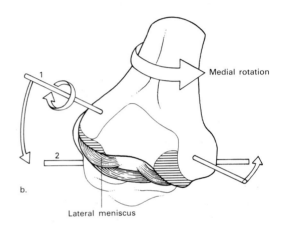

FIGURE 22·10 *The locking mechanism. (a) Inferior view of the human femoral condyles. The medial condyle is longer than the lateral condyle. (b) As the knee approaches extension the anterior border of the shorter lateral femoral condyle wedges against the anterior edge of the lateral meniscus. As the knee continues to extend the femur rotates medially, the lateral femoral condyle (with its meniscus) slides forward and the medial condyle (and meniscus) backward bringing the axis of the femoral condyles (1) in line with the axis of the tibial condyles (2). (c) Superior view of the fully extended knee. After Kapandji (1987).*

cruciate becomes tense. At this point, the femoral attachment of the anterior cruciate acts as the pivot around which the femur medially rotates. At the same time the lateral condyle (with its meniscus) slides forward as the medial condyle both slides backward and continues in extension for another 10–15°. (The pivotal point on the lateral condyle can be seen to be the centre of the curve described by the medial condyle). The movements of the menisci are held in check by their various attachments to the tibial plateau, and (in the case of the medial meniscus) the tibial collateral ligament and (in the case of the lateral meniscus) popliteus.

The hyperextended locked knee ensures stability by: (1) putting both the medial and lateral tibial and femoral condyles in their most stable, close-packed position; (2) tensing the anterior cruciate ligament; and (3) tensing the collateral ligaments.

Chimpanzees lack both the habitual use of the knee in an extended posture and the features of the knee which produce locking. Three features of the chimpanzee knee are inconsistent with the locking mechanism. Firstly, the articular surfaces on the femoral condyles are more similar in length than they are in humans. Secondly, the curvature of both condyles is similar. Thirdly, tension in the posterior cruciate ligament increases as

the joint extends as it does in humans, but it pulls the whole femur back counteracting any tendency for the anterior cruciate to act as a pivot for medial rotation. This is due to the more centrally and anteriorly placed femoral attachment of the posterior cruciate ligament in the ape (Corbridge, 1987).

Movement and condylar shape

In humans and apes there is a considerable difference in the size of the lateral tibial condyle in relation to the medial tibial condyle (Fig. 22.7). In humans the tibial condyles are sub-equal in size (lateral condyle 92% of the area of the medial condyle) while in the large apes, the medial condyle is much larger than the lateral condyle (lateral condyle 84% of the area of the medial condyle) (Spencer, 1989). This size difference reflects the difference in weight transfer through the lower limb in humans and apes and mirrors a similar difference in the size of the femoral condyles.

There is also a marked difference in shape of the lateral tibial condyle in humans and apes as there is with the lateral femoral condyle. In humans the shape of the lateral tibial condyle in sagittal section can vary from slightly concave to flat to convex (Fig. 22.11). The smaller ape lateral condyle normally has a more marked convexity than

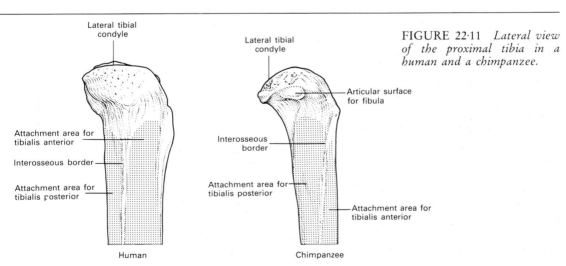

FIGURE 22·11 *Lateral view of the proximal tibia in a human and a chimpanzee.*

491

found in humans (Thomson, 1889; Martin and Saller, 1959). In both apes and humans (with the convex shape) the convexity is greatest on the posterior half of the condyle where the joint surface wraps over its posterior edge giving it a rounded rim. This contrasts with the sharp posterior rim of the medial condyle.

This morphology can be associated with the potential for more extreme flexion of the knee joint. In both humans and apes, flexion of the knee joint is associated with medial rotation of the tibia (or lateral rotation of the femur). As this rotation occurs the lateral femoral condyle and its meniscus slide onto the posterior border of the lateral tibial condyle (Brantigan and Voshell, 1941; Trinkaus, 1975). This produces the rounding (or convexity) of the posterior surface of the lateral condyle.

A convex lateral condyle in humans has been associated with hyperflexion in squatting postures (e.g. Thompson, 1889). However, displacement of the lateral meniscus onto the posterior border of the condyle may also be associated with other movements of the knee (Smith, 1956). For example, in apes the high degree of convexity may well be associated with the greater rotational capacity of the knee joint and the more mobile lateral meniscus.

Joint congruity and rotational ability in the knee

Although the rounded femoral condyles and the basically flat tibial condyles result in an incongruous joint, there is more congruity, or a greater area of contact, in the human joint than in the ape joint (Fig. 22.12). The edges of the two tibial condyles slope up on the sides of the two tibial spines marking the border of the non-articular intercondylar area. When the human joint approaches extension the slope of these edges fits the curvature of the femoral condyles and provides a 'guide' for the femur as it medially rotates and locks. In the chimpanzee joint, good congruence exists only between the medial femoral and tibial condyles. As the knee rotates the lateral femoral condyle and its meniscus have a large freedom of movement on the convex lateral tibial plateau.

Tardieu (1981) assessed the congruence and rotary capability of the knee joint by comparing the distance between the two tibial spines and the width of the intercondylar notch (Fig. 22.13). Humans have a narrow intercondylar space in relation to the distance between the tibial spines as might be expected from the high degree of congruence and limited rotary capability characteristic of the human knee. Next to humans come gorillas then orang-utans, chimpanzees and gibbons

FIGURE 22·12 *Posterior view of the knee in (a) a human and (b) a chimpanzee.*

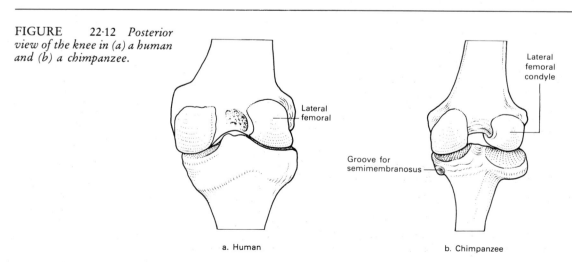

Lateral femoral

Lateral femoral condyle

Groove for semimembranosus

a. Human

b. Chimpanzee

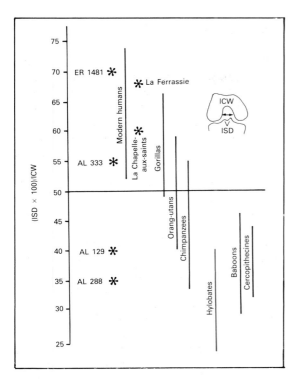

FIGURE 22·13 *The relationship between the interspinal distance (ISD) of the tibia and the width of the intercondylar notch (ICW) of the femur [(ISD × 100)/ICW] in humans, selected non-human primates and fossil hominids. After Tardieu (1981).*

and cercopithecine monkeys. This progression suggests a size related effect on congruence and rotary ability at the knee.

The fossil hominids also seem to follow this size-related series. The smallest knees assigned to *Australopithecus afarensis* (AL 129 and AL 288) have the least congruence and greatest rotary capability while the larger *A. afarensis* knee (AL 333) and the *Homo* knee (KNM-ER 1481) fall within the human range of variation.

Other features of the lower leg in humans and the apes

There are a number of other features of the tibia and fibula that distinguish modern humans from apes. These features reflect differences in both muscle attachments and in the structure of the ankle joint in these species.

The proximal tibia

One of the most important of these additional features is the **retroversion** of the tibial plateau. This refers to the angle that the tibial plateau makes with the long axis of the tibial shaft (**retroversion angle**) (Fig. 22.14) or alternatively the angle that it makes with the mechanical axis of the tibia defined as the line connecting the centre of the medial

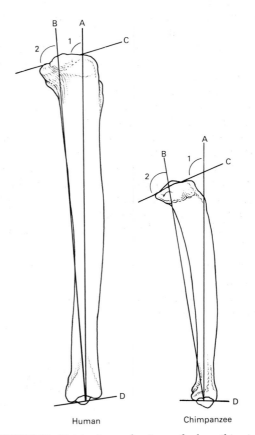

FIGURE 22·14 *Lateral view of the tibia in a human and a chimpanzee. A = long axis of the shaft; B = mechanical axis of the shaft; C = tangent of the medial tibial condyle; D = axis of the tibiotalar joint surface; angle 1 = retroversion angle; angle 2 = inclination angle.*

493

condyle with the centre of the distal articular surface (**inclination angle**) (Fig. 22.14). There is a wide range of variation in the magnitude of both these angles in human populations and what data are available suggest that the range of variation in the apes may overlap the upper human range of variation (Martin and Saller, 1959; Povinelli and Sterling, 1988).

Highly angled tibial plateaus are characteristic of human foetuses and are generally associated with the flexed-knee foetal position assumed in the womb. In adults highly angled plateaus have been explained by habitual squatting especially during the period of growth and development (Charles, 1893; and references cited in Trinkaus, 1975). This association is supported by the observation that the angle tends to be the highest among those human populations that commonly adopt a squatting posture when at rest or work and the least among those populations that use chairs, stools, etc. Furthermore, it is well known that tension accelerates epiphyseal growth while compression retards it (Heuter–Volkmann law) (Heuter, 1862; Volkmann, 1862). If squatting were a frequently adopted posture during the period of growth, the patellar ligament would habitually put the anterior border of the tibia under tension and the weight of the femoral condyles would put the posterior part under compression, resulting over time in a highly angled plateau.

However, Trinkaus (1975) pointed out that tibial retroversion does not always correlate with other features that indicate squatting in human populations (e.g. a rounded lateral tibial condyle and anterior rounding of the distal articulation, see below) and argued that squatting may not be the only explanation for this feature. Noting that during locomotion the forces at the knee are at their maximum level and are many times body weight when the knee is in partial flexion (between 10° and 20°), he suggested that tibial retroversion orients the main force acting on the knee perpendicular to the plane of the tibial condyles and thereby minimizes the anteroposterior shear stress at the knee. High activity levels and corresponding high levels of mechanical stress in the knee during locomotion, therefore may be an additional, and perhaps more universal, explanation for highly angled tibial plateaus in some human populations.

Non-human primates do habitually move with a flexed knee and are generally characterized by a more angled tibial plateau than is common in humans. However, a highly angled plateau in human fossils need not necessarily indicate, as once thought, a locomotor pattern involving a less than fully extended knee. In order to arrive at such a conclusion it would be necessary to have other anatomical evidence that could be exclusively associated with such a flexed-knee mode of progression.

There are three other important morphological features of the proximal tibia that distinguish humans from apes. Firstly, in humans the origin for semimembranosus is a horizontal groove on the posteromedial margin of the epicondyle immediately below the medial condyle (Fig. 22.12). In the African apes, and particularly in the chimpanzee, the origin is found in the same position but frequently takes the shape of a marked circular depression tailing off posteriorly in a tear-drop fashion. Secondly, in humans there is a distinctive horizontally oriented curved groove that separates the tibial tuberosity from the intracapsular (within the joint capsule) bone above it (Fig. 22.15). This groove is lacking in its human form in the apes. Thirdly, in apes the proximal shaft lateral to the tuberosity looks as if it has been hollowed out (Fig. 22.15). This gives the lateral side of the tuberosity a sharper edge than the medial side and also accentuates the (proximodistally) narrow lateral tibial epicondyle. In fact both tibial epicondyles in the apes are much narrower than in humans giving the tibial plateau a shelf-like appearance. In humans the epicondyles are more massive and the shaft fans out in a gentle curve to blend with them.

FIGURE 22·15 *Anterior, and posterior views of the human and chimpanzee proximal tibia. Shaded areas are the attachments of the indicated muscles. The arrow indicates the hollowed out lateral surface of the chimpanzee tibial shaft.*

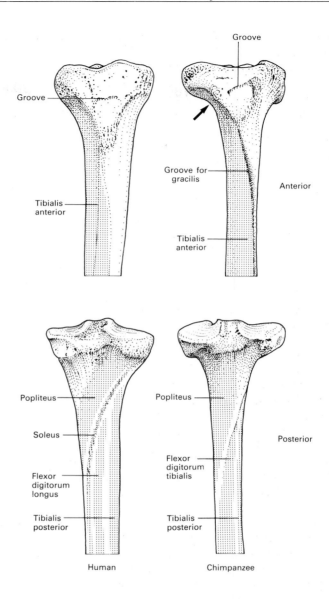

The shaft of the tibia

The hollowed-out appearance of the proximal part of the lateral side of the tibial shaft in the apes is associated with a different pattern of muscle attachment to that found in humans. In both humans and apes, the interosseous border is a variably developed ridge that provides the attachment for the interosseous membrane that binds the tibia and the fibula together. In humans this ridge runs straight up the lateral side of the shaft stopping about two-fingers' breadth short of the lateral condyle, whereas in apes the ridge slants forward to join the lateral side of the tibial tuberosity (Fig. 22.11). There are also consequent differences in the attachment areas of tibialis anterior and tibialis posterior in humans and the apes. Tibialis anterior originates anterior to the interosseous ridge and in humans it occupies the lateral side of

495

the tibial shaft extending from the tuberosity (and the sharp anterior border of the tibia) to the interosseous ridge. Tibialis posterior then occupies the lateral aspect of the posterior part of the shaft. In the apes, on the other hand, it is tibialis posterior that occupies the lateral side of the shaft in the region of the tibial tuberosity while tibialis anterior attaches across the more rounded anterior border of the shaft. In the apes this rounded anterior border has a variably sharp medial edge in the proximal third of the shaft. This edge marks the anterior extent of a groove for the origin of gracilis (Fig. 22.15). In the apes this muscle is much more strongly developed than it is in humans.

The proximal third of the posterior surface of the shaft also has a different topography in the African apes than it does in humans (Fig. 22.15). The African apes lack the roughened soleal line which in humans arches from the edge of the shaft below the lateral condyle to the medial border of the shaft. This soleal line separates the attachment area of popliteus above from that of tibialis posterior and flexor digitorum longus below and provides the tibial part of the origin for soleus. In the African apes the origin of soleus is normally confined to the fibula. The ridge that mimics the soleal line in the African apes is in reality the posterior border of tibialis posterior. The upper part of this ridge separates the attachment areas of tibialis posterior and popliteus and the lower part separates tibialis posterior from flexor digitorum longus (tibialis) and is thereby homologous with the human vertical line.

There is also a marked difference in the curvature of the long axis of the shaft in humans and apes. In the majority of humans the shaft is laterally concave in its upper half and laterally convex in its lower half (Fig. 22.16). This curvature can be easily seen because the anterior border of the human tibial shaft clearly follows a similar concave/convex course down the front of the shaft. This curvature ensures that the plane of the tibiotalar joint surface is both perpendicular

to the mechanical axis of the shaft and parallel to the plane of the tibial condyles. The tibial shaft in the African apes, on the other hand, is laterally concave throughout its entire length. This positions the tibiotalar joint surface at an angle to both the mechanical axis of the shaft and to the plane of the tibial condyles.

In addition to these differences, the African ape tibial shaft is more robust, or thicker in

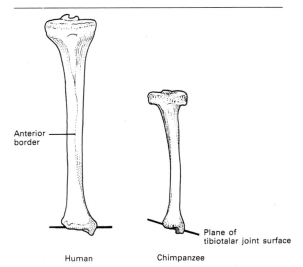

FIGURE 22·16 *Anterior view of the tibia of a human and a chimpanzee. Note the differences in the curvature of the long axis of the tibial shaft.*

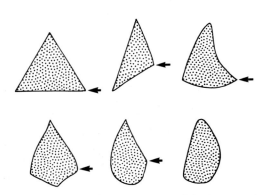

FIGURE 22·17 *The cross-section of the human tibia at mid-shaft can assume a variety of different shapes. The arrow indicates the interosseous border. After Martin and Saller (1959).*

relation to its length, than is the human shaft. Furthermore, it tends to have a different cross-sectional shape from that normally found in humans. In humans the shaft cross-section can vary in shape from an equilateral triangle to an elongated oval (Fig. 22.17). However, the anterior border is generally quite sharp giving the cross-section an almond-shaped, or amygdaloid (amygdale (Gk) = almond), appearance. In the African apes the tibia rarely has the sharp anterior margin of the human triangular cross-section and resembles a more-or-less elongate oval.

A mediolaterally flattened tibia is described as **platycnemic** (platy (Gk) = flat; kneme (Gk) = knee). In platycnemia the rounded posterior surface in some cases gives the appearance of a pillar extending down the back of the shaft. Lovejoy et al. (1976) have determined that the more platycnemic tibiae are stronger in anteroposterior bending and in torsion but also have a reduced mediolateral bending strength in comparison to less flattened **eurycnemic** (eu (Gk) = well) tibiae. Among modern human populations eurycnemic tibiae are frequently found in Blacks and Caucasians, while the most platycnemic tibiae are found in populations such as the Ainu (Japan), the Vedda (India) and the North American Indians (Martin and Saller, 1959).

The distal tibia

The form of the distal tibia, and particularly of the tibiotalar articular surface, reflects the requirements of the ankle joint in either human bipedal locomotion or ape quadrupedal locomotion. Two main features of the distal tibia separate humans from the African apes.

Firstly, in humans the distal joint surface of the tibia is oriented perpendicularly to the long axis of the tibial shaft while in the African apes it has a lateral inclination (Fig. 22.16). The human orientation minimizes potentially damaging shear stress on the articular cartilage and is also related to the anteroposterior or sagittal plane of motion of the human leg during normal gait (Latimer et al., 1987) (see Chapter 23 and Fig. 23.13).

Secondly, relative to the plane of the tibial plateau, the distal joint surface shows a greater degree of lateral torsion in humans than it does in the African apes or in any other studied primate (Fig. 22.18). The degree of human tibial torsion may be directly related to the modifications of the human foot that are adaptations to striding bipedal locomotion (Chapter 23). In particular, Lewis (1981) has argued that the human foot has been remodelled around the subtalar axis (see Fig. 23.16). Because this axis is oriented medially to the sagittal plane of the foot,

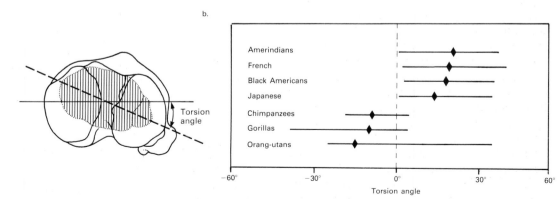

FIGURE 22·18 *(a) Mediolateral axis of the tibial plateau (solid line) and of the (shaded) tibiotalar articular surface (dashed line). (b) Angle of tibial torsion in selected human populations and the large apes. Data after Martin and Saller (1959).*

human feet would automatically become pigeon-toed without modifications to the ankle joint that would reorient the subtalar axis in a more sagittal direction. One of these modifications was most probably the reorientation of the talar trochlea on the body of the talus (Elftman and Manter, 1935b; Lewis, 1981a) and another could well have been the lateral torsion of the distal joint surface of the tibia.

An additional characteristic has also been claimed to provide an important distinction between human bipedal and the ape quadrupedal distal tibiae. Stern and Susman (1983) and Susman et al. (1984) have argued that in the human tibia the distal joint surface has an anterior tilt in the sagittal plane while in the apes it has a posterior tilt (Fig. 22.14) (see also Davis, 1964). Although the majority of human tibiae do have an anterior tilt of the distal articular surface (Stern and Susman, 1983), Latimer et al. (1987) argued that this trait is both highly variable in the African apes and has no functional significance. Analyses of motion at the ankle joint have shown that the human condition neither enhances the range of dorsiflexion of the foot (which some argue is an important feature of bipedalism) nor does the ape condition increase the degree of plantarflexion (Latimer et al., 1987). Rather than being directly related to the requirements of locomotion, it is most probable that this feature is an allometric consequence of body size.

Squatting facets

Accessory articular facets are frequently found on the anterior surface of the distal tibia. These facets are formed when the tibiotalar articular surface wraps round onto the front of the bone (Fig. 22.19). In hyperdorsiflexion of the foot at the ankle joint these facets articulate with opposing accessory facets on the neck of the talus. The tibial facets are variable in form and may be expressed as well-marked facets extending onto the anterior surface of the bone or simply as an eversion of the anterior margin

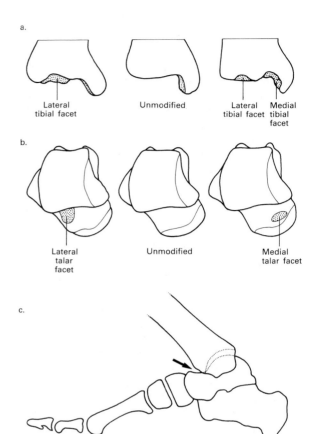

FIGURE 22.19 *Squatting facets on (a) the distal tibia and (b) the talus. After Barnett (1954) and Singh (1959). (c) In hyperdorsiflexion of the ankle joint the accessory facets on the distal tibial articulate with those on the neck of the talus (arrow).*

of the tibiotalar surface (Thomson 1889).

On both the tibia and the talus facets can occur on either the medial side or the lateral side or both sides of the bone (Fig. 22.19). In human populations lateral facets occur at a much higher frequency than do medial facets (Thomson, 1889; Trinkaus, 1975). In normal locomotion, tension in the ligaments limits the degree of dorsiflexion possible at the ankle joint and prohibits contact between the tibia and talar neck (Trinkaus, 1975). The posterior part of the deltoid ligament and the tibiofibular ligaments stretch in dorsiflexion as the distal tibia and fibula are pushed apart by the wedge-shaped talar trochlea (see Chapter 23) (Barnett and Napier, 1952; Hicks, 1953; Trinkaus, 1975). Habitual squatting, particularly from childhood, has the effect of stretching these ligaments and thereby permitting contact between the tibia and the talar neck and the formation of the accessory facets (Trinkaus, 1975).

In modern humans, squatting facets are less frequent in Europeans and more frequent in populations such as Asian Indians (Singh, 1959) and Australians (Rao, 1966) that habitually engage in a squatting posture (Trinkaus, 1975). However, even in Europeans there is a much higher frequency of squatting facets in foetuses than there is in adults (Barnett, 1954). The lack of persistent dorsiflexion of the foot during childhood and adulthood results in the early obliteration of the foetal facets (Barnett, 1954). Furthermore, the presence of squatting facets correlates well with other indicators of a squatting posture, such as a convex lateral tibial condyle and extreme platycnemia (Thomson, 1889; Trinkaus, 1975).

Squatting facets on both the tibia and the talus occur at a high frequency in non-human primates. A small sample of five gorillas, eight orang-utans and three baboons were reported to have both tibial and talar facets but out of six chimpanzees only a single tibial facet was found (Thomson, 1889). Squatting has not been reported for non-human primates, and these accessory facets are most probably associated with hyperdorsiflexion of the foot in locomotion (Thomson, 1889), particularly, while climbing.

The proximal fibula and shaft

The most noticeable difference between the human and ape fibula is robusticity. In comparison to total shaft length, the ape fibula has a uniformly larger circumference (and therefore a greater robusticity index) than does the human fibula. The form of the shaft is highly variable in both humans and the apes. However, the neck of the human fibula is frequently considerably more slender than is the body of the shaft (Sprecher, 1932; Martin and Saller, 1959; Susman and Stern, 1982). In addition, the human fibula is generally either straight or anteriorly concave, while the ape fibula is anteriorly convex (Martin and Saller, 1959). The surface for the origin of peroneus brevis also tends to be convex in the apes rather than concave (or flat) as it is in humans (Susman and Stern, 1982).

There is considerable difference between the head of the fibula in humans and apes. The characteristic human styloid process is missing from the ape fibula which articulates neatly under the overhanging ape lateral epicondyle. Viewed from the top, the head of the ape fibula is relatively flat and the tibiofibular articular surface is well defined. This contrasts with the condition in humans where the fibula articulates more against the side of the lateral epicondyle than underneath it and where there is considerable variation in the form of this articulation (Martin and Saller, 1959). There is also a corresponding variation in the form of the fibular articular surface in humans where it can vary from slightly concave to convex.

The distal fibula

There are a number of distinctions between the human and ape distal fibulae that directly reflect differences in the movement capabilities of the ankle joint. One of the most marked differences is the orientation of the articular facet (Figs 22.20 and 22.21). In humans this facet faces medially while in the African apes it faces inferiomedially (Stern and Susman, 1983). In the majority of humans the angle made by this surface with the vertical is less than 5° while in the African apes it ranges between 20° and 35°. This difference reflects the fact that the human fibula does not transmit any significant degree of weight to the foot, but rather acts to stabilize the ankle joint on the lateral side. In the African apes, however, where the

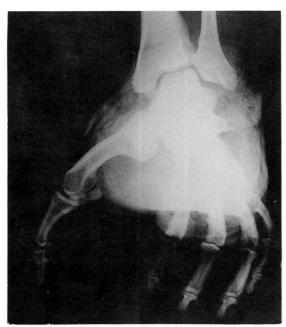

FIGURE 22·20 *Radiograph of a chimpanzee ankle joint from the anterior aspect. Compare with Figure 23.9. Courtesy of Theya Molleson.*

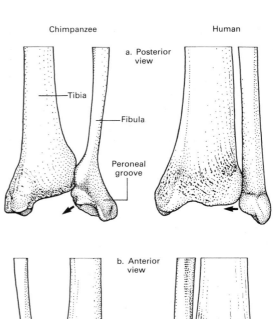

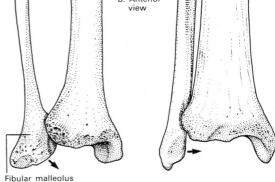

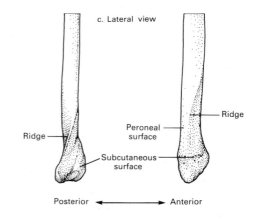

tibial shaft is mediolaterally curved and the tibiotalar joint is not parallel to the ground, the fibula would be expected to transmit a considerable amount of weight.

The difference in the orientation of the distal articular facet is also one of the factors that is related to the marked difference in the shape of the malleolus in humans and the apes. In the apes, rather than pointing downward in line with the shaft and with the vertically oriented articular facet, the malleolus looks as if it has been pulled out laterally (Figs 22.20 and 22.21). This difference is reflected in the different orientation of the ridge separating the subcutaneous surface of the fibula from the distal part of the peroneal surface. In humans this ridge travels in a gentle arc from the anterior border of the shaft downward to the tip of the malleolus. In the apes it is laterally concave, and swings in a sharp ridge from the anterior border of the shaft to the tip of the laterally extended malleolus.

FIGURE 22·21 *The human and chimpanzee distal tibia and fibula in (a) posterior view, and (b) anterior view. Lateral view of (c) the human and (d) the chimpanzee distal fibula. The arrow indicates the orientation of the articular facet on the fibula for the talus.*

Together with the inferomedially oriented articular surface in the apes, the wide peroneal groove in these primates may also be related to their laterally extended malleolar morphology. Both these features would expand the fibular malleolus. From electromyographical analyses Stern and Susman (1983) argued that large peroneal muscles (reflected in the large ape peroneal groove) function to stabilize the foot in arboreal locomotion.

Humans also differ from the apes in the shape of the proximal border of the distal articular facet of the fibula (Fig. 22.22). In humans it is perpendicular to the long axis of the shaft and parallel with the ground surface, while in apes this border is oblique to the long axis of the shaft (Stern and Susman, 1983). Stern and Susman (1983) have argued that this difference is directly related to the degree of plantarflexion and dorsiflexion that is characteristic of bipedalism on the one hand and quadrupedalism on the other. In particular, they argued that the ape condition is compatible with a greater degree of plantarflexion. However, Latimer et al. (1987) have questioned whether the range of joint motion can be accurately inferred from this feature and whether the ape pattern necessarily precludes human bipedal locomotion. These authors argued that there is no difference between humans and the African apes in the actual range of dorsiflexion

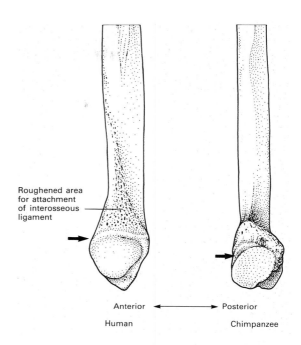

FIGURE 22·22 *Medial view of the human and the chimpanzee fibulae. Note the difference in the shape of the proximal border of the articular facet (arrows).*

and plantarflexion possible at the ankle joint irrespective of the orientation of this surface.

The remaining significant feature of the distal fibula is the length of the fibular malleolus relative to the length of the tibial malleolus (Fig. 22.23). The fibular malleolus

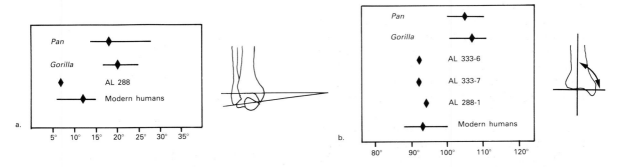

FIGURE 22·23 *(a) Variation in the angle between the plane of the supratalar joint and the tips of the tibial and fibular malleoli in modern humans, the African apes and* Australopithecus afarensis *(AL 288-*

1). (b) Variation in the angle between the axis of the tibial shaft and the orientation of the tibiotalar joint surface in the mediolateral plane. After Latimer et al. *(1987).*

is long in relation to the tibial malleolus in apes and short in humans. Latimer et al. (1987) have demonstrated that a line drawn between the tips of the tibial and fibular malleoli corresponds to the axis of rotation of the ankle joint. In humans the shorter fibular malleolus, relative both to the plane of the tibial joint and to the tibial malleolus, results in a rotational axis that is more parallel to the ground than is the case in the African apes with a longer fibular malleolus. This together with the previously mentioned features of the ankle joint restricts the motion of the human leg in normal gait to the sagittal plane (Latimer et al., 1987).

The knee joint and lower leg in hominid evolution

Lower leg bones are surprisingly rare in the fossil record. Discussion of the lower leg has centred on the three hominid taxa for which material is available, *Australopithecus afarensis*, *Homo habilis*, and Neanderthal.

Australopithecus afarensis

As with other parts of the *Australopithecus afarensis* skeleton, the tibia and fibula show a mosaic morphology. Many features of both bones have their greatest similarity with the African apes while aspects of the distal tibia, in particular, are consistent with the requirements of bipedal locomotion.

THE PROXIMAL TIBIA IN AUSTRALOPITHECUS AFARENSIS

The *Australopithecus afarensis* tibial plateau is chimpanzee like in morphology (Senut and Tardieu, 1985) in that it has the skeletal features which would suggest that the lateral meniscus was circular with a single attachment to the non-articular central stripe anterior to the tibial spine. In particular, the three fossils studied (AL 129-1b, AL 288-1aq and AL 333x-26) have an ape-like short and continuous posterior border to the lateral condyle (Fig. 22.24). They uniformly lack the longer posterior border that is notched to receive

the second (posterior) tibial attachment of the uniquely human semicircular lateral meniscus (compare Fig. 22.7). The ape-like morphology of all three *A. afarensis* specimens contrasts with other functional conclusions drawn for these hominids (Tardieu, 1981). Both in the proportions of the femoral condyles and in the rotational ability of the knee, the AL 129 and AL 288 distal femora and proximal tibiae are pongid like while the AL 333 bones are at the lower limit of the variation found in modern humans (e.g. Fig 22.13) (Tardieu, 1981; Senut and Tardieu, 1985).

Other ape-like features of the proximal

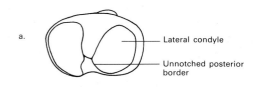

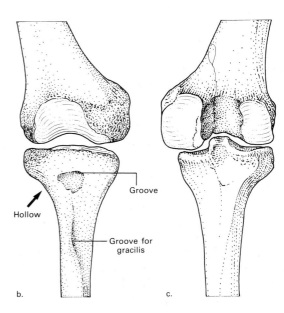

FIGURE 22·24 *The* Australopithecus afarensis *(AL 129-1) knee. (a) superior view of the tibial plateau, (b) the anterior view and (c) the posterior view of the articulated tibia and femur. After Tardieu (1983).*

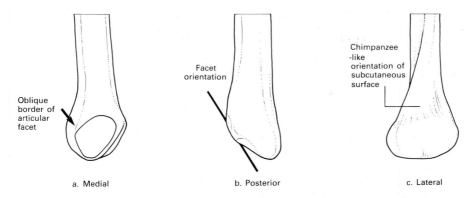

Oblique border of articular facet

Facet orientation

Chimpanzee -like orientation of subcutaneous surface

a. Medial

b. Posterior

c. Lateral

FIGURE 22·25 _The_ Australopithecus afarensis _(AL 288-1) distal fibula from the medial (a), posterior (b) and lateral (c) aspects._

tibia are the tear-drop shaped pit on the medial epicondyle for the attachment of semimembranosus and the elongate groove on the medial side of the shaft below the tibial tuberosity for the insertion of gracilis (Fig. 22.24). Furthermore, in the smaller _A. afarensis_ fossils (AL 288 and AL 129) the lateral side of the shaft immediately under the epicondyle has the hollowed-out appearance that is common in the apes where this morphology is associated with the attachment of tibialis posterior on the lateral aspect of the shaft rather than the posterior.

THE DISTAL TIBIA AND FIBULA IN AUSTRALOPITHECUS AFARENSIS

There are two main features of the _Australopithecus afarensis_ distal tibia and fibula (tibiae AL 288-1ar, AL 333-6 and AL 333-7; fibulae AL 288-1at, AL 333-9A, AL 333-9B, AL 333-85 and AL 333w-37) that are consistent with the requirements of bipedal locomotion and different from the condition seen in the apes. Firstly, the distal articular surface of the tibia is perpendicular to the long axis of the tibial shaft rather than inclined in a lateral direction (Fig. 22.23). Secondly, the axis of rotation of the ankle joint as measured by a line connecting the tips of the medial and lateral malleoli is more parallel to both the ground surface and the distal articular surface

of the tibia than is the case in the African apes (Fig. 22.23).

On the other hand, the distal fibula has a number of ape-like features (Fig. 22.25). In all of the known _A. afarensis_ fibular specimens the proximal border of the distal articular surface is oblique to the axis of the shaft of the fibula rather than perpendicular as in humans. Furthermore, the distal articular surface faces distally as well as medially when the shaft is vertically oriented, while in humans the surface faces medially. The _A. afarensis_ fibulae also have a deep peroneal groove bordered by a well-developed medial lip and the subcutaneous surfaces of all specimens (except AL 333-85) face anteriorly rather than laterally as in humans.

Stern and Susman (1983) suggested that the orientation of the subcutaneous surface in _A. afarensis_ may simply be related to the large peroneal groove. These authors offered two possible functional explanations for large peroneal muscles in _A. afarensis_. Either the _A. afarensis_ foot was used extensively in arboreal locomotion or a greater amount of muscle activity was required to stabilize the foot during the stance phase of bipedal locomotion than is the case in modern humans. Although Stern and Susman (1983) favoured the first explanation, the bipedal features of the distal tibia together with

503

significant features of the foot (Chapter 23) provide some support for the second explanation. The other ape-like features may be explained by an increased range of joint movement, by greater weight transfer through the fibula, or simply as retentions from a quadrupedal ancestor.

The final trait of the *A. afarensis* distal tibia is equivocal. In the small AL 288-1 (Lucy) tibia the distal joint surface is angled posteriorly (as in many apes), while in the other available larger specimens it is angled anteriorly (as in humans). Stern and Susman (1983) have suggested that the ape-like orientation in AL 288-1 reflects a greater degree of arboreal locomotion in this individual, while the human-like orientation in the other specimens reflects a greater degree of bipedal locomotion in these individuals. However, Latimer et al. (1987) argued that differences in anteroposterior orientation of the distal tibial joint have no relationship to the range of movement of this joint and that the orientation is variable even within modern human populations.

The tibia and fibula of Homo habilis

Discussion concerning the morphology of the tibia and fibula of *Homo habilis* has centred on the fragmentary left tibia and fibula found in the early 1960s at the FLK site at Olduvai Gorge (OH 35) (Fig. 22.26) (Davis, 1964; Day, 1976, 1978; Lovejoy, 1975; Susman and Stern, 1982).

There is general agreement that the distal tibia and fibula have a human-like morphology. Not only is the tibiotalar articular surface perpendicular to the tibial shaft but also the extent and degree of development of the distal interosseous ligamentous areas are human like and not ape like (compare Fig. 22.22). In humans these interosseous ligaments are both strong and extensive, leaving rugose markings that extend proximally further up the tibial shaft. Furthermore, unlike *Australopithecus afarensis*, there are no ape-like characteristics of the distal fibula. However, there has been controversy over

the functional significance of the morphology of the OH 35 tibial shaft (Davis, 1964; Susman and Stern, 1982). When first described, Davis (1964) noted that the shaft curved convex medially in its upper part and convex laterally in its lower third and was therefore human like in form. However, he also pointed out a number of non-human characteristics. Firstly, the shaft has a rounded anterior border typical of apes and different from the normally sharp border of the human tibia. Secondly, at the level of midshaft, the area of insertion of tibialis posterior is more extensive than that of flexor digitorum longus; the reverse is true in humans.

FIGURE 22·26 *The* Homo habilis *(OH 35) tibia and fibula. Courtesy of M.H. Day.*

In addition, Davis argued that the form of the soleal attachment of OH 35 was intermediate between humans and apes; in apes soleus is confined to the fibula, whereas in humans it has an extensive tibial attachment (Chapter 19). In OH 35 the tibial attachment, is not as extensive as in humans in that the soleal line does not reach the lateral border. Furthermore, he argued that the markings for popliteus on the fossil were different from both humans and apes, suggesting a nearly vertical pull of the muscles and a knee joint that was not fully adapted to bipedalism. Davis (1964) concluded that OH 35 was habitually bipedal but had a gait that differed from that of modern humans.

More recent authors have suggested that Davis misinterpreted both the soleal line and the nature of the popliteal surface and that in these, and in most other features, the OH 35 tibia falls within the range of variation of modern human tibiae (Lovejoy, 1975, 1978; Day, 1978; Susman and Stern, 1982). Susman and Stern (1982) follow Lovejoy (1975, 1978) in noting that a weakly developed interosseous ridge and the marked platycnemia of the shaft (associated with a well-developed posterior pilaster and a laterally directed origin for tibialis posterior) can be matched in modern human populations. They also pointed out characteristics of the fibula that may be more common in apes but are also occasionally found in human material. These features include: (1) a convex rather than concave surface for the origin of peroneus brevis on the lateral side of the shaft; and (2) a marked ridge between the area on the posterior surface for the origin of flexor hallucis longus and the region on the medial surface for the origin of the same muscle. The consensus of these authors is that the shafts of both the tibia and the fibula are within the range of modern human variation and only differ from the human condition in one feature, namely the particularly marked development of the crest between the origins of the tibialis posterior and flexor digitorum longus.

An additional *Homo habilis* tibia, the recently reported OH 62 tibia from the FLK site at Olduvai (Johanson et al., 1987), is similar to some of the *A. afarensis* tibiae (AL 288-1 and AL 129-1b) with respect to the morphology of the tibial tuberosity and the adjacent shaft. However, it differs from *A. afarensis* tibiae in lacking the groove on the medial side of the shaft associated with the medial attachment of tibialis anterior and the attachment of gracilis that these *A. afarensis* specimens share with the apes.

The Neanderthal lower leg

The Neanderthal tibia and fibula are both very robust bones, but in other respects they fall well within the range of variation found in modern humans (Trinkaus, 1983a). The robusticity of the fibula (midshaft circumference $\times 100$/maximum length) overlaps the highest range for modern humans and approaches the robusticity found in the living apes (Trinkaus, 1983a). Furthermore, the Neanderthals appear to have had moderate hypertrophy of their malleoli which are broad in relation to both malleolar thickness and maximum fibular length. Trinkaus (1983a) suggested that because the lateral malleolus helps to maintain the stability of the ankle joint, its hypertrophy would be consistent with the many other features that suggest lower limb robusticity in these hominids.

The tibia is also robust in relation to modern humans. Mechanical analyses of the strength of the Neanderthal tibia suggest that it is far stronger in comparison to modern human tibiae than even its external measurements suggest. The amygdaloid (almond shaped) cross-section that is typical of Neanderthal tibiae together with the magnitude and distribution of the cortical bone in the cross-section combine to give Neanderthals twice the tibial strength in bending and torsion (when scaled against the length of the tibia) than found in a comparative modern human sample (Lovejoy and Trinkaus, 1980).

Furthermore, there are also a number of features that suggest that the quadriceps

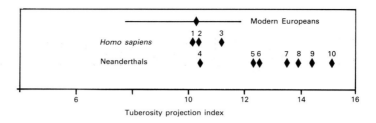

FIGURE 22·27 *The distance between the tibial tuberosity and the mechanical axis of the shaft in relation to the length of the tibia in modern Europeans, archaic* Homo sapiens *and Neanderthals.* Skhūl 4 (1), Cro Magnon (2), Kabwe (3), Tabūn (4), La Ferrassie (5), Shanidar 5 (6), Shanidar 2 (7), Kiik-Koba (8), La Chapelle (9), and Spy 2 (10). Data after Trinkaus (1983a).

femoris muscle had a greater relative strength in Neanderthals than it has in modern humans. Firstly, the tibial tuberosity in the Neanderthals projects more anteriorly than it does in modern humans (Fig. 22.27). Secondly, the tibial condyles are located in a more posterior position in relation to the axis of the tibial shaft than they are in modern humans. Thirdly, the patella is both absolutely and relatively thicker than it is in humans. All these features combine to move the insertion of quadriceps femoris further anterior to the axis of rotation of the knee joint, thereby lengthening the moment arm of the muscle and increasing its strength (Trinkaus, 1983a).

CHAPTER TWENTY-THREE

THE HOMINOID FOOT

The most obvious difference between the ape foot and the human foot is the opposability of the great toe (hallux) in the apes and its absence in humans (Fig. 23.1). However, there are a number of more subtle differences that are important for human bipedal loco-motion. Similarly, there are a number of features of the ape foot that are important to its function as a grasping organ. These features are best understood in the context of the kinematics (movement) of the foot in locomotion.

The human and ape foot in bipedal locomotion

In human bipedal locomotion, the first part of the foot to come into contact with the ground at the beginning of the stance phase is the lateral margin of the heel (Fig. 23.2). In many individuals the next part of the foot to contact the ground is the lateral (or fifth) metatarsal followed by the more medial metatarsals. In other individuals the heads of the five metatarsals may strike the ground simultaneously (Reeser et al., 1983). In both cases the big toe (hallux) is the last part of the foot to contact the ground. As the body

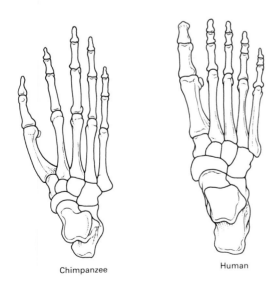

FIGURE 23·1 *A chimpanzee and a human foot from the dorsal aspect. After Schultz (1963).*

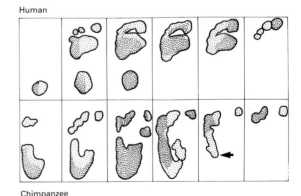

FIGURE 23·2 *Distribution of pressure in the human foot and chimpanzee foot during bipedal walking. The arrow indicates the midtarsal break in the chimpanzee foot. After Elftman and Manter (1935a).*

weight moves forward over the foot, it also rolls medially across the ball of the foot. The heel is the first part of the foot to leave the ground followed in many individuals by the fifth metatarsal and then by the remaining metatarsals. In other individuals the metatarsals may leave the ground simultaneously. In both cases, however, the big toe (hallux) is the last part of the foot to leave the ground at the terminal part of the stance phase of the stride known as toe-off.

Depending on the speed of movement, variations in this pattern may occur (Bojsen-Møller, 1979). However, there are two features that are invariant. Firstly, during the midstance phase of the stride, the body weight is always transferred from the lateral side of the foot to the medial side, or ball, of the foot in preparation for toe-off. Secondly, the midtarsal region of the foot behaves as a rigid lever both at heel-strike and at toe-off.

It is precisely in these two features that the kinematics of the chimpanzee foot differs from the kinematics of the human foot in bipedal walking (Elftman and Manter, 1935a). In chimpanzee walking there is no medial weight transfer, rather the weight is transferred in a straight line from the posterior to the anterior part of the foot. At heel strike the lateral border of the foot comes into contact with the substrate at about the same time as the heel followed by the medial border and then by full foot contact. The second and third toes are normally the last to leave the ground rather than the great toe as is the case in humans. The midtarsal region of the foot is also much more mobile. As the heel leaves the ground, the foot bends, or pivots at the calcaneocuboid joint and part of the lateral side of the foot stays in contact with the ground (Fig. 23.3). This is called the 'midtarsal break' and is a direct result of the mobility of the calcaneocuboid joint in the ape foot.

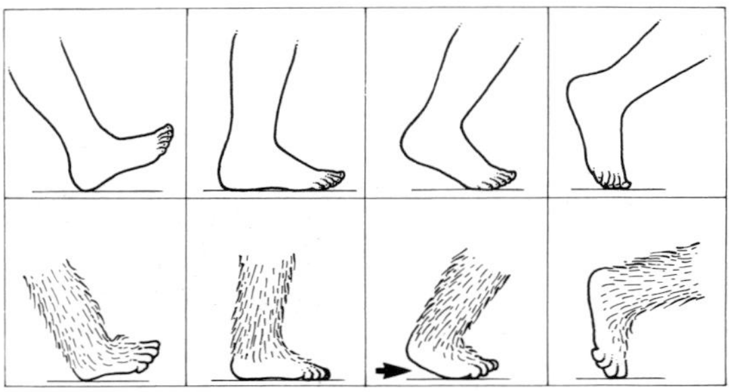

Major differences in the structure of the human and ape foot

In its adaptation to bipedal locomotion the human foot is among the most specialized of all primate feet. Two of the major specializations of the human foot, the longitudinal arch and the unique structure of the human calcaneocuboid joint, are directly related to the differences in the kinematics of the ape and human foot. The human longitudinal arch permits the unique medial weight transfer during the midstance phase of the stride while at the same time distributes the weight of the body over the sole of the foot in standing and acts as a shock-absorber to cushion the body from the rather considerable forces generated by walking and running. Furthermore, the unique form of the human calcaneocuboid joint prevents the midtarsal break in human walking and stabilizes the midtarsal region of the foot. Other specializations of the human foot that are directly related to bipedal locomotion are: (1) the relative proportions of the major parts of the foot that allow it to work as an effective bipedal lever; (2) the form of the ankle joint permitting the leg to move back and forth in a straight line over the foot; and (3) the structure of the first tarsometatarsal joint that prohibits the opposability of the great toe.

Human

Chimpanzee

FIGURE 23.3 *Lateral view of foot contact during bipedal walking in a human and a chimpanzee. The arrow indicates the midtarsal break in the chimpanzee foot. After Susman (1983).*

The human arch and the calcaneocuboid joint

The human foot is unique in being arched or, more precisely, being shaped like a half dome with its hollow surfaces facing both

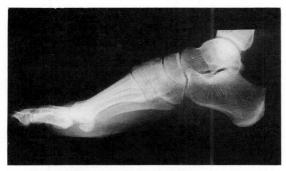

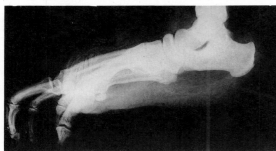

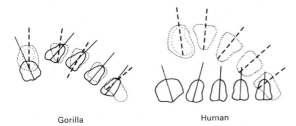

Gorilla   Human

FIGURE 23·4 *Medial radiographs of a human foot and a chimpanzee foot. Courtesy of Peter Abrahams and Theya Molleson. Transverse sections through the metatarsals (instep) of a gorilla foot and a human foot. After Morton (1922). In the gorilla foot both the metatarsal bases (dotted outlines) and heads (solid outlines) are elevated in line with the transverse arch of the foot. In the human foot the metatarsal bases are elevated but the metatarsal heads are on the ground. The axes of the metatarsal heads (solid lines) have rotated in relation to those of the gorilla to permit the heads of the metatarsal to lie squarely on the ground.*

downward and medially. The dome shape results from the presence of both a longitudinal arch as well as a transverse arch. Other primates have only the transverse (mediolateral or side-to-side) arch, their feet being flat in the longitudinal direction (Fig. 23.4).

In normal standing posture muscle activity is not required to maintain the human arch (Basmajian and De Luca, 1985). Rather the shapes of the bones themselves and, more importantly, the ligaments that bind the bones together and the **plantar aponeurosis** are enough to maintain the integrity of the structure. Muscles reinforce the human arch only when it is subjected to the larger forces of movement and locomotion.

There are three major ligaments that are important to arch support and to the dynamics of the foot in locomotion, the **plantar calcaneonavicular ligament** (commonly called the **spring ligament**), the **short plantar ligament** and the **long plantar ligament** (Fig. 23.5). The spring ligament spans the

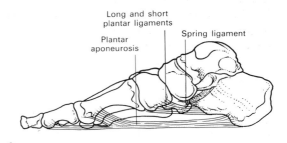

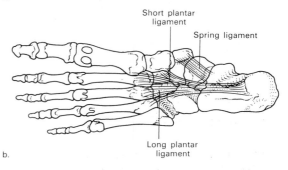

FIGURE 23·5 *Ligaments of the human foot from the medial (a) and plantar (b) aspects.*

509

distance between the navicular and the sustentaculum tali. It forms the floor of the socket supporting the head of the talus. The short plantar ligament extends from the plantar surface of the anterior part of the calcaneus to the large area just posterior to the cuboid ridge. And the long plantar ligament spans the short ligament from the calcaneus to the bases of the second to the fifth metatarsals. In addition to these a large number of smaller plantar interosseous ligaments bind each bone to its neighbours and provide further reinforcement both to the longitudinal and to the transverse arches.

The **plantar aponeurosis** is also important to the maintenance of the human arch. This is a strong band of collagenous tissue that acts as a truss to bind the foot together (Fig. 23.5). It runs the length of the foot from the calcaneus to the phalanges of the five toes. Experiments have shown that the plantar aponeurosis takes up about 60% of the stress imposed on the foot in normal standing while the beam action of the medial four metatarsals takes up 25% (Hicks, 1954, 1955). The remaining 15% of stress could be borne by the fifth metatarsal which was not measured in these experiments. When the toes are extended as they are during the later part of the stance phase of the stride the plantar aponeurosis acts as a windlass, takes proportionally more stress and heightens the arch.

The human arched, or domed, foot is not rigid but is a highly mobile structure that has been likened to a twisted plate (Fig. 23.6a) (MacConaill, 1945). One end of this hypothetical rectangular plate is analogous to the heads of the five metatarsals and maintains contact with the substrate. The other end which represents the heel is turned on end giving a twist to the structure. This structure can be further twisted or untwisted around the transverse tarsal joint (Chapter 19) and these movements are essential to both standing on uneven substrates and to the dynamics of bipedal locomotion.

The medial side of the foot, which is the highest part of the arch in standing, shows

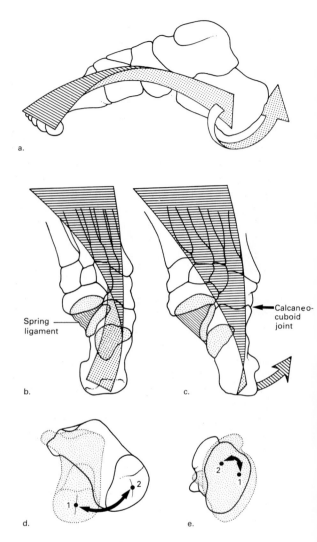

FIGURE 23·6 *Dynamics of the human foot in walking. (a) The human arch as a 'twisted plate'. After MacConaill and Basmajian (1969). (b) Dorsal aspect of the human foot during the early part of the stance phase of the bipedal stride. Note the lax spring ligament. (c) The foot during the later part of the stance phase. The human heel exorotates and swings laterally at the calcaneocuboid joint (small arrow) untwisting the foot plate and tightening the spring ligament. The range of movement of (d) the human heel and (e) the chimpanzee heel from full endorotation (1) to full exorotation (2). Note the absence of the lateral swing in the chimpanzee. Views (b)–(e) after Lewis (1980b).*

the greatest mobility during walking. During the middle part of the stance phase when the full foot is in contact with the substrate the leg and talus rotate medially as a unit in relation to the foot plate (at the peritalar joint) (Chapter 19). This moves the body weight to the medial side of the foot and simultaneously drives the head of the talus into the navicular. Because the subtalar joint acts like a screw this also causes calcaneus to move posteriorly in relation to the talus and navicular, stretching the calcaneonavicular (or spring) ligament and putting the medial side of the foot into a rigid, close-packed position (Fig. 23.6). At the same time the heel exorotates and swings laterally at the calcaneocuboid joint. This untwists the foot plate and wedges the calcaneus against the cuboid in the close-packed position. This gives firm support to the foot which is thereby transformed into a rigid lever as the heel is lifted off the ground and the foot moves into the toe-off stage of the stride. The untwisting and twisting of the foot plate and the resulting stretching and relaxing of the ligaments during the different stages of the stride also helps to absorb the considerable forces to which the foot is subjected during locomotion. These forces are approximately 1.2 times body weight during walking, 2 times body weight during running and up to 5 times body weight when landing on the feet from a height (Bowden, 1967).

The human calcaneocuboid joint is unique among primates in permitting not only rotation of the calcaneus but also the lateral swing that wedges the calcaneus against the cuboid. In the apes (chimpanzee and gorilla) the calcaneocuboid joint is so arranged that it lacks the lateral swing but permits a considerably greater degree of rotary movement than it does in humans (Fig. 23.6). When the ape foot is used in prehension it is normally inverted while the anterior part of the foot rotates at the calcaneocuboid joint and causes the navicular to ride up on the talar head, impacting the calcaneus against the navicular at the calcaneonavicular articulation

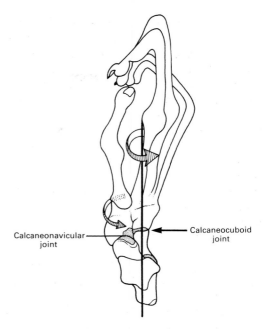

FIGURE 23·7 *The foot of a South American monkey* (Pithecia monachus) *illustrating the increased rotatory movement at the calcaneocuboid joint in the non-human primate foot. The vertical line represents the axis of rotation. After Lewis (1980b).*

(Fig. 23.7). The joints are in a close-packed condition, stabilizing the foot in a posture where it is ready to grasp a rounded branch firmly.

Proportions of the foot

One of the most obvious differences in proportions between the human and the ape foot is the length of the toes (or phalanges). In relation to total foot length, the length of the phalanges of the third digit makes up only 18% of total foot length in humans while it accounts for 33% in gorillas and 35% in chimpanzees (Fig. 23.8) (Keith, 1929). Schultz (1963) has demonstrated that the relatively short human lateral toes are not the result of an unusually long great toe. The lengths of the phalanges of the great toe in relation to metatarsal length are about the same in humans and apes (Table 23.1). The

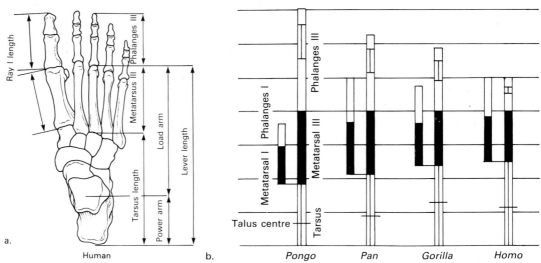

FIGURE 23·8 *Foot proportions. (a) Bone and lever lengths. (b) Average relative lengths measured on the foot skeletons of selected higher primates reduced to the same lever length. After Schultz (1963).*

TABLE 23·1 Average proportions of the skeleton of the foot in humans and apes. After Schultz (1963)[1]

	Specimens	Ray I/ Ray III	Phalanges I/ Metatarsal I	Phalanges III/ Metatarsal III	*Foot length/ Trunk length	*Lever length/ Trunk length	Power arm/ Load arm
Homo	25 (18)*	101.8	92.5	65.9	43.8	35.4	39.4
Gorilla	20 (8)	67.5	90.4	118.1	46.1	31.2	46.1
Pan	20 (16)	70.0	90.6	121.8	47.0	30.2	28.2
Pongo	9 (5)	35.0	69.7	145.9	59.3	33.7	19.6
Symphalangus	5 (2)	67.0	85.0	133.3	51.3	30.8	22.7
Hylobates	18 (10)	66.8	82.4	135.7	51.6	30.0	18.2

*Sample sizes for these ratios are in parentheses.

short lateral toes have resulted primarily from a reduction in length of all three phalanges of the lateral four digits and particularly from the reduction of the intermediate phalanges of these digits in relation to metatarsal length. Indeed, the lengths of the intermediate phalanges of each toe are always shorter relative to the lengths of their corresponding proximal phalanges in humans than they are in the apes (Stern and Susman, 1983).

The lengths of the individual toes in relation to each other, as well as the lengths of the individual phalanges in relation to each other, also distinguish humans from apes. In humans the sequence of total toe length (as measured by the summed length of the phalanges of each digit) is normally I>II>III>IV>V (Martin and Saller, 1959) while in the apes the first toe is normally the shortest while the third is normally the longest. In humans the sequence of proximal phalange (PP) length is normally the same as that of total toe length; however some slight variation can occur. Notwithstanding this variation, PPII is always longer than both PPIII and PPIV in humans while it is shorter than both these proximal phalanges in the apes (Stern and Susman, 1983).

Humans also differ from the apes in having a shorter total foot length in relation to trunk

TABLE 23·2 Metatarsal robusticity in humans, and African apes. After Archibald et al. (1972).

	Most robust(%)	2nd most robust(%)	3rd most robust(%)	4th most robust(%)	Least robust(%)
Humans (N = 50)					
1st metatarsal	(100)				
2nd metatarsal			8	22	(70)
3rd metatarsal			10	(68)	22
4th metatarsal		8	(76)	8	8
5th metatarsal		(92)	6	2	
Chimpanzees (N = 10) and gorillas (N = 10)					
1st metatarsal	(100)				
2nd metatarsal		30	(40)	30	
3rd metatarsal		(65)	35		
4th metatarsal		5	25	(65)	(5)
5th metatarsal				5	95

Robusticity = [(midshaft diameter × 100)/length]. Numbers represent the percentage of bones in each robusticity class. Circled percentages highlight the most common robusticity sequence.

length and a longer lever length of the foot in relation to trunk length (Table 23.1). However, gorillas have a higher ratio between the power arm of the foot (taken to be the distance between the centre of the trochlea of the talus to the calcaneal tuberosity) and the load arm (the distance between the centre of the talar trochlea to the head of the third metatarsal) of their feet (Schultz, 1963).

There are also differences in the robusticity of the metatarsals as measured by the midshaft circumference of the bone divided by its length (Archibald et al., 1972). In the great majority of cases the human fifth metatarsal is second only to the first metatarsal in robusticity (Table 23.2). This reflects the fact that in bipedal locomotion the lateral side of the foot carries a large proportion of body weight during the midstance phase of the locomotor cycle. In apes the fifth metatarsal is generally the least robust of the metatarsals. In these primates the more medial metatarsals are subjected to an increased force when the foot is used in prehension.

The ankle joint

The ankle joint (or talocrural joint) resembles a carpenter's mortise and tenon joint when viewed in the coronal plane (Fig. 23.9). The tibial and fibular malleoli and the lower articular surface of the tibia are the mortise, and the talus the tenon. The joint is held together by two sets of ligaments, the **medial (deltoid) ligament** and the **lateral ligament**. The direction of the ligaments, the orientation of the malleoli and the wedge-shape of the talar trochlea prevent backward displacement of the talus in relation to the leg (Fig. 23.10).

In humans, and more so in the African apes, the talar trochlea is narrower posteriorly than it is anteriorly (Fig. 23.11). As a result, the malleoli of the tibia and fibula do not articulate as firmly with the talus in plantarflexion as in dorsiflexion. Because the ape trochlea is more wedge-shaped than the human trochlea, there is a greater latitude for movement in the ape ankle joint when it is in plantarflexion. However, when the foot is

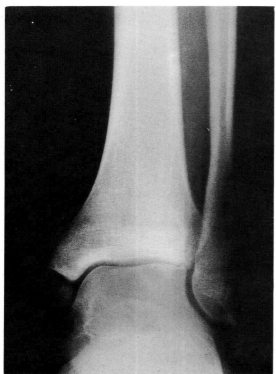

FIGURE 23·9 *Radiograph of a human ankle joint from the anterior aspect. Compare with Figure 20.22. Courtesy of Peter Abrahams.*

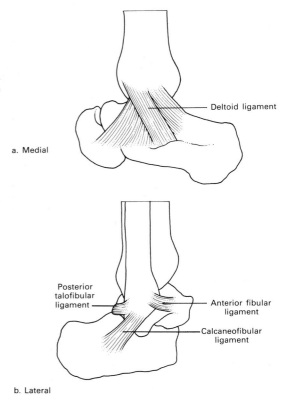

a. Medial — Deltoid ligament

b. Lateral — Posterior talofibular ligament / Anterior fibular ligament / Calcaneofibular ligament

c. Superior — Lateral malleolus / Trochlea / Medial malleolus

FIGURE 23·10 *Ligaments of the ankle from the medial (a), lateral (b), and superior (c) aspects. Note that the orientation of the ligaments and the wedge shape of the trochlea prevent backward displacement of the talus in relation to the leg.*

dorsiflexed the tibial and fibular malleoli articulate more firmly with the wider anterior trochlea and the ligaments of the ankle are tensed. This is true of the distal **tibiofibular** ligament that binds the two bones of the leg together and also particularly true for the **tibiocalcaneal** and **fibulocalcaneal** ligaments that tense and produce a downward pressure on the leg at the joint. The dorsiflexed joint in both humans and apes is in its close-packed position and is thereby much more stable than it is in plantarflexion.

In both humans and apes the trochlear surface is shaped like the surface of a cone with the smallest radius of curvature on the medial side and the largest on the lateral side (Fig. 23.12). However apes differ from humans in having a much larger apical angle to this cone. This means that the medial border of the trochlea has a much smaller radius of curvature than the lateral border in the apes and that it is also much lower than the lateral border. As a result in plantarflexion the anterior part of the ape foot is automatically displaced medially and everted and in dorsiflexion it is displaced laterally and inverted (Lewis, 1980a). When the foot is

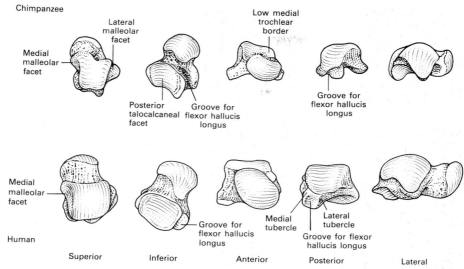

FIGURE 23·11   *The chimpanzee and human talus.*

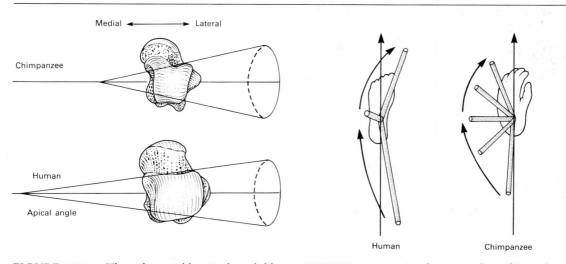

FIGURE 23·12   *The talar trochlea is shaped like the surface of a cone in both chimpanzees and humans. See text for discussion. After Latimer* et al. *(1987).*

FIGURE 23·13   *In humans the tibia takes a straighter path over the foot during the stance phase of the stride than it does in chimpanzees. After Latimer* et al. *(1987).*

fixed and the tibia mobile this cone shaped trochlea causes the tibia to rotate around its long axis medially in dorsiflexion and laterally in plantarflexion. Together with the form of the distal tibia it also forces the leg to trace an arcuate path as it moves forward and backward over the foot (Fig. 23.13).

The degree of longitudinal rotation of the tibia over the stationary foot (or of medial/lateral displacement of the foot on the stationary leg) is dependent not only on the size of the apical angle of the cone of which the trochlear surface is a part but also on the range of joint motion. The larger the range of motion the more rotation will occur. Latimer et al. (1987) demonstrated that the human tibia rotates through an average of 11.5° as it moves from extreme dorsiflexion to extreme plantarflexion. This figure contrasts with an average rotation of over 17° in the chimpanzee and the gorilla.

The ankle joint of the African apes differs from the human joint in another way. In the African apes the articular surfaces on the talus for the medial and lateral malleoli have a greater lateral flare than they do in humans (Fig. 23.11). During dorsiflexion, when the ape foot is automatically inverted the medial malleolus is received into the cup-like articular surface produced by the flaring medial malleolar facet on the talus and the fibular malleolus into the flaring lateral malleolar facet. With the foot inverted much of the weight of the body is carried through the medial malleolus to the medial side of the talus. The foot is habitually dorsiflexed during climbing and is also frequently used in an inverted posture during locomotion on flat surfaces (Lewis, 1980a).

The first tarsometatarsal joint

The form of the joint surfaces between the medial cuneiform and the first metatarsal largely determines the prehensile capability of the great toe (Fig. 23.14). In both humans and apes, the axis of this joint lies vertically in the dorsoplantar plane. However humans differ from the apes in having a much less convex joint surface on the medial cuneiform that is also oriented in a more anterior direction. What convexity still exists on the human medial cuneiform articular surface (and concavity on the articular surface of the first metatarsal) is so arranged as to screw the first metatarsal into a close-packed articulation with the cuneiform as the body weight passes over the great toe at the toe-off phase of the locomotor cycle. The details of these joint surfaces are discussed in a later section.

The subtalar joint and the transformation of the ape foot into a human foot

The human foot is more than an ape foot placed flat on the ground with a longitudinal arch and an hallux adducted in line with the other toes. It has been suggested that the major differences between the human foot and the ape foot have resulted from the realignment of the human foot around the subtalar axis of the primitive prehensile primate foot (Keith, 1929; Lewis, 1981).

The subtalar axis is the axis of the subtalar joint, or the joint between the talus and the calcaneus (Fig. 23.15). This joint has

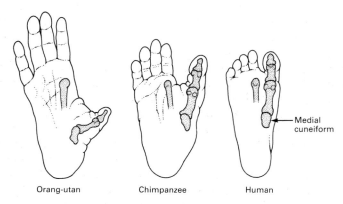

Orang-utan   Chimpanzee   Human

FIGURE 23·14 *The first tarsometatarsal joint in an orang-utan, a chimpanzee and a human. After Schultz (1950).*

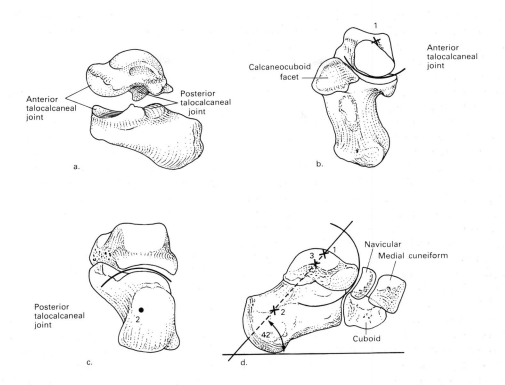

FIGURE 23·15 *The subtalar axis. The subtalar joint (a) comprises two reciprocal concave and convex articular surfaces, (b) the anterior talocalcaneal joint and (c) the posterior talocalcaneal joint. The subtalar axis passes through the centres of the circles (1 and 2) formed by the curved surfaces of the two joints.*

(d) In the medial view the human subtalar axis also passes through the centre of an additional circle (3) formed by the anterior talocalcaneal joint and is inclined by about 42° from the horizontal. After Shephard (1951).

reciprocal concave and convex articular surfaces at either end of the talus. The convex head of the talus articulates with the concave anterior (and medial) facets on the calcaneus. At the same time the concave posterior facet on the talus articulates with the convex posterior facet on the calcaneus. The axis of rotation of the joint is the line passing through the centres of the circles of which the joints are arcs. It is this axis around which the foot inverts and everts.

The human and ape subtalar joints are very similar, the major differences being that the axis of the subtalar joint is dorsoplantarly steeper in humans than it is in apes. This results from the disproportionally large radius of curvature of the subtalar joint surfaces in the human foot. The larger radius of curvature also limits the degree of inversion and eversion in the human foot. Furthermore, the axis of the subtalar joint makes a more acute angle with the long axis of the foot in humans than it does in the apes (Fig. 23.16).

Lewis (1981) has suggested that the reason why the human subtalar axis makes a more acute angle with the long axis of the foot is that the entire foot has been realigned around this axis (Fig. 23.16). In particular, he suggested that the adducted human great toe has been brought in line with the remaining

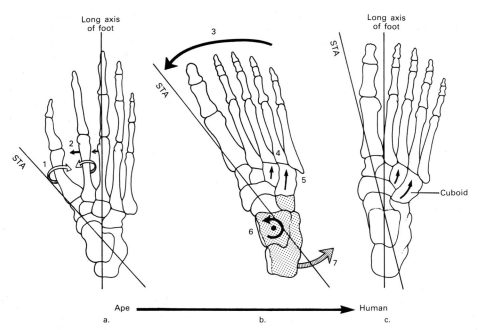

FIGURE 23·16 *Transformation of an ape foot into a human foot. (a) The ape metatarsals rotate so that their heads sit firmly on the ground (1) and the lateral metatarsals move medially (2) toward the subtalar axis (STA). (b) The medial realignment of the lateral four metatarsals (3) is associated with their angled bases (4) and with the elongated and medially bent cuboid and lateral cuneiform (5). The talar trochlear rotates medially on the talus (6) and the posterior part of the calcaneus moves laterally (7) to bring the long axis of the foot closer to the subtalar axis. This realignment results in a closer approximation between the subtalar and long axes in the human foot (c) than in the ape foot (a).*

toes as the result of the realignment of the lateral four metatarsals toward the first metatarsal and the oblique subtalar axis rather than in the realignment of the first metatarsal towards the lateral four metatarsals and away from the subtalar axis. It is claimed that this realignment is associated with the following skeletal modifications. Firstly, the cuboid and the lateral cuneiform in the human foot have a noticeable medial bend in their long axes. This orients the third, fourth and fifth tarsometatarsal articulations medially towards the great toe. Secondly, the bases of the lateral four metatarsals are angled in relation to the long axes of the metatarsal shafts. This also has the effect of orienting the shafts medially toward the great toe. Thirdly, the

shafts of the lateral four metatarsals have a medial twist or torsion. This orients the heads of these metatarsals so that they are in square contact with the flat substrate.

This medial realignment toward the subtalar axis and the great toe may not seem intuitively logical. However, the ape great toe is in its close-packed and most stable position when it is abducted and lies in line with the subtalar axis (Lewis, 1972c, 1980b, c). It would be logical to retain this stability on the medial side of the human foot by realigning the lateral toes towards the great toe rather than breaking the stability by realigning the great toe toward the lateral side of the foot. Furthermore, the morphology of the metatarsals and of the anterior tarsal bones is consistent with this medial realignment.

Other modifications to the primitive foot that would be necessary to achieve the human bipedal foot under this scenario would include the following. Firstly, the first metatarsal would rotate laterally to the point where it sits firmly on the substrate rather than facing the lateral toes as it does in the apes (Fig. 23.16). Secondly, if the toes were aligned medially towards the subtalar axis and the great toe the incipient human foot would be seriously pigeon-toed. A change in the orientation of the trochlea of the talus would be needed to counteract this. A medial rotation of the trochlea would reduce the angle between the talar neck and the trochlea and bring the long axis of the trochlea, and of the foot, more in line with the subtalar axis. A final alteration would involve a lateral reorientation of the calcaneus to bring its long axis in line with the subtalar axis.

Significant features of the bones of the foot

Many of the significant differences between the bones of the foot in apes and humans have already been mentioned in the previous discussions. However, there are other features that are also important in distinguishing ape foot bones from human foot bones.

The talus

The important distinguishing features of the talus that have already been discussed are the morphology of the trochlear surface and the flare of the tibial and fibular trochlear facets (in the context of the ankle joint) and the radius of curvature of the posterior talocalcaneal joint (in the context of the subtalar axis). Other important features are the following.

ANGLE OF TORSION OF THE TALAR HEAD

The angle of torsion of the talar head (Lisowski, 1967) or of the talar neck (Day

and Wood, 1968) has been defined as the angle made by the long axis of the head of the talus and the plane of the trochlea when viewed from the front (Fig. 23.17). The African apes have considerably smaller torsion angles than do humans. However, torsion angles are better measures of differences in trochlear orientation between these species than they are of head (or neck) torsion. When the tali are aligned with the axes of their subtalar joints in parallel the long axes of the heads of both human and

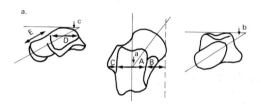

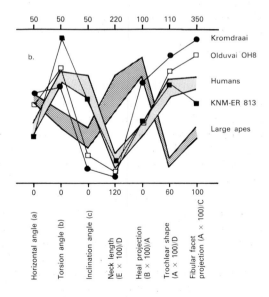

FIGURE 23·17   (a) Selected talar measurements. (b) Talar measurements in modern humans, chimpanzees and gorillas and selected fossil hominids. After Day and Wood (1968) and Wood (1974a).

African ape tali have similar orientations (Fig. 22.11) (Elftman and Manter, 1935b). However, the orientations of trochlear surfaces are very different.

THE ANGLE OF INCLINATION OF THE TALAR NECK

Day and Wood (1968) defined this angle as that enclosed by a line drawn as a tangent to the trochlea at the midpoint of the lateral malleolar surface and a second line drawn from the midpoint of the articular surface of the head through a point midway between the upper and lower borders of the neck (Fig. 23.17). The unique large angle of inclination of the human talar neck is directly related to the equally unique presence of a longitudinal arch in the human foot.

LENGTH OF THE TALAR NECK

The neck length index is defined by Day and Wood (1968) as the length of the neck from the trochlea to a line tangent to the most distal (anterior) extent of the head divided by the length of the fibular facet (Fig. 23.17). Humans are characterized by a shorter neck than are the African apes. The short and relatively stout human talar neck goes against the trend towards relatively elongated tarsal bones in the human bipedal foot and may well act to resist the increased stress on the talar neck resulting from the pattern of weight transfer to the medial side of the foot characteristic of bipedal walking.

THE GROOVE FOR FLEXOR HALLUCIS LONGUS

The groove for flexor hallucis longus is located on the posterior aspect of the talus (Fig. 22.11). Its upper margins are located immediately posterior to the trochlea and are defined by the medial and lateral talar tubercles. In humans both these tubercles contact the distal margin of the tibia when the foot is in plantarflexion and the groove is vertically oriented with parallel margins. In the African apes only the medial tubercle contacts the distal margin of the tibia in

plantarflexion and the groove is trapezoidal in shape. Because tendons align along their line of action, the oblique orientation of this groove in the apes reflects the arcuate path traced by the leg as it moves from a position of plantarflexion to one of dorsiflexion, while the more vertical orientation of the human groove reflects the more sagittal movement of the human leg (Latimer et al., 1987) (Fig. 23.13).

LATERAL MALLEOLAR GROOVE

In the African apes there is normally a very noticeable groove on the lateral malleolar surface that receives the tensed posterior talofibular liagment in dorsiflexion. This groove is normally absent in humans.

The calcaneus

The calcaneus, or heel bone, is the largest bone in the pedal skeleton and is the only tarsal bone with a epiphysis. The human calcaneus is more robust than that of the African apes. It is mediolaterally wider in relation to its dorsoplantar dimension and its longitudinal axis is straighter than it is in the apes (Fig. 23.18). The ape calcaneus appears in both the dorsal and the plantar view to be angled with reference to the very well-developed **peroneal trochlea** (see below). It has already been suggested that the straight human calcaneal axis could well have arisen from an ancestral ape-like angled calcaneus by the lateral movement of the posterior part of the calcaneus. This would bring the long axis of the calcaneus more in line with the subtalar axis.

There are also a number of other features of the human calcaneus that directly reflect its role in bipedal locomotion; these features are discussed below.

PRESENCE OF THE LATERAL PLANTAR TUBERCLE

The lateral plantar tubercle is a highly variable feature on the lateral plantar surface of the human calcaneal tuberosity (Fig. 23.19). In some individuals it is completely fused to

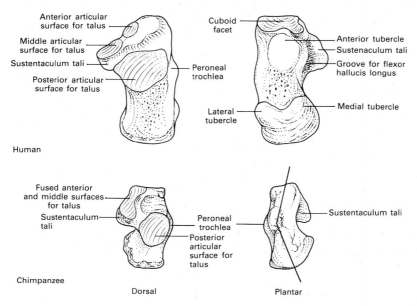

FIGURE 23·18 *The human and the chimpanzee calcaneus from the plantar and dorsal aspects. Note the angled axis of the chimpanzee calcaneus.*

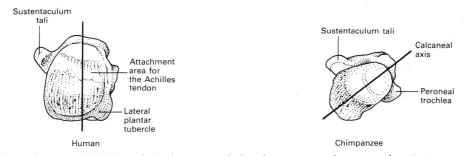

FIGURE 23·19 *Posterior view of the human and the chimpanzee calcaneus. After Latimer and Lovejoy (1989).*

the calcaneal tuberosity. In these cases the tuberosity has a quadrate form with its long axis roughly perpendicular to the ground. In other individuals the lateral plantar tubercle is largely separated from the calcaneal tuberosity, connected to it by a 'bridge' of bone. In these individuals the calcaneal tuberosity is oval with its axis oblique to the ground and resembles the condition found in the apes which totally lack a lateral plantar tubercle (Weidenreich, 1940).

MORPHOLOGY OF THE CALCANEAL TUBEROSITY

In humans the calcaneal tuberosity is clearly divided into three areas (Fig. 23.19). The most inferior area is vertically striated and ends above in an irregular ridge which marks the most inferior area of attachment for the Achilles tendon. The area immediately above this ridge is the area of main attachment for the tendon and above this is a smooth area which in life is covered by a bursa lying between bone and tendon. Although this morphology is related to body size it also reflects the role of the triceps surae (and the Achilles tendon) in lifting the heel and propelling the body forward in bipedal loco-motion (Chapter 19).

ORIENTATION OF THE SUSTENTACULUM TALI

The **sustentaculum tali** (sustentaculum (L) = prop or support) is the bracket-like support that projects medially from the body of the calcaneus and supports the talar head (Fig. 23.19). There is considerable confusion in the literature over the orientation of the sustentaculum in humans and the apes. In humans the sustentaculum is oriented perpendicular to the long axis of the calcaneal tuberosity and parallel to the ground surface. In the apes it is oriented at an acute angle to the long axis of the tuberosity (and to the ground when the long axis of the tuberosity is perpendicular to the ground). However, there is no significant difference in sustentaculum orientation between chimpanzees and humans when the bones are oriented according to their subtalar axes (Elftman and Manter, 1935b; Latimer and Lovejoy, 1989) (Fig. 23.19). Whatever differences exist between chimpanzees and humans in the orientation of the sustentaculum to the ground during locomotion result from differences in the position of the entire calcaneus and not from any change in the sustentaculum alone.

PERONEAL TROCHLEA

The peroneal trochlea is located on the lateral side of the calcaneus and separates the tendons of peroneus longus (which runs below it) and peroneus brevis (which runs above it) (Fig. 23.20). Weidenreich (1940) argued that the lateral plantar tubercle developed out of the peroneal trochlea during the course of the evolution of the human foot. This interpretation has recently been criticized by Latimer and Lovejoy (1989). However, it is true that apes, which lack a lateral plantar tubercle,

have a very large peroneal trochlea, while humans, with a lateral plantar tubercle, have a peroneal trochlea which is either small or absent.

CALCANEONAVICULAR ARTICULATION

The African apes retain the calcaneonavicular articulation as a narrow and elongated facet on the dorsal side of the calcaneous adjacent to the calcaneocuboid facet (Fig. 23.7). This morphology is similar to that found in other non-human primates and is important to the stability of the foot when it is in an inverted posture as is common in climbing. The absence of this facet in modern humans is associated with the fact that the human foot is not capable of the degree of endorotation (medial rotation of the forefoot in relation to the hindfoot) found in non-human primates.

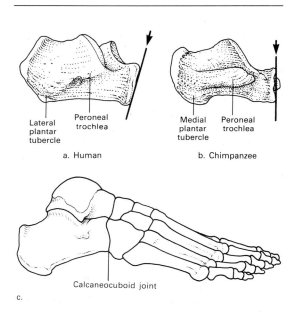

FIGURE 23·20 *Lateral view of the human (a) and the chimpanzee (b) calcaneus. Note the difference in the angles of the calcaneocuboid joint surfaces (arrows). (c) Human foot from the lateral aspect. Note the vertical position of the calcaneocuboid joint when the human calcaneus is in its proper anatomical position.*

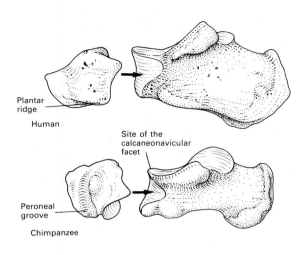

Plantar
ridge

Human

Site of the
calcaneonavicular
facet

Peroneal
groove

Chimpanzee

FIGURE 23·21 *Medial view of the human and the chimpanzee calcaneus and cuboid.*

CUBOID ARTICULAR SURFACE

The cuboid articular surface on the calcaneus is one of the most distinctive features of the human foot (Lewis, 1980b, 1981). In relation to the long axis of the calcaneus the superior margin of human cuboid facet projects further anteriorly than does the inferior margin (Fig. 23.20). The facet only assumes a vertical orientation in relation to the ground when the anterior aspect of the calcaneus is raised as it would be in a normally arched living foot. This inclination of the facet positions the cuboid in the 'keystone' position of the lateral (low) arch of the foot and, as such, directly reflects the presence of a longitudinal arch.

In humans the cuboid facet on the calcaneus is normally highly asymmetrical (Fig. 23.21). There is generally a relatively deep concavity on the medial aspect of the facet which articulates with the beak-like projection on the articular surface of the cuboid. From this medial concavity, the calcaneal articular facet flares out laterally in a moderately convex surface that has been termed the anterolateral articular process of the calcaneus (Elftman

and Manter, 1935b). The complex shape of this joint surface (together with the reciprocal shape of the cuboid articular surface) allows the distinctive lateral swing of the human calcaneocuboid joint (Fig. 23.6). This brings the joint into a close-packed position during the stance phase of the bipedal locomotor cycle and allows the mid-tarsal region of the foot to function as a rigid lever (see the previous discussion on the human arch and calcaneocuboid joint).

The apes lack the anterolateral process and in the chimpanzees the concavity in the calcaneal articular surface for the cuboid beak is more symmetrically located on the medial edge of the facet than it is in humans. In gorillas the joint surface is normally almost flat (Lewis, 1980b, 1981), however in rare instances it may resemble the chimpanzee morphology. Both these morphologies, and particularly the chimpanzee morphology, permit rotary movement at the calcaneocuboid joint, but lack the locking mechanism which is so important to human bipedal locomotion. The absence of this locking mechanism results in the 'midtarsal break' in the stance phase of the ape locomotor cycle.

The cuboid

The cuboid forms the 'keystone' of the lateral arch of the human foot (Fig. 23.20). In addition to the elongation of the human cuboid there are a number of other features that distinguish it from an ape cuboid.

THE CALCANEOCUBOID ARTICULAR FACET

The morphology of the cuboid articular facet for the calcaneus is a mirror image of the morphology of the calcaneal facet (Fig. 23.21). In humans there is a prominent projecting beak on the medial side of the facet and a laterally extending concavity that is the reciprocal of the convex anterolateral process of the calcaneus. In the chimpanzee the beak may be more centrally located and blends into the remainder of the joint surface which is either flat or slightly concave

(Bojsen-Møller, 1979; Lewis, 1980b). The lateral extension is absent. In gorillas the surface is generally flat with only the slightest indication of an elevation (beak).

THE CUBOID–METATARSAL FACETS

Anteriorly the cuboid articulates with both the fourth and the fifth metatarsals. In apes the articular facet for the fifth metatarsal is generally convex in shape (however, concave surfaces are known) (Fig. 23.22) (Lewis, 1980b). In humans the facet for the fifth metatarsal is normally either convex or sellar (saddle shaped); however, concave and flat surfaces have also been observed (Day, pers. comm.).

THE PLANTAR (INFERIOR) SURFACE

There is a well-developed plantar ridge on the inferior surface of the human cuboid that reflects the importance of the strongly developed long and short plantar ligaments in maintaining the foot in a rigid, close-packed posture during the mid-stance phase of the bipedal locomotor cycle (Fig. 23.21). This ridge is not as well developed in the ape cuboid but the adjacent peroneal groove is wider in the apes than it is in humans (Susman, 1983).

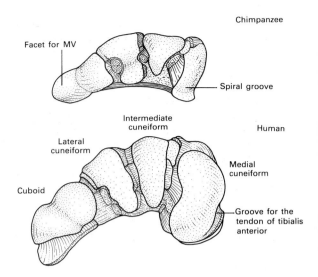

FIGURE 23·22   *Cuboid and cuneiform metatarsal joint surfaces in a chimpanzee and a human from the anterior aspect. After Lewis (1980b).*

The navicular

Of all of the tarsal bones, the navicular is perhaps the least indicative of locomotor pattern. The human navicular is distinctive only in relative size (Fig. 23.23) and in the fact that it does not articulate with the cuboid as it does in the apes. It is long

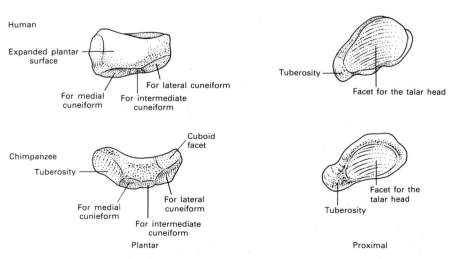

FIGURE 23·23   *Plantar and posterior views of a human and a chimpanzee navicular.*

(anteroposteriorly) in relation to the combined length of the tarsals and metatarsals as are the cuboid and the cuneiforms (Elftman and Manter, 1935b). The human navicular is also deeper (in the dorsoplantar plane) and is expanded in the area where the cubonavicular and plantar calcaneonavicular ligaments insert (Susman, 1983). Because of the more robust nature of the human navicular the tuberosity for the insertion of tibialis posterior appears proportionally smaller than it does in the African apes. The articular surface on the navicular for the medial cuneiform is also normally flat to concave in humans and convex in the apes (Susman and Stern, 1982).

The cuneiforms

All three of the human cuneiforms are elongated in relation to the tarsal and metatarsal skeleton in comparison to the ape cuneiforms and have morphological features that reflect bipedal locomotion. The medial orientation of the lateral cuneiform (and of the cuboid) has already been discussed in the context of the reorientation of the human foot around the subtalar axis. The medial orientation of these tarsal bones also affects the articulations between them. In the African apes the cuboid and the lateral cuneiform

and the lateral cuneiform and intermediate cuneiform generally articulate with each other at both their anterior and posterior ends (Fig. 23.24). Humans normally lack the anterior articulations which have been replaced by strong interosseous ligaments binding these bones together (Lewis, 1980b). It should be noted, however, that there is variation in these articulations and both humans and the African apes may have the pattern normally found in the other (Lewis, 1980b).

Of all the cuneiforms the medial cuneiform, which articulates with the first metatarsal, shows the most obvious differences in morphology between humans and apes. Specifically, the morphology and orientation of the articular facet for the first metatarsal on the anterior surface of the human medial cuneiform is markedly different from that of the apes with prehensile great toes (Fig. 23.22).

MORPHOLOGY

In prehensile-toed African apes the first tarsometatarsal articular facet is cylindrical about a vertical axis in its upper, or dorsal, half (Fig. 23.22). This convexity blends into the concave lower, or plantar, part of the joint which takes the form of a concave spiral

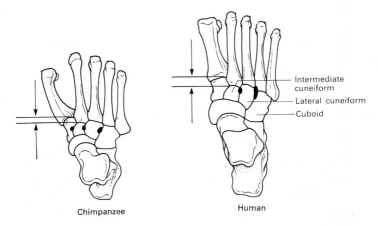

Chimpanzee Human

Intermediate cuneiform
Lateral cuneiform
Cuboid

FIGURE 23·24 *Dorsal view of a chimpanzee and a human foot. Note that in the human foot the cuboid and lateral cuneiform and the lateral and intermediate cuneiforms do not have anterior articulations. Arrows indicate the degree of indentation of the second metatarsal into the anterior tarsal row. After Lewis (1981) and Schultz (1930).*

525

groove that courses obliquely across the lower half of the joint surface and is bounded both medially and on the plantar margin by a ridge. As the ape hallux moves medially into abduction the first metatarsal not only moves medially around the axis of the cuneiform convexity but also rotates around its lower axis to the point where its lower medial margin lies in an oblique groove (Lewis, 1972c). This locks the joint into its maximally congruent, close-packed position. This morphology is found in both chimpanzees and gorillas. However, occasionally in gorillas (as in some humans) the joint surface is divided into upper and lower halves by a narrow non-articular band of bone.

In humans the first tarsometatarsal joint is normally almost flat over all of its surface and it is bordered medially by a distinct groove which guides the metatarsal tendon of tibialis anterior to its insertion on the base of the first metatarsal (Fig. 23.22). In some individuals this groove may extend onto the joint surface producing a spiral concavity similar to, but shallower than, that found in the African apes (Lewis, 1972c). In these individuals the upper, or dorsal, part of the joint may also be rather convex.

ORIENTATION
The human first tarsometatarsal joint lies at an angle of approximately 50° to the morphological axis of the foot and is thereby more anteriorly directed than is the corresponding joint in the apes (Fig. 23.25). In the African apes, and particularly in the orang-utan, the joint makes a sharp angle with the morphological axis of the foot. The mountain gorilla (*Gorilla gorilla beringei*) approaches the human condition most closely.

THE RELATIONSHIP BETWEEN THE MEDIAL AND THE INTERMEDIATE CUNEIFORM
The first tarsometatarsal articular surface (on the medial cuneiform) also extends further anteriorly than does the second tarsometatar-

sal articular surface (on the intermediate cuneiform) (Fig. 23.24). In humans the anteroposterior distance between these joint surfaces is 6.4% of total foot length, in the mountain gorilla it is 5.0%, in the lowland it

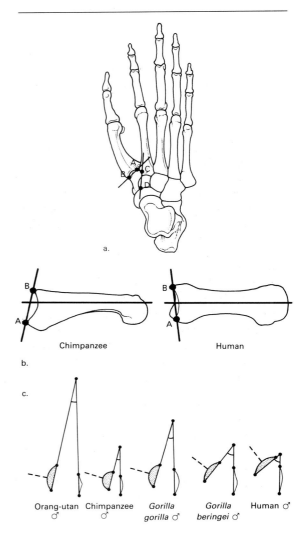

FIGURE 23·25 *Orientation and curvature of the anterior medial cuneiform joint surface (first tarsometatarsal joint) in humans and the larger apes. (a) The measurements on the medial cuneiform. (b) Dorsal view of a chimpanzee and a human first metatarsal showing the difference in the orientation of the tarsometatarsal joint surface. (c) Differences in the angle and the curvature of the first tarsometatarsal joint. After Schultz (1930).*

526

is 1.3% and in the chimpanzee it is between 1% and 2% (Schultz, 1930). The 'indentation' of the second metatarsal further into the anterior tarsal row is most probably an adaptation for stability in terrestrial locomotion in the human (and mountain gorilla) foot.

The first metatarsal
The first metatarsal has a number of morphological features that reflect the different functions of the great toe in humans and in the apes.

ORIENTATION OF THE JOINT FOR THE MEDIAL CUNEIFORM
In humans this joint lies in a plane opposite to that characteristic of the apes (Fig. 23.25). The medial edge of the human joint projects further posteriorly than does the lateral edge. This unique human pattern compensates for the anteromedial orientation of the first metatarsal articular surface on the medial cuneiform. It explains why this joint can lie at an angle to the morphological axis of the foot and the first metatarsal can still lie parallel to the second metatarsal. The opposite condition characteristic of the apes accentuates the medial projection, or abduction, of the great toe in these primates.

SHAPE AND TORSION OF THE HEAD
In humans the head of the first metatarsal has become flattened and also has rotated in relation to its base to lie squarely on the ground (see Fig. 23.4). In apes the head is both more spherical in shape and oriented more towards the other digits.

ROBUSTICITY
The human first metatarsal is also more robust than is the ape first metatarsal in relation to metatarsal length. This is particularly true of its dorsoplantar diameter where it far exceeds the robusticity found in any of the apes (Fig. 23.26). The stout nature of the human bone provides good resistance to the bending stresses particularly during the toe-

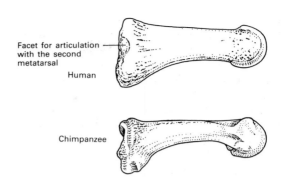

FIGURE 23·26 *Lateral view of a human and a chimpanzee first metatarsal.*

off phase of bipedal walking.

The human first metatarsal is also distinctive in normally having a facet near its base on the lateral surface for articulation with the second metatarsal (Fig. 23.26). Such facets are generally accepted as indications of a non-opposable great toe where the first metatarsal lies close to and parallel to the second metatarsal (Day and Napier, 1964; Day, 1978). However, Lewis (1980b) warned that ligamentous tuberosities are found on gorilla metatarsals and sometimes can be confused with true articular facets. It must be said, however, that gorillas are also the apes which next to humans have the least opposable great toe.

The lateral four metatarsals
The remaining metatarsals also have a number of features that reflect either human bipedalism or ape quadrupedalism. The torsion of these metatarsals together with the angled orientation of their shafts have already been discussed in the context of the reorientation of the human foot around the subtalar axis. Other distinguishing features concern the morphology of the fifth metatarsal and the shape of the metatarsal heads.

THE MORPHOLOGY OF THE FIFTH METATARSAL
The base of the human fifth metatarsal is expanded and the lateral border of this bone

traces a gentle curve as it passes from the proximal end to the shaft. In the apes the lateral border of this bone is straighter. In addition, the articular surface at the base of the fifth metatarsal is set at a more acute angle to the axis of the shaft than is the case in the apes.

MORPHOLOGY OF THE METATARSAL HEADS

In humans the articular surfaces on the heads of the metatarsals are separated from the epicondyles by a greater distance than is the case in the apes (Fig. 23.27). There is also a resulting sulcus, or depression, between the head and the shaft. This morphology is lacking in the apes. In humans it relates to an increased potential for dorsiflexion at the metatarsophalangeal joints. Such dorsiflexion is essential to human bipedalism where the posterior part of the foot 'rolls' over the toes during the toe-off phase of the stride.

The phalanges

The differences between humans and apes in the relative lengths of the pedal (foot) phalanges have already been discussed with the proportions of the foot. As with the phalanges of the hand, human pedal (foot) phalanges are also straighter than are ape phalanges (Fig. 23.28) (Stern and Susman,

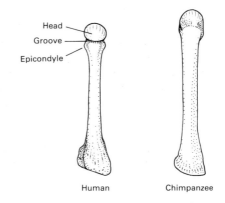

FIGURE 23·27 *The third metatarsals (MIII) of a human and a chimpanzee from the dorsal aspect.*

1983). Furthermore, in humans PPV is less curved than PPIV while in the apes the reverse is true. Other distinguishing features are found in the morphology of the proximal phalanges of all toes and distal phalanx of the great toe.

MORPHOLOGY OF THE PROXIMAL PHALANGES

The proximal phalanges of apes are characterized by broad bases (adjacent to the metatarsophalangeal joints) and well-developed flexor ridges. As in the proximal phalanges of the hand, these ridges provide the area of

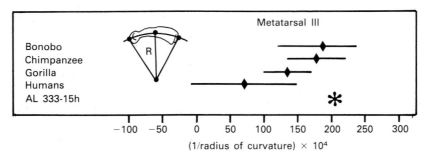

FIGURE 23·28 *Curvature of the proximal phalanges of the toes in humans, African apes and Australopithecus afarensis (AL 333). Radius of curvature = [(H-D/2)² +(L/2)²]/2(H-D/2) where L = interarticular length of the proximal phalanx, D = dorsoplantar midshaft diameter, H = height of the dorsal surface of the bone above a line connecting the centres of the anterior and posterior joint surfaces. After Stern and Susman (1983) and Susman et al. (1984).*

attachment of the flexor sheaths which are strong bands of tissue that hold the flexor tendons next to the plantar surface of the phalanges. Well-developed ridges indicate strong flexor tendons and, together with the curved phalanges, are adaptations to arboreal locomotion. In humans both the broad phalangeal bases and the well-developed flexor ridges are absent.

Human proximal phalanges also differ from ape proximal phalanges in that their anterior (distal) joint surfaces subtend smaller angles (are flatter) than is the case in apes. Furthermore, the dorsal rim of the basal articular surface exhibits a marked concavity from side to side (Stern and Susman, 1983).

THE DISTAL PHALANX OF THE GREAT TOE

The distal phalanx of the human great toe has two unique features. Firstly, it has a marked lateral deflection (towards the fibular side of the foot) of approximately 14° in relation to the plane of the interphalangeal joint (Fig. 23.29) (Wilkinson, 1954; Barnett,

1962). This deflection is completely absent in the apes. Secondly, the human distal phalanx is characterized by a marked degree of axial torsion (Fig. 23.29) (Day and Napier, 1966; Day, 1967). In relation to the plane of the interphalangeal joint, the distal (anterior) end of the phalanx is rotated laterally. As with the lateral deflection of the bone, this torsion is also absent in the apes.

Both these features of the distal phalanx are related to bipedal locomotion. During toe-off at the end of the stride phase in the bipedal locomotor cycle, the foot is angled laterally and the medial side of the distal phalanx of the great toe frequently takes the weight of the body. The lateral deflection of the bone is found not only in adults but also in human foetuses (Wilkinson, 1954), thereby providing some evidence that this feature is an inherited trait rather than a phenotypic response to bipedal locomotion.

The Australopithecus afarensis foot

The large collection of foot bones from the Hadar, which includes an associated ankle joint (AL 288-1) and 13 phalanges from a single foot, provides by far the best evidence for the pedal morphology of any species of _Australopithecus_ or _Paranthropus_ (Johanson et al., 1982; Latimer et al., 1982). Other pedal bones relating to these genera that have been discussed in the literature are the TM 1517 _Paranthropus robustus_ talus from Kromdraai, South Africa (Broom and Schepers, 1946; Le Gros Clark, 1947a,b; Day and Wood, 1968; Robinson, 1972) and the recently described _P. robustus_ pedal bones from Swartkrans, South Africa (Susman and Brain, 1988; Susman 1988b). McHenry and Temerin (1979) and Day (1978) also give brief mention of other australopithecine and paranthropine foot bones.

The _Australopithecus afarensis_ foot bones show a mosaic of features. Some are more similar to those found in the human bipedal foot while others are reminiscent of the ape foot.

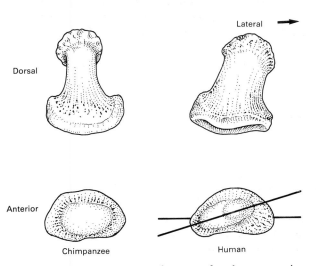

FIGURE 23·29 _Dorsal view of a human and a chimpanzee first distal phalanx. Note the lateral deflection of the human bone. Anterior view of a human and a chimpanzee first distal phalanx. Note the axial torsion in the human bone._

529

Features that suggest bipedal locomotion

MORPHOLOGY OF THE ANKLE JOINT

Although the *Australopithecus afarensis* talus gives the general appearance of being squat and ape-like it shows a number of features which, together with the associated distal tibia and fibula, suggest a human-like ankle joint (Latimer et al., 1987) (Chapter 22). The most important of these features are: (1) the shape of the talar trochlea; (2) the orientation of the axis of the ankle joint; and (3) the orientation of the groove for the flexor hallucis longus (Fig. 23.30).

As in humans the radius of curvature of the medial border of the talar trochlea is closer to that of the lateral border than is the case in apes. This morphology, together with the increased height of the medial border of the trochlea, reduces both the rotation of, and arc inscribed by, the tibia as it moves from a position of plantarflexion to one of dorsiflexion. The result is that the *A. afarensis* leg, like the human leg, would move in a more sagittal plane in relation to the foot. These features are also related to the axis of rotation of the ankle joint which has been experimentally determined by Latimer et al. (1987) to have been more parallel to the plane of the trochlea than is the case in the apes.

The position of the tibia vertically over the talus is also related to the human-like vertical orientation of the groove for flexor hallucis longus which is in marked contrast to the oblique groove found in the tali of the African apes (see Fig. 23.11).

ABILITY TO DORSIFLEX THE TOES

The heads of the second to the fifth metatarsals show a human-like pattern where the joint surfaces are well separated from the epicondyles by a sulcus (Fig. 23.30). This allows a greater degree of dorsiflexion at the metatarsophalangeal joints than in the apes and is consistent with bipedalism. The fifth metatarsal is also human-like in its expanded

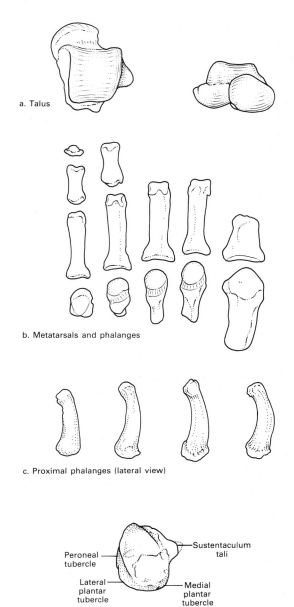

a. Talus

b. Metatarsals and phalanges

c. Proximal phalanges (lateral view)

Peroneal tubercle

Sustentaculum tali

Lateral plantar tubercle

Medial plantar tubercle

d. Calcaneus (posterior view)

FIGURE 23·30 Australopithecus afarensis *foot bones. (a) Dorsal and anterior views of the talus (AL 288-1as), (b) dorsal view of the metatarsals and phalanges (AL 333-115), (c) lateral view of the proximal phalanges (AL 333-115), (d) posterior view of the calcaneus (AL 333-55). After Johanson* et al. *(1982).*

base but is reminiscent of the apes in its straight, rather than curved, lateral edge.

DEVELOPMENT OF THE LONGITUDINAL ARCH

Two features have been suggested by Stern and Susman (1983) to indicate at least an incipient development of the longitudinal arch. Firstly, although the *Australopithecus afarensis* navicular bones are narrow in the dorsoplantar plane and, therefore, are reminiscent of ape naviculars, they have very well-developed insertions for the plantar cubonavicular (spring) ligament and a human-like groove for the calcaneonavicular ligaments. Both these features are important in the support of the human longitudinal arch. Secondly, the lateral cuneiform is elongated in a human-like fashion, suggesting that the anterior tarsal skeleton has been remodelled in a manner consistent with bipedal locomotion.

Primitive features of the Australopithecus afarensis *foot*

THE PHALANGES

Stern and Susman (1983) described the phalanges of the second to fifth digits as being strikingly pongid-like in morphology (Fig. 23.30). In relation to the size of the femoral head the phalanges are longer than would be expected in humans and they also show a greater radius of curvature (although PPV is less curved than the others as is the case in humans). Both these features can be associated with arboreal locomotion. In addition, the second proximal phalanx resembles the apes in being shorter than the third proximal phalanx. Furthermore, the middle phalanges are longer in relation to the proximal phalanges than is the case in humans. Indeed, all the proximal phalanges have a distinct ape-like morphology. They have expanded bases, well-developed flexor ridges and distal (anterior) trochlea that are larger in their dorsoplantar dimension than is the case in humans.

THE CALCANEUS

Although the calcaneus has a wide tuberosity (human like), it also has features that are similar to those seen in ape calcanei (Fig. 23.30). These include an oval orientation of the tuberosity, a massive peroneal tubercle, and a ridge which runs obliquely from the large peroneal trochlea to an indistinct lateral plantar tubercle (Stern and Susman, 1983; Latimer and Lovejoy, 1989).

THE FIRST TARSOMETATARSAL ARTICULATION

The articular facet on the medial cuneiform for the first metatarsal is damaged but has been described by Latimer et al. (1982) as 'markedly convex' as is the case in apes with an opposable great toe. In addition, although the lateral cuneiform is elongated (human like) its plantar tuberosity has a hook-like form which is typical of apes rather than running the anteroposterior length of the bone as is the case in humans (Stern and Susman, 1983).

THE FIRST METATARSAL

The first metatarsal has an ape-like rounded head rather than a human-like flattened head (Stern and Susman, 1983; Susman, 1983).

The foot of Paranthropus robustus

The foot of *Paranthropus robustus* is represented by a few foot bones from the site of Swartkrans, southern Africa (Susman, 1988b). The most informative of these is a complete left first metatarsal (SKX 5017) from Bed I (Susman and Brain, 1988). This bone has a number of distinctly human features (Fig. 23.31). Firstly, the posterior (proximal) articular surface is flat suggesting that the hallux was adducted in line with the remaining toes rather than abducted as it is in the living apes. Secondly, the metatarsal base is expanded inferiorly indicating the presence of well-developed plantar ligaments as in the human foot. Thirdly, the bone is

531

more robust than non-human first metatarsals indicating that it supported a greater amount of weight in locomotion than the ape MI. Fourthly, the bone lacks the degree of torsion found in ape first metatarsals. Moreover, the articular surface on the head wraps around onto the back, or dorsum, of the bone. This would have allowed the human-like dorsiflexion of the great toe during bipedal locomotion.

Although these features are consistent with a striding bipedal gait, there is an additional ape-like feature that suggests that bipedalism in *P. robustus* would have lacked the human-like toe-off mechanism at the end of the stance phase of the bipedal walking cycle. The head of the *P. robustus* MI has an ape-like outline being narrower dorsally than the human metatarsal head. Susman and Brain (1988) argue that this would have precluded close packing of the metatarsophalangeal joint when the joint was in extreme dorsiflexion. They argue further that without the stability of the close-packed joint it would have been unlikely that weight was shifted to the medial side of the foot and fully onto the great toe during the stance phase of the cycle. It therefore seems probable that, although *P. robustus* was bipedal, bipedalism in this hominid involved a different pattern of weight transfer than it does in modern humans.

The foot of Homo habilis

The best known and most complete set of foot bones that have been referred to the taxon *Homo habilis* are the OH 8 bones from Olduvai Gorge (Fig. 23.32). OH 8 represents the bones of a left foot that were found in Bed I at the site of FLK NN in 1960. These bones were originally described as adult (Day and Napier, 1964). However, since then they have been interpreted as juvenile (Susman and Stern, 1982). All the tarsal bones and metatarsals are present; however, the phalanges are missing along with the posterior part of the calcaneus, the metatarsal heads and the styloid process of the fifth metatarsal.

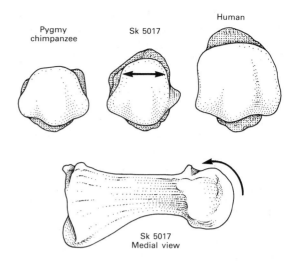

FIGURE 23·31 *(a) Anterior view of the head of the first metatarsal in a pygmy chimpanzee,* Paranthropus robustus *(SK 5017) and a human. The arrow indicates the dorsally narrow head on SK 5017. (b) Medial view of the* Paranthropus robustus *(SK 5017) first metatarsal. The arrow indicates the dorsal extent of the anterior articular surface. After Susman and Brain (1988).*

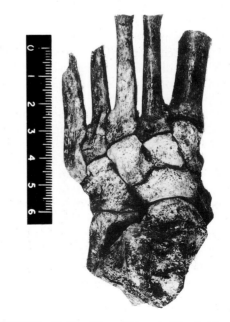

FIGURE 23·32 *The Olduvai Hominid 8 foot. Courtesy of M.H. Day.*

Since the initial description (Day and Napier, 1964), the OH 8 foot has generated a substantial amount of controversy. Some have agreed with Day and Napier (1964) and argued that its principal affinities are with *Homo sapiens* (Day and Wood, 1968; Susman, 1983), while others have argued that it is more similar to the feet of monkeys and/or apes (Lisowski, 1967; Oxnard, 1972a,b; Lisowski et al., 1974, 1976; Lewis, 1980c, 1981). The protagonists agree that some of the features of OH 8 are consistent with bipedalism, while others are strikingly conservative, the consensus opinion being that although the foot is from a bipedal individual the bipedalism of this individual differed to a greater or lesser degree from that of modern humans.

Some have also argued that the foot is so different from that of modern bipeds that it should be excluded from the genus *Homo* and referred to the genus *Australopithecus* (Wood, 1974a,b; Oxnard, 1975; Lewis, 1980c, 1981). The foot was not directly associated with *H. habilis* cranial or dental material, but was initially assigned to *Homo* largely on the strength of the later debated conclusion that its closest affinities were with modern humans (Day and Napier, 1964). Important to the question of taxonomic affinity of the OH 8 foot is the morphology of its talus (see below) and the existence of the contemporary KNM-ER 813 talus from Koobi Fora. The KNM-ER 813 talus has been assigned to the genus *Homo* (Leakey, 1972). Both anatomical (Leakey and Wood, 1973) and metrical analyses (Fig. 23.17) (Wood, 1974a,b) have shown that this fossil is much more similar to modern human tali than is the talus of the OH 8 foot and, therefore, has a 'better claim to belong to a *Homo* foot' than does the OH 8 talus (Wood, 1974a: 375).

Affinities of the OH 8 foot

METATARSAL ROBUSTICITY
The least disputed evidence for the affinity between OH 8 and modern humans comes from the anterior part of the foot and particularly from the morphology of the metatarsals. The robusticity pattern of the metatarsals is similar to humans in that the fifth metatarsal is the second most robust after the first metatarsal. In apes the fifth metatarsal is the least robust (Table 23.2) (Day and Napier, 1964; Susman, 1983). The metatarsal robusticity pattern of OH 8 (1>5>3>4>2) differs slightly from the normal human pattern (1>5>4>3>2); however, Archibald et al. (1972) have shown that the OH 8 pattern is found in a small percentage of modern human feet.

RELATIVE FOOT LENGTH
The relative length of the OH 8 foot is similar to the relative length of the human foot and much shorter than the relative length of the ape foot (Susman and Stern, 1982). In this comparison, foot length (from the anterior surface of the navicular to the epicondyle of the second metatarsal) is relative to the length of the tibia. The OH 8 foot is compared with the OH 35 tibia which is considered by Susman and Stern (but not by Day (personal communication)) to have belonged to the same juvenile individual as the OH 8 foot.

ADDUCTION OF THE HALLUX
There is strong evidence that the OH 8 hallux (big toe) was adducted and did not show the degree of opposability found in ape feet (Fig. 23.33). There are three pieces of evidence to support this. Firstly, there is a contact facet between the first and second metatarsals near their bases that indicates the presence of at least a rudimentary synovial joint between the two approximated bones (Day and Napier, 1964; Day, 1978). Secondly, the articular surface for the first metatarsal on the medial cuneiform is both less convex and more anterior facing than is the case in the apes (Susman, 1983). Thirdly, there is a single articulation between the medial and intermediate cuneiforms as in humans rather than

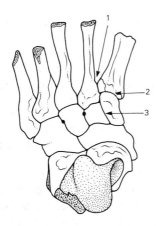

FIGURE 23·33 *Dorsal view of the Olduvai Homi-nid 8 foot. Note (1) the contact between the first and second metatarsals, (2) the relatively flat first tarsometatarsal joint, and (3) the single articulation between the medial and intermediate cuneiforms. Compare with Figure 23.24. After Lewis (1981).*

the dual articulations of the apes (Lewis, 1980c).

However, there is also evidence that the hallux may have had a certain degree of grasping function over and above what is present in modern humans. The articular surface on the medial cuneiform for the first metatarsal is similar in structure to the apes and different from humans. This similarity lies in the convex upper part of the joint which is confluent with the concave medial part of the joint. This joint shape allows the first metatarsal to be 'screwed' into a close-packed articulation with the medial cuneiform which in a grasping foot would stabilize the joint (Lewis, 1972c). When this joint is in the close-packed position there is some divergence of the hallux that 'is quite in accord with some residual grasping function' (Lewis, 1980c: 293).

TORSION OF THE METATARSALS
There has also been some dispute over the significance of the torsion of the first metatarsal. This refers to the lateral rotation of the head of the metatarsal in relation to the base to face more towards the ground

rather than towards the second digit (Elftman and Manter, 1935b). Susman (1983) argued that the OH 8 first metatarsal has been rotated to the extent that it occupies a more dorsoplantar neutral position than is found in the apes. The OH 8 first metatarsal shows a degree of rotation that is possibly less than that found in the gorilla and certainly not of the same magnitude as that found in humans. Lewis (1980c) concluded that this would be expected if the OH 8 metatarsal still retained a degree of grasping function as indicated by the form of the anterior medial cuneiform joint surface.

TORSION OF THE LATERAL FOUR METATARSALS
Whilst the direction of torsion is the same in the human and ape first metatarsal it is different in the more lateral metatarsals. In the apes the second metatarsal (and to a lesser extent the remaining metatarsals) are rotated to face the first metatarsal, while in humans they are clearly rotated in the opposite direction to face the ground (Fig. 23.4) (Lewis, 1980c). The human pattern ensures that the heads of the metatarsals contact the ground in spite of the well-developed transverse arch of the human foot. While it is not possible to determine the torsion of the OH 8 fourth and fifth metatarsals, the second and third ones show the human direction and degree of torsion.

THE ANTERIOR INTERTARSAL AND TARSOMETATARSAL JOINTS
The second metatarsal shows a human articulation in the OH 8 foot. Its base is wedged between the lateral and the medial cuneiforms and has facets for the articulation with these bones (Fig. 23.33). However, the remaining anterior intertarsal and tarsometatarsal joints of the OH 8 foot are conservative in form and are most similar to those found in the apes (Lewis, 1980c). In particular, the anterior part of the articulation between the intermedi-ate cuneiform and lateral cuneiform and between the lateral cuneiform and cuboid are

present as they are in the apes. In humans these articulations are absent and interosseous ligaments appear in their places (Lewis, 1980b,c) (compare Fig. 23.24). Furthermore, the cuboid articulation for the fifth metatarsal is concave. In humans this joint is normally either convex or saddle shaped. In apes it is normally convex; however, concave surfaces have been observed (Lewis, 1980b,c). This facet is also at a sharper angle to the MIV facet than is normally the case in humans.

THE LONGITUDINAL ARCH

The presence or absence of a longitudinal arch in the OH 8 foot is controversial. Day and Napier (1964) argued for the presence of both a transverse and a longitudinal arch primarily on the basis of well-developed markings for the ligaments and tendons that are associated with the static and dynamic support of the tarsal arches. Furthermore, they noted that the robustness of the lateral metatarsals indicate that they were subject to a greater amount of weight than is the case in other primates.

Susman (1983) has reaffirmed this interpretation on the basis of the morphology of the inferior surface of the navicular which is inflated at the point of insertion of the plantar calcaneonavicular and cubonavicular ligaments, both of which are important in humans in the maintenance of the longitudinal arch. However, the morphology of the cuboid apparently does not support the existence of a human-type longitudinal arch. Lewis (1981) argued that this bone is bent dorsally in OH 8 as it is in apes. In humans it has a distinct plantar bend that is consistent with its position as the 'keystone' of the lateral part of the human longitudinal arch. Moreover, there is a distinct articular surface for the navicular. This surface is present in apes and normally absent in humans. It is matched in OH 8 by an equivalent articular surface on the navicular for the cuboid.

Furthermore, Oxnard and Lisowski (1980) argued that the reconstructed arch in the generally available casts of the OH 8 foot is reconstructed artificially high and is, in fact, anatomically incorrect (Fig. 23.34). They asserted that when the tarsal bones are reconstructed in what they believe to be the correct anatomical positions, the transverse arch closely resembles the condition in the ape. These authors implied that the resemblance between the OH 8 foot and the ape foot comes from the fact that both lacked the longitudinal arch characteristic of the human foot.

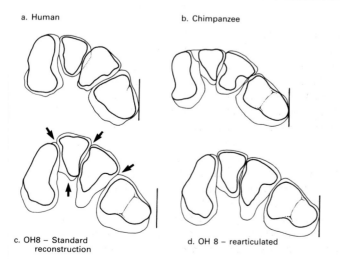

a. Human   b. Chimpanzee

c. OH8 – Standard reconstruction   d. OH 8 – rearticulated

FIGURE 23·34 *The transverse arch in (a) a human and (b) a chimpanzee. The standard reconstruction of the OH 8 transverse arch (c) is claimed to be anatomically incorrect in the articulations of the intermediate and lateral cuneiforms (arrows). When the bones are rearticulated the transverse arch resembles the lower chimpanzee arch rather than the higher human arch. After Oxnard and Lisowski (1980).*

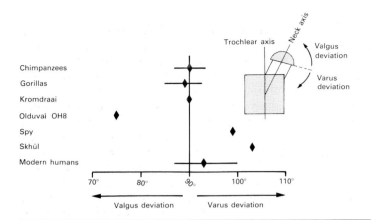

FIGURE 23·35 *The orientation of the talar head in relation to the talar neck in modern humans, the African apes and selected fossil hominids. After Day and Wood (1968).*

THE CALCANEOCUBOID JOINT

The OH 8 cuboid resembles the human bone in having a flange on the inferomedial side of the bone that articulates with an opposing concavity on the anterior face of the calcaneus. The flange in the apes is located in a more medial position. Lewis (1980c) argued that this similarity between OH 8 and modern humans is superficial and that the movement possible at this joint is quite different. In particular, the lateral expansion of the OH 8 joint is less than in humans and the joint rotates into the close-packed position without allowing the lateral deviation of the heel that is so important in human bipedalism.

Lewis (1980c) emphasized that the medial position of the cuboid flange gives the OH 8 joint a much greater stability than is possible in the ape joint where the flange is more centrally located. The important difference is that the eccentric OH 8 flange wedges itself under the sustentaculum tali. Therefore, the OH 8 joint resembles the human one in producing a stable calcaneocuboid joint (Susman, 1983) but differs from the human condition in not also allowing the calcaneus to swing laterally and thereby tense the plantar ligaments and provide additional support for a longitudinal arch. This is another feature that suggests that the OH 8 foot did not possess a well-developed longitudinal arch and did not function in a manner identical to the human foot in bipedal locomotion.

THE TALUS

The talus is the single part of the OH 8 foot that has caused the most doubt in relation to its close affinity with modern human bipeds. It resembles the ape talus in being squat and foreshortened (Lewis, 1981). Its neck and torsion angles are also similar to the apes and to the paranthropine talus from Kromdraai and quite different from the human condition (Lisowski, 1967). The wide neck angle suggests that the trochlea was not medially rotated as it is in humans and the torsion angle suggests that the medial border of the trochlea may have been low in relation to the lateral border as it is in the apes (see above). In addition, the subtalar axis is relatively flat and at a large angle to the axis of the foot as a whole, while the head of the talus is unique among modern primate tali in not sitting symmetrically on the talar neck but being rotated laterally (Fig. 23.35). A metric comparison of the OH 8 talus, the Kromdraai talus and tali of bipeds and quadrupeds underscores the unique morphology of the OH 8 talus (Day and Wood, 1968).

Summary

There is no doubt that many of the features of the OH 8 foot are consistent with the requirements of bipedal locomotion. These features include the metatarsal robusticity pattern, the length of the foot relative to the length of the tibia and the morphology of the MI that suggests that the hallux was adducted in line with the other toes (at least to a greater extent than that found in most living apes). Most importantly, they also include the morphology of the calcaneocuboid joint which would have prevented the mid-tarsal break during the stance phase of the walking cycle. However, other features of the OH 8 foot are much more ape-like in morphology. There is evidence that the subtalar axis was at an ape-like large angle to the long axis of the foot (dorsally) and that the calcaneocuboid joint would not have allowed the calcaneus to swing laterally during the mid-stance phase of the walking cycle and thereby tense the plantar ligaments supporting a longitudinal arch. Moreover, the anterior intertarsal articulations are ape-like in lacking the strong interosseous ligaments that bind these bones together in humans and the talus is squat and foreshortened with neck and torsion angles that are most similar to those found in the apes. These features suggest that the longitudinal arch may have been absent and that the weight transfer through the OH 8 foot in both standing and walking was different than it is in the modern human foot.

Susman and Brain (1988) argued on the basis of the morphology of the first metatarsal that *Homo habilis* (as represented by the OH 8 foot) had reached the same grade of bipedalism as had *Paranthropus robustus* (as represented by the SKX 5017 first metatarsal) (see above). This stage of bipedalism lacked the transfer of weight to the medial side of the foot and fully onto the great toe during the final half of the stance phase of the walking cycle. This interpretation (based on the MI) is consistent with the functional conclusions that can be derived from the other bones of the OH 8 foot.

The Neanderthal foot

The most complete Neanderthal foot fossils are La Ferrassie (1 and 2), Kiik-Koba 1, Shanidar 1 (Fig. 23.36) and Tabun C1 (Trinkaus, 1983a). Analysis of these fossils, together with a much larger number of less complete fossils from both Europe and south western Asia, suggests that the Neanderthal foot was fully adapted to bipedal locomotion (Trinkaus, 1983a,b). Neanderthal feet differ from modern human feet only in their greater robusticity and in the relative proportions of the phalanges of their great toes. This interpretation is contrary to earlier opinions which suggested that Neanderthals may have had at least a partially opposable great toe (Boule, 1911–13; Morton, 1926) and that the Neanderthal talus differs significantly from that of modern humans (Oxnard, 1972a, 1973a,b; Lisowski et al., 1974; but see Rhoads and Trinkaus, 1977).

The robusticity of the Neanderthal foot

Trinkaus (1983a) has highlighted three main features that suggest a greater degree of robusticity in the Neanderthal foot than is commonly found in the modern human foot. Firstly, many areas of attachment for the plantar muscles and ligaments are more strongly developed in the Neanderthal foot than they are in the modern human foot. These areas include the medial process of the calcaneus, the navicular tuberosity, the

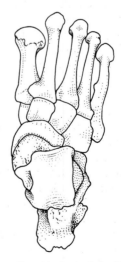

FIGURE 23.36 *Dorsal view of the Shanidar I right foot. After Trinkaus (1983a).*

537

tuberosity of the lateral cuneiform and the areas on the bases of the proximal phalanges for the attachment of both muscles and ligaments. Secondly, the articular surfaces on both the talus and the calcaneus are expanded. In particular there is a high frequency of fusion of the medial and anterior talocalcaneal facets which is associated with expansion of these surfaces (Trinkaus, 1983a). Thirdly, the first metatarsals and all five proximal phalanges are more robust than is the norm for modern humans.

In addition, the Neanderthal proximal phalanges are generally wider than they are high, while in modern humans they are higher than they are wide. Trinkaus (1983a) suggested that this is related to a more extreme mediolateral bending stress than is common in modern humans.

Relative proportions of the phalanges of the great toe

The Neanderthal first proximal phalanx is short relative both to medial subtalar length (calcaneal tuberosity to the head of the first metatarsal) and to the length of the distal phalanx. A short proximal phalanx is also characteristic of the Neanderthal thumb (Chapter 18).

Opposability of the great toe

The joint surface for the first metatarsal on the medial cuneiform is generally convex in Neanderthals. This has led some authors to suggest that the great toe in Neanderthals had a degree of opposability intermediate between the living apes and modern humans (Boule, 1911–13; Morton, 1926). However, Trinkaus (1983a) has pointed out the following features that suggest that the Neanderthal great toe was fully adducted, lacking any

significant degree of opposability. Firstly, the morphology of the first tarsometatarsal joint is highly variable in modern humans. Secondly, the concavoconvex joint surface would act as an effective buttress against the torsional forces generated in bipedal locomotion. The specific morphology of the first metatarsal where the concavity results from an expansion of the dorsomedial and plantolateral articular margins would result in a close-packed joint with tightened ligaments during the toe-off phase of the stride. Thirdly, there is a high frequency of articular facets between the bases of the first and second metatarsals; and lastly, the angle of torsion of the first metatarsal is indistinguishable from modern humans. Trinkaus (1983a) argued that the greater degree of first metatarsal torsion in apes gives the metatarsophalangeal joint a more mediolateral axis. This results in the great toe opposing the phalanges of the other digits rather than lying plane to the ground.

The talus

The main feature distinguishing the Neanderthal talus is an apparently relatively short neck length (Trinkaus, 1983a). However, neck length is measured from the edge of the trochlea. Rhoads and Trinkaus (1977) argued that the apparent short neck length in Neanderthals is a direct result of the hypertrophy of the trochlear surface. It is not an absolute reduction and has no functional significance. Other features that distinguish the Neanderthal talus are an expansion of the lateral malleolar surface and a slightly greater height of the bone (Rhoads and Trinkaus, 1977). Both these features reflect the greater overall robustness of the Neanderthal postcranial skeleton (Rhoads and Trinkaus, 1977; Trinkaus, 1983a).

REFERENCES

Abitbol, M. M. (1987a). Evolution of the lumbos-acral angle. *American Journal of Physical Anthropology* **72**: 361–372.

Abitbol, M. M. (1987b). Evolution of the sacrum in hominoids. *American Journal of Physical Anthropology* **74**: 65–81.

Abitbol, M. M. (1988a) Effect of posture and locomotion on energy expenditure. *American Journal of Physical Anthropology* **77**: 191–200.

Abitbol, M. M. (1988b). Evolution of the ischial spine and of the pelvic floor in the Hominoidea. *American Journal of Physical Anthropology* **75**: 53–67.

Adams, L. M. and Moore, W. J. (1975). Biomechanical appraisal of some skeletal features associated with head balance and posture in the hominoidea. *Acta Anatomica* **92**: 580–594.

Aiello, L. C. (1981a). On analysis of shape and strength in the long bones of higher primates. Ph.D. thesis, University of London.

Aiello, L. C. (1981b). The allometry of primate body proportions. In: *Vertebrate Locomotion* (M. H. Day, Ed.), *Symposia of the Zoological Society of London*, pp. 331–358. London: Academic Press.

Aiello, L. C. (1981c). Locomotion in the miocene hominoidea. In: *Aspects of Human Evolution* (C. B. Stringer, Ed.), pp. 63–98. London: Taylor & Francis.

Aitchison, J. (1960). The unilateral distribution of epipteric bones. *British Dental Journal* **109**: 55–59.

Alexander, R. McN. (1980). Optimum walking techniques for quadrupeds and bipeds. *Journal of Zoology* **192**: 97–117.

Alexander, R. McN. (1984a). Walking and running. *American Scientist* **72**: 348–354.

Alexander, R. McN. (1984b). Stride length and speed for adults, children, and fossil hominids. *American Journal of Physical Anthropology* **63**: 23–27.

Andrews, P. (1984). An alternative interpretation of characters used to define *Homo erectus*. In: *The Early Evolution of Man* (P. Andrews and Frazen, Eds), *Courier Forschungs Institut Senckenberg* **69**: 167–175.

Andrews, P. (1985). Family group systematics and evolution among catarrhine primates. In: *Ancestors: The Hard Evidence* (E. Delson, Ed.), pp. 14–22. New York: Alan R. Liss.

Andrews, P. (1986a). Fossil evidence on human origins and dispersal. *Cold Spring Harbor Symposia on Quantitative Biology* **51**: 419–428.

Andrews, P. (1986b). Molecular evidence for catarrhine evolution. In: *Major Topics in Primate and Human Evolution* (B. Wood, L. Martin, and P. Andrews, Eds.), pp. 107–129. Cambridge: Cambridge University Press.

Andrews, P. and Cronin, J. E. (1982). The relationships of *Sivapithecus* and *Ramapithecus* and the evolution of the orang-utan. *Nature* **297**: 541–546.

Andrews, P. and Martin, L. (1987). Cladistic relationships of extant and fossil hominoids. *Journal of Human Evolution* **16**: 101–118.

Ankel, F. (1967). Morphologie von Wirbelsäule und Brustkorb. *Primatologia* **4**: 1–120.

Ankel, F. (1972). Vertebral morphology of fossil and extant primates. In: *The Functional and Evolutionary Biology of the Primates* (R. H. Tuttle, Ed.), pp. 223–240. Chicago: Aldine-Atherton.

Anton, S. C. (1989). Intentional cranial vault deformation and induced changes of the cranial base and face. *American Journal of Physical Anthropology* **79**: 253–267.

Anon. (1976). hominid remains from Hadar, Ethiopia. *Nature* **260**: 389.

Archibald, J. D., Lovejoy, C. O. and Heiple, K. G. (1972). Implications of relative robusticity in the Olduvai metatarsus. *American Journal of Physical Anthropology* **37**: 93–96.

Arensburg, B., Tillier, A. M., Vandermeersch, B., Duday, H., Schepartz and Rak, Y. (1989). A middle palaeolithic human hyoid bone. *Nature* **338**: 758–760.

Armstrong, E. and Onge, M. (1981). Number of neurones in the anterior thalamic complex of the primate limbic system. *American Journal of Physical Anthropology* **54**: 197 Abstract.

Asfaw, B. (1985). Proximal femur articulation in Pliocene hominids. *American Journal of Physical Anthropology* **68**: 535–538.

Ashcroft, M. T., Heneage, P. and Lovell, H. G. (1966). Heights and weights of Jamaican schoolchildren of various ethnic groups. *American Journal of Physical Anthropology* **24**: 35–44.

Ashley-Montagu, M. F. (1933). The anthropological significance of the pterion in primates. *American Journal of Physical Anthropology* **18**: 159–336.

Ashley-Montagu, M. F. (1943). The mesethmoid–presphenoid relationships in primates. *American Journal of Physical Anthropology* **1**: 129–141.

Ashton, E. H. (1957). Age changes in the basicranial axis of the anthropoidea. *Proceedings of the Zoological Society of London* **129**: 61–74.

Ashton, E. H. and Oxnard, C. E. (1963). The musculature of the primate shoulder. *Transactions of the Zoological Society of London* **29**: 553–650.

Ashton, E. H. and Oxnard, C. E. (1964). Functional Adaptations in the Primate Shoulder Girdle. *Proceedings of the Zoological Society of London* **142**: 49–66.

Ashton, E. H. and Spence, T. F. (1958). Age changes in the cranial capacity and foramen magnum of hominoids. *Proceedings of the Zoological Society of London* **130**: 169–181.

Ashton, E. H. and Zuckerman, S. (1952). Age changes in the position of the occipital condyles in the chimpanzee and gorilla. *American Journal of Physical Anthropology* **10**: 227–288.

Ashton, E. H. and Zuckerman, S. (1956). Age changes in the position of the foramen magnum in hominoids. *Proceedings of the Zoological Society of London* **126**: 315–325.

Ashton, E. H. and Zuckerman, S. (1954). The anatomy of the articular fossa (fossa mandibularis) in man and apes. *American Journal of Physical Anthropology* **12**: 29–61.

Ashton, E. H., Flinn, R. M., Oxnard, C. E. and Spence, T. F. (1976). The adaptive and classificatory significance of certain quantitative features of the forelimb in primates. *Journal of Zoology* **17**: 515–550.

Baker, M. A. (1982). Brain cooling in endotherms in heat and exercise. *Annual Reviews in Physiology* **44**: 85–96.

Baldwin, H. (1989). The thumb in human evolution. B.Sc. Thesis, University College London.

Barnett, C. H. (1953). Locking at the knee joint. *Journal of Anatomy* **87**: 91–95.

Barnett, C. H. (1954). Squatting facets on the European talus. *Journal of Anatomy* **88**: 509–513.

Barnett, C. H. (1962). Valgus deviation of the distal phalanx of the great toe. *Journal of Anatomy* **96**: 171–177.

Barnett, C. H. and Napier, J. (1952). The axis of rotation at the ankle joint in man. Its influence upon the form of the talus and the mobility of the fibula. *Journal of Anatomy* **86**: 1–9.

Barnett, C. H. and Napier, J. R. (1953). The rotary mobility of the fibula in eutherian mammals. *Journal of Anatomy* **87**: 11–21.

Bartstra, G. J. (1982). River-laid strata near Trinil, Java. *Modern Quaternary Research in South East Asia* **7**: 97–130.

Bar-Yosef, O., Vandermeersch, B., Arensberg, B., Goldberg, P., Laville, H., Meignen, L., Rak, Y., Tchernov, E. and Tillier, A.-M. (1986). New data on the origin of modern man in the Levant. *Current Anthropology* **27**: 63–64.

Basmajian, J. V. (1975). *Grant's Method of Anatomy*, 9th edition. Baltimore: The Williams and Wilkins Company.

Basmajian, J. V. and De Luca, C. J. (1985). *Muscles Alive*, 5th edition. Baltimore: The Williams and Wilkins Company.

Basmajian, J. V. and Latif, A. (1957). Integrated actions and functions of the chief flexors of the elbow: A detailed electromyographic analysis. *Journal of Bone and Joint Surgery* **39A**: 1106–1118.

Batson, O. V. (1953). The temporalis muscle. *Oral Surgery, Oral Medicine and Oral Pathology* **6**: 40–57.

Baycroft, S. (1988). Morphological features that influence movement in the temporomandibular joint of the extant and fossil hominids and the great apes. M.Sc. Thesis, University College London.

Beaumont, P. B., de Villiers, H. and Vogel, J. C. (1978). Modern man in sub-Saharan African prior to 49,000 BP,; a review and evaluation with particular reference to Border Cave. *South African Journal of Science* **74**: 409–419.

Behrensmeyer, A. K. and Laporte, L. F. (1981). Footprints of a Pleistocene hominid in northern Kenya. *Nature* **289**: 167–169.

Bellugue, P. (1962). *Introduction a l'Etude de la Forme Humaine: Anatomie Plastique et Méchanique*. Paris: Edition Posthume à Compte d'Auteur.

Ben-Itzhak, S., Smith, P. and Bloom, R. A. (1988). Radiographic study of the humerus in Neanderthals and *Homo sapiens sapiens*. *American Journal of Physical Anthropology* **77**: 231–242.

Berge, C. and Kazmierczak, J.-B. (1986). Effects of size and locomotor adaptations on the hominid pelvis: evaluation of australopithecine bipedality with a new multivariate method. *Folia Primatologia* **46**: 185–204.

Berge, C., Orban-Segebarth, R. and Schmid, P. (1984). Obstetrical interpretation of the australopithecine pelvic cavity. *Journal of Human Evolution* **13**: 573–587.

Berkowitz, B. K. B., Holland, G. R. and Moxham, B. J. (1978). *A Colour Atlas of Oral Anatomy*. London: Wolfe Medical Publications.

Beynon, A. D. and Dean, M. C. (1987). Crown formation time of a fossil hominid premolar tooth. *Archives of Oral Biology* **32**: 773–780.

Beynon, A. D. and Dean, M. C. (1988). Distinct dental development patterns in early fossil hominids. *Nature* **335**: 509–514.

Beynon, A. D. and Reid, D. J. (1987). Relationships between perikymata counts and crown formation times in the human permanent dentition. *Journal of Dental Research* **66**: 889–890.

Beynon, A. D. and Wood, B. A. (1986). Variations in enamel thickness and structure in East African hominids. *American Journal of Physical Anthropology* **70**: 177–195.

Beynon, A. D. and Wood, B. A. (1987). Patterns and rates of molar crown formation times in East African fossil hominds. *Nature* **326**: 493–496.

Biegert, J. (1957). Der formuandel des primatenschädels und seine Beziehungen zur ontogenctischen Entwicklung und den phylogenetischen Spezialisationen der Kopforgane. *Gengenbaurs Morphologisches Jahrbuch* **98**: 77–199.

Biegert, J. (1963). The evaluation of characters of the skull hands and feet for primate taxonomy. In: *Classification and Human Evolution* (S. L. Washburn, Ed.), pp. 116–145. Chicago: Aldine.

Biegert, J. and Maurer, R. (1972). Rumpfskelettlänge, Allometrien und Körperproportionen bei catarrhinen Primaten. *Folia primatologica* **17**: 142–156.

Bilsborough, A. and Wood, B. A. (1988). Cranial morphometry of early hominids: facial region. *American Journal of Physical Anthropology* **76**: 61–86.

Bjork, A. (1955). Cranial base development. *American Journal of Orthodontics* **41**: 198–225.

Bjork, A. and Kuroda, T. (1968). Congenital bilateral hypoplasia of the mandibular condyles associated with congenital bilateral ptosis. A radiographic analysis of the craniofacial growth by the implant method in one case. *Acta Morphol. Neerl. – Scand.* **10**: 49–58.

Blake, M. L. (1967). The quantitative myology of the hind limb of primates with special reference to their locomotor adaptations. Ph.D. Thesis, Magdalene College, University of Cambridge.

Blaney, S. P. A. (1986). An allometric study of the frontal sinus in *Gorilla, Pan* and *Pongo*. *Folia Primatologica* **47**: 81–96.

Blumberg, J. E., Hylander, W. L. and Goepp, R. A. (1971). Taurodontism; A biometric study. *American Journal of Physical Anthropology* **34**: 243–255.

Blumenberg, B. (1985). Population characteristics of extinct hominid endocranial volume. *American Journal of Physical Anthropology* **68**: 269–279.

Bluntschli, H. (1929). Die Kaumuskulature des Orang Utan und ihre Bedeutung für die

Formung des Schädels. *Gengenbaurs Morphologisches Jahrbuch* **63**: 531–606.

Boaz, N. T. (1988). Status of *Australopithecus afarensis*. *Yearbook of Physical Anthropology* **31**: 85–113.

Bock, W. J. (1977). Foundations and methods of evolutionary classification. In: *Major Patterns in Vertebrate Evolution* (M. K. Hecht, P. C. Goody, and B. M. Hecht, Eds), pp. 851–895. New York: Plenum.

Bojsen-Møller, F. (1979). The calcaneocuboid joint and stability of the longitudinal arch at high and low gear push off. *Journal of Anatomy* **129**: 165–176.

Bolk, L. (1909) On the position and displacement of the foramen magnum in the Primates. *Verk. Akad. Wet. Amst.* **12**: 362–377.

Bonč-Osmolovsky, G. A. (1941). Kist'iskopaemogo cheloveka iz grota Kiik-Koba (The hand of the fossil man from Kiik-Koba). *Paleolit Kryma* **2**.

Bonde, N. (1977). Cladistic classification as applied to vertebrates. In: *Major Patterns in Vertebrate Evolution* (M. K. Hecht, P. C. Goody, and B. M. Hecht, Eds), pp. 741–804. New York: Plenum.

Bonde, N. (1981). Problems of species concepts in paleontology. In: *Concepts and Methods in Palaeontology* (J. Martinelli, Ed.), pp. 19–34. Barcelona: University of Barcelona.

Bornstein, M. N. and Bornstein, H. G. (1976). The pace of life. *Nature* **259**: 557–558.

Boule, M. (1911–13). L'homme fossile de la Chapelle-aux-Saints. *Annales de Paleontologie* **6**: 111–172; **7**: 21–56, 85–192; **8**: 1–70.

Boule, M. and Vallois, H. V. (1912). *Fossil man*. London: Thames and Hudson.

Bowen, W. H. and Koch, G. (1970). Determination of age in monkeys (*Macaca irus*) on the basis of dental development. *Laboratory Animals* **4**: 113–123.

Bowden, R. E. M. (1967). The functional anatomy of the foot. *Physiotherapy* **53**: 120–126.

Boyde, A. (1963). Estimation of age at death of young human skeletal remains from incremental lines in dental enamel. *Third International Meeting in Forensic Immunology, Medicine, Pathology and Toxicology* (16–24 April 1963) London.

Boyde, A. (1964). The structure and development of mammalian enamel. Ph.D. thesis, University of London.

Boyde, A. (1969). Correlation of ameloblast size with enamel prism pattern; use of scanning electron microscope to make surface area measurements. *Z. Zellforsch* **93**: 583–593.

Boyde, A. (1976). Amelogenesis and the development of teeth. In: *Scientific Foundations of Dentistry* (B. Cohen and I. R. H. Kramer, Eds). pp. 335–352. London: Heinemann.

Boyde, A. (1989). Enamel. In: *Handbook of Microscopic Anatomy*, Vol. V/6, *Teeth*, pp. 309–473. Springer-Verlag, Berlin.

Boyde, A. and Hobdell, M. H. (1969). Scanning electron microscopy of primary membrane bone. *Z. Zellforsch.* **99**: 98–108.

Boyde, A. and Martin, L. B. (1984). The microstructure of primate dental enamel. In: *Food Acquisition and Processing in Primates* (D. J. Chivers, B. A. Wood and A. Billsborough, Eds), pp. 341–367. New York: Plenum Press.

Boyde, A. and Martin, L. B. (1987). Tandem scanning reflected light microscopy of primate enamel. *Scanning Microscopy* **1**; 1935–1948.

Boyde, A., Hendel, P., Hendel, R., Maconnachie, E. and Jones, S. J. (1990). Human cranial bone structure and the healing of cranial bone grafts: A study using backscattered electron imaging and confocal microscopy. *Anatomy and Embryology* **181**: 235–251.

Brain, C. K., Vrba, E. S. and Robinson, J. T. (1974). A new hominid innominate bone from Swartkrans. *Annals of the Transvaal Museum* **29**: 55–66.

Brandes, R. (1932) Über den kehlkopf des orangutan in verschiedenen Alterstadien mit besonderer Berücksichtigung der Kehlsackfrage. *Gengenbaurs Morphologisches Jahrbuch* **6**: 1–61.

Brantigan and Voshell (1941). The mechanisms of the ligaments and menisci of the knee joint. *Journal of Bone and Joint Surgery* **23**: 44–46.

Broecker, W. and Bender, M. (1972). Age determination on marine strandlines. In: *Calibration of Hominid Evolution* (W. W. Bishop and J. A. Miller, Eds). Edinburgh: Scottish Academic Press.

Bromage, T. G. (1985). Taung Facial remodelling; a growth and development study. In: *Hominid Evolution, Past, Present and Future* (P. V. T. Tobias, Ed.), pp. 239–246. New York: Alan Liss.

Bromage, T. G. (1986). A comparative scanning electron microscope study of facial remodelling in *Homo sapiens*, *Pan*, sp. and Plio-Pleistocene

Hominidae. Ph.D. thesis, University of Toronto, Canada.

Bromage, T. G. (1989). Ontogeny of the early hominid face. *Journal of Human Evolution* **18**: 751–773.

Bromage, T. G. and Dean, M. C. (1985). Re-evaluation of the age at death of Plio-Pleistocene fossil hominids. *Nature* **317**: 525–528.

Bronowski, J. and Long, W. M. (1952). Statistics of descrimination in anthropology. *American Journal of Physical Anthropology* **10**: 385–394.

Broom, R. (1938a). The Pleistocene anthropoid apes of South Africa. *Nature* **142**: 377–379.

Broom, R. (1938b). Further evidence on the structure of the South African Pleistocene anthropoids. *Nature* **142**: 897–899.

Broom, R. (1943). An ankle-bone of the ape-man *Paranthropus robustus*. *Nature* **152**: 689–690.

Broom, R. (1950). The genera and species of the South African Fossil ape-man. *American Journal of Physical Anthropology* **8**: 1–14.

Broom, R. and Robinson, J. T. (1947). Further remains of the Sterkfontein ape-man, *Plesianthropus*. *Nature* **160**, 430–431.

Broom, R. and Schepers, G. W. H. (1946). The South African fossil ape-men—the Australopithecinae. *Transvaal Museum Memoir No. 2.* Pretoria: South Africa.

Broom, R., Robinson, J. T. and Schepers, G. W. H. (1950). Sterkfontein ape-man, *Plesianthropus*. *Transvaal Museum Memoir No. 4.* Pretoria: South Africa.

Brown, F. H., Harris, J., Leakey, R. and Walker, A. (1985). Early *Homo erectus* from West Lake Turkana, Kenya. *Nature* **316**: 788–792.

Browning, H. (1953). The confluence of dural venous sinuses. *American Journal of Anatomy* **93**: 307–329.

Brues, A. M. (1977). *People and Races.* New York: Macmillan.

Bull, J. W. D. (1969). Tentorium cerebelli. *Proceedings of the Royal Society of Medicine* **62**: 1301–1310.

Bullion, S. K. (1987). The biological application of teeth in archaeology. Ph.D. thesis, University of Lancaster.

Bush, M. E., Lovejoy, C. O., Johanson, D. C. and Coppens, Y. (1982). Hominid carpal, metacarpal, and phalangeal bones recovered from the Hadar formation: 1974–1977 collections. *American Journal of Physical Anthropology* **57**: 651–678.

Butler, H. (1949). A rare suture in the anterior cranial fossa of the human skull. *Man* **49**: 25–27.

Butler, P. (1981). Dentition in function. In: *Dental Anatomy and Embryology, a Companion to Dental Studies* (J. W. Osborn, Ed.), pp. 329–356. Oxford: Blackwell Scientific Publications.

Butler, P. M. (1986). Problems of dental evolution in the higher primates. In: *Major Topics in Primate and Human Evolution* (B. A. Wood, L. B. Martin and P. A. Andrews, Eds), pp. 89–106. Cambridge University Press.

Butzer, K. (1979). Comment. *Current Anthropology* **20**: 28.

Butzer, K., Brown, R. and Thurber, D. (1969). Horizontal sediments of the Lower Omo valley: the Kibish formation. *Quaternaria* **11**: 15–39.

Byrne, R. and Whiten, A. (1987). The thinking primates guide to deception. New Scientist **1589**: 54–57.

Cabanac, M. (1986). Keeping a cool head. *News in Physiological Sciences* **1**: 41–44.

Cabanac, M. and Caputa, M. (1979a). Natural selective cooling of the human brain: evidence of its occurrence and magnitude. *Journal of Physiology, London* **286**: 255–264.

Cabanac, M. and Caputa, M. (1979b). Open loop increase in trunk temperature produced by face cooling in working humans. *Journal of Physiology* **289**: 163–174.

Cachel, S. (1984). Growth and allometry in primate masticatory muscles. *Archives of Oral Biology* **29**: 287–293.

Campbell, B. G. (1966). *Human Evolution.* Chicago: Aldine.

Carlson, D.S. (1977). Condylar translation and the function of the superficial masseter muscle in the Rhesus monkey (*Macaca mulata*). *American Journal of Physical Anthropology* **47**: 53–64.

Carlsoo, S. (1972). *How man moves: kinesiological studies and methods.* London: Heinemann.

Cartmill, M. and Milton, K. (1977). The Lorisiform wrist joint and the evolution of 'brachiating' adaptations in the hominoidea. *American Journal of Physical Anthropology* **47**: 249–272.

Cave, A. J. E. (1961). The frontal sinus of the gorilla. *Proceedings of The Zoological Society of London* **136**: 359–373.

Cave, A. J. E. and Haines, R. W. (1940).

The paranasal sinuses of the anthropoid apes. *Journal of Anatomy* **74**: 493–523.

Chamberlain, A. and Wood, B. A. (1985). A reappraisal of variation in hominid mandibular corpus dimensions. *American Journal of Physical Anthropology* **66**: 399–405.

Chandler, S. B. and Derezinski, C. F. (1935). The variations of the middle meningeal artery within the middle cranial fossa. *Anatomical Record* **62**: 309–319.

Charles, R. H. (1893). The influence of function, as exemplified in the morphology of the lower extremity of the Punjabi. *Journal of Anatomy and Physiology* **28**: 271–280.

Charteris, J., Wall, J. C. and Nottrodt, J. W. (1981). Functional reconstruction of gait from the Pliocene hominid foot prints at Laetoli, northern Tanzania. *Nature* **290**: 496–498.

Charteris, J., Wall, J. C. and Nottrodt, J. W. (1982). Pliocene hominid gait: new interpretations based on available footprint data from Laetoli. *American Journal of Physical Anthropology* **58**: 133–144.

Chavaillon, J., Chavaillon, N., Coppens, Y. and Senut, B. (1977). Présence d'hominides dans le site Oldowayen de Gomboré I à Melka Kunturé, Ethiopie. *Comptes Rendus Hébdomadaires des Séances de l'Académie des Sciences* **285D**: 961–964.

Ciochon, R. L. and Corruccini, R. S. (1976). Shoulder joint of Sterkfontein *Australopithecus. South African Journal of Science* **72**: 80–82.

Ciochon, R. L. and Corruccini, R. S. (1977). The coraco-acromial ligament and projection index in man and other anthropoid primates. *Journal of Anatomy* **124**: 627–632.

Clarke, R. J. (1977). The cranium of the Swartkrans hominid SK 847 and its relevance to human origins. Ph.D. Thesis, University of Witwatersrand, Johannesburg, South Africa.

Conroy, G. C. and Vannier, M. W. (1987). Dental development of the Taung skull from computerised tomography. *Nature* **329**: 625–627.

Cook, D. C., Buikstra, J. E., DeRousseau, C. J. and Johanson, D. C. (1983). Vertebral pathology in the Afar australopithecines. *American Journal of Physical Anthropology* **60**: 83–102.

Coolridge, H. J. (1933). *Pan paniscus.* Pigmy chimpanzee from South of the Congo River. *American Journal of Physical Anthropology* **28**: 1–57.

Corbridge, R. (1987). *The knee joint: a functional analysis.* B.Sc. Thesis, University College London.

Corruccini, R. S. and Henderson, A. M. (1978). Multivariate dental allometry in primates. *American Journal of Physical Anthropology* **48**: 203–208.

Corruccini, R. S. and McHenry, H. M. (1978). Relative femoral head size in early hominids. *American Journal of Physical Anthropology* **49**: 145–148.

Cousin, R. P. and Fenart, R. (1971). Etude Ontogénétique des Eléménts Sagittaux du Fossé Cérébrale Antérieure chez l'Homme Orientation Vestibulaire. *Archives of Anatomy and Pathology* **9**: 383–395.

Cramer, D. L. (1977). Craniofacial morphology of *Pan paniscus:* a morphometric and evolutionary appraisal. *Contributions to Primatology*, Vol. 10. Basel: S. Karger.

Critchley, E. M. R. (1985). The human face. *British Medical Journal* **291**: 1222–1223.

Cunningham, D. J. (1886). The lumbar curve in man and apes. *Royal Irish Academy Cunningham Memoire No. 2.*

Dart, R. A. (1925). *Australopithecus africanus:* the ape-man of South Africa. *Nature* **115**: 195–197.

Dart, R. A. (1926). Taung and its significance. *Natural History* **26**: 315–327.

Davis, P. R. (1964). Hominid fossils from Bed I, Olduvai Gorge, Tanganyika. *Nature* **201**: 967–970.

Day, M. H. (1967). Olduvai Hominid 10: a multivariate analysis. *Nature* **215**: 323–324.

Day, M. H. (1969a). Femoral fragment of a robust australopithecine from Olduvai Gorge, Tanzania. *Nature* **221**: 230–233.

Day, M. H. (1969b). Omo human skeletal remains. *Nature* **222**: 1135–1138.

Day, M. H. (1971). Post cranial remains of *Homo erectus* from Bed IV, Olduvai Gorge, Tanzania. *Nature* **232**: 383–387.

Day, M. H. (1972). The Omo human skeletal remains. In: *The Origin of the Genus Homo* (F. Bordes, Ed.), pp. 35–36. Paris: UNESCO.

Day, M. H. (1973). Locomotor features of the lower limb in hominids. *Symposia of the Zoological Society of London* **33**: 29–51.

Day, M. H. (1974). The interpolation of isolated fossil foot bones into a discriminant function analysis: a reply. *American Journal of Physical Anthropology* **41**: 233–236.

Day, M. H. (1976). Hominid postcranial material from Bed I, Olduvai Gorge. In: *Human Origins* (G. Isaac and E. McCown, Eds), pp. 363–374. California: W. A. Benjamin.

Day, M. H. (1978). Functional interpretations of the morphology of postcranial remains of early African hominids. In: *Early Hominids of Africa* (C. J. Jolly, Ed.), pp. 311–345. London: Duckworth.

Day, M. H. (1982). The *Homo erectus* pelvis: punctuation or gradualism? *Première Congres Internationale Paleontologie Humaine* (Nice, October 1982), Pretirage, CNRS, Vol. 1, pp. 411–421.

Day, M. H. (1984). The postcranial remains of *Homo erectus* from Africa, Asia and possibly Europe. In: *The Early Evolution of Man* (P. Andrews and F. Fraanzen, Eds), *Courier forschungs Institut Senckenberg* **69**: 113–121.

Day, M. H. (1986). *Guide to Fossil Man*. London: Cassells.

Day, M. H. and Leakey, R. E. F. (1973). New evidence for the genus *Homo* from East Rudolf, Kenya (I). *American Journal of Physical Anthropology* **39**: 341–354.

Day, M. H. and Leakey, R. E. F. (1974). New evidence for the genus *Homo* from East Rudolf, Kenya (III). *American Journal of Physical Anthropology* **41**: 367–380.

Day, M. H. and Molleson, T. (1973). The Trinil femora. In: *Human Evolution (Symposia of the Society for the Study of Human Biology. 11)* (M. H. Day, Ed.), pp. 127–154. London: Society for Human Biology.

Day, M. H. and Molleson, T. I. (1976). The puzzle from JK2—a femur and a tibial fragment (OH 34) from Olduvai Gorge, Tanzania. *Journal of Human Evolution* **5**: 455–465.

Day, M. H. and Napier, J. R. (1961). The two heads of flexor pollicis brevis. *Journal of Anatomy* **95**: 123–130.

Day, M. H. and Napier, J. R. (1963). Functional significance of the deep head of flexor pollicis brevis in primates. *Folia Primatologica* **1**: 122–134.

Day, M. H. and Napier, J. R. (1964). Hominid fossils from Bed I, Olduvai Gorge, Tanganyika. Fossil foot bones. *Nature* **201**: 967–970.

Day, M. H. and Napier, J. R. (1966). A hominid toe bone from bed I, Olduvai Gorge, Tanzania. *Nature* **211**: 929–930.

Day, M. H. and Stringer, C. (1982). A reconsideration of the Omo Kibish remains and the *erectus–sapiens* transition. In: *L'Homo Erectus et la Place de l'Homme de Tautavel Parmi les Hominides Fossiles. Première Congress Internationale de Paléontologie Humaine.* (Nice, October 1982). Prétirage; CNRS, Vol. 1, pp. 814–846.

Day, M. H. and Wood, J. R. (1968). Functional affinities of the Olduvai hominid 8 talus. *Man* **3**: 440–455.

Day, M. H., Leakey, R. E. F., Walker, A. C. and Wood, B. A. (1976a). New hominids from East Turkana, Kenya, *American Journal of Physical Anthropology* **45**: 369–436.

Day, M. H., Leakey, R. E. F., Walker, A. C. and Wood, B. A. (1976b). New hominids from East Rudolf, Kenya (I), *American Journal of Physical Anthropology* **42**: 461–476.

Dean, M. C. (1983). The comparative anatomy of the hominoid cranial base. Ph.D. Thesis, University of London.

Dean, M. C. (1985a). The comparative myology of the hominoid cranial base. II: The muscles of the prevertebral and upper pharyngeal region. *Folia Primatologica* **44**: 40–51.

Dean, M. C. (1985b). Variation in the developing root cone angle of the permanent mandibular teeth of modern man and certain fossil hominids. *American Journal of Physical Anthropology* **68**: 233–238.

Dean, M. C. (1985c). The comparative myology of the hominoid cranial base. I: The muscular relations and bony attachments of the digastric muscle. *Folia Primatologica* **43**: 157–180.

Dean, M. C. (1985d). The eruption pattern of the permanent incisors and first permanent molars in *Australopithecus (Paranthropus) robustus. American Journal of Physical Anthropology* **67**: 251–257.

Dean, M. C. (1987a). Growth layers and incremental markings in hard tissues: a review of the literature and some preliminary observations about enamel structure in *Paranthropus boisei. Journal of Human Evolution* **16**: 157–172.

Dean, M. C. (1987b). The dental developmental status of six east African juvenile fossil hominids. *Journal of Human Evolution* **16**: 197–313.

Dean, M. C. (1988a). Growth processes in the cranial base of hominoids and their bearing on morphological similarities that exist in the cranial base of *Homo* and *Paranthropus*. In *Evolutionary History of the 'Robust' Australopithecines* (F.E. Grine, Ed.), pp. 107–112. New

545

York: Aldine de Gruyter.

Dean, M. C. (1988b). Another look at the nose and the functional significance of the face and the nasal mucous membrane for cooling the brain in fossil hominds. *Journal of Human Evolution* 17: 715–718.

Dean, M. C. and Wood, B. A. (1981a). Developing pongid dentition and its use for ageing individual crania in comparative cross-sectional growth studies. *Folia Primatologica* 36: 111–127.

Dean, M. C. and Wood, B. A. (1981b). Metrical analysis of the basicranium of extant hominoids and Australopithecus. *American Journal of Physical Anthropology* 59: 53–71.

Dean, M. C. and Wood, B. A. (1982). Basicranial anatomy of Plio-Pleistocene hominids from East and South Africa. *American Journal of Physical Anthropology* 59: 157–174.

Dean, M. C., Stringer, C. B. and Bromage, T. G. (1986). A new age at death for the Neanderthal child from Devil's Tower, Gibraltar and the implications for studies of general growth and development in Neanderthals. *American Journal of Physical Anthropology* 70: 301–309.

Demes, B. (1987). Another look at an old face; biomechanics of the neanderthal facial skeleton reconsidered. *Journal of Human Evolution* 16: 297–305.

Demes, B. and Creel, N. (1988). Bite force, diet and cranial morphology of fossil hominids. *Journal of Human Evolution* 17: 657–670.

Demirjian, A. (1978). Dentition. In: *Human Growth* (L. Falkner and J. M. Tanner, Eds), Vol. 2, pp. 413–444. New York: Plenum Press.

Demirjian, A., Tanner, J. M. and Goldstein, H. (1973). A new system of dental age assessment. *Human Biology* 45: 211.

Dobzhansky, T. (1967). *The Biology of Ultimate Concern*, p. 152. New York: New American library.

Downs, W. R. (1952). The role of cephalometrics in orthodontic case analysis. *American Journal of Orthodontics* 38: 162–182.

Dubois, E. (1894). *Pithecanthropus erectus, eine menschenähnliche Übergansform aus Java.* Batavia: Landesdruckerei.

Dubois, E. (1926). On the principal characters of the femur of *Pithecantropus erectus. Proc. K. ned. Akad. Wet.* 29: 730–743.

Dubois, E. (1927). Figures of the femur of *Pithecanthropus erectus. Proceedings K. ned.*

Akad. Wet. 29: 1275–1277.

Dubosset, J. F. (1981). Finger rotation during prehension. In: *The Hand* (R. Tubiana, Ed.), Vol. I, pp. 202–206. Philadelphia: W. G. Saunders.

DuBrul, E. L. (1950). Posture, locomotion and the skull in Lagomorpha. *American Journal of Anatomy* 87: 277–313.

DuBrul, E. L. (1977). Early hominid feeding mechanisms. *American Journal of Physical Anthropology* 47: 305–320.

DuBrul, E. L. (1979). Origin and adaptations of the hominid jaw joint. In: *The Temporomandibular Joint*, 3rd edn. (B. G. Sarnat and D. M. Laskin, Eds), pp. 5–34. Springfield, Illinois: Thomas.

DuBrul, E. L. (1980). *Sicher's Oral Anatomy.* St. Louis: Mosby.

DuBrul, E. L. and Sicher, H. (1954). *The Adaptive Chin.* Springfield: Thomas.

Duchin, L. E. (1990). The evolution of articulate speech; comparative anatomy of the oral cavity in *Pan* and *Homo. Journal of Human Evolution* 19: 684–695

Duckworth, W. L. H. (1904). *Morphology and Anthropology.* Cambridge: Cambridge University Press.

Dunbar, R. I. M. (1990). Co-evolution of cognitive capacity, group size and social grooming in primates; implications for the evolution of language. *Journal of Human Evolution*, in press.

Eberhart, H. D., Inman, V. T. and Bresler, B. (1954). The principal elements in human locomotion. In: *Human Limbs and Their Substitutes* (P. E. Klopsteg and P. D. Wilson, Eds), pp. 437–471. New York: McGraw-Hill.

Eccles, J. C. (1984). *The Human Mystery. The Gifford Lectures,* University of Edinburgh 1977–78. London: Routledge & Kegan Paul. Boston.

Eckenhoff, J. E. (1970). The physiologic significance of the vertebral venous plexus. *Journal of Surgery Gynecology and Obstetrics* 131: 72–78.

Eckhardt, R. B. (1987). Hominoid nasal region polymorphism and its phylogenetic significance. *Nature* 328: 333–335.

Eisenberg, N. and Brodie, A. (1965). Antagonism of temporal fascia to masseteric contraction. *Anatomical Record* 152: 185–192.

Eldredge, N. and Cracraft, J. (1980). *Phylogenetic Patterns and the Evolutionary Process.* New

York: Columbia University Press.

Elftman, H. and Manter, J. (1935a). Chimpanzee and human feet in bipedal walking. *American Journal of Physical Anthropology* **20**: 69–79.

Elftman, H. and Manter, J. (1935b). The evolution of the human foot, with especial reference to the joints. *Journal of Anatomy* **70**: 56–67.

Endo, B. and Kimura, T. (1970). Postcranial skeleton of the Amud Man. In: *The Amud Man and His Cave Site* (H. Suzuki and F. Takai, Eds), pp. 231–406. Tokyo: Academic Press.

Enlow, D. H. (1966). A comparative study of facial growth in *Homo* and *Macaca*. *American Journal of Physical Anthropology* **24**: 293–307.

Enlow, D. H. (1968). *The Human Face: An Account of Postnatal Growth and Development of the Craniofacial System*. New York: Harper and Row.

Epstein, H. M., Linde, H. W., Crampton, A. R., Ciric, I. S. and Eckenhoff, J. E. (1970). The vertebral venous plexus as a major cerebral venous outflow tract. *Anaesthesiology* **32**: 332–337.

Eriksen, E. F. (1986). Normal and pathological remodelling of human trabecular bone: three dimensional reconstruction of the remodelling sequence in normals and in metabolic bone disease. *Endocrine Reviews* **7**: 379–408.

Evans, F. G. and Krahl, V. E. (1945). The torsion of the humerus: a phylogenetic study from fish to man. *American Journal of Anatomy* **76**: 303–337.

Falk, D. (1975). Comparative anatomy of the larynx in man and the chimpanzee: implications for language in Neanderthals. *American Journal of Physical Anthropology* **43**: 123–132.

Falk, D. (1980). A reanalysis of the South African australopithecine natural endocasts. *American Journal of Physical Anthropology* **53**: 525–539.

Falk, D. (1983a). The Taung endocasts: a reply to Holloway. *American Journal of Physical Anthropology* **60**: 479–489.

Falk, D. (1983b). Cerebral cortices of East African early hominids. *Science* **221**: 1072–1074.

Falk, D. (1985). Hadar AL 162-28 endocast as evidence that brain enlargement preceded cortical reorganisation in hominid evolution. *Nature* **313**: 45–47.

Falk, D. (1986a). Endocranial casts and their significance for primate brain evolution. In: *Comparative Primate Biology. Vol. 1: Systematics, Evolution and Anatomy*, pp. 477–490.

New York: A. Liss.

Falk, D. (1986b) Evolution of cranial blood drainage in hominoids; enlarged occipital–marginal sinuses and emissary foramina. *American Journal of Physical Anthropology* **70**: 311–324.

Falk, D. and Conroy, E. G. (1984). The cranial venous sinus system in *Australopithecus afarensis*. *Nature* **306**: 779–781.

Falk, D. and Kasinga, S. (1983). Cranial capacity of a female robust australopithecine (KNM ER 407) from Kenya. *Journal of Human Evolution* **12**: 515–518.

Fanning, E. A. and Moorrees, C. F. A. (1969). A comparison of permanent mandibular molar formation in Australian Aborigines and Caucasoids. *Archives of Oral Biology* **14**: 999–1006.

Fedak, M. A., Pinshow, B. and Schmidt-Nielsen, K. (1974). Energetic cost of bipedal running. *American Journal of Physiology* **227**: 1038–1044.

Fedak, M. A. and Seeherman, A. J. (1979) Reappraisal of energetics of locomotion shows identical cost in bipeds and quadrupeds including ostrich and horse. *Nature* **282**: 713–716.

Feibel, C. S., Brown, F. H. and McDougall, I. (1989). Stratigraphic context of fossil hominids from the Omo group deposits: northern Turkana Basin, Kenya and Ethiopia. *American Journal of Physical Anthropology* **78**: 595–622.

Feldesman, M. R. (1979). Further morphometric studies of the ulna from the Omo Basin, Ethiopia. *American Journal of Physical Anthropology* **51**: 409–416.

Feldesman, M. R. and Lundy, J. K. (1988). Stature estimates for some African Plio-Pleistocene fossil hominids. *Journal of Human Evolution* **17**: 583–596.

Fenart, R. and Deblock, R. (1973). *Pan paniscus* et *Pan troglodytes*—craniométrie—Etude comparative et ontogénique selon les méthodes classique et vestibulaire. Tome 1. *Musée Royale de l'Afrique Centrale—Tervuren, Belgique Annales, Series, 1N*, 8° Sciences Zoologiques, No. 204.

Fenart, R. and Empereur-Buisson, R. (1970). Application de la méthode 'vestibulaire' d'orientation au crânes du Pech de l'Azé et comparaison avec d'autres crânes néanderthaliens. *Archives de l'Institut de Paléontologie Mémoires Humains* **33**: 89–148.

Field, E. J. and Harrison, R. J. (1968). *Anatomical Terms, their Origin and Derivation*, 3rd edn,

p. 212, Cambridge: Heffer.

Fischer, E. (1906). Die Variationen am Radius und Ulna des Menschen. *Zeitschrift für Morphologie und Anthropologie* **9**: 147–247.

Fleagle, J. G. (1976). Locomotion and posture of the Malayan Siamang and implications for hominoid evolution. *Folia Primatologica* **26**: 245–269.

Fleagle, J. G., Stern, J. T. Jr., Jungers, W. L., Susman, R. L., Vangor, A. K. and Wells, J. P. (1981). Climbing: a biomechanical link with brachiation and with bipedalism. In: *Vertebrate Locomotion* (M. H. Day, Ed.) *Symposium of the Zoological Society of London* **48**: 359–375.

Fletcher, A. M. (1985). Ethnic variations in sagittal condylar angles. *Journal of Dentistry* **13**: 304–310.

Fransiscus, R. G. and Trinkaus, E. (1988). Nasal morphology and the emergence of *Homo erectus*. *American Journal of Physical Anthropology* **75**: 517–527.

Fraser, J. E. (1920). *The Anatomy of the Human Skeleton*. London: Churchill.

Garn, S. M. (1970). *The Earlier Gain and the Later Loss of Cortical Bone in Nutritional Perspective*. Springfield: Thomas.

Garn, S. M. and Lewis, A. B. (1963). Phylogenetic and intra-specific variations in tooth sequence polymorphism. In: *Dental Anthropology* (D. R. Brothwell, Ed.), pp. 53–73. Oxford: Pergamon Press.

Garn, S. M., Lewis, A. B. and Polacheck, D. L. (1959). Variability of tooth formation. *Journal of Dental Research* **38**: 135–148.

Garn, S. M., Negy, J. M. and Sandusky, S. T. (1972). Differential sexual dimorphism in bone diameters of subjects of european and African ancestry. *American Journal of Physical Anthropology* **37**: 127–130.

Geissmann, T. (1986a). Estimation of australopithecine stature from long bones: AL 288-1 as a test case. *Folia Primatologica* **47**: 119–127.

Geissmann, T. (1986b). Length estimate for KNM ER 736, a hominid femur from the Lower Pleistocene of East Africa. *Human Evolution* **1**: 481–493.

Gentry, A. W. and Gentry, A. (1978). The Bovidae (Mammalia) of Olduvai Gorge, Tanzania, 1 and 2. *Bulletin of the British Museum (Natural History), Geology* **29**: 289–446; **30**: 1–83.

Geschwind, N. (1965). Disconnection syndromes in animals and man. *Brain* **88**: 237–294,

585–644.

Geschwind, N. (1972). Language and the brain. *Scientific American* **226**: 76–83.

Gill, H. I. (1971). Neuromuscular spindles in the human lateral pterygoid muscles. *Journal of Anatomy* **109**: 157–167.

Girgis, F. G., Marshall, J. L. and Al Monajen, A. R. S. (1975). The cruciate ligaments of the knee joint. *Clinical Orthopaedics and Related Research* **106**: 216–231.

Gleiser, I. and Hunt, E. E. (1955). The permanent mandibular first molar; its calcification, eruption and decay. *American Journal of Physical Anthropology* **13**: 253–284.

Gordon, A. M., Huxley, A. F. and Julian, F. J. (1966). The variation in isometric tension with sarcomere length in vertebrate muscle fibres. *Journal of Physiology* **184**: 170–192.

Gorjanović-Kramberger, D. (1906). *Der diluviale Mensch von Krapina in Kroatien: Ein Beitrag zur Paläoanthropologie*. Wiesbaden: C. W. Kreidel's Verlag.

Gorjanović-Kramberger, D. (1914). Der Axillarrand des Schulterblattes des Menschen von Krapina. *Glasnik hrv. prirod. der Zagreb* **26**: 231–257.

Granados, J. I. (1979). The influence of the loss of teeth and attrition on the articular eminence. *Journal of Prosthetic Dentistry* **42**: 78–85.

Grausz, H. M., Leakey, R. E., Walker, A. C. and Ward, C. V. (1988). Associated cranial and postcranial bones of *Australopithecus boisei*. In: *Evolutionary History of the 'Robust' Australopithecines* (F. E. Grine, Ed.), pp. 127–132. New York: Aldine de Grugter.

Gray, E. G. and Basmajian, J. V. (1968). Electromyography and cinematography of leg and foot ('normal' and flat) during walking. *Anatomical Record* **161**: 1–5.

Green, J. H. and Silver, P. H. S. (1981). *An Introduction to Human Anatomy*. Oxford: Oxford University Press.

Greene and Sibley (1986). Neanderthal pubic morphology and gestation length revisited. *Current Anthropology* **27**: 517–518.

Gregory, W. K. (1912). Notes on the principles of quadrupedal locomotion. *Annals of the New York Academy of Science* **22**: 267–294.

Grine, F. E. (1985a). Australopithecine evolution: the deciduous dental evidence. In: *Ancestors: The Hard Evidence* (E. A. Delson, Ed.), pp. 153–167. New York: A. Liss.

Grine, F. E. (1985b). Dental morphology and

the systematic affinities of the Taung fossil hominid. In: *Hominid Evolution Past, Present and Future* (P. V. Tobias, Ed.), pp. 247–253.

Grine, F. E. (1986). Dental evidence for dietary differences in *Australopithecus* and *Paranthropus*; a quantitative analysis of permanent molar microwear. *Journal of Human Evolution* **15**: 783–822.

Grine, F. E. (1988). New craniodental fossils of *Paranthropus* from the Swartkrans formation and their significance in 'robust' australopithecine evolution. In: *Evolutionary History of the 'Robust' Australopithecines* (F. E. Grine, Ed.) New York: Aldine de Gruyter.

Grine, F. E. and Martin, L. B. (1988). Enamel thickness and development in *Australopithecus* and *Paranthropus*. In: *Evolutionary History of the 'Robust' Australopithecines* (F. E. Grine, Ed.), pp. 3–42. New York: Aldine de Gruyter.

Gustafson, G. and Koch, G. (1974). Age estimation up to 16 years of age based upon dental development. *Odontologisk Revy* **25**: 297–306.

Haines, D. E. (1986). The primate cerebellum. In: *Comparative Primate Biology, Vol. 1: Systematics, Evolution and Anatomy*, pp. 491–535. New York: A. Liss.

Harvey, P. H., Martin, R. D. and Clutton-Brock, T. H. (1986). Life histories in comparative perspective. In: *Primate Societies* (B. B. Smuts, D. L. Cheney, R. M. Seyfarth, R. W. Wrandham and T. T. Struhsaker, Eds), pp. 181–196. Chicago: The University of Chicago Press.

Haughton, S. (1873). *Principles of Animal Mechanics*. London: Longman Green.

Hawkes, J. (1965). *Prehistory in the History of Mankind. Cultural and Scientific Development.* Vol. 1, Part 1, Unesco. London: New English Library.

Haxton, H. (1945). The functions of the patella and the effects of its excision. *Surgery, Gynecology and Obstetrics* **80**: 389–395.

Hayward, J. N. (1967). Cerebral cooling during increased cerebral blood flow in the monkey. *Proceedings of the Society of Experimental Biology* **124**: 555–557.

Hayward, J. N. and Baker, M. A. (1968). Role of cerebral arterial blood in the regulation of brain temperature in the monkey. *American Journal of Physiology* **215**: 389–403.

Heim, J.-L. (1972). Les Néandertaliens adultes de la Ferrassie (Dordogne). Études anthropologiques et comparatives. Thèse de Doctorat d'État, Université de Paris VI.

Heim, J. L. (1974). Les hommes fossiles de La Ferrassie (Dordogne) et le problème de la définition des Néanderthaliens Classiques. *L'Anthropologie* **78**: 81–112, 321–378.

Heim, J. L. (1983). Les variations du squelette post-cranien. Des hommes de Neandertal suivant le sexe. *Anthropologie Journal* **87**: 379–416.

Heiple, K. G. and Lovejoy, C. L. (1971). The distal femoral anatomy of *Australopithecus. American Journal of Physical Anthropology* **35**: 75–84.

Hennig, W. (1966). *Phylogenetic systematics.* Urbana: University of Illinois Press.

Hess, A. F., Lewis, J. M. and Roman, B. (1932). A radiographic study of calcification of the teeth from birth to adolescence. *Dental Cosmos* **74**: 1053–1061.

Heuter, C. (1862). Anatomische Studien an der Extremitätengelenk Neugeborener und Erwaschener. *Virchows Archiv* **25**: 572–599.

Hicks, J. H. (1953). The mechanics of the foot. I. The joints. *Journal of Anatomy* **87**: 345–357.

Hicks, J. H. (1954). The mechanics of the foot. II. The plantar aponeurosis and the arch. *Journal of Anatomy* **88**: 25–30.

Hicks, J. H. (1955). The foot as a support. *Acta Anatomica* **25**: 34–45.

Hiiemae, K. M. (1978). Mammalian mastication; a review of the activity of the jaw muscles and the movements they produce. In: *Development, Function and Evolution of Teeth* (P. M. Butler and K. A. Joysey, Eds), pp. 359–398. New York: Academic press.

Hill, A. and Ward, S. (1988). Origin of the hominidae: The record of African large hominoid evolution between 14 my and 4 my. *Yearbook of Physical Anthropology* **31**: 49–83.

Hollingshead, W. H. (1982). *Anatomy for Surgeons, Vol. 1: The Head and Neck*, 3rd edn. Philadelphia: Harper and Row.

Holloway, R. L. (1970). New endocranial values for the Australopithecines. *Nature* **227**: 199–200.

Holloway, R. L. (1972a). New australopithecine endocast, SK 1585, from Swartkrans, South Africa. *American Journal of Physical Anthropology* **37**: 173–186.

Holloway, R. L. (1972b). Australopithecine endocasts, brain evolution in the hominoidea, and a model of hominid evolution. In: *The Functional and Evolutionary Biology of Primates* (R. Tuttle, Ed.). Atherton: Aldine.

Holloway, R. L. (1973a). Endocranial volumes

of early African hominids and the role of the brain in human mosaic evolution. *Journal of Human Evolution* 2: 449–458.

Holloway, R. L. (1973b). New endocranial values for the East African early hominids. *Nature* 243: 97–99.

Holloway, R. L. (1974). The casts of fossil hominid brains. *Scientific American* 231: 106–115.

Holloway, R. L. (1976). Some problems of hominid brain endocast reconstruction, allometry, and neural reorganization. In: *Colliquium VI of the XI Congress of the UISPP* (Nice, 1976), (P. V. Tobias and Y. Coppens, Eds), pp. 69–119. Pretriage.

Holloway, R. L. (1978). Problems of brain endocast interpretation and African hominid evolution. In: *Early hominids of Africa* (C. J. Jolly, Ed.). London: Duckworth.

Holloway, R. L. (1980a). The OH 7 (Olduvai Gorge, Tanzania) hominid partial brain endocast revisited. *American Journal of Physical Anthropology* 53: 267–274.

Holloway, R. L. (1980b). Indonesian 'Solo' (Ngandong) endocranial reconstructions; some preliminary observations and comparisons with Neanderthal and *Homo erectus* groups. *American Journal of Physical Anthropology* 53: 285–295.

Holloway, R. L. (1981a). Exploring the dorsal surface of hominid brain endocasts by stereoplotter and discriminant analysis. *Philosophical Transactions of The Royal Society of London, Series B* 292: 155–166.

Holloway, R. L. (1981b). Volumetric and asymmetry determinations on recent hominid endocasts; Spy I and II, Djebel Irhoud I, and the Salé *Homo erectus* specimens with some notes on Neanderthal brain size. *American Journal of Physical Anthropology* 55: 385–393.

Holloway, R. L. (1981c). The Indonesian *Homo erectus* brain endocasts revisited. *American Journal of Physical Anthropology* 55: 503–521.

Holloway, R. L. (1981d). The endocast of the Omo L338y-6 Juvenile hominid: 'Gracile' or 'robust' *Australopithecus*? *American Journal of Physical Anthropology* 54: 109–118.

Holloway, R. L. (1981e). Revisiting the Taung australopithecine endocast; the position of the lunate sulcus as determined by the stereoplotting technique. *American Journal of Physical Anthropology* 56: 43–58.

Holloway, R. L. (1983). Cerebral brain endocast pattern of *Australopithecus afarensis* hominid. *Nature* 303: 420–422.

Holloway, R. L. (1985). The poor brain of *Homo sapiens neanderthalensis*; see what you please. In: *Ancestors; The Hard Evidence* (E. Delson, Ed.), pp. 319–324. New York: A. Liss.

Holloway, R. L. (1988). 'Robust' australopithecine brain endocasts; some preliminary observations. In: *Evolutionary History of the 'Robust' Australopithecines* (F. E. Grine, Ed.), pp. 97–106. New York: Aldine de Gruyter.

Holloway, R. L. and De La Coste-Lareymondie, M. C. (1982). Brain endocast asymmetry in pongids and hominids; some preliminary findings on the paleontology of cerebral dominance. *American Journal of Physical Anthropology* 58: 101–110.

Holloway, R. L. and Post, D. C. (1982). The relativity of relative brain measures and hominid mosaic evolution. In: *Primate Brains; Evolution, Methods and Concepts* (E. Armstrong and D. Falk, Eds), pp. 57–76. New York: Plenum.

Howell, F. C. (1978). Hominidae. In: *Evolution of African Mammals* (V. J. Maglio, and H. B. S. Cooke, Eds.), pp. 154–248. Cambridge: Harvard University Press.

Howell, P. G. T. (1987). Sexual dimorphism in mastication and speech. Or do men and women eat and talk differently. *Australian Prosthodontic Journal* 1: 9–17.

Howell, F. C. and Wood, B. A. (1974). Early hominid Ulna from the Omo Basin, Ethiopia. *Nature* 249: 174–176.

Hrdlicka, A. (1934a). Contributions to the study of the femur: the crista aspera and the pilaster. *American Journal of Physical Anthropology* 19: 17–37.

Hrdlicka, A. (1934b). The hypotrochanteric fossa of the femur. *Smithsonian Miscellaneous Collections* 92: 1–49.

Hrdlicka, A. (1937). The gluteal ridge and gluteal tuberosities (3rd trochanters). *American Journal of Physical Anthropology* 23: 129–198.

Hughes, A. R. and Tobias, P. V. (1977). A fossil skull probably of the genus *Homo* from Sterkfontein, Transvaal. *Nature* 265: 310–312.

Humphrey, N. (1986). *The Inner Eye*. London: Faber & Faber.

Huxley, T. H. (1863). *Evidence as to Man's Place in Nature*, London: Williams and Norgate.

Huxley, T. H. (1867). On two widely contrasted forms of the human cranium. *Journal of*

Anatomy and Physiology, London **1**: 60–77.

Hylander, W. L. (1975). The human mandible 'lever or link'. *American Journal of Physical Anthropology* **43**: 227–242.

Hylander, W. L. (1979a). An experimental analysis of temporomandibular joint reaction force in macaques. *American Journal of Physical Anthropology* **51**: 433–456.

Hylander, W. L. (1979b). The functional significance of primate mandibular form. *Journal of Morphology* **160**: 223–240.

Hylander, W. L. (1979c). Functional anatomy. In: *The Temporomandibular Joint*. (B. E. Sarnat and D. M. Laskin, Eds), pp. 85–113.

Hylander, W. L. (1983). Posterior temporalis function in macaques and humans. *American Journal of Physical Anthropology* **60**: 208 (Abstr.).

Hylander, W. L. (1984). Stress and strain in the mandibular symphysis of primates: a test of competing hypotheses. *American Journal of Physical Anthropology* **64**: 1–46.

Hylander, W. L. (1988). Implications of *in vivo* experiments for interpreting the functional significance of 'Robust' Australopithecine jaws. In: *Evolutionary History of the 'Robust' Australopithecines* (F. E. Grine, Ed.), pp. 55–83. New York: Aldine de Gruyter.

Inman, V. T., Saunders, J. and Abbott, L. C. (1944). Observations on the function of the shoulder joint. *Journal of Bone and Joint Surgery* **26**: 1–29.

Ishida, H., Kimura, T. and Okada, M. (1975). Patterns of bipedal walking in anthropoid primates. In: *Symposium of the 5th Congress of the International Primatological Society (1974)* (S. Kondo, M. Kawai, A. Ehara and S. Kawamura, Eds.) Tokyo: Japan Science Press, pp. 287–301.

Ishida, H., Kimura, T., Okada, M. and Yamazaki, N. (1985a). Kinesiological aspects of bipedal walking in gibbons. In: *Primate Morphophysiology: Locomotor Analysis and Human Bipedalism* (S. Kondo, Ed.). Tokyo: University of Tokyo Press.

Ishida, H., Kumakura, H. and Kondo, S. (1985b). Primate bipedalism and quadrupedalism: comparative electromyography. In: *Primate Morphophysiology, Locomotor Analysis and Human Bipedalism* (S. Kondo, Ed.) pp. 59–80. Tokyo: University of Tokyo Press.

Jaspers, M. T. and Witkop, C. J. (1980). Taurodontism, and isolated trait associated with syndromes and X-chromosomal aneuploidy. *American Journal of Human Genetics* **32**: 396–413.

Jenkins, F. (1972). Chimpanzee bipedalism: cineradiographic analysis and implications for the evolution of gait. *Science* **178**: 877–879.

Jenkins, F. A. and Camazin, S. M. (1977). Hip structure and locomotion in ambulatory and cursorial carnivores. *Journal of Zoology* **181**: 351–370.

Jenkins, F. A. and Fleagle, J. G. (1975). Knuckle walking and the functional anatomy of the wrist in living apes. In: *Primate Functional Morphology and Evolution* (R. H. Tuttle, Ed.), pp. 213–227. The Hague: Mouton.

Jerison, H. J. (1973). *Evolution of the Brain and Intelligence*. London: Academic Press.

Johanson, D. C. (1985). The most primitive australopithecines. In: *Hominid Evolution; Past, Present and Future* (P. V. Tobias, Ed.), pp. 203–212.

Johanson, D. C. and Taieb, M. (1976). Plio-Pleistocene hominid discoveries in Hadar, Ethiopia. *Nature* **260**: 293–297.

Johanson, D. C. and White, T. D. (1979). A systematic assessment of early African hominids. *Science* **203**, 321–330.

Johanson, D. C. and White, T. D. (1985). *Science* **203**: 321–330.

Johanson, D. C., White, T. D. and Coppens, Y. (1978). A new species of the genus *Australopithecus* (Primates; Hominidae) from the Pliocene of Eastern Africa. *Kirtlandia* **28**: 1–14.

Johanson, D. C., Lovejoy, C. O., Kimbel, W. H., White, T. D., Ward, S. C., Bush, M. E., Latimer, B. M. and Coppens, Y. (1982). Morphology of the Pliocene Partial hominid skeleton (AL 288-1) from the Hadar formation, Ethiopia. *American Journal of Physical Anthropology* **57**: 403–452.

Johanson, D. C., Masao, F. T., Eck, G. G., White, T. D., Walter, R. C., Kimbel, W. H., Asfaw, B., Manega, P., Nolessokia, P. and Suwa, G. (1987). New partial skeleton of *Homo habilis* from Olduvai Gorge, Tanzania. *Nature* **327**: 205–209.

Jungers, W. L. (1982). Lucy's limbs: skeletal allometry and locomotion in *Australopithecus afarensis*. *Nature* **297**: 676–678.

Jungers, W. L. (1984). Aspects of size and scaling in primate biology with special reference to the locomotor skeleton. *Yearbook of Physical Anthropology* **27**: 73–97.

Jungers, W. L. (1985). Body size and scaling of limb proportions in primates. In: *Size and Scaling in Primate Biology* (W. L. Jungers, Ed.), pp. 345–381. New York: Plenum Press.

Jungers, W. L. (1988a). Relative joint size and hominoid locomotor adaptations with implications for the evolution of hominid bipedalism. *Journal of Human Evolution* 17: 247–265.

Jungers, W. L. (1988b). Lucy's length: stature reconstruction in *Australopithecus afarensis* (AL 288-1) with implications for other small-bodied hominids. *American Journal of Physical Anthropology,* 76: 227–231.

Jungers, W. L. (1988c). New estimates of body size in australopithecines. In: *Evolutionary History of the 'Robust' Australopithecines* (F. E. Grine, Ed.), pp. 115–126. New York: Aldine de Gruyter.

Jungers, W. L. and Grine, F. E. (1986). Dental trends in the australopithecines; the allometry of mandibular molar dimensions. In: *Major Topics in Primate and Human Evolution* (B. A. Wood, L. B. Martin and P. Andrews, Eds), pp. 205–219. Cambridge: Cambridge University Press.

Jungers, W. L. and Stern, J. T., Jr. (1983). Body proportions, skeletal allometry and locomotion in the Hadar hominids: a reply to Wolpoff. *Journal of Human Evolution* 12: 673–684.

Jungers, W. L., Stern, J. T., Jr., and Jouffroy, F. K. (1983). Functional morphology of the *quadriceps femoris* in primates: a comparative anatomical and experimental analysis. *Annales des Sciences Naturelles, Zoologie (Paris)* 5: 101–116.

Juniper, R. P. (1981). The superior pterygoid muscle? *British Journal of Oral Surgery* 19: 121–128.

Kapandji, I. A. (1974) *The physiology of the joints. Volume 3: The trunk and vertebral column.* Edinburgh: Churchill Livingstone.

Kapandji, I. A. (1982). *The Physiology of the Joints. Vol. 1: Upper Limb*, 5th edn. Edinburgh: Churchill Livingstone.

Kapandji, I. A. (1987). *The Physiology of the Joints, Volume 2: Lower Limb*, 5th edn. Edinburgh: Churchill Livingstone.

Kay, R. F. (1973). Humerus of robust *Australopithecus*. *Science* 182: 396.

Kay, R. F. and Hiiemae, K. M. (1974). Jaw movement and tooth use in recent and fossil primates. *American Journal of Physical Anthropology* 40: 227–256.

Keith, A. (1899). On the chimpanzees and their relations to the gorilla. *Proceedings of the Zoological Society of London* 1899: 296–312.

Keith, A. (1913). Problems relating to the earlier forms of prehistoric man. *Proceedings of The Royal Society of Medicine (Odontology)* 6: 103–119.

Keith, A. (1915). *The Antiquity of Man*. London: Williams and Norgate.

Keith, A. (1926). *The Engines of the Human Body*. Philadelphia: J. B. Lippincott.

Keith, A. (1929). The history of the human foot and its bearing on orthopaedic practice. *Journal of Bone and Joint Surgery* 11: 10–32.

Keleman, G. (1969). Anatomy of the larynx and the anatomical basis of vocal performance. In: *The Chimpanzee, Anatomy, Behavior, and Diseases of Chimpanzees* (G. H. Bourne, Ed.), pp. 165–186. Basel: S. Karger.

Kennedy, G. (1983a). A morphometric and taxonomic assessment of a hominine femur from the lower member, Koobi Fora, Lake Turkana. *American Journal of Physical Anthropology* 61: 429–436.

Kennedy, G. (1983b). Some aspects of femoral morphology in *Homo erectus*. *Journal of Human Evolution* 12: 587–616.

Kennedy, G. (1984). The emergence of *Homo sapiens*: the postcranial evidence. *Man* 19: 94–110.

Kern, H. H. and Straus, W. L. (1949). The femur of *Plesianthropus transvaalensis*. *American Journal of Physical Anthropology* 7: 53–77.

Kimbel, W. H. (1984). Variation in the pattern of cranial venous sinuses and hominid phylogeny. *American Journal of Physical Anthropology* 63: 243–263.

Kimbel, W. H. and Rak, Y. (1985). Functional morphology of the asterionic region in extant hominoids and fossil hominids. *American Journal of Physical Anthropology* 66: 31–54.

Kimbel, W. H., White, T. D. and Johanson, D. C. (1984). Cranial morphology of *Australopithecus afarensis*: A comparative study based on a composite reconstruction of the adult skull. *American Journal of Physical Anthropology* 64: 337–388.

Kimbel, W. H., White, T. D. and Johanson, D. C. (1985). Craniodental morphology of the hominids from Hadar and Laetoli; evidence of 'Paranthropus' and *Homo* in the mid-Pliocene of Eastern Africa. In: *Ancestors: The Hard Evidence* (E. Delson, Ed.), pp. 120–137. New

York: A. Liss.

Kimura, T. (1976). Correction to the metacarpal I of the Amud Man. A new description especially on the insertion area of m. opponens pollicis. *Journal of the Anthropological Society of Nippon* **84**: 48–54.

Kimura, T. (1985). Bipedal and quadrupedal walking of primates: comparative dynamics. In: *Primate Morphophysiology, Locomotor Analysis and Human Bipedalism* (S. Kondo, Ed.), pp. 81–104. Tokyo: University of Tokyo Press.

Kimura, T., Okada, M. and Ishida, H. (1979). Kinesiological characteristics of primate walking: its significance in human walking. In: *Environment, Behavior and Morphology: Dynamic Interactions in Primates* (M. E. Morbeck, H. Preuschoft and N. Gomberg, Eds), pp. 297–311. New York: Gustav Fischer.

Knowles, F. H. S. (1915). The glenoid fossa in the skull of the Eskimo. *Canada Geological Survey Bulletin No. 9. Anthropological series No. 4*, pp. 1–25.

Knussmann, R. (1967). *Humerus, Ulna und Radius der Simiae. Bibliotheca Primatologica*, Vol. 5. Basel: S. Karger.

Kollar, E. J. and Baird, G. E. (1969). The influence of the dental papilla on the development of tooth shape in embryonic mouse tooth germs. *Journal of Embryology and Experimental Morphology* **21**: 131–148.

Kovacs, Z. (1971). A systematic description of dental roots. In: *Dental Morphology and Evolution*, (A. A. Dahlberg, Ed.), pp. 211–256. Chicago: Chicago University Press.

Kozam, G. (1985). An anatomists view of the temporomandibular joint. *Journal of the New Jersey Dental Association* **56**: 64–66.

Kramer, A. (1986). Hominid–pongid distinctiveness in the Miocene–Pliocene fossil record: the Lothagam mandible. *American Journal of Physical Anthropology* **70**: 457–475.

Krantz, G. S. (1963). The functional significance of the mastoid processes in man. *American Journal of Physical Anthropology* **21**: 591–593.

Krogman, W. M. (1931a). Studies in growth changes in the skull and face of anthropoids: growth changes in the skull and face of the gorilla. *American Journal of Anatomy* **46**: 325–342.

Krogman, W. M. (1931b). Studies in growth changes in the skull and face of anthropoids: growth changes in the skull and face of the chimpanzee. *American Journal of Anatomy* **47**: 325–342.

Krogman, W. M. (1931c). Studies in growth changes in the skull and face of anthropoids: growth changes in the skull and face of the orang utan. *American Journal of Anatomy* **47**: 343–365.

Krogman, W. M. and Iscan, M. Y. (1986). *The Human Skeleton in Forensic Medicine*. Springfield: Thomas.

Kummer, B. K. F. (1975). Functional adaptation to posture in the pelvis of man and other primates. In: *Primate Functional Morphology and Evolution* (R. Tuttle, Ed.), pp. 281–290. The Hague: Mouton.

Laitman, J. T. (1977). The ontogenetic and phylogenetic development of the upper respiratory system and basicranium in man. Ph.D. Dissertation, Yale University.

Laitman, J. T., Heimubuch, R. C. and Crelin, E. S. (1978). Developmental change in a basicranial line and its relationship to the upper respiratory system in living primates. *American Journal of Anatomy* **152**: 467–483.

Laitman, J. T., Heimbuch, R. C. and Crelin, E. S. (1979). The basicranium of fossil hominids as an indicator of their upper respiratory systems. *American Journal of Physical Anthropology* **51**: 15–34.

Laitman, J. T. and Heimbuch, R. C. (1982). The basicranium of Plio-Pleistocene hominids as an indicator of their upper respiratory systems. *American Journal of Physical Anthropology* **59**: 323–343.

Lanier, R. R. (1939). The presacral vertebrae of American white and negro males. *American Journal of Physical Anthropology* **25**: 341–420.

Larson, S. G. (1988). Subscapularis function in gibbons and chimpanzees: implications for interpretation of humeral head torsion in hominoids. *American Journal of Physical Anthropology* **76**: 449–462.

Larson, S. G. and Stern, J. T. (1986). EMG of scapulohumeral muscles in the chimpanzee during reaching and 'arboreal' locomotion. *American Journal of Anatomy* **176**: 171–190.

Larson, S. G. and Stern, J. T. (1989). Role of supraspinatus in the quadrupedal locomotion of vervets (*Cercopithecus aethiops*); implications for interpretation of humeral morphology. *American Journal of Physical Anthropology* **79**: 369–377.

Last, R. J. (1955). The muscles of the head and neck: a review. *British Dental Journal* **5**:

338–354.

Latham, R. A. (1972). The sella point and postnatal growth of the cranial base. *American Journal of Orthodontics* **61**: 156–162.

Latimer, B. and Lovejoy, C. O. (1989). The calcaneus of *Australopithecus afarensis* and its implications for the evolution of bipedality. *American Journal of Physical Anthropology* **78**: 369–386.

Latimer, B., Lovejoy, C. O., Johanson, D. C. and Coppens, Y. (1982). Hominid tarsal, metatarsal and phalangeal bones recovered from the Hadar formation: 1974–1977 collections. *American Journal of Physical Anthropology* **57**: 701–719.

Latimer, B., Ohman, J. C. and Lovejoy, C. O. (1987). Talocrural joint in African hominoids: implications for *Australopithecus afarensis*. *American Journal of Physical Anthropology* **74**: 155–175.

Leakey, L. S. B., Tobias, P. V. and Napier, J. R. (1964). A new species of the genus *Homo* from Olduvai Gorge. *Nature* **202**: 7–9.

Leakey, M. D. (1979). Footprints in the ashes of time. *National Geographic* **155**: 446–457.

Leakey, M. D. and Hay, R. L. (1979). Pliocene footprints in the Laetolil beds at Laetoli, north Tanzania. *Nature* **278**: 317–323.

Leakey, M. G. and Leakey R. E. (1978). *Koobi Fora Research Project volume 1: The Fossil Hominids and an Introduction to their Context 1968–1974*. Oxford: Clarendon Press.

Leakey, R. E. F. (1971). Further evidence of Lower Pleistocene hominids from East Rudolf, North Kenya. *Nature* **231**: 241–245.

Leakey, R. E. F. (1972). Further evidence of Lower Pleistocene hominids from East Rudolf, North Kenya. *Nature* **237**: 264–269.

Leakey, R. E. F. (1973). Further evidence of Lower Pleistocene hominids from East Rudolf, Northern Kenya, 1972. *Nature* **242**: 170–173.

Leakey, R. E. F. and Walker, A. (1988). New *Australopithecus boisei* specimens from East and West Turkana, Kenya. *American Journal of Physical Anthropology* **76**: 1–24.

Leakey, R. E. F. and Wood, B. A. (1973). New evidence of the genus *Homo*, East Rudolf, Kenya. *American Journal of Physical Anthropology* **39**: 355–368.

Leakey, R. E. F., Mungai, J. M. and Walker, A. C. (1971). New australopithecines from East Rudolf, Kenya. *American Journal of Physical Anthropology* **35**: 175–186.

Le Gros Clark, W. E. (1947a). The importance of the fossil Australopithecinae in the study of human evolution. *Science Progress* **35**: 377–395.

Le Gros Clark, W. E. (1947b). Observations on the anatomy of the fossil Australopithecinae. *Journal of Anatomy* **81**: 300–333.

Le Gros Clark, W. E. (1947c). Note on the palaeontology of the lemuroid brain. *Journal of Anatomy* **79**: 123–126.

Le Gros Clark, W. E. (1950). Hominid characters of the australopithecine dentition. *Journal of The Royal Anthropological Institute (Great Britain and Northern Ireland)* **80**: 37–54.

Le Gros Clark, W. E. (1964). *The Fossil Evidence for Human Evolution*, 2nd edn. Chicago: University of Chicago Press.

Le Gros Clark, W. E. (1971). *The Antecedents of Man*, 3rd edn. Edinburgh: Edinburgh University Press.

Le Gros Clark, W. E. (1972). *The Fossil Evidence for Human Evolution*, 2nd edn. Chicago: Chicago University Press.

LeMay, M. (1976). Morphological cerebral asymmetry of modern man, fossil man and non-human primates. *Annals of The New York Academy of Science* **280**: 348–366.

Leutenegger, W. (1972). Newborn size and pelvic dimensions of *Australopithecus*. *Nature* **240**: 568–569.

Leutenegger, W. (1974). Functional aspects of pelvic morphology in simian primates. *Journal of Human Evolution* **3**: 207–222.

Leutenegger, W. (1977). A functional interpretation of the sacrum of *Australopithecus africanus*. *South African Journal of Science* **73**: 308–310.

Leutenegger, W. (1982). Encephalization and obstetrics in primates with particular reference to human evolution. In: *Evolution: Methods and Concepts* (E. Armstrong and D. Falk, Eds), pp. 85–95. New York: Plenum Press.

Leutenegger, W. (1987). Neonatal brain size and neurocranial dimensions in Pliocene hominids: implications for obstetrics. *Journal of Human Evolution* **16**: 291–296.

Lewin, R. (1987). The earliest 'humans' were more like apes. *Science* **236**: 1061–1063.

Lewis, O. J. (1962). The comparative morphology of m. flexor accessorious and the associated long flexor tendons. *Journal of Anatomy* **96**: 321–333.

Lewis, O. J. (1964). The tibialis posterior tendon in the primate foot. *Journal of Anatomy* **98**:

209–218.

Lewis, O. J. (1972a). Evolution of the hominoid wrist. In: *The Functional and Evolutionary Biology of the Primates* (R. H. Tuttle, Ed.), pp. 207–222. Chicago: Aldine-Atherton.

Lewis, O. J. (1972b). Osteological features characterizing the wrists of monkeys and apes, with a reconsideration of this region in *Dryopithecus (Proconsul) africanus*. *American Journal of Physical Anthropology* **36**: 45–58.

Lewis, O. J. (1972c). The evolution of the hallucial tarsometatarsal joint in the Anthropoidea. *American Journal of Physical Anthropology* **37**: 13–34.

Lewis, O. J. (1973). The hominoid os capitatum with special reference to the fossil bones from Sterkfontein and Olduvai Gorge. *Journal of Human Evolution* **2**: 1–11.

Lewis, O. J. (1974). The wrist articulations of the Anthropoidea. In: *Primate Locomotion* (F. A. Jenkins, Ed.), pp. 143–169. New York: Academic Press.

Lewis, O. J. (1977). Joint remodelling and the evolution of the human hand. *Journal of Anatomy* **123**: 157–201.

Lewis, O. J. (1980a). The joints of the evolving foot. Part I. The ankle joint. *Journal of Anatomy* **130**: 527–543.

Lewis, O. J. (1980b). The joints of the evolving foot. Part II. The intrinsic joints. *Journal of Anatomy* **130**: 833–857.

Lewis, O. J. (1980c). The joints of the evolving foot. Part III. The fossil evidence. *Journal of Anatomy* **131**: 275–298.

Lewis, O. J. (1981). Functional morphology of the joints of the evolving foot. *Symposia of the Zoological Society of London* **46**: 169–188.

Lewis, A. B. and Garn, S. M. (1960). The relationship between tooth formation and other maturational factors. *Angle Orthodontist* **30**: 70–77.

Lieberman, P. and Crelin, E. S. (1971). On the speech of Neanderthal man. *Linguistic Inquiry* **2**: 203–222.

Lightoller, G. S. (1929). The facial muscles of three orang-utans and two cercopithecidae. *Journal of Anatomy* **63**: 19–81.

Lisowski, F. P. (1967). Angular growth changes and comparisons in the primate talus. *Folia Primatologica* **7**: 81–97.

Lisowski, F. P., Albrecht, G. H. and Oxnard, C. E. (1974). The form of the talus in some higher primates: a multivariate study. *American Journal of Physical Anthropology* **41**: 191–216.

Lisowski, F. P., Albrecht, G. H. and Oxnard, C. E. (1976). African fossil tali: further multivariate morphometric studies. *American Journal of Physical Anthropology* **45**: 5–18.

Loth, E. (1938). Beiträge zur Kenntnis der Weichteilanatomie des Neanderthalers. *Zeitschrift für Rassenkunde* **7**: 13–35.

Lovejoy, C. O. (1974). The gait of australopithecines. *Yearbook of Physical Anthropology* **17**: 147–161.

Lovejoy, C. O. (1975). Biomechanical perspectives on the lower limb of early hominids. In: *Primate Functional Morphology and Evolution* (R. H. Tuttle, Ed.), pp. 291–326. The Hague: Mouton.

Lovejoy, C. O. (1978). A biomechanical review of the locomotor diversity of early hominids. In: *Early Hominids of Africa* (C. J. Jolly, Ed.), pp. 403–429. New York: St Martins Press.

Lovejoy, C. O. (1988). Evolution of human walking. *Scientific American* **259**: 82–89.

Lovejoy, C. O. and Heiple, K. G. (1970). A reconstruction of the femur of *Australopithecus africanus*. *American Journal of Physical Anthropology* **32**: 33–40.

Lovejoy, C. O. and Heiple, K. G. (1972). Proximal femoral anatomy of *Australopithecus*. *Nature* **235**: 175–176.

Lovejoy, C. O. and Trinkaus, E. (1980). Strength and robusticity of the Neanderthal tibia. *American Journal of Physical Anthropology* **53**: 465–470.

Lovejoy, C. O., Heiple, K. G. and Burstein, A. H. (1973). The gait of *Australopithecus*. *American Journal of Physical Anthropology* **38**: 757–779.

Lovejoy, C. O., Burstein, A. H. and Heiple, K. G. (1976). The biomechanical analysis of bone strength: a method and its applications to platycnemia. *American Journal of Physical Anthropology* **44**: 489–506.

Lovejoy, C. O., Johanson, D. C. and Coppens, Y. (1982a). Hominid lower limb bones recovered from the Hadar Formation: 1974–1977 collections. *American Journal of Physical Anthropology* **57**: 679–700.

Lovejoy, C. O., Johanson, D. C. and Coppens, Y. (1982b). Hominid upper limb bones recovered for the Hadar formation: 1974–1977 collections. *American Journal of Physical Anthropology* **57**: 637–650.

Lovejoy, C. O., Johanson, D. C. and Coppens,

(1982c). Elements of the axial skeleton recovered from the Hadar Formation: 1974–1977 collections. *American Journal of Physical Anthropology* **57**: 631–636.

Lovell, N. C. (1989). Test of Phenice's technique for determine sex from the os pubis. *American Journal of Physical Anthropology* **79**: 117–120.

MacConaill, M. A. (1945). The postural mechanism of the human foot. *Proceedings of the Royal Irish Academy* **50B**: 265–278.

MacConaill, M. A. and Basmajian, J. V. (1969). *Muscles and Movements.* Baltimore: The Williams & Wilkins Company.

MacNamara, J. A. (1973). The independent functions of the two heads of the lateral pterygoid muscle. *American Journal of Anatomy* **138**: 197–205.

MacNamara, J. A. (1974). An electromyographic study of mastication in the rhesus monkey (*Macaca mulatta*). *Archives of Oral Biology* **19**: 821–823.

Maier, N. and Nkini, A. (1984). Olduvai hominid 9: new results of investigation. In: *The Early Evolution of Man* (P. Andrews and J. L. Franzen, Eds). *Cour. Forsch. Inst. Senckenberg* **69**: 69–82.

Manato, S., Baron, G., Stephan, H. and Frahm, H. D. (1985a). Volume comparisons in the cerebellar complex of primates (11 cerebellar nuclei). *Folia Primatologica* **44**: 182–203.

Manato, S., Stephan, H., and Baron, G. (1985b). Volume comparisons in the cerebellar complex of primates. (1 ventral pons). *Folia Primatologica* **44**: 171–181.

Mann, R. and Inman, V. T. (1964). Phasic activity of intrinsic muscles of the foot. *Journal of Bone and Joint Surgery* **46A**: 469–481.

Marshall, J. B. (1986). Mandibular symphysis (medial suture) closure in modern *Homo sapiens*: preliminary evidence from archaological populations. *American Journal of Physical Anthropology* **69**: 499–502.

Martin, C. P. (1932). Some variations in the lower end of the femur which are especially prevalent in the bones of primitive people. *Journal of Anatomy* **66**: 371–383.

Martin, L. B. (1985). Significance of enamel thickness in hominoid evolution. *Nature* **314**: 260–263.

Martin, R. (1928). *Lehrbuch der Anthropologie,* Vols 1–3, 2nd edn. Jena: Gustav Fischer.

Martin, R. and Saller, K. (1957). *Lehrbuch der Anthropologie,* Vol. I. 3rd edn. Stuttgart: Fischer.

Martin, R. and Saller, K. (1959). *Lehrbuch der Anthropologie,* Vol. 2, 3rd edn. Stuttgart: Gustav Fischer.

Martin, R. D. (1982). Allometric approaches to the evolution of the primate nervous system. In: *Primate Brain Evolution: Methods and Concepts* (E. Armstrong and D. Falk, Eds), pp. 39–56. New York: Plenum Press.

Martin, R. D. (1983). *Human Brain Evolution in an Ecological Context.* 52nd James Arthur Lecture on The Evolution of the Human Brain 1982. New York: American Museum of Natural History.

Marzke, M. W. (1983). Joint function and grips of the *Australopithecus afarensis* hand, with special reference to the region of the capitate. *Journal of Human Evolution* **12**: 197–211.

Marzke, W. M. and Marzke, R. F. (1987). The third metacarpal styloid process in humans: origin and functions. *American Journal of Physical Anthropology* **73**: 415–432.

Marzke, W. M. and Shackley, M. S. (1986). Hominid hand use in the Pliocene and Pleistocene: evidence from experimental archaeology and comparative morphology. *Journal of Human Evolution* **15**: 439–460.

Marzke, W. M., Longhill, J. M. and Rasmussen, S. A. (1988). Gluteus maximus muscle function and the origin of hominid bipedality. *American Journal of Physical Anthropology* **77**: 519–528.

Matsushima, T., Rhoton, A. L., de Oliveira, E. and Peace, D. (1983). Microsurgical anatomy of the veins of the posterior fossa. *Journal of Neurosurgery* **59**: 63–105.

Mayr, E. (1940). Speciation phenomena in birds. *American Naturalist* **74**: 249–278.

Mayr, E. (1969). *Principles of Systematic Zoology.* New York: McGraw-Hill.

McCormick, W. F. (1981). Sternal foramina in man. *American Journal of Forensic Medicine and Pathology* **2**: 249–252.

McCown, T. D. and Keith, A. (1939). *The Stone Age of Mount Carmel. 2, The Fossil Human Remains from the Levalloiso-Mousterian.* Oxford: Clarendon Press.

McHenry, H. M. (1973). Early hominid humerus from East Rudolf, Kenya. *Science* **180**: 739–741.

McHenry, H. M. (1974). How large were the Australopithecines? *American Journal of Physical Anthropology* **40**: 329–340.

McHenry, H. M. (1975a). Biomechanical interpretations of the early hominid hip. *Journal of Human Evolution* **4**: 343–355.

McHenry, H. M. (1975b). Fossil hominid body weight and brain size. *Nature* **254**: 686–688.

McHenry, H. M. (1975c). A new pelvic fragment from Swartkrans and the relationship between the robust and gracile australopithecines. *American Journal of Physical Anthropology* **43**: 245–262.

McHenry, H. M. (1975d). Fossils and the mosaic nature of human evolution. *Science* **190**: 425–431.

McHenry, H. M. (1975e). The ischium and hip extensor mechanism in human evolution. *American Journal of Physical Anthropology* **43**: 39–46.

McHenry, H. M. (1976). Early hominid body weight and encephalization. *American Journal of Physical Anthropology* **45**: 77–84.

McHenry, H. M. (1978). Fore- and hind-limb proportions in Plio-Pleistocene hominids. *American Journal of Physical Anthropology* **49**: 15–22.

McHenry, H. M. (1982). The pattern of human evolution: studies on bipedalism, mastication and encephalisation. *Annual Review of Anthropology* **11**: 151–173.

McHenry, H. M. (1983). The capitate of *Australopithecus afarensis* and *A. africanus*. *American Journal of Physical Anthropology* **62**: 187–198.

McHenry, H. M. (1986). The first bipeds: a comparison of the *A. afarensis* and *A. africanus* postcranium and implications for the evolution of bipedalism. *Journal of Human Evolution* **15**: 177–191.

McHenry, H. M. (1988). New estimates of body weight in early hominids and their significance to encephalization and megadontia in 'robust' australopithecines, In: *Evolutionary History of the 'Robust' Australopithecines* (F. E. Grine, Ed.), pp. 133–148. New York: Aldine de Gruyter.

McHenry, H. M. and Corruccini, R. S. (1975). Distal humerus in hominoid evolution. *Folia Primatologia* **23**: 227–244.

McHenry, H. M. and Corruccini, R. S. (1976a). Fossil hominid femora and the evolution of walking. *Nature* **259**: 657–658.

McHenry, H. M. and Corruccini, R. S. (1976b). Fossil hominid femora (Reply). *Nature* **264**: 813.

McHenry, H. M. and Corruccini, R. S. (1978). The femur in early human evolution. *American Journal of Physical Anthropology* **49**: 473–488.

McHenry, H. M. and Temerin, L. A. (1979). The evolution of hominid bipedalism: evidence from the fossil record. *Yearbook of Physical Anthropology* **22**: 105–131.

McHenry, H. M., Corruccini, R. S. and Howell, F. C. (1976). Analysis of an early hominid ulna from the Omo Basin, Ethiopia. *American Journal of Physical Anthropology* **44**: 295–304.

Miles, A. E. W. (1963). The dentition in the assessment of individual age in skeletal material. In: *Dental Anthropology* (D. R. Brothwell, Ed.), pp. 191–209. Oxford: Pergamon Press.

Miller, R. A. (1932). Evolution of the pectoral girdle and forelimb in the primates. *American Journal of Physical Anthropology* **17**: 1–56.

Mobb, G. E. and Wood, B. A. (1977). Allometry and sexual dimorphism in the primate innominate bone. *American Journal of Anatomy* **150**: 531–538.

Moore, W. J., Adams, L. M. and Lavelle, C. L. B. (1973). Head posture in the hominoidea. *Journal of The Zoological Society, London* **169**: 409–416.

Moore, W. J. and Lavelle, C. B. L. (1974). *Growth of the Facial Skeleton in the Hominoidea*. London: Academic Press.

Morrison, J. B. (1970). The mechanics of the knee joint in relation to normal walking. *Journal of Biomechanics* **3**: 51–61.

Morton, D. J. (1922). Evolution of the human foot. *American Journal of Physical Anthropology* **5**: 305–325.

Morton, D. J. (1926). Significant characteristics of the Neanderthal foot. *Natural History* **26**: 310–314.

Moss, M. L. (1958). The pathogenesis of artificial cranial deformation. *American Journal of Physical Anthropology* **16**: 269–285.

Moss, M. L. (1963). Morphological variations of the crista galli and medial orbital margin. *American Journal of Physical Anthropology* **21**: 259–264.

Moss, M. L. and Young, R. W. (1960). A functional approach to craniology. *American Journal of Physical Anthropology* **18**: 281–292.

Moss, M. L., Norback, C. B. and Robertson, G. G. (1956). Growth of certain human fetal cranial bones. *American Journal of Anatomy* **97**: 155–176.

Murphy, T. (1955). The spheno-ethmoidal articulation in the anterior cranial fossa of the Australian aborigine. *American Journal of Physical Anthropology* **13**: 285–300.

Musgrave, J. H. (1969). A comparative study of the hand bones of Neanderthal man. *Human Biology* **41**: 587–588.

Musgrave, J. H. (1970). *An Anatomical Study of the Hands of Pleistocene and Recent Man.* Ph.D. Thesis, Churchill College, University of Cambridge.

Musgrave, J. H. (1971). How dextrous was Neanderthal man? *Nature* **233**: 538–541.

Musgrave, J. H. (1973). The phalanges of Neanderthal and Upper Paleolithic hands. In: *Human Evolution* (M. H. Day, Ed.), *Symposia of the Society for the Study of Human Biology* **11**: 59–85.

Napier, J. R. (1959). *Fossil Metacarpals from Swartkrans. Fossil Mammals of Africa, No. 17* London: British Museum (Natural History).

Napier, J. R. (1960). Studies of the hands of living primates. *Proceedings of the Zoological Society of London* **134**: 647–657.

Napier, J. R. (1962). Fossil hand bones from Olduvai Gorge. *Nature* **196**: 409–411.

Napier, J. (1963). Brachiation and brachiators. In: *The Primates* (J. Napier and N. A. Barnicot, Eds.), *Symposia of the Zoological Society of London* **10**: 183–196.

Napier, J. R. (1964). The evolution of bipedal walking in the hominids. *Archives de Biologie (Liège)* **75**: 673–708.

Napier, J. R. (1980). *Hands.* London: George Allen & Unwin.

Napier, J. R. and Davis, P. R. (1959). The forelimb and associated remains of *Proconsul africanus. Fossil Mammals of Africa* **16**: 1–70.

Newell-Morris, L. and Fahrenbuch, C. E. (1985). Practical considerations for use of the nonhuman primate model in prenatal research. In: *Nonhuman Primate Models for Human Growth and Development* (E. S. Watts, Ed.), pp. 9–40. New York: A. Liss.

Nissen, H. W. and Reisen, A. H. (1964). The eruption of the permanent dentition in the chimpanzee. *American Journal of Physical Anthropology* **22**: 285–294.

O'Conner, B. L. and Rarey, K. E. (1979). Normal amplitudes of radioulnar pronation and supination in several genera of anthropoid primate. *American Journal of Physical Anthropology* **51**: 39–44.

Ogawa, T., Kamiya, T., Sakai, S. and Hosokawa, H. (1970). Some observations on the endocranial cast of the Amud man. In: *The Amud Man and His Cave Site* (H. Suzuki and F. Takai, Eds). Tokyo: Academic Press of Japan.

Ogden, G. R. (1988). The significance of taurodontism in dental surgery. *Dental Update* **15**: 32–34.

Ohman, J. C. (1986). The first rib of hominoids. *American Journal of Physical Anthropology* **70**: 209–230.

Ohtsuki, F. (1977). Developmental changes of cranial bone thickness in the human fetal period. *American Journal of Physical Anthropology* **46**: 141–154.

Okada, M. (1943). Hard tissues of animal body. Highly interesting details of Nippon studies in periodic patterns of hard tissues are described. *The Shanghai Evening Post (Special Edition) Health, Recreation and Medical Progress,* pp. 15–31.

Okada, M. (1985). Primate bipedal walking: comparative kinematics. In: *Primate Morphophysiology, Locomotor Analysis and Human Bipedalism* (S. Kondo, Ed.), pp. 47–58. Tokyo: University of Tokyo Press.

Oliver, G. (1976). The stature of *Australopithecus. Journal of Human Evolution* **5**: 529–534.

Olivier, G. and Tissier, H. (1975). Determination of cranial capacity in fossil men. *American Journal of Physical Anthropology* **43**: 353–362.

Olson, T. R. (1985). Cranial morphology and systematics of the Hadar formation hominids and 'Australopithecus' africanus. In: *Ancestors; The Hard Evidence* (E. Delson, Ed.), pp. 102–119. New York: A. Liss.

Orchardson, R. and MacFarlane, S. H. (1980). The effect of local anaesthesia on the maximum biting force achieved by human subjects. *Archives of Oral Biology* **25**: 799–804.

Owen, R. (1831). On the anatomy of the orangutan. *Proceedings of The Zoological Society of London* **1**: 28–29.

Owen, R. (1859). On the orang, chimpanzee and gorilla. In: *On the Classification and Geographical Distribution of the Mammalia,* Appendix B, pp. 64–103. (Reade Lecture, Cambridge, May 1859). London: Parker.

Oxnard, C. E. (1963). Locomotor adaptations in the primate forelimb. *Symposia of the Zoological Society of London* **10**: 165–182.

Oxnard, C. E. (1967). The functional morphology of the primate shoulder as revealed by comparative anatomical, osteometric and discriminant function techniques. *American Journal of Physical Anthropology* **26**: 219–240.

Oxnard, C. E. (1968). Note on the fragmentary

Sterkfontein scapula. *American Journal of Physical Anthropology* **28**: 213–218.

Oxnard, C. E. (1969). Evolution of the human shoulder: some possible pathways. *American Journal of Physical Anthropology* **30**: 319–331.

Oxnard, C. E. (1972a). Some African fossil foot bones: a note on the interpolation of fossils into a matrix of extant species. *American Journal of Physical Anthropology* **37**: 3–12.

Oxnard, C. E. (1972b). Functional morphology of primates: some mathematical and physical methods. In: *Functional and Evolutionary Biology of Primates* (R. H. Tuttle, Ed.), pp. 305–336. Chicago: Aldine-Atherton.

Oxnard, C. E. (1973a). *Form and Pattern in Human Evolution*. Chicago: University of Chicago Press.

Oxnard, C. E. (1973b). Functional inferences from morphometrics: problems posed by uniqueness and diversity among the primates. *Systematic Zoology* **22**: 409–424.

Oxnard, C. E. (1975). The place of the australopithecines in human evolution: grounds for doubt? *Nature* **258**: 389–395.

Oxnard, C. E. (1984). *The Order of Man*. New Haven, Yale University Press.

Oxnard, C. E. and Lisowski, F. P. (1980). Functional articulation of some hominoid foot bones: implications for the Olduvai (Hominid 8) foot. *American Journal of Physical Anthropology* **52**: 107–117.

Parsons, F. G. (1898). The muscles of mammals with special reference to human myology. *Journal of Anatomy and Physiology, London* **32**: 428–450.

Patterson, B. and Howells, W. W. (1967). Hominid humeral fragment from Early Pleistocene of northwestern Kenya. *Science* **156**: 64–66.

Pauly, J. E., Rushing, J. L. and Scheving, L. E. (1967). An electromyographic study of some muscles crossing the elbow joint. *Anatomical Record* **159**: 47–54.

Pearson, K. and Bell, J. (1919). A study of the long bones of the English skeleton. I. The femur. *Drapers Company Memoirs*, Biometric Series 10, pp. 1–224. Cambridge: Cambridge University Press.

Pegington, J. (1985). *Clinical Anatomy in Action*. Edinburgh: Churchill Livingstone.

Penning, L. (1988). Functional significance of the uncovertebral joints. *Annals of The Royal College of Surgeons of England* **70**: 164.

Phenice, T. W. (1969). A newly developed visual method of sexing the os pubis. *American Journal of Physical Anthropology* **30**: 297–302.

Pickford, M., Johanson, D. C., Lovejoy, C. O., White, T. D. and Aronson, J. L. (1983). A hominoid humeral fragment from the Pliocene of Kenya. *American Journal of Physical Anthropology* **60**: 337–346.

Picton, D. C. A. (1962). Distortion of the jaws during biting. *Archives of Oral Biology* **7**: 573–580.

Pilbeam, D. (1972). *The Ascent of Man*. New York: MacMillan.

Popper, K. R. and Eccles, J. C. (1977). *The Self and Its Brain*. London: Routledge & Kegan Paul.

Povinelli, D. and Sterling, E. (1988). *Retroversion of the tibial plateau in humans and apes*. Yale University; unpublished manuscript.

Prejzner-Morawska, P. and Urbanowicz, M. (1981). Morphology of some of the lower limb muscles in primates. In: *Primate Evolutionary Biology* (A. B. Chlarelli and R. S. Corruccini, Eds), pp. 60–67. Berlin: Springer Verlag.

Preuschoft, H. (1970). Functional anatomy of the lower extremity. In: *The Chimpanzee*, Vol. 3. (G. H. Bourne, Ed.), pp. 221–294. Basel: Karger.

Preuschoft, H. (1973a). Body posture and locomotion in some East African miocene Dryopithecinae. In: *Human Evolution* (M. H. Day, Ed.), *Symposia of the Society for the Study of Human Biology* **11**: 13–46.

Preuschoft, H. (1973b). Functional anatomy of the upper extremity. In: *The Chimpanzee*, Vol. 6, (G. H. Bourne, Ed.), pp. 34–120. Basel: Karger.

Prost, J. H. (1980). Origin of bipedalism. *American Journal of Physical Anthropology* **52**: 175–189.

Raisz, L. G. (1988). Local and systemic factors in the pathogenesis of osteoporosis. *The New England Journal of Medicine*, 818–828.

Rak, Y. (1978). The functional significance of the squamosal suture in *Australopithecus boisei*. *American Journal of Physical Anthropology* **49**: 71–78.

Rak, Y. (1983). *The Australopithecine Face*. London: Academic Press.

Rak, Y. (1985). Australopithecine taxonomy and phylogeny in light of facial morphology. *American Journal of Physical Anthropology* **66**: 281–287.

Rak, Y. (1987). The Neanderthal; a new look at

an old face. *Journal of Human Evolution* **15**: 151–164.

Rak, Y. (1988). On variation in the masticatory system of *Australopithecus boisei*. In: *Evolutionary History of the 'Robust' Australopithecines*, pp. 193–198. New York: Aldine de Gruyter.

Rak, Y. and Arensburg, B. (1987). Kebara 2 Neanderthal pelvis: first look at a complete inlet. *American Journal of Physical Anthropology* **73**: 227–231.

Rak, Y. and Howell, F. C. (1978). Cranium of a juvenile *Australopithecus boisei* from the lower Omo basin, Ethiopia. *American Journal of Physical Anthropology* **48**: 345–366.

Ralston, H. J. (1976). Energetics of human walking. In: *Neural Control of Locomotion* (R. M. Herman, S. Grillner, P. Stern and D. Stuart, Eds) New York: Plenum Press.

Rao, P. D. P. (1966). Squatting facets on the talus and tibia in Australian Aborigines. *Archaeology and Physical Anthropology in Oceania* **1**: 51–56.

Raven, H. (1950). *The Anatomy of the Gorilla* (W. Gregory, Ed.). New York: Colombia University Press.

Rees, L. A. (1954). The structure and function of the mandibular joint. *British Dental Journal* **96**: 125–133.

Reeser, L. A., Susman, R. L. and Stern, J. T. (1983). Electromyographic studies of the human foot: experimental approaches to hominid evolution. *Foot and Ankle* **3**: 391–407.

Reynolds, T. R. (1985). Mechanics of increased support of weight by the hindlimbs in primates. *American Journal of Physical Anthropology* **67**: 335–349.

Reynolds, T. R. (1987). Stride length and its determinants in humans, early hominids, primates, and mammals. *American Journal of Physical Anthropology* **72**: 101–115.

Reynolds, V. (1976). *The Biology of Human Action*. Reading, San Francisco: Freeman.

Rhoads, J. G. and Trinkaus, E. (1977). Morphometrics of the Neanderthal talus. *American Journal of Physical Anthropology* **46**: 29–44.

Rightmire, G. P. (1983). The Lake Ndutu cranium and early *Homo sapiens* in Africa. *American Journal of Physical Anthropology* **61**: 245–254.

Roberts, D. (1974). Structure and function of the primate scapula. In: *Primate Locomotion* (F. A. Jenkins, Jr., Ed.), pp. 171–200. New York: Academic Press.

Robinson, J. T. (1954). The genera and species of the australopithecinae. *American Journal of Physical Anthropology* **12**: 181–200.

Robinson, J. T. (1956). The dentition of the Australopithecinae. *Transvaal Museum Memoirs No. 9*, pp. 1–179. Pretoria, South Africa.

Robinson, J. T. (1965). *Homo habilis* and the australopithecines. *Nature* **205**: 121–124.

Robinson, J. T. (1966). On the distinctiveness of *Homo habilis*. *Nature* **209**: 957–960.

Robinson, J. T. (1972). *Early Hominid Posture and Locomotion*. Chicago: University of Chicago Press.

Robinson, J. T. (1978). Evidence for locomotor difference between gracile and robust early hominids from South Africa. In: *Early Hominids of Africa* (C. J. Jolly, Ed.) pp. 441–457. New York: St. Martin.

Rodenstein, D. O., Perlmutter, N. and Stanescu, D. C. (1985). Infants are not obligatory nasal breathers. *American Review of Respiratory Disease* **131**: 343–347.

Rodman, P. S. and McHenry, H. M. (1980). Bioenergetics and the origin of hominid bipedalism. *American Journal of Physical Anthropology* **52**: 103–106.

Romanes, G. J. (1964). *Cunningham's Textbook of Anatomy*. London: Oxford University Press.

Rose, M. D. (1975). Functional proportions of primate lumbar vertebral bodies. *Journal of Human Evolution* **4**: 21–38.

Rose, M. D. (1984). A hominine hip bone, KNM-ER 3228, from East lake Turkana, Kenya. *American Journal of Physical Anthropology* **63**: 371–378.

Rosenberg, K. R. (1986). Sexual dimorphism in the Neanderthal pelvis. *American Journal of Physical Anthropology* **69**: 257 (Abstr.).

Rouvier, H. (1927). *Anatomie Humaine; Descriptive et Topographique*, Vol. 1. Paris: Masson and Company.

Ruff, R. (1988). Hindlimb articular surface allometry in Hominoidea and *Macaca*, with comparisons to diaphyseal scaling. *Journal of Human Evolution* **17**: 687–714.

Saban, R. (1977). Les impressions vasculaires pariétales endocraniennes dans la lignée des Hominides. *C.R. Acad. des Sci. Paris* **284**: 803–806.

Saban, R. (1986) Veines meningées et hominisation. Fossil Man, new facts, new ideas. *Anthropos, (Brno)* **23**: 15–33.

Sakka, M. (1972). Anatomie comparée de l'écaille de l'occipital (squama occipitalis P.N.A.) et des muscles de la nuque chez l'Homme et les pongides. 1: Ostéologie. *Mammalia* **36**: 696–750.

Sakka, M. (1984). Cranial morphology and masticatory adaptations. In: *Food Acquisition and Processing in Primates* (D. J. Chivers, B. A. Wood and A. Billsborough, Eds), pp. 415–427. New York: Plenum Press.

Santa Luca, A. P. (1980). The Ngandong fossil hominids: a comparative study of a far eastern *Homo erectus* group. *Yale University Publications in Anthropology, No. 78.*

Sarasin, F. (1932). Die Variation im Bau des Handskeletts verschiedener Menschenformen. *Zeitschrift für Morphologie und Anthropologie* **30**: 252–316.

Saunders, J. B. de C. M., Inman, V. T. and Eberhart, H. D. (1953). The major determinants in normal and pathological gait. *Journal of Bone and Joint Surgery* **35A**: 543–558.

Schepers, G. W. H. (1946). In: The South African fossil ape-men the Australopithecinae (R. Broom and G. W. H. Schepers, Eds), *Transvaal Museum Memoirs* **2**: 165–272.

Schepers, G. W. H. (1950). Sterkfontein ape-man Plesianthropus. Part 2. *Transvaal Museum Memoirs No. 4.*

Schmid, P. (1983) Eine Rekonstrucktion des Skelettes von A. L. 288-1 (Hadar) und deren Konsequenzen. *Folia Primatologica* **40**: 283–306.

Schmidt, R. F. (1978). *Fundamentals of Neurophysiology.* New York: Springer-Verlag.

Schmidt-Nielsen, K. (1979). *Animal Physiology, Adaptation and Environment,* 2nd edn. Cambridge: Cambridge University Press.

Schultz, A. H. (1930). The skeleton of the trunk and limbs of higher primates. *Human Biology* **2**: 303–438.

Schultz, A. H. (1935). Eruption and decay of the permanent teeth in primates. *American Journal of Physical Anthropology* **19**: 489–581.

Schultz, A. H. (1937). Proportions, variability, and asymmetries of the long bones of the limbs and the clavicles in man and apes. *Human Biology* **9**: 281–328.

Schultz, A. H. (1940). The size of the orbit and of the eye in primates. *American Journal of Physical Anthropology* **26**: 389–408.

Schultz, A. H. (1941). Growth and development of the orang-utan. *Contributions to Embryology* **29**: 57–111.

Schultz, A. H. (1942). Conditions for balancing the head in primates. *American Journal of*

Physical Anthropology **29**: 483–497.

Schultz, A. H. (1949). Sex differences in the pelves of primates. *American Journal of Physical Anthropology* **7**: 401–423.

Schultz, A. H. (1950). The physical distinction of man. *Proceedings of the American Philosophical Society* **94**: 428–449.

Schultz, A. H. (1953). The relative thickness of the long bones and the vertebrae in primates. *American Journal of Physical Anthropology* **11**: 277–310.

Schultz, A. H. (1955). The position of the occipital condyles and of the face relative to the skull base in primates. *American Journal of Physical Anthropology* **13**: 97–120.

Schultz, A. H. (1956). Postembryonic age changes. *Primatologia* **1**: 887–964.

Schultz, A. H. (1961). Vertebral column and thorax. *Primatologia* **4**: 1–66.

Schultz, A. H. (1963). The relative lengths of the foot skeleton and its main parts in primates. *Symposia of the Zoological Society of London* **10**: 199–206.

Schultz, A. H. (1969a). *The Life of the Primates.* London: Weidenfeld and Nicolson.

Schultz, A. H. (1969b). Observations on the acetabulum of primates. *Folia primatologica* **11**: 181–199.

Schwartz, J. H. (1983). Palatine fenestrae, the orang-utan and hominoid evolution. *Primates* **24**: 231–240.

Scott, J. H. (1954). Heat regulating function of the nasal mucous membrane. *Journal of Laryngology and Otology* **68**: 308–317.

Scott, J. H. (1957). The form of the dental arch. *Journal of Dental Research* **36**: 996–1003.

Scott, J. H. (1967). Dento-facial development and growth. Oxford: Pergamon Press.

Senut, B. (1981a). Outlines of the distal humerus in hominoid primates: application to some Plio-Pleistocene hominids. In: *Primate Evolutionary Biology* (A. B. Chiarelli and R. S. Corruccini, Eds), pp. 81–92. Berlin: Springer Verlag.

Senut, B. (1981b). Humeral outlines in some hominoid primates and in Plio-Pleistocene hominids. *American Journal of Physical Anthropology* **56**: 275–283.

Senut, B. and Tardieu, C. (1985). Functional aspects of Plio-Pleistocene Hominid limb bones: implications for taxonomy and phylogeny. In: *Ancestors: The Hard Evidence* (E. Delson, Ed.), pp. 193–201. New York: A. Liss.

Shapiro, L. J. and Jungers, W. L. (1988). Back muscle function during bipedal walking in chimpanzee and gibbon: implications for the evolution of human locomotion. *American Journal of Physical Anthropology* 77: 201–212.

Shea, B. T. (1985). On aspects of skull form in African apes and orang utans, with implications for hominoid evolution. *American Journal of Physical Anthropology* 68: 329–342.

Shellis, R. P. (1984). Variations in growth of the enamel crown in human teeth and a possible relationship between growth and enamel structure. *Archives of Oral Biology* 29: 697–705.

Shenkin, H. A., Harmel, M. H. and Kety, S. S. (1948). Dynamic anatomy of the cerebral circulation. *Archives of Neurology and Psychiatry* 60: 240–252.

Shephard, E. (1951). Tarsal movements. *Journal of Bone and Joint Surgery* 33B: 258–263.

Shipman, P., Walker, A. and Birchell, D. (1985). *The Human Skeleton.* Cambridge: Harvard University Press.

Shore, L. R. (1930). A report on the spinous processes of the cervical vertebrae in the native races of South Africa. *Journal of Anatomy* 64: 482–505.

Sibley, C. G. and Ahlquist, J. E. (1984). The phylogeny of hominoid primates, as indicated by DNA–DNA hybridization. *Journal of Molecular Evolution* 20: 2–15.

Sicher, H. (1937). Phylogenesis of human temporo-mandibular articulation. *Journal of Dental Research* 16: 339–340. (Abstr.).

Siddiqi, M. A. H. (1934). Variations in the lower end of the femur from Indians. *Journal of Anatomy* 68: 331–337.

Sigmon, B. A. (1974). A functional analysis of pongid hip and thigh musculature. *Journal of Human Evolution* 3: 161–185.

Sigmon, B. A. (1975). Functions and evolution of hominid hip and thigh musculature. In: *Primate Functional Morphology and Evolution* (R. H. Tuttle, Ed.), pp. 235–252. The Hague: Mouton.

Sigmon, B. A. (1982). Comparative morphology of the locomotor skeleton of *Homo erectus* and the other fossil hominids, with special reference to the Tautavel innominate and femora. *Première Congress Internationale Paleontologie Humaine*, (Nice), Vol. 1, pp. 422–446. Pretinage, CNRS.

Sigmon, B. A. and Farslow, D. L. (1986). The primate hindlimb. In: *Comparative Primate Biology, Vol. 1: Systematics, Evolution, and Anatomy* (D. R. Swindler and J. Erwin, Eds), pp. 671–718. New York: A. Liss.

Simpson, G. G. (1961). *Principles of Animal Taxonomy*, New York: Columbia University Press.

Simpson, G. G. (1963). The meaning of taxonomic statements. In: *Classification and Human Evolution* (S. L. Washburn, Ed.), pp. 1–31. Chicago: Aldine.

Singh, I. (1959). Squatting facets on the talus and tibia in Indians. *Journal of Anatomy* 93: 540–550.

Smith, B. H. (1986a) Dental development in *Australopithecus* and early *Homo. Nature* 323: 327–330.

Smith, B. H. (1986a). Development and evolution of the helicoidal plane of dental occlusion. *American Journal of Physical Anthropology* 69: 21–35.

Smith, B. H. (1989). Dental development as a measure of life history in primates. *Evolution* 43: 683–688.

Smith, B. H. and Garn, S. M. (1987). Polymorphisms in eruption sequence of permanent teeth in American children. *American Journal of Physical Anthropology* 74: 289–304.

Smith, J. M. and Savage, R. J. G. (1955). Some locomotor adaptations in mammals. *Journal of the Linnean Society (Zoology)* 42: 603–622.

Smith, J. W. (1956). Observations on the postural mechanism of the human knee joint. *Journal of Anatomy* 90: 236–260.

Smith, P. (1986) A quantitative study of Australopithecine dental attrition. In: *Teeth Revisited.* Proceedings of the VIIth International Symposium on Dental Morphology, Paris, 1986. (D. E. Russell, J.-P. Santoro and D. Sigogneau-Russell, Eds). *Mém. Mus. Natn. Hist. Nat.*, Paris, (Série C) 53: 389–398.

Smith, R. J. (1978). Mandibular biomechanics and temporomandibular joint function in primates. *American Journal of Physical Anthropology* 49: 341–349.

Smith, R. J. (1984a). Comparative functional morphology of maximum mandibular opening (gape) in primates. In: *Food Acquisition and Processing in Primates* (D. J. Chivers, B. A. Wood, and A. Billsborough, Eds), pp. 231–255. New York: Plenum Press.

Smith, R. J. (1984b). Allometric scaling in comparative biology: problems of concept and method. *American Journal of Physiology*

(*Regulatory Integrative Comparative Physiology 15*) **246**: R152–R160.

Smith, R. J. (1985). Functions of condylar translation in human mandibular movement. *American Journal of Orthodontics* **88**: 191–202.

Sokolof, S. (1972). The muscular anatomy of the chimpanzee foot. *Gegenbaurs Morphologische Jahrbuch, Leipzig* **119**: 86–125.

Sonntag, C. F. (1923). On the anatomy, physiology and pathology of the chimpanzee. *Proceedings of the Zoological Society of London* **1**: 323–429.

Sonntag, C. F. (1924). *The Morphology and Evolution of the Apes and Man*. London: J. Bale, Sons and Danielsson.

Soriano, M. (1970). The fluoritic origin of the bone lesion in the *Pithecanthropus erectus* femur. *American Journal of Physical Anthropology* **32**: 49–58.

Spencer, J. (1989). *A morphological analysis of the proximal tibia in humans and African apes*. B.Sc. Thesis, University College London.

Sperber, G. H. (1981). *Craniofacial Embryology*, 3rd edn, p. 173. Bristol: J. Wright and Sons.

Sprecher, H. (1932). *Morphologische Untersuchungen an der Fibula des Menschen unter Berücksichtigung anderer Primaten*. Zürich: Anthropologische Institute der Universitat Zürich.

Sprinz, R. and Kaufman, M. H. (1987). The sphenoidal canal. *Journal of Anatomy* **153**: 47–54.

Steele-Russell, I. (1979). Brain size and intelligence: a comparative perspective. In: *Brain, Behaviour and Evolution* (D. A. Oakley and H. C. Plotkin, Eds), pp. 126–153. London: Methuen.

Stephan, H., Bauchot, R. and Andy, O. J. (1970). Data on size of the brain and various brain parts in insectivors and primates. In: *The Primate Brain* (C. R. Norback and W. Montagna, Eds), pp. 289–297. New York: Appleton-Centuary-crofts.

Stern, J. T. (1971). Functional myology of the hip and thigh of cebid monkeys and its implications for the evolution of erect posture. *Bibliotheca primatologica* **14**: 1–318.

Stern, J. T. (1988). *Essentials of Gross Anatomy*. Philadelphia: F. A. Davis Co.

Stern, J. T. and Jungers, W. L. (1987). Absence of a relationship between the capitular joint of the first rib and locomotor use of the upper limb. *American Journal of Physical Anthropology* **72**: 257 (Abstr.).

Stern, J. T. and Susman, R. L. (1981). Electyromyography of the gluteal muscles in *Hylobates*, *Pongo* and *Pan*: application for the evolution of hominid bipedality. *American Journal of Physical Anthropology* **60**: 279–317.

Stern, J. T. and Susman, R. L. (1983). The locomotor anatomy of *Australopithecus afarensis*. *American Journal of Physical Anthropology* **60**: 279–317.

Stern, J. T., Wells, J. P., Vangor, A. K. and Fleagle, J. G. (1977). Electromyography of some muscles of the upper limb in *Ateles* and *Lagothrix*. *Yearbook of Physical Anthropology* **20**: 498–507.

Stewart, T. D. (1933). The tympanic plate and external auditory meatus in the Eskimo. *American Journal of Physical Anthropology* **17**: 481–496.

Stewart, T. D. (1960). Form of the pubic bone in Neanderthal man. *Science* **131**: 1437–1438.

Stewart, T. D. (1962). Neanderthal scapulae with special attention to the Shanidar Neanderthals from Iraq. *Anthropos* **57**: 779–800.

Straus, W. L. (1929). Studies on primate ilia. *American Journal of Anatomy* **43**: 403–460.

Straus, W. L. (1942). Rudimentary digits in primates. *Quarterly Review of Biology* **17**: 228–243.

Straus, W. L. (1948). The humerus of *Paranthropus robustus*. *American Journal of Physical Anthropology* **6**: 285–311.

Streudel, K. (1981). Sexual dimorphism and allometry in primate ossa coxae. *American Journal of Physical Anthropology* **55**: 209–215.

Stringer, C. B. (1974). Population relationships of Later Pleistocene hominids. *Journal of Archaeological Sciences* **1**: 317–342.

Stringer, C. B. (1984a). The definition of *Homo erectus* and the existence of the species in Africa and Europe. In: *The Early Evolution of Man* (P. Andrews and Frazen, Eds), *Cour. Forsch. Inst. Senckenberg* **69**: 131–143.

Stringer, C. B. (1984b). Human evolution and biological adaptation in the Pleistocene. In: *Hominid Evolution and Community Ecology* (R. Foley, Ed.), pp. 55–83. London: Academic Press.

Stringer, C. B. (1986). An archaic character in the Broken Hill innominate E719. *American Journal of Physical Anthropology* **71**: 115–120.

Stringer, C. B., Dean M. C. and Martin, R. D. (1990). *A Comparative Study of Cranial and*

Dental Development within a Recent British Sample and among Neanderthals. Wenner Gren Symposium. New York: A. Liss, in press.

Sunderland, E. P., Smith, C. J. and Sunderland, R. (1987). A histological study of the chronology of initial mineralization in the human deciduous dentition *Archives of Oral Biology* **32**: 167–174.

Susman, R. L. (1979). The comparative and functional morphology of hominoid fingers. *American Journal of Physical Anthropology* **50**: 215–236.

Susman, R. L. (1983). Evolution of the human foot: evidence from Plio-Pleistocene hominids. *Foot and Ankle* **3**: 365–376.

Susman, R. L. (1988a). Hand of *Paranthropus. Science* **240**: 781–784.

Susman, R. L. (1988b). New postcranial remains from Swartkrans and their bearing on the functional morphology and behavior of *Paranthropus robustus.* In: *Evolutionary History of the 'Robust' Australopithecines* (F. E Grine, Ed.), pp. 149–172. New York: Aldine de Gruyter.

Susman, R. L. and Brain, T. M. (1988). New first metatarsal (SKX 5017) from Swartkrans and the gait of *Paranthropus robustus. American Journal of Physical Anthropology* **77**: 7–16.

Susman, R. L. and Creel, N. (1979). Functional and morphological affinities of the subadult hand (O.H.7) from Olduvai Gorge. *American Journal of Physical Anthropology* **51**: 311–332.

Susman, R. L. and Stern, J. T., Jr. (1979). Telemetered electromyography of flexor digitorum profundus and flexor digitorum superficialis in *Pan troglodytes* and implications for interpretation of the O.H. 7 hand. *American Journal of Physical Anthropology* **50**: 565–574.

Susman, R. L. and Stern, J. T. (1982). Functional morphology of *Homo habilis. Science* **217**: 931–934.

Susman, R. L., Stern, J. T. Jr. and Jungers, W. L. (1984). Arboreality and bipedality in Hadar hominids. *Folia Primatologica* **43**: 113–156.

Suzuki, R. (1985). Human adult walking. In: *Primate Morphophysiology, Locomotor Analysis and Human Bipedalism* (S. Kondo, Ed.), pp. 3–24. Tokyo: University of Tokyo Press.

Swindler, D. R. (1976). *Dentition of Living Primates.* London: Academic Press.

Swindler, D. R. (1985). Nonhuman primate dental development and its relationship to human dental development. In: *Nonhuman Primate*

Models for Human Growth and Development (E. S. Watts, Ed.), pp. 67–94. New York: A. Liss.

Tague, R. G. and Lovejoy, C. O. (1986). The obstetric pelvis of A.L. 288-1 (Lucy). *Journal of Human Evolution* **15**: 237–255.

Tardieu, C. (1981). Morpho-functional analysis of the articular surfaces of the knee-joint in primates. In: *Primate Evolutionary Biology* (A. B. Chiarelli and R. S. Corruccini, Eds), pp. 68–80. Berlin: Springer Verlag.

Tardieu, C. (1983). *L'Articulation du genou.* France: Editions du CNRS.

Tattersall, I. (1986). Species recognition in human paleontology. *Journal of Human Evolution* **15**: 165–175.

Tattersall, I. and Eldredge, N. (1977). Fact, theory and fantasy in human paleontology. *American Scientist* **65**: 204–211.

Taxman, R. M. (1963). Incidence and size of the juxtamastoid eminence in modern crania. *American Journal of Physical Anthropology* **21**: 153–157.

Taylor, C. R. and Rowntree, V. J. (1973). Running on two or four legs: which consumes more energy: *Science* **179**: 186–187.

Taylor, C. R., Schmidt-Nielsen, K. and Raab, J. L. (1970). Scaling of energetic cost of running to body size in mammals. *American Journal of Physiology* **219**: 1104–1107.

Ten Cate, A. R. (1985). *Oral Histology: Development, Structure and Function,* 2nd edn. Princeton, NJ: The C.V. Mosby Co.

Thoma, A. (1981). The position of the Vertesszollos find in relation to *Homo erectus; Papers in Honor of Davidson Black* (B. A. Sigmon and J. S. Cybulski, Eds), pp. 105–114. Toronto: University of Toronto.

Thomson, A. (1889). The influence of posture on the form of the articular surfaces of the tibia and astragalus in the different races of man and the higher apes. *Journal of Anatomy* **23**: 616–639.

Thompson, D. D. (1979). The core technique in determination of age at death in skeletons. *Journal of Forensic Science* **24**: 902–915.

Thompson, D. D. and Trinkaus, E. (1987). Femoral diaphyseal histomorphometric age determinations for the Shanidar III, IV, V and VI Neanderthals and Neanderthal longevity. *American Journal of Physical Anthropology* **72**: 123–130.

Tillier, A. M. (1984). L'enfant *Homo* II de Qafzeh

(Israel) et son apport à la compréhension des modalités de la croissance de squelettes Moustériens. *Paleorient* 10: 7–48.

Tobias, P. V. (1966). On the distinctiveness of *Homo habilis. Nature* 209: 953–957.

Tobias, P. V. (1967). *Olduvai Gorge. Vol. 2. The Cranium and Maxillary Dentition of Australopithecus (Zinjanthropus) boisei.* Cambridge: Cambridge University Press.

Tobias, P. V. (1974). In: *Perspectives in Palaeoanthropology* (A. K. Ghosh, Ed.), pp. 9–17. Calcutta.

Tobias, P. V. (1980). The natural history of the helicoidal occlusal plane and its evolution in early *Homo. American Journal of Physical Anthropology* 53: 173–187.

Tobias, P. V. (1987). The brain of *Homo habilis*: a new level of organisation in cerebral evolution. *Journal of Human Evolution* 16: 741–761.

Todd, T. W. (1923). Age changes in the pubic sysphysis. VII. The anthropoid strain in human pubic symphyses of the third decade. *Journal of Anatomy* 27: 274–294.

Tredgold, A. F. (1897). Variations of ribs in the primates, with especial reference to the number of sternal ribs in man. *Journal of Anatomy and Physiology* 31: 288–302.

Trevor, J. C. (1963). The history of the word 'brachiator' and a problem of authorship in primate nomenclature. In: *The Primates* (J. Napier and N. A. Barnicot, Eds) *Symposia of the Zoological Society of London* 10: 197–198.

Trinkaus, E. (1975). Squatting among the Neandertals: a problem in the behavioral interpretation of skeletal morphology. *Journal of Archaeological Sciences* 2: 327–351.

Trinkaus, E. (1976a). The morphology of European and Southwest Asian Neandertal pubic bones. *American Journal of Physical Anthropology* 44: 95–104.

Trinkaus, E. (1976b). The evolution of the hominid femoral diaphysis during the Upper Pleistocene in Europe and the Near East. *Zeitschrift fur Morphologie und Anthropologie* 67: 291–319.

Trinkaus, E. (1977). A functional interpretation of the axillary border of the Neandertal scapula. *Journal of Human Evolution* 6: 231–234.

Trinkaus, E. (1980). Sexual differences in Neandertal limb bones. *Journal of Human Evolution* 9: 377–397.

Trinkaus, E. (1981). Neanderthal limb proportions and cold adaptation. In: *Aspects of Human Evolution* (C. B. Stringer, Ed.), pp. 187–224. *Symposia for the Society for the Study of Human Biology*, Vol. 21. London: Taylor & Francis.

Trinkaus, E. (1982). The Shanidar 3 Neandertal. *American Journal of Physical Anthropology* 57: 37–60.

Trinkaus, E. (1983a). *The Shanidar Neanderthals.* New York: Academic Press.

Trinkaus, E. (1983b). Functional aspects of Neandertal pedal remains. *Foot and Ankle* 3: 377–390.

Trinkaus, E. (1984). Neandertal pubic morphology and gestation length. *Current Anthropology* 25: 509–514.

Trinkaus, E. (1989). Oludavi hominid 7 trapezial metacarpal 1 articular morphology: contrasts with recent humans. *American Journal of Physical Anthropology* 80: 411–416.

Trinkaus, E. and Churchill, S. E. (1988). Neandertal radial tuberosity orientation. *American Journal of Physical Anthropology* 75: 15–21.

Trinkaus, E. and Howells, W. W. (1979). The Neandertals. *Scientific American* 241: 94–105.

Trotter, M. L. and Gleser, G. (1952). Estimation of stature from long bones of American whites and negroes. *American Journal of Physical Anthropology* 10: 463–514.

Trotter, M. L. and Gleser, G. (1958). A re-evaluation of stature based on measurements taken during life and of long bones after death. *American Journal of Physical Anthropology* 16: 79–123.

Turner, A. and Chamberlain, A. (1989). Speciation, morphological change and the status of African *Homo erectus. Journal of Human Evolution* 18: 115–130.

Tuttle, R. H. (1967). Knuckle-walking and the evolution of hominid hands. *American Journal of Physical Anthropology* 26: 171–206.

Tuttle, R. (1969a). Knuckle-walking and the problems of human origins. *Science* 166: 953–961.

Tuttle, R. H. (1969b). Quantitative and functional studies on the hands of the anthropoidea. *Journal of Morphology* 128: 309–364.

Tuttle, R. (1974a). Electromyography of brachial muscles in *Pan gorilla* and hominoid evolution. *American Journal of Physical Anthropology* 41: 71–90.

Tuttle, R. (1974b). Darwin's apes, dental apes, and the descent of man: normal science in evolutionary anthropology. *Current Anthro-*

pology **15**: 389–398.

Tuttle, R. (1975). Parallelism, brachiation, and hominoid phylogeny. In: *Phylogeny of the Primates, a Multidisciplinary Approach* (W. P. Luckett and F. S. Szalay, Eds), pp. 447–480. New York: Plenum Press.

Tuttle, R. (1981). Evolution of hominid bipedalism and prehensile capabilities. *Philosophical Transactions of the Royal Society, London, Ser. B* **292**: 89–94.

Tuttle, R. H. (1984). Bear facts and Laetoli impressions. *American Journal of Physical Anthropology* **63**: 230.

Tuttle, R. H. and Basmajian, J. V. (1974a). Electromyography of brachial muscles in *Pan gorilla* and hominoid evolution. *American Journal of Physical Anthropology* **41**: 71–90.

Tuttle, R. H. and Basmajian, J. V. (1974b). Electromyography of forearm musculature in *Gorilla* and problems related to knuckle-walking. In: *Primate Locomotion* (F. A. Jenkins, Jr., Ed.), pp. 293–347. New York: Academic Press.

Tuttle, R. H. and Basmajian, J. V. (1977). Electromyography of pongid shoulder muscles and hominoid evolution. I. Retractors of the humerus and 'rotators' of the scapula. *Yearbook of Physical Anthropology* **20**: 491–497.

Tuttle, R. H. and Basmajian, J. V. (1978a). Electromyography of pongid shoulder muscles. II: Deltoid, rhomboid and 'rotator' cuff. *American Journal of Physical Anthropology* **49**: 47–56.

Tuttle, R. H. and Basmajian, J. V. (1978b). Electromyography of pongid shoulder muscles. III: Quadrupedal positional behaviour. *American Journal of Physical Anthropology* **49**: 57–78.

Tuttle, R. H., Basmajian, J. V. and Ishida, H. (1975). Electromyography of the gluteus maximus muscle in gorilla and the evolution of bipedalism. In: *Primate Functional Morphology and Evolution* (R. H. Tuttle, Ed.), pp. 253–269. The Hague: Mouton.

Tuttle, R. H., Basmajian, J. V., and Ishida, H. (1978). Electromyography of pongid gluteal muscles and hominid evolution. In: *Recent Advances in Primatology*, Vol. 3 (D. J. Chivers and K. A. Joysey, Eds), pp. 463–468.

Tuttle, R. H., Basmajian, J. V., and Ishida, H. (1979a). Activities of pongid thigh muscles during bipedal behavior. *American Journal of Physical Anthropology* **50**: 123–136.

Tuttle, R. H., Cortright, G. W. and Buxhoeveden, D. P. (1979b). Anthropology on the move: progress in experimental studies of nonhuman primate positional behavior. *Yearbook of Physical Anthropology* **22**: 187–214.

Vallois, H. V. (1928–46). L'Omoplate Humaine: Etude anatomique et anthropologique. *Bulletins et Mémoires de la Société d'Anthropologie de Paris, Série 7* **9**: 129–168; **10**: 110–191; *Série 8* **3**: 3–153; *Série 9* **7**: 16–100.

Van Gerven, D. P. (1972). The contribution of size and shape variation to patterns of sexual dimorphism of the human femur. *American Journal of Physical Anthropology* **37**: 49–60.

Vilensky, J. A., Van Hoesen, G. W. and Damasio, A. R. (1982). The limbic system and human evolution. *Journal of Human Evolution* **11**: 447–460.

Virchow, H. (1929). Das Os Centrale Carpi des Menschen. *Gengenbaurs Morphologische Jahrbuch* **63**: 480–531.

Vlček, E. (1975). Morphology of the first metacarpal of Neanderthal individuals from the Crimea. *Bulletins et Mémoires de la Société d'Anthropologie de Paris, Serie 13* **2**: 257–276.

Volkmann, R. (1862). Chirurgische Erfahrungen über Konchenverbregungen und Knochenwachsthum. *Archive für Pathologische Anatomie* **24**: 512–540.

Vrba, E. S. (1979). A new study of the scapula of *Australopithecus africanus* from Sterkfontein. *American Journal of Physical Anthropology* **51**: 117–130.

Vrba, E. S. (1980). Evolution, species and fossils: how does life evolve? *South African Journal of Science* **76**: 61–84.

Vrba, E. (1982). Biostratigraphy and chronology, based particularly on bovidae, of southern hominid-associated assemblages: Makapansgat, Sterkfontein, Taung, Kromdraai, Swartkrans and also Elandsfontein (Saldanha), Broken Hill (now Kabwe) and Cave of Hearths. In: *L'Homo Erectus et Alpalce de l'Homme de Tautavel Parmi les Hominides Fossils. 1st Congress Internationale Paléontologie Humaine* (Nice, October, 1982) Pretirage, CNRS, Vol. 1, pp. 707–752.

Walensky, N. A. (1964). A re-evaluation of the mastoid region of contemporary and fossil man. *Anatomical Record* **149**: 67–72.

Walker, A. C. (1973). New *Australopithecus* femora from East Rudolf, Kenya. *Journal of Human Evolution* **2**: 545–555.

Walker, A. C. (1978). Functional anatomy of oral tissues: mastication and deglutition. In: *Textbook of Oral Biology* (J. H. Shaw, E. A. Sweeney, C. C. Cappuccino and S. M. Meller, Eds), pp. 277–296.

Walker, A. C. and Leakey, R. E. (1986). *Homo erectus* skeleton from West Lake Turkana, Kenya. *American Journal of Physical Anthropology* **69**: 275 (Abstr.).

Walker, A. C. and Leakey, R. E. F. (1988). The evolution of *Australopithecus boisei*. In: *Evolutionary History of the 'Robust' Australopithecines* (F. E. Grine, Ed.), pp. 247–258. New York: Aldine de Gruyter.

Walker, A., Falk, D., Smith, R. and Pickford, M. (1983). The skull of *Proconsul africanus*: reconstruction and cranial capacity. *Nature* **305**: 525–527.

Walker, A., Leakey, R. E., Harris, J. M. and Brown, F. H. (1986). 2.5-myr *Australopithecus boisei* from west of Lake Turkana, Kenya. *Nature* **322**: 517–522.

Walker, P. S. and Hajek, J. V. (1972). The load-bearing area in the knee joint. *Journal of Biomechanics* **5**: 581–589.

Walmsley, T. (1933). The vertical axes of the femur and their relations. A contribution to the study of the erect posture. *Journal of Anatomy* **67**: 284–300.

Wanner, J. A. (1977). Variations in the anterior patellar groove of the human femur. *American Journal of Physical Anthropology* **47**: 99–102.

Ward, S. C. and Kimbel, W. H. (1983). Subnasal alveolar morphology and systematic position of *Sivapithecus*. *American Journal of Physical Anthropology* **61**: 157–171.

Warick-James, W. (1960). *The Jaws and Teeth of Primates*. London: Pitman Medical.

Waterman, H. C. (1929). Studies on the evolution of the pelvis of man and other primates. *Bulletin of the American Museum of Natural History* **58**: 585–641.

Weidenreich, F. (1913). Über das Huftbein und das Becken der Primaten und ihre Umformung durch den aufrechten Gang. *Anatomischer Anzieger* **44**: 497–513.

Weidenreich, F. (1936). The mandibles of *Sinanthropus pekinensis*; a comparative study. *Palaeontoligia Sinica, Series D, No. 7, Fast 4*, pp. 1–162. Peking: The Geological Survey of China.

Weidenreich, F. (1937). The dentition of *Sinanthropus pekinensis*; a comparative odontography of the Hominids. *Palaeontologica Sinica. N.S. D No. 1*. Peking: The Geological Survey of China.

Weidenreich, F. (1940). The external tubercle of the human tuber calcanei. *American Journal of Physical Anthropology* **23**: 473–487.

Weidenreich, F. (1941). The extremity bones of *Sinanthropus pekinensis*. *Paleontologica Sinica (N.S. D) No. 5*, pp. 373–383. Peking: Geological Survey of China.

Weidenreich, F. (1943). The skull of *Sinanthropus pekinensis*: a comparative study on a primitive hominid skull. *Palaeontologia Sinica, New Series D No. 10 (Whole series No. 127)*. Peking: The Geological Survey of China.

Weidenreich, F. (1951). Morphology of solo man. *Anthropological Papers of the American Museum of Natural History New York* **43, Part 3**: 205–290.

Weinert, H. (1928). *Pithecanthropus erectus*. *Zeitschrift für die Gesamte Anatomie*. **87**: 429–547.

Weinert, C. R., McMaster, J. H. and Ferguson, R. J. (1973). Dynamic function of the human fibula. *American Journal of Anatomy* **138**: 145–150.

Weisl, H. (1954). The articular surfaces of the sacroiliac joint and their relation to the movements of the sacrum. *Acta Anatomica (Basel)* **22**: 1–14.

Wells, L. H. (1958). A reconsideration of some mandibular profiles. *South African Journal of Science* **54**: 55–58.

White, T. D. (1980). Evolutionary implications of Pliocene hominid footprints. *Science* **208**: 175–176.

White, T. D. (1981). New fossil hominids from Laetoli, Tanzania. *American Journal of Physical Anthropology* **46**: 197–230.

White, T. D. and Suwa, G. (1987). Hominid footprints at Laetoli: facts and interpretations. *American Journal of Physical Anthropology* **72**: 485–514.

White, T. D., Johanson, D. C. and Kimbel, W. H. (1981). *Australopithecus africanus*: its phyletic position reconsidered. *South African Journal of Science* **77**: 445–470.

Wiley, E. O. (1978). The evolutionary species concept reconsidered. *Systematic Zoology* **27**: 17–26.

Wilkinson, J. L. (1954). The terminal phalanx of the great toe. *Journal of Anatomy* **88**: 537–541.

Willoughby, D. P. (1978). *All about Gorillas*.

New Jersey: Barnes.

Winkler, L. A. (1988). Variation in sub-occipital anatomy. In: *Orang Utan Biology* (J. H. Schwartz, Ed.), pp. 191–199. Oxford: Oxford University Press.

Winkler, L. A. (1989). Morphology and relationships of the orang utan fatty cheek pads. *American Journal of Primatology* **17**: 305–319.

Wolpoff, M. H. (1968). Climatic influence on the skeletal nasal aperture. *American Journal of Physical Anthropology* **29**: 405–423.

Wolpoff, M. H. (1973). Posterior tooth size, body size, and diet in South African gracile australopithecines. *American Journal of Physical Anthropology* **39**: 375–394.

Wolpoff, M. H. (1976). Fossil hominid femora. *Nature* **264**: 812–813.

Wolpoff, M. H. (1978). Some aspects of canine size in the australopithecines. *Journal of Human Evolution* **7**: 115–126.

Wolpoff, M. H. (1979). The Krapina dental remains. *American Journal of Physical Anthropology* **50**: 67–114.

Wolpoff, M. H. (1983a). Lucy's little legs. *Journal of Human Evolution* **12**: 443–453.

Wolpoff, M. H. (1983b). Lucy's lower limbs: long enough for Lucy to be fully bipedal? *Nature* **304**: 59–61.

Woo, J. K. (1966). The hominid skull of Lantian. *Shensi Vertebrate Palasiat* **10**: 14–22.

Wood, B. A. (1974a). Olduvai Bed I post-cranial fossils: a reassessment. *Journal of Human Evolution* **3**: 373–378.

Wood, B. A. (1974b). A *Homo* talus from East Rudolf, Kenya. *Journal of Anatomy* **117**: 203–204.

Wood, B. A. (1979). On analysis of tooth and body size relationships in five primate taxa. *Folia Primatologica* **31**: 187–211.

Wood, B. A. (1984). The origin of *Homo erectus*. In: *The Early Evolution of Man* (P. Andrews and J. L. Frazen, Eds). *Courier Forschungsinstitut Senckenberg* **69**: 99–111.

Wood, B. A. and Chamberlain, A. T. (1986). The primate pelvis: allometry or sexual dimorphism? *Journal of Human Evolution* **15**: 257–269.

Wood, B. A. and Engleman, C. A. (1988). Analysis of the dental morphology of Plio-Pleistocene hominids. Part 5: Maxillary postcanine tooth morphology. *Journal of Anatomy* **161**: 1–35.

Wood, B. A. and Stack, C. G. (1980). Does allometry explain the differences between 'gracile' and 'robust' australopithecines? *American Journal of Physical Anthropology* **52**: 55–62.

Wood, B. A. and Uytterschaut, H. (1987). Analysis of the dental morphology of Plio-Pleistocene hominids. Part 3: Mandibular premolar crowns. *Journal of Anatomy* **154**: 121–156.

Wood, B. A., Abbott, S. A. and Graham, S. H. (1983). Analysis of the dental morphology of Plio-Pleistocene hominids. Part 2: Mandibular molars—study of cusp areas, fissure pattern and cross-sectional shape of the crown. *Journal of Anatomy* **137**: 287–314.

Wood, B. A., Abbott, S. A. and Uytterschaut, H. (1988). Analysis of the dental morphology of Plio-Pleistocene hominids. Part 4: Mandibular postcanine root morphology. *Journal of Anatomy* (in press).

Wood-Jones, F. (1948). *Hallmarks of Mankind.* London: Baillière Tindall.

Wood-Jones, F. (1949). *Structure and Function as Seen in the Foot, 2nd edn.* London: Baillière-Tindall.

Woodhall, B. (1936). Variations of the cranial venous sinuses in the region of the torcular herophili. *Archives of Surgery* **33**: 297–314.

Woodhall, B. (1939). Anatomy of the cranial blood sinuses with particular reference to the lateral. *Laryngoscope* **49**: 966–1010.

Wu, X. (1981). A well preserved cranium of an archaic type of early *Homo sapiens* from Dali China. *Scienta Sinica* **24**: 530–539.

Yamazaki, N. (1985). Primate bipedal walking: computer simulation. In: *Primate morphophysiology, locomotor analysis and human bipedalism* (S. Kondo, Ed.) pp. 59–80. Japan: University of Tokyo Press.

Yilmaz, S., Newman, H. N. and Poole, D. F. G. (1977). Diurnal periodicity of von Ebner growth lines in pig dentine. *Archives of Oral Biology* **22**: 511–513.

Zapfe, H. (1960). Die Primatenfunde aus der miozänen Spaltenfüllung von Neudorf an der March (Děvínska Nová Ves), Tschechoslowakei. *Schweizerische Palaeontologische Abhandlungen* **78**: 1–293.

Zihlman, A. L. (1984). Body build and tissue composition in *Pan paniscus* and *Pan troglodytes*, with comparisons to other hominoids. In: *The Pygmy Chimpanzee* (R. S. Susman, Ed.), pp. 179–200.

Zihlman, A. and Brunker, L. (1979). Hominid

bipedalism: then and now. *Yearbook of Physical Anthropology* **22**: 132–162.

Zihlman, A. L. and Cramer, D. L. (1978). A skeletal comparison between pygmy (*Pan paniscus*) and common chimpanzees (*Pan troglodytes*). *Folia Primatologica* **29**: 86–94.

Zouaoui, A. and Hidden, G. (1989). The cervical vertebral venous plexus, a drainage route for the brain. *Surgical and Radiologic Anatomy*, Vol. II, pp. 79–81. Berlin: Springer-Verlag.

INDEX

Notes

1. References to illustrations are indicated by **bold** figures. These are limited to specific detailed features and not general topics.
2. There are usually textual references on the same pages as illustrations. These are subsumed under the emboldened figures.